CONFORMAL METHODS IN GENERAL RELATIVITY

This book is a systematic exposition of conformal methods and how they can be used to study the global properties of solutions to the equations of Einstein's theory of gravity. It shows that combining these ideas with techniques of the theory of partial differential equations can elucidate the stability of the basic solutions of the theory. Introducing the differential geometric, spinorial and PDE background required to gain a deep understanding of conformal methods, this text provides an accessible account of key results in mathematical relativity over the last 30 years, including the stability of de Sitter and Minkowski spacetimes.

For graduate students and researchers, this self-contained account includes useful visual models to help the reader grasp abstract concepts and a list of further reading, making this the perfect reference companion on the topic.

This title, first published in 2017, has been reissued as an Open Access publication on Cambridge Core.

JUAN A. VALIENTE KROON is a Reader in Applied Mathematics at Queen Mary University of London. He was a Lise Meitner fellow of the Austrian Science Fund (FWF), an Engineering and Physical Sciences (EPSRC) Advanced Research fellow and he specialises in various aspects of mathematical general relativity.

CAMBRIDGE MONOGRAPHS ON MATHEMATICAL PHYSICS

General Editors: P. V. Landshoff, D. R. Nelson, S. Weinberg

S. J. Aarseth *Gravitational N-Body Simulations: Tools and Algorithms*†
J. Ambjørn, B. Durhuus and T. Jonsson *Quantum Geometry: A Statistical Field Theory Approach*†
A. M. Anile *Relativistic Fluids and Magneto-fluids: With Applications in Astrophysics and Plasma Physics*
J. A. de Azcárraga and J. M. Izquierdo *Lie Groups, Lie Algebras, Cohomology and Some Applications in Physics*†
O. Babelon, D. Bernard and M. Talon *Introduction to Classical Integrable Systems*†
F. Bastianelli and P. van Nieuwenhuizen *Path Integrals and Anomalies in Curved Space*†
D. Baumann and L. McAllister *Inflation and String Theory*
V. Belinski and E. Verdaguer *Gravitational Solitons*†
J. Bernstein *Kinetic Theory in the Expanding Universe*†
G. F. Bertsch and R. A. Broglia *Oscillations in Finite Quantum Systems*†
N. D. Birrell and P. C. W. Davies *Quantum Fields in Curved Space*†
K. Bolejko, A. Krasiński, C. Hellaby and M-N. Célérier *Structures in the Universe by Exact Methods: Formation, Evolution, Interactions*
D. M. Brink *Semi-Classical Methods for Nucleus-Nucleus Scattering*†
M. Burgess *Classical Covariant Fields*†
E. A. Calzetta and B.-L. B. Hu *Nonequilibrium Quantum Field Theory*
S. Carlip *Quantum Gravity in 2+1 Dimensions*†
P. Cartier and C. DeWitt-Morette *Functional Integration: Action and Symmetries*†
J. C. Collins *Renormalization: An Introduction to Renormalization, the Renormalization Group and the Operator-Product Expansion*†
P. D. B. Collins *An Introduction to Regge Theory and High Energy Physics*†
M. Creutz *Quarks, Gluons and Lattices*†
P. D. D'Eath *Supersymmetric Quantum Cosmology*†
J. Dereziński and C. Gérard *Mathematics of Quantization and Quantum Fields*
F. de Felice and D. Bini *Classical Measurements in Curved Space-Times*
F. de Felice and C. J. S Clarke *Relativity on Curved Manifolds*†
B. DeWitt *Supermanifolds, 2nd edition*†
P. G. O. Freund *Introduction to Supersymmetry*†
F. G. Friedlander *The Wave Equation on a Curved Space-Time*†
J. L. Friedman and N. Stergioulas *Rotating Relativistic Stars*
Y. Frishman and J. Sonnenschein *Non-Perturbative Field Theory: From Two Dimensional Conformal Field Theory to QCD in Four Dimensions*
J. A. Fuchs *Affine Lie Algebras and Quantum Groups: An Introduction, with Applications in Conformal Field Theory*†
J. Fuchs and C. Schweigert *Symmetries, Lie Algebras and Representations: A Graduate Course for Physicists*†
Y. Fujii and K. Maeda *The Scalar-Tensor Theory of Gravitation*†
J. A. H. Futterman, F. A. Handler, R. A. Matzner *Scattering from Black Holes*†
A. S. Galperin, E. A. Ivanov, V. I. Ogievetsky and E. S. Sokatchev *Harmonic Superspace*†
R. Gambini and J. Pullin *Loops, Knots, Gauge Theories and Quantum Gravity*†
T. Gannon *Moonshine beyond the Monster: The Bridge Connecting Algebra, Modular Forms and Physics*†
M. Göckeler and T. Schücker *Differential Geometry, Gauge Theories and Gravity*†
C. Gómez, M. Ruiz-Altaba and G. Sierra *Quantum Groups in Two-Dimensional Physics*†
M. B. Green, J. H. Schwarz and E. Witten *Superstring Theory Volume 1: Introduction*
M. B. Green, J. H. Schwarz and E. Witten *Superstring Theory Volume 2: Loop Amplitudes, Anomalies and Phenomenology*
V. N. Gribov *The Theory of Complex Angular Momenta: Gribov Lectures on Theoretical Physics*†
J. B. Griffiths and J. Podolský *Exact Space-Times in Einstein's General Relativity*†
S. W. Hawking and G. F. R. Ellis *The Large Scale Structure of Space-Time*†
F. Iachello and A. Arima *The Interacting Boson Model*†
F. Iachello and P. van Isacker *The Interacting Boson-Fermion Model*†
C. Itzykson and J. M. Drouffe *Statistical Field Theory Volume 1: From Brownian Motion to Renormalization and Lattice Gauge Theory*†
C. Itzykson and J. M. Drouffe *Statistical Field Theory Volume 2: Strong Coupling, Monte Carlo Methods, Conformal Field Theory and Random Systems*†
G. Jaroszkiewicz *Principles of Discrete Time Mechanics*
C. V. Johnson *D-Branes*†
P. S. Joshi *Gravitational Collapse and Spacetime Singularities*†

J. I. Kapusta and C. Gale *Finite-Temperature Field Theory: Principles and Applications, 2nd edition*†
V. E. Korepin, N. M. Bogoliubov and A. G. Izergin *Quantum Inverse Scattering Method and Correlation Functions*†
M. Le Bellac *Thermal Field Theory*†
Y. Makeenko *Methods of Contemporary Gauge Theory*†
N. Manton and P. Sutcliffe *Topological Solitons*†
N. H. March *Liquid Metals: Concepts and Theory*†
I. Montvay and G. Münster *Quantum Fields on a Lattice*†
L. O'Raifeartaigh *Group Structure of Gauge Theories*†
T. Ortín *Gravity and Strings, 2nd edition*
A. M. Ozorio de Almeida *Hamiltonian Systems: Chaos and Quantization*†
L. Parker and D. Toms *Quantum Field Theory in Curved Spacetime: Quantized Fields and Gravity*
R. Penrose and W. Rindler *Spinors and Space-Time Volume 1: Two-Spinor Calculus and Relativistic Fields*†
R. Penrose and W. Rindler *Spinors and Space-Time Volume 2: Spinor and Twistor Methods in Space-Time Geometry*†
S. Pokorski *Gauge Field Theories, 2nd edition*†
J. Polchinski *String Theory Volume 1: An Introduction to the Bosonic String*†
J. Polchinski *String Theory Volume 2: Superstring Theory and Beyond*†
J. C. Polkinghorne *Models of High Energy Processes*†
V. N. Popov *Functional Integrals and Collective Excitations*†
L. V. Prokhorov and S. V. Shabanov *Hamiltonian Mechanics of Gauge Systems*
A. Recknagel and V. Schiomerus *Boundary Conformal Field Theory and the Worldsheet Approach to D-Branes*
R. J. Rivers *Path Integral Methods in Quantum Field Theory*†
R. G. Roberts *The Structure of the Proton: Deep Inelastic Scattering*†
C. Rovelli *Quantum Gravity*†
W. C. Saslaw *Gravitational Physics of Stellar and Galactic Systems*†
R. N. Sen *Causality, Measurement Theory and the Differentiable Structure of Space-Time*
M. Shifman and A. Yung *Supersymmetric Solitons*
H. Stephani, D. Kramer, M. MacCallum, C. Hoenselaers and E. Herlt *Exact Solutions of Einstein's Field Equations, 2nd edition*†
J. Stewart *Advanced General Relativity*†
J. C. Taylor *Gauge Theories of Weak Interactions*†
T. Thiemann *Modern Canonical Quantum General Relativity*†
D. J. Toms *The Schwinger Action Principle and Effective Action*†
A. Vilenkin and E. P. S. Shellard *Cosmic Strings and Other Topological Defects*†
R. S. Ward and R. O. Wells, Jr *Twistor Geometry and Field Theory*†
E. J. Weinberg *Classical Solutions in Quantum Field Theory: Solitons and Instantons in High Energy Physics*
J. R. Wilson and G. J. Mathews *Relativistic Numerical Hydrodynamics*†
T. Ortín *Gravity and Strings*
D. Baumann and L. McAllister *Inflation and String Theory*
S. Raychaudhuri and K. Sridhar *Particle Physics of Brane Worlds and Extra Dimensions*

† Available in paperback

Conformal Methods in General Relativity

JUAN A. VALIENTE KROON
Queen Mary University of London

CAMBRIDGE
UNIVERSITY PRESS

CAMBRIDGE
UNIVERSITY PRESS

Shaftesbury Road, Cambridge CB2 8EA, United Kingdom

One Liberty Plaza, 20th Floor, New York, NY 10006, USA

477 Williamstown Road, Port Melbourne, VIC 3207, Australia

314–321, 3rd Floor, Plot 3, Splendor Forum, Jasola District Centre, New Delhi – 110025, India

103 Penang Road, #05–06/07, Visioncrest Commercial, Singapore 238467

Cambridge University Press is part of Cambridge University Press & Assessment, a department of the University of Cambridge.

We share the University's mission to contribute to society through the pursuit of education, learning and research at the highest international levels of excellence.

www.cambridge.org
Information on this title: www.cambridge.org/9781009291347

DOI: 10.1017/9781009291309

First published 2017
Reissued as OA 2022

A catalogue record for this publication is available from the British Library.

ISBN 978-1-009-29134-7 Hardback
ISBN 978-1-009-29133-0 Paperback

ἀγεωμέτρητος μηδεὶς εἰσίτω (Let no one untrained in geometry enter)

– *Epigram at the Academy of Plato*

Contents

Preface

This book discusses an approach to the study of global properties of solutions to the equations of general relativity, the Einstein field equations, in which the notion of conformal transformation plays a central role. The use of conformal transformations in differential geometry dates back, at least, to the work of *Hermann Weyl* in the 1920s.[1] Their application to global questions in general relativity, as presented in this book, stems from the seminal work of *Roger Penrose* in the 1960s in which the close connection between the global causal structure of the solutions to the equations of general relativity and conformal geometry was established.[2] Penrose's key insights are that the close relation between the propagation of the gravitational field and the structure of light cones which holds locally in a spacetime is also preserved in the case of large scales and that the asymptotic behaviour of the gravitational field can be conveniently analysed in terms of conformal extensions of the spacetime. In the following decade Penrose's ideas were polished, extended and absorbed into the mainstream research of general relativity by a considerable number of researchers[3] – finally leading to the influential notion of *asymptotic simplicity.* The subject reached its maturity when this *formal* theory was combined with the methods of the theory of partial differential equations (PDEs). This breakthrough is mainly due to the work of *Helmut Friedrich* in the early 1980s, who – through the *conformal Einstein field equations*[4] – showed that ideas of conformal geometry can be used to establish the existence of large classes of solutions to the Einstein field equations satisfying Penrose's notion of asymptotic simplicity. As a result of this work it is now clear that Penrose's original insights hold for large classes of spacetimes and not only for special explicitly known solutions.

This book develops the theory of the conformal Einstein field equations from the ground up and discusses their applications to the study of asymptotically simple spacetimes. Special attention is paid to results concerning the existence and stability of *de Sitter-like spacetimes*, the semiglobal existence and stability of *Minkowski-like spacetimes* using hyperboloidal Cauchy problems and the

[1] See Weyl (1968).
[2] See Penrose (1963, 1964).
[3] See e.g. Hawking and Ellis (1973); Geroch (1976).
[4] See Friedrich (1981a,b, 1983).

construction of *anti-de Sitter-like spacetimes* from initial boundary value problems. These results belong to the canon of modern mathematical relativity. In addition to their mathematical interest, they are of great physical relevance as they express, among other things, the internal consistency of general relativity and provide an approach for the global evaluation of spacetimes by means of numerical methods.

Why a book on the subject? The applications of conformal methods in general relativity constitute a mature subject with a number of *core results* which will withstand the pass of time. Still, it provides a number of challenging open questions whose resolution will strengthen its connections with other research strands in general relativity. This book aims at making the subject accessible to physicists and mathematicians alike who want to make use of conformal methods to analyse the global structure and properties of spacetimes. Hopefully, this book will provide an alternative to the use of original references while learning the subject or doing research.

Anyone who wants to engage with the subject of this book faces a number of challenges. To begin with, one has a vast literature spreading over more than 50 years. As it is to be expected from a living subject, the perspectives change through time, the importance of certain problems rise and wane and it is sometimes hard to differentiate the fundamental from the subsidiary. The combination of results from various references is often hindered by changing notation and conflicting conventions. Moreover, to appreciate and understand the results of the theory one requires a considerable amount of background material: conformal geometry, spinors, PDE theory, causal theory, etc. These methods are an essential part of the toolkit of a modern mathematical relativist. This book endeavours to bring together in a single volume all the required background material in a concise and coherent manner.

As a cautionary note, it should be mentioned what this book is not intended to be. This book is not an introductory book to general relativity. A certain familiarity with the subject is assumed from the outset – ideally at the level of Part I of R. Wald's book *General Relativity*.[5] This is also not a book on the applications of the theory of PDEs in general relativity. For this, there are other books available.[6] Also, although the Cauchy problem in general relativity is a leading theme, this book should not be viewed as a monograph on the topic – for this, I refer the interested reader to H. Ringström's monograph.[7]

I have endeavoured to write a book which not only serves as an *introduction to the subject* but also is a *tool for research.* With this idea in mind, I have striven to provide as much detail as possible of the arguments and calculations. However, at some stages supplying further details is neither possible nor desirable. Indeed, quoting the preface of J. L. Synge's classical book on general relativity: "There

[5] See Wald (1984).
[6] See e.g. Choquet-Bruhat (2008); Rendall (2008).
[7] See Ringström (2009).

are heavy calculations in the book, but there are places where the reader will find me sitting on the fence, whistling, instead of rushing into the fray"; see Synge (1960).[8] In an attempt to keep the readability and the length of the text under control, I have not endeavoured to provide completely general or optimal theorems – the attentive reader will realise this and is referred to the literature for further details, if required. As a picture is better than a thousand words, I have complemented the text with a considerable number of figures and diagrams which, I hope, will help to explain the content of the main text and provide useful visual models.

In writing this book, I have assumed the reader to have a certain mathematical maturity. Some basic knowledge of topology is needed – Appendix A in Wald's book contains the required background – as well as familiarity with basic tensorial calculus. I have, however, not assumed any prior knowledge of 2-spinors. The necessary toolkit is developed in the course of two chapters. Readers looking for a supplementary source on the topic are referred to J. Stewart's book.[9] The applications of conformal methods discussed in this book require certain knowledge of the theory of PDEs. I provide all the required material in a chapter of its own – nevertheless, some previous exposure to the basic ideas of the theory of PDEs is an advantage. Some arguments in the book make use of very concrete results of analysis. In these cases, I have included the necessary ideas in appendices to the various chapters.

[8] I am thankful to R. Beig for bringing my attention to this quote.

[9] See Stewart (1991).

Acknowledgements

First and foremost, I would like to acknowledge the influence that Helmut Friedrich had in the development of my understanding of the theory of the conformal Einstein field equations. He has been an inexhaustible source of knowledge and insights. I also would like to acknowledge the role that Robert Beig, Sergio Dain and Malcolm MacCallum have played in shaping complementary perspectives on the subject. I am also grateful for the extensive discussions with my collaborators Christian Lübbe and Jose Luis Jaramillo which have allowed to form, define and refine some of the points of view and concerns presented in this book.

I want to thank my students (Diego Carranza, Michael Cole, Edgar Gasperín, Adem Hursit and Jarrod Williams), collaborators and colleagues (Artur Alho, Robert Beig, Alberto Carrasco, Daniela Pugliese, Alfonso García-Parrado, David Hilditch, Jose Luis Jaramillo, Malcolm MacCallum, Belgin Seymenoglu and Claire Sommé) who have carefully read various parts of this book, thus providing valuable feedback and a seemingly never-ending list of typos. Of course, I am fully responsible for any mistakes or omissions left in the text. Finally, I would also like to thank Jean-Phillipe Nicolas for useful conversations on some of the topics covered in this book.

Although all the calculations presented in this book have been carried out by hand, the system `xAct` for `Mathematica` has been a valuable resource in case of trouble. I am thankful to Thomas Bäckdahl and Alfonso García-Parrado for their help at various stages with this system. I also thank Peter Hübner and Anil Zenginoglu for allowing me to use figures and diagrams from their publications.

I would like to thank my home department, the School of Mathematical Sciences of Queen Mary University of London, for the support granted during the course of the project and for the sabbatical term in spring 2014 when a considerable portion of this text took form. The primal notes which eventually turned into this book stem from a minicourse I held at the Centre of Mathematics of the University of Minho in Portugal in April 2011; further notes come from a minicourse held at the Institute of Mathematics and System Sciences of the Chinese Academy of Sciences in June 2013. I thank these institutions and particularly my hosts (Filipe Mena and Xiaoning Wu, respectively) for the invitation and hospitality. I also thank the Gravitational Physics Group of the

Faculty of Physics of the University of Vienna for their hospitality at various times during the last years.

Finally, I would like to express my deepest gratitude to my wife, Christiane Maria Losert-Valiente Kroon, for her constant support during the course of this project and for a careful and critical reading of the whole manuscript and for spotting an incredible amount of typos and mismatching indices in the equations. The conclusion of this book would have taken much longer without her help! Finally I would like to thank my father, Antonio Valiente, for instilling in me the desire of writing a book and my in-laws, Linde and Peter Losert, for the interest they have taken in this project.

Juan A. Valiente Kroon

Symbols

d^*_{ij} alternative description of the components of the magnetic part of the rescaled Weyl tensor, page 264

$(e_a, \Gamma_a{}^b{}_c, \Xi, s, L_{ab}, d^a{}_{bcd}, T_{ab})$ unknowns in the frame version of the standard conformal field equations, page 196

$(e_a, \hat{\Gamma}_a{}^b{}_c, \hat{L}_{ab}, d^a{}_{bcd}, T_{ab}, \Xi, d_a)$ unknowns in the frame version of the extended conformal field equations, page 205

$(e_{AA'}, \Gamma_{AA'BC}, \Xi, s, L_{AA'BB'}, \phi_{ABCD}, T_{AA'BB'})$ unknowns in the spinorial version of the standard conformal field equations, page 199

$(e_{AA'}, \hat{\Gamma}_{AA'BC}, \hat{L}_{AA'BB'}, \phi_{ABCD}, T_{AA'BB'}, \Xi, d_{AA'})$ unknowns in the spinorial version of the extended conformal field equations, page 208

$(\boldsymbol{o}, \boldsymbol{\iota})$ spin basis in index-free notation, page 66

$(\hat{\Sigma}_{AA'BB'}, \hat{\Xi}^C{}_{DAA'BB'}, \hat{\Delta}_{CC'DD'BB'}, \hat{\Lambda}_{BB'CD}, \delta_{AA'}, \varsigma_{AA'BB'}, \gamma_{AA'BB'})$ zero quantities in the spinorial extended conformal field equations, page 208

$(\hat{\Sigma}_{ab}, \hat{\Xi}^c{}_{dab}, \hat{\Delta}_{cdb}, \Lambda_{bcd}, \delta_a, \gamma_{ab}, \varsigma_{ab})$ zero quantities in the frame extended conformal field equations, page 205

$(\mathcal{M}, \boldsymbol{g})$ generic spacetime, page 45

$(\mathcal{U}, \varphi)$ coordinate chart, page 28

$(\Sigma_{AA'BB'}, \Xi^C{}_{DAA'BB'}, Z_{AA'BB'}, Z_{AA'}, \Delta_{CDBB'}, \Lambda_{BB'CD}, Z, M_{AA'})$ zero quantities in the spinorial version of the standard conformal equations, page 199

$(\Sigma_{ab}, \Xi^c{}_{dab}, Z_{ab}, Z_a, \Delta_{cdb}, \Lambda_{bcd}, Z, M_a)$ zero quantities in the frame version of the standard conformal field equations, page 196

$(g_{ab}, \Xi, s, L_{ab}, d^a{}_{bcd}, T_{ab})$ unknowns in the metric standard conformal field equations, page 191

$(h_{ij}, s_{ij}, \zeta, \varsigma)$ unknowns in the conformal static equations, page 511

(o^A, ι^A) spin basis in abstract index notation, page 71

$(u, r, \theta^{\mathcal{A}})$ Bondi coordinates, page 236

$(x(\tau), \boldsymbol{\beta}(\tau))$ conformal geodesic with parameter τ, page 127

(x^μ) local coordinates in a four-dimensional manifold, page 28

$[\nabla_a, \nabla_b]$ commutator of covariant derivatives, page 39

$[[\boldsymbol{\xi}, \boldsymbol{\eta}]]$ antisymmetric product of $\boldsymbol{\xi}$, $\boldsymbol{\eta} \in \mathfrak{S}$, page 65

$[\boldsymbol{g}]$ conformal class of the metric $\boldsymbol{g}$, page 113

$[\boldsymbol{u}, \boldsymbol{v}]$ commutator of the vector fields $\boldsymbol{u}$ and $\boldsymbol{v}$, page 34

$\alpha_{\boldsymbol{a}}$, $\beta_{\boldsymbol{a}}$, $\omega_{\boldsymbol{a}}, \ldots$ components of the covectors $\boldsymbol{\alpha}$, $\boldsymbol{\beta}$, $\boldsymbol{\omega}, \ldots$ with respect to the frame $\{\boldsymbol{e_a}\}$, page 51

α_a, β_a, $\omega_a, \ldots$ generic covectors in abstract index notation

$\approx$ diffeomorphism between sets, page 27

$\bar{\xi}^{A'}$, $\bar{\eta}_{A'}, \ldots$ complex conjugates of the spinors ξ^A, $\eta_B, \ldots$, page 72

β^2 norm of the covector $\tilde{\boldsymbol{\beta}}$, page 134

β_a covector associated to a conformal geodesic in abstract index notation, page 127

$\boldsymbol{\alpha}$, $\boldsymbol{\beta}$, $\boldsymbol{\omega}, \ldots$ generic covectors in index-free notation

$\boldsymbol{\beta}$ covector associated to a conformal geodesic in index-free notation, page 127

$\boldsymbol{\chi}$ Weingarten map, page 56

$\boldsymbol{D}$ generic three-dimensional connection in index-free notation

$\boldsymbol{d}$ rescaled conformal geodesic covector, page 201

$\boldsymbol{\delta}$ Euclidean metric on $\mathbb{R}^3$, page 143

$\boldsymbol{e}$, $\boldsymbol{e_{AB}}$ space spinor irreducible components of the frame vector $\boldsymbol{e_{AA'}}$, page 104

$\boldsymbol{\ell}$ three-dimensional Lorentzian metric on the conformal boundary of an anti-de Sitter-like spacetime, page 456

$\boldsymbol{f}$ covector defining a Weyl connection in index-free notation, page 119

$\boldsymbol{f}$ unphysical conformal geodesic covector, page 201

$\boldsymbol{g}$ generic four-dimensional Lorentzian metric tensor in index-free notation

$\boldsymbol{g}^\sharp$ generic contravariant four-dimensional Lorentzian metric tensor in index-free notation

$\boldsymbol{g}_{\mathscr{E}}$ standard metric on the Einstein cylinder, page 144

$\boldsymbol{\gamma}$ metric in the quotient manifold, page 141

$\boldsymbol{h}$ generic (negative definite) Riemannian three-dimensional metric

$\boldsymbol{\hbar}$ standard metric on the unit 3-sphere, page 142

$\boldsymbol{K}$ extrinsic curvature tensor of a hypersurface in index-free notation, page 61

$\boldsymbol{k}$ intrinsic metric of compact two-dimensional surfaces

$\boldsymbol{M}$, $\boldsymbol{N}, \ldots$ generic higher rank tensors in index-free notation

$\boldsymbol{N}$ tangent to the generators of null infinity

$\boldsymbol{\nabla}$, $\bar{\boldsymbol{\nabla}}$ generic linear connections in index-free notation, page 38

$\boldsymbol{\nu}$ unit normal to a hypersurface $\mathcal{S}$, page 54

$\boldsymbol{\omega}$, $\boldsymbol{\omega^{AB}}$ space spinor irreducible components of the frame covector $\boldsymbol{\omega^{AA'}}$, page 104

$\boldsymbol{\partial}_\mu$ coordinate basis vector

$\boldsymbol{Q}$ transition tensor between connections in index-free notation, page 42

$\boldsymbol{q}$ intrinsic metric of null infinity

$\boldsymbol{\Sigma}$ torsion tensor of a connection $\boldsymbol{\nabla}$ in index-free notation, page 39

$\boldsymbol{\sigma}$ standard metric on the unit 2-sphere

$\sigma_{\mathbf{L}}(\boldsymbol{\xi})$ symbol of a differential operator $\mathbf{L}$, page 252

$\boldsymbol{t}$ vector field generating a timelike congruence

$\boldsymbol{\tau}$ vector counterpart of the spinor $\tau_{AA'}$, page 102

$\boldsymbol{v}$, $\boldsymbol{u}$, $\boldsymbol{w}, \ldots$ generic vectors in index-free notation

ς shear tensor, page 226
$\boldsymbol{z}$, $\boldsymbol{\zeta}$ deviation vector and covector, respectively, page 135
$\boldsymbol{\mathcal{L}_h}$ conformal Killing operator of the metric $\boldsymbol{h}$, page 257
$\breve{\mathbf{u}}$ perturbation quantity in an evolution system
$\chi_{(AB)CD}$ spinorial counterpart of the Weingarten tensor
χ_{AB} spinorial counterpart of the acceleration vector
$\circ$ composition of functions, page 36
$\coprod$ disjoint union of sets
$\delta(i)$ Dirac's delta, page 279
$\Delta_{\boldsymbol{h}}$ Laplacian operator of the Riemannian metric $\boldsymbol{h}$
$\delta_\mu{}^\nu$, $\delta_a{}^b$, $\delta_i{}^j$, $\delta_\alpha{}^\beta$, $\delta_A{}^B$, $\delta_{\boldsymbol{A}}{}^{\boldsymbol{B}}$, $\delta_{\boldsymbol{a}}{}^{\boldsymbol{b}}$, $\delta_{\boldsymbol{i}}{}^{\boldsymbol{j}}$ Kronecker's delta
$\delta_{\alpha\beta}$ components of the three-dimensional Euclidean metric in Cartesian coordinates, page 47
$\delta_{\boldsymbol{AB}}$ Sen connection on a timelike conformal boundary, page 471
$\delta_{\boldsymbol{ij}}$ components of a three-dimensional Riemannian metric with respect to an orthonormal basis, page 45
$\dot{\gamma}(\mathrm{s})$ tangent vector to a curve, page 30
$\dot{\boldsymbol{x}}(\mathrm{s})$ alternative notation for the tangent vector to a curve, page 30
$\epsilon = \pm 1$ encodes the causal character of a hypersurface, page 54
$\epsilon_{\boldsymbol{abcd}}$ components of the volume form with respect to an orthonormal basis
$\epsilon_{\boldsymbol{AB}}$, $\epsilon^{\boldsymbol{AB}}$ components of the spinors ϵ_{AB}, ϵ^{AB} with respect to a spin basis, page 71
$\epsilon_{A'B'}$, $\epsilon^{A'B'}$ complex conjugates of the spinors ϵ_{AB}, ϵ^{AB}
$\epsilon_{AA'BB'CC'DD'}$ spinorial counterpart of the volume form, page 78
ϵ_{ABCDEF} spinorial counterpart of the three-dimensional volume form, page 105
ϵ_{abcd} volume form of a metric g_{ab}, page 49
ϵ_{AB}, ϵ^{AB} antisymmetric spinors, page 67
$\equiv$ definition
$\eta_{\boldsymbol{ABCD}}$ components of the electric part of the Weyl spinor, page 373
$\eta_{\boldsymbol{ab}}$ components of a four-dimensional Lorentzian metric with respect to an orthonormal basis, page 45
$\eta_{\mu\nu}$ components of the Minkowski metric tensor in Cartesian coordinates, page 47
ð, $\bar{\text{ð}}$ eth and eth-bar operators, page 241
exp exponential map, page 275
Γ geodesic distance, page 276
$\gamma(\mathrm{s})$ curve in a manifold with parameter s, page 30
$\Gamma_{\boldsymbol{a}}{}^{\boldsymbol{c}}{}_{\boldsymbol{b}}$ connection coefficients of $\boldsymbol{\nabla}$ with respect to $\{\boldsymbol{e_a}\}$
$\gamma_{\boldsymbol{i}}{}^{\boldsymbol{j}}{}_{\boldsymbol{k}}$ connection coefficients of the three-dimensional connection $\boldsymbol{D}$ with respect to the frame $\{\boldsymbol{e_i}\}$, page 59
$\Gamma_\mu{}^\nu{}_\lambda$ Christoffel symbols of the metric $\boldsymbol{g}$ in the coordinates (x^μ)
$\Gamma_{\boldsymbol{A'A'}}{}^{\boldsymbol{BB'}}{}_{\boldsymbol{CC'}}$ spinorial counterpart of the connection coefficients $\Gamma_{\boldsymbol{a}}{}^{\boldsymbol{b}}{}_{\boldsymbol{c}}$, page 82
$\Gamma_{\boldsymbol{AA'}}{}^{\boldsymbol{B}}{}_{\boldsymbol{C}}$ reduced spin connection coefficients, page 82

Γ_{ABCD} space spinor counterpart of the reduced spin connection coefficients $\Gamma_{AA'CD}$, page 107
$\gamma_{AB}{}^{C}{}_{D}$ reduced spatial spin connection coefficients, page 109
$\gamma_{AB}{}^{CD}{}_{EF}$ spinorial counterpart of the three-dimensional connection coefficients $\gamma_i{}^j{}_k$, page 109
$\grave{\Omega}$, $\check{\Omega}$ massless and, respectively, massive part of the conformal factor associated to Euclidean initial data sets, page 529
$\hat{\boldsymbol{\nabla}}$ generic Weyl connection in index-free notation, page 119
$\hat{\Gamma}_a{}^b{}_c$ connection coefficients of a Weyl connection $\hat{\boldsymbol{\nabla}}$, page 119
$\hat{\Gamma}_{AA'}{}^{B}{}_{C}$ reduced Weyl connection spin coefficients, page 206
$\hat{\nabla}_a$ generic Weyl connection in abstract index notation, page 119
$\hat{\rho}^c{}_{dab}$ Weyl connection algebraic curvature, page 203
$\hat{\rho}_{ABCC'DD'}$ Weyl connection reduced spinorial algebraic curvature, page 207
$\hat{P}^c{}_{dab}$ Weyl connection geometric curvature, page 203
$\hat{P}_{ABCC'DD'}$ Weyl connection reduced spinorial geometric curvature, page 207
κ conformal factor associated to the construction of the cylinder at spatial infinity, page 541
Λ Newman-Penrose Ricci scalar, page 87
λ cosmological constant, page 2
$\Lambda_{(ABCD)}$, Λ_{AB} irreducible components of the spinorial Bianchi equation, page 351
$\langle \boldsymbol{\omega}, \boldsymbol{v} \rangle$ action of the covector $\boldsymbol{\omega}$ on the vector $\boldsymbol{v}$
$\langle \boldsymbol{t} \rangle^{\perp}|_p$ subspace orthogonal to $\boldsymbol{t}$, page 55
$\langle \boldsymbol{t} \rangle$ one-dimensional subspace spanned by $\boldsymbol{t}$, page 55
$\langle\langle \boldsymbol{\xi}, \boldsymbol{\eta} \rangle\rangle$ Hermitian product of $\boldsymbol{\xi}, \boldsymbol{\eta} \in \mathfrak{S}$, page 94
$[\![\nabla_a, \nabla_b]\!]$ modified commutator of covariant derivatives, page 40
$\mathbb{H}^n$ n-dimensional half Euclidean space, page 29
$\mathbb{R}^+$ non-negative real numbers
$\mathbb{R}^2$ Euclidean plane
$\mathbb{R}^n$ n-dimensional Euclidean space
$\mathbb{S}^2$ 2-sphere
$\mathbb{S}^3$ three-dimensional unit sphere, page 142
$\mathbf{A}^*$ transpose of the complex conjugate of the matrix $\mathbf{A}$
$\mathbf{A}^3$ normal matrix in an initial boundary value problem, page 314
$\mathbf{A}^\mu$ symmetric matrices in a symmetric hyperbolic system, page 294
$\mathbf{d}$ exterior derivative (differential), page 31
$\mathbf{d}x^\mu$ coordinate basis covector
$\mathbf{L}$ generic differential operator
$\mathbf{L}^*$ formal adjoint of the differential operator $\mathbf{L}$
$\mathbf{L}_h$ Yamabe operator, page 256
$\mathbf{T}$ map associated to the prescription of boundary conditions in an initial boundary value problem, page 314
$\mathbf{u}, \mathbf{v}, \mathbf{w}, \ldots$ $\mathbb{C}^N$-valued functions

$\mathcal{B}_a(p)$ ball of radius $a > 0$ centred at the point p
$\mathcal{C}_p$ null cone at a point $p \in \mathcal{M}$, page 45
$\mathcal{C}_p^+$, $\mathcal{C}_p^-$ future and, respectively, past null cone at a point $p \in \mathcal{M}$, page 45
$\mathcal{D}$ a generic derivation, page 30
$\mathcal{D}_{AB}$ Sen connection of $\nabla_{AA'}$ induced by $\tau_{AA'}$, page 105
$\mathcal{E}$ corner in an initial boundary value problem, page 314
$\mathcal{G}$ generic lens-shaped domain, page 301
$\mathcal{H}_k$ standard hyperboloids, page 154
$\mathcal{I}$ cylinder at spatial infinity, page 542
$\mathcal{I}^0$ intersection of the cylinder at spatial infinity with a Cauchy initial hypersurface, page 542
$\mathcal{I}^\pm$ critical sets where null infinity touches spatial infinity, page 542
$\mathcal{M}$, $\mathcal{N}$ generic (unphysical) spacetime manifolds
$\mathcal{N}$, $\mathcal{N}'$ initial null hypersurfaces in a characteristic problem, page 320
$\mathcal{N}_i$ complex null cone at i, page 522
$\mathcal{N}_{\mathbb{C}}(i)$ complexification of the null cone through i, page 532
$\mathcal{P}$ covariant derivative in the direction of $\tau_{AA'}$, page 105
$\mathcal{Q}$ generic quotient manifold, page 141
$\mathcal{R}$ generic subset of a hypersurface $\mathcal{S}$
$\mathcal{S}$ generic hypersurface on a manifold $\mathcal{M}$
$\mathcal{T}$ timelike boundary, page 314
$\mathcal{U}$, $\mathcal{V}$ generic open subsets of a manifold or $\mathbb{R}^n$
$\mathcal{U}_{\mathbb{C}}$ complexification of a neighbourhood $\mathcal{U}$ of the point at infinity, page 532
$\mathcal{Z}$ intersection of initial null hypersurfaces in a characteristic problem, page 320
$\mathfrak{S}$ complex vector space, page 65
$\mathfrak{S}(\mathcal{M})$ spin structure (spin bundle) over $\mathcal{M}$, page 81
$\mathfrak{S}(\mathcal{S})$ space spinor structure over a three-dimensional manifold $\mathcal{S}$, page 101
$\mathfrak{S}^*$ dual of the complex vector space $\mathfrak{S}$, page 65
$\mathfrak{S}^\bullet(\mathcal{M})$, $\mathfrak{S}_A(\mathcal{M})$, $\mathfrak{S}^A(\mathcal{M})$, $\mathfrak{S}_{AA'}{}^B(\mathcal{M}), \ldots$ various spin bundles over $\mathcal{M}$
$\mathfrak{S}^\bullet$ spin algebra, page 66
$\mathfrak{S}^A$, $\mathfrak{S}_A$, ... alternative notation for the vector spaces $\mathfrak{S}$, $\mathfrak{S}^*, \ldots$, page 66
$\mathfrak{S}^{A'}$, $\mathfrak{S}_{A'B'}$, ... complex conjugates of the spaces $\mathfrak{S}^A$, $\mathfrak{S}_{AB}, \ldots$, page 72
$\mathfrak{T}^\bullet(\mathcal{M})$ tensor bundle over $\mathcal{M}$, page 34
$\mathfrak{T}^a(\mathcal{M})$ alternative notation for the tangent bundle over $\mathcal{M}$, page 36
$\mathfrak{T}^{a_1 \cdots a_k}{}_{b_1 \cdots b_l}(\mathcal{M})$ alternative notation for the tensor bundle over $\mathcal{M}$, page 36
$\mathfrak{T}_a(\mathcal{M})$ alternative notation for the cotangent bundle over $\mathcal{M}$, page 36
$\mathfrak{X}(\mathcal{M})$ set of of scalar fields over $\mathcal{M}$, page 30
$\mathring{\mathbf{u}}$ background quantity in an evolution system
$\mathscr{C}$ generic cut of null infinity
$\mathscr{C}_\star$ fiduciary cut of null infinity
$\mathscr{E}$ extension operator of functions between Sobolev spaces, page 308
$\mathscr{I}$ part of the conformal boundary that is a hypersurface, page 178
$\mathscr{I}^\pm$ future and, respectively, past null infinity

$\mathscr{N}_i^+$, $\mathscr{N}_i^-$ null cones generated by the null geodesics through i, page 531
$\mathscr{N}_u$ outgoing null hypersurface associated to the retarded time u
$\mathscr{R}_{\boldsymbol{h}}$ linearised Ricci operator, page 289
$\mathscr{Z}$ generic intersection of null infinity with a null hypersurface
$\operatorname{int}\mathcal{A}$ topological interior of the set $\mathcal{A}$, page 397
i square root of -1
$\mu_{\boldsymbol{ABCD}}$ components of the magnetic part of the Weyl spinor, page 373
$\nabla_{\boldsymbol{a}}$ covariant directional derivative in the direction of $\boldsymbol{e_a}$, page 51
$\nabla_{\boldsymbol{u}}\boldsymbol{v}$ covariant derivative of $\boldsymbol{v}$ with respect to $\boldsymbol{u}$, page 38
∇_a, $\bar{\nabla}_a$ generic linear connections in abstract index notation, page 38
$\nabla_{\boldsymbol{AA'}}$ directional spinorial covariant derivative, page 82
$\nabla_{AA'}$, $\tilde{\nabla}_{AA'}, \ldots$ spinor covariant derivatives, page 81
∇_{AB} space spinor counterpart of $\nabla_{AA'}$, page 105
Ω generic three-dimensional conformal factor
$\oplus$ direct sum
$\otimes$ tensor product between tensors or tensor spaces
$\overline{\mathcal{A}}$ topological closure of the set $\mathcal{A}$, page 394
$\| \mathbf{u} \|_{\mathcal{S},m}$ Sobolev norm of order m of a function over $\mathcal{S}$, page 306
$\partial\mathbb{H}^n$ boundary of the n-dimensional half Euclidean space, page 29
$\partial\mathcal{M}$ boundary of $\mathcal{M}$
ϕ unphysical conformally coupled scalar field, page 216
ϕ_0 radiation field in the asymptotic characteristic problem on a cone, page 500
$\Phi_{ABA'B'}$ spinorial counterpart of the trace-free Ricci tensor, page 89
Φ_{ab} trace-free Ricci tensor of a connection ∇_a in abstract index notation, page 48
ϕ_{AB} unphysical Maxwell spinor, page 215
Π generic distribution, page 55
$\Pi\,|_p$ hyperplane induced by a distribution at a point $p \in \mathcal{M}$, page 55
$\pounds_{\boldsymbol{v}}$ Lie derivative in the direction of $\boldsymbol{v}$, page 37
Ψ_{ABCD} Weyl spinor, page 87
ρ boundary-defining function, page 285
ρ polar radial coordinate, page 514
ρ^α three-dimensional unit position vector, page 514
$\rho^{\boldsymbol{C}}{}_{\boldsymbol{DAA'BB'}}$ reduced spinorial algebraic curvature, page 198
$\rho^{\boldsymbol{c}}{}_{\boldsymbol{dab}}$ components of the algebraic curvature, page 195
$\rho^{AA'}$ spatial spinor used to introduce a $1+1+2$ spinor formalism, page 464
Ric, ***Ric***$[\boldsymbol{g}]$ Ricci tensor of a connection $\boldsymbol{\nabla}$ in index-free notation, page 48
Riem Riemann curvature tensor of a connection $\boldsymbol{\nabla}$ in index-free notation, page 40
Schouten, ***Schouten***$[\boldsymbol{g}]$ Schouten tensor of a connection $\boldsymbol{\nabla}$ in index-free notation, page 48
σ Newman-Penrose spin connection coefficient corresponding to $\Gamma_{\boldsymbol{01'00}}$
$\sigma^{\boldsymbol{a}}{}_{\boldsymbol{AA'}}$, $\sigma_{\boldsymbol{a}}{}^{\boldsymbol{AA'}}$ spacetime Infeld-van der Waerden symbols, page 74

$\Sigma_{\boldsymbol{a}}{}^{\boldsymbol{c}}{}_{\boldsymbol{b}}$ components of the torsion tensor with respect to an orthonormal frame, page 53

$\sigma_{\boldsymbol{i}}{}^{\boldsymbol{k}}{}_{\boldsymbol{j}}$, $\Pi^{\boldsymbol{k}}{}_{\boldsymbol{lij}}$, $\pi_{\boldsymbol{klij}}$ components of the three-dimensional torsion, geometric and algebraic curvatures, page 264

$\sigma_{\boldsymbol{i}}{}^{\boldsymbol{AB}}$, $\sigma^{\boldsymbol{i}}{}_{\boldsymbol{AB}}$ spatial Infeld-van der Waerden symbols, page 99

$\Sigma_a{}^c{}_b$ torsion tensor of a connection ∇_a in abstract index notation, page 39

$\simeq$ equality at the conformal boundary

$\square$ D'Alembertian operator, page 89

$\square_{AB}$ box commutator, page 89

$\overset{\star}{\simeq}$ equality at a fiduciary cut of null infinity

$\tau_{AA'}$ privileged timelike spinor inducing a space spinor formalism, page 102

Θ conformal factor associated to a conformal geodesic, page 132

$\theta = (\theta^{\mathcal{A}})$ local coordinates on $\mathbb{S}^2$

$\Theta_{\boldsymbol{ABCD}}$ space spinor counterpart of the components of the Schouten tensor of a Weyl connection, page 373

$\tilde{\boldsymbol{\eta}}$ Minkowski metric

$\tilde{\boldsymbol{g}}_{\mathscr{E}}$ metric of the anti-de Sitter spacetime, page 159

$\tilde{\boldsymbol{g}}_{\mathscr{S}}$ metric of the Schwarzschild spacetime, page 163

$\tilde{\boldsymbol{g}}_{dS}$ metric of the de Sitter spacetime, page 155

$\tilde{\mathcal{E}}_k$ asymptotic ends of asymptotically Euclidean manifold $\tilde{\mathcal{S}}$, page 272

$\tilde{\mathcal{F}}_{ab}$ self-dual Faraday tensor, page 213

$\tilde{\mathcal{M}}$ generic (physical) spacetime manifold

$\tilde{\phi}$ physical conformally coupled scalar field, page 216

$\tilde{\phi}_{AB}$ physical Maxwell spinor, page 215

$\tilde{\varrho}$ density of a perfect fluid, page 219

$\tilde{\varrho}$ energy density, page 254

$\tilde{F}_{ab}$ physical Faraday tensor, page 213

$\tilde{j}_k$ energy flux vector, page 254

$\tilde{p}$ pressure of a perfect fluid, page 219

$\tilde{T}_{ab}$ physical energy-momentum tensor

$\tilde{u}^a$ physical 4-velocity of a perfect fluid, page 219

$\underline{x}$ spatial coordinates (x^1, x^2, x^3)

Υ_a logarithmic gradient of a conformal factor, page 116

$\Upsilon_{AA'}$ spinorial counterpart of the logarithmic gradient of a conformal factor, page 123

φ^* pull-back, page 36

φ_* push-forward, page 36

$\varpi_{\boldsymbol{AA'}}$ components of $\varpi_{AA'}$ with respect to a spin basis, page 95

$\varpi_{AA'}$ Hermitian spinor assocated to a Hermitian inner product, page 95

ϱ conformally rescaled density of a perfect fluid, page 220

ϱ unphysical energy density, page 255

$\boldsymbol{Weyl}$, $\boldsymbol{Weyl}[\boldsymbol{g}]$ Weyl tensor of a connection $\boldsymbol{\nabla}$ in index-free notation, page 48

$\xi^{\boldsymbol{A}}$, $\eta_{\boldsymbol{A}}, \ldots$ components of the spinors ξ^A, $\eta_A, \ldots$ with respect to a spin basis

ξ^A, $\eta_A, \ldots$ generic spinors in abstract-index notation
ξ_{ABCC}, χ_{ABCD} real and imaginary parts of Γ_{ABCD}, page 107
Ξ_{ij}, S_i, S_{ij}, H_{kij} zero quantities associated to the conformal static field equations, page 511
$\zeta_0, \ldots \zeta_4$ components of the spin-2 zero-rest mass field ζ_{ABCD}, page 551
ζ_{ABCD} spin-2 zero-rest mass field, page 551
$\{c_i\}$ global orthonormal frame on $\mathbb{S}^3$, page 142
$\{e_a\}$ vector basis in index-free notation, page 31
$\{\omega^a\}$ covector basis in index-free notation, page 31
$\{\mathcal{S}_t\}_{t\in\mathbb{R}}$ foliation of $\mathcal{M}$, page 54
$\{e_i\}$ three-dimensional vector basis in index-free notation, page 59
$\{e_{AA'}\}$ alternative index-free notation for the Newman-Penrose null tetrad, page 79
$\{e_{AB}\}$, $\{\omega^{AB}\}$ three-dimensional basis and cobasis with spin frame indices, page 109
$\{l, n, m, \bar{m}\}$ Newman-Penrose null tetrad in index-free notation, page 77
$\{\omega^i\}$ three-dimensional covector basis in index-free notation, page 59
$\{\omega^{AA'}\}$ soldering form, page 79
$\{\epsilon_A{}^A\}$, $\{\epsilon^A{}_A\}$ alternative abstract index notation for a spin basis and its dual, page 71
$\{\omega^a{}_a\}$ covector basis in abstract index notation, page 36
$\{\omega^i{}_i\}$ three-dimensional covector basis in index-free notation, page 59
$\{e_a{}^a\}$ vector basis in abstract index notation, page 36
$\{e_i{}^i\}$ three-dimensional vector basis in abstract index notation, page 59
$\{l^a, n^a, m^a, \bar{m}^a\}$ Newman-Penrose null tetrad in abstract index notation, page 77
$\{m, m_\alpha, m_{\alpha_1\alpha_2}, \ldots\}$ sequence of multipole moments of a static spacetime, page 519
b_{ABCD} Cotton spinor, page 512
C^*_p characteristic set of a symmetric hyperbolic system at the point p, page 297
C^∞ class of infinitely differentiable (smooth) functions
$C^\infty(\mathbb{R}^3, \mathbb{C}^N)$ space of smooth functions from $\mathbb{R}^3$ to $\mathbb{C}^N$, page 306
$C^c{}_{dab}$ Weyl tensor of a connection ∇_a in abstract index notation, page 48
C^k class of k-times differentiable functions
$C^k(\mathbb{R}^3, \mathbb{C}^N)$ set of C^k functions from $\mathbb{R}^3$ to $\mathbb{C}^N$, page 307
$C^k\big([0,T]; H^m(\mathbb{R}^3, \mathbb{C}^N)\big)$ set of C^k functions from $[0,T]$ to $H^m(\mathbb{R}^3, \mathbb{C}^N)$, page 307
D bounded open subset of $H^m(\mathbb{R}^3, \mathbb{C}^N)$ such that for $\mathbf{w} \in D$ the matrix $\mathbf{A}^0(0, \underline{x}, \mathbf{w})$ is positive definite bounded away from zero by $\delta > 0$ for all $p \in \mathbb{R}^3$, page 309
$D(\mathcal{R})$ domain of dependence of $\mathcal{R}$, page 304
D, Δ, δ, $\bar{\delta}$ Newman-Penrose directional covariant derivatives, page 92
$D^\pm(\mathcal{A})$, $D(\mathcal{A})$ future/past and total domain of dependence of a set $\mathcal{A}$, page 392

$d^a{}_{bcd}$ rescaled Weyl tensor, page 188
$d_{\boldsymbol{a}}$ components of the rescaled physical conformal geodesics covector, page 203
$D_{\boldsymbol{i}}$ three-dimensional directional covariant derivative in the direction of $\boldsymbol{e_i}$, page 59
D_i generic three-dimensional connection in abstract index notation
$D_{\boldsymbol{AB}}$ three-dimensional covariant directional derivative, page 109
$d_{\boldsymbol{ij}}$, $d_{\boldsymbol{ijk}}$ components of the electric and magnetic parts of the rescaled Weyl tensor, page 261
D_{AB} spinorial counterpart of a three-dimensional Levi-Civita connection $\boldsymbol{D}$, page 106
$F^{\boldsymbol{a}}(x)$, $F^{\mu}(x)$ coordinate gauge source functions, page 339
$f_{\boldsymbol{a}}$ components of the unphysical conformal geodesics covector, page 203
f_a covector defining a Weyl connection in abstract index notation, page 119
$F_{\boldsymbol{AB}}(x)$ frame gauge source functions, page 345
F_{ab} unphysical Faraday tensor, page 214
g^{ab} generic contravariant four-dimensional Lorentzian metric tensor in abstract index notation
G_{ab} Einstein tensor of a metric g_{ab}
g_{ab} generic four-dimensional Lorentzian metric tensor in abstract index notation
$H^{\pm}(\mathcal{A})$, $H(\mathcal{A})$ future/past and total Cauchy horizons of the set $\mathcal{A}$, page 394
$H^m(\mathbb{R}^3, \mathbb{C}^N)$ Sobolev space of order m of functions from $\mathbb{R}^3$ to $\mathbb{C}^N$, page 307
$h_a{}^b$ projector associated to a distribution Π, page 55
$h_{\boldsymbol{ABCD}}$ components of h_{ABCD} with respect to a spin frame $\{\epsilon_{\boldsymbol{A}}{}^A\}$, page 99
$h_{AA'}{}^{BB'}$ spinorial counterpart of the projector $h_{\boldsymbol{a}}{}^{\boldsymbol{b}}$, page 98
h_{ABCD} space spinor counterpart of $h_{AA'}{}^{BB'}$ and of a three-dimensional Riemannian metric, page 98
I generic interval in $\mathbb{R}$
i^0 spatial infinity
$i^{\pm}$ future and, respectively, past timelike infinity
$I^{\pm}(\mathcal{U})$ chronological future and, respectively, past of a set $\mathcal{U}$, page 391
$J^+(o, \mathcal{M}')$ set consisting of o and all points of $\mathcal{M}'$ which can be joined to o by a causal curve in $\mathcal{M}'$, page 497
$J^{\pm}(\mathcal{U})$ causal future and, respectively, past of a set $\mathcal{U}$, page 391
j_k unphysical flux vector, page 255
$J_{\boldsymbol{jk}}$, $J_{\boldsymbol{j}}$ normal components of the rescaled Cotton tensor, page 262
K_{ij} extrinsic curvature tensor of a hypersurface in abstract index notation, page 61
L_{ab} Schouten tensor of a connection ∇_a in abstract index notation, page 48
l_{ij} three-dimensional Schouten tensor, page 60
$p \prec\!\prec q$ timelike related points, page 391
p conformally rescaled pressure of a perfect fluid, page 220
$p \prec q$ strictly causally related points, page 391

$p \preceq q$ causally related points, page 391
$P^{\boldsymbol{C}}{}_{\boldsymbol{DAA'BB'}}$ reduced spinorial geometric curvature, page 198
$P^{\boldsymbol{c}}{}_{\boldsymbol{dab}}$ components of the geometric curvature, page 194
$P^{\boldsymbol{CC'}}{}_{\boldsymbol{DD'AA'BB'}}$ spinorial geometric curvature, page 197
$P_n^{(\alpha,\beta)}(\tau)$ Jacobi polynomial of degree n with parameters (α,β), page 553
$Q_a{}^b{}_c$ transition tensor between connections in abstract index notation, page 42
r three-dimensional Ricci scalar, page 60
$R(x)$ conformal gauge source function, page 348
R, $R[\boldsymbol{g}]$ Ricci scalar of a connection ∇_a, page 48
$R^{\boldsymbol{c}}{}_{\boldsymbol{dab}}$ components of the Riemann tensor with respect to an orthonormal frame, page 53
$R^d{}_{cab}$ Riemann curvature tensor of a connection ∇_a in abstract index notation, page 40
$r^k{}_{lij}$ three-dimensional Riemann curvature tensor in abstract index notation, page 60
$r_{ABCDEFGH}$ spinorial counterpart of the three-dimensional Riemann curvature tensor, page 110
R_{ab} Ricci tensor of a connection ∇_a in abstract index notation, page 48
r_{ACEFGH}, r_{ABCE} reduced three-dimensional curvature spinors, page 110
$R_{CC'DD'AA'BB'}$ spinorial counterpart of the Riemann curvature tensor, page 86
$R_{CDAA'BB'}$ reduced Riemann curvature spinor, page 86
r_{ij} three-dimensional Ricci tensor in abstract index notation, page 60
s the Friedrich scalar, page 186
s_{ABCD} spinorial counterpart of the three-dimensional trace-free Ricci tensor, page 110
s_{ij} three-dimensional trace-free Ricci tensor, page 60
$SO(3)$ three-dimensional special orthogonal group
$T(\mathcal{M})$ tangent bundle over $\mathcal{M}$, page 34
$T\,|_p\,(\mathcal{M})$ tangent space at a point $p \in \mathcal{M}$, page 31
$T^*(\mathcal{M})$ cotangent bundle over $\mathcal{M}$, page 34
$T^*\,|_p\,(\mathcal{M})$ cotangent space at a point $p \in \mathcal{M}$, page 31
$T^\bullet\,|_p\,(\mathcal{M})$ tensor algebra at $p \in \mathcal{M}$, page 33
$T_l^k\,|_p\,(\mathcal{M})$ space of (k,l)-tensors at the point $p \in \mathcal{M}$, page 33
$T^{a_1\cdots a_k}{}_{b_1\cdots b_l}$ arbitrary (k,l)-tensor in abstract index notation
T_{ab} unphysical energy-momentum tensor
T_{cdb} rescaled Cotton tensor, page 189
u retarded time
U, $X^{\mathcal{A}}$, ω, $\xi^{\mathcal{A}}$ components of an adapted frame in the asymptotic characteristic problem, page 482
u, v retarded and, respectively, advanced time coordinates
$u^{\boldsymbol{a}}$, $v^{\boldsymbol{a}}$, $w^{\boldsymbol{a}}$,... components of the vectors $\boldsymbol{u}$, $\boldsymbol{v}$, $\boldsymbol{w}$ with respect to the coframe $\{\boldsymbol{\omega}^{\boldsymbol{a}}\}$, page 51

u^a unphysical 4-velocity of a perfect fluid, page 220
u^a, v^a, w^a, ... generic vectors in abstract index notation
v norm of a static Killing vector, page 504
$x(\mathrm{s})$ alternative notation for a curve with parameter s, page 30
X_{CDAB}, $Y_{CDA'B'}$ curvature spinors, page 86
$Y[\boldsymbol{h}]$ Yamabe invariant, page 280
Y_{abc} four-dimensional Cotton tensor, page 116
y_{ijk} three-dimensional Cotton tensor, page 118
y_{ij} three-dimensional Cotton-York tensor, page 118
$z_{\boldsymbol{AA'}}$, z, $z_{(AB)}$ spacetime and space spinor components of the spinorial counterpart of the deviation vector of a congruence of conformal geodesics, page 383
${}^*F_{ab}$ Hodge dual of an antisymmetric tensor F_{ab}, page 50
${}^*R_{abcd}$, R^*_{abcd} left and, respectively, right duals of the tensor R_{abcd}, page 50
$+$ Hermitian conjugation, page 96
$\dagger$, $\ddagger$ generalised dualisation operations, page 50
$\sharp$, $\flat$ musical operators, page 44
$\alpha, \beta, \gamma, \ldots$ spatial coordinate indices
$\boldsymbol{A}, \boldsymbol{B}, \boldsymbol{C}, \ldots$ spinor frame indices, page 74
$\boldsymbol{a}, \boldsymbol{b}, \ldots$ spacetime frame indices ranging $\mathbf{0}, \ldots, \mathbf{3}$
$\boldsymbol{i}, \boldsymbol{j}, \boldsymbol{k}, \ldots$ frame indices ranging either $\mathbf{0}$, $\mathbf{1}$, $\mathbf{2}$ or $\mathbf{1}$, $\mathbf{2}$, $\mathbf{3}$
$\perp$ perpendicular component
$\mu, \nu, \lambda, \ldots$ spacetime coordinate indices
$A, B, C, \ldots$ abstract spinor indices, page 66
$a, b, c \ldots$ abstract spacetime indices
$i, j, k, \ldots$ abstract spatial indices
${}_sY_{lm}$ spin-weighted spherical harmonics
$(a_1 \cdots a_l)$ symmetrisation over the indices $a_1 \cdots a_l$, page 36
$[a_1 \cdots a_l]$ antisymmetrisation over the indices $a_1 \cdots a_l$, page 36
$\mathcal{A}, \mathcal{B}, \ldots$ arbitrary string of indices
$\{a_1 \cdots a_l\}$ symmetric trace-free part over the indices $a_1 \cdots a_l$, page 47

1
Introduction

This book discusses an approach to the analysis of asymptotic and global properties of solutions to the equations of Einstein's theory of general relativity (the Einstein field equations) based on ideas arising in conformal geometry. This approach allows a geometric and rigorous formulation of problems and notions of great physical relevance in the context of general relativity. At the same time, it provides valuable insights into the properties of the Einstein field equations under optimal regularity conditions.

Before entering into the subject, it is useful to discuss the motivation behind this type of endeavour. Accordingly, a brief account of certain aspects of what can be called *mathematical general relativity* is necessary.

1.1 On the Einstein field equations

Einstein's theory of general relativity is the best theory of gravity we have. It is a relativistic theory of gravity which considers four-dimensional differentiable, orientable manifolds $\tilde{\mathcal{M}}$ endowed with a Lorentzian metric $\tilde{\boldsymbol{g}}$; a discussion of these differential geometric notions is provided in Chapter 2. The pair $(\tilde{\mathcal{M}}, \tilde{\boldsymbol{g}})$ is called a ***spacetime***. Here, and in the rest of this book, quantities associated to the spacetime $(\tilde{\mathcal{M}}, \tilde{\boldsymbol{g}})$ will be distinguished by a tilde (˜); the motivation behind this notation will become clear in the following. The gravitational field is described in general relativity as a manifestation of the curvature of spacetime.

The fundamental equations of general relativity, the ***Einstein field equations***, describe how matter produces the curvature of spacetime. They are given, in the abstract index notation discussed in Section 2.2.6, by

$$\tilde{R}_{ab} - \frac{1}{2}\tilde{R}\tilde{g}_{ab} + \lambda\tilde{g}_{ab} = \tilde{T}_{ab}, \tag{1.1}$$

where $\tilde{g}_{ab}$ is the abstract index version of $\tilde{\boldsymbol{g}}$, and where $\tilde{R}_{ab}$ and $\tilde{R}$ denote, respectively, the Ricci tensor and Ricci scalar of the metric $\tilde{\boldsymbol{g}}$. Moreover, λ is the so-called ***cosmological constant*** and $\tilde{T}_{ab}$ denotes the energy–momentum tensor of the matter in the spacetime. Precise definitions and conventions for the curvature tensors are provided in Chapter 2, while a discussion of the energy–momentum tensors for a range of matter models is provided in Chapter 9. The energy–momentum tensor satisfies the ***conservation equation***

$$\tilde{\nabla}^a \tilde{T}_{ab} = 0,$$

where $\tilde{\nabla}_a$ denotes the covariant derivative of the metric $\tilde{\boldsymbol{g}}$. The ***Bianchi identity*** satisfied by the Riemann curvature tensor $\tilde{R}^a{}_{bcd}$ of the metric $\tilde{\boldsymbol{g}}$ ensures the consistency between the conservation equation and the Einstein field equations. A ***solution to the Einstein field equations*** is a pair $(\tilde{\mathcal{M}}, \tilde{\boldsymbol{g}})$, together with a $\tilde{\boldsymbol{g}}$-divergence-free tensor $\tilde{T}_{ab}$ such that Equation (1.1) holds. In suitable open subsets of $\tilde{\mathcal{M}}$ the metric $\tilde{\boldsymbol{g}}$ is expressed, using some *local coordinates* (x^μ), in terms of its components $(\tilde{g}_{\mu\nu})$; here and in what follows, Greek indices are used as *coordinate indices*. In general, several coordinate charts will be needed to cover the spacetime manifold $\tilde{\mathcal{M}}$. Two metrics $\tilde{\boldsymbol{g}}$ and $\bar{\boldsymbol{g}}$ over $\tilde{\mathcal{M}}$ are said to be ***isometric*** if they are related, everywhere on $\tilde{\mathcal{M}}$, by some coordinate transformation.

In the cases where $\tilde{T}_{ab} = 0$, a direct computation shows that Equation (1.1) implies

$$\tilde{R}_{ab} = \lambda \tilde{g}_{ab}. \tag{1.2}$$

In what follows, the latter will be known as the ***vacuum Einstein field equations*** and a solution thereof as an ***Einstein spacetime***. The full curvature of a four-dimensional manifold is described by the tensor $\tilde{R}^a{}_{bcd}$. This tensor has 20 independent components. By contrast, the Ricci tensor appearing in the Einstein field Equations (1.1) and (1.2) has only 10 independent components. Hence, even in the absence of a cosmological constant, where the vacuum field Equations (1.2) reduce to

$$\tilde{R}_{ab} = 0, \tag{1.3}$$

it is possible to have solutions with a non-vanishing Riemann tensor. As a consequence, solutions to the vacuum field equations play a special role in general relativity, as they describe *pure gravitational configurations*. Vacuum spacetimes are often deemed more fundamental, as they exclude potential pathologies which may arise from the choice of a particular matter model.

General relativity has two main domains of applicability: *cosmology* and *isolated systems*. To make use of the Einstein field Equations (1.1) within these two domains, one requires a number of idealisations. On the one hand, in cosmology it is usually assumed that the matter content of the universe can be described by a perfect fluid with an equation of state which depends on a particular cosmological era. It is a convention in mathematical relativity to refer to spacetimes with compact spacelike sections as ***cosmological spacetimes***. On the other hand, ***isolated systems*** are convenient idealisations of astrophysical objects for which it is assumed that the cosmological expansion has no influence. The transition between the regime of isolated systems and the cosmological one is a topic of fundamental relevance for the understanding of the physical content of the Einstein field equations; see, for example, Ellis (1984, 2002).

The validity of general relativity has been verified in a number of experiments covering a wide range of scenarios ranging from the dynamics of the solar system to cosmological scales; see, for example, Will (2014) for a discussion of the subject. Surveys of the physical content of general relativity and its various domains of applicability can be found, for example, in Poisson and Will (2015) and Shapiro (1999).

Note. *In the remainder of this chapter, in order to simplify the presentation, the discussion will be restricted to Einstein spaces, that is, solutions to the vacuum Equations* (1.2). *The inclusion of matter very often requires a case-by-case analysis.*

1.2 Exact solutions

A natural first step to developing an understanding of the properties of solutions to the Einstein field equations is the construction of ***exact solutions***, that is, explicit solutions written in terms of *elementary functions* of some coordinates. The first non-trivial exact solution to the Einstein field equations ever obtained is the Schwarzschild solution. It describes a static spherically symmetric vacuum configuration; see Schwarzschild (1916), an English translation of which can be found in Schwarzschild (2003). Remarkably, despite the complexity of the field equations, the literature contains a vast number of exact solutions to the equations of general relativity; see, for example, Stephani et al. (2003) for a monograph on the subject. The number of solutions with a physical or geometric significance is, arguably, much smaller; see, for example, Bičák (2000) and Griffiths and Podolský (2009).

1.2.1 Construction of exact solutions

The construction of exact solutions to the Einstein field equations requires a number of assumptions concerning the nature of the solutions. The most natural assumptions involve the presence of continuous symmetries (*Killing vectors*) of some type in the solution, for example, spherical symmetry, axial symmetry, stationarity (including staticity) and homogeneity. Other types of assumptions involve the algebraic structure of the curvature tensors of the spacetime (e.g. the Petrov type of the Weyl tensor). These types of assumptions are harder to justify on a physical basis.

Exact solutions are usually constructed in a coordinate system adapted to the assumptions being made. Very often, these *natural coordinates* cover only a portion of the whole spacetime manifold. Thus, one needs to find new coordinate systems (charts) for the exact solution which allow one to uncover a full *maximal analytic extension* of the spacetime. This maximal extension usually paves the way to the interpretation of the exact solution and gives access to its global properties.

1.2.2 The limitations of exact solutions

Several of the well-known consequences of general relativity have been developed through the analysis of exact solutions, for example, the notion of a black hole. Thus, the study of exact solutions to the Einstein field equations helps to develop a physical and geometric intuition which, in turn, can lead to questions concerning more generic solutions. However, despite the valuable insights they provide, the construction of exact solutions is not a systematic approach to explore the *space of solutions of the theory.* In particular, this approach leaves open the question of whether certain properties of a solution are *generic*, that is, satisfied by a broader class of spacetimes. Moreover, exact solutions do not lend themselves to the analysis of dynamic situations such as, for example, the description of the gravitational radiation produced by an isolated system. Thus, it is not possible to address issues involving *stability* just by means of exact solutions. In order to analyse the above issues one has to consider whether it is possible to formulate an *initial value problem for the Einstein field equations* by means of which large classes of solutions can be constructed.

1.3 The Cauchy problem in general relativity

As in the case of many other physical theories, general relativity admits the formulation of an *initial value problem* (*Cauchy problem*). This aspect of the theory is obscured by both the *tensorial character of the Einstein field equations* and the *absence of a background geometry in the theory*; it is a priori not clear that the field equations give rise to a system of partial differential equations (PDEs) of a recognisable type.

Classical physical theories are expected to satisfy a ***causality principle:*** *the future of an event in spacetime cannot influence its past, and, moreover, signals must propagate at finite speed*. Among the three main types of PDEs (elliptic, hyperbolic and parabolic), *hyperbolic differential equations* are the only ones compatible with the causality principle. This observation suggests it should be possible to extract from the Einstein field equations a system of evolution equations with *hyperbolic properties.*

1.3.1 Hyperbolic reductions

The seminal work of Fourès-Bruhat (1952) has shown that the hyperbolic properties of the Einstein field equations can be made manifest by means of a suitable choice of coordinates. Following modern terminology, a choice of coordinates is a particular example of *gauge choice*. Indeed, by choosing the spacetime coordinates (x^μ) in such a way that they satisfy the wave equation associated with the metric $\tilde{\boldsymbol{g}}$, the Einstein field equations can be shown to imply a *system of quasilinear wave equations* for the components $(\tilde{g}_{\mu\nu})$ of the (a priori unknown) metric $\tilde{\boldsymbol{g}}$ with respect to the *wave coordinates.* For quasilinear wave differential equations there exists a developed theory which allows the formulation of a *well-posed Cauchy problem.* The use of wave coordinates is not the only way of bringing to the fore the hyperbolic aspects of the Einstein field equations. In this book, it will be shown that the Einstein field equations can be reformulated in such a way that after a suitable gauge choice they imply a so-called (first order) *symmetric hyperbolic evolution system* – a class of PDEs with properties similar to those of wave equations and for which a comparable theory is available. The procedure of extracting suitable hyperbolic *evolution equations* through a particular reformulation of the Einstein field equations and a suitable gauge choice is known as a ***hyperbolic reduction***; hyperbolic reductions are further discussed in Chapter 13. Besides its natural relevance in mathematical relativity, the construction of hyperbolic reductions for the Einstein field equations is of fundamental importance for numerical relativity; see, for example, Alcubierre (2008) and Baumgarte and Shapiro (2010).

In the same way that the Einstein field equations are geometric in nature, a proper formulation of the Cauchy problem in general relativity must also be done in a geometric way; see, for example, Choquet-Bruhat (2007). This idea is, in principle, in conflict with the discussion of hyperbolicity properties of the Einstein field equations, as the associated procedure of gauge fixing breaks the *spacetime covariance* of the field equations. As will be seen in the following, this tension can be resolved in a satisfactory manner.

1.3.2 Initial data and the constraint equations

The formulation of an initial value problem for the Einstein field equations requires the prescription of suitable initial data for the evolution equations on a three-dimensional manifold $\tilde{\mathcal{S}}$. This manifold will be later interpreted as a hypersurface of the spacetime $(\tilde{\mathcal{M}}, \tilde{\boldsymbol{g}})$. An important feature of general relativity is that the initial data for the evolution equations implied by the Einstein field equations are constrained. The ***constraint equations of general relativity*** (*Einstein constraints*) can be formulated as a set of equations intrinsic to the initial hypersurface $\tilde{\mathcal{S}}$ for a pair of symmetric tensors $\tilde{\boldsymbol{h}}$ and $\tilde{\boldsymbol{K}}$ describing, respectively, the intrinsic geometry of the hypersurface (***intrinsic metric*** or ***first fundamental form***) and the way the initial hypersurface is curved within the spacetime $(\tilde{\mathcal{M}}, \tilde{\boldsymbol{g}})$ – the so-called ***extrinsic curvature*** or ***second fundamental form***. A priori, it is not clear what the *freely specifiable data* for these constraint equations consist of, or whether, given a particular choice of free data, the equations can be solved. The systematic analysis of the constraint equations has shown that under suitable assumptions, they can be recast as a set of *elliptic partial differential equations*; see, for example, Bartnik and Isenberg

(2004). For this type of equation a theory is available to discuss the existence and uniqueness of solutions.

The constraint equations play a fundamental role in the theory and ensure that the solution of the evolution equations is, in fact, a solution to the Einstein field equations; this type of analysis is often called the *propagation of the constraints.* The constraint equations of general relativity will be discussed in Chapter 11.

1.3.3 The well-posedness of the Cauchy problem in general relativity

The formulation of the Cauchy problem in general relativity ensures, at least locally, the existence of a solution to the Einstein field equations which is consistent with the prescribed initial data. More precisely, one has the following result first proven in Fourès-Bruhat (1952).

Theorem 1.1 (***local existence of solutions to the initial value problem***) *Given a solution* $(\tilde{\boldsymbol{h}}, \tilde{\boldsymbol{K}})$ *to the Einstein constraint equations on a three-dimensional manifold* $\tilde{\mathcal{S}}$ *there exists a vacuum spacetime* $(\tilde{\mathcal{M}}, \tilde{\boldsymbol{g}})$ *such that* $\tilde{\mathcal{S}}$ *is a spacelike hypersurface of* $\tilde{\mathcal{M}}$, $\tilde{\boldsymbol{h}}$ *is the intrinsic metric induced by* $\tilde{\boldsymbol{g}}$ *on* $\tilde{\mathcal{S}}$ *and* $\tilde{\boldsymbol{K}}$ *is the associated extrinsic curvature.*

The spacetime $(\tilde{\mathcal{M}}, \tilde{\boldsymbol{g}})$ obtained as a result of Theorem 1.1 is called a ***development of the initial data set*** $(\tilde{\mathcal{S}}, \tilde{\boldsymbol{h}}, \tilde{\boldsymbol{K}})$. Not every spacetime can be *globally* constructed from an initial value problem. Those which can be constructed in this way are said to be ***globally hyperbolic.*** There are important examples of spacetimes which do not possess this property – most noticeably, the *anti-de Sitter spacetime.* A general result concerning globally hyperbolic spacetimes states that their topology is that of $\mathbb{R} \times \tilde{\mathcal{S}}$ with each slice $\tilde{\mathcal{S}}_t \equiv \{t\} \times \tilde{\mathcal{S}}$ being intersected only once by each timelike curve in the spacetime. The slices $\tilde{\mathcal{S}}_t$ are known as ***Cauchy surfaces.*** The above points will be further discussed in Chapter 14.

The Cauchy problem for the Einstein field equations provides an appropriate setting for the discussion of dynamics. In particular, it allows one to investigate whether a given solution of the Einstein field equations is *stable*, that is, whether its essential features are retained if the initial data set is perturbed. Moreover, it also allows one to analyse whether a given property of a solution is *generic*, that is, whether the property holds for all solutions in an open set in the *space of initial data.*

1.3.4 Geometric uniqueness and the maximal globally hyperbolic development

An important observation concerning Theorem 1.1 is that it does not ensure the uniqueness of the development $(\tilde{\mathcal{M}}, \tilde{\boldsymbol{g}})$ of the initial data set $(\tilde{\mathcal{S}}, \tilde{\boldsymbol{h}}, \tilde{\boldsymbol{K}})$: a different hyperbolic reduction procedure will, in general, give rise to an alternative development $(\tilde{\mathcal{M}}', \tilde{\boldsymbol{g}}')$. From the point of view of the Cauchy problem of general relativity, the solution manifold is not known a priori. Instead, it is obtained as a part of the evolution process.

Given that an initial data set for the Einstein field equations gives rise to an infinite number of developments (one for each *reasonable* gauge choice), it is natural to ask whether it is possible to combine these various developments to obtain a *maximal development.* This question is answered in the positive by the following fundamental result; see Choquet-Bruhat and Geroch (1969).

Theorem 1.2 (***existence of a maximal development***) *Given an initial data set for the Einstein field equations* $(\tilde{\mathcal{S}}, \tilde{\boldsymbol{h}}, \tilde{\boldsymbol{K}})$, *there exists a unique maximal development* $(\tilde{\mathcal{M}}, \tilde{\boldsymbol{g}})$, *that is, a development such that if* $(\tilde{\mathcal{M}}', \tilde{\boldsymbol{g}}')$ *is another development, then* $\tilde{\mathcal{M}}' \subseteq \tilde{\mathcal{M}}$ *and on* $\tilde{\mathcal{M}}'$ *the metrics* $\tilde{\boldsymbol{g}}$ *and* $\tilde{\boldsymbol{g}}'$ *are isometric.*

The *maximal development* $(\tilde{\mathcal{M}}, \tilde{\boldsymbol{g}})$ is also known as the *maximal globally hyperbolic development* of the data $(\tilde{\mathcal{S}}, \tilde{\boldsymbol{h}}, \tilde{\boldsymbol{K}})$. Theorem 1.2 clarifies the sense in which one can expect uniqueness from the Cauchy problem in general relativity; this idea is known as *geometric uniqueness.*

One can think of the maximal development of an initial data set as the largest spacetime that can be uniquely constructed out of an initial value problem. The boundary of this maximal development, if any at all, sets the limits of predictability of the data – accordingly, one has a close link with the notion of *classical determinism*. In certain spacetimes, it is possible to extend the maximal development of a hypersurface to obtain a *maximal extension.* Accordingly, in general, maximal developments and maximal extensions do not coincide. A further discussion of the Cauchy problem in general relativity is provided in Chapter 14.

1.3.5 Construction of maximal developments and global existence of solutions

Given some initial data set $(\tilde{\mathcal{S}}, \tilde{\boldsymbol{h}}, \tilde{\boldsymbol{K}})$, it is natural to ask, How can one construct its maximal development $(\tilde{\mathcal{M}}, \tilde{\boldsymbol{g}})$? In general, this is a very difficult task, as it requires controlling the evolution dictated by the Einstein field equations under very general circumstances – something for which the required mathematical technology is not yet available. There are, nevertheless, some conjectures concerning the global behaviour of maximal developments. The origin of these conjectures goes back to Penrose (1969) – see Penrose (2002) for a reprint – and are usually known by the name *cosmic censorship*. In particular, the so-called *strong cosmic censorship* states that the maximal development of generic initial data for the Einstein field equations cannot be extended as a Lorentzian manifold.

Given an exact solution to the Einstein equations, if one knows its maximal extension, one can determine the maximal development $(\tilde{\mathcal{M}}, \tilde{\boldsymbol{g}})$ of one of its (Cauchy) hypersurfaces, say, $\tilde{\mathcal{S}}$. In what follows, let $(\tilde{\boldsymbol{h}}, \tilde{\boldsymbol{K}})$ denote the initial data implied on $\tilde{\mathcal{S}}$ by the spacetime metric $\tilde{\boldsymbol{g}}$. The explicit knowledge of the maximal development allows one to provide a physical interpretation of the solution and to analyse its global structure in some detail. One can now ask whether certain aspects of $(\tilde{\mathcal{M}}, \tilde{\boldsymbol{g}})$ – say, its basic global structure – are shared by a wider class of solutions to the Einstein field equations. A strategy to address this question within the framework of the Cauchy problem in general relativity is to consider initial data sets $(\tilde{\mathcal{S}}, \bar{\boldsymbol{h}}, \bar{\boldsymbol{K}})$ which are, in some sense, close to the initial data for the exact solution. One can then try to show that the associated maximal globally hyperbolic development $(\bar{\mathcal{M}}, \bar{\boldsymbol{g}})$ has the desired global properties. If this is the case, one has obtained a statement about the *stability* of the solution and the *genericity* of the property one is interested in. The standard convention, to be used in this book, is to call $(\tilde{\mathcal{M}}, \tilde{\boldsymbol{g}})$ and $(\tilde{\mathcal{S}}, \tilde{\boldsymbol{h}}, \tilde{\boldsymbol{K}})$, respectively, the ***background spacetime*** and the ***background initial data set*** and $(\bar{\mathcal{M}}, \bar{\boldsymbol{g}})$ and $(\tilde{\mathcal{S}}, \bar{\boldsymbol{h}}, \bar{\boldsymbol{K}})$ the ***perturbed spacetime*** and ***perturbed initial data set***, respectively. In practice, the notion of closeness between initial data sets is dictated by the requirements of the PDE theory used to prove the existence of solutions to the evolution equations. In the previous discussion it has been assumed that the 3-manifolds on which the background and perturbed initial data are prescribed are the same. The stability analysis allows one to conclude that the spacetime manifolds $\tilde{\mathcal{M}}$ and $\bar{\mathcal{M}}$ are the same – they are, however, endowed with different metrics.

In analysing the stability of the background solution $(\tilde{\mathcal{M}}, \tilde{\boldsymbol{g}})$ one needs to show that the solutions to the evolution equations with perturbed initial data exist as long as the background solution. The expectation is that the assumption of having initial data close to data for an exact solution whose global structure is well understood will ease this task. In the following sections a strategy to exploit this assumption will be discussed.

1.4 Conformal geometry and general relativity

Special relativity provides a framework for the discussion of the notion of ***causality*** – that is, the relation between cause and effect – which is consistent with the *principle of relativity.* The *causal structure* of special relativity is determined by the light cones associated with the Minkowski metric $\tilde{\boldsymbol{\eta}}$. It allows the determination of whether a signal travelling not faster than the speed of light can be sent between two events – if this is the case, then the two events are said to be ***causally related***. More generally, one can talk of *Lorentzian causality*: any Lorentzian metric $\tilde{\boldsymbol{g}}$ gives rise to a causal structure determined by the light cones associated to $\tilde{\boldsymbol{g}}$. Thus, general relativity provides a natural generalisation of the notions of causality of special relativity – one in which the light cones vary from event to event in spacetime. Crucially, however, in general relativity the causal structure is a basic unknown of the theory.

The theory of hyperbolic differential equations provides notions of causality which, in principle, are independent from the notions of Lorentzian causality. It is, nevertheless, a remarkable feature of general relativity that locally, the propagation of fields dictated by the Einstein field equations is governed by the structure of the light cones of the solutions – the so-called *characteristic surfaces* of the evolution equations. Thus, the notions of Lorentzian and PDE causality coincide. This aspect of the Einstein field equations is further discussed in Chapter 14.

1.4.1 Conformal transformations and conformal geometry

Locally, a light cone can be described (away from its vertex) in terms of a condition of the form $\phi(x^\mu) = constant$ where $\phi : \tilde{\mathcal{M}} \to \mathbb{R}$ is such that

$$\tilde{g}^{\mu\nu}\partial_\mu\phi\partial_\nu\phi = 0. \tag{1.4}$$

The structure of the light cones of a spacetime $(\tilde{\mathcal{M}}, \tilde{\boldsymbol{g}})$ is preserved by ***conformal rescalings***, that is, transformations of the spacetime metric of the form

$$\tilde{\boldsymbol{g}} \mapsto \boldsymbol{g} \equiv \Xi^2\tilde{\boldsymbol{g}}, \qquad \Xi > 0 \tag{1.5}$$

where Ξ is a smooth function on $\tilde{\mathcal{M}}$ – the so-called ***conformal factor***. Throughout this book, the metrics $\tilde{\boldsymbol{g}}$ and $\boldsymbol{g}$ will be called the ***physical metric*** and the ***unphysical metric***, respectively. The rescaling (1.5) gives rise to a ***conformal transformation*** of $(\tilde{\mathcal{M}}, \tilde{\boldsymbol{g}})$ to $(\tilde{\mathcal{M}}, \boldsymbol{g})$. Precise definitions and further discussion of these notions are provided in Chapter 5. In elementary geometry, conformal transformations are usually described as transformations preserving the angle between vectors. In Lorentzian geometry, they preserve the light cones; from (1.4) it follows that $g^{\mu\nu}\partial_\mu\phi\partial_\nu\phi = 0$, so that the condition $\phi(x^\mu) = constant$ also describes the light cones of the metric $\boldsymbol{g}$.

One key aspect of conformal rescalings is that they allow one to introduce *conformal extensions* of the spacetime $(\tilde{\mathcal{M}}, \tilde{\boldsymbol{g}})$; see Figure 1.1. In a Riemannian setting, the most basic example of conformal extensions of manifolds is the so-called *conformal completion* of the Euclidean plane $\mathbb{R}^2$ into the 2-sphere $\mathbb{S}^2$ by

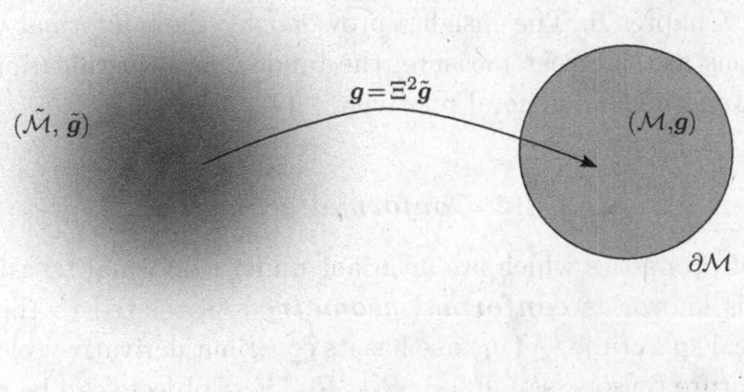

Figure 1.1 Schematic representation of the conformal extension of a manifold. The *physical* manifold $(\tilde{\mathcal{M}}, \tilde{\boldsymbol{g}})$ has infinite extension, while the *unphysical* (extended) manifold $(\mathcal{M}, \boldsymbol{g})$ is compact with boundary $\partial\mathcal{M}$. The boundary $\partial\mathcal{M}$ corresponds to the points for which $\Xi = 0$. Further details can be found in Chapter 5. Adapted from Penrose (1964).

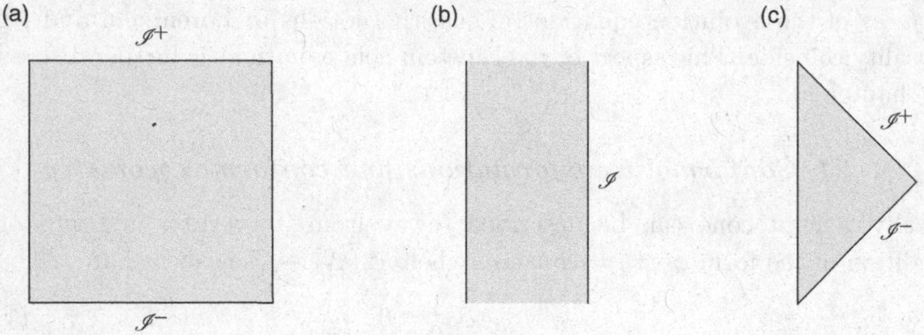

Figure 1.2 Penrose diagrams of the three spacetimes of constant curvature: (a) the de Sitter spacetime; (b) the anti-de Sitter spacetime; (c) the Minkowski spacetime. Details of these constructions can be found in Chapter 6.

means of stereographic coordinates. By suitably choosing the conformal factor Ξ, the metric $\boldsymbol{g}$ given by the rescaling (1.5) may be well defined even at the points where $\Xi = 0$. If this is the case, it can be verified that the set of points $\partial\mathcal{M}$ for which $\Xi = 0$ corresponds to *ideal points at infinity* for the spacetime $(\tilde{\mathcal{M}}, \tilde{\boldsymbol{g}})$ and is called the ***conformal boundary***. The pair $(\mathcal{M}, \boldsymbol{g})$ where $\mathcal{M}$ is the extended manifold obtained from attaching to $\tilde{\mathcal{M}}$ its conformal boundary is usually known as the ***unphysical spacetime***. Of particular interest are the portions of the conformal boundary which are hypersurfaces of the manifold $\mathcal{M}$ – these sets are characterised by the additional requirement of $\mathbf{d}\Xi \neq 0$, so that they have a well-defined normal. This part of the conformal boundary is denoted by $\mathscr{I}$.

Explicit calculations show that the three spacetimes of *constant curvature* – the Minkowski, de Sitter and anti-de Sitter spacetimes – can be conformally extended. The details of these constructions are described in Chapter 6. These conformal extensions are conveniently represented in terms of *Penrose diagrams*; see Figure 1.2. A discussion of the construction of Penrose diagrams can also be found in Chapter 6. The insights provided by the conformal extensions of these solutions are, in great measure, the fundamental justification for the use of conformal methods in general relativity.

1.4.2 Conformal geometry

The study of properties which are invariant under conformal transformations of a manifold is known as ***conformal geometry***. Associated to the metric $\boldsymbol{g}$ of the unphysical spacetime $(\mathcal{M}, \boldsymbol{g})$ one has its covariant derivative (connection) ∇_a and its curvature tensors, say, $R^a{}_{bcd}$, R_{ab}, R. These objects can be related to the corresponding objects associated to the physical metric $\tilde{\boldsymbol{g}}$ ($\tilde{\nabla}_a$, $\tilde{R}^a{}_{bcd}$, $\tilde{R}_{ab}$ and $\tilde{R}$) and the conformal factor Ξ and its derivatives. Their transformation laws show, in particular, that the Riemann tensor, the Ricci tensor and the Ricci scalar are not conformal invariants. There is, however, another part of the curvature which is conformally invariant. It is described by the *Weyl tensor*, for which it holds that

$$\tilde{C}^a{}_{bcd} = C^a{}_{bcd}, \qquad \text{on } \tilde{\mathcal{M}}.$$

In view of the above, one can regard the Weyl tensor as a property of the collection of metrics conformally related to $\tilde{\boldsymbol{g}}$ – the ***conformal class*** $[\tilde{\boldsymbol{g}}]$. If the vacuum Einstein field Equations (1.3) hold, the Bianchi identities imply that

$$\tilde{\nabla}_a \tilde{C}^a{}_{bcd} = 0 \tag{1.6}$$

irrespectively of the value of the cosmological constant.

1.4.3 Conformal invariance of equations of physics

A number of equations in physics have nice conformal properties. The prototypical example is given by the source-free Maxwell equations

$$\tilde{\nabla}^a \tilde{F}_{ab} = 0, \qquad \tilde{\nabla}_{[a}\tilde{F}_{bc]} = 0, \tag{1.7}$$

where $\tilde{F}_{ab}$ denotes the ***Faraday tensor***. One can introduce an *unphysical Faraday tensor* F_{ab} by requiring it to coincide with $\tilde{F}_{ab}$ on $\tilde{\mathcal{M}}$. Using the transformation properties relating the covariant derivatives $\tilde{\nabla}_a$ and ∇_a, it follows that the Maxwell equations are ***conformally invariant***; that is, one has that

$$\nabla^a F_{ab} = 0, \qquad \nabla_{[a}F_{bc]} = 0.$$

The above equations are well defined everywhere on the unphysical spacetime manifold $\mathcal{M}$, in particular at the conformal boundary. These equations allow the extension of the definition of the unphysical field F_{ab} to the conformal boundary $\partial\mathcal{M}$.

In contrast to the Maxwell equations, the vacuum Einstein field Equations (1.2) are not conformally invariant. The transformation law for the Ricci tensor under the rescaling (1.5) implies the equation

$$R_{ab} = -\frac{2}{\Xi}\nabla_a\nabla_b\Xi - g_{ab}g^{cd}\left(\frac{1}{\Xi}\nabla_c\nabla_d\Xi - \frac{3}{\Xi^2}\nabla_c\Xi\nabla_d\Xi\right). \tag{1.8}$$

The above equation is, at least formally, singular at the points where $\Xi = 0$. Thus, it does not provide a good equation for the analysis of the evolution of the unphysical metric $\boldsymbol{g}$ on $\mathcal{M}$. Nevertheless, as pointed out by Penrose (1963) the Bianchi identity (1.6) has a nice *conformal covariance* property. More precisely, one has that

$$\nabla_a\left(\Xi^{-1}\tilde{C}^a{}_{bcd}\right) = 0.$$

The above equation suggests defining the ***rescaled Weyl tensor*** $d^a{}_{bcd} \equiv \Xi^{-1}\tilde{C}^a{}_{bcd}$. Under certain assumptions, the Weyl tensor can be shown to vanish at $\mathscr{I}$ so that the rescaled Weyl tensor is well defined at this portion of the conformal boundary – this important result is analysed in detail in Chapter 10. The rescaled

Weyl tensor is not a conformal invariant; it transforms in a homogeneous fashion under the rescaling (1.5). The above discussion leads to the equation

$$\tilde{\nabla}_a d^a{}_{bcd} = 0, \tag{1.9}$$

the so-called ***Bianchi equation***. In addition, in view of the symmetries of the Weyl tensor it can be shown that

$$\tilde{\nabla}_{[e} d^a{}_{|b|cd]} = 0. \tag{1.10}$$

Note the similarity between Equations (1.9) and (1.10) and the Maxwell Equations (1.7). In particular, the equations are regular even at the conformal boundary. These equations are full of physical significance, as the Weyl tensor can be thought of as describing the *free gravitational field*, that is, a gravitational analogue of the Faraday tensor. Chapter 8 provides a detailed derivation and discussion of the equations presented in this section.

1.4.4 Asymptotics of the gravitational field and asymptotic simplicity

One of the basic predictions of general relativity is the existence of gravitational waves propagating at the speed of light across the fabric of spacetime. As a dynamical process governed by the Einstein field equations, gravitational radiation is closely related to the structure of the light cones of spacetime – thus, if one wants to analyse gravitational radiation one has to examine the propagation of the gravitational field along null directions. This analysis is complicated by the absence of a background geometry so that, a priori, it is not clear what the asymptotic behaviour of the gravitational field should be. This concern lies at the heart of the subject of the ***asymptotics of spacetime*** – that is, the study of the limit behaviour of fields at large distances and large times and the characterisation of spacetimes by data obtained by taking such limits.

In theories which describe fields on a given background, one can discuss limits *at infinity* in a meaningful way in terms of the background geometry. The situation is radically different in general relativity, where the spacetime $(\tilde{\mathcal{M}}, \tilde{\boldsymbol{g}})$ – with respect to which the limits of fields derived from $\tilde{\boldsymbol{g}}$ are to be formulated – is the central objects of study. Accordingly, making sense of limiting procedures in general relativity is a delicate process and requires a careful analysis of the geometry and the way it is determined by the Einstein field equations. An approach to this analysis is provided by Penrose's suggestion that the close relation between the propagation of the gravitational field and the structure of null cones which holds locally is also preserved at large scales and that the asymptotic behaviour of the gravitational field can be conveniently analysed in terms of conformal extensions of the spacetime; see Penrose (1963, 1964) and Penrose (2011) for a reprint of the latter reference. With this idea in mind, Penrose introduced the notion of ***asymptotically simple spacetimes***, namely, spacetimes admitting a *smooth* conformal extension which is similar to that of one of the three constant curvature spacetimes. Proceeding in this manner, one attempts to single out a class of *sufficiently well-behaved spacetimes* for which it is possible to relate the structure of the light cones in spacetime to the structure of the field equations and the large-scale behaviour of their solutions. For asymptotically simple spacetimes the causal character of $\mathscr{I}$ is determined by the sign of the cosmological constant; moreover, as already seen, the Weyl tensor vanishes on the conformal boundary – the latter is the basic observation in a collection of results known generically as ***peeling***.

Minkowski-like spacetimes, that is, those asymptotically simple spacetimes for which $\lambda = 0$, are of particular relevance in the study of asymptotics with regard to their connection to the notion of isolated systems in general relativity; compare the discussion at the end of Section 1.1. For this type of spacetime $\mathscr{I}$ is a null hypersurface describing idealised *observers at infinity.* Penrose's original insight was to use the notion of asymptotic simplicity as a way of characterising isolated systems in general relativity – this idea has been called ***Penrose's proposal*** by Friedrich (2002). One of the appealing features of this approach to the study of isolated systems is that it provides a general framework in which notions of physical interest such as gravitational radiation and the associated mass/momentum-loss can be rigorously formulated and analysed. A substantial amount of work has been invested in pursuing these ideas, as attested by the sprawling literature on the subject. An exposition of the notion of asymptotic simplicity, some of its basic consequences and Penrose's proposal is given in Chapters 7 and 10.

1.4.5 The conformal Einstein field equations

In view of Penrose's ideas on the relation between general relativity and conformal geometry one can ask: *to what extent is it possible to draw conclusions about the global structure of spacetimes from an analysis of the behaviour, under conformal rescalings, of the Einstein field equations?* As will be seen in this book, by considering this question one is led to *analyse the behaviour of solutions to the Einstein field equations under optimal regularity conditions.* To address the above question one needs a suitable set of equations to work with. As already observed, the direct transcription of the Einstein field equations as an equation for the unphysical metric $\boldsymbol{g}$ does not provide a set of equations which are adequate from the point of view of PDE theory.

An alternative set of field equations, the so-called ***conformal Einstein field equations***, has been constructed in the seminal work by Friedrich (1981a,b, 1983). The construction of this conformal representation of the equations begins with a revised reading of the *singular Equation* (1.8) not as an equation for the unphysical metric (or alternatively, its Ricci tensor) but for the derivatives of the conformal factor Ξ. To complete this alternative point of view one upgrades the curvature tensors to the level of unknowns and, accordingly, provides equations for them. The required equations are supplied by the Bianchi identities in a way which is consistent with the Einstein field equations satisfied by the physical metric $\tilde{\boldsymbol{g}}$. The resulting system consists of equations for the conformal factor and its first- and second-order derivatives, the unphysical metric $\boldsymbol{g}$ (through the definition of its Ricci tensor), the unphysical Ricci tensor R_{ab} and the rescaled Weyl tensor $d^a{}_{bcd}$ – the equation for the latter field is Equation (1.9). The equations derived by Friedrich have two key properties: (i) they are *formally regular* even at the points where $\Xi = 0$ and (ii) whenever $\Xi \neq 0$, they imply a solution to the Einstein field equations. The considerations leading to the conformal Einstein field equations will be discussed in Chapter 8.

The equations described in the previous paragraph are usually known as the *metric conformal field equations*. One can extend the basic construction to incorporate more *gauge freedom* so as to obtain a more flexible set of equations. A natural first step in this direction consists of rewriting the field equations in a frame formalism. This leads, in turn, in an almost direct way to the spinorial version of the equations; see below. A more extreme generalisation consists of a reformulation of the field equations in terms of a covariant derivative $\hat{\nabla}_a$ which is not the Levi-Civita connection of a metric, but which nevertheless respects the structure of the conformal class $[\tilde{\boldsymbol{g}}]$, a so-called *Weyl connection*. The resulting equations are known as the *extended conformal Einstein field equations*. As will be seen below, this particular formulation of the equations allows the use of gauges with conformally privileged properties.

Friedrich's conformal Einstein equations are not the only possible type of conformal representation of the Einstein field equations; see, for example, Mason (1995) and Anderson (2005a). In any case, they are the ones which have been studied in a more systematic manner in the literature.

1.4.6 Gauge conditions and conformal geodesics

As already mentioned, the procedure of hyperbolic reduction requires the specification of a gauge in terms of which the evolution equations are to be expressed. Earlier in this chapter, the notion of a gauge choice had been restricted to a specification of coordinates. For the conformal field equations, the gauge specification involves three aspects: a coordinate, a frame and a conformal aspect. The precise choice of these three aspects of the gauge depends on the particulars of the problem at hand. A discussion of the gauge freedom contained in the conformal field equations is given in Chapter 13.

The presence of a *conformal gauge freedom* – that is, the freedom to specify the representative in the conformal class one wants to work with – is one of the most attractive aspects of the conformal field equations. Given the bewildering freedom one has in this respect, the use of conformal gauges related to conformal invariants is a natural choice. *Conformal geodesics* are a good example of the type of invariants one can consider. These curves are defined through a set of equations which are invariant under conformal rescalings. In general, the conformal class $[\tilde{\boldsymbol{g}}]$ does not contain a metric for which the conformal geodesics can be recast as *standard (metric) geodesics*. However, there is always a *Weyl connection* for which they are affine geodesics. Conformal geodesics can be used to construct *conformal Gaussian gauge systems* for which coordinates and an adapted frame are propagated off an initial hypersurface. Conformal geodesics allow one to specify a *privileged unphysical metric* $\boldsymbol{g} = \Theta^2\tilde{\boldsymbol{g}}$ where Θ is a conformal factor determined through the conformal geodesic equations. Crucially, for solutions to the vacuum field equations (1.2), the conformal factor Θ can be determined explicitly from the initial data for a congruence of these curves – it turns out to be a quadratic polynomial of a suitable parameter of the curves in the congruence. To fully exploit the advantages provided by conformal Gaussian systems, it is necessary to express the conformal field equations in terms of Weyl connections – these considerations lead to the already mentioned extended conformal field equations. Conformal geodesics and their properties are analysed in Chapter 5.

1.4.7 Spinors

This book adopts an approach to the extraction of information from the conformal Einstein field equations which makes systematic use of a formalism based on the so-called *2-spinors*. The use of spinors to carry out this analysis is not essential to the purposes of the book, but it has the advantage of simplifying certain algebraic aspects of the discussion.

Spinors are the most basic objects subject to Lorentz transformations. To every tensor and tensorial operation there exists a spinorial counterpart. More precisely, to every tensor of rank k there corresponds a spinor of rank $2k$. In some particular cases – for example, null vectors or the Weyl tensor – by exploiting symmetries one can associate to the tensor a spinor of the same rank k.

Spinors are well adapted to the discussion of the geometry of null hypersurfaces. Thus, it is not surprising that they are a valuable tool in the discussion of the Einstein field equations. In this book, spinorial representations of the conformal field equations are systematically used as a part of the hyperbolic reduction procedure. In particular, a 2-spinor formalism usually known as the *space spinor formalism*, which can be regarded as a spinorial analogue of the *1+3 formalism for tensors*, provides an almost completely algorithmic approach to the decomposition of the field equations into (symmetric hyperbolic) evolution equations and constraint equations. The basic spinorial formalism used in this book is described in Chapter 3, while the space spinor formalism is dealt with in Chapter 4.

1.5 Existence of asymptotically simple spacetimes

The conformal field equations provide a powerful tool for the analysis and construction of asymptotically simple spacetimes. In broad terms, they allow the reformulation of problems involving unbounded domains in the physical spacetime $(\tilde{\mathcal{M}}, \tilde{\boldsymbol{g}})$ as problems on bounded domains of the unphysical spacetime $(\mathcal{M}, \boldsymbol{g})$. From the point of view of PDE theory, problems involving a finite existence time are simpler to analyse than global existence questions. Under the appropriate conditions, the existence of solutions to hyperbolic differential equations on a fixed finite time interval can be shown by invoking the property of *Cauchy stability*; this and other basic notions of PDE theory are discussed in Chapter 12 where a brief account of basic existence results for symmetric hyperbolic systems is given. Prior to its use with the conformal Einstein field equations, the technique for the analysis of evolution equations based on a combination of conformal techniques and Cauchy stability had been used to show the existence of global solutions of the Yang-Mills equations on the Minkowski and de Sitter spacetimes; see Choquet-Bruhat and Christodoulou (1981).

The remainder of this section provides a brief survey of some of the existence results for asymptotically simple spacetimes which have been obtained using the conformal Einstein equations. These results will be elaborated in Part IV of this book.

1.5.1 Characteristic initial value problems

Characteristic problems are a particular type of initial value problem where data are prescribed on *null initial hypersurfaces*. Typically, these data are prescribed on two intersecting null hypersurfaces $\mathscr{N}_1$ and $\mathscr{N}_2$. The relevant PDE theory then allows one to conclude the existence and uniqueness of solutions on neighbourhoods of $\mathscr{N}_1 \cap \mathscr{N}_2$ which are either to the future or to the past of their intersection. In a different type of characteristic problem one prescribes initial data on a null cone $\mathscr{N}$, including its vertex, and one endeavours to obtain a solution inside the cone – at least in a neighbourhood of the vertex. Conformal methods allow the formulation of characteristic problems for which initial data are prescribed on a null conformal boundary – in this case one talks of an ***asymptotic characteristic initial value problem***; see Friedrich (1981a,b, 1982, 1986c). An attractive feature of characteristic initial value problems is that the field equations, expressed in an adapted gauge, have structural properties which simplify their analysis. In particular, the constraint equations on the initial null hypersurfaces reduce to ordinary differential equations.

Asymptotic characteristic problems allow the aspects of the theory of the asymptotics of isolated systems to be set on a rigorous footing. The basic theory of characteristic problems for hyperbolic equations is discussed in Chapter 12. Applications of this theory to the conformal field equations are given in Chapter 18.

1.5.2 De Sitter-like spacetimes

The simplest type of standard (i.e. non-characteristic) initial value problem for the conformal Einstein field equations involves the construction of de Sitter-like spacetimes. In this case one considers compact initial hypersurfaces $\mathcal{S}$ which are diffeomorphic to the 3-sphere $\mathbb{S}^3$. One has the following concise statement first proved in Friedrich (1986b).

Theorem 1.3 (***global existence and stability of de Sitter-like spacetimes***) *Solutions to the Einstein field Equations* (1.2) *with a de Sitter-like value of the cosmological constant arising from Cauchy initial data close to data for the de Sitter spacetime are asymptotically simple.*

The proof of this result relies on the fact that a conformal representation of the *exact* de Sitter spacetime can be recast as a solution of the conformal Einstein field equations which extends beyond the conformal boundary. It follows from the general theory of hyperbolic equations that the solution of the evolution equations for an initial data set which is close to initial data for the background solution will give rise, in its development, to a spacelike hypersurface on which the conformal factor vanishes. This hypersurface can then be interpreted as the conformal boundary of the perturbed spacetime. Thus, the resulting perturbed spacetime has the same global structure as the de Sitter spacetime, and one can say that, in this case, the notion of asymptotic simplicity is *stable*. Remarkably, a variation of Theorem 1.3 allows for the possibility of prescribing initial data on the conformal boundary.

Theorem 1.3 can be extended to include the coupling of the gravitational field with various types of *trace-free matter*. A detailed discussion of the proof of Theorem 1.3 is given in Chapter 15.

1.5.3 Anti-de Sitter-like spacetimes

As already mentioned, the anti-de Sitter spacetime provides one of the basic examples of non-globally hyperbolic spacetimes. This peculiarity of the spacetime can be attributed to the timelike nature of its conformal boundary; this is further discussed in Chapter 14. As a consequence of the above, spacetimes with a global structure which is similar to that of the anti-de Sitter spacetime cannot be constructed using a standard initial value problem, and the initial data have to be supplemented by suitable boundary data on the hypothetic conformal boundary. This type of setting was first analysed in Friedrich (1995) and requires the identification of initial data which can be described as *anti-de Sitter-like* and appropriate boundary data for the conformal Einstein field equations on a timelike hypersurface representing the conformal boundary. It turns out that initial data sets $(\tilde{\mathcal{S}}, \tilde{\boldsymbol{h}}, \tilde{\boldsymbol{K}})$ for anti-de Sitter-like spacetimes are characterised by the fact that they admit a conformal extension $(\mathcal{S}, \boldsymbol{h}, \boldsymbol{K})$ such that $\mathcal{S}$ has a boundary $\partial\mathcal{S}$ with the topology of the 2-sphere $\mathbb{S}^2$. Based on the example of the exact anti-de Sitter spacetime one expects the conformal boundary to intersect $\mathcal{S}$ on $\partial\mathcal{S}$ and be of the form $\mathscr{I}_c = (-c, c) \times \partial\mathcal{S}$ for some $c > 0$. A detailed analysis of the conformal evolution equations on $\mathscr{I}_c$ reveals that suitable boundary data for the conformal field equations consists of a three-dimensional Lorentzian metric $\boldsymbol{\ell}$. In order to ensure the smoothness of solutions, the underlying PDE theory requires certain compatibility conditions (*corner conditions*) between the initial and the boundary data which are implied by the conformal field equations. Taking into account the above observations one has the following.

Theorem 1.4 (***local existence of anti-de Sitter-like spacetimes***) *Consider an anti-de Sitter-like initial data set* $(\tilde{\mathcal{S}}, \tilde{\boldsymbol{h}}, \tilde{\boldsymbol{K}})$ *for the Einstein field equations and a Lorentzian three-dimensional metric* $\boldsymbol{\ell}$ *on* $\mathscr{I}_c$. *Assume that the above data satisfy suitable corner conditions. Then, there exists a solution to the Einstein field equations* $(\tilde{\mathcal{M}}, \tilde{\boldsymbol{g}})$ *with anti-de Sitter-like cosmological constant and an associated conformal extension* $(\mathcal{M}, \boldsymbol{g})$ *such that* $\tilde{\mathcal{S}}$ *is a spacelike hypersurface of* $(\tilde{\mathcal{M}}, \tilde{\boldsymbol{g}})$ *and so that* $(\tilde{\boldsymbol{h}}, \tilde{\boldsymbol{K}})$ *coincides with the intrinsic metric and extrinsic curvature implied by* $(\tilde{\mathcal{M}}, \tilde{\boldsymbol{g}})$ *on* $\tilde{\mathcal{S}}$. *Furthermore,* $\mathscr{I}_c$ *is the conformal boundary of* $(\mathcal{M}, \boldsymbol{g})$ *and the intrinsic metric of* $\mathscr{I}_c$ *implied by* $\boldsymbol{g}$ *belongs to the conformal class of* $\boldsymbol{\ell}$.

The proof of the above theorem is described in Chapter 17. The above theorem ensures only local existence of anti-de Sitter-like spacetimes, that is, the existence of a solution close to $\tilde{\mathcal{S}}$. It says nothing about the global existence or stability of solutions. Accordingly, it does not require assumptions on the smallness of the data. At the time of writing, the question of the stability (or lack thereof) is an open problem.

1.5.4 Minkowski-like spacetimes

The analysis of Minkowski-like spacetimes gives rise to some of the most challenging open problems in the application of conformal methods in general relativity.

In principle, one would like to construct Minkowski-like spacetimes by prescribing suitable *asymptotically Euclidean* initial data on a three-dimensional manifold $\tilde{\mathcal{S}}$ which is a Cauchy hypersurface of the hypothetic spacetime. However, it turns out that a simpler problem consists of the specification of initial data on a 3-manifold $\tilde{\mathcal{H}}$ describing a hypersurface of $\tilde{\mathcal{M}}$ which in the conformal extension intersects $\mathscr{I}$ —a so-called ***hyperboloid***. Hyperboloidal initial data sets $(\tilde{\mathcal{H}}, \tilde{\boldsymbol{h}}, \tilde{\boldsymbol{K}})$ admit conformal extensions $(\mathcal{H}, \boldsymbol{h}, \boldsymbol{K})$ for which $\mathcal{H}$ is a manifold with boundary $\partial\mathcal{H}$ which has the topology of the 2-sphere $\mathbb{S}^2$ – this boundary corresponds to the intersection of the hyperboloid with $\mathscr{I}$. Hyperboloidal initial data sets are similar in structure to anti-de Sitter-like initial data. There is, in fact, a correspondence between the two; this relation is explored in Chapter 11. An important feature of hyperboloids is that they are not Cauchy hypersurfaces; that is, they do not allow the reconstruction of a whole Minkowski-like spacetime. Despite this shortcoming, one has the following *semi-global* existence and stability result first proved in Friedrich (1986b).

Theorem 1.5 (***semi-global existence and stability of the hyperboloidal initial value problem***) *Solutions to the hyperboloidal initial value problem for the Einstein Equation* (1.3) *with initial data* $(\tilde{\mathcal{H}}, \tilde{\boldsymbol{h}}, \tilde{\boldsymbol{K}})$ *which are suitable perturbations of Minkowski hyperboloidal data are asymptotically simple to the future of* $\tilde{\mathcal{H}}$ *and have a conformal boundary with the same global structure as the conformal boundary of Minkowski spacetime.*

A detailed account of this result is given in Chapter 16. Aside from some technical details, the key ideas of the proof of this result are similar to those of Theorem 1.3 for de Sitter-like spacetimes. Again, a conformal point of view allows one to provide a global existence result for the Einstein field equations in terms of a problem involving a finite existence time. A proof of the non-linear stability of the Minkowski spacetime making use of initial data prescribed on a Cauchy initial hypersurface has been given in the work by Christodoulou and Klainerman (1993). This proof relies on a detailed analysis of the decay of the gravitational field using carefully constructed estimates. Remarkably, the main result of this work does not provide enough regularity at infinity for us to conclude that the spacetime obtained is asymptotically simple.

Time-independent solutions

An important source of intuition on the behaviour of general Minkowski-like spacetimes is provided by the analysis of *time-independent spacetimes*, that is, spacetimes possessing a continuous symmetry which (at least) in the asymptotic region is timelike. If the Killing vector of a time-independent solution is hypersurface orthogonal, then one speaks of a *static spacetime*. Otherwise, one has a *stationary solution*. In the vacuum case, static and stationary solutions can be thought of as describing the exterior gravitational field of some compact matter configuration. In addition, the Schwarzschild and Kerr spacetimes describe time-independent black holes. From the point of view of conformal geometry, their relevance lies in that they allow a detailed analysis of *spatial infinity*, that is, the portion of the conformal boundary intersecting the conformal extension $\mathcal{S}$ of a Cauchy hypersurface $\tilde{\mathcal{S}}$. Vacuum time-independent spacetimes can be shown to admit conformal extensions which are as smooth as one can expect.

Time-independent spacetimes are described by equations which, in a suitable gauge, are elliptic. This feature of this class of solutions explains many of their rigidity and uniqueness properties – in particular, they are characterised through a sequence of *multipole moments*. The analysis of these expansions and other asymptotic properties of static and stationary solutions can be performed in a very convenient manner through conformal methods. In addition, and quite remarkably, static spacetimes can be shown to have a close relation to spacetimes constructed from an asymptotic characteristic initial value problem on a light cone. These and further aspects of static solutions are discussed in Chapter 19.

Spatial infinity

The *asymptotic region* of Cauchy hypersurfaces of Minkowski-like spacetimes can be conformally extended to include a further point – the *point at infinity*. In these conformal extensions, domains in the asymptotic region are transformed into suitable neighbourhoods of the point at infinity. This *point compactification* procedure is a generalisation of the compactification of $\mathbb{R}^2$ into $\mathbb{S}^2$. From a spacetime perspective, the point at infinity gives rise to *spatial infinity* i^0. In this picture, i^0 can be thought of as the vertex of the light cone of $\mathscr{I}$, and the Minkowski-like spacetime corresponds to the exterior of the cone; this construction is analysed in Chapters 19 and 20.

The construction of Minkowski-like asymptotically simple spacetimes from Cauchy initial data requires a precise understanding of the behaviour of the gravitational field in a neighbourhood of spatial infinity. It was first observed by Penrose (1965) that for spacetimes with non-vanishing mass the conformal structure becomes singular at spatial infinity. As a consequence, the initial data implied by the Bianchi Equation (1.9) – which, as already discussed, is one of the key constituents of the conformal field equations – blows up at spatial infinity. The resulting singularity makes the analysis of solutions to the conformal field equations in this region of spacetime particularly challenging. This observation explains, to some extent, why the first results on the existence of Minkowski-like spacetimes were restricted to the developments of hyperboloidal initial data. Early attempts to analyse this situation – see, for example, Beig and Schmidt (1982), Beig (1984) and Friedrich (1988) – reached an impasse due to the lack of a suitable representation of spatial infinity. A breakthrough in this direction was given in Friedrich (1998c) where a representation of spatial infinity based on the properties of conformal geodesics, the so-called *cylinder at spatial infinity*, allows one to formulate a *regular* finite initial value problem for the conformal field equations at spatial infinity. In recent years, a considerable amount of work has been devoted to exploring the implications of this construction. The picture that has progressively emerged is that the conditions required to ensure the existence of asymptotically simple developments out of asymptotically Euclidean initial data are much more restrictive than what one would first expect.

The analysis of the structure of spatial infinity has been informed by developments in the construction of solutions to the constraint equations of general relativity. The exterior asymptotic gluing constructions introduced in Corvino (2000) and Corvino and Schoen (2006) allow one to *glue* static and stationary asymptotic regions to otherwise completely general asymptotically Euclidean initial data sets, the basic ideas of the exterior asymptotic gluing construction are briefly discussed in Chapter 11. As already observed, time-independent solutions to the Einstein field equations are well behaved in a neighbourhood of spatial infinity. Chruściel and Delay (2002) have shown that it is possible to combine this observation with Theorem 1.5 to obtain complete Minkowski-like asymptotically simple spacetimes. The spacetimes obtained in this manner are very special, as they are exactly static, or, more generally, stationary in a neighbourhood of spatial infinity – nevertheless, radiation is registered at null infinity. It is natural to ask whether it is possible to relax this *rigid* behaviour so as to obtain more general types of asymptotically simple spacetimes. The analysis of the *problem of spatial infinity* remains a challenging open area of research; an introductory discussion to the problem of spatial infinity is provided in Chapter 20.

1.6 Perspectives

At the time of writing, the use of conformal methods to analyse the global existence and stability of solutions to the Einstein field equations has been mainly restricted to asymptotically simple spacetimes. One of the motivations behind this book is to encourage researchers interested in the open problems of mathematical relativity to further extend the available conformal methods so as to make them suitable for the analysis of more complicated spacetimes – for example, black holes. From the author's point of view, the realisation of this vision requires the development of not only analytic tools, but also a computational framework which allows one to perform *numerical relativity* using the conformal field equations. Some ideas in this direction are put forward in the concluding Chapter 21.

1.7 Structure of this book

This book is divided in four parts. Throughout, a combination of abstract index notation and index-free notation has been used. An index-free notation has been preferred whenever it simplifies the presentation and emphasises structural aspects of an equation, while abstract indices are used, mostly, in detailed calculations. The spinorial conventions follow those in the monograph of Penrose and Rindler (1984). In view of the systematic use of spinors, this book adopts a $(+---)$ convention for the signature of Lorentzian metrics. As a consequence of this convention the sign of the cosmological constant in the de Sitter spacetime is negative, while for the anti-de Sitter spacetime it is positive. In order to avoid confusion – inasmuch as it is possible – with other sources, a negative cosmological constant will be described as being *de Sitter-like* and a positive one as being *anti-de Sitter-like*. Further details on conventions can be found in Chapters 2, 3 and 4.

Throughout this book ***bold italics*** are systematically used to denote that a given concept is being defined, while *italics* are used to highlight an idea; the attentive reader will realise that sometimes the distinction between these two is blurry.

The content of the four parts of this book can be briefly described as follows.

Part I (Geometric tools) provides a self-contained discussion of the differential geometric and spinorial notions that will be used throughout the book. The presentation and selection of material is tailored to the needs of the discussion in Parts II and III and the applications in Part IV. Chapter 2 gives a brief account of the required notions of differential geometry. The purpose of the chapter is not only to serve as a quick reference in later parts of the book but also to elaborate certain ideas which are not readily available elsewhere in the literature. Chapter 3 provides an account of 2-spinors, while Chapter 4 develops the so-called space spinor formalism. Chapter 5 provides an introduction to conformal geometry which covers not only the transformation formulae for the connection and curvature but also not so well-known topics such as Weyl connections and conformal geodesics – two key notions which will be further developed in Parts II and III.

Part II (General relativity and conformal geometry) provides an introduction to the use of conformal methods in general relativity. It also develops a toolkit of other mathematical methods which will be used to extract information from the Einstein field equations. Chapter 6 provides a brief survey of the construction of conformal extensions of basic solutions to the Einstein field equations – the Minkowski, de Sitter, anti-de Sitter and Schwarzschild spacetimes – as well as a general framework for the construction of Penrose diagrams of spherically symmetric static spacetimes. Chapter 7 provides a discussion of one of the leading themes of this book, the concept of asymptotically simple spacetimes and a formulation of the so-called Penrose's proposal. Chapter 8 gives a derivation and detailed discussion of the main tool of this book, the conformal Einstein field equations. Several versions of the equations are considered – metric, frame, spinorial and in terms of Weyl connections. Chapter 9 complements Chapter 8 and describes matter models amenable to treatment by means of conformal methods. Several of the main results of this book for the vacuum case can be generalised by including these matter models. Chapter 10 provides a brief discussion of the *formal* theory of the asymptotics of spacetime – sometimes also called *asymptopia*. This is a vast topic with a sprawling literature. It is thus impossible to do full justice to the subject in a concise chapter. Accordingly, the decision has been made to restrict the material to aspects of the subject which motivate the later parts of the book.

Part III (Methods of PDE theory) provides an account of PDE and spinorial methods that will be used systematically in Part IV to obtain statements about the existence of various types of solutions to the Einstein field equations. Chapter 11 provides a discussion of the constraint equations implied by the conformal Einstein field equations on spacelike and timelike hypersurfaces – the so-called conformal constraint equations. The proper discussion of this material requires the introduction of certain notions of elliptic PDE theory. This is done at various places in the chapter. Chapter 12 provides a discussion of the methods of the theory of hyperbolic PDEs which will be used in the latter parts of the book. This chapter has been written with the applications in Part IV in mind and covers basic local existence and uniqueness results for initial value, boundary value and characteristic initial value problems. Chapter 13 discusses in detail various hyperbolic reduction procedures for the conformal Einstein field equations by means of spinorial methods. The analysis is not restricted to the evolution systems, but also considers the subsidiary evolution equations required to prove the *propagation of the constraints*. Part III of the book concludes with Chapter

14 where a brief discussion of Lorentzian causality and key aspects of the Cauchy problem in general relativity are given.

Part IV (Applications) is concerned with applications of the conformal Einstein field equations to the analysis of the existence of asymptotically simple spacetimes. Chapter 15 analyses the global existence and stability of de Sitter-like spacetimes; see Theorem 1.3. Two different proofs are provided: the first one makes use of the standard conformal field equations and gauge source functions, and the second one relies on the extended conformal field equations and conformal Gaussian systems. Chapter 16 provides a proof of the semiglobal existence and stability result for hyperboloidal initial data for the Minkowski spacetime and a detailed analysis of the structure of the conformal boundary of the resulting spacetimes; see Theorem 1.5. Chapter 17 provides a discussion of the construction of anti-de Sitter-like spacetimes by means of an initial boundary value problem; see Theorem 1.5. Chapter 18 discusses a different setting for the construction of solutions to the conformal field equations, that of asymptotic characteristic initial value problems either on intersecting null hypersurfaces (one of them representing null infinity) or on a cone (representing past null infinity). Chapter 19 analyses the properties of static solutions by means of conformal methods. The main purpose of this chapter is to pave the way for the discussion of the problem of spatial infinity, which is analysed in Chapter 20. In particular, a discussion of the construction of the so-called cylinder at spatial infinity is provided.

The book concludes with Chapter 21, which provides a subjective selection of open problems in mathematical general relativity where it is felt that the use of conformal methods can provide fresh insights.

Further reading sections. Each chapter provides a brief literature survey. The purpose of this is to provide the interested reader a convenient point of entry into the literature in case more details or an alternative perspective on the subject are required.

Part I

Geometric tools

2

Differential geometry

The language of general relativity is differential geometry. The present chapter provides a brief review of the ideas and notions of differential geometry that will be used in this book. It also serves the purpose of setting the notation and conventions. The chapter assumes a prior knowledge of the subject at the level, say, of the first chapter of Choquet-Bruhat (2008) or Stewart (1991), or chapters 2 and 3 of Wald (1984). In view of the applications in later parts of this book, some topics which may not be regarded as belonging to the standard baggage of a relativist are discussed in some detail – for example, general (i.e. non-Levi-Civita) connections, the so-called $1+3$ split of tensors – that is, a split based on a congruence of timelike curves, rather than on a foliation, as in the usual $3+1$ – and the analysis of the geometry of submanifolds using a frame formalism.

2.1 Manifolds

The basic objects of study in differential geometry are differentiable manifolds. Intuitively, a manifold is a space that, locally, looks like $\mathbb{R}^n$ for some $n \in \mathbb{N}$. Despite this simplicity at a small scale, the global structure of a manifold can be much more complicated and leads to considerations of differential topology.

2.1.1 On the definition of a manifold

A differentiable function f between open sets $\mathcal{U}, \mathcal{V} \subset \mathbb{R}^n$, $f : \mathcal{U} \to \mathcal{V}$, is called a ***diffeomorphism*** if it is bijective and if its inverse $f^{-1} : \mathcal{V} \to \mathcal{U}$ is differentiable. If f and f^{-1} are C^k functions, then one has a C^k-***diffeomorphism***. Furthermore, if f and f^{-1} are C^∞ functions, one speaks of a ***smooth diffeomorphism*** and one writes $\mathcal{U} \approx \mathcal{V}$. Throughout this book, the word ***smooth*** will be used as a synonym for C^∞. The words *function*, *map* and *mapping* will be used as synonyms of each other.

A ***topological space*** is a set with a well-defined notion of *open* and *closed* sets. Given some topological space $\mathcal{M}$, a ***chart*** on $\mathcal{M}$ is a pair $(\mathcal{U}, \varphi)$, with

$\mathcal{U} \subset \mathcal{M}$ and φ a bijection from $\mathcal{U}$ to an open set $\varphi(\mathcal{U}) \subset \mathbb{R}^n$ such that given $p \in \mathcal{U}$

$$\varphi(p) \equiv (x^1, \ldots, x^n).$$

The entries $x^1, \ldots, x^n$ are called ***local coordinates*** of the point $p \in \mathcal{U}$. The set $\mathcal{U}$ is called the ***domain*** of the chart. Two charts $(\mathcal{U}_1, \varphi_1)$ and $(\mathcal{U}_2, \varphi_2)$ are said to be ***C^k-related*** if the map

$$\varphi_2 \circ \varphi_1^{-1} : \varphi_1(\mathcal{U}_1 \cap \mathcal{U}_2) \to \varphi_2(\mathcal{U}_1 \cap \mathcal{U}_2)$$

and its inverse are C^k. The map $\varphi_1 \circ \varphi_2^{-1}$ defines changes of local coordinates $(x^\mu) = (x^1, \ldots, x^n) \mapsto (y^\mu) = (y^1, \ldots, y^n)$ in the intersection $\mathcal{U}_1 \cap \mathcal{U}_2$; see Figure 2.1. Thus, one can regard the coordinates (y^μ) as functions of the coordinates (x^μ). *All throughout this book the Greek letters μ, $\nu, \ldots$ will be used to denote coordinate indices.* The functions $y^\mu(x^1, \ldots, x^n)$ are C^k and, moreover, the ***Jacobian*** $\det(\partial y^\mu / \partial x^\nu)$ is different from zero.

A ***C^k-atlas*** on $\mathcal{M}$ is a collection of charts whose domains cover the set $\mathcal{M}$. The collection of all C^k-related charts is called a ***maximal atlas***. The pair consisting of the space $\mathcal{M}$ together with its maximal C^k-atlas is called a ***C^k-differentiable manifold***. If the charts are C^∞-related, one speaks of a ***smooth differentiable manifold***. If for each φ in the atlas, the map $\varphi : \mathcal{U} \to \mathbb{R}^n$ has the same n, then the manifold is said to have ***dimension*** n. *In what follows, the discussion will be restricted to manifolds of dimension* 3 *and* 4.

Remark. In introductory discussions of differential geometry one generally considers smooth structures. However, as will be seen in later chapters, when one looks at general relativity from the perspective of conformal geometry, the

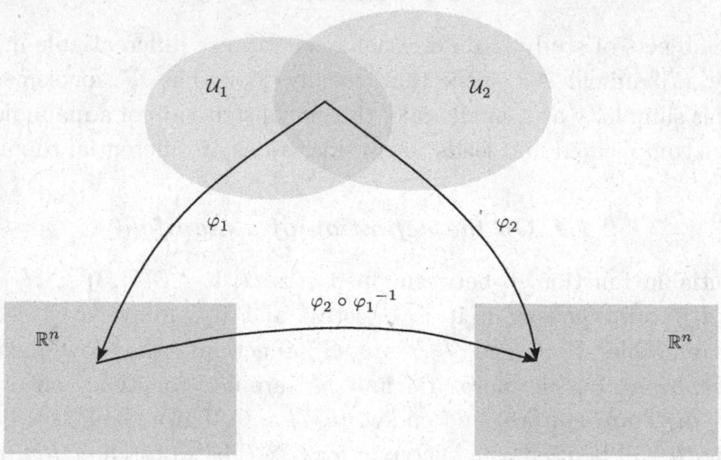

Figure 2.1 Schematic representation of the change of coordinates between charts – see the main text for further details. The figure is adapted from Stewart (1991).

smoothness (or lack thereof) encodes important physical content. Accordingly, one is led to consider the more general class of C^k-differentiable manifolds.

The differentiable manifolds used in general relativity are generally assumed to be Hausdorff and paracompact. A differentiable manifold is ***Hausdorff*** if every two points in it admit non-intersecting open neighbourhoods. The reason for requiring the Hausdorff condition is to ensure that a convergent sequence of points cannot have more than one limit point. If $\mathcal{M}$ is ***paracompact***, then there exists a countable basis of open sets. Paracompactness is used in several basic constructions in differential geometry. In particular, it is required to show that every Riemannian manifold admits a metric. *In what follows, all differentiable manifolds to be considered will be assumed to be Hausdorff and paracompact.* Accordingly, in the rest of the book Hausdorff, paracompact differentiable manifolds will be simply called ***manifolds***.

Orientability

An open set of $\mathbb{R}^n$ is naturally oriented by the order of the coordinates $(x^\mu) = (x^1, \ldots, x^n)$. Hence, a chart $(\mathcal{U}, \varphi)$ inherits an orientation from its image in $\mathbb{R}^n$. In an orientable manifold the orientation of these charts matches together properly. More precisely, a manifold is said to be ***orientable*** if its maximal atlas is such that the Jacobian of the coordinate transformation for each pair of overlapping charts is positive.

An alternative description of the notion of orientability in terms of orthonormal frames will be given in Section 2.5.3. Orientability is a necessary and sufficient condition for the existence of a spinorial structure on $\mathcal{M}$; see, for example, Chapter 3.

2.1.2 Manifolds with boundary

Manifolds with boundary arise naturally when discussing general relativity from the perspective of conformal geometry. In order to introduce this concept one requires the following subsets of $\mathbb{R}^n$:

$$\mathbb{H}^n \equiv \{(x^1, \cdots, x^n) \in \mathbb{R}^n \mid x^n \geq 0\},$$
$$\partial\mathbb{H}^n \equiv \{(x^1, \cdots, x^n) \in \mathbb{R}^n \mid x^n = 0\}.$$

One says that $\mathcal{M}$ is a ***manifold with boundary*** if it can be covered with charts mapping open subsets of $\mathcal{M}$ either to open sets of $\mathbb{R}^n$ or to open subsets of $\mathbb{H}^n$. The ***boundary of*** $\mathcal{M}$, $\partial\mathcal{M}$, is the set of points $p \in \mathcal{M}$ for which there is a chart $(\mathcal{U}, \varphi)$ with $p \in \mathcal{U}$ such that $\varphi(\mathcal{U}) \subset \mathbb{H}^n$ and $\varphi(p) \in \partial\mathbb{H}^n$. The boundary $\partial\mathcal{M}$ is an $(n-1)$-dimensional differentiable manifold in its own right. Hence, it is a *submanifold* of $\mathcal{M}$ – see Section 2.7.1.

2.2 Vectors and tensors on a manifold

In order to probe the geometric properties of a manifold one needs vectors and, more generally, tensors. This section provides a brief discussion of these fundamental notions.

2.2.1 *Some ancillary notions*

Derivations

Denote by $\mathfrak{X}(\mathcal{M})$ the set of ***scalar fields (i.e. functions) over*** $\mathcal{M}$; that is, smooth functions $f : \mathcal{M} \to \mathbb{R}$.

Definition 2.1 (***derivations***) *A **derivation** is a map* $\mathcal{D} : \mathfrak{X}(\mathcal{M}) \to \mathfrak{X}(\mathcal{M})$ *such that:*

(i) ***Action on constants.*** *For all constant fields c,* $\mathcal{D}(c) = 0$.
(ii) ***Linearity.*** *For all* $f, g \in \mathfrak{X}(\mathcal{M})$, $\mathcal{D}(f + g) = \mathcal{D}(f) + \mathcal{D}(g)$.
(iii) ***Leibnitz rule.*** *For all* $f, g \in \mathfrak{X}(\mathcal{M})$, $\mathcal{D}(fg) = \mathcal{D}(f)g + f\mathcal{D}(g)$.

The connection between derivations and covariant derivatives is discussed in Section 2.4.1.

Curves

The notion of a vector is intimately related to that of a curve. Given an open interval $I = (a, b) \subset \mathbb{R}$ where either or both of a, b can be infinite, a smooth ***curve*** on $\mathcal{M}$ is a map $\gamma : I \to \mathcal{M}$ such that for any chart $(\mathcal{U}, \varphi)$, the composition $\varphi \circ \gamma : I \to \mathbb{R}^n$ is a smooth map. One often speaks of the curve $\gamma(\mathrm{s})$ with $\mathrm{s} \in (a, b)$; s is called the ***parameter of the curve.*** If the domain (a, b) of a curve can be extended to, say, $[a, b]$ while keeping $\gamma(\mathrm{s})$ smooth, one has an ***extendible curve.*** A curve which is not extendible is called ***inextendible.***

A ***tangent vector to a curve*** $\gamma(\mathrm{s})$ at a point $p \in \mathcal{M}$, to be denoted as $\dot\gamma(p)$, is the map defined by

$$\dot\gamma(p) : f \mapsto \frac{\mathrm{d}}{\mathrm{ds}}(f \circ \gamma)\big|_p = \dot\gamma(f)\big|_p, \qquad f \in \mathfrak{X}(\mathcal{M}).$$

Given a chart $(\mathcal{U}, \varphi)$ with local coordinates (x^μ), the components of $\dot\gamma(p)$ with respect to the chart are given by

$$\dot x^\mu(p) \equiv \frac{\mathrm{d}}{\mathrm{ds}} x^\mu(\gamma(\mathrm{s}))\big|_p.$$

In a slight abuse of notation the points of the curve γ will often be denoted by $x(\mathrm{s}) \in \mathcal{M}$ and its tangent vector by $\dot{\boldsymbol{x}}(\mathrm{s})$.

2.2.2 Tangent vectors and covectors

To each point $p \in \mathcal{M}$, one can associate a vector space $T|_p(\mathcal{M})$, the ***tangent space at*** p consisting of all the tangent vectors at p. In what follows, the elements of this space will be simply known as ***vectors***. All throughout, vectors will mostly be denoted with lowercase bold Latin letters: $\boldsymbol{v}$, $\boldsymbol{u}$, $\boldsymbol{w}, \ldots$ Abstract index notation will also be used to denote vectors; see Section 2.2.6. The tangent space $T|_p(\mathcal{M})$ can be characterised either as *the set of derivations at p of smooth functions on $\mathcal{M}$* or as *the set of equivalence classes of curves through p under a suitable equivalence relation.* With the first characterisation one considers the vectors as *directional derivatives*, while with the second one they are considered as *velocities.* If the dimension of the manifold $\mathcal{M}$ is n, then $T|_p(\mathcal{M})$ is a vector space of dimension n. Local coordinates (x^μ) in a neighbourhood of the point p give a basis of $T|_p(\mathcal{M})$ consisting of the partial derivative operators $\{\boldsymbol{\partial}/\partial x^\mu\}$; where no confusion arises about which coordinates are meant, one simply writes $\{\boldsymbol{\partial}_\mu\}$. In particular, for the vector tangent to a curve one has that $\dot{\boldsymbol{x}}(\mathrm{s}) = \dot{x}^\mu(\mathrm{s})\boldsymbol{\partial}_\mu$. In this last expression and in what follows, ***Einstein's summation convention*** has been adopted – that is, repeated up and down coordinate indices indicate summation for all values of the range of the index. That is,

$$\dot{x}^\mu(\mathrm{s})\boldsymbol{\partial}_\mu \equiv \sum_{\mu=1}^{n} \dot{x}^\mu(\mathrm{s})\boldsymbol{\partial}_\mu.$$

Covectors

The ***dual space*** $T^*|_p(\mathcal{M})$, the ***cotangent space at*** p, is the vector space of linear maps $\boldsymbol{\omega} : T|_p(\mathcal{M}) \to \mathbb{R}$. Generic elements of $T^*|_p(\mathcal{M})$ will be denoted by lowercase bold Greek letters: $\boldsymbol{\alpha}$, $\boldsymbol{\beta}$, $\boldsymbol{\omega}$, ... Being dual to $T|_p(\mathcal{M})$, the space $T^*|_p(\mathcal{M})$ has also dimension n, and its elements are called ***covectors***. If $\boldsymbol{\omega}$ acts on $\boldsymbol{v} \in T|_p(\mathcal{M})$, then one writes $\langle \boldsymbol{\omega}, \boldsymbol{v} \rangle \in \mathbb{R}$.

Given $f \in \mathfrak{X}(\mathcal{M})$, for each $\boldsymbol{v} \in T|_p(\mathcal{M})$, one has that $\boldsymbol{v}(f)$ is a scalar. Hence, f defines a map, the ***differential*** of f, $\mathbf{d}f : T|_p(\mathcal{M}) \to \mathbb{R}$ via

$$\mathbf{d}f(\boldsymbol{v}) = \boldsymbol{v}(f).$$

As a consequence of the linearity of $\boldsymbol{v}$ one has that $\mathbf{d}f$ is linear, and thus $\mathbf{d}f \in T^*|_p(\mathcal{M})$. Given a chart $(\mathcal{U}, \varphi)$ with coordinates (x^μ), the coordinate differentials $\mathbf{d}x^\mu$ form a basis for $T^*|_p(\mathcal{M})$, the so-called ***dual basis***. The dual basis satisfies $\langle \mathbf{d}x^\mu, \boldsymbol{\partial}_\nu \rangle = \delta_\nu{}^\mu$, where $\delta_\nu{}^\mu$ is the so-called ***Kronecker's delta***. It follows that every covector $\boldsymbol{\omega}$ at $p \in \mathcal{M}$ can be written as $\boldsymbol{\omega} = \langle \boldsymbol{\omega}, \boldsymbol{\partial}_\mu \rangle \mathbf{d}x^\mu$.

Bases

The previous discussion is extended in a natural way to more general bases. Given any basis $\{\boldsymbol{e}_a\}$ of $T|_p(\mathcal{M})$, its ***dual basis*** $\{\boldsymbol{\omega}^b\}$ of $T^*|_p(\mathcal{M})$ is defined by the condition $\langle \boldsymbol{\omega}^{\boldsymbol{b}}, \boldsymbol{e}_{\boldsymbol{a}} \rangle = \delta_{\boldsymbol{a}}{}^{\boldsymbol{b}}$. In the rest of the book, lowercase bold indices such as $\boldsymbol{a}$, $\boldsymbol{b}, \ldots$ denote ***spacetime frame indices*** ranging $\mathbf{0}, \ldots, \mathbf{3}$. These will be used when working with four-dimensional manifolds. The lowercase bold Latin letters $\boldsymbol{i}$, $\boldsymbol{j}$, $\boldsymbol{k}, \ldots$ will range, depending on the context, over either $\mathbf{0}$, $\mathbf{1}$, $\mathbf{2}$ or $\mathbf{1}$, $\mathbf{2}$, $\mathbf{3}$. *For simplicity of presentation, and unless explicitly stated, a four-dimensional manifold will be assumed in the subsequent discussion.*

Given another pair of bases $\{\tilde{\boldsymbol{e}}_{\boldsymbol{a}}\}$ and $\{\tilde{\boldsymbol{\omega}}^{\boldsymbol{b}}\}$ of $T|_p(\mathcal{M})$ and $T^*|_p(\mathcal{M})$, respectively, these are related to the bases $\{\boldsymbol{e}_{\boldsymbol{a}}\}$ and $\{\boldsymbol{\omega}^{\boldsymbol{b}}\}$ by non-singular matrices $(A_{\boldsymbol{a}}{}^{\boldsymbol{b}})$ and $(A^{\boldsymbol{a}}{}_{\boldsymbol{b}})$ such that

$$\tilde{\boldsymbol{e}}_{\boldsymbol{a}} = A_{\boldsymbol{a}}{}^{\boldsymbol{b}} \boldsymbol{e}_{\boldsymbol{b}}, \qquad \tilde{\boldsymbol{\omega}}^{\boldsymbol{a}} = A^{\boldsymbol{a}}{}_{\boldsymbol{b}} \boldsymbol{\omega}^{\boldsymbol{b}}, \tag{2.1}$$

satisfying $A^{\boldsymbol{a}}{}_{\boldsymbol{b}} A^{\boldsymbol{b}}{}_{\boldsymbol{c}} = \delta_{\boldsymbol{c}}{}^{\boldsymbol{a}}$ so that $(A_{\boldsymbol{a}}{}^{\boldsymbol{b}})$ and $(A^{\boldsymbol{a}}{}_{\boldsymbol{b}})$ are inverses of each other. In these last expressions and in what follows, ***Einstein's summation convention*** for repeated contravariant and covariant *frame* indices has been adopted so that a sum from $\boldsymbol{b} = \mathbf{0}$ to $\boldsymbol{b} = \mathbf{3}$ is implied.

Condition (2.1) ensures that the new bases $\{\tilde{\boldsymbol{e}}_{\boldsymbol{a}}\}$ and $\{\tilde{\boldsymbol{\omega}}^{\boldsymbol{b}}\}$ are dual to each other; that is, $\langle \tilde{\boldsymbol{\omega}}^{\boldsymbol{b}}, \tilde{\boldsymbol{e}}_{\boldsymbol{a}} \rangle = \delta_{\boldsymbol{a}}{}^{\boldsymbol{b}}$. Given $\boldsymbol{v} \in T|_p(\mathcal{M})$, $\boldsymbol{\alpha} \in T^*|_p(\mathcal{M})$, the above transformation rules for the bases imply

$$\boldsymbol{v} = v^{\boldsymbol{a}} \boldsymbol{e}_{\boldsymbol{a}} = \tilde{v}^{\boldsymbol{a}} \tilde{\boldsymbol{e}}_{\boldsymbol{a}} = (\tilde{v}^{\boldsymbol{a}} A_{\boldsymbol{a}}{}^{\boldsymbol{b}}) \boldsymbol{e}_{\boldsymbol{b}},$$
$$\boldsymbol{\alpha} = \alpha_{\boldsymbol{a}} \boldsymbol{\omega}^{\boldsymbol{a}} = \tilde{\alpha}_{\boldsymbol{a}} \tilde{\boldsymbol{\omega}}^{\boldsymbol{a}} = (\tilde{\alpha}_{\boldsymbol{a}} A^{\boldsymbol{a}}{}_{\boldsymbol{b}}) \boldsymbol{\omega}^{\boldsymbol{b}}.$$

The two bases are said to have the ***same orientation*** if $\det(A_{\boldsymbol{a}}{}^{\boldsymbol{b}}) > 0$.

2.2.3 Higher rank tensors

Higher rank tensors can be constructed using elements of $T|_p(\mathcal{M})$ and $T^*|_p(\mathcal{M})$ as basic building blocks. A ***contravariant tensor of rank*** k at the point p is a multilinear map

$$\boldsymbol{M} : \underbrace{T^*|_p(\mathcal{M}) \times \cdots \times T^*|_p(\mathcal{M})}_{k \text{ terms}} \longrightarrow \mathbb{R},$$

that is, a function taking k covectors as arguments. Similarly, a ***covariant tensor of rank*** l at the point p is a multilinear map

$$\boldsymbol{N} : \underbrace{T|_p(\mathcal{M}) \times \cdots \times T|_p(\mathcal{M})}_{l \text{ terms}} \longrightarrow \mathbb{R},$$

that is, a function taking l vectors as arguments. More generally, one can also have ***tensors of mixed type***: a (k,l) tensor at p is a multilinear map

$$\boldsymbol{T} : \underbrace{T^*|_p(\mathcal{M}) \times \cdots \times T^*|_p(\mathcal{M})}_{k \text{ terms}} \times \underbrace{T|_p(\mathcal{M}) \times \cdots \times T|_p(\mathcal{M})}_{l \text{ terms}} \longrightarrow \mathbb{R},$$

so that $\boldsymbol{T}$ takes k covectors and l vectors as arguments. In particular, a $(k,0)$-tensor corresponds to a contravariant tensor of rank k, while a $(0,l)$-tensor is a covariant tensor of rank l. The ***space of (k,l)-tensors at the point*** p will be denoted by $T_l^k|_p(\mathcal{M})$. In particular, one has the identifications $T^1|_p(\mathcal{M}) = T|_p(\mathcal{M})$ and $T_1|_p(\mathcal{M}) = T^*|_p(\mathcal{M})$. Formally, the space $T_l^k|_p(\mathcal{M})$ is obtained as the *tensor product* of k copies of $T^*|_p(\mathcal{M})$ and l copies of $T|_p(\mathcal{M})$. That is, one has that

$$T_l^k|_p(\mathcal{M}) = \underbrace{T|_p(\mathcal{M}) \otimes \cdots \otimes T|_p(\mathcal{M})}_{k \text{ terms}} \otimes \underbrace{T^*|_p(\mathcal{M}) \otimes \cdots \otimes T^*|_p(\mathcal{M})}_{l \text{ terms}}.$$

The ordering given in the previous expression is known as the ***standard order***. Notice, however, that an arbitrary tensor does not need not to have its arguments in standard order.

As an example of the previous discussion consider $\boldsymbol{v} \in T|_p(\mathcal{M})$ and $\boldsymbol{\alpha} \in T^*|_p(\mathcal{M})$. Their ***tensor product*** $\boldsymbol{v} \otimes \boldsymbol{\alpha}$ is then defined by

$$(\boldsymbol{v} \otimes \boldsymbol{\alpha})(\boldsymbol{u}, \boldsymbol{\beta}) = \langle \boldsymbol{\beta}, \boldsymbol{v} \rangle \langle \boldsymbol{\alpha}, \boldsymbol{u} \rangle, \qquad \boldsymbol{u} \in T|_p(\mathcal{M}), \qquad \boldsymbol{\beta} \in T^*|_p(\mathcal{M}). \tag{2.2}$$

One readily sees that $\boldsymbol{v} \otimes \boldsymbol{\alpha}$ is a bilinear map and thus a $(1,1)$-tensor at $p \in \mathcal{M}$. The action of the tensor product given in Equation (2.2) can be extended directly to an arbitrary (finite) number of tensors and covectors. If $\{\boldsymbol{e}_a\}$ and $\{\boldsymbol{\omega}^b\}$ denote, respectively, bases of $T|_p(\mathcal{M})$ and $T^*|_p(\mathcal{M})$, then a basis of $T_l^k|_p(\mathcal{M})$ is given by

$$\{\boldsymbol{e}_{b_1} \otimes \cdots \otimes \boldsymbol{e}_{b_k} \otimes \boldsymbol{\omega}^{a_1} \otimes \cdots \otimes \boldsymbol{\omega}^{a_l}\}.$$

The collection of all the tensor spaces of the form $T_l^k|_p(\mathcal{M})$ is called the ***tensor algebra*** at p and will be denoted by $T^\bullet|_p(\mathcal{M})$. The tensor algebra is defined by means of a *direct sum*.

Symmetries of tensors

A covariant tensor of rank l, say, $\boldsymbol{S}$, is said to be ***symmetric*** with respect to its ith and jth arguments if

$$\boldsymbol{S}(\boldsymbol{v}_1, \ldots, \boldsymbol{v}_i, \ldots, \boldsymbol{v}_j, \ldots, \boldsymbol{v}_l) = \boldsymbol{S}(\boldsymbol{v}_1, \ldots, \boldsymbol{v}_j, \ldots, \boldsymbol{v}_i, \ldots, \boldsymbol{v}_l). \tag{2.3}$$

Similarly, $\boldsymbol{A}$ it is said to be ***antisymmetric*** if

$$\boldsymbol{A}(\boldsymbol{v}_1, \ldots, \boldsymbol{v}_i, \ldots, \boldsymbol{v}_j, \ldots, \boldsymbol{v}_l) = -\boldsymbol{A}(\boldsymbol{v}_1, \ldots, \boldsymbol{v}_j, \ldots, \boldsymbol{v}_i, \ldots, \boldsymbol{v}_l). \tag{2.4}$$

If the properties (2.3) and (2.4) hold under interchange of any arbitrary pair of indices, one says that $\boldsymbol{S}$ is ***totally symmetric*** and $\boldsymbol{A}$ is ***totally antisymmetric***, respectively. The above definitions can be extended to contravariant tensors of arbitrary rank. A totally antisymmetric covariant tensor of rank l is also called an ***l-form***. Symmetry properties of tensors are best expressed in terms of abstract index notation.

2.2.4 Tensor fields

The discussion in the previous subsections concerned the notion of a tensor at a point $p \in \mathcal{M}$. The ***tensor bundle over*** $\mathcal{M}$, $\mathfrak{T}^\bullet(\mathcal{M})$, is the *disjoint union* of the tensor algebras $T^\bullet|_p(\mathcal{M})$ for all $p \in \mathcal{M}$:

$$\mathfrak{T}^\bullet(\mathcal{M}) \equiv \coprod_{p\in\mathcal{M}} T^\bullet|_p(\mathcal{M}).$$

The disjoint union emphasises that although for $p, q \in \mathcal{M}$, $p \neq q$, the spaces $T^\bullet|_p(\mathcal{M})$ and $T^\bullet|_q(\mathcal{M})$ are *isomorphic*; they are regarded as different sets. Important subsets of the tensor bundles are the ***tangent bundle*** and the ***cotangent bundle*** given, respectively, by

$$T(\mathcal{M}) \equiv \coprod_{p\in\mathcal{M}} T|_p(\mathcal{M}), \qquad T^*(\mathcal{M}) \equiv \coprod_{p\in\mathcal{M}} T^*|_p(\mathcal{M}).$$

A ***smooth tensor field*** over $\mathcal{M}$ is a prescription of a tensor $\boldsymbol{T} \in T^\bullet|_p(\mathcal{M})$ at each $p \in \mathcal{M}$ such that when $\boldsymbol{T}$ is represented locally in a system of coordinates around p, the corresponding components are smooth functions on the local chart and, more generally, across the atlas. This idea can be naturally extended to consider tensor fields which are not smooth but just C^k for some positive integer k. An important property of tensor fields is that they are multilinear over $\mathfrak{X}(\mathcal{M})$. This property is often referred to as ***$\mathfrak{X}$-linearity***. It can be used to characterise tensors. More precisely, one has the following lemma which will be used repeatedly (see Penrose and Rindler (1984) for a proof):

Lemma 2.1 (***characterisation of tensors***) *A map*

$$\boldsymbol{T} : T^*(\mathcal{M}) \times \cdots \times T^*(\mathcal{M}) \times T(\mathcal{M}) \times \cdots \times T(\mathcal{M}) \to \mathfrak{X}(\mathcal{M})$$

is induced by a (k, l)-tensor field if and only if it is multilinear over $\mathfrak{X}(\mathcal{M})$.

The discussion of tensor fields and the tensor bundle is naturally carried out using the language of *fibre bundles*; see, for example, Kobayashi and Nomizu (2009). This point of view will, however, not be used in this book.

2.2.5 The commutator of vector fields

Given $\boldsymbol{u}, \boldsymbol{v} \in T(\mathcal{M})$, their ***commutator*** $[\boldsymbol{u}, \boldsymbol{v}] \in T(\mathcal{M})$ is the vector field defined by

$$[\boldsymbol{u}, \boldsymbol{v}]f \equiv \boldsymbol{u}(\boldsymbol{v}(f)) - \boldsymbol{v}(\boldsymbol{u}(f)),$$

for $f \in \mathfrak{X}(\mathcal{M})$. Given a basis $\{\boldsymbol{e_a}\}$ one has that the components of the commutator with respect to this basis are given by

$$[\boldsymbol{u}, \boldsymbol{v}]^{\boldsymbol{a}} = \boldsymbol{u}(v^{\boldsymbol{a}}) - \boldsymbol{v}(u^{\boldsymbol{a}}), \qquad u^{\boldsymbol{a}} \equiv \langle \boldsymbol{\omega}^{\boldsymbol{a}}, \boldsymbol{u} \rangle, \quad v^{\boldsymbol{a}} \equiv \langle \boldsymbol{\omega}^{\boldsymbol{a}}, \boldsymbol{v} \rangle.$$

One can readily verify that

$$[\boldsymbol{u}, \boldsymbol{v}] = -[\boldsymbol{v}, \boldsymbol{u}],$$
$$[\boldsymbol{u} + \boldsymbol{v}, \boldsymbol{w}] = [\boldsymbol{u}, \boldsymbol{w}] + [\boldsymbol{v}, \boldsymbol{w}],$$
$$[[\boldsymbol{u}, \boldsymbol{v}], \boldsymbol{w}] + [[\boldsymbol{v}, \boldsymbol{w}], \boldsymbol{u}] + [[\boldsymbol{w}, \boldsymbol{u}], \boldsymbol{v}] = 0.$$

The last identity is known as the ***Jacobi identity*** – not to be confused with the Jacobi identity for spinors, to be discussed in Chapter 3.

2.2.6 Abstract index notation for tensors

The presentation of tensors in this section has so far used an ***index-free notation***. In the sequel, the so-called ***abstract index notation*** will also be used where convenient; see Penrose and Rindler (1984). To this end, lowercase Latin indices will be employed. Accordingly, a vector field $\boldsymbol{v} \in T(\mathcal{M})$ will also be written as v^a. Similarly, for $\boldsymbol{\alpha} \in T^*(\mathcal{M})$ one writes α_a. More generally, a (k, l)-tensor $\boldsymbol{T}$ will be denoted by $T^{a_1 \cdots a_k}{}_{b_1 \cdots b_l}$. *It is important to stress that the indices in these expressions do not represent components with respect to some coordinates or frame.* These components are denoted, respectively, by Greek indices and bold lowercase Latin indices such as in v^μ and $v^{\boldsymbol{a}}$. The role of the abstract indices is to specify in a simple way the nature of the object under consideration and to describe in a convenient fashion operations between tensors. In particular, the action $\langle \boldsymbol{\alpha}, \boldsymbol{v} \rangle$ of a 1-form on a vector is denoted in abstract index notation by $\alpha_a v^a$, while its tensor product $\boldsymbol{\alpha} \otimes \boldsymbol{v}$ is written as $\alpha_a v^b$. Similarly, the operation defined in Equation (2.2) is expressed as $\alpha_a u^a \beta_b v^b$.

The idea behind the use of abstract indices is to have a notation for tensorial expressions that mirrors the expressions for their basis components (had a basis been introduced). Using the index notation one can write only tensorial expressions since no basis has been specified; see, for example, Wald (1984) for a further discussion on this subject.

Each type of notation has its own advantages. In particular, the index-free notation is better to describe conceptual and structural aspects, while the abstract index notation is useful in explicit computations. In particular, the abstract index notation allows the expression, in a convenient way, of tensors whose arguments are not given in standard order as in $F_{ab}{}^c{}_d$.

An operation which has a particularly convenient description in terms of abstract indices is the ***contraction*** between a contravariant and a covariant index. For example, given $F_{ab}{}^c{}_d$, the contraction between the contravariant index ^{c} and, say, the covariant index ${}_d$ is denoted by $F_{ab}{}^c{}_c$. Following the convention that repeated indices are *dummy* indices one has, for example, that $F_{ab}{}^c{}_c = F_{ab}{}^d{}_d$. Given a basis $\{\boldsymbol{e_a}\}$ and a cobasis $\{\boldsymbol{\omega^a}\}$, their elements are denoted, using abstract index notation, as $e_{\boldsymbol{a}}{}^a$ and $\omega^{\boldsymbol{a}}{}_a$, respectively. If $F_{\boldsymbol{ab}}{}^{\boldsymbol{c}}{}_{\boldsymbol{d}} \equiv F_{ab}{}^c{}_d e_{\boldsymbol{a}}{}^a e_{\boldsymbol{b}}{}^b \omega^{\boldsymbol{c}}{}_c e_{\boldsymbol{d}}{}^d$ denotes the components of $F_{ab}{}^c{}_d$ with respect to a basis $\{\boldsymbol{e_a}\}$ and its associated cobasis $\{\boldsymbol{\omega^a}\}$, then the components of the contraction $F_{ab}{}^c{}_c$

are given by $F_{\boldsymbol{ab}}{}^{\boldsymbol{c}}{}_{\boldsymbol{c}}$. *Following Einstein's summation convention, a sum on the index* $\boldsymbol{c}$ *is understood.* Although this definition is given in terms of components with respect to a basis, the contraction is a *geometric* (i.e. coordinate- and base-independent) operation transforming a tensor of rank (k, l) into a tensor of rank $(k-1, l-1)$.

Symmetries of tensors are expressed in a convenient fashion using abstract index notation. For example, if S_{ab} and A_{ab} denote, respectively, symmetric and antisymmetric covariant tensors of rank 2, then $S_{ab} = S_{ba}$ and $A_{ab} = -A_{ba}$. More generally, given M_{ab}, its ***symmetric*** and ***antisymmetric parts*** are defined, respectively, by the expressions

$$M_{(ab)} \equiv \frac{1}{2}(M_{ab} + M_{ba}), \qquad M_{[ab]} \equiv \frac{1}{2}(M_{ab} - M_{ba}).$$

The operations of symmetrisation and antisymmetrisation can be extended to higher rank tensors. In particular, it is noticed that for a rank 3 covariant tensor T_{abc} one has

$$T_{[abc]} \equiv \frac{1}{3!}(T_{abc} + T_{bca} + T_{cab} - T_{acb} - T_{cba} - T_{bac}).$$

If a tensor $S_{a_1 \cdots a_l}$ is symmetric with respect to the indices $a_1, \ldots, a_l$, then one writes $S_{a_1 \cdots a_l} = S_{(a_1 \cdots a_l)}$. Similarly, if $A_{a_1 \cdots a_l}$ is antisymmetric with respect to $a_1, \ldots, a_l$, one writes $A_{a_1 \cdots a_l} = A_{[a_1 \cdots a_l]}$ and $A_{a_1 \cdots a_l}$ is said to be an ***l-form***.

Consistent with the abstract index notation for tensors, it is convenient to introduce a similar convention to denote the various tensor spaces. Accordingly the bundle $\mathfrak{T}^k_l(\mathcal{M})$ will, in the following, be denoted by $\mathfrak{T}^{a_1 \cdots a_k}{}_{b_1 \cdots b_l}(\mathcal{M})$. In particular, in this notation the tangent bundle $T(\mathcal{M})$ is denoted by $\mathfrak{T}^a(\mathcal{M})$, while the cotangent bundle $T^*(\mathcal{M})$ is given by $\mathfrak{T}_a(\mathcal{M})$.

A further discussion of the abstract index notation with specific remarks in the treatment of spinors is given in Section 3.1.4.

2.3 Maps between manifolds

This section discusses maps between manifolds. In what follows let $\mathcal{M}$ and $\mathcal{N}$ denote two manifolds. These manifolds could be the same.

2.3.1 Push-forwards and pull-backs

A map $\varphi : \mathcal{N} \to \mathcal{M}$ is said to be smooth (C^∞) if for every smooth function $f \in \mathfrak{X}(\mathcal{M})$, the composition $\varphi^* f \equiv f \circ \varphi : \mathcal{N} \to \mathbb{R}$ is also smooth. Given $p \in \mathcal{N}$, let $T|_p(\mathcal{N})$, $T|_{\varphi(p)}(\mathcal{M})$ denote, respectively, the tangent spaces at $p \in \mathcal{N}$ and $\varphi(p) \in \mathcal{M}$. The map $\varphi : \mathcal{N} \to \mathcal{M}$ induces a map $\varphi_* : T|_p(\mathcal{N}) \to T|_{\varphi(p)}(\mathcal{M})$, the ***push-forward***, through the formula

$$(\varphi_* \boldsymbol{v}) f(p) \equiv \boldsymbol{v}(f \circ \varphi)(p), \qquad \boldsymbol{v} \in T|_p(\mathcal{N}).$$

It can be readily verified that φ_* so defined is a $\mathfrak{X}$-linear map; that is, given $\boldsymbol{v}$, $\boldsymbol{u} \in T|_p(\mathcal{N})$ and a function $f \in \mathfrak{X}(\mathcal{M})$ one has $\varphi_*(f\boldsymbol{v} + \boldsymbol{u}) = f\varphi_*\boldsymbol{v} + \varphi_*\boldsymbol{u}$.

Note that the above definition is made in a point-wise manner. *Smooth vector fields do not, in general, push forward to smooth vector fields, except in the case of diffeomorphisms.* For example, if φ is not surjective, then there is no way of deciding which vector to assign to a point not on the image of φ. If φ is not injective, then for some points of $\mathcal{M}$, there may be several different vectors obtained as push-forwards of a vector on $\mathcal{N}$. However, given $\varphi : \mathcal{N} \to \mathcal{M}$ a diffeomorphism, for every $\boldsymbol{v} \in T(\mathcal{N})$ there exists a unique vector field on $T(\mathcal{M})$ obtained as the pull-back of $\boldsymbol{v}$; see Lee (2002).

The push-forward $\varphi_* : T(\mathcal{N}) \to T(\mathcal{M})$ can be used, in turn, to define a map $\varphi^* : T^*(\mathcal{M}) \to T^*(\mathcal{N})$, the ***pull-back***, as

$$\langle \varphi^* \boldsymbol{\omega}, \boldsymbol{v} \rangle \equiv \langle \boldsymbol{\omega}, \varphi_* \boldsymbol{v} \rangle, \qquad \boldsymbol{\omega} \in T^*(\mathcal{M}), \qquad \boldsymbol{v} \in T(\mathcal{N}).$$

Again, it can be readily verified that φ^* so defined is $\mathfrak{X}$-linear: $\varphi^*(f\boldsymbol{\omega} + \boldsymbol{\zeta}) = f^* \varphi^* \boldsymbol{\omega} + \varphi^* \boldsymbol{\zeta}$ for $\boldsymbol{\omega}$, $\boldsymbol{\zeta} \in T^*(\mathcal{M})$. The pull-back commutes with the differential $\mathbf{d}$; that is, $\varphi^*(\mathbf{d}f) = \mathbf{d}(\varphi^* f)$. *Contrary to the case of push-forwards, pull-backs of smooth covector fields always lead to smooth covector fields. There is no ambiguity in the construction.* In the case that $\varphi : \mathcal{N} \to \mathcal{M}$ is a diffeomorphism, then the inverse pull-back $(\varphi^*)^{-1}$ is well defined so that covectors can be pulled back from $T^*(\mathcal{N})$ to $T^*(\mathcal{M})$.

The operations of push-forward and pull-back can be extended in a natural way, respectively, to arbitrary contravariant and covariant tensors. The case of most relevance for the subsequent discussion is that of a covariant tensor of rank 2, $\boldsymbol{g} \in \mathfrak{T}_2(\mathcal{M})$. Its pull-back $\varphi^* \boldsymbol{g} \in \mathfrak{T}_2(\mathcal{N})$ satisfies

$$(\varphi^* \boldsymbol{g})(\boldsymbol{u}, \boldsymbol{v}) = \boldsymbol{g}(\varphi_* \boldsymbol{u}, \varphi_* \boldsymbol{v}), \qquad \boldsymbol{u}, \boldsymbol{v} \in T(\mathcal{N}).$$

2.3.2 Lie derivatives

Smooth maps of the manifold into itself, $\varphi : \mathcal{M} \to \mathcal{M}$, lead to the notion of the Lie derivative. Given a vector $\boldsymbol{v}$, the ***Lie derivative*** $\pounds_{\boldsymbol{v}}$ measures the change of a tensor field along the integral curves of $\boldsymbol{v}$.

In what follows, let $f \in \mathfrak{X}(\mathcal{M})$ denote a smooth function and $\boldsymbol{u}$, $\boldsymbol{v} \in T(\mathcal{M})$, $\boldsymbol{\alpha} \in T^*(\mathcal{M})$. The action of $\pounds_{\boldsymbol{v}}$ on functions and vectors is given by

$$\pounds_{\boldsymbol{v}} f \equiv \boldsymbol{v}(f), \qquad \pounds_{\boldsymbol{v}} \boldsymbol{u} \equiv [\boldsymbol{v}, \boldsymbol{u}].$$

The Lie derivative can be extended to act on covectors by requiring the *Leibnitz rule*

$$\pounds_{\boldsymbol{v}} \langle \boldsymbol{\alpha}, \boldsymbol{u} \rangle = \langle \pounds_{\boldsymbol{v}} \boldsymbol{\alpha}, \boldsymbol{u} \rangle + \langle \boldsymbol{\alpha}, \pounds_{\boldsymbol{v}} \boldsymbol{u} \rangle.$$

A coordinate expression can be obtained from the latter. The action of $\pounds_{\boldsymbol{v}}$ can be extended to arbitrary tensor fields by means of the Leibnitz rule

$$\pounds_{\boldsymbol{v}} (\boldsymbol{S} \otimes \boldsymbol{T}) = \pounds_{\boldsymbol{v}} \boldsymbol{S} \otimes \boldsymbol{T} + \boldsymbol{S} \otimes \pounds_{\boldsymbol{v}} \boldsymbol{T}.$$

The reader interested in the derivation of the above expressions and their precise relation to the notions of push-forward and pull-back of tensor fields is referred to, for example, Stewart (1991) where a list of coordinate expressions for the computation of the derivatives is also provided.

2.4 Connections, torsion and curvature

This section discusses the further structure required on a manifold to describe the geometric notion of curvature – a key ingredient of the equations of general relativity.

2.4.1 Covariant derivatives and connections

The notion of linear connection allows one to relate tensors at different points of the manifold $\mathcal{M}$.

Definition 2.2 (***linear connection***) *A* ***linear connection (connection*** *for short) is a map* $\boldsymbol{\nabla} : \mathfrak{T}^1(\mathcal{M}) \times \mathfrak{T}^1(\mathcal{M}) \rightarrow \mathfrak{T}^1(\mathcal{M})$ *sending the pair of vector fields* $(\boldsymbol{u}, \boldsymbol{v})$ *to a vector field* $\nabla_{\boldsymbol{v}}\boldsymbol{u}$ *satisfying:*

(i) $\nabla_{\boldsymbol{u}+\boldsymbol{v}}\boldsymbol{w} = \nabla_{\boldsymbol{u}}\boldsymbol{w} + \nabla_{\boldsymbol{v}}\boldsymbol{w}$
(ii) $\nabla_{\boldsymbol{u}}(\boldsymbol{v}+\boldsymbol{w}) = \nabla_{\boldsymbol{u}}\boldsymbol{v} + \nabla_{\boldsymbol{u}}\boldsymbol{w}$
(iii) $\nabla_{f\boldsymbol{u}}\boldsymbol{v} = f\nabla_{\boldsymbol{u}}\boldsymbol{v}$
(iv) $\nabla_{\boldsymbol{u}}(f\boldsymbol{v}) = \boldsymbol{u}(f)\boldsymbol{v} + f\nabla_{\boldsymbol{u}}\boldsymbol{v}$

for $f \in \mathfrak{X}(\mathcal{M})$*. The vector* $\nabla_{\boldsymbol{u}}\boldsymbol{v}$ *is called the* ***covariant derivative of*** $\boldsymbol{v}$ ***with respect to*** $\boldsymbol{u}$*.*

Any manifold admits a connection. In four dimensions this can be shown through the specification of 4^3 functions on the spacetime manifold $\mathcal{M}$; see, for example, Willmore (1993). The reason behind this result becomes more transparent once the so-called *connection coefficients* have been introduced; see Section 2.6.

As a consequence of the requirement (iv) $\nabla_{\boldsymbol{u}}\boldsymbol{v}$ is not $\mathfrak{X}$-linear in $\boldsymbol{v}$; however, it is $\mathfrak{X}$-linear in $\boldsymbol{u}$. Thus, using Lemma 2.1 for a fixed second argument it defines a mixed $(1,1)$-tensor. Using abstract index notation the latter is denoted by $\nabla_a v^b$, so that $\nabla_a v^b \in \mathfrak{T}_a{}^b(\mathcal{M})$.

From the discussion in the previous paragraph it follows that one can regard the connection $\boldsymbol{\nabla}$ as a map $\nabla_a : \mathfrak{T}^b(\mathcal{M}) \rightarrow \mathfrak{T}_a{}^b(\mathcal{M})$. Moreover, a connection $\boldsymbol{\nabla}$ induces a map $\nabla_a : \mathfrak{T}_b(\mathcal{M}) \rightarrow \mathfrak{T}_{ab}(\mathcal{M})$ via

$$(\nabla_a \omega_b) v^b = \nabla_a(\omega_b v^b) - \omega_b(\nabla_a v^b).$$

This map is fixed if one requires the *Leibnitz rule* to hold between the product of a vector and a covector. To extend the covariant derivative to arbitrary tensors one uses again the Leibnitz rule. For example, from

$$
\begin{aligned}
\nabla_e(\omega_a T^a{}_{bcd} u^b v^c w^d) = (\nabla_e \omega_a) T^a{}_{bcd} u^b v^c w^d + \omega_a (\nabla_e T^a{}_{bcd}) u^b v^c w^d \\
+ \omega_a T^a{}_{bcd} (\nabla_e u^b) v^c w^d + \omega_a T^a{}_{bcd} u^b (\nabla_e v^c) w^d \\
+ \omega_a T^a{}_{bcd} u^b v^c (\nabla_e w^d),
\end{aligned}
$$

it follows that

$$
\begin{aligned}
(\nabla_e T^a{}_{bcd}) \omega_a u^b v^c w^d = \nabla_e(\omega_a T^a{}_{bcd} u^b v^c w^d) - (\nabla_e \omega_a) T^a{}_{bcd} u^b v^c w^d \\
- \omega_a T^a{}_{bcd} (\nabla_e u^b) v^c w^d - \omega_a T^a{}_{bcd} u^b (\nabla_e v^c) w^d \\
- \omega_a T^a{}_{bcd} u^b v^c (\nabla_e w^d),
\end{aligned}
$$

so that one obtains a $\mathfrak{X}$-linear map $\mathfrak{T}^a{}_{bcd}(\mathcal{M}) \to \mathfrak{T}_e{}^a{}_{bcd}(\mathcal{M})$.

The subsequent discussion will make use of the ***commutator of covariant derivatives***. This is defined as

$$
[\nabla_a, \nabla_b] \equiv 2\nabla_{[a}\nabla_{b]}.
$$

One has that

$$
\begin{aligned}
[\nabla_a, \nabla_b](T_{\mathcal{A}} + S_{\mathcal{A}}) &= [\nabla_a, \nabla_b] T_{\mathcal{A}} + [\nabla_a, \nabla_b] S_{\mathcal{A}}, \\
[\nabla_a, \nabla_b](T_{\mathcal{A}} R_{\mathcal{B}}) &= ([\nabla_a, \nabla_b] T_{\mathcal{A}}) R_{\mathcal{B}} + T_{\mathcal{A}}([\nabla_a, \nabla_b] R_{\mathcal{B}}),
\end{aligned}
$$

where ${}_{\mathcal{A}}$ and ${}_{\mathcal{B}}$ denote an arbitrary string of (covariant and contravariant) indices.

Covariant derivatives and derivations on a manifold are related in a natural way: given a derivation $\mathcal{D}$ and a connection $\boldsymbol{\nabla}$ on $\mathcal{M}$ there exists a unique $\boldsymbol{v} \in T(\mathcal{M})$ such that $\mathcal{D}f = v^a \nabla_a f$ for any $f \in \mathfrak{X}(\mathcal{M})$; see, for example, O'Neill (1983).

2.4.2 Torsion of a connection

The notion of ***torsion*** arises from the analysis of the action of the commutator of covariant derivatives on scalar fields. For convenience the abstract index notation is used. Consider $x^{ab} \in \mathfrak{T}^{ab}(\mathcal{M})$ and $f,\ g \in \mathfrak{X}(\mathcal{M})$. One readily has that

$$
\begin{aligned}
x^{ab}[\nabla_a, \nabla_b](f+g) &= x^{ab}[\nabla_a, \nabla_b] f + x^{ab}[\nabla_a, \nabla_b] g, \\
x^{ab}[\nabla_a, \nabla_b](fg) &= (x^{ab}[\nabla_a, \nabla_b] f) g + f(x^{ab}[\nabla_a, \nabla_b] g).
\end{aligned}
$$

It follows from the latter that the operator $x^{ab}[\nabla_a, \nabla_b]$ *must be a derivation*; see Definition 2.1. Thus, there exists $u^a \in \mathfrak{T}^a(\mathcal{M})$ such that

$$
x^{ab}[\nabla_a, \nabla_b] = u^a \nabla_a. \tag{2.5}
$$

The map $x^{ab} \mapsto u^a \nabla_a$ defined by Equation (2.5) is $\mathfrak{X}$-linear. It defines a tensor $\boldsymbol{\Sigma}$, the ***torsion tensor*** of the connection $\boldsymbol{\nabla}$, via $u^c = x^{ab}\Sigma_a{}^c{}_b$. Hence,

$$
\nabla_a \nabla_b f - \nabla_b \nabla_a f = \Sigma_a{}^c{}_b \nabla_c f, \qquad f \in \mathfrak{X}(\mathcal{M}). \tag{2.6}
$$

One readily sees that

$$\Sigma_a{}^c{}_b = -\Sigma_b{}^c{}_a.$$

That is, the torsion is an antisymmetric tensor. If a connection $\boldsymbol{\nabla}$ is such that $\Sigma_a{}^c{}_b = 0$, then it is said to be ***torsion-free***.

Remark. Alternatively, one could have defined the torsion via the relation

$$\boldsymbol{\Sigma}(\boldsymbol{u}, \boldsymbol{v}) = \nabla_{\boldsymbol{u}}\boldsymbol{v} - \nabla_{\boldsymbol{v}}\boldsymbol{u} - [\boldsymbol{u}, \boldsymbol{v}], \qquad \boldsymbol{u},\, \boldsymbol{v} \in T(\mathcal{M}). \tag{2.7}$$

2.4.3 Curvature of a connection

In order to discuss the notion of curvature of a connection it is convenient to define the ***modified commutator*** of covariant derivatives

$$[\![\nabla_a, \nabla_b]\!] \equiv [\nabla_a, \nabla_b] - \Sigma_a{}^c{}_b \nabla_c.$$

Clearly, one has that $[\![\nabla_a, \nabla_b]\!] f = 0$ for $f \in \mathfrak{X}(\mathcal{M})$ so that

$$[\![\nabla_a, \nabla_b]\!](f T_{\mathcal{A}}) = f [\![\nabla_a, \nabla_b]\!] T_{\mathcal{A}},$$

for ${}_{\mathcal{A}}$ denoting an arbitrary string of covariant or contravariant indices. In particular, one has that

$$[\![\nabla_a, \nabla_b]\!](f u^c) = f [\![\nabla_a, \nabla_b]\!] u^c,$$
$$[\![\nabla_a, \nabla_b]\!](u^c + v^c) = [\![\nabla_a, \nabla_b]\!] u^c + [\![\nabla_a, \nabla_b]\!] v^c.$$

From the previous expressions one concludes that the map $u^d \mapsto [\![\nabla_a, \nabla_b]\!] u^d$ is $\mathfrak{X}$-linear. Thus, using Lemma 2.1 it defines a tensor field $R^d{}_{cab}$, the ***Riemann curvature tensor*** of the connection $\boldsymbol{\nabla}$. One writes

$$[\![\nabla_a, \nabla_b]\!] u^d = \left([\nabla_a, \nabla_b] - \Sigma_a{}^c{}_b \nabla_c\right) u^d = R^d{}_{cab} u^c. \tag{2.8}$$

Alternatively, one has that

$$\left(\nabla_a \nabla_b - \nabla_b \nabla_a\right) u^d = R^d{}_{cab} u^c + \Sigma_a{}^c{}_b \nabla_c u^d.$$

The antisymmetry of $[\![\nabla_a, \nabla_b]\!]$ on the indices ${}_a$ and ${}_b$ is inherited by the Riemann curvature tensor, so that

$$R^d{}_{cab} = -R^d{}_{cba}.$$

The action of the commutator of covariant derivatives can be extended to other tensors using the Leibniz rule. For example, from

$$[\![\nabla_a, \nabla_b]\!](\omega_d v^d) = \left([\![\nabla_a, \nabla_b]\!] \omega_d\right) v^d + \omega_d [\![\nabla_a, \nabla_b]\!] v^d,$$

one can conclude that

$$\left(\nabla_a \nabla_b - \nabla_b \nabla_a\right) \omega_d = -R^c{}_{dab} \omega_c + \Sigma_a{}^c{}_b \nabla_c \omega_d.$$

Similarly, evaluating $[\![\nabla_a, \nabla_b]\!](S^d{}_{ef}\omega_d u^e v^f)$, one concludes that

$$(\nabla_a \nabla_b - \nabla_b \nabla_a)\, S^d{}_{ef} = R^d{}_{cab} S^c{}_{ef} - R^c{}_{eab} S^d{}_{cf} - R^c{}_{fab} S^d{}_{ec} + \Sigma_a{}^c{}_b \nabla_c S^d{}_{ef}.$$

Remark. The curvature can be defined in an alternative way via the relation

$$\boldsymbol{Riem}(\boldsymbol{u},\boldsymbol{v})\boldsymbol{w} = \nabla_{\boldsymbol{u}}\nabla_{\boldsymbol{v}}\boldsymbol{w} - \nabla_{\boldsymbol{v}}\nabla_{\boldsymbol{u}}\boldsymbol{w} - \nabla_{[\boldsymbol{u},\boldsymbol{v}]}\boldsymbol{w}, \qquad \boldsymbol{u},\,\boldsymbol{v},\,\boldsymbol{w} \in T(\mathcal{M}), \quad (2.9)$$

where the expression $\boldsymbol{Riem}(\boldsymbol{u},\boldsymbol{v})\boldsymbol{w}$ corresponds to $R^d{}_{cab} w^c u^a v^b$ in abstract index notation.

Bianchi identities

In order to investigate further symmetries of the curvature tensor, consider the triple derivative $\nabla_{[a}\nabla_b\nabla_{c]} f$ of $f \in \mathfrak{X}(\mathcal{M})$. A computation shows, on the one hand, that

$$\begin{aligned} 2\nabla_{[a}\nabla_b\nabla_{c]} f &= 2\nabla_{[[a}\nabla_{b]}\nabla_{c]} f = [\nabla_{[a}, \nabla_b]\nabla_{c]} f \\ &= \Sigma_{[a}{}^d{}_b \nabla_{|d|}\nabla_{c]} f - R^d{}_{[cab]}\nabla_d f, \end{aligned}$$

and on the other hand that

$$\begin{aligned} 2\nabla_{[a}\nabla_b\nabla_{c]} f &= 2\nabla_{[a}\nabla_{[b}\nabla_{c]]} f \\ &= \nabla_{[a}[\nabla_b, \nabla_{c]}] f = \nabla_{[a}\left(\Sigma_b{}^d{}_{c]}\nabla_d f\right) \\ &= \nabla_{[a}\Sigma_b{}^d{}_{c]}\nabla_d f + \Sigma_{[b}{}^d{}_c \nabla_{a]}\nabla_d f. \end{aligned}$$

Putting these two computations together and using the definition of the torsion tensor, Equation (2.6), one concludes that

$$\nabla_{[a}\Sigma_b{}^d{}_{c]}\nabla_d f + R^d{}_{[cab]}\nabla_d f + \Sigma_{[a}{}^d{}_b \Sigma_{c]}{}^e{}_d \nabla_e f = 0.$$

As the scalar field f is arbitrary, one concludes that

$$R^d{}_{[cab]} + \nabla_{[a}\Sigma_b{}^d{}_{c]} + \Sigma_{[a}{}^e{}_b \Sigma_{c]}{}^d{}_e = 0. \qquad (2.10)$$

This is the so-called ***first Bianchi identity***. In the case of a *torsion-free connection* it takes the familiar form

$$R^d{}_{[cab]} = 0.$$

As a consequence of the antisymmetry in the last two indices, the latter can be written as

$$R^d{}_{cab} + R^d{}_{abc} + R^d{}_{bca} = 0.$$

Next, consider the action of $\nabla_{[a}\nabla_b\nabla_{c]}$ on a vector field v^d. As in the case of the first Bianchi identity, one can compute this object in two different ways. On the one hand, one has that

$$\begin{aligned}
2\nabla_{[a}\nabla_b\nabla_{c]}v^d &= 2\nabla_{[[a}\nabla_{b]}\nabla_{c]}v^d \\
&= [\nabla_{[a},\nabla_b]\nabla_{c]}v^d \\
&= [\![\nabla_{[a},\nabla_b]\!]\nabla_{c]}v^d + \Sigma_{[a}{}^e{}_b\nabla_{|e|}\nabla_{c]}v^d \\
&= -R^e{}_{[cab]}\nabla_e v^d + R^d{}_{e[ab}\nabla_{c]}v^e + \Sigma_{[a}{}^e{}_b\nabla_{|e|}\nabla_{c]}v^d,
\end{aligned}$$

and on the other hand that

$$\begin{aligned}
2\nabla_{[a}\nabla_b\nabla_{c]}v^d &= 2\nabla_{[a}\nabla_{[b}\nabla_{c]]}v^d \\
&= 2\nabla_{[a}[\![\nabla_b,\nabla_{c]}]\!]v^d + \nabla_{[a}\left(\Sigma_b{}^e{}_{c]}\nabla_e v^d\right) \\
&= \nabla_{[a}R^d{}_{|e|bc]}v^e + R^d{}_{e[bc}\nabla_{a]}v^e + \nabla_{[a}\Sigma_b{}^e{}_{c]}\nabla_e v^d + \Sigma_{[b}{}^e{}_c\nabla_{a]}\nabla_e v^d.
\end{aligned}$$

Equating the two expressions for $2\nabla_{[a}\nabla_b\nabla_{c]}v^d$ and using the first Bianchi identity, Equation (2.10), to eliminate covariant derivatives of the torsion tensor one concludes that

$$\nabla_{[a}R^d{}_{|e|bc]} + \Sigma_{[a}{}^f{}_b R^d{}_{|e|c]f} = 0. \tag{2.11}$$

This is the so-called ***second Bianchi identity***. For a *torsion-free connection* one obtains the well-known expression

$$\nabla_{[a}R^d{}_{|e|bc]} = 0. \tag{2.12}$$

2.4.4 Change of connection

Consider two connections $\boldsymbol{\nabla}$ and $\bar{\boldsymbol{\nabla}}$ on the manifold $\mathcal{M}$. A natural question to be asked is whether there is any relation between these connections and their associated torsion and curvature tensors. By definition one has that

$$(\bar{\nabla}_a - \nabla_a)f = 0, \qquad f \in \mathfrak{X}(\mathcal{M}).$$

Moreover, one also has that

$$(\bar{\nabla}_a - \nabla_a)(fv^a) = f(\bar{\nabla}_a - \nabla_a)v^a.$$

It follows that the map $v^b \mapsto (\bar{\nabla}_a - \nabla_a)v^a$ is $\mathfrak{X}$-linear, so that, invoking Lemma 2.1, there exists a tensor field, the ***transition tensor*** $Q_a{}^b{}_c$, such that

$$(\bar{\nabla}_a - \nabla_a)v^b = Q_a{}^b{}_c v^c. \tag{2.13}$$

Now, from

$$(\bar{\nabla}_a - \nabla_a)(\omega_b v^b) = 0,$$

one readily concludes that

$$(\bar{\nabla}_a - \nabla_a)\omega_b = -Q_a{}^c{}_b\omega_c. \tag{2.14}$$

A different choice of covariant derivatives gives rise to a different choice of transition tensor. The set of connections over a manifold $\mathcal{M}$ is an *affine space*: given a connection $\boldsymbol{\nabla}$ on the manifold, any other connection can be obtained by a suitable choice of transition tensor. If $\boldsymbol{Q}$ denotes the index-free version of the tensor $Q_a{}^b{}_c$, then the relation between the connection $\boldsymbol{\nabla}$ and $\bar{\boldsymbol{\nabla}}$ will be denoted, in a schematic way, as

$$\bar{\boldsymbol{\nabla}} - \boldsymbol{\nabla} = \boldsymbol{Q}.$$

In Chapter 5 specific forms for the transition tensor will be investigated.

Transformation of the torsion and the curvature

A direct computation using Equations (2.6) and (2.13) renders the following relation between the torsion tensors of the connections $\bar{\boldsymbol{\nabla}}$ and $\boldsymbol{\nabla}$:

$$\bar{\Sigma}_a{}^c{}_b - \Sigma_a{}^c{}_b = -2Q_{[a}{}^c{}_{b]}. \tag{2.15}$$

In particular, it follows that if $Q_a{}^c{}_b = \frac{1}{2}\Sigma_a{}^c{}_b$, then $\bar{\Sigma}_a{}^c{}_b = 0$. That is, *it is always possible to construct a connection which is torsion-free.*

An analogous, albeit lengthier computation using Equations (2.6) and (2.8) renders the following relation between the respective curvature tensors:

$$\bar{R}^c{}_{dab} - R^c{}_{dab} = 2\nabla_{[a}Q_{b]}{}^c{}_d - \Sigma_a{}^e{}_b Q_e{}^c{}_d + 2Q_{[a}{}^c{}_{|e|}Q_{b]}{}^e{}_d. \tag{2.16}$$

2.4.5 The geodesic and geodesic deviation equations

Given a covariant derivative $\boldsymbol{\nabla}$, one can introduce the notion of ***parallel propagation***. Given $\boldsymbol{u}, \boldsymbol{v} \in T(\mathcal{M})$, then $\boldsymbol{u}$ is said to be parallely propagated in the direction of $\boldsymbol{v}$ if it satisfies the equation $\nabla_{\boldsymbol{v}}\boldsymbol{u} = 0$.

A ***geodesic*** $\gamma \subset \mathcal{M}$ is a curve whose tangent vector is parallely propagated along itself. Following the convention of Section 2.2.1, let $\dot{\boldsymbol{x}}$ denote the tangent vector to γ. One has that

$$\nabla_{\dot{\boldsymbol{x}}}\dot{\boldsymbol{x}} = 0. \tag{2.17}$$

A ***congruence of geodesics*** is the set of integral curves of a vector field $\dot{\boldsymbol{x}}$ satisfying Equation (2.17). Any vector $\boldsymbol{z}$ such that $[\dot{\boldsymbol{x}}, \boldsymbol{z}] = 0$ is called a ***deviation vector*** of the congruence of geodesics. *Assuming that the connection* $\boldsymbol{\nabla}$ *is torsion-free* so that $\nabla_{\dot{\boldsymbol{x}}}\boldsymbol{z} = \nabla_{\boldsymbol{z}}\dot{\boldsymbol{x}}$, a computation shows that $\boldsymbol{z}$ satisfies the ***geodesic deviation equation***

$$\nabla_{\dot{\boldsymbol{x}}}\nabla_{\dot{\boldsymbol{x}}}\boldsymbol{z} = \boldsymbol{Riem}(\dot{\boldsymbol{x}}, \boldsymbol{z})\dot{\boldsymbol{x}}.$$

Remark. The set of geodesics emanating from a point $p \in \mathcal{M}$ allows one to define a diffeomorphism between a neighbourhood of the origin of $T|_p(\mathcal{M})$ and a suitably small neighbourhood $\mathcal{U}$ of p, the so-called ***exponential map***. A precise definition of the exponential map is given in Section 11.6.2. Further properties and applications are given in Sections 14.2 and 18.4.1.

2.5 Metric tensors

A ***metric*** on the manifold $\mathcal{M}$ is a symmetric rank 2 covariant tensor field $\boldsymbol{g}$ – to be denoted by g_{ab} in abstract index notation. The metric tensor $\boldsymbol{g}$ is said to be ***non-degenerate*** if $\boldsymbol{g}(\boldsymbol{u}, \boldsymbol{v}) = 0$ for all $\boldsymbol{u}$ if and only if $\boldsymbol{v} = 0$. In the sequel, and unless otherwise explicitly stated, it is assumed that all the metrics under consideration are non-degenerate. If $\boldsymbol{g}(\boldsymbol{u}, \boldsymbol{v}) = 0$, then the vectors $\boldsymbol{u}$ and $\boldsymbol{v}$ are said to be ***orthogonal***. Pointwise, the components $g_{\boldsymbol{ab}} \equiv \boldsymbol{g}(\boldsymbol{e_a}, \boldsymbol{e_b})$ with respect to a basis $\{\boldsymbol{e_a}\}$ define a symmetric $(n \times n)$-matrix $(g_{\boldsymbol{ab}})$. As this matrix is symmetric, it has n real eigenvalues. The ***signature*** of $\boldsymbol{g}$ is the difference between the number of positive and negative eigenvalues. If the signature is n or $-n$, then $\boldsymbol{g}$ is said to be a ***Riemannian metric***. If the signature is $\pm(n-2)$, then $\boldsymbol{g}$ is a ***Lorentzian metric***.

From the non-degeneracy of $\boldsymbol{g}$ it follows that there exists a unique contravariant rank 2 tensor to be denoted by either $\boldsymbol{g}^\sharp$ or g^{ab} such that

$$g_{ab}g^{bc} = \delta_a{}^c.$$

In terms of components with respect to a basis this means that the matrices $(g_{\boldsymbol{ab}})$ and $(g^{\boldsymbol{ab}})$ are inverses of each other. Accordingly, $\boldsymbol{g}^\sharp$ is also non-degenerate and one obtains an isomorphism between the vector spaces $T|_p(\mathcal{M})$ and $T^*|_p(\mathcal{M})$. More precisely, given $\boldsymbol{v} \in T|_p(\mathcal{M})$, then $\boldsymbol{v}^\flat \equiv \boldsymbol{g}(\boldsymbol{v}, \cdot) \in T^*|_p(\mathcal{M})$ as $\boldsymbol{g}(\boldsymbol{u}, \boldsymbol{v}) \in \mathbb{R}$ for any $\boldsymbol{u} \in T|_p(\mathcal{M})$. Similarly, given $\boldsymbol{\omega} \in T^*|_p(\mathcal{M})$, one has that $\boldsymbol{\omega}^\sharp \equiv \boldsymbol{g}^\sharp(\boldsymbol{\omega}, \cdot) \in T|_p(\mathcal{M})$. In terms of abstract indices, the operations $^\flat$ (*flat*) and $^\sharp$ (*sharp*) correspond to the operations of ***lowering and raising of indices*** by means of g_{ab} and g^{ab}:

$$v_a \equiv g_{ab}v^b, \qquad \omega^a \equiv g^{ab}\omega_b.$$

The operations $^\flat$ and $^\sharp$ are inverses of each other. They can be extended in a natural way to tensors of arbitrary rank.

Given two manifolds $\mathcal{M}$ and $\bar{\mathcal{M}}$ with metrics $\boldsymbol{g}$ and $\bar{\boldsymbol{g}}$, respectively, a diffeomorphism $\varphi : \mathcal{M} \to \bar{\mathcal{M}}$ is called an ***isometry*** if $\varphi^*\bar{\boldsymbol{g}} = \boldsymbol{g}$. If an isometry exists, then the pairs $(\mathcal{M}, \boldsymbol{g})$ and $(\bar{\mathcal{M}}, \bar{\boldsymbol{g}})$ are said to be ***isometric***. If $\mathcal{M} = \bar{\mathcal{M}}$ and $\boldsymbol{g} = \bar{\boldsymbol{g}}$, one speaks of an isometry of $\mathcal{M}$.

Remark. Most of the Lorentzian metrics to be considered in this book will be associated to four-dimensional manifolds. These Lorentzian metrics will be assumed to have signature -2. This convention leads one to consider three-dimensional ***negative-definite*** Riemannian metrics, that is, metrics with signature -3. In this book, only three-dimensional Riemannian manifolds will be considered. In the sequel, the symbol $\boldsymbol{g}$ will be used to denote a generic Lorentzian metric, while $\boldsymbol{h}$ will be used for a generic negative-definite Riemannian metric.

Specifics for Lorentzian metrics

Following the standard terminology of general relativity, a pair $(\mathcal{M}, \boldsymbol{g})$ consisting of a four-dimensional manifold and a Lorentzian metric will be called a ***spacetime***. The metric $\boldsymbol{g}$ can be used to classify vectors in a pointwise manner as ***timelike***, ***null*** or ***spacelike*** depending on whether $\boldsymbol{g}(\boldsymbol{v}, \boldsymbol{v}) > 0$, $\boldsymbol{g}(\boldsymbol{v}, \boldsymbol{v}) = 0$ or $\boldsymbol{g}(\boldsymbol{v}, \boldsymbol{v}) < 0$, respectively. A basis $\{\boldsymbol{e}_a\}$ is said to be ***orthonormal*** if

$$\boldsymbol{g}(\boldsymbol{e}_a, \boldsymbol{e}_b) = \eta_{ab}, \qquad \eta_{ab} \equiv \mathrm{diag}(1, -1, -1, -1).$$

It follows that $\boldsymbol{g}$ can be written as

$$\boldsymbol{g} = \eta_{ab}\boldsymbol{\omega}^a \otimes \boldsymbol{\omega}^b, \tag{2.18}$$

where $\{\boldsymbol{\omega}^a\}$ denotes the coframe dual to $\{\boldsymbol{e}_a\}$. A change of basis, as given by Equation (2.1), preserving Equation (2.18), is called a ***Lorentz transformation***. A calculation readily shows that for a Lorentz transformation one has that

$$\eta_{ab}A^a{}_c A^b{}_d = \eta_{cd}.$$

Further aspects of Lorentz transformations are discussed in Sections 3.1.9, 3.1.12 and 5.1.1.

The set of null vectors at a point $p \in \mathcal{M}$ is called the ***null cone at*** p and will be denoted by $\mathcal{C}_p$. By definition timelike vectors lie inside the null cone, while spacelike ones lie outside it. The null cone is made of two half cones. If one of these half cones can be singled out and called the ***future half cone*** $\mathcal{C}_p^+$ and the other the ***past half cone*** $\mathcal{C}_p^-$, then $T|_p(\mathcal{M})$ is said to be ***time oriented***. A timelike vector inside $\mathcal{C}_p^+$ is said to be ***future directed***; similarly a timelike vector inside $\mathcal{C}_p^-$ is called ***past directed***. If $T(\mathcal{M})$ can be time oriented in a continuous manner for all $p \in \mathcal{M}$, then $(\mathcal{M}, \boldsymbol{g})$ is said to be a ***time-oriented spacetime***. A curve $\gamma \subset \mathcal{M}$ with a *timelike, future-oriented* tangent vector $\dot{\boldsymbol{x}}$ is said to be ***parametrised by its proper time*** if $\boldsymbol{g}(\dot{\boldsymbol{x}}, \dot{\boldsymbol{x}}) = 1$.

Specifics for Riemannian metrics

A Riemannian metric $\boldsymbol{h}$ endows the tangent spaces of the manifold with an inner product. Because of the signature conventions, this inner product is negative definite. A basic result of Riemannian geometry is that every differential manifold admits a Riemannian metric. The proof of this argument relies heavily on the paracompactness of the manifold; see, for example, Choquet-Bruhat et al. (1982).

In the case of a Riemannian metric $\boldsymbol{h}$, a basis $\{\boldsymbol{e}_i\}$ is said to be ***orthonormal*** if

$$\boldsymbol{h}(\boldsymbol{e}_i, \boldsymbol{e}_j) = -\delta_{ij}, \qquad \delta_{ij} \equiv \mathrm{diag}(1, 1, 1).$$

Thus, using the associated coframe basis $\{\boldsymbol{\omega}^i\}$ one can write

$$\boldsymbol{h} = -\delta_{ij}\boldsymbol{\omega}^i \otimes \boldsymbol{\omega}^j.$$

2.5.1 Metric connections and Levi-Civita connections

Two further conditions which are usually required from a connection are metric compatibility and torsion-freeness. In this section the consequences of these assumptions are briefly reviewed.

Metric connections

A connection ∇ on $\mathcal{M}$ is said to be ***metric with respect to g*** if $\nabla g = 0$ (i.e. $\nabla_a g_{bc} = 0$). The Riemann curvature tensor of the connection ∇ acquires, by virtue of the metricity condition, a further symmetry. This can be better seen by applying the modified commutator $[\![\nabla_a, \nabla_b]\!]$ to the metric g_{ab}. On the one hand, by the assumption of metricity one has $[\![\nabla_a, \nabla_b]\!] g_{cd} = 0$, while on the other hand

$$[\![\nabla_a, \nabla_b]\!] g_{cd} = -R^e{}_{cab} g_{ed} - R^e{}_{dab} g_{ce} = -R_{dcab} - R_{cdab},$$

where $R_{dcab} \equiv g_{de} R^e{}_{cab}$. Hence, one concludes that

$$R_{cdab} = -R_{dcab}. \tag{2.19}$$

The Levi-Civita connection

A connection ∇ is said to be the ***Levi-Civita connection of the metric g*** if ∇ is *torsion-free* and *metric* with respect to g. The *Fundamental Theorem of Riemannian Geometry* (also valid in the Lorentzian case) ensures that the Levi-Civita connection of a metric g is unique. The proof of this result is well known and readily available in most books on Riemannian geometry; see, for example, Choquet-Bruhat et al. (1982). The Levi-Civita connection ∇ of the metric g is characterised by the so-called ***Koszul formula***

$$\begin{aligned} 2g(\nabla_v u, w) &= v(g(u,w)) + u(g(w,v)) - w(g(v,u)) \\ &\quad - g(v,[u,w]) + g(u,[w,v]) + g(w,[v,u]). \end{aligned} \tag{2.20}$$

Of particular interest are the further symmetries that the Riemann tensor of a Levi-Civita connection possesses. First of all, because of the metricity, the curvature tensor has the symmetry given in Equation (2.19). Furthermore, as the connection is torsion-free, the first Bianchi identity implies $R_{c[dab]} = 0$. From the latter one readily has that

$$\begin{aligned} 2R_{cdab} &= R_{cdab} + R_{dcba} \\ &= -R_{cabd} - R_{cbda} - R_{dbac} - R_{dacb} \\ &= -R_{acdb} - R_{bcad} - R_{bdca} - R_{adbc} \\ &= R_{abcd} + R_{badc}. \end{aligned}$$

Hence, one recovers the well-known symmetry of interchange of pairs

$$R_{cdab} = R_{abcd}.$$

Characterisation of flatness

An open subset $\mathcal{U} \subset \mathcal{M}$ of a spacetime $(\mathcal{M}, \boldsymbol{g})$ is said to be flat if the metric $\boldsymbol{g}$ on $\mathcal{U}$ is isometric to the ***Minkowski metric***

$$\boldsymbol{\eta} \equiv \eta_{\mu\nu}\mathbf{d}x^\mu \otimes \mathbf{d}x^\nu, \qquad (\eta_{\mu\nu}) \equiv \mathrm{diag}(1, -1, -1, -1).$$

In the case of a three-dimensional Riemannian manifold $(\mathcal{S}, \boldsymbol{h})$, flatness implies a local isometry with the three-dimensional ***Euclidean metric***

$$\boldsymbol{\delta} \equiv -\delta_{\alpha\beta}\mathbf{d}x^\alpha \otimes \mathbf{d}x^\beta, \qquad (\delta_{\alpha\beta}) \equiv \mathrm{diag}(1, 1, 1).$$

The Riemann tensor of a Levi-Civita connection provides a local characterisation of the flatness of a manifold. More precisely, a metric is flat on $\mathcal{U}$ if and only if its Riemann tensor vanishes on $\mathcal{U}$. The *if* part of the result follows by direct evaluation of the Riemann tensor. The *only if* part is more complicated; see, for example, Choquet-Bruhat et al. (1982), page 310 for a proof.

Traces

A metric $\boldsymbol{g}$ on a manifold $\mathcal{M}$ allows one to introduce a further operation on tensors which reduces their rank by 2 – the ***trace*** with respect to $\boldsymbol{g}$. Given $\boldsymbol{T} \in \mathfrak{T}_2(\mathcal{M})$, its trace, $\mathbf{tr}_{\boldsymbol{g}}\boldsymbol{T}$, is the scalar described in abstract index notation by $g^{ab}T_{ab}$. Observing that $g^{ab}T_{ab} = T^a{}_a$, one sees that taking the trace of a tensor is a generalisation of the operation of contraction. The operation of taking the trace can be generalised to any pair of indices of the same type in an arbitrary tensor – for example, $g^{ac}M_{abcd}$ and $g^{bc}M_{abcd}$ denote the traces of M_{abcd} with respect to the first and third arguments and the second and third ones, respectively.

Given a symmetric tensor on a four-dimensional manifold $\mathcal{M}$, $T_{ab} = T_{(ab)} \in \mathfrak{T}_{ab}(\mathcal{M})$, its ***trace-free part*** $T_{\{ab\}}$ is given by

$$T_{\{ab\}} \equiv T_{ab} - \frac{1}{4}g_{ab}g^{cd}T_{cd}.$$

In the case of a three-dimensional manifold $\mathcal{S}$ with metric $\boldsymbol{h}$, the above definition has to be modified to

$$T_{\{ij\}} \equiv T_{ij} - \frac{1}{3}h_{ij}h^{kl}T_{kl},$$

for a symmetric tensor $T_{ij} \in \mathfrak{T}_{ij}(\mathcal{S})$. The operation of taking the trace-free part of a tensor can be extended to tensors of arbitrary rank. Unfortunately, the expressions to compute them become increasingly cumbersome. A more efficient approach to describe this operation is in terms of spinors; see Chapters 3 and 4. A tensor $M_{a_1\cdots a_k}$ is said to be ***trace-free*** if $M_{a_1\cdots a_k} = M_{\{a_1\cdots a_k\}}$.

2.5.2 Decomposition of the Riemann tensor

In what follows, consider a spacetime $(\mathcal{M}, \boldsymbol{g})$ and a connection $\bar{\boldsymbol{\nabla}}$ on $\mathcal{M}$ – not necessarily the Levi-Civita connection of the metric $\boldsymbol{g}$. Let $\bar{R}^a{}_{bcd}$ denote the Riemann curvature tensor of the connection $\bar{\boldsymbol{\nabla}}$. A ***concomitant*** of $\bar{R}^a{}_{bcd}$ is any tensorial object which can be constructed from the curvature tensor by means of the operations of covariant differentiation and contraction with g_{ab} and g^{ab}. The basic concomitant of $\bar{R}^a{}_{bcd}$ is the ***Ricci tensor*** $\bar{R}_{cd}$ defined by the contraction

$$\bar{R}_{bd} \equiv \bar{R}^a{}_{bad}.$$

When working in index-free notation the Ricci tensor will be denoted by ***Ric***. Using the contravariant metric g^{ab} one can define a further concomitant, the ***Ricci scalar*** relative to the metric $\boldsymbol{g}$, $\bar{R}$, as

$$\bar{R} \equiv g^{bd}\bar{R}_{bd}.$$

A concomitant of $\bar{R}^a{}_{bcd}$ which will appear recurrently in this book is the ***Schouten tensor*** relative to $\boldsymbol{g}$, $\bar{L}_{ab}$. In four dimensions it is defined as

$$\bar{L}_{ab} = \frac{1}{2}\bar{R}_{ab} - \frac{1}{12}\bar{R}g_{ab}.$$

The definition of the Schouten tensor is dimension dependent. The definition for three dimensions will be discussed in Section 2.7. When working in index-free notation the Schouten tensor will be denoted by ***Schouten***. In the discussion of spinors in Chapter 3 a further concomitant arises in a natural way: the ***trace-free Ricci tensor*** $\bar{\Phi}_{ab}$. In four dimensions one has that

$$\bar{\Phi}_{ab} \equiv \frac{1}{2}\bar{R}_{\{ab\}} = \frac{1}{2}\left(\bar{R}_{(ab)} - \frac{1}{4}\bar{R}g_{ab}\right),$$

where the overall factor of $\frac{1}{2}$ is conventional. *It is important to observe that the tensors $\bar{R}_{ab}$ and $\bar{L}_{ab}$ are not symmetric unless $\bar{\nabla}$ is a Levi-Civita connection.*

Finally, one can define the ***Weyl tensor*** of $\bar{\boldsymbol{\nabla}}$ relative to $\boldsymbol{g}$, $\bar{C}^a{}_{bcd}$, as the fully trace-free part of $\bar{R}^a{}_{bcd}$. When working in index-free notation the Weyl tensor will be denoted by ***Weyl***.

The case of a Levi-Civita connection

If $\bar{\boldsymbol{\nabla}}$ is the Levi-Civita connection of the metric $\boldsymbol{g}$, so that $\bar{\boldsymbol{\nabla}} = \boldsymbol{\nabla}$, it can be shown that

$$R^c{}_{dab} = C^c{}_{dab} + 2(\delta^c{}_{[a}L_{b]d} - g_{d[a}L_{b]}{}^c), \qquad (2.21a)$$
$$= C^c{}_{dab} + 2S_{d[a}{}^{ce}L_{b]e}, \qquad (2.21b)$$

where

$$S_{ab}{}^{cd} = \delta_a{}^c\delta_b{}^d + \delta_a{}^d\delta_b{}^c - g_{ab}g^{cd}.$$

This tensor will play a special role in the context of conformal geometry; see Chapter 5. A spinorial derivation of this decomposition is provided in Chapter 3.

Remark. The decomposition given by Equations (2.21a) and (2.21b) is unique; that is, the Rieman tensor cannot be reconstructed from any other combination of the Schouten and Weyl tensors. Moreover, if $C^c{}_{dab} = 0$ and $L_{ab} = 0$, then necessarily $R^c{}_{dab} = 0$. These remarks also hold for the generalisations of the decomposition to Weyl connections; see Section 5.3 and, in particular, Equation (5.28a).

The Einstein tensor

An important concomitant of the Riemann tensor of a Levi-Civita connection $\boldsymbol{\nabla}$ is the ***Einstein tensor*** $\boldsymbol{G}$ defined in four dimensions by

$$G_{ab} \equiv R_{ab} - \frac{1}{2}R g_{ab}.$$

Starting from the second Bianchi identity, Equation (2.12), contracting the indices ^{d} and $_b$ and then contracting the resulting expression with g^{ae} yields

$$\nabla^a R_{ab} = \frac{1}{2}\nabla_b R, \qquad \text{that is,} \qquad \nabla^a G_{ab} = 0.$$

That is, the Einstein tensor is divergence-free.

2.5.3 Volume forms and Hodge duals

The spacetime ***volume form*** of the metric $\boldsymbol{g}$, ϵ_{abcd}, is defined by the conditions

$$\epsilon_{abcd} = \epsilon_{[abcd]}, \qquad \epsilon_{abcd}\epsilon^{abcd} = -24,$$

and

$$\epsilon_{abcd}\boldsymbol{e_0}{}^a\boldsymbol{e_1}{}^b\boldsymbol{e_2}{}^c\boldsymbol{e_3}{}^d = 1,$$

where $\{\boldsymbol{e_a}\}$ is a $\boldsymbol{g}$-orthonormal frame. A spacetime $(\mathcal{M}, \boldsymbol{g})$ has a non-vanishing volume element if and only if $\mathcal{M}$ is orientable; see, for example, O'Neill (1983); Willmore (1993). The following properties can be directly verified:

$$\epsilon_{abcd}\epsilon^{pqrs} = -24\delta_a{}^{[p}\delta_b{}^q\delta_c{}^r\delta_d{}^{s]}, \tag{2.22a}$$
$$\epsilon_{abcd}\epsilon^{pqrd} = -6\delta_a{}^{[p}\delta_b{}^q\delta_c{}^{r]}, \tag{2.22b}$$
$$\epsilon_{abcd}\epsilon^{pqcd} = -4\delta_a{}^{[p}\delta_b{}^{q]}, \tag{2.22c}$$
$$\epsilon_{abcd}\epsilon^{pbcd} = -6\delta_a{}^p; \tag{2.22d}$$

see, for example, Penrose and Rindler (1984). If $\boldsymbol{\nabla}$ denotes the Levi-Civita covariant derivative of the metric $\boldsymbol{g}$, one can then readily verify that $\nabla_a\epsilon_{bcde} = 0$. That is, the volume form is compatible with the Levi-Civita connection of the metric $\boldsymbol{g}$.

The Hodge duals

Given an antisymmetric tensor $F_{ab} = F_{[ab]}$, one can use the volume form to define its ***Hodge dual*** $^*F_{ab}$ as

$$^*F_{ab} \equiv -\frac{1}{2}\epsilon_{ab}{}^{cd}F_{cd}.$$

This definition can be naturally extended to any tensor with a pair of antisymmetric indices. Using the identity (2.22c) one readily finds that

$$^{**}F_{ab} = -F_{ab}.$$

Of special relevance are the Hodge duals of the Riemann and Weyl tensors. If R_{abcd} denotes the Riemann curvature of the Levi-Civita connection $\boldsymbol{\nabla}$, then one can define a ***left dual*** and a ***right dual***, respectively, by

$$^*R_{abcd} \equiv -\frac{1}{2}\epsilon_{ab}{}^{pq}R_{pqcd}, \qquad R^*_{abcd} \equiv -\frac{1}{2}\epsilon_{cd}{}^{pq}R_{abpq}.$$

The Hodge dual can be used to recast the Bianchi identities in an alternative way. More precisely, one has that

$$R_{a[bcd]} = \delta_{[b}{}^p\delta_c{}^q\delta_{d]}{}^r R_{apqr} = -\frac{1}{6}\epsilon_{sbcd}\left(\epsilon^{spqr}R_{apqr}\right) = \frac{1}{3}\epsilon_{sbcd}R^*{}_{ap}{}^{sp}.$$

Thus, the first Bianchi identity $R_{a[bcd]} = 0$ is equivalent to

$$R^*_{ab}{}^{cb} = 0. \tag{2.23}$$

Furthermore,

$$\frac{1}{2}\epsilon_f{}^{abc}\nabla_{[a}R^d{}_{|e|bc]} = \nabla_a\left(\frac{1}{2}\epsilon_f{}^{abc}R^d{}_{ebc}\right) = -\nabla_a R^{*d}{}_{ef}{}^a.$$

Thus, one has that

$$\nabla^a R^*{}_{abcd} = 0.$$

Finally, it is noticed that the duals of the Weyl tensor satisfy

$$^*C_{abcd} = C^*_{abcd}.$$

Sometimes it is convenient to make use of operations of dualisation on one or three indices. Given an arbitrary tensor J_a and another tensor K_{abc} antisymmetric in ${}_{abc}$ one defines

$$^\dagger J_{abc} \equiv \epsilon_{abc}{}^d J_d, \qquad ^\ddagger K_a \equiv \frac{1}{6}\epsilon_a{}^{bcd}K_{bcd}. \tag{2.24}$$

Using the properties of contractions of the volume form, it can be shown that

$$^{\ddagger\dagger}J_a = J_a, \qquad ^{\dagger\ddagger}K_{abc} = K_{abc}.$$

Further details on the calculations required to obtain all of the properties discussed in this section can be found in Penrose and Rindler (1984).

2.6 Frame formalisms

Frame formalisms have been used in many areas of relativity to analyse the properties of the Einstein field equations and their solutions; see, for example, Ellis and van Elst (1998); Ellis et al. (2012); Wald (1984). One of the advantages of frame formalisms is that they lead to consider scalar objects and equations, which are, in general, simpler to manipulate than their tensorial counterparts. A further advantage of frames is that they lead to a straight forward transcription of tensorial expressions into spinors; see Chapter 3.

The purpose of this section is to develop and fix the conventions of a frame formalism used in Friedrich (2004).

2.6.1 Basic definitions and conventions

Given a spacetime $(\mathcal{M}, \boldsymbol{g})$, let $\{\boldsymbol{e_a}\}$ denote a frame and let $\{\boldsymbol{\omega^b}\}$ denote its dual coframe basis. *For the time being, this frame is not assumed to be* $\boldsymbol{g}$*-orthogonal.* By definition one has that

$$\langle \boldsymbol{\omega^b}, \boldsymbol{e_a} \rangle = \delta_{\boldsymbol{a}}{}^{\boldsymbol{b}}. \tag{2.25}$$

In what follows, it will be assumed one has a connection $\boldsymbol{\nabla}$ *which, for the time being, is assumed to be general*; that is, it is not necessarily metric or torsion-free. The ***connection coefficients*** of $\boldsymbol{\nabla}$ with respect to the frame $\{\boldsymbol{e_a}\}$, to be denoted by $\Gamma_{\boldsymbol{a}}{}^{\boldsymbol{b}}{}_{\boldsymbol{c}}$, are defined via

$$\nabla_{\boldsymbol{a}} \boldsymbol{e_b} = \Gamma_{\boldsymbol{a}}{}^{\boldsymbol{c}}{}_{\boldsymbol{b}} \boldsymbol{e_c}, \tag{2.26}$$

where $\nabla_{\boldsymbol{a}} \equiv e_{\boldsymbol{a}}{}^a \nabla_a$ denotes the ***covariant directional derivative*** in the direction of $\boldsymbol{e_a}$. As $\nabla_{\boldsymbol{a}} \boldsymbol{e_b}$ is a vector, it follows that

$$\langle \boldsymbol{\omega^c}, \nabla_{\boldsymbol{a}} \boldsymbol{e_b} \rangle = \langle \boldsymbol{\omega^c}, \Gamma_{\boldsymbol{a}}{}^{\boldsymbol{d}}{}_{\boldsymbol{b}} \boldsymbol{e_d} \rangle = \Gamma_{\boldsymbol{a}}{}^{\boldsymbol{d}}{}_{\boldsymbol{b}} \langle \boldsymbol{\omega^c}, \boldsymbol{e_d} \rangle = \Gamma_{\boldsymbol{a}}{}^{\boldsymbol{c}}{}_{\boldsymbol{b}}.$$

This expression could have been used, alternatively, as a definition of the connection coefficients. In order to carry out computations one also needs an expression for $\nabla_{\boldsymbol{a}} \boldsymbol{\omega^b}$. By analogy with Equation (2.26) one can write $\nabla_{\boldsymbol{a}} \boldsymbol{\omega^b} = \mho_{\boldsymbol{a}}{}^{\boldsymbol{b}}{}_{\boldsymbol{c}} \boldsymbol{\omega^c}$. The coefficients $\mho_{\boldsymbol{a}}{}^{\boldsymbol{b}}{}_{\boldsymbol{c}}$ can be expressed in terms of the connection coefficients $\Gamma_{\boldsymbol{a}}{}^{\boldsymbol{c}}{}_{\boldsymbol{b}}$ by differentiating Equation (2.25) with respect to $\nabla_{\boldsymbol{d}}$. Noting that $\delta_{\boldsymbol{a}}{}^{\boldsymbol{b}}$ is a constant scalar one has, on the one hand, that

$$\nabla_{\boldsymbol{d}}(\langle \boldsymbol{\omega^b}, \boldsymbol{e_a} \rangle) = \boldsymbol{e_d}(\langle \boldsymbol{\omega^b}, \boldsymbol{e_a} \rangle) = \boldsymbol{e_d}(\delta_{\boldsymbol{a}}{}^{\boldsymbol{b}}) = 0,$$

while, on the other hand, one has

$$\nabla_{\boldsymbol{d}}(\langle \boldsymbol{\omega^b}, \boldsymbol{e_a} \rangle) = \langle \nabla_{\boldsymbol{d}} \boldsymbol{\omega^b}, \boldsymbol{e_a} \rangle + \langle \boldsymbol{\omega^b}, \nabla_{\boldsymbol{d}} \boldsymbol{e_a} \rangle = \left(\mho_{\boldsymbol{d}}{}^{\boldsymbol{b}}{}_{\boldsymbol{c}} + \Gamma_{\boldsymbol{d}}{}^{\boldsymbol{b}}{}_{\boldsymbol{c}} \right) \langle \boldsymbol{\omega^c}, \boldsymbol{e_a} \rangle,$$

so that $\mho_{\boldsymbol{d}}{}^{\boldsymbol{b}}{}_{\boldsymbol{c}} = -\Gamma_{\boldsymbol{d}}{}^{\boldsymbol{b}}{}_{\boldsymbol{c}}$. Consequently, one has

$$\nabla_{\boldsymbol{a}} \boldsymbol{\omega^b} = -\Gamma_{\boldsymbol{a}}{}^{\boldsymbol{b}}{}_{\boldsymbol{c}} \boldsymbol{\omega^c}. \tag{2.27}$$

It is observed that the specification of the 4^3 connection coefficients $\Gamma_{\boldsymbol{a}}{}^{\boldsymbol{b}}{}_{\boldsymbol{c}}$ fully determines the connection $\boldsymbol{\nabla}$; a generalisation of this argument shows that every manifold admits a connection; see, for example, Willmore (1993).

Consider now $\boldsymbol{v} \in T(\mathcal{M})$ and $\boldsymbol{\alpha} \in T^*(\mathcal{M})$. Writing the above in terms of the frame and coframe, respectively, one has

$$\boldsymbol{v} = v^{\boldsymbol{a}} \boldsymbol{e}_{\boldsymbol{a}}, \qquad v^{\boldsymbol{a}} \equiv \langle \boldsymbol{\omega}^{\boldsymbol{a}}, \boldsymbol{v} \rangle,$$
$$\boldsymbol{\alpha} = \alpha_{\boldsymbol{a}} \boldsymbol{\omega}^{\boldsymbol{a}}, \qquad \alpha_{\boldsymbol{a}} \equiv \langle \boldsymbol{\alpha}, \boldsymbol{e}_{\boldsymbol{a}} \rangle.$$

In order to further develop the frame formalism it will be convenient to define

$$\nabla_{\boldsymbol{a}} v^{\boldsymbol{b}} \equiv \langle \boldsymbol{\omega}^{\boldsymbol{b}}, \nabla_{\boldsymbol{a}} \boldsymbol{v} \rangle, \qquad \nabla_{\boldsymbol{a}} \alpha_{\boldsymbol{b}} \equiv \langle \nabla_{\boldsymbol{a}} \boldsymbol{\alpha}, \boldsymbol{e}_{\boldsymbol{b}} \rangle.$$

It follows from Equations (2.26) and (2.27) that

$$\nabla_{\boldsymbol{a}} v^{\boldsymbol{b}} = \boldsymbol{e}_{\boldsymbol{a}}(v^{\boldsymbol{b}}) + \Gamma_{\boldsymbol{a}}{}^{\boldsymbol{b}}{}_{\boldsymbol{c}} v^{\boldsymbol{c}}, \qquad \nabla_{\boldsymbol{a}} \alpha_{\boldsymbol{b}} = \boldsymbol{e}_{\boldsymbol{a}}(\alpha_{\boldsymbol{b}}) - \Gamma_{\boldsymbol{a}}{}^{\boldsymbol{c}}{}_{\boldsymbol{b}} \alpha_{\boldsymbol{c}}. \tag{2.28}$$

The above expressions extend in the obvious way to higher rank components. Notice, in particular, that

$$\nabla_{\boldsymbol{a}} \delta_{\boldsymbol{b}}{}^{\boldsymbol{c}} = -\Gamma_{\boldsymbol{a}}{}^{\boldsymbol{d}}{}_{\boldsymbol{b}} \delta_{\boldsymbol{d}}{}^{\boldsymbol{c}} - \Gamma_{\boldsymbol{a}}{}^{\boldsymbol{c}}{}_{\boldsymbol{d}} \delta_{\boldsymbol{b}}{}^{\boldsymbol{d}} = -\Gamma_{\boldsymbol{a}}{}^{\boldsymbol{c}}{}_{\boldsymbol{b}} + \Gamma_{\boldsymbol{a}}{}^{\boldsymbol{c}}{}_{\boldsymbol{b}} = 0.$$

Metric connections

Now assume that the connection $\boldsymbol{\nabla}$ is $\boldsymbol{g}$-compatible (i.e. $\boldsymbol{\nabla g} = 0$) and that the frame $\{\boldsymbol{e}_{\boldsymbol{a}}\}$ is $\boldsymbol{g}$-orthogonal; that is, $\boldsymbol{g}(\boldsymbol{e}_{\boldsymbol{a}}, \boldsymbol{e}_{\boldsymbol{b}}) = \eta_{\boldsymbol{ab}}$. It follows then that

$$\nabla_{\boldsymbol{a}} \left(\boldsymbol{g}(\boldsymbol{e}_{\boldsymbol{b}}, \boldsymbol{e}_{\boldsymbol{c}}) \right) = \boldsymbol{e}_{\boldsymbol{a}}(\eta_{\boldsymbol{bc}}) = 0$$

and that

$$\nabla_{\boldsymbol{a}} \boldsymbol{g}(\boldsymbol{e}_{\boldsymbol{b}}, \boldsymbol{e}_{\boldsymbol{c}}) = \boldsymbol{g}(\nabla_{\boldsymbol{a}} \boldsymbol{e}_{\boldsymbol{b}}, \boldsymbol{e}_{\boldsymbol{c}}) + \boldsymbol{g}(\boldsymbol{e}_{\boldsymbol{b}}, \nabla_{\boldsymbol{a}} \boldsymbol{e}_{\boldsymbol{c}}).$$

Thus, using Equation (2.26) one concludes that

$$\Gamma_{\boldsymbol{a}}{}^{\boldsymbol{d}}{}_{\boldsymbol{b}} \eta_{\boldsymbol{dc}} + \Gamma_{\boldsymbol{a}}{}^{\boldsymbol{d}}{}_{\boldsymbol{c}} \eta_{\boldsymbol{bd}} = 0. \tag{2.29}$$

Finally, in the case of a Levi-Civita connection and with the choice of a coordinate basis $\{\boldsymbol{\partial}_\mu\}$, the Koszul formula, Equation (2.20), shows that the connection coefficients reduce to the classical expression for the Christoffel symbols:

$$\Gamma_\mu{}^\nu{}_\lambda = \frac{1}{2} g^{\nu\rho} (\partial_\mu g_{\rho\lambda} + \partial_\lambda g_{\mu\rho} - \partial_\rho g_{\mu\lambda}).$$

2.6.2 Frame description of the torsion and curvature

Following the spirit of the previous subsections, let

$$\Sigma_a{}^c{}_b \equiv e_a{}^a e_b{}^b \omega^c{}_c \Sigma_a{}^c{}_b$$

denote the ***components of the torsion tensor*** $\Sigma_a{}^c{}_b$ with respect to $\{e_a\}$ and $\{\omega^a\}$. Given $f \in \mathfrak{X}(\mathcal{M})$, a short computation shows that

$$\begin{aligned}
\Sigma_a{}^c{}_b e_c(f) &= \nabla_a e_b(f) - \nabla_b e_a(f) \\
&= (e_a e_b(f) - \Gamma_a{}^c{}_b e_c(f)) - (e_b e_a(f) - \Gamma_b{}^c{}_a e_c(f)) \\
&= [e_a, e_b](f) - (\Gamma_a{}^c{}_b - \Gamma_b{}^c{}_a)\, e_c(f),
\end{aligned}$$

where it has been used that $\nabla_a f = e_a(f)$. Thus, one obtains that

$$\Sigma_a{}^c{}_b e_c = [e_a, e_b] - (\Gamma_a{}^c{}_b - \Gamma_b{}^c{}_a)\, e_c. \tag{2.30}$$

To obtain a frame description of the Riemann curvature tensor one makes use of Equation (2.8) with $u^c = e_d{}^c$, and contracts with $e_a{}^a e_b{}^b \omega^c{}_d$. One then has that

$$R^c{}_{dab} \equiv e_a{}^a e_b{}^b e_d{}^d \omega^c{}_c R^c{}_{dab}.$$

Furthermore, one can compute

$$\begin{aligned}
e_a{}^a e_b{}^b \omega^c{}_c \nabla_a \nabla_b e_d{}^c &= \omega^c{}_c \nabla_a (\nabla_b e_d{}^c) - \omega^c{}_c (\nabla_a e_b{}^b)(\nabla_b e_d{}^c), \\
&= \omega^c{}_c \nabla_a (\Gamma_b{}^f{}_d e_f{}^c) - \omega^c{}_c \Gamma_a{}^f{}_b \nabla_f e_d{}^c \\
&= \omega^c{}_c e_a (\Gamma_b{}^f{}_d) e_f{}^c + \omega^c{}_c \Gamma_b{}^f{}_d \nabla_a e_f{}^c - \Gamma_a{}^f{}_b \Gamma_f{}^c{}_d \\
&= e_a(\Gamma_b{}^c{}_d) + \Gamma_b{}^f{}_d \Gamma_a{}^c{}_f - \Gamma_a{}^f{}_b \Gamma_f{}^c{}_d.
\end{aligned}$$

A similar computation can be carried out for $e_a{}^a e_b{}^b \omega^c{}_c \nabla_b \nabla_a e_d{}^c$ so that one obtains

$$\begin{aligned}
R^c{}_{dab} &= e_a(\Gamma_b{}^c{}_d) - e_b(\Gamma_a{}^c{}_d) + \Gamma_f{}^c{}_d(\Gamma_b{}^f{}_a - \Gamma_a{}^f{}_b) \\
&\quad + \Gamma_b{}^f{}_d \Gamma_a{}^c{}_f - \Gamma_a{}^f{}_d \Gamma_b{}^c{}_f - \Sigma_a{}^f{}_b \Gamma_f{}^c{}_d.
\end{aligned} \tag{2.31}$$

Remark. Equations (2.30) and (2.31) are sometimes known as the ***(Cartan) structure equations***. They can be conveniently expressed in the language of differential forms; see, for example, Frankel (2003); Wald (1984).

2.7 Congruences and submanifolds

The formulation of an initial value problem in general relativity requires the decomposition of tensorial objects in terms of *temporal* and *spatial components*. This decomposition requires, in turn, an understanding of the way geometric structures of the spacetime are inherited by suitable subsets thereof. For concreteness, in what follows a spacetime $(\mathcal{M}, \boldsymbol{g})$ is assumed. Hence $\mathcal{M}$ is a four-dimensional manifold and $\boldsymbol{g}$ denotes a Lorentzian metric.

2.7.1 Basic notions

Submanifolds

Intuitively, a ***submanifold*** of $\mathcal{M}$ is a set $\mathcal{N} \subset \mathcal{M}$ which inherits a manifold structure from $\mathcal{M}$. A more precise definition of submanifolds requires the concept of embedding. Given two smooth manifolds $\mathcal{M}$ and $\mathcal{N}$, an ***embedding*** is a map $\varphi : \mathcal{N} \to \mathcal{M}$ such that:

(a) The push-forward $\varphi_* : T|_p(\mathcal{N}) \to T|_{\varphi(p)}(\mathcal{M})$ is injective for every point $p \in \mathcal{N}$.
(b) The manifold $\mathcal{N}$ is diffeomorphic to the image $\varphi(\mathcal{N})$.

In terms of the above, one defines a ***submanifold*** $\mathcal{N}$ of $\mathcal{M}$ as the image, $\varphi(\mathcal{S}) \subset \mathcal{M}$, of a k-dimensional manifold $\mathcal{S}$ $(k < 4)$ by an embedding $\varphi : \mathcal{S} \to \mathcal{M}$. Often it is convenient to identify $\mathcal{N}$ with $\varphi(\mathcal{S})$ and denote, in an abuse of notation, both manifolds by $\mathcal{N}$. A three-dimensional submanifold of $\mathcal{M}$ is called a ***hypersurface***. In what follows, a generic hypersurface will be denoted by $\mathcal{S}$. As a consequence of its manifold structure, one can associate to $\mathcal{S}$ tangent and cotangent bundles, $T(\mathcal{S})$ and $T^*(\mathcal{S})$ and, more generally, a tensor bundle $\mathfrak{T}^\bullet(\mathcal{S})$.

A vector $\boldsymbol{u}$ (u^i) on $\mathcal{S}$ can be associated to a vector of $\mathcal{M}$ by the push-forward $\varphi_*\boldsymbol{u}$. A vector on $\boldsymbol{v} \in T(\mathcal{M})$ is said to be ***normal*** to $\mathcal{S}$ if $\boldsymbol{g}(\boldsymbol{v}, \varphi_*\boldsymbol{u}) = 0$ for all $\boldsymbol{u} \in T(\mathcal{S})$. If $\epsilon \equiv \boldsymbol{g}(\boldsymbol{v}, \boldsymbol{v}) = \pm 1$, one speaks of a ***unit normal vector*** – in this case the surface is said to be ***timelike*** if $\epsilon = -1$ and ***spacelike*** if $\epsilon = 1$. A hypersurface $\mathcal{S}$ of a Lorentzian manifold $\mathcal{M}$ is orientable if and only if there exists a unique smooth normal vector field on $\mathcal{S}$; see, for example, O'Neill (1983).

A natural way of specifying a hypersurface is as the level surface of some function $f \in \mathfrak{X}(\mathcal{M})$. In this case one has that the gradient $\mathbf{d}f \in T^*(\mathcal{M})$ gives rise to a normal vector $(\mathbf{d}f)^\sharp \in T(\mathcal{M})$. The ***unit normal*** of $\mathcal{S}$, $\boldsymbol{\nu}$ (ν_a), is then defined as a unit 1-form in the direction of $\mathbf{d}f$; that is, $\boldsymbol{g}^\sharp(\boldsymbol{\nu}, \boldsymbol{\nu}) = \epsilon$. The normal of $\mathcal{S}$ is defined in the restriction to $\mathcal{S}$ of the cotangent bundle $T^*(\mathcal{M})$. In the case of a spacelike hypersurface, the normal constructed in this way is taken, conventionally, to be future pointing.

Foliations

A ***foliation*** of a spacetime $(\mathcal{M}, \boldsymbol{g})$ is a family, $\{\mathcal{S}_t\}_{t\in\mathbb{R}}$, of *spacelike* hypersurfaces $\mathcal{S}_t$, such that

$$\bigcup_{t\in\mathbb{R}} \mathcal{S}_t = \mathcal{M}, \qquad \mathcal{S}_{t_1} \cap \mathcal{S}_{t_2} = \emptyset \quad \text{for} \quad t_1 \neq t_2.$$

The hypersurfaces $\mathcal{S}_t$ are called the ***leaves*** or ***slices*** of the foliation. The foliation $\{\mathcal{S}_t\}_{t\in\mathbb{R}}$ can be defined in terms of a scalar field $f \in \mathfrak{X}(\mathcal{M})$ such that the leaves of the foliation are level surfaces of f. That is, given $p \in \mathcal{S}_t$, then $f(p) = t$. The scalar field f is said to be a ***time function***. In what follows, it will be convenient to identify f and t. The ***normal of a foliation*** is a normalised vector field $\boldsymbol{\nu}$ orthogonal to each leaf of a foliation. The gradient $\mathbf{d}t$ provides a further 1-form normal to the leaves. In general, one has that

$$\boldsymbol{\nu} = N\mathbf{d}t.$$

The proportionality factor N is called the ***lapse*** of the foliation.

Distributions

A ***distribution*** Π is an assignment at each $p \in \mathcal{M}$ of a k-dimensional subspace $\Pi|_p$ of the tangent space $T|_p(\mathcal{M})$. The vector spaces $\Pi|_p$ are called ***hyperplanes*** if their dimension is one less than that of $\mathcal{M}$. A submanifold $\mathcal{N}$ of $\mathcal{M}$ such that $\Pi|_p = T|_p(\mathcal{N})$ for all $p \in \mathcal{N}$ is said to be an ***integrable manifold*** of Π. If for every $p \in \mathcal{M}$ there is an integrable manifold, then Π is said to be ***integrable***. One has the following result (see e.g. Choquet-Bruhat et al. (1982) for details):

Theorem 2.1 (***Frobenius theorem***) *A distribution Π on $\mathcal{M}$ is integrable if and only if for $\boldsymbol{u}$, $\boldsymbol{v} \in \Pi$, one has $[\boldsymbol{u}, \boldsymbol{v}] \in \Pi$.*

The ***projector associated to the distribution*** Π is a tensor field $h_a{}^b$ satisfying $h_a{}^b h_b{}^c = \delta_a{}^c$ such that for $v^a \in \mathfrak{X}(\mathcal{M})$ one has that $h_a{}^b v^a \in \Pi$.

2.7.2 Geometry of congruences

Integral curves

A curve $\gamma : I \to \mathcal{M}$ is the ***integral curve*** of a vector $\boldsymbol{v}$ if the tangent vector of the curve γ coincides with $\boldsymbol{v}$. Standard theorems of the theory of ordinary differential equations – see, for example, Hartman (1987) – ensure that, given $\boldsymbol{v} \in T(\mathcal{M})$, for all $p \in \mathcal{M}$ there exists an interval $I \ni 0$ and a unique integral curve $\gamma : I \to \mathcal{M}$ of $\boldsymbol{v}$ such that $\gamma(0) = p$. If the domain of an integral curve is $\mathbb{R}$, then the integral curve is said to be ***complete***.

Congruences

The notion of a *congruence of geodesics* has been discussed in Section 2.4.5. More generally, a ***congruence of curves*** is the set of integral curves of a (nowhere vanishing) vector field $\boldsymbol{v}$ on $\mathcal{M}$. In the remaining part of this section *it will be assumed that the curves of a congruence are non-intersecting and timelike.* This will be the case of most relevance in this book. In what follows, $\boldsymbol{t}$ will denote the vector field generating a timelike congruence. Without loss of generality it is assumed that $\boldsymbol{g}(\boldsymbol{t}, \boldsymbol{t}) = 1$.

As in previous sections let $\{\boldsymbol{e}_a\}$ denote a $\boldsymbol{g}$-orthonormal frame. The orthonormal frame can be adapted to the congruence defined by the vector field $\boldsymbol{t}$ by setting $\boldsymbol{e}_0 = \boldsymbol{t}$. Given a point $p \in \mathcal{M}$, the tangent space $T|_p(\mathcal{M})$ is naturally split in a part tangential to $\boldsymbol{t}$, to be denoted by $\langle \boldsymbol{t}\rangle|_p$ (the one-dimensional subspace spanned by $\boldsymbol{t}$), and a part orthogonal to it which will be denoted by $\langle \boldsymbol{t}\rangle^\perp|_p = \langle \boldsymbol{e}_i\rangle|_p$ (the three-dimensional subspace generated by $\{\boldsymbol{e}_i\}$ with $\boldsymbol{i} = \mathbf{1}, \mathbf{2}, \mathbf{3}$). The space $\langle \boldsymbol{t}\rangle^\perp|_p$ is an example of a hyperplane. One writes then

$$T|_p(\mathcal{M}) = \langle \boldsymbol{t}\rangle|_p \oplus \langle \boldsymbol{t}\rangle^\perp|_p, \tag{2.32}$$

where $\oplus$ denotes the ***direct sum*** of vectorial spaces – that is, any vector in $T|_p(\mathcal{M})$ can be written in a unique way as the sum of an element in $\langle \boldsymbol{t}\rangle|_p$ and an element in $\langle \boldsymbol{t}\rangle^\perp|_p$. Hence, one sees that *the congruence generated by* $\boldsymbol{t}$ *gives rise to a three-dimensional distribution* Π. At every point $p \in \mathcal{M}$, the subspace $\Pi_p \subset T|_p(\mathcal{M})$ corresponds to $\langle \boldsymbol{e}_i\rangle|_p$; that is, $\{\boldsymbol{e}_i\}$ is a basis of Π_p. In the sequel, $\langle \boldsymbol{t}\rangle$ and $\langle \boldsymbol{t}\rangle^\perp$ will denote, respectively, the *disjoint union* of all the spaces $\langle \boldsymbol{t}\rangle|_p$ and $\langle \boldsymbol{t}\rangle|_p^\perp$, $p \in \mathcal{M}$, and one has that $\Pi = \langle \boldsymbol{t}\rangle^\perp$. The Frobenius theorem, Theorem 2.1, gives the necessary and sufficient conditions for the distribution defined by $\langle \boldsymbol{t}\rangle|_p^\perp$ to be integrable; that is, for the vector $\boldsymbol{t}$ to be the unit normal of a foliation $\{\mathcal{S}_t\}$ of the spacetime.

Making use of $\boldsymbol{g}^\sharp$ one obtains an analogous decomposition for the cotangent space. Namely, one has that

$$T^*|_p(\mathcal{M}) = \langle \boldsymbol{t}^\flat\rangle|_p \oplus \langle \boldsymbol{t}^\flat\rangle^\perp|_p, \tag{2.33}$$

with $\langle \boldsymbol{t}^\flat\rangle^\perp|_p = \langle \boldsymbol{\omega}_i\rangle|_p$. The decompositions (2.32) and (2.33) can be extended in a natural way to higher rank tensors by considering tensor products. Given a tensor T_{ab} with components with respect to the frame $\{\boldsymbol{e}_a\}$ given by $T_{\boldsymbol{ab}}$, one has that $T_{\boldsymbol{ij}} \equiv e_{\boldsymbol{i}}{}^a e_{\boldsymbol{j}}{}^b T_{ab}$ and $T_{\mathbf{00}} \equiv t^a t^b T_{ab}$ correspond, respectively, to the components of T_{ab} ***transversal*** and ***longitudinal*** to $\boldsymbol{t}$; finally, $T_{\mathbf{0}\boldsymbol{i}} \equiv t^a e_{\boldsymbol{i}}{}^b T_{ab}$ and $T_{\boldsymbol{i}\mathbf{0}} \equiv e_{\boldsymbol{i}}{}^a t^b T_{ab}$ are ***mixed transversal-longitudinal*** components.

The covariant derivative of $\boldsymbol{t}$

To further discuss the geometry of the congruence generated by the timelike vector $\boldsymbol{t}$ it is convenient to introduce the ***Weingarten map*** $\boldsymbol{\chi} : \langle \boldsymbol{t}\rangle^\perp \to \langle \boldsymbol{t}\rangle^\perp$ defined by

$$\boldsymbol{\chi}(\boldsymbol{u}) \equiv \nabla_{\boldsymbol{u}}\boldsymbol{t}, \qquad \boldsymbol{u} \in \langle \boldsymbol{t}\rangle^\perp.$$

One can readily verify that

$$\boldsymbol{g}(\boldsymbol{t}, \boldsymbol{\chi}(\boldsymbol{u})) = \boldsymbol{g}(\boldsymbol{t}, \nabla_{\boldsymbol{u}}\boldsymbol{t}) = \frac{1}{2}\nabla_{\boldsymbol{u}}(\boldsymbol{g}(\boldsymbol{t}, \boldsymbol{t})) = 0, \tag{2.34}$$

so that indeed $\boldsymbol{\chi}(\boldsymbol{u}) \in \langle \boldsymbol{t}\rangle^\perp$. Hence, it is enough to consider the Weingarten map evaluated on a basis $\{\boldsymbol{e}_i\}$ of $\langle \boldsymbol{t}\rangle^\perp$. Accordingly, one defines

$$\boldsymbol{\chi}_{\boldsymbol{i}} \equiv \boldsymbol{\chi}(\boldsymbol{e}_{\boldsymbol{i}}) = \chi_{\boldsymbol{i}}{}^{\boldsymbol{j}} \boldsymbol{e}_{\boldsymbol{j}}, \qquad \chi_{\boldsymbol{i}}{}^{\boldsymbol{j}} \equiv \langle \boldsymbol{\omega}^{\boldsymbol{j}}, \boldsymbol{\chi}_{\boldsymbol{i}}\rangle.$$

In the following, it will be more convenient to work with $\chi_{ij} \equiv \eta_{jk}\chi_i{}^k$. The scalars χ_{ij} can be considered as the components of a rank 2 covariant tensor on $\boldsymbol{\chi} \in \langle \boldsymbol{t}\rangle^\perp \otimes \langle \boldsymbol{t}\rangle^\perp$ – the ***Weingarten (or shape) tensor*** of the congruence. The symmetric part $\theta_{ij} \equiv \chi_{(ij)}$ and the antisymmetric part $\omega_{ij} \equiv \chi_{[ij]}$ are called the ***expansion*** and the ***twist*** of the congruence, respectively. From $\boldsymbol{g}(\boldsymbol{t}, \boldsymbol{e}_i) = 0$ it follows that $\boldsymbol{g}(\nabla_j \boldsymbol{t}, \boldsymbol{e}_i) = -\boldsymbol{g}(\boldsymbol{t}, \nabla_j \boldsymbol{e}_i)$. Hence, one can compute

$$\begin{aligned}
\chi_{ij} &= \boldsymbol{g}(\boldsymbol{e}_i, \boldsymbol{\chi}_j) = \boldsymbol{g}(\boldsymbol{e}_i, \nabla_j \boldsymbol{t}) = -\boldsymbol{g}(\boldsymbol{t}, \nabla_j \boldsymbol{e}_i) \\
&= -\boldsymbol{g}(\boldsymbol{t}, \nabla_i \boldsymbol{e}_j - [\boldsymbol{e}_i, \boldsymbol{e}_j]) = \boldsymbol{g}(\nabla_i \boldsymbol{t}, \boldsymbol{e}_j) + \boldsymbol{g}(\boldsymbol{t}, [\boldsymbol{e}_i, \boldsymbol{e}_j]) \\
&= \boldsymbol{g}(\boldsymbol{\chi}_i, \boldsymbol{e}_j) + \boldsymbol{g}(\boldsymbol{t}, [\boldsymbol{e}_i, \boldsymbol{e}_j]) \\
&= \chi_{ji} + \boldsymbol{g}(\boldsymbol{t}, [\boldsymbol{e}_i, \boldsymbol{e}_j]),
\end{aligned}$$

where in the third line it has been used that $\nabla_i \boldsymbol{e}_i - \nabla_j \boldsymbol{e}_i = [\boldsymbol{e}_i, \boldsymbol{e}_j]$ as $\boldsymbol{\nabla}$ is torsion-free. Hence, by the Frobenius theorem, Theorem 2.1, the symmetry relation $\chi_{ij} = \chi_{ji}$ holds if and only if the distribution $\langle \boldsymbol{t}\rangle^\perp$ is integrable. The components χ_{ij} are related to the connection coefficients of $\boldsymbol{\nabla}$ as can be seen from

$$\chi_i{}^j = \langle \boldsymbol{\omega}^j, \boldsymbol{\chi}_i\rangle = \langle \boldsymbol{\omega}^j, \nabla_i \boldsymbol{e}_0\rangle = \langle \boldsymbol{\omega}^j, \Gamma_i{}^b{}_0 \boldsymbol{e}_b\rangle = \Gamma_i{}^j{}_0.$$

Alternatively, one has that

$$\chi_{ij} = \Gamma_i{}^c{}_0 \eta_{cj} = -\Gamma_i{}^c{}_j \eta_{c0} = -\Gamma_i{}^0{}_j,$$

where the last two equalities follow from the metricity of the connection; see Equation (2.29). Now, from $\boldsymbol{g}(\boldsymbol{t}, \boldsymbol{t}) = 1$, it readily follows that $\boldsymbol{g}(\nabla_a \boldsymbol{t}, \boldsymbol{t}) = 0$. Consequently, one has that the ***acceleration*** of the congruence, $\boldsymbol{a} \equiv \nabla_0 \boldsymbol{t} = \nabla_0 \boldsymbol{e}_0$, if non-vanishing, must be spatial; that is, $\boldsymbol{g}(\boldsymbol{a}, \boldsymbol{t}) = 0$ so that $\boldsymbol{a} \in \langle \boldsymbol{t}\rangle^\perp$. Using the definition of connection coefficients of the connection $\boldsymbol{\nabla}$ it follows that

$$a^i \equiv \langle \boldsymbol{\omega}^i, \boldsymbol{a}\rangle = \Gamma_0{}^i{}_0.$$

2.7.3 Geometry of hypersurfaces

Given a spacetime $(\mathcal{M}, \boldsymbol{g})$ and a hypersurface thereof, $\mathcal{S}$, the embedding $\varphi : \mathcal{S} \to \mathcal{M}$ induces on $\mathcal{S}$ a rank 2 covariant tensor $\boldsymbol{h}$, the ***intrinsic metric*** or ***first fundamental form*** of $\mathcal{S}$ via the pull-back of $\boldsymbol{g}$ to $\mathcal{S}$:

$$\boldsymbol{h} \equiv \varphi^* \boldsymbol{g}.$$

As a consequence of the definition of an embedding, the intrinsic metric $\boldsymbol{h}$ will be non-degenerate if the hypersurface $\mathcal{S}$ is timelike or spacelike. Its signature will be $(+,-,-)$ in the former case and $(-,-,-)$ in the latter. The (unique) Levi-Civita connection of $\boldsymbol{h}$ will be denoted by $\boldsymbol{D}$. Alternatively, one can define the ***pull-back connection***

$$\varphi^* \boldsymbol{\nabla} : T(\mathcal{S}) \times T(\mathcal{S}) \to T(\mathcal{S})$$

via

$$\varphi_*\big((\varphi^*\nabla)_{\boldsymbol{v}}\boldsymbol{u}\big) \equiv \nabla_{\varphi_*\boldsymbol{v}}(\varphi_*\boldsymbol{u}), \qquad \boldsymbol{u},\, \boldsymbol{v} \in T(\mathcal{S}). \tag{2.35}$$

It can be verified that $\varphi^*\boldsymbol{\nabla}$ as defined above is indeed a linear connection. Given a function $f \in \mathfrak{X}(\mathcal{M})$, the action of $\varphi^*\boldsymbol{\nabla}$ on the pull-back $\varphi^* f$ is defined by

$$(\varphi^*\nabla)_{\boldsymbol{v}}(\varphi^* f) \equiv \varphi^*(\nabla_{\varphi_*\boldsymbol{v}} f) \in T(\mathcal{M}). \tag{2.36}$$

In order to define the action of $\varphi^*\boldsymbol{\nabla}$ on covectors, one requires the Leibnitz rule

$$(\varphi^*\nabla)_{\boldsymbol{v}}\langle\varphi^*\boldsymbol{\omega}, \boldsymbol{u}\rangle = \langle(\varphi^*\nabla)_{\boldsymbol{v}}(\varphi^*\boldsymbol{\omega}), \boldsymbol{u}\rangle + \langle\varphi^*\boldsymbol{\omega}, (\varphi^*\nabla)_{\boldsymbol{v}}\boldsymbol{u}\rangle,$$

for $\boldsymbol{\omega} \in T^*(\mathcal{M})$ and $\boldsymbol{u},\, \boldsymbol{v} \in T(\mathcal{S})$. A calculation using this expression with the definitions (2.35) and (2.36) shows that for $\boldsymbol{\omega} \in T^*(\mathcal{M})$ one has

$$(\varphi^*\nabla)_{\boldsymbol{v}}\varphi^*\boldsymbol{\omega} \equiv \varphi^*(\nabla_{\varphi_*\boldsymbol{v}}\boldsymbol{\omega}).$$

In a natural way, the embedding $\varphi : \mathcal{S} \to \mathcal{M}$ takes the connection $\boldsymbol{\nabla}$ to the connection $\boldsymbol{D}$. More precisely, one has the following result:

Lemma 2.2 *Given* $\boldsymbol{u},\, \boldsymbol{w} \in T(\mathcal{S})$

$$\varphi_*(D_{\boldsymbol{w}}\boldsymbol{u}) = \nabla_{\varphi_*\boldsymbol{w}}(\varphi_*\boldsymbol{u}). \tag{2.37}$$

Proof Given a function $f \in \mathfrak{X}(\mathcal{M})$ one has that

$$\begin{aligned}
(\varphi^*\nabla)_{\boldsymbol{u}}\left((\varphi^*\nabla)_{\boldsymbol{v}}(\varphi^* f)\right) &= (\varphi^*\nabla)_{\boldsymbol{u}}\left(\varphi^*(\nabla_{\varphi_*\boldsymbol{v}} f)\right) \\
&= \varphi^*\left(\nabla_{\varphi_*\boldsymbol{u}}\nabla_{\varphi_*\boldsymbol{v}} f\right) \\
&= \varphi^*\left(\nabla_{\varphi_*\boldsymbol{v}}\nabla_{\varphi_*\boldsymbol{u}} f\right) \\
&= (\varphi^*\nabla)_{\boldsymbol{v}}\left((\varphi^*\nabla)_{\boldsymbol{u}}(\varphi^* f)\right),
\end{aligned}$$

where to pass from the second to the third line it has been used that the connection $\boldsymbol{\nabla}$ is torsion-free. One thus concludes that the connection $\varphi^*\boldsymbol{\nabla}$ is indeed torsion free. Finally, it can be readily verified that one has compatibility with the metric $\boldsymbol{h}$. Indeed,

$$(\varphi^*\nabla)_{\boldsymbol{v}}\boldsymbol{h} = (\varphi^*\nabla)_{\boldsymbol{v}}(\varphi^*\boldsymbol{g}) = \varphi^*(\nabla_{\varphi_*\boldsymbol{v}}\boldsymbol{g}) = 0,$$

where the last equality follows from the $\boldsymbol{g}$-compatibility of the connection $\boldsymbol{\nabla}$. As $\varphi^*\boldsymbol{\nabla}$ is torsion-free and $\boldsymbol{h}$-compatible, it follows from the fundamental theorem of Riemannian geometry that it must coincide with the connection $\boldsymbol{D}$. In other words, one has that $\varphi^*\boldsymbol{\nabla} = \boldsymbol{D}$, as given in Equation (2.37). □

A frame formalism on hypersurfaces

The present discussion of the geometry of hypersurfaces is valid for both the spacelike and timelike case. To accommodate these two possibilities, all throughout, the following conventions concerning frame indices will be used: if the hypersurface is *timelike* so that $\epsilon = 1$, the frame indices $\boldsymbol{i}, \boldsymbol{j}, \boldsymbol{k}, \ldots$ take the values $\mathbf{1}, \mathbf{2}, \mathbf{3}$; if the hypersurface is *spacelike* so that $\epsilon = -1$, the indices $\boldsymbol{i}, \boldsymbol{j}, \boldsymbol{k}, \ldots$ take the values $\mathbf{0}, \mathbf{1}, \mathbf{2}$.

Following the conventions given in the previous paragraph, let $\{\boldsymbol{e_i}\} \subset T(\mathcal{S})$ denote a triad of $\boldsymbol{h}$-orthogonal vectors. If $\mathcal{S}$ is spacelike one has that $\boldsymbol{h}(\boldsymbol{e_i}, \boldsymbol{e_j}) = -\delta_{\boldsymbol{ij}}$, while in the timelike case $\boldsymbol{h}(\boldsymbol{e_i}, \boldsymbol{e_j}) = \mathrm{diag}(1, -1, -1)$. Using the push-forward $\varphi_* : T(\mathcal{S}) \to T(\mathcal{M})$ one obtains the vectors $\varphi_* \boldsymbol{e_i}$ defined on the restriction of $T(\mathcal{M})$ to $\mathcal{S}$. The triad $\{\boldsymbol{e_i}\}$ can be naturally extended to a tetrad $\{\boldsymbol{e_a}\}$ on the restriction of $T(\mathcal{M})$ to $\mathcal{S}$ by setting $\boldsymbol{e_0} = \boldsymbol{\nu}^\sharp$ in the spacelike case and $\boldsymbol{e_3} = \boldsymbol{\nu}^\sharp$ in the timelike case. In order to discuss these two cases simultaneously, the notation $\boldsymbol{e}_\perp$ will be used. Similarly, the notation $\boldsymbol{\omega}^\perp$ will be used to denote the normal element of the coframe, that is, $\boldsymbol{\omega^0}$ or $\boldsymbol{\omega^3}$. Given $\boldsymbol{v}$ and $\boldsymbol{\alpha}$ on the restriction of $T(\mathcal{M})$ and $T^*(\mathcal{M})$ to $\mathcal{S}$, their components along the normal will be denoted by $v^\perp$ and $\alpha_\perp$, respectively.

To simplify the presentation, the notation $\boldsymbol{e_i}$ will often be used to denote both the vectors of $T(\mathcal{S})$ and their push-forward to $T(\mathcal{M})$. The appropriate point of view should be clear from the context. In the cases where confusion may arise, it is convenient to make use of abstract index notation: given $\boldsymbol{e_i} \in T(\mathcal{S})$, we shall write $e_{\boldsymbol{i}}{}^i$; its push-forward $\varphi_* \boldsymbol{e_i} \in T(\mathcal{M})$ will be denoted by $e_{\boldsymbol{i}}{}^a$. Similarly, $\boldsymbol{\omega^i} \in T^*(\mathcal{M})$ will be written as $\omega^{\boldsymbol{i}}{}_a$, while the pull-back $\varphi^* \boldsymbol{\omega^i} \in T^*(\mathcal{S})$ will be denoted by $\omega^{\boldsymbol{i}}{}_i$. Given $\boldsymbol{u} \in T(\mathcal{S})$, one has that $u^{\boldsymbol{i}} \equiv \langle \varphi^* \boldsymbol{\omega^i}, \boldsymbol{u} \rangle = \langle \boldsymbol{\omega^i}, \varphi_* \boldsymbol{u} \rangle$. Written in index notation $u^a \omega^{\boldsymbol{i}}{}_a = u^i \omega^{\boldsymbol{i}}{}_i$; that is, the (spatial) components of $\boldsymbol{u}$ and its push-forward $\varphi_* \boldsymbol{u}$ coincide.

As a consequence of the existence of two covariant derivatives, one also has two sets of directional covariant derivatives. Firstly, acting on spacetime objects, $\nabla_{\boldsymbol{a}} = e_{\boldsymbol{a}}{}^a \nabla_a$, so that in particular $\nabla_{\boldsymbol{i}} = e_{\boldsymbol{i}}{}^a \nabla_a$. Secondly, acting on hypersurface-defined objects, one has $D_{\boldsymbol{i}} = e_{\boldsymbol{i}}{}^i D_i$. The connection coefficients of $\boldsymbol{D}$ with respect to $\{\boldsymbol{e_i}\}$ are given by $\gamma_{\boldsymbol{i}}{}^{\boldsymbol{j}}{}_{\boldsymbol{k}} \equiv \langle \boldsymbol{\omega^j}, D_{\boldsymbol{i}} \boldsymbol{e_k} \rangle$. Now, given $\boldsymbol{u} \in T(\mathcal{S})$ and $\boldsymbol{\alpha} \in T^*(\mathcal{S})$ and defining

$$D_{\boldsymbol{i}} u^{\boldsymbol{j}} \equiv \langle \boldsymbol{\omega^j}, D_{\boldsymbol{i}} \boldsymbol{u} \rangle, \qquad D_{\boldsymbol{i}} \alpha_{\boldsymbol{j}} \equiv \langle D_{\boldsymbol{i}} \boldsymbol{\alpha}, \boldsymbol{e_j} \rangle,$$

one has, by analogy to Equation (2.28), that

$$D_{\boldsymbol{i}} u^{\boldsymbol{j}} = \boldsymbol{e_i}(u^{\boldsymbol{j}}) + \gamma_{\boldsymbol{i}}{}^{\boldsymbol{j}}{}_{\boldsymbol{k}} u^{\boldsymbol{k}}, \qquad D_{\boldsymbol{i}} \alpha_{\boldsymbol{j}} = \boldsymbol{e_i}(\alpha_{\boldsymbol{j}}) - \gamma_{\boldsymbol{i}}{}^{\boldsymbol{k}}{}_{\boldsymbol{j}} \alpha_{\boldsymbol{k}}.$$

To investigate relations between the directional covariant derivatives $\nabla_{\boldsymbol{i}}$ and $D_{\boldsymbol{i}}$ one makes use of the formula (2.37) with $\boldsymbol{w} = \boldsymbol{e_i}$, $\boldsymbol{u} = \boldsymbol{e_j}$ so that $\varphi_*(D_{\boldsymbol{i}} \boldsymbol{e_j}) = \nabla_{\boldsymbol{i}}(\varphi_* \boldsymbol{e_j}) = \nabla_{\boldsymbol{i}} \boldsymbol{e_j}$ – the last equality given in a slight abuse of notation as, strictly speaking, $\nabla_{\boldsymbol{i}}$ acts on spacetime objects. From the definition of connection coefficients one has that

$$\Gamma_i{}^j{}_k = \langle \omega^j, \nabla_i e_k \rangle = \langle \omega^j, \varphi_*(D_i e_k) \rangle$$
$$= \langle \varphi^* \omega^j, D_i e_k \rangle = \gamma_i{}^j{}_k. \tag{2.38}$$

Given a *spatial* vector $u \in T(\mathcal{M})$ (i.e. $u^\perp \equiv n_a u^a = 0$) and recalling that $\nabla_a u^b \equiv e_a{}^a \omega^b{}_b \nabla_a u^b$, using Equation (2.28) one has that

$$\nabla_a u^b = e_a(u^b) + \Gamma_a{}^b{}_k u^k. \tag{2.39}$$

Restricting the free frame indices in the above expression and using (2.38) one finds

$$\nabla_i u^j = e_i(u^j) + \Gamma_i{}^j{}_k u^k$$
$$= e_i(u^j) + \gamma_i{}^j{}_k u^k = D_i u^j.$$

The intrinsic curvature tensors on the hypersurface

In order to describe the intrinsic curvature of the submanifold $\mathcal{S}$, one considers the ***three-dimensional Riemann curvature tensor*** $r^k{}_{lij}$ of the Levi-Civita connection $\boldsymbol{D}$ of the intrinsic metric $\boldsymbol{h}$. Given $v \in T(\mathcal{S})$, and recalling that $\boldsymbol{D}$ is torsion-free, one has by analogy to Equation (2.8) that

$$D_i D_j v^k - D_j D_i v^k = r^k{}_{lij} v^l.$$

As $r^k{}_{lij}$ is the Riemann tensor of a Levi-Civita connection one has the symmetries

$$r_{klij} = r_{[kl]ij} = r_{kl[ij]} = r_{[kl][ij]},$$
$$r_{klij} = r_{ijkl}, \qquad r_{k[lij]} = 0.$$

In what follows, let $r_{lj} \equiv r^k{}_{lkj}$ and $r \equiv h^{lj} r_{lj}$ denote, respectively, the Ricci tensors and scalars of $\boldsymbol{D}$. It is convenient to also consider the ***trace-free part of the three-dimensional Ricci tensor*** s_{ij} and the ***three-dimensional Schouten tensor*** l_{ij} given by

$$s_{ij} \equiv r_{\{ij\}} = r_{ij} - \frac{1}{3} r h_{ij}, \qquad l_{ij} \equiv s_{ij} + \frac{1}{12} r h_{ij}.$$

The three-dimensionality of the submanifold $\mathcal{S}$ leads to the decomposition

$$r_{klij} = 2h_{k[i}l_{j]l} + 2h_{l[j}l_{i]k}. \tag{2.40}$$

A computation using the above expressions shows that the second Bianchi identity $D_{[i} r_{jk]lm} = 0$ takes, in this case, the form

$$D^i s_{ij} = \frac{1}{6} D_j r.$$

Given the $\boldsymbol{h}$-orthogonal triad $\{\boldsymbol{e}_i\}$ and its associated coframe basis $\{\boldsymbol{\omega}^i\}$ one defines the components $r^k{}_{lij} \equiv e_i{}^i e_j{}^j \omega^k{}_k e_l{}^l r^k{}_{lij}$. A computation similar to that leading to Equation (2.31) yields

$$
\begin{aligned}
r^k{}_{lij} &= \boldsymbol{e}_i(\gamma_j{}^k{}_l) - \boldsymbol{e}_j(\gamma_i{}^k{}_l) + \gamma_m{}^k{}_l(\gamma_j{}^m{}_i - \gamma_i{}^m{}_j) \\
&\quad + \gamma_j{}^m{}_l \gamma_i{}^k{}_m - \gamma_i{}^m{}_l \gamma_j{}^k{}_m. \qquad (2.41)
\end{aligned}
$$

Moreover, the definition of the torsion tensor implies:

$$
[\boldsymbol{e}_i, \boldsymbol{e}_j] = (\gamma_i{}^k{}_j - \gamma_j{}^k{}_i)\boldsymbol{e}_k. \qquad (2.42)
$$

Remark. Equations (2.41) and (2.42) are the *three-dimensional analogue of the (Cartan) structure Equations* (2.30) and (2.31).

Extrinsic curvature

The discussion in Section 2.7.2 concerning the Weingarten map can be specialised to the case of the tangent space of a hypersurface. This leads to the notion of ***extrinsic curvature*** or ***second fundamental form*** of the hypersurface $\mathcal{S}$. The latter is defined via the map $\boldsymbol{K} : T(\mathcal{S}) \times T(\mathcal{S}) \to \mathbb{R}$ given by

$$
\boldsymbol{K}(\boldsymbol{u}, \boldsymbol{v}) \equiv \langle \nabla_{\boldsymbol{u}} \boldsymbol{\nu}, \boldsymbol{v} \rangle = \boldsymbol{g}(\nabla_{\boldsymbol{u}} \boldsymbol{\nu}^\sharp, \boldsymbol{v}). \qquad (2.43)
$$

From the discussion of the Weingarten map it follows that $\boldsymbol{K}$ as defined above is a symmetric three-dimensional tensor. In abstract index notation the latter will be written as K_{ij}.

Now, given an orthonormal frame $\{\boldsymbol{e}_i\}$ on $\mathcal{S}$ and choosing $\boldsymbol{v} = \boldsymbol{e}_i$ and $\boldsymbol{u} = \boldsymbol{e}_j$ in formula (2.43) one finds that the components K_{ij} are given by

$$
K_{ij} = \nabla_i \nu_j \equiv \langle \nabla_j \boldsymbol{\nu}, \boldsymbol{e}_i \rangle = \langle \nabla_j \boldsymbol{\omega}^\perp, \boldsymbol{e}_i \rangle \qquad (2.44)
$$

so that, comparing Equation (2.44) with the definition of the connection coefficients one finds that

$$
\begin{aligned}
K_{ij} &= \Gamma_i{}^a{}_\perp \eta_{aj} \\
&= -\Gamma_i{}^a{}_j \eta_{a\perp} = -\epsilon \Gamma_i{}^\perp{}_j. \qquad (2.45)
\end{aligned}
$$

Now, looking again at Equation (2.39) and setting $\boldsymbol{a} \mapsto \boldsymbol{i}$, $\boldsymbol{b} \mapsto \perp$ one obtains

$$
\begin{aligned}
\nabla_i v^\perp &= \boldsymbol{e}_i(v^\perp) + \Gamma_i{}^\perp{}_k v^k \\
&= \Gamma_i{}^\perp{}_k v^k = -\epsilon K_{ik} v^k, \qquad (2.46)
\end{aligned}
$$

as $\boldsymbol{v} \in T(\mathcal{S})$ so that $v^\perp = 0$.

The Gauss-Codazzi and Codazzi-Mainardi equations

The curvature tensors of the connections $\boldsymbol{\nabla}$ and $\boldsymbol{D}$ are related to each other by means of the ***Gauss-Codazzi equation***

$$R_{ijkl} = r_{ijkl} + K_{ik}K_{jl} - K_{il}K_{jk}, \tag{2.47}$$

and the ***Codazzi-Mainardi equation***

$$R_{i\perp jk} = D_j K_{ki} - D_k K_{ji}. \tag{2.48}$$

The proof of the Gauss-Codazzi equation follows by considering the commutator of $\boldsymbol{\nabla}$, Equation (2.8), acting on the frame vectors $\boldsymbol{e}_l$:

$$\nabla_a \nabla_b e_l{}^c - \nabla_b \nabla_a e_l{}^c = R^c{}_{dab} e_l{}^d \equiv R^c{}_{lab}.$$

Contracting the previous equation with $e_i{}^a e_j{}^b \omega^k{}_c$, and using

$$\nabla_b e_l{}^c = \omega^b{}_b \Gamma_b{}^a{}_l e_a{}^c,$$

together with formulae (2.38) and (2.46) and the expression for the components of the three-dimensional Riemann tensor in terms of the connection coefficients, Equation (2.41), yields (2.47). The proof of the Codazzi-Mainardi Equation (2.48) involves less computation. In this case one evaluates the commutator of covariant derivatives on the covector $\boldsymbol{\nu}$. Contracting with $e_i{}^a e_j{}^b e_k{}^c$ one readily finds that

$$\nabla_i \nabla_j \nu_k - \nabla_j \nabla_i \nu_k = -R^{\perp}{}_{kij},$$

where $R^{\perp}{}_{kij} \equiv R^d{}_{cab}\nu_d e_k{}^c e_i{}^a e_j{}^b$. Now, using Equation (2.44) one finds that

$$\nabla_i K_{jk} - \nabla_j K_{ik} = -R^{\perp}{}_{kij}.$$

Formula (2.48) follows from the above expression by noticing that $\nabla_i K_{jk} = D_i K_{jk}$ as K_{jk} corresponds to the spatial components of a spatial tensor.

A remark concerning foliations

The discussion in the previous subsections was restricted to a single hypersurface $\mathcal{S}$. However, it can be readily extended to a foliation $\{\mathcal{S}_t\}$. In this case the contravariant version of the normal $\boldsymbol{\nu}^\sharp$ and the unit vector $\boldsymbol{t}$ generating the congruence coincide. Moreover, one has a distribution which is integrable so that the Weingarten tensor $\boldsymbol{\chi} \in \big(\langle \boldsymbol{t}\rangle^\perp \otimes \langle \boldsymbol{t}\rangle^\perp\big)|_p$, for $p \in \mathcal{M}$ can be identified with the second fundamental form $\boldsymbol{K} \in T|_p(\mathcal{S}_{t(p)}) \otimes T|_p(\mathcal{S}_{t(p)})$ where $t(p) \in \mathbb{R}$ is the only value of the time function such that $p \in \mathcal{S}_{t(p)}$. In particular one has that $\chi_{ij} = \chi_{(ij)}$.

2.8 Further reading

There is a vast choice of books on differential geometry ranging from introductory texts to comprehensive monographs. An introductory discussion geared towards applications in general relativity can be found in the first chapter of Stewart (1991) or the second and third chapters of Wald (1984). A more extensive introduction with broader applications in physics is Frankel (2003). A more advanced discussion, again aimed at applications in physics, is the classical textbook by Choquet-Bruhat et al. (1982). A systematic and coherent discussion of the theory from a modern mathematical point of view covering topological manifolds, smooth manifolds and differential geometry can be found in Lee (1997, 2000, 2002). A more concise alternative to the latter three books is given in Willmore (1993). A monograph on Lorentzian geometry with applications to general relativity is O'Neill (1983). Readers who like the style of this reference will also find the brief summary of differential geometry given in the first chapter of O'Neill (1995) useful. The present discussion of differential geometry has avoided the use of the language of fibre bundles. Readers interested in the latter are referred to Taubes (2011).

Books on numerical relativity like Baumgarte and Shapiro (2010) and Alcubierre (2008) also provide introductions to the 3+1 decomposition of general relativity. In these references, the reader will encounter an approach to this topic based on the so-called *projection formalism*. A more detailed discussion, also aimed at numerical relativity, can be found in Gourgoulhon (2012).

3

Spacetime spinors

The notion of spinors arises naturally in the construction of a relativistic first-order equation for a quantum wave function – the so-called *Dirac equation*. Spinors are the most basic objects to which one can apply a Lorentz transformation. The seminal work in Penrose (1960) has shown that spinors constitute a powerful tool to analyse the structure of the Einstein field equations and their solutions. Most applications of spinors in general relativity make use not of the *Dirac spinors* but of the so-called *2-spinors*. The latter are more elementary objects, and indeed, the whole theory of the Dirac equation can be reformulated in terms of 2-spinors. In the sequel, 2-spinors will be very often simply called *spinors*.

The purpose of this chapter is to develop the basic formalism of spinors in a spacetime. Accordingly, one speaks of *spacetime spinors*, sometimes also called $SL(2,\mathbb{C})$ spinors; see, for example, Ashtekar (1991). A discussion of spinors in the presence of a singled-out timelike direction, the so-called *space spinor formalism*, is given in Chapter 4. One of the motivations for the use of spinors in general relativity is that they provide a simple representation of null vectors and of several tensorial operations. Although spinors will be used systematically in this book, they are not essential for the analysis. All the key arguments could be carried out in a tensorial way at the expense of lengthier and less transparent computations.

The presentation in this chapter differs sligthly in focus and content from that given in other texts; see, for example, Penrose and Rindler (1984); Stewart (1991); O'Donnell (2003). For reasons to be discussed in the main text, a systematic use of the so-called *Newman-Penrose formalism* will be avoided – although the basic notational conventions of Penrose and Rindler (1984), the authoritative work on the subject, are retained.

3.1 Algebra of 2-spinors

In what follows let $(\mathcal{M}, \boldsymbol{g})$ be a spacetime. The present discussion begins by analysing spinorial structures at a given point p of the spacetime manifold $\mathcal{M}$. The concept of a spinor is closely related to the representation theory of the group $SL(2, \mathbb{C})$. This group has two inequivalent representations in terms of two-dimensional complex vector spaces which are complex conjugates of each other; for a discussion of this aspect of the theory, see, for example, Carmeli (1977); Sexl and Urbantke (2000). Thus, the discussion of this chapter starts with a brief discussion of complex vector spaces.

3.1.1 Complex vector spaces

By a ***complex vector space*** it will be understood a vector space over the field of the complex numbers, $\mathbb{C}$. In what follows let $\mathfrak{S}$ denote a complex vector space, and let $\mathfrak{S}^*$ denote its dual, that is, the complex vector space of all linear maps from $\mathfrak{S}$ to $\mathbb{C}$. As in the case of real vector spaces, given $\boldsymbol{\varsigma} \in \mathfrak{S}$ and $\boldsymbol{\zeta} \in \mathfrak{S}^*$, the application of $\boldsymbol{\zeta}$ on $\boldsymbol{\varsigma}$ will be denoted by $\langle \boldsymbol{\zeta}, \boldsymbol{\varsigma} \rangle$. Notice, however, that in this case $\langle \boldsymbol{\zeta}, \boldsymbol{\varsigma} \rangle \in \mathbb{C}$.

Given $\mathfrak{S}$, it is natural to define an operation of complex conjugation over $\mathfrak{S}$: given $\boldsymbol{\varsigma} \in \mathfrak{S}$, its complex conjugate $\bar{\boldsymbol{\varsigma}}$ is defined via

$$\langle \boldsymbol{\zeta}, \bar{\boldsymbol{\varsigma}} \rangle \equiv \overline{\langle \boldsymbol{\zeta}, \boldsymbol{\varsigma} \rangle}, \qquad \boldsymbol{\zeta} \in \mathfrak{S}^*.$$

The operation of complex conjugation from $\mathfrak{S}$ to $\mathfrak{S}^*$ can be defined in an analogous way: given $\boldsymbol{\zeta} \in \mathfrak{S}^*$, its complex conjugate $\bar{\boldsymbol{\zeta}}$ satisfies

$$\langle \bar{\boldsymbol{\zeta}}, \boldsymbol{\varsigma} \rangle \equiv \overline{\langle \boldsymbol{\zeta}, \boldsymbol{\varsigma} \rangle}, \qquad \boldsymbol{\varsigma} \in \mathfrak{S}.$$

Given $\boldsymbol{\xi}$, $\boldsymbol{\zeta} \in \mathfrak{S}$ and $z \in \mathbb{C}$, the complex conjugate of the linear combination $\boldsymbol{\xi}+z\boldsymbol{\zeta}$ is $\bar{\boldsymbol{\xi}}+\bar{z}\bar{\boldsymbol{\zeta}}$. Thus, the operation of complex conjugation is not an isomorphism between $\mathfrak{S}$ and itself, but an *anti-isomorphism* between $\mathfrak{S}$ and the vector space $\bar{\mathfrak{S}}$, the ***complex conjugate*** of $\mathfrak{S}$. Similarly, the complex conjugation defines an anti-isomorphism between $\mathfrak{S}^*$ and the space, $\overline{\mathfrak{S}^*}$, the complex conjugate of $\mathfrak{S}^*$. If one considers the complex conjugate of the spaces $\overline{\mathfrak{S}}$ and $\overline{\mathfrak{S}^*}$, one recovers the spaces $\mathfrak{S}$ and $\mathfrak{S}^*$, respectively. Moreover, because of the way the complex conjugate operation has been defined, one has that $\overline{\mathfrak{S}^*} = \bar{\mathfrak{S}}^*$, so that $\overline{\mathfrak{S}}$ and $\overline{\mathfrak{S}^*}$ are duals of each other.

The vector spaces $\mathfrak{S}$, $\mathfrak{S}^*$, $\overline{\mathfrak{S}}$ and $\overline{\mathfrak{S}^*}$ will be regarded as the elementary building blocks in the construction of a spinorial formalism. As in the case of real vector spaces one can construct *higher rank* objects by considering arbitrary tensor products of these vector spaces. This will be discussed later in the chapter once further structure and an *abstract index notation* for spinors has been introduced.

3.1.2 Simplectic vector spaces

Key to the notion of spinors is the definition of a ***symplectic vector space***.

Definition 3.1 (*simplectic vector space*) *A* ***simplectic vector space*** *consists of an even-dimensional vector space $\mathfrak{S}$ endowed with a function $[[\cdot,\cdot]] : \mathfrak{S}\times\mathfrak{S}\to\mathbb{C}$ which is:*

(i) ***antisymmetric (skew)****; that is, given $\boldsymbol{\xi}$, $\boldsymbol{\eta}\in\mathfrak{S}$*

$$[[\boldsymbol{\xi},\boldsymbol{\eta}]] = -[[\boldsymbol{\eta},\boldsymbol{\xi}]]$$

(ii) ***bilinear****; that is,*

$$[[\boldsymbol{\xi}+z\boldsymbol{\zeta},\boldsymbol{\eta}]] = [[\boldsymbol{\xi},\boldsymbol{\eta}]] + z[[\boldsymbol{\zeta},\boldsymbol{\eta}]], \qquad [[\boldsymbol{\xi},\boldsymbol{\eta}+z\boldsymbol{\zeta}]] = [[\boldsymbol{\xi},\boldsymbol{\eta}]] + z[[\boldsymbol{\xi},\boldsymbol{\zeta}]]$$

(iii) ***non-degenerate****; that is, if $[[\boldsymbol{\xi},\boldsymbol{\eta}]]=0$ for all $\boldsymbol{\eta}$ then $\boldsymbol{\xi}=0$.*

The antisymmetric product $[[\cdot,\cdot]]$ defines in a canonical way an isomorphism between $\mathfrak{S}$ and $\mathfrak{S}^*$: to $\boldsymbol{\xi}\in\mathfrak{S}$ one associates $\boldsymbol{\xi}^\flat \equiv [[\boldsymbol{\xi},\cdot]]\in\mathfrak{S}^*$. A transformation $\mathbf{Q}:\mathfrak{S}\to\mathfrak{S}$ satisfying $[[\mathbf{Q}\boldsymbol{\xi},\mathbf{Q}\boldsymbol{\eta}]] = [[\boldsymbol{\xi},\boldsymbol{\eta}]]$ is called a ***symplectic transformation***.

Remark. The rest of this book will be concerned only with the case where the dimension of $\mathfrak{S}$ is 2.

3.1.3 Spin bases

From the definition of a symplectic vector space it follows directly that given non-zero $\boldsymbol{\xi}$, $\boldsymbol{\eta}\in\mathfrak{S}$ such that $[[\boldsymbol{\xi},\boldsymbol{\eta}]]=0$, there exists $z\in\mathbb{C}$, $z\neq 0$ such that $\boldsymbol{\xi}=z\boldsymbol{\eta}$. Alternatively, given $\boldsymbol{\xi}$, $\boldsymbol{\eta}\in\mathfrak{S}$, they are linearly independent if and only if $[[\boldsymbol{\xi},\boldsymbol{\eta}]]\neq 0$. This observation leads to the idea of a ***spin basis***.

Definition 3.2 (*spin basis*) *Given non-zero $\boldsymbol{o}$, $\boldsymbol{\iota}\in\mathfrak{S}$, the pair $\{\boldsymbol{o},\boldsymbol{\iota}\}$ is said to be a* ***spin basis*** *for $\mathfrak{S}$ if $[[\boldsymbol{o},\boldsymbol{\iota}]]=1$.*

Now, given $\boldsymbol{\xi}\in\mathfrak{S}$, the components of $\boldsymbol{\xi}$ with respect to the basis $\{\boldsymbol{o},\boldsymbol{\iota}\}$ are defined by the equation

$$\boldsymbol{\xi} = \xi^{\mathbf{0}}\boldsymbol{o} + \xi^{\mathbf{1}}\boldsymbol{\iota},$$

where

$$\xi^{\mathbf{0}} \equiv [[\boldsymbol{\xi},\boldsymbol{\iota}]], \qquad \xi^{\mathbf{1}} \equiv -[[\boldsymbol{\xi},\boldsymbol{o}]].$$

3.1.4 Abstract index notation for spinors

The discussion of spinors in this book makes use of a combination of index-free and abstract index notations. Following the general discussion on abstract index notation given in Penrose and Rindler (1984), an element $\boldsymbol{\xi}\in\mathfrak{S}$ will also be denoted by ξ^A, where the abstract superindex ^{A} provides information about the vector space to which the object belongs – in this case $\mathfrak{S}$. Similarly given $\boldsymbol{\eta}\in\mathfrak{S}^*$, it will often be written as η_A. This notation of abstract sub- and superindices will also be extended to the vector spaces themselves; thus, the symbols $\mathfrak{S}^A$ and $\mathfrak{S}_A$ will be used, respectively, instead of $\mathfrak{S}$ and $\mathfrak{S}^*$. Furthermore, given $\xi^A \in \mathfrak{S}^A$, then ξ_A will denote $\boldsymbol{\xi}^\flat$, the dual of $\boldsymbol{\xi}$ under the antisymmetric product in $\mathfrak{S}$. Following this notation, the product $[[\boldsymbol{\eta}, \boldsymbol{\xi}]] = \langle \boldsymbol{\eta}^\sharp, \boldsymbol{\xi} \rangle$ will be written as $\eta_A \xi^A$.

In order to extend the formalism, one introduces an infinite number of copies (realisations) of the spaces $\mathfrak{S}$ and $\mathfrak{S}^*$: $\mathfrak{S}^A$, $\mathfrak{S}^B, \ldots$ and $\mathfrak{S}_A$, $\mathfrak{S}_B$, ... The different realisations are connected to each other by a ***sameness map*** such that ξ^A and ξ^B correspond to two different copies of the same object $\boldsymbol{\xi}$ belonging to different realisations of $\mathfrak{S}$, that is, $\mathfrak{S}^A$ and $\mathfrak{S}^B$. A peculiarity of the abstract index notation is that although ξ^A and ξ^B describe the same object, expressions like $\xi^A = \xi^B$ are not allowed – *the indices in an equation must be balanced.*

Objects like ξ^A and η_B are called ***valence 1 spinors***. Following the terminology used for tensors, ξ^A is said to be ***contravariant***, while η_A is said to be ***covariant***. ***Higher valence spinors*** can be introduced using the tensorial product $\otimes$ of the basic vector spaces $\mathfrak{S}$ and $\mathfrak{S}^*$. The use of the abstract index notation simplifies the underlying discussion of these tensorial products. For example, a valence 3 spinor $\chi_{AB}{}^C$ is defined through a multilinear map $\boldsymbol{\chi} : \mathfrak{S}^A \times \mathfrak{S}^B \times \mathfrak{S}_C \to \mathbb{C}$. As a consequence of the $\mathfrak{S}$-linearity of this mapping, there exists a spinor $\chi_{AB}{}^C \in \mathfrak{S}_{AB}{}^C$. The space $\mathfrak{S}_{AB}{}^C$ is a vector space. This procedure extends in a natural way to higher valence spinors with arbitrary combinations of covariant and contravariant indices. The collection of all the spaces of the form $\mathfrak{S}_{A\cdots C}{}^{D\cdots F}$ is called the ***spin algebra*** and is denoted by $\mathfrak{S}^\bullet$. The spin algebra ensures that the multiplication of spinors renders a spinor. The operation of addition in $\mathfrak{S}^\bullet$ is defined only between spinors of the same type, that is, the same rank and same combination of covariant and contravariant indices.

3.1.5 The spinor ϵ_{AB}

As the antisymmetric 2-form $[[\cdot, \cdot]]$ is a function from $\mathfrak{S} \otimes \mathfrak{S}$ to $\mathbb{C}$, it follows that there exists a valence 2 spinor $\epsilon_{AB} \in \mathfrak{S}_{AB}$ such that

$$[[\boldsymbol{\xi}, \boldsymbol{\eta}]] = \epsilon_{AB} \xi^A \eta^B.$$

The spinor ϵ_{AB} is called the ***ϵ-spinor***. Now, as $[[\boldsymbol{\xi}, \boldsymbol{\eta}]] = -[[\boldsymbol{\eta}, \boldsymbol{\xi}]]$, it follows that $\epsilon_{AB} = -\epsilon_{BA}$; that is, ϵ_{AB} is *antisymmetric*. It has already been shown that $[[\boldsymbol{\xi}, \boldsymbol{\eta}]]$ can be written as $\xi_A \eta^A$; thus, it follows that

$$\xi_B = \epsilon_{AB} \xi^A = \xi^A \epsilon_{AB}. \tag{3.1}$$

That is, ϵ_{AB} can be regarded as an *index lowering object*. In other words, the spinor ϵ_{AB} provides a convenient way to express the duality between the spaces $\mathfrak{S}$ and $\mathfrak{S}^*$. This duality is a bijection, so that it follows that there must exist a further spinor, $(\epsilon^{-1})^{AB} \in \mathfrak{S}^{AB}$, by means of which one can raise back the index of the spinor ξ_A; that is, $\xi^A = (\epsilon^{-1})^{CA}\xi_C$. In order to simplify the appearance of the above expressions it is convenient to define a further spinor $\epsilon^{AB} \in \mathfrak{S}^{AB}$ via

$$\epsilon^{AB} \equiv -(\epsilon^{-1})^{AB}, \tag{3.2}$$

so that one obtains

$$\xi^A = -\epsilon^{CA}\xi_C. \tag{3.3}$$

Combining Equations (3.1) and (3.3) one obtains $\xi_B = -\epsilon_{AB}\epsilon^{CA}\xi_C$, which together with the requirement that ϵ_{AB} and $(\epsilon^{-1})^{AB}$ represent inverse operations, implies

$$\delta_B{}^C = -\epsilon_{AB}\epsilon^{CA},$$

with $\delta_B{}^C$ the ***two-dimensional Kronecker's delta***. The spinor ϵ^{AB} is also *antisymmetric*. This can be seen from

$$[[\boldsymbol{\xi}, \boldsymbol{\eta}]] = \xi_B\eta^B = \xi_B\delta_C{}^B\eta^C = -\xi_B(\epsilon_{DC}\epsilon^{BD})\eta^C$$
$$= \epsilon^{BD}\xi_B(\epsilon_{CD}\eta^C) = \epsilon^{BD}\xi_B\eta_D.$$

A similar computation shows that $[[\boldsymbol{\eta}, \boldsymbol{\xi}]] = \epsilon^{DB}\eta_D\xi_B$. Finally, as $[[\boldsymbol{\xi}, \boldsymbol{\eta}]] = -[[\boldsymbol{\eta}, \boldsymbol{\xi}]]$ one concludes that $\epsilon^{AB} = -\epsilon^{BA}$ as claimed.

If $\epsilon^A{}_C$ and $\epsilon_A{}^C$ denote the spinors in $\mathfrak{S}^\bullet$ obtained by raising the first and second index of ϵ_{AB}, respectively, it follows from the above calculations that

$$\epsilon_C{}^A = -\epsilon^A{}_C = \delta_C{}^A, \qquad \epsilon_{AB}\epsilon^{AB} = \epsilon_A{}^A = 2.$$

The above formulae lead to the so-called ***see-saw rule***. Given a spinor $\chi^{P\cdots QA}$ one has that

$$\chi^{P\cdots QA} = \epsilon^{AB}\chi^{P\cdots Q}{}_B = -\chi^{P\cdots Q}{}_B\epsilon^{BA} = \chi^{P\cdots QB}\epsilon_B{}^A, \tag{3.4a}$$
$$\chi^{P\cdots Q}{}_A = -\epsilon_{AB}\chi^{P\cdots QB} = \chi^{P\cdots QB}\epsilon_{BA} = -\chi^{P\cdots Q}{}_B\epsilon^B{}_A. \tag{3.4b}$$

Comparing the above expressions one concludes that

$$\chi^{P\cdots Q}{}_A{}^A = -\chi^{P\cdots QA}{}_A.$$

3.1.6 The Jacobi identity and decompositions in irreducible components

As $\mathfrak{S}$ is a vector space of dimension 2, it follows that any antisymmetrisation over a set of three or more spinorial indices must vanish. In particular, one obtains what is known as the ***Jacobi identity***:

$$\epsilon_{A[B}\epsilon_{CD]} = \epsilon_{AB}\epsilon_{CD} + \epsilon_{AC}\epsilon_{DB} + \epsilon_{AD}\epsilon_{BC} = 0. \tag{3.5}$$

A direct consequence of the Jacobi identity is the following lemma:

Lemma 3.1 (***irreducible decomposition of a pair of indices***) *Consider the spinor* $\zeta_{\cdots AB\cdots}$. *Then*

$$\zeta_{\cdots AB\cdots} = \zeta_{\cdots(AB)\cdots} + \frac{1}{2}\epsilon_{AB}\zeta_{\cdots C}{}^{C}{}_{\cdots}.$$

Proof Consider the Jacobi identity rewritten in the form

$$\epsilon_A{}^C\epsilon_B{}^D - \epsilon_B{}^C\epsilon_A{}^D = \epsilon_{AB}\epsilon^{CD},$$

and multiply it by $\zeta_{\cdots CD\cdots}$. One readily obtains

$$2\zeta_{\cdots[AB]\cdots} = \epsilon_{AB}\zeta_{\cdots C}{}^{C}{}_{\cdots}.$$

Finally, combining the latter with the identity

$$\zeta_{\cdots AB\cdots} = \zeta_{\cdots(AB)\cdots} + \zeta_{\cdots[AB]\cdots},$$

one obtains the required result. □

The previous result can be used to interchange the order of two spinorial indices. In this case Lemma 3.1 directly yields

$$\zeta_{\cdots BA\cdots} = \zeta_{\cdots AB\cdots} - \epsilon_{AB}\zeta_{\cdots P}{}^{P}{}_{\cdots}. \tag{3.6}$$

The above lemma leads to the following result:

Proposition 3.1 (***irreducible decomposition of spinors***) *Any spinor* $\zeta_{A\cdots F}$ *can be decomposed as the sum of the spinor* $\zeta_{(A\cdots F)}$ *and products of* ϵ*-spinors with symmetrised contractions of* $\zeta_{A\cdots F}$.

Proof Assume $\zeta_{ABC\cdots F}$ to have valence n. In the following argument, the symbol $\sim$ between two spinors indicates that their difference is a linear combination of the outer product of ϵ-spinors and spinors of lower valence. The key idea of the decomposition is to show that

$$\zeta_{ABC\cdots EF} \sim \zeta_{(ABC\cdots EF)}.$$

To this end, one first notices that

$$n\zeta_{(ABC\cdots EF)} = \zeta_{A(BC\cdots EF)} + \zeta_{B(AC\cdots EF)} + \zeta_{C(AB\cdots EF)} + \cdots + \zeta_{F(AB\cdots E)}. \tag{3.7}$$

Now, one looks at the terms in the right-hand side of the above equation and considers the difference between the first and the second term, the first and the third term and so on. Using Lemma 3.1, these differences can be rewritten as

$$\begin{aligned}
\zeta_{A(BC\cdots EF)} - \zeta_{B(AC\cdots EF)} &= -\zeta^X{}_{(XC\cdots EF)}\epsilon_{AB},\\
\zeta_{A(BC\cdots EF)} - \zeta_{C(AB\cdots EF)} &= -\zeta^X{}_{(XB\cdots EF)}\epsilon_{AC},\\
&\ \vdots\\
\zeta_{A(BC\cdots EF)} - \zeta_{F(ABC\cdots E)} &= -\zeta^X{}_{(XBC\cdots E)}\epsilon_{AF}.
\end{aligned}$$

The above expressions can be used in Equation (3.7) to eliminate the terms

$$\zeta_{B(AC\cdots EF)}, \qquad \zeta_{B(AC\cdots EF)}, \qquad \cdots \qquad \zeta_{F(ABC\cdots E)}.$$

One obtains

$$\zeta_{(ABC\cdots EF)} = \zeta_{A(BC\cdots EF)} + \frac{1}{n}\zeta^{X}{}_{(XC\cdots EF)}\epsilon_{AB} + \cdots + \frac{1}{n}\zeta^{X}{}_{(XBC\cdots E)}\epsilon_{AF}.$$

That is,

$$\zeta_{(ABC\cdots EF)} \sim \zeta_{A(BC\cdots EF)}.$$

The procedure described above can be repeated for each of the terms

$$\zeta^{X}{}_{(XC\cdots EF)}, \qquad \cdots \qquad \zeta^{X}{}_{(XB\cdots E)},$$

to obtain

$$\zeta_{(ABC\cdots EF)} \sim \zeta_{A(BC\cdots EF)} \sim \zeta_{AB(C\cdots EF)} \sim \cdots \sim \zeta_{ABC\cdots(EF)} \sim \zeta_{ABC\cdots EF}.$$

□

Remark. If one has a spinor with a set of contravariant indices, these can be lowered so that Proposition 3.1 applies.

The type of decompositions of spinors provided by Proposition 3.1 will be used systematically in the rest of the book. A particularly useful example is given by

$$\begin{aligned}\chi_{ABCD} = \chi_{(ABCD)} &+ \frac{1}{2}\chi_{(AB)P}{}^{P}\epsilon_{CD} + \frac{1}{2}\chi_{P}{}^{P}{}_{(CD)}\epsilon_{AB} + \frac{1}{4}\chi_{P}{}^{P}{}_{Q}{}^{Q}\epsilon_{AB}\epsilon_{CD} \\ &+ \frac{1}{2}\epsilon_{A(C}\chi_{D)B} + \frac{1}{2}\epsilon_{B(C}\chi_{D)A} - \frac{1}{3}\epsilon_{A(C}\epsilon_{D)B}\chi, \end{aligned} \tag{3.8}$$

with

$$\chi_{AB} \equiv \chi_{Q(AB)}{}^{Q}, \qquad \chi \equiv \chi_{PQ}{}^{PQ}.$$

A decomposition like the one given in Equation (3.8) will be called a ***decomposition in irreducible components***. The spinors $\chi_{(ABCD)}, \chi_{(AB)P}{}^{P}, \ldots, \chi$ are *independent* in the sense that $\chi_{ABCD} = 0$ if and only if

$$\chi_{(ABCD)} = 0, \quad \chi_{(AB)P}{}^{P} = 0, \quad \cdots \quad \chi = 0.$$

The latter fact will be used repeatedly in the following. Finally, it is observed that the number of independent components an arbitrary symmetric spinor can have is given by the following proposition; see Penrose and Rindler (1984).

Proposition 3.2 (***number of independent components***) *If* $\zeta_{A\cdots C} = \zeta_{(A\cdots C)}$ *is of valence* p*, then it has* $(p+1)$ *independent components.*

In conjunction with Proposition 3.1 the latter result can be used to count the total number of independent components of an arbitrary spinor.

3.1.7 Components with respect to a basis

As in the case of tensors, it is often convenient to discuss spinors in terms of a specific basis. To express this idea, it is convenient to introduce bold indices $\boldsymbol{A}, \boldsymbol{B}, \ldots$ ranging over $\mathbf{0}$ and $\mathbf{1}$. Thus, $\xi^{\boldsymbol{A}}$ and $\eta_{\boldsymbol{A}}$ represent the components of ξ^A and η_B with respect to a specific basis. This idea extends in a natural way to higher valence spinors.

Given a spin basis $\{\boldsymbol{o}, \boldsymbol{\iota}\}$, one often requires a notation to describe the basis in a more systematic manner. This will be done by means of the symbol $\epsilon_{\boldsymbol{A}}{}^A$ where

$$\epsilon_{\mathbf{0}}{}^A \equiv o^A, \qquad \epsilon_{\mathbf{1}}{}^A \equiv \iota^A. \tag{3.9}$$

Similarly, the dual cobasis of $\epsilon_{\boldsymbol{A}}{}^A$ will be denoted collectively by $\epsilon^{\boldsymbol{A}}{}_A$. By definition one has that

$$\epsilon_{\boldsymbol{A}}{}^A \epsilon^{\boldsymbol{B}}{}_A = \delta_{\boldsymbol{A}}{}^{\boldsymbol{B}}.$$

It follows from Equation (3.9) and the previous condition that

$$\epsilon^{\mathbf{0}}{}_A = -\iota_A, \qquad \epsilon^{\mathbf{1}}{}_A = o_A.$$

Using this notation and given two spinors ξ^A and η_B, one can write

$$\xi^A = \xi^{\boldsymbol{A}} \epsilon_{\boldsymbol{A}}{}^A, \qquad \eta_B = \eta_{\boldsymbol{B}} \epsilon^{\boldsymbol{B}}{}_B,$$

where

$$\xi^{\boldsymbol{A}} \equiv \xi^A \epsilon^{\boldsymbol{A}}{}_A, \qquad \eta_{\boldsymbol{B}} \equiv \eta_B \epsilon_{\boldsymbol{B}}{}^B.$$

Hence

$$[[\eta, \xi]] = \eta_A \xi^A = \left(\eta_{\boldsymbol{P}} \epsilon^{\boldsymbol{P}}{}_A\right)\left(\xi^{\boldsymbol{Q}} \epsilon_{\boldsymbol{Q}}{}^A\right) = \eta_{\boldsymbol{P}} \xi^{\boldsymbol{P}}.$$

The components $\epsilon_{\boldsymbol{AB}}$ of the antisymmetric spinor ϵ_{AB} with respect to the basis $\epsilon_{\boldsymbol{A}}{}^A$ are given by

$$(\epsilon_{\boldsymbol{AB}}) \equiv \left(\epsilon_{AB} \epsilon_{\boldsymbol{A}}{}^A \epsilon_{\boldsymbol{B}}{}^B\right) = \begin{pmatrix} o_A o^A & o_A \iota^A \\ \iota_A o^A & \iota_A \iota^A \end{pmatrix} = \begin{pmatrix} 0 & 1 \\ -1 & 0 \end{pmatrix}. \tag{3.10}$$

Now, a direct computation shows that

$$\begin{pmatrix} 0 & 1 \\ -1 & 0 \end{pmatrix}^{-1} = \begin{pmatrix} 0 & -1 \\ 1 & 0 \end{pmatrix}.$$

Hence, consistent with Equation (3.2) one has that

$$(\epsilon^{\boldsymbol{AB}}) \equiv \left(\epsilon^{AB} \epsilon^{\boldsymbol{A}}{}_A \epsilon^{\boldsymbol{B}}{}_B\right) = \begin{pmatrix} 0 & 1 \\ -1 & 0 \end{pmatrix}.$$

An alternative way of rewriting the previous discussion is

$$\delta_A{}^B = \epsilon_A{}^{\boldsymbol{A}} \epsilon_{\boldsymbol{A}}{}^B, \qquad \epsilon_{AB} = \epsilon_{\boldsymbol{AB}} \epsilon_A{}^{\boldsymbol{A}} \epsilon_B{}^{\boldsymbol{B}}, \qquad \epsilon^{AB} = \epsilon^{\boldsymbol{AB}} \epsilon_{\boldsymbol{A}}{}^A \epsilon_{\boldsymbol{B}}{}^B.$$

From the latter it follows that

$$\delta_A{}^B = o_A \iota^B - \iota_A o^B, \tag{3.11a}$$

$$\epsilon_{AB} = o_A \iota_B - \iota_A o_B, \tag{3.11b}$$

$$\epsilon^{AB} = o^A \iota^B - \iota^A o^B. \tag{3.11c}$$

3.1.8 Complex conjugation of spinors

In order to relate spinors with tensors one has to consider the operation of complex conjugation discussed in Section 3.1.1. The convention to denote the operation of complex conjugation in the abstract index notation is to add a bar to the kernel symbol and a prime to each of the indices. For example, one has that

$$\overline{\zeta^A} = \bar{\zeta}^{A'} \in \mathfrak{S}^{A'}.$$

The operation of complex conjugation is *idempotent* – given $\zeta \in \mathfrak{S}$, then $\bar{\bar{\zeta}} = \zeta$. Using abstract index notation one writes the latter as $\overline{\bar{\zeta}^{A'}} = \zeta^A$.

A spinor $\xi^{A\cdots CS'\cdots U'}{}_{D\cdots EW'\cdots Y'}$ with, say, p unprimed contravariant indices, r primed contravariant indices, q unprimed covariant indices and s primed covariant indices describes the most general type of spinors. It is obtained from the $\mathfrak{S}$-linear map

$$\boldsymbol{\xi} : \underbrace{\mathfrak{S}_A \times \cdots \times \mathfrak{S}_C}_{p \text{ times}} \times \underbrace{\mathfrak{S}_{S'} \times \cdots \times \mathfrak{S}_{U'}}_{r \text{ times}} \times \underbrace{\mathfrak{S}^D \times \cdots \times \mathfrak{S}^E}_{q \text{ times}} \times \underbrace{\mathfrak{S}^{W'} \times \cdots \times \mathfrak{S}^{Y'}}_{s \text{ times}} \to \mathbb{C}.$$

The algebra $\mathfrak{S}^\bullet$ is then extended to accommodate this more general type of spinors with unprimed and primed indices.

An important consequence of the fact that the spaces $\mathfrak{S}$ and $\bar{\mathfrak{S}}$ are not isomorphic is that it is not possible to single out 2-spinors which are intrinsically real or imaginary unless one assumes further structure on $\mathfrak{S}^\bullet$. From a notational point of view, as $\mathfrak{S}$ and $\bar{\mathfrak{S}}$ are not isomorphic, the relative position of primed and unprimed indices is irrelevant. Thus, one can write expressions like $\zeta_{AA'} = \zeta_{A'A}$. Notice, in contrast, that the reordering of groups of primed indices or groups of unprimed indices is not allowed unless the spinor possesses special symmetries.

The rules for the raising and lowering of indices of valence 1 spinors are extended to higher valence spinors in a natural way. Primed indices are raised and lowered using the spinors $\epsilon^{A'B'} \in \mathfrak{S}^{A'B'}$ and $\epsilon_{A'B'} \in \mathfrak{S}_{A'B'}$ which are related, respectively, to ϵ^{AB} and ϵ_{AB} by complex conjugation. That is,

$$\bar{\epsilon}_{A'B'} \equiv \overline{\epsilon_{AB}}, \qquad \bar{\epsilon}^{A'B'} \equiv \overline{\epsilon^{AB}}.$$

It is conventional to write $\epsilon_{A'B'}$, $\epsilon^{A'B'}$ *instead of* $\bar{\epsilon}_{A'B'}$ *and* $\bar{\epsilon}^{A'B'}$.

Finally, note that the discussion of Section 3.1.6 concerning the decomposition of spinors in irreducible components, and in particular Lemma 3.1 and Proposition 3.1, can be directly extended to the case of spinors containing primed indices or combinations of primed or unprimed indices. In particular, one has the following decomposition of a spinor with two unprimed and two primed indices:

$$\eta_{AA'BB'} = \eta_{(AB)(A'B')} + \frac{1}{2}\eta_{P}{}^{P}{}_{(A'B')}\epsilon_{AB} + \frac{1}{2}\eta_{(AB)Q'}{}^{Q'}\epsilon_{A'B'} + \frac{1}{4}\epsilon_{AB}\epsilon_{A'B'}\eta_{Q}{}^{Q}{}_{Q'}{}^{Q'}. \tag{3.12}$$

A particular case of the above decomposition is when $\zeta_{AA'BB'}$ is the spinorial counterpart of an antisymmetric rank-2 tensor $\zeta_{ab} = -\zeta_{ba}$. In this case one has that

$$\zeta_{AA'BB'} = \zeta_{AB}\epsilon_{A'B'} + \bar{\zeta}_{A'B'}\epsilon_{AB}, \tag{3.13}$$

where $\zeta_{AB} \equiv \frac{1}{2}\zeta_{AP'B}{}^{P'}$, and one has that $\zeta_{AB} = \zeta_{(AB)}$.

3.1.9 The relation between spinors and tensors

Spinors provide a simple representation of several tensorial operations. Although every four-dimensional tensor (***world tensor***) can be represented in terms of spinors, the converse is not true. There are spinors which admit no discussion in terms of tensors. This observation is based on the fact that 2-spinors are related to representations of the group of (2×2) complex matrices with unit determinant, $SL(2,\mathbb{C})$, while tensors are related to the *Lorentz group*. These groups are not isomorphic to each other. The group $SL(2,\mathbb{C})$ covers the Lorentz group in a $2:1$ way; see, for example, Carmeli (1977); Sexl and Urbantke (2000) for further discussions on this issue.

Hermitian spinors

The key property to relate 2-spinors to world tensors is *hermicity.* A spinor $\boldsymbol{\xi} \in \mathfrak{S}^{\bullet}$ is said to be ***Hermitian*** if and only if $\boldsymbol{\xi} = \bar{\boldsymbol{\xi}}$, that is, if the spinor is equal to its complex conjugate. For this to be the case, $\boldsymbol{\xi}$ needs to have the same number of unprimed and primed indices. By raising and lowering the indices as necessary one can, without loss of generality, assume that the spinor has the same number of unprimed and primed contravariant indices and the same number of unprimed and primed covariant indices, for example, $\xi_{AA'\cdots DD'}{}^{EE'\cdots HH'}$. In this case the hermicity condition reads

$$\xi_{AA'\cdots DD'}{}^{EE'\cdots HH'} = \bar{\xi}_{AA'\cdots DD'}{}^{EE'\cdots HH'},$$

where on the right-hand side it has been used that the position of primed and unprimed indices can be interchanged.

Consider now $\xi^{AA'} \in \mathfrak{S}^{AA'}$. If $\{\boldsymbol{o}, \boldsymbol{\iota}\}$ and $\{\bar{\boldsymbol{o}}, \bar{\boldsymbol{\iota}}\}$ are, respectively, spin bases of $\mathfrak{S}$ and $\bar{\mathfrak{S}}$, one can write

$$\xi^{AA'} = ao^{A}\bar{o}^{A'} + b\iota^{A}\bar{\iota}^{A'} + co^{A}\bar{\iota}^{A'} + d\iota^{A}\bar{o}^{A'}, \tag{3.14}$$

for some $a,\ b,\ c,\ d \in \mathbb{C}$. In other words, a pair ${}^{AA'}$ of indices is associated to four complex components. If one assumes, in addition, $\xi^{AA'}$ to be Hermitian, then it follows that $a,\ b \in \mathbb{R}$ and $c = \bar{d}$. Thus, the hermicity condition reduces the number of independent components to four real ones. Consequently, one can think of the Hermitian spinor $\xi^{AA'} \in \mathfrak{S}^{AA'}$ as describing a four-dimensional vector (***world-vector***) ξ^a.

The argument described in the previous paragraph can be extended in a natural fashion to higher valence Hermitian spinors, $\xi_{AA'\cdots DD'}{}^{EE'\cdots HH'}$, so that one can regard each pair of unprimed-primed indices (i.e. ${}_{AA'},\ {}^{EE'},\cdots$) as associated to a tensorial index (i.e. ${}_a,\ {}^e,\cdots$).

In what follows let

$$g_{AA'BB'} \equiv \epsilon_{AB}\epsilon_{A'B'}. \tag{3.15}$$

A computation then shows that $\bar{g}_{AA'BB'} = g_{AA'BB'}$ and, in addition, that

$$\begin{aligned} g^{AA'BB'} &= \epsilon^{AB}\epsilon^{A'B'}, \\ g_{AA'BB'}g^{BB'CC'} &= g_{AA'}{}^{CC'} \equiv \delta_A{}^C\delta_{A'}{}^{C'}, \\ g_{AA'BB'}g^{AA'BB'} &= 4, \\ g_{AA'BB'} &= g_{BB'AA'}. \end{aligned}$$

Furthermore, given $v_{AA'} \in \mathfrak{S}_{AA'}$ it can be readily verified that

$$v_{AA'}g^{AA'BB'} = v^{BB'}, \qquad v^{AA'}g_{AA'BB'} = v_{BB'}.$$

Hence, the spinor $g_{AA'BB'}$ has all the properties of a spinorial counterpart of the metric tensor. These ideas will now be put in more precise terms.

The Infeld-van der Waerden symbols

In order to describe explicitly the correspondence between spinors and tensors at a point $p \in \mathcal{M}$, consider a basis $\{\boldsymbol{e_a}\} \subset T|_p(\mathcal{M})$ and let $g_{\boldsymbol{ab}} \equiv \boldsymbol{g}(\boldsymbol{e_a}, \boldsymbol{e_b})$ denote the components of the metric $\boldsymbol{g}$ with respect to this basis. Let also $\{\boldsymbol{\omega^a}\} \subset T^*|_p(\mathcal{M})$ denote the dual basis to $\{\boldsymbol{e_a}\}$ so that $\langle \boldsymbol{\omega^b}, \boldsymbol{e_a}\rangle = \delta_{\boldsymbol{a}}{}^{\boldsymbol{b}}$. *It is conventional to assume that the basis is* $\boldsymbol{g}$*-orthogonal*; that is, $g_{\boldsymbol{ab}} = \eta_{\boldsymbol{ab}}$. Finally, let $\{\epsilon_{\boldsymbol{A}}\} \subset \mathfrak{S}$ denote a spin basis, and let $\epsilon_{\boldsymbol{AB}}$ denote the components of the spinor ϵ_{AB} with respect to the latter basis. The scalars $g_{\boldsymbol{ab}}$ and $\epsilon_{\boldsymbol{AB}}$ can be put in correspondence with each other via an equation of the form

$$\epsilon_{\boldsymbol{AB}}\epsilon_{\boldsymbol{A'B'}} = \sigma^{\boldsymbol{a}}{}_{\boldsymbol{AA'}}\sigma^{\boldsymbol{b}}{}_{\boldsymbol{BB'}}\eta_{\boldsymbol{ab}}, \tag{3.16}$$

where $\sigma^{\boldsymbol{a}}{}_{\boldsymbol{AA'}}$ are the so-called ***Infeld-van der Waerden symbols***. These can be regarded as the entries of four (2×2) matrices $(\sigma^{\boldsymbol{a}}{}_{\boldsymbol{AA'}})$, $\boldsymbol{a} = \boldsymbol{0}, \ldots, \boldsymbol{3}$. Unprimed indices denote the rows and the primed indices the columns of the matrix. Given $\sigma^{\boldsymbol{a}}{}_{\boldsymbol{AA'}}$, one defines the inverse symbol $\sigma_{\boldsymbol{b}}{}^{\boldsymbol{BB'}}$ via the relations

$$\sigma_{\boldsymbol{a}}{}^{\boldsymbol{AA'}}\sigma^{\boldsymbol{b}}{}_{\boldsymbol{AA'}} = \delta_{\boldsymbol{a}}{}^{\boldsymbol{b}}, \qquad \sigma_{\boldsymbol{a}}{}^{\boldsymbol{AA'}}\sigma^{\boldsymbol{a}}{}_{\boldsymbol{BB'}} = \delta_{\boldsymbol{B}}{}^{\boldsymbol{A}}\delta_{\boldsymbol{B'}}{}^{\boldsymbol{A'}}. \tag{3.17}$$

From these expressions it follows that the correspondence (3.16) can be inverted to yield

$$\eta_{ab} = \sigma_a{}^{AA'} \sigma_b{}^{BB'} \epsilon_{AB}\epsilon_{A'B'}. \tag{3.18}$$

Using Equation (3.18) and observing that $\eta_{ab} = \overline{\eta_{ab}}$, it follows that

$$\sigma_a{}^{AA'} = \overline{\sigma_a{}^{AA'}}. \tag{3.19}$$

Hence, $(\sigma_a{}^{AA'})$ and $(\sigma^a{}_{AA'})$ describe *Hermitian matrices*. An explicit computation shows that the matrices

$$(\sigma_0{}^{AA'}) \equiv \frac{1}{\sqrt{2}}\begin{pmatrix} 1 & 0 \\ 0 & 1 \end{pmatrix}, \qquad (\sigma_1{}^{AA'}) \equiv \frac{1}{\sqrt{2}}\begin{pmatrix} 0 & 1 \\ 1 & 0 \end{pmatrix},$$

$$(\sigma_2{}^{AA'}) \equiv \frac{1}{\sqrt{2}}\begin{pmatrix} 0 & \mathrm{i} \\ -\mathrm{i} & 0 \end{pmatrix}, \qquad (\sigma_3{}^{AA'}) \equiv \frac{1}{\sqrt{2}}\begin{pmatrix} 1 & 0 \\ 0 & -1 \end{pmatrix},$$

and

$$(\sigma^0{}_{AA'}) \equiv \frac{1}{\sqrt{2}}\begin{pmatrix} 1 & 0 \\ 0 & 1 \end{pmatrix}, (\sigma^1{}_{AA'}) \equiv \frac{1}{\sqrt{2}}\begin{pmatrix} 0 & 1 \\ 1 & 0 \end{pmatrix},$$

$$(\sigma^2{}_{AA'}) \equiv \frac{1}{\sqrt{2}}\begin{pmatrix} 0 & -\mathrm{i} \\ \mathrm{i} & 1 \end{pmatrix}, \qquad (\sigma^3{}_{AA'}) \equiv \frac{1}{\sqrt{2}}\begin{pmatrix} 1 & 0 \\ 0 & -1 \end{pmatrix},$$

satisfy the relations (3.16), (3.17), (3.18) and (3.19). The above matrices correspond, up to a normalisation factor, to the so-called *Pauli matrices*.

Now, consider arbitrary $\boldsymbol{v} \in T|_p(\mathcal{M})$ and $\boldsymbol{\alpha} \in T^*|_p(\mathcal{M})$. In terms of the bases $\{\boldsymbol{e_a}\}$ and $\{\boldsymbol{\omega^a}\}$, $\boldsymbol{v}$ and $\boldsymbol{\alpha}$ can be written as

$$\boldsymbol{v} = v^a \boldsymbol{e_a}, \qquad v^a \equiv \langle \boldsymbol{\omega^a}, \boldsymbol{v}\rangle,$$
$$\boldsymbol{\alpha} = \alpha_a \boldsymbol{\omega^a}, \qquad \alpha_a \equiv \langle \boldsymbol{\alpha}, \boldsymbol{e_a}\rangle.$$

The components v^a and α_a can be put in correspondence with Hermitian spinors using the Infeld-van der Waerden symbols via the rules

$$v^a \mapsto v^{AA'} = v^a \sigma_a{}^{AA'} \tag{3.20a}$$

$$\alpha_a \mapsto \alpha_{AA'} = \alpha_a \sigma^a{}_{AA'}. \tag{3.20b}$$

In terms of arrays of explicit components and matrices one has

$$(v^0, v^1, v^2, v^3) \mapsto \frac{1}{\sqrt{2}}\begin{pmatrix} v^0 + v^3 & v^1 + \mathrm{i}v^2 \\ v^1 - \mathrm{i}v^2 & v^0 - v^3 \end{pmatrix},$$

$$(\alpha_0, \alpha_1, \alpha_2, \alpha_3) \mapsto \frac{1}{\sqrt{2}}\begin{pmatrix} \alpha_0 + \alpha_3 & \alpha_1 - \mathrm{i}\alpha_2 \\ \alpha_1 + \mathrm{i}\alpha_2 & \alpha_0 - \alpha_3 \end{pmatrix}.$$

A quick computation shows that

$$\begin{aligned}\langle \boldsymbol{\alpha}, \boldsymbol{v}\rangle = v^a\alpha_a &= v^{AA'}\alpha_{AA'} \\ &= v^{00'}\alpha_{00'} + v^{01'}\alpha_{01'} + v^{10'}\alpha_{10'} + v^{11'}\alpha_{11'} \\ &= v^0\alpha_0 - v^1\alpha_1 - v^2\alpha_2 - v^3\alpha_3.\end{aligned}$$

Thus, one has that the assignments defined in (3.20a) and (3.20b) are consistent with the inner product defined on $T|_p(\mathcal{M})$ by the metric $\boldsymbol{g}$.

The assignment given by (3.20a) and (3.20b) can be extended to tensors of arbitrary rank. For example, given the tensor $T_{ab}{}^c$, denote its components with respect to $\{\boldsymbol{e_a}\}$ and $\{\boldsymbol{\omega^b}\}$ by $T_{\boldsymbol{ab}}{}^{\boldsymbol{c}}$. One then has the assignment

$$T_{\boldsymbol{ab}}{}^{\boldsymbol{c}} \mapsto T_{\boldsymbol{AA'BB'}}{}^{\boldsymbol{CC'}} \equiv \sigma^{\boldsymbol{a}}{}_{\boldsymbol{AA'}}\sigma^{\boldsymbol{b}}{}_{\boldsymbol{BB'}}\sigma_{\boldsymbol{c}}{}^{\boldsymbol{CC'}}T_{\boldsymbol{ab}}{}^{\boldsymbol{c}}.$$

The object $T_{\boldsymbol{AA'BB'}}{}^{\boldsymbol{CC'}}$ will be called the ***spinorial counterpart*** of the tensor components $T_{\boldsymbol{ab}}{}^{\boldsymbol{c}}$.

3.1.10 The spinorial representation of null vectors

As already mentioned in the introduction to this chapter, one of the key advantages of the use of spinors is the convenient representation of null vectors they provide. More precisely, one has the following result:

Proposition 3.3 (***spinorial counterpart of null vectors***) *The spinorial counterpart of a non-vanishing real null vector k^a can be written as*

$$k^{AA'} = \pm\kappa^A\bar{\kappa}^{A'}, \tag{3.21}$$

for some valence 1 spinor κ^A.

Proof A direct computation shows that $k^{AA'}$ as given by Equation (3.21) is indeed the spinorial counterpart of a null vector. Conversely, a computation yields

$$\begin{aligned} \boldsymbol{g}(\boldsymbol{k},\boldsymbol{k}) &= \epsilon_{AB}\epsilon_{A'B'}k^{AA'}k^{BB'} \\ &= 2(k^{00'}k^{11'} - k^{01'}k^{10'}) = \det(k^{AA'}). \end{aligned}$$

Thus, the requirement $\boldsymbol{g}(\boldsymbol{k},\boldsymbol{k}) = 0$ implies that $k^{AA'}$, regarded as a (2×2) matrix, has rows/columns which are linearly dependent. Accordingly, there exist valence 1 spinors κ^A and λ^B such that $k^{AA'} = \kappa^A\bar{\lambda}^{A'}$. As, $\boldsymbol{k}$ is non-zero, it follows that κ_A, $\lambda_B \neq 0$. From the reality of $\boldsymbol{k}$, it follows that its spinor counterpart $k^{AA'}$ must be Hermitian; that is, $k^{AA'} = \bar{k}^{AA'}$. Hence, $\kappa^A\bar{\lambda}^{A'} = \bar{\kappa}^{A'}\lambda^A$. Contracting the latter with κ_A one has that $\kappa_A\lambda^A = 0$, so that κ^A and λ^A must be proportional to each other. The proportionality factor can be absorbed into κ^A by means of a redefinition of the spinor. The sign in Equation (3.21) is that of the proportionality constant. □

Remark. A null vector constructed using the *positive sign* in Equation (3.21) will be said to be ***future pointing***, while one using the *negative* sign will be called ***past pointing***.

From Proposition 3.3 it follows that every valence 1 spinor κ^A defines a null vector $\boldsymbol{k}$. However, this is not a one-to-one correspondence. More precisely, a spinor differing from κ^A by a complex phase, that is, $e^{i\vartheta}\kappa^A$, with $\vartheta \in \mathbb{R}$ will give rise to the same null vector. The phase change is said to be ***right-handed*** if $\vartheta > 0$. This phase does not affect the construction of the vector $\boldsymbol{k}$. Nevertheless, it contains some geometric information. To see this, consider a further spinor μ^A such that $\kappa_A\mu^A = 1$ so that $\{\kappa^A, \mu^A\}$ constitute a spin basis. Now, one can readily verify that

$$s^{AA'} \equiv \frac{1}{\sqrt{2}}(\kappa^A\bar{\mu}^{A'} + \mu^A\bar{\kappa}^{A'}), \qquad t^{AA'} = \frac{\mathrm{i}}{\sqrt{2}}(\kappa^A\bar{\mu}^{A'} - \mu^A\bar{\kappa}^{A'}),$$

are the spinorial counterparts of two unit spacelike vectors $\boldsymbol{s}$ and $\boldsymbol{t}$ and that they are both orthogonal to $\boldsymbol{k}$. At each point $p \in \mathcal{M}$, $\boldsymbol{s}$ and $\boldsymbol{t}$ span a subspace of $T|_p(\mathcal{M})$ which is orthogonal to $\boldsymbol{k}$. This subspace is called the ***flag*** of the spinor κ^A; the ***pole of the flag*** is the vector $\boldsymbol{k}$.

Now, suppose κ^A is subject to a phase change such that

$$\kappa^A \mapsto e^{i\vartheta}\kappa^A. \tag{3.22}$$

In order to retain the normalisation $\kappa_A\mu^A = 1$, the transformation (3.22) implies the transformation $\mu^A \mapsto e^{-i\vartheta}\mu^A$. Furthermore, one has that

$$\boldsymbol{s} \mapsto \cos 2\vartheta\boldsymbol{s} + \sin 2\vartheta\boldsymbol{t}, \qquad \boldsymbol{t} \mapsto -\sin 2\vartheta\boldsymbol{s} + \cos 2\vartheta\boldsymbol{t},$$

so that a phase change of ϑ in κ^A implies a change of 2ϑ in its flag; the flagpole, however, remains unchanged.

3.1.11 Null tetrads

Inspection of Equation (3.14) shows that every spin basis $\{\boldsymbol{o}, \boldsymbol{\iota}\}$ gives rise to an associated vector basis consisting of null vectors. This ***null tetrad*** has the peculiarity of consisting of two *real* null vectors and two *complex* null vectors which are the complex conjugates of each other. In order to analyse this further, let

$$l^{AA'} \equiv o^A\bar{o}^{A'}, \qquad n^{AA'} \equiv \iota^A\bar{\iota}^{A'}, \qquad m^{AA'} \equiv o^A\bar{\iota}^{A'}, \qquad \bar{m}^{AA'} \equiv \iota^A\bar{o}^{A'}.$$

Furthermore, let l^a, n^a, m^a and $\bar{m}^a$ (or $\boldsymbol{l}$, $\boldsymbol{n}$, $\boldsymbol{m}$, $\bar{\boldsymbol{m}}$) denote the tensorial counterparts of the above spinors. Using the above definitions one can verify that

$$l_a n^a = -m_a\bar{m}^a = 1, \tag{3.23}$$

while all the other remaining contractions vanish. Using relations (3.11a)–(3.11c) it can be readily shown that

$$g_{ab} = 2l_{(a}n_{b)} - 2m_{(a}\bar{m}_{b)}, \qquad g^{ab} = 2l^{(a}n^{b)} - 2m^{(a}\bar{m}^{b)}.$$

An orthonormal tetrad $\{\boldsymbol{e_a}\}$ can be readily obtained from the null tetrad $\{\boldsymbol{l}, \boldsymbol{n}, \boldsymbol{m}, \bar{\boldsymbol{m}}\}$. Namely, let

$$\boldsymbol{e_0} = \frac{1}{\sqrt{2}}(\boldsymbol{l} + \boldsymbol{n}), \tag{3.24a}$$

$$\boldsymbol{e_1} = \frac{1}{\sqrt{2}}(\boldsymbol{m} + \bar{\boldsymbol{m}}), \tag{3.24b}$$

$$\boldsymbol{e_2} = \frac{\mathrm{i}}{\sqrt{2}}(\boldsymbol{m} - \bar{\boldsymbol{m}}), \tag{3.24c}$$

$$\boldsymbol{e_3} = \frac{1}{\sqrt{2}}(\boldsymbol{l} - \boldsymbol{n}). \tag{3.24d}$$

Using the relations in (3.23) it can be verified that the latter vectors indeed constitute an orthonormal tetrad. Furthermore, it can be readily checked that $\boldsymbol{e_0}$ is timelike while $\boldsymbol{e_1}$, $\boldsymbol{e_2}$ and $\boldsymbol{e_3}$ are spacelike. The vector $\boldsymbol{e_0}$ is said to be *future pointing* as both $\boldsymbol{l}$ and $\boldsymbol{n}$ are future pointing in the sense of Section 3.1.10. Moreover, a right-handed phase change (i.e. $\vartheta > 0$) in the spin basis of the form $o^A \mapsto e^{i\vartheta} o^A$, $\iota^A \mapsto e^{-i\vartheta} \iota^A$ leads to the *right-handed* rotations

$$\boldsymbol{e_1} \mapsto \cos 2\vartheta \boldsymbol{e_1} + \sin 2\vartheta \boldsymbol{e_2}, \qquad \boldsymbol{e_2} \mapsto -\sin 2\vartheta \boldsymbol{e_1} + \cos 2\vartheta \boldsymbol{e_2},$$

while at the same time leaving $\boldsymbol{e_0}$ and $\boldsymbol{e_3}$ unchanged. Accordingly, the triad of spacelike vectors $\{\boldsymbol{e_1}, \boldsymbol{e_2}, \boldsymbol{e_3}\}$ defined by (3.24b)–(3.24d) is said to be *right-handed.* The inverse relations to (3.24a)–(3.24d) are given by

$$\boldsymbol{l} = \frac{1}{\sqrt{2}}(\boldsymbol{e_0} + \boldsymbol{e_3}), \qquad \boldsymbol{n} = \frac{1}{\sqrt{2}}(\boldsymbol{e_0} - \boldsymbol{e_3}),$$

$$\boldsymbol{m} = \frac{1}{\sqrt{2}}(\boldsymbol{e_1} - \mathrm{i}\boldsymbol{e_2}), \qquad \bar{\boldsymbol{m}} = \frac{1}{\sqrt{2}}(\boldsymbol{e_1} + \mathrm{i}\boldsymbol{e_2}).$$

The spinorial counterpart of the volume form

The ***spinorial counterpart of the volume 4-form*** ϵ_{abcd} is given by

$$\epsilon_{AA'BB'CC'DD'} = \mathrm{i}(\epsilon_{AB}\epsilon_{CD}\epsilon_{A'C'}\epsilon_{B'D'} - \epsilon_{AC}\epsilon_{BD}\epsilon_{A'B'}\epsilon_{C'D'}). \tag{3.25}$$

Using the Jacobi identity (3.5) it can be verified that the above expression is indeed totally antisymmetric under interchange of the pairs $_{AA'}$, $_{BB'}$, $_{CC'}$ and $_{DD'}$. Moreover, one has

$$\epsilon_{AA'BB'CC'DD'}\epsilon^{AA'BB'CC'DD'} = 24,$$

and

$$\sigma_{\boldsymbol{0}}{}^{\boldsymbol{AA'}}\sigma_{\boldsymbol{1}}{}^{\boldsymbol{BB'}}\sigma_{\boldsymbol{2}}{}^{\boldsymbol{CC'}}\sigma_{\boldsymbol{3}}{}^{\boldsymbol{DD'}}\epsilon_{\boldsymbol{AA'BB'CC'DD'}} = 1;$$

compare Section 2.5.3. The expression (3.25) can be deduced applying a decomposition in irreducible components to $\epsilon_{AA'BB'CC'DD'}$ and exploiting its antisymmetry properties.

3.1.12 Changes of basis and $SL(2,\mathbb{C})$ transformations

Let $\{\epsilon_{\boldsymbol{A}}{}^{A}\}$ and $\{\tilde{\epsilon}_{\boldsymbol{A}}{}^{A}\}$ denote two spin bases for $\mathfrak{S}$. The spinors of one basis can be expressed as linear combinations of the spinors of the other basis. This can be conveniently be written as

$$\tilde{\epsilon}_{\boldsymbol{A}}{}^{A} = \Lambda_{\boldsymbol{A}}{}^{\boldsymbol{P}}\,\epsilon_{\boldsymbol{P}}{}^{A}, \tag{3.26}$$

where $(\Lambda_{\boldsymbol{A}}{}^{\boldsymbol{P}})$ denotes an invertible (2×2) matrix. The associated spinor cobases $\{\epsilon^{\boldsymbol{A}}{}_{A}\}$ and $\{\tilde{\epsilon}^{\boldsymbol{A}}{}_{A}\}$ are related in a similar way:

$$\tilde{\epsilon}^{\boldsymbol{A}}{}_{A} = \Lambda^{\boldsymbol{A}}{}_{\boldsymbol{P}}\epsilon^{\boldsymbol{P}}{}_{A}; \tag{3.27}$$

where $(\Lambda^{\boldsymbol{A}}{}_{\boldsymbol{P}})$ is another invertible (2×2) matrix. Now, one has that

$$\begin{aligned}\delta_{\boldsymbol{A}}{}^{\boldsymbol{B}} &= \tilde{\epsilon}_{\boldsymbol{A}}{}^{P}\tilde{\epsilon}^{\boldsymbol{B}}{}_{P} = (\Lambda_{\boldsymbol{A}}{}^{\boldsymbol{P}}\epsilon_{\boldsymbol{P}}{}^{Q})(\Lambda^{\boldsymbol{B}}{}_{\boldsymbol{Q}}\epsilon^{\boldsymbol{Q}}{}_{Q}) = (\Lambda_{\boldsymbol{A}}{}^{\boldsymbol{P}}\Lambda^{\boldsymbol{B}}{}_{\boldsymbol{Q}})\,\epsilon_{\boldsymbol{P}}{}^{Q}\epsilon^{\boldsymbol{Q}}{}_{Q}\\ &= \Lambda_{\boldsymbol{A}}{}^{\boldsymbol{P}}\Lambda^{\boldsymbol{B}}{}_{\boldsymbol{Q}}\delta_{\boldsymbol{P}}{}^{\boldsymbol{Q}} = \Lambda_{\boldsymbol{A}}{}^{\boldsymbol{P}}\Lambda^{\boldsymbol{B}}{}_{\boldsymbol{P}}.\end{aligned}$$

Hence, the matrices $(\Lambda_{\boldsymbol{A}}{}^{\boldsymbol{P}})$ and $(\Lambda^{\boldsymbol{A}}{}_{\boldsymbol{P}})$ are inverses of each other.

Now, given a contravariant valence 1 spinor κ^{A}, one can expand it in terms of the bases $\{\epsilon_{\boldsymbol{A}}{}^{A}\}$ and $\{\tilde{\epsilon}_{\boldsymbol{A}}{}^{A}\}$ as

$$\kappa^{A} = \kappa^{\boldsymbol{A}}\epsilon_{\boldsymbol{A}}{}^{A} = \tilde{\kappa}^{\boldsymbol{A}}\tilde{\epsilon}_{\boldsymbol{A}}{}^{A}.$$

As a consequence of the change of basis (3.26), the coefficients $\kappa_{\boldsymbol{A}}$ and $\tilde{\kappa}_{\boldsymbol{A}}$ are related to each other via

$$\tilde{\kappa}^{\boldsymbol{A}} = \Lambda^{\boldsymbol{A}}{}_{\boldsymbol{P}}\kappa^{\boldsymbol{P}}.$$

Similarly, from the transformation rule (3.27), the components $\mu_{\boldsymbol{A}}$ and $\tilde{\mu}_{\boldsymbol{A}}$ of a valence 1 covariant spinor μ_{A} with respect to the spin cobasis $\{\epsilon^{\boldsymbol{A}}{}_{A}\}$ and $\{\tilde{\epsilon}^{\boldsymbol{A}}{}_{A}\}$ can be found to be related via

$$\tilde{\mu}_{\boldsymbol{A}} = \Lambda_{\boldsymbol{A}}{}^{\boldsymbol{P}}\mu_{\boldsymbol{P}}.$$

The transformation rules given in the previous paragraph can be extended in a natural way to higher valence spinors and to spinors with primed indices. For example, if $v^{\boldsymbol{AA'}}$ and $\tilde{v}^{\boldsymbol{AA'}}$ denote the components of the spinor $v^{AA'}$ with respect to the two different sets of bases, one has that

$$\tilde{v}^{\boldsymbol{AA'}} = \Lambda^{\boldsymbol{A}}{}_{\boldsymbol{P}}\bar{\Lambda}^{\boldsymbol{A'}}{}_{\boldsymbol{P'}}v^{\boldsymbol{PP'}}.$$

A case of special importance is that of the antisymmetric spinor ϵ_{AB} for which the transformation rule between bases is given by

$$\tilde{\epsilon}_{\boldsymbol{AB}} = \Lambda_{\boldsymbol{A}}{}^{\boldsymbol{P}}\Lambda_{\boldsymbol{B}}{}^{\boldsymbol{Q}}\epsilon_{\boldsymbol{PQ}}. \tag{3.28}$$

Earlier in the chapter, the notion of simplectic transformations was introduced. The properties of these transformations can be investigated from Equation (3.28). As a consequence of the discussion of Section 3.1.7 the matrices

(ϵ_{AB}) and $(\tilde{\epsilon}_{AB})$ both have the form given by Equation (3.10). It follows from Equation (3.28) that

$$\det(\tilde{\epsilon}_{AB}) = \left(\det\left(\Lambda_A{}^B\right)\right)^2 \det(\epsilon_{AB}).$$

Furthermore as $\det(\tilde{\epsilon}_{AB}) = \det(\epsilon_{AB}) = 1$, one concludes that $\det\left(\Lambda_A{}^B\right) = \pm 1$. Hence, if one restricts attention to the transformations with positive determinant, one finds that *the set of transformations that preserve the antisymmetric product* $[[\cdot,\cdot]]$ *is given by the group* $SL(2,\mathbb{C})$.

Relation to the Lorentz transformations

Following the discussion of the previous paragraphs, the components $g_{AA'BB'}$ of the spinorial counterpart of the metric transform under a change of spin basis as

$$\tilde{g}_{AA'BB'} \equiv \tilde{\epsilon}_{AB}\tilde{\epsilon}_{A'B'} = \Lambda_A{}^P \bar{\Lambda}_{A'}{}^{P'} \Lambda_B{}^Q \bar{\Lambda}_{B'}{}^{Q'} \epsilon_{PQ}\epsilon_{P'Q'}.$$

Using the Infeld-van der Waerden symbols, the latter can be rewritten as

$$\tilde{\eta}_{ab} = \Lambda_a{}^c \Lambda_b{}^d \eta_{cd},$$

with

$$\Lambda_a{}^c \equiv \sigma_a{}^{AA'} \sigma^c{}_{PP'} \Lambda_A{}^P \bar{\Lambda}_{A'}{}^{P'}.$$

The above expression provides the relation between $SL(2,\mathbb{C})$ and Lorentz transformations; see, for example, Sexl and Urbantke (2000) for more details.

3.1.13 Soldering forms

The connection between spinors and world tensors has been implemented in terms of the components with respect to some vector and spin bases. There is a different perspective of this translation in terms of so-called soldering forms.

The metric tensor $\boldsymbol{g}$ can be written in terms of the orthonormal cobasis $\{\boldsymbol{\omega}^a\}$ as

$$\boldsymbol{g} = \eta_{ab}\boldsymbol{\omega}^a \otimes \boldsymbol{\omega}^b.$$

This last expression can be rewritten, using the correspondence (3.18), as

$$\boldsymbol{g} = \epsilon_{AB}\epsilon_{A'B'}\sigma_a{}^{AA'}\sigma_b{}^{BB'}\boldsymbol{\omega}^a \otimes \boldsymbol{\omega}^b = \epsilon_{AB}\epsilon_{A'B'}\boldsymbol{\omega}^{AA'} \otimes \boldsymbol{\omega}^{BB'}, \tag{3.29}$$

where $\boldsymbol{\omega}^{AA'} \equiv \sigma_a{}^{AA'}\boldsymbol{\omega}^a$. The four covectors $\{\boldsymbol{\omega}^{AA'}\}$ are called the ***soldering forms*** . In terms of abstract index notation one writes the soldering form as $\omega^{AA'}{}_a$. A similar discussion can be made with the contravariant metric $\boldsymbol{g}^\sharp$. From $\boldsymbol{g}^\sharp = \eta^{ab}\boldsymbol{e}_a \otimes \boldsymbol{e}_b$, together with (3.16), one can write

$$\boldsymbol{g}^\sharp = \epsilon^{AB}\epsilon^{A'B'}\boldsymbol{e}_{AA'} \otimes \boldsymbol{e}_{BB'}, \tag{3.30}$$

where $\boldsymbol{e}_{AA'} \equiv \sigma^a{}_{AA'}\boldsymbol{e}_a$. In abstract index notation one would write $e_{AA'}{}^a$ instead of $\boldsymbol{e}_{AA'}$. In view of the above, given a vector $\boldsymbol{v} \in T|_p(\mathcal{M})$ and a covector $\boldsymbol{\alpha} \in T^*|_p(\mathcal{M})$, one can write

$$\boldsymbol{v} = v^{AA'}\boldsymbol{e}_{AA'}, \qquad \boldsymbol{\alpha} = \alpha_{AA'}\boldsymbol{\omega}^{AA'}.$$

As a final remark concerning the connection between spinors and world tensors, it is observed that $\boldsymbol{e}_a = \delta_a{}^b\boldsymbol{e}_b$. Thus, $\delta_a{}^b$ can be interpreted as the components $e_a{}^b$ of the frame vector $\boldsymbol{e}_a$ with respect to the frame $\{\boldsymbol{e}_a\}$. Contracting $e_a{}^b$ with $\sigma_b{}^{BB'}$ one finds

$$e_a{}^{BB'} \equiv e_a{}^b\sigma_b{}^{BB'} = \sigma_a{}^{BB'}.$$

3.2 Calculus of spacetime spinors

The discussion of the previous section has been restricted to spinors at a given point of the spacetime manifold $\mathcal{M}$. It is now assumed that a spinorial structure can be constructed in a consistent way on the whole of $\mathcal{M}$ – the conditions ensuring this are discussed in Section 3.3, and essentially amount to requiring the spacetime to be orientable. The spinorial structure over $\mathcal{M}$ (also called a ***spin bundle***) will be denoted by $\mathfrak{S}(\mathcal{M})$. Consistent with this notation, the spinorial structure at a point $p \in \mathcal{M}$ will be denoted by $\mathfrak{S}|_p(\mathcal{M})$.

As is the case with tensors, the idea of relating spinors defined at different points of the spacetime manifold requires the use of the notion of a *connection* and its associated *covariant derivative*. Thus, it is necessary to extend the notion of a connection in such a way that it applies to ***spinor fields***. In what follows, by a spinor field it is understood a *smooth* assignment of a spinor, say, $\xi_{A\cdots CD'\cdots F'}{}^{G\cdots LP'\cdots N'}$, to each point of the spacetime manifold. The sets of spinorial fields over $\mathcal{M}$ will be denoted in a similar manner to the sets of spinors at a point, that is, $\mathfrak{S}^\bullet(\mathcal{M})$, $\mathfrak{S}_A(\mathcal{M})$, $\mathfrak{S}^A(\mathcal{M})$, $\mathfrak{S}_{AA'}{}^B(\mathcal{M})$, and so on.

3.2.1 The spinorial covariant derivative

A ***spinor covariant derivative*** $\nabla_{AA'}$ is a map

$$\nabla_{AA'} : \mathfrak{S}^{B\cdots C'}{}_{D\cdots E'}(\mathcal{M}) \to \mathfrak{S}^{B\cdots C'}{}_{AD\cdots A'E'}(\mathcal{M}).$$

Given an arbitrary spinor $\zeta^{B\cdots C'}{}_{D\cdots E'}$, its spinorial covariant derivative will be denoted by $\nabla_{AA'}\zeta^{B\cdots C'}{}_{D\cdots E'}$. The mapping defined by $\nabla_{AA'}$ is required to satisfy the following properties:

(i) ***Linearity.*** Given $\zeta^{B\cdots C'}{}_{D\cdots E'}, \eta^{B\cdots C'}{}_{D\cdots E'} \in \mathfrak{S}^{B\cdots C'}{}_{D\cdots E'}(\mathcal{M})$,

$$\nabla_{AA'}(\zeta^{B\cdots C'}{}_{D\cdots E'} + \eta^{B\cdots C'}{}_{D\cdots E'}) = \nabla_{AA'}\zeta^{B\cdots C'}{}_{D\cdots E'} + \nabla_{AA'}\eta^{B\cdots C'}{}_{D\cdots E'}.$$

(ii) ***Leibnitz rule.*** Given fields $\zeta^{B\cdots C'}{}_{D\cdots E'} \in \mathfrak{S}^{B\cdots C'}{}_{D\cdots E'}(\mathcal{M})$ and $\xi^{F\cdots G'}{}_{H\cdots I'} \in \mathfrak{S}^{F\cdots G'}{}_{H\cdots I'}(\mathcal{M})$,

$$\nabla_{AA'}(\zeta^{B\cdots C'}{}_{D\cdots E'}\xi^{F\cdots G'}{}_{H\cdots I'}) = \xi^{F\cdots G'}{}_{H\cdots I'}\nabla_{AA'}\zeta^{B\cdots C'}{}_{D\cdots E'} + \zeta^{B\cdots C'}{}_{D\cdots E'}\nabla_{AA'}\xi^{F\cdots G'}{}_{H\cdots I'}.$$

(iii) ***Hermicity.*** Given $\zeta^{B\cdots C'}{}_{D\cdots E'} \in \mathfrak{S}^{B\cdots C'}{}_{D\cdots E'}(\mathcal{M})$,

$$\overline{\nabla_{AA'}\zeta^{B\cdots C'}{}_{D\cdots E'}} = \nabla_{AA'}\bar{\zeta}^{B'\cdots C}{}_{D'\cdots E}.$$

(iv) ***Action on scalars.*** Given a scalar ϕ, then $\nabla_{AA'}\phi$ is the spinorial counterpart of $\nabla_a\phi$.

(v) ***Representation of derivations.*** Given a derivation $\mathcal{D}$ on spinor fields, there exists a spinor $\xi^{AA'}$ such that

$$\mathcal{D}\zeta^{B\cdots C'}{}_{D\cdots E'} = \xi^{AA'}\nabla_{AA'}\zeta^{B\cdots C'}{}_{D\cdots E'},$$

for all $\zeta^{B\cdots C'}{}_{D\cdots E'} \in \mathfrak{S}^{\bullet}(\mathcal{M})$.

Remark. The above list of properties is more general than the ones given in, say, Penrose and Rindler (1984) and Stewart (1991), as the present discussion does not assume that the spinor covariant derivative is compatible with the ϵ-spinor; that is, $\nabla_{AA'}\epsilon_{BC} = 0$.

For completeness, the following result proved in Penrose and Rindler (1984) is recalled:

Theorem 3.1 (***existence of the spinorial covariant derivative***) *Every covariant derivative* $\boldsymbol{\nabla}$ *over* $\mathcal{M}$ *has a spinorial counterpart* $\nabla_{AA'}$.

3.2.2 Spin connection coefficients

In specific computations, given a spin basis $\{\epsilon_{\boldsymbol{A}}{}^{A}\}$, it is convenient to introduce the notion of the ***spin connection coefficients*** associated to a certain connection. The direct spinorial counterparts of the connection coefficients $\Gamma_{\boldsymbol{a}}{}^{\boldsymbol{c}}{}_{\boldsymbol{b}}$ are given after suitable contraction with the Infeld-van der Waerden symbols by the spinor components

$$\Gamma_{\boldsymbol{AA'}}{}^{\boldsymbol{BB'}}{}_{\boldsymbol{CC'}} \equiv \omega^{\boldsymbol{BB'}}{}_{BB'}\nabla_{\boldsymbol{AA'}}e_{\boldsymbol{CC'}}{}^{BB'}, \tag{3.31}$$

where $\nabla_{\boldsymbol{AA'}} \equiv e_{\boldsymbol{AA'}}{}^{AA'}\nabla_{AA'}$ denotes the ***directional covariant derivative in the direction of*** $\boldsymbol{e_{AA'}}$. Now, using that

$$\omega^{\boldsymbol{BB'}}{}_{BB'} = \epsilon^{\boldsymbol{B}}{}_{B}\bar{\epsilon}^{\boldsymbol{B'}}{}_{B'}, \qquad e_{\boldsymbol{CC'}}{}^{CC'} = \epsilon_{\boldsymbol{C}}{}^{C}\bar{\epsilon}_{\boldsymbol{C'}}{}^{C'},$$

it follows that

$$\begin{aligned}\Gamma_{\boldsymbol{AA'}}{}^{\boldsymbol{BB'}}{}_{\boldsymbol{CC'}} &= \epsilon^{\boldsymbol{B}}{}_{B}\bar{\epsilon}^{\boldsymbol{B'}}{}_{B'}\bar{\epsilon}_{\boldsymbol{C'}}{}^{B'}\nabla_{\boldsymbol{AA'}}\epsilon_{\boldsymbol{C}}{}^{B} + \epsilon^{\boldsymbol{B}}{}_{B}\bar{\epsilon}^{\boldsymbol{B'}}{}_{B'}\epsilon_{\boldsymbol{C}}{}^{B}\nabla_{\boldsymbol{AA'}}\bar{\epsilon}_{\boldsymbol{C'}}{}^{B'} \\ &= \epsilon^{\boldsymbol{B}}{}_{B}\delta_{\boldsymbol{C'}}{}^{\boldsymbol{B'}}\nabla_{\boldsymbol{AA'}}\epsilon_{\boldsymbol{C}}{}^{B} + \bar{\epsilon}^{\boldsymbol{B'}}{}_{B'}\delta_{\boldsymbol{C}}{}^{\boldsymbol{B}}\nabla_{\boldsymbol{AA'}}\bar{\epsilon}_{\boldsymbol{C'}}{}^{B'}.\end{aligned}$$

Hence, defining the ***spin connection coefficients***

$$\Gamma_{AA'}{}^{B}{}_{C} \equiv \epsilon^{B}{}_{B}\nabla_{AA'}\epsilon_{C}{}^{B}, \tag{3.32}$$

one obtains

$$\Gamma_{AA'}{}^{BB'}{}_{CC'} = \Gamma_{AA'}{}^{B}{}_{C}\delta_{C'}{}^{B'} + \bar{\Gamma}_{AA'}{}^{B'}{}_{C'}\delta_{C}{}^{B}. \tag{3.33}$$

Using $\delta_{C}{}^{B} = \epsilon_{C}{}^{Q}\epsilon^{B}{}_{Q}$, the definition of $\Gamma_{AA'}{}^{B}{}_{C}$ and requiring that

$$\nabla_{AA'}\delta_{C}{}^{B} = 0$$

one also has that

$$\Gamma_{AA'}{}^{B}{}_{C} = -\epsilon_{C}{}^{Q}\nabla_{AA'}\epsilon^{B}{}_{Q}.$$

The spin connection coefficients provide a way of *computing the covariant derivative of spinors without a tensorial counterpart.* Given $\kappa_{A} = \kappa_{\boldsymbol{A}}\epsilon_{A}{}^{\boldsymbol{A}} \in \mathfrak{S}_{A}(\mathcal{M})$ one has that

$$\begin{aligned}
\nabla_{AA'}\kappa_{B} &\equiv \epsilon_{B}{}^{Q}\nabla_{AA'}\kappa_{Q} \\
&= \epsilon_{B}{}^{Q}\nabla_{AA'}(\kappa_{P}\epsilon^{P}{}_{Q}) \\
&= \epsilon_{B}{}^{Q}\left(e_{AA'}(\kappa_{P})\epsilon^{P}{}_{Q} + \kappa_{P}\nabla_{AA'}\epsilon^{P}{}_{Q}\right) \\
&= e_{AA'}(\kappa_{B}) - \Gamma_{AA'}{}^{P}{}_{B}\kappa_{P}.
\end{aligned}$$

Similar computations show, for example, that

$$\begin{aligned}
\nabla_{AA'}\zeta^{B} &= e_{AA'}(\zeta^{B}) + \Gamma_{AA'}{}^{B}{}_{P}\zeta^{P}, \\
\nabla_{AA'}\xi_{B'}{}^{CC'} &= e_{AA'}(\xi_{B'}{}^{CC'}) - \bar{\Gamma}_{AA'}{}^{Q'}{}_{B'}\xi_{Q'}{}^{CC'} \\
&\quad + \Gamma_{AA'}{}^{C}{}_{Q}\xi_{B'}{}^{QC'} + \bar{\Gamma}_{AA'}{}^{C'}{}_{Q'}\xi_{B'}{}^{CQ'}.
\end{aligned}$$

The generalisation to spinors of arbitrary valence and number of primed indices can be readily obtained from the above examples.

Metric and Levi-Civita spin connection coefficients

So far, the discussion of the spin connection coefficients has been completely general. *In the present section it is assumed that the connection is metric.*

The spinorial counterpart of the metric compatibility condition $\nabla_{a}g_{bc} = 0$ is given by

$$\nabla_{AA'}(\epsilon_{BC}\epsilon_{B'C'}) = \epsilon_{B'C'}\nabla_{AA'}\epsilon_{BC} + \epsilon_{BC}\nabla_{AA'}\epsilon_{B'C'} = 0.$$

Regarding the second equality as a (partial) decomposition in irreducible terms, one has that

$$\nabla_{AA'}\epsilon_{BC} = 0, \qquad \nabla_{AA'}\epsilon_{B'C'} = 0.$$

In order to investigate the implications of a metric connection on its associated spin connection coefficients, it is convenient to compute

$$\begin{aligned}\nabla_{\boldsymbol{AA'}}\epsilon_{\boldsymbol{BC}} &= e_{\boldsymbol{AA'}}(\epsilon_{\boldsymbol{BC}}) - \Gamma_{\boldsymbol{AA'}}{}^{\boldsymbol{Q}}{}_{\boldsymbol{B}}\epsilon_{\boldsymbol{QC}} - \Gamma_{\boldsymbol{AA'}}{}^{\boldsymbol{Q}}{}_{\boldsymbol{C}}\epsilon_{\boldsymbol{BQ}} \\ &= -\Gamma_{\boldsymbol{A'A'CB}} + \Gamma_{\boldsymbol{AA'BC}} = 0\end{aligned}$$

as $e_{\boldsymbol{AA'}}(\epsilon_{\boldsymbol{BC}}) = 0$; again, the components $\epsilon_{\boldsymbol{BC}}$ are constants. Hence, one concludes that

$$\Gamma_{\boldsymbol{AA'BC}} = \Gamma_{\boldsymbol{AA'(BC)}}.$$

3.2.3 Spinorial curvature

The spinorial counterpart of the curvature tensors can be introduced in a natural way by looking at the commutator of spinorial covariant derivatives. More precisely, one can write

$$[\![\nabla_{AA'}, \nabla_{BB'}]\!]\xi^{CC'} = R^{CC'}{}_{PP'AA'BB'}\xi^{PP'} \tag{3.34}$$

with

$$[\![\nabla_{AA'}, \nabla_{BB'}]\!] \equiv \nabla_{AA'}\nabla_{BB'} - \nabla_{BB'}\nabla_{AA'} - \Sigma_{AA'}{}^{PP'}{}_{BB'}\nabla_{PP'},$$

consistent with the notation of Section 2.4.3 and with $\Sigma_{AA'}{}^{CC'}{}_{BB'}$ representing the ***spinorial counterpart of the torsion tensor*** of $\boldsymbol{\nabla}$. The spinor $R^{CC'}{}_{DD'AA'BB'}$ is the ***spinorial counterpart of the Riemann curvature tensor*** $R^c{}_{dab}$. In the following discussion it is assumed that the connection $\boldsymbol{\nabla}$ is completely general – in particular, it could have torsion and be non-metric, so that $\nabla_{AA'}\epsilon_{BC} \neq 0$. As a consequence, the curvature spinor has only the symmetry

$$R^{CC'}{}_{DD'AA'BB'} = -R^{CC'}{}_{DD'BB'AA'}.$$

The curvature spinor in terms of the spin connection coefficients

In order to obtain a simpler representation of the curvature spinor it is convenient to look first at its expression in terms of spin connection coefficients. To this end, one can consider the frame expression (2.31) for the Riemann tensor, and contract it with the Infeld-van der Waerden symbols. One readily obtains

$$\begin{aligned}R^{\boldsymbol{CC'}}{}_{\boldsymbol{DD'AA'BB'}} &= e_{\boldsymbol{AA'}}(\Gamma_{\boldsymbol{BB'}}{}^{\boldsymbol{CC'}}{}_{\boldsymbol{DD'}}) - e_{\boldsymbol{BB'}}(\Gamma_{\boldsymbol{AA'}}{}^{\boldsymbol{CC'}}{}_{\boldsymbol{DD'}}) \\ &\quad + \Gamma_{\boldsymbol{FF'}}{}^{\boldsymbol{CC'}}{}_{\boldsymbol{DD'}}\Gamma_{\boldsymbol{BB'}}{}^{\boldsymbol{FF'}}{}_{\boldsymbol{AA'}} - \Gamma_{\boldsymbol{FF'}}{}^{\boldsymbol{CC'}}{}_{\boldsymbol{DD'}}\Gamma_{\boldsymbol{AA'}}{}^{\boldsymbol{FF'}}{}_{\boldsymbol{BB'}} \\ &\quad + \Gamma_{\boldsymbol{BB'}}{}^{\boldsymbol{FF'}}{}_{\boldsymbol{DD'}}\Gamma_{\boldsymbol{AA'}}{}^{\boldsymbol{CC'}}{}_{\boldsymbol{FF'}} - \Gamma_{\boldsymbol{AA'}}{}^{\boldsymbol{FF'}}{}_{\boldsymbol{DD'}}\Gamma_{\boldsymbol{BB'}}{}^{\boldsymbol{CC'}}{}_{\boldsymbol{FF'}} \\ &\quad - \Sigma_{\boldsymbol{AA'}}{}^{\boldsymbol{FF'}}{}_{\boldsymbol{BB'}}\Gamma_{\boldsymbol{FF'}}{}^{\boldsymbol{CC'}}{}_{\boldsymbol{DD'}}.\end{aligned}$$

Now, making use of the decomposition (3.32) for the spin connection coefficients, one obtains after a lengthy, but straightforward calculation that

$$R^{\boldsymbol{CC'}}{}_{\boldsymbol{DD'AA'BB'}} = R^{\boldsymbol{C}}{}_{\boldsymbol{DAA'BB'}}\delta_{\boldsymbol{D'}}{}^{\boldsymbol{C'}} + \bar{R}^{\boldsymbol{C'}}{}_{\boldsymbol{D'AA'BB'}}\delta_{\boldsymbol{D}}{}^{\boldsymbol{C}}, \qquad (3.35)$$

where

$$\begin{aligned} R^{\boldsymbol{C}}{}_{\boldsymbol{DAA'BB'}} \equiv\ & e_{\boldsymbol{AA'}}(\Gamma_{\boldsymbol{BB'}}{}^{\boldsymbol{C}}{}_{\boldsymbol{D}}) - e_{\boldsymbol{BB'}}(\Gamma_{\boldsymbol{AA'}}{}^{\boldsymbol{C}}{}_{\boldsymbol{D}}) \\ & -\Gamma_{\boldsymbol{FB'}}{}^{\boldsymbol{C}}{}_{\boldsymbol{D}}\Gamma_{\boldsymbol{AA'}}{}^{\boldsymbol{F}}{}_{\boldsymbol{B}} - \Gamma_{\boldsymbol{BF'}}{}^{\boldsymbol{C}}{}_{\boldsymbol{D}}\bar{\Gamma}_{\boldsymbol{AA'}}{}^{\boldsymbol{F'}}{}_{\boldsymbol{B'}} + \Gamma_{\boldsymbol{FA'}}{}^{\boldsymbol{C}}{}_{\boldsymbol{D}}\Gamma_{\boldsymbol{BB'}}{}^{\boldsymbol{F}}{}_{\boldsymbol{A}} \\ & +\Gamma_{\boldsymbol{AF'}}{}^{\boldsymbol{C}}{}_{\boldsymbol{D}}\bar{\Gamma}_{\boldsymbol{BB'}}{}^{\boldsymbol{F'}}{}_{\boldsymbol{A'}} + \Gamma_{\boldsymbol{AA'}}{}^{\boldsymbol{C}}{}_{\boldsymbol{F}}\Gamma_{\boldsymbol{BB'}}{}^{\boldsymbol{F}}{}_{\boldsymbol{D}} - \Gamma_{\boldsymbol{BB'}}{}^{\boldsymbol{C}}{}_{\boldsymbol{F}}\Gamma_{\boldsymbol{AA'}}{}^{\boldsymbol{F}}{}_{\boldsymbol{D}} \\ & -\Sigma_{\boldsymbol{AA'}}{}^{\boldsymbol{FF'}}{}_{\boldsymbol{BB'}}\Gamma_{\boldsymbol{FF'}}{}^{\boldsymbol{C}}{}_{\boldsymbol{D}}. \end{aligned}$$

This last expression can be regarded as the *spinorial counterpart of the first Cartan structure equation*; see Equation (2.31).

The commutator of covariant derivatives on arbitrary spinors

The commutator expression (3.34) applies only to spinors arising from a tensorial counterpart. In this section this commutator expression is applied to arbitrary valence spinors. In order to do this, observe that Equation (3.35) also holds if expressed in terms of abstract spinorial indices. More precisely, one has that

$$R^{CC'}{}_{DD'AA'BB'} = R^{C}{}_{DAA'BB'}\delta_{D'}{}^{C'} + \bar{R}^{C'}{}_{D'AA'BB'}\delta_{D}{}^{C}, \qquad (3.36)$$

where, in general $R_{CDAA'BB'} \neq R_{(CD)AA'BB'}$.

Applying the commutator (3.34) to the particular case when $\xi^{\boldsymbol{CC'}} = \epsilon_{\boldsymbol{D}}{}^{\boldsymbol{C}}\epsilon_{\boldsymbol{D'}}{}^{\boldsymbol{C'}}$ one obtains, after taking into account the split (3.36), that

$$\begin{aligned} \epsilon_{\boldsymbol{D'}}{}^{C'}[\![\nabla_{AA'},\nabla_{BB'}]\!]\epsilon_{\boldsymbol{D}}{}^{C} + \epsilon_{\boldsymbol{D}}{}^{C}[\![\nabla_{AA'},\nabla_{BB'}]\!]\epsilon_{\boldsymbol{D'}}{}^{C'} \\ = \epsilon_{\boldsymbol{D'}}{}^{C'}R^{C}{}_{DAA'BB'}\epsilon_{\boldsymbol{D}}{}^{D} + \epsilon_{\boldsymbol{D}}{}^{C}\bar{R}^{C'}{}_{D'AA'BB'}\epsilon_{\boldsymbol{D'}}{}^{D'}. \end{aligned}$$

From the latter one can conclude that

$$\begin{aligned} [\![\nabla_{AA'},\nabla_{BB'}]\!]\epsilon_{\boldsymbol{D}}{}^{C} &= R^{C}{}_{QAA'BB'}\epsilon_{\boldsymbol{D}}{}^{Q}, \\ [\![\nabla_{AA'},\nabla_{BB'}]\!]\epsilon_{\boldsymbol{D'}}{}^{C'} &= \bar{R}^{C'}{}_{Q'AA'BB'}\epsilon_{\boldsymbol{D'}}{}^{Q'}. \end{aligned}$$

Now, using that $\epsilon_{\boldsymbol{P}}{}^{C}\epsilon^{\boldsymbol{Q}}{}_{C} = \delta_{\boldsymbol{P}}{}^{\boldsymbol{Q}}$, and that $[\![\nabla_{AA'},\nabla_{BB'}]\!]\delta_{\boldsymbol{P}}{}^{\boldsymbol{Q}} = 0$, one finds that

$$\begin{aligned} \epsilon_{\boldsymbol{P}}{}^{C}[\![\nabla_{AA'},\nabla_{BB'}]\!]\epsilon^{\boldsymbol{Q}}{}_{C} &= -\epsilon^{\boldsymbol{Q}}{}_{C}[\![\nabla_{AA'},\nabla_{BB'}]\!]\epsilon_{\boldsymbol{P}}{}^{C} && (3.37\text{a}) \\ &= -\epsilon^{\boldsymbol{Q}}{}_{C}R^{C}{}_{DAA'BB'}\epsilon_{\boldsymbol{P}}{}^{D}. && (3.37\text{b}) \end{aligned}$$

Multiplying the previous expression by $\epsilon^{\boldsymbol{P}}{}_{D}$ and using that $\epsilon^{\boldsymbol{P}}{}_{D}\epsilon_{\boldsymbol{P}}{}^{C} = \delta_{D}{}^{C}$ one obtains

$$\delta_{D}{}^{C}[\![\nabla_{AA'},\nabla_{BB'}]\!]\epsilon^{\boldsymbol{Q}}{}_{C} = -\epsilon^{\boldsymbol{Q}}{}_{C}R^{C}{}_{QAA'BB'}\delta_{D}{}^{Q}.$$

Finally, using that $[\![\nabla_{AA'}, \nabla_{BB'}]\!]\delta_D{}^C = 0$, one concludes that

$$[\![\nabla_{AA'}, \nabla_{BB'}]\!]\epsilon^{\boldsymbol{Q}}{}_D = -R^C{}_{DAA'BB'}\epsilon^{\boldsymbol{Q}}{}_C. \tag{3.38}$$

A similar argument applied to primed basis spinors yields

$$[\![\nabla_{AA'}, \nabla_{BB'}]\!]\epsilon^{\boldsymbol{Q}'}{}_{D'} = -\bar{R}^{C'}{}_{D'AA'BB'}\epsilon^{\boldsymbol{Q}'}{}_{C'}. \tag{3.39}$$

Now, using that $[\![\nabla_{AA'}, \nabla_{BB'}]\!]$ applied to a scalar is zero, one has that Equations (3.37a), (3.37b), (3.38) and (3.39) render the following formulae for arbitrary valence 1 spinors:

$$[\![\nabla_{AA'}, \nabla_{BB'}]\!]\mu^C = R^C{}_{QAA'BB'}\mu^Q, \tag{3.40a}$$

$$[\![\nabla_{AA'}, \nabla_{BB'}]\!]\bar{\lambda}^{C'} = \bar{R}^{C'}{}_{Q'AA'BB'}\bar{\lambda}^{Q'}, \tag{3.40b}$$

$$[\![\nabla_{AA'}, \nabla_{BB'}]\!]\kappa_C = -R^Q{}_{CAA'BB'}\kappa_Q, \tag{3.40c}$$

$$[\![\nabla_{AA'}, \nabla_{BB'}]\!]\bar{\nu}_{C'} = -\bar{R}^{Q'}{}_{C'AA'BB'}\bar{\nu}_{Q'}. \tag{3.40d}$$

The extension to higher valence spinors follows from the Leibniz rule. For example, one has that

$$\begin{aligned}[\![\nabla_{AA'}, \nabla_{BB'}]\!]\xi_{CD}{}^{E'} &= -R^Q{}_{CAA'BB'}\xi_{QD}{}^{E'} - R^Q{}_{DAA'BB'}\xi_{CQ}{}^{E'} \\ &\quad + \bar{R}^{E'}{}_{Q'AA'BB'}\xi_{CD}{}^{Q'}.\end{aligned}$$

3.2.4 Decomposition of a general curvature spinor

Expression (3.36) is a convenient starting point to analyse the decomposition of the curvature spinor in terms of irreducible components. Lowering the index pair CC' using the ϵ-spinor one obtains:

$$\begin{aligned}R_{CC'DD'AA'BB'} &= R_{CDAA'BB'}\epsilon_{D'C'} + \bar{R}_{C'D'AA'BB'}\epsilon_{DC} \\ &= -R_{CDAA'BB'}\epsilon_{C'D'} - \bar{R}_{C'D'AA'BB'}\epsilon_{CD}.\end{aligned} \tag{3.41}$$

For the curvature spinor of a general connection one has that $R_{CDAA'BB'} \neq R_{(CD)AA'BB'}$. However, one still has that

$$R_{CDAA'BB'} = -R_{CDBB'AA'}.$$

This antisymmetry can be exploited using the split (3.13) in such a way that the indices CD are not touched. Accordingly, one obtains

$$R_{CDAA'BB'} = X_{CDAB}\epsilon_{A'B'} + Y_{CDA'B'}\epsilon_{AB}, \tag{3.42}$$

where

$$X_{CDAB} = X_{CD(AB)} \equiv \frac{1}{2}R_{CDAQ'B}{}^{Q'},$$

$$Y_{CDA'B'} = Y_{CD(A'B')} \equiv \frac{1}{2}R_{CDA'QB'}{}^{Q}.$$

To complete the decomposition of the curvature spinor in irreducible components one can apply the decomposition formulae (3.8) and (3.12) for valence 4 spinors to X_{CDAB} and $Y_{CDA'B'}$. This idea will not be pursued any further here. However, it will be convenient to single out certain components of the decomposition in irreducible terms of X_{CDAB}. It is conventional to set

$$\Psi_{ABCD} \equiv X_{(ABCD)}, \qquad \Lambda \equiv \frac{1}{6} X_{PQ}{}^{PQ}.$$

Let $C_{CC'DD'AA'BB'}$ denote the spinor obtained from the split (3.41) of the curvature spinor by setting $X_{ABCD} = X_{(ABCD)}$ and $Y_{CDA'B'} = 0$. One has that

$$C_{CC'DD'AA'BB'} = -\Psi_{ABCD}\epsilon_{A'B'}\epsilon_{C'D'} - \bar{\Psi}_{A'B'C'D'}\epsilon_{AB}\epsilon_{CD}. \tag{3.43}$$

As a consequence of the total symmetry of Ψ_{ABCD} it can be readily verified that $C_{CC'DD'AA'BB'}$ is the spinorial counterpart of a trace-free tensor. Following the discussion in Section 2.5.2, it must be the ***spinorial counterpart of the Weyl tensor*** C_{cdab}.

Decomposition of the curvature spinor of a torsion-free connection

The decomposition of the curvature spinor is now particularised to the case of a torsion-free connection. In this case, the Riemann curvature tensor has the cyclic symmetry of the Bianchi identity. The latter is best exploited using the alternative expression of the identity given by Equation (2.23) involving the right-dual of the Riemann tensor. Using the spinorial counterpart of the volume form given by Equation (3.25) one has that

$$\begin{aligned} R^*_{AA'BB'CC'DD'} &= \frac{\mathrm{i}}{2}(\delta_C{}^E\delta_D{}^F\delta_{C'}{}^{F'}\delta_{D'}{}^{E'} - \delta_C{}^F\delta_D{}^E\delta_{C'}{}^{E'}\delta_{D'}{}^{F'})R_{AA'BB'EE'FF'} \\ &= \mathrm{i}R_{AA'BB'CD'DC'}, \end{aligned}$$

so that the spinorial counterpart of Equation (2.23) is given by

$$R_{CC'QQ'A}{}^{Q'Q}{}_{A'} = 0.$$

A direct evaluation of the above condition using the splits (3.41) and (3.42) shows that

$$X_{CQA}{}^{Q}\epsilon_{C'A'} - \bar{X}_{C'Q'A'}{}^{Q'}\epsilon_{CA} + Y_{CAC'A'} - \bar{Y}_{C'A'AC} = 0,$$

so that

$$X_{PQ}{}^{PQ} = \bar{X}_{PQ}{}^{PQ}, \qquad \bar{Y}_{C'A'AC} = Y_{CAC'A'}.$$

Hence, one has that $X_{PQ}{}^{PQ}$ (i.e. Λ) is a real scalar, while $Y_{ABA'B'}$ is a Hermitian tensor, and, thus, it is the spinorial counterpart of a rank 2 tensor.

Decomposition on the curvature spinor of a metric connection

As already seen, a connection which is compatible with a metric $\boldsymbol{g}$ satisfies $\nabla_{AA'}\epsilon_{CD} = 0$. It follows then that $[\![\nabla_{AA'}, \nabla_{BB'}]\!]\epsilon_{CD} = 0$. However, one also has that

$$[\![\nabla_{AA'}, \nabla_{BB'}]\!]\epsilon_{CD} = -R^{Q}{}_{CAA'BB'}\epsilon_{QD} + R^{Q}{}_{DAA'BB'}\epsilon_{CQ},$$

from which one concludes that

$$R_{CDAA'BB'} = R_{(CD)AA'BB'}.$$

The latter can be reexpressed in terms of the following symmetries of the spinors X_{ABCD} and $Y_{ABA'B'}$:

$$X_{ABCD} = X_{(AB)CD}, \qquad Y_{ABA'B'} = Y_{(AB)A'B'}.$$

Decomposition of the curvature spinor of a Levi-Civita connection

Finally, one can collect the results of the previous subsections to obtain the well-known irreducible decomposition of the spinorial counterpart of the Riemann tensor of a Levi-Civita connection. As the Levi-Civita connection associated to the metric $\boldsymbol{g}$ is both torsion-free and metric, it follows then that

$$X_{ABCD} = X_{(AB)(CD)}, \qquad X_{CQA}{}^{Q} = 0.$$

It follows from (3.8) that $X_{ABCD} = X_{CDAB}$ and that

$$\begin{aligned} X_{ABCD} &= X_{(ABCD)} - \frac{1}{3}\epsilon_{A(C}\epsilon_{D)B}X_{PQ}{}^{PQ} \\ &= \Psi_{ABCD} + \Lambda(\epsilon_{DB}\epsilon_{CA} + \epsilon_{CB}\epsilon_{DA}). \end{aligned}$$

Similarly, for $Y_{ABA'B'}$ one has that

$$Y_{ABA'B'} = Y_{(AB)(A'B')},$$

so that according to the general split (3.12) $Y_{ABA'B'}$ corresponds to a trace-free rank 2 tensor.

To conclude the analysis, it is convenient to compute the Ricci tensor and scalar in terms of the spinors X_{ABCD} and $Y_{ABA'B'}$. From Equations (3.41) and (3.42) it follows directly that

$$\begin{aligned} &R_{AA'BB'} = -X_{QA}{}^{Q}{}_{B}\epsilon_{A'B'} - \bar{X}_{Q'A'}{}^{Q'}{}_{B'}\epsilon_{AB} + 2Y_{ABA'B'} \\ &R = -4X_{PQ}{}^{PQ}, \end{aligned}$$

where $R_{AA'BB'}$ denotes the ***spinorial counterpart of the Ricci tensor*** R_{ab} and it has been used that for a Levi-Civita connection $Y_{ABA'B'} = \bar{Y}_{A'B'AB}$ and $X_{PQ}{}^{PQ} = \bar{X}_{P'Q'}{}^{P'Q'}$. In particular, one has that

$$R = -24\Lambda.$$

As $Y_{ABA'B'}$ is trace-free, it has to be related to Φ_{ab}, the symmetric trace-free part of the Ricci tensor. Indeed, a calculation for its spinorial counterpart shows that

$$\begin{aligned} 2\Phi_{ABA'B'} &\equiv R_{AA'BB'} - \frac{1}{4}R\epsilon_{AB}\epsilon_{A'B'} \\ &= 2Y_{ABA'B'}. \end{aligned}$$

It can be verified that $\Phi_{ABA'B'}$ satisfies the symmetries

$$\Phi_{ABA'B'} = \Phi_{BAA'B'} = \Phi_{ABB'A'} = \Phi_{BAB'A'}. \tag{3.44}$$

Putting together the discussion of this section, one finds that the spinor counterpart of the Riemann curvature tensor of a Levi-Civita connection can be decomposed as

$$\begin{aligned} R_{AA'BB'CC'DD'} &= -\epsilon_{A'B'}\epsilon_{C'D'}(\Psi_{ABCD} + 2\Lambda\epsilon_{A(C}\epsilon_{D)B}) \\ &\quad -\epsilon_{AB}\epsilon_{CD}(\bar{\Psi}_{A'B'C'D'} + 2\Lambda\epsilon_{A'(C'}\epsilon_{D')B'}) \\ &\quad +\epsilon_{A'B'}\epsilon_{CD}\Phi_{ABC'D'} + \epsilon_{AB}\epsilon_{C'D'}\Phi_{CDA'B'}. \end{aligned}$$

Working back from this expression one can recover the decomposition of the Riemann tensor in terms of the Weyl and Schouten tensor given in Equations (2.21a) and (2.21b).

3.2.5 The $\square_{AB}$-operator

In some applications it is convenient to have a more explicit expression for the commutator of spinorial covariant derivatives. *In the remainder of this section it is assumed that $\nabla_{AA'}$ is the spinorial counterpart of a Levi-Civita connection.*

Exploiting the antisymmetry of Equation (3.40a) with respect to the pairs $_{AA'}$ and $_{BB'}$ one can rewrite it as

$$(\epsilon_{A'B'}\square_{AB} + \epsilon_{AB}\square_{A'B'})\mu^C = R^C{}_{QAA'BB'}\mu^Q, \tag{3.45}$$

where

$$\square_{AB} \equiv \nabla_{Q'(A}\nabla_{B)}{}^{Q'}, \qquad \square_{A'B'} \equiv \nabla_{Q(A'}\nabla^Q{}_{B')}.$$

It can be verified that both $\square_{AB}$ and $\square_{A'B'}$ are linear and satisfy the Leibnitz rule – one has, for example, that

$$\square_{AB}(\mu_C\lambda^D) = (\square_{AB}\mu_C)\lambda^D + \mu_C(\square_{AB}\lambda^D).$$

Defining the ***D'Alembertian operator*** as $\square \equiv \nabla_{PP'}\nabla^{PP'}$, one obtains the decomposition

$$\nabla_{AQ'}\nabla_B{}^{Q'} = \frac{1}{2}\epsilon_{AB}\square + \square_{AB}.$$

Now, contracting indices suitably in Equation (3.45) one readily obtains

$$\square_{AB}\mu^C = X^C{}_{QAB}\mu^Q, \qquad \square_{A'B'}\mu^C = Y^C{}_{QA'B'}\mu^Q.$$

Using the explicit expressions for the curvature spinors X_{ABCD} and $Y_{ABA'B'}$ for a Levi-Civita connection, as given in Section 3.2.4 one concludes that

$$\square_{AB}\mu_C = \Psi_{ABCD}\mu^D - 2\Lambda\mu_{(A}\epsilon_{B)C}, \qquad \square_{A'B'}\mu_C = \Phi_{CDA'B'}\mu^D. \tag{3.46}$$

The above expressions can be extended to higher order valence spinors by means of the Leibnitz rule.

The expressions in (3.46) can be extended to the case of connections with torsion; see Penrose (1983) for the general theory and Gasperín and Valiente Kroon (2015) for explicit expressions and applications.

3.3 Global considerations

The discussion on null vectors and their flagpoles in Section 3.1.10 makes a natural connection with the notion of orientability and the assumptions needed to ensure the existence of spinorial structures on a region of spacetime.

As seen in Proposition 3.3, every non-vanishing null vector is either future pointing or past pointing, in accordance with the choice of sign made in Equation (3.21). Thus, the existence of spinors on a region of spacetime provides a way to define a time orientation. In a similar way, the idea of a right-handed phase change of a triad of orthonormal vectors $\{\boldsymbol{e}_1, \boldsymbol{e}_2, \boldsymbol{e}_3\}$, as discussed in Section 3.1.10, can be used to define a notion of space orientation. Thus, at least at an intuitive level, the existence of a spinorial structure over a spacetime seems to imply that the spacetime is time orientable and space orientable. It turns out that the converse is also true: time and space orientability ensure the existence of a spinorial structure. More precisely, one has the following result proved in Geroch (1968):

Theorem 3.2 (***orientability and the existence of a spinor structure***) *A non-compact spacetime $(\mathcal{M}, \boldsymbol{g})$ has a spinor structure if and only if there exists on $\mathcal{M}$ a global system of orthonormal tetrads.*

Part IV of this book will be concerned with the construction of spacetimes from suitably posed initial value problems. Thus, it is convenient to have a criterion to encode the existence of a spinorial structure in an initial value problem. An example of this is the following result in Geroch (1970c):

Proposition 3.4 (***global hyperbolicity and the existence of a spinor structure***) *Every globally hyperbolic spacetime has a spinor structure.*

The notion of global hyperbolicity is discussed in Section 14.1.

An orientable spacetime may have several spinorial structures. One can ensure uniqueness of the spinorial structure if one restricts further the topology of the spacetime. More precisely, one has that (see Geroch (1968)):

Proposition 3.5 (***uniqueness of the spinorial structure***) *The spinorial structure of a spacetime is unique if and only if $\mathcal{M}$ is simply connected.*

3.4 Further reading

Further details on the various topics covered in the present chapter can be found in Penrose and Rindler (1984), Stewart (1991) and O'Donnell (2003). The discussion in these references leads, in a natural way, to the *Newman-Penrose formalism* and applications like the *Petrov algebraic classification of the Weyl tensor*. Some discussion on the use of spinors in the construction and analysis of exact solutions to the Einstein field equations can be found in Stephani et al. (2003) and Griffiths and Podolský (2009). The relation between Dirac spinors and 2-spinors is presented in Penrose and Rindler (1984) and Stewart (1991). A pure mathematics perspective can be found, for example, in Petersen (1991); see also Choquet-Bruhat et al. (1982).

A more general perspective of the discussion of the present chapter can be obtained by making use of the notion of *fibre bundles*; see, for example, Ashtekar et al. (1982). In terms of this language, the spinorial structure arises as a *principal fibre bundle* over the spacetime manifold $\mathcal{M}$ with structure group $SL(2,\mathbb{C})$. This point of view is convenient for computer algebra implementations; see, for example, Martín-García (2014). The fibre bundles are useful in analyses that require the *blowing up* of particular points of spacetime – as in the analysis of caustics in Friedrich and Stewart (1983) or the so-called problem of spatial infinity of Friedrich (1998c).

Appendix: the Newman-Penrose formalism

The idea of a spinor-based null tetrad formalism was introduced in the seminal article by Newman and Penrose (1962); see also Newman and Penrose (1963). This so-called ***Newman-Penrose (NP) formalism*** was first used as a way of analysing the asymptotics of gravitational radiation. The potential of the formalism to obtain exact solutions to the Einstein field equations, in particular, ones having an algebraically special Weyl tensor, was quickly realised; see, for example, Stephani et al. (2003) for an entry point to the literature of exact solutions. Refinements of the formalism which are adapted to specific configurations or types of problems are available in the literature, most noticeably Geroch et al. (1973); see also Machado and Vickers (1995, 1996).

The key aspects of a generic spinor-based null tetrad formalism have already been covered in this book. One of the peculiarities of the formalism, as introduced in Newman and Penrose (1962), is the use of specific symbols to denote directional derivatives and the spin coefficients. This notation will not be used in this book as the Newman-Penrose (NP) formalism assumes, from the onset, a Levi-Civita connection. However, the discussion in this book will very often use more general connections. Hence, one has more independent spin coefficients. Moreover, the labelling of spin coefficients through indices lends itself better for a systematic analysis of the properties of the relevant equations. Additional difficulties with the NP formalism arise with the space spinor formalism; see the next chapter.

The purpose of this appendix is to provide a guide to the translation, whenever possible, between NP objects and the ones used in this book.

The directional derivatives

Let $\{\boldsymbol{o}, \boldsymbol{\iota}\}$ denote, as usual, a spin basis. Also, let $\{\boldsymbol{l}, \boldsymbol{n}, \boldsymbol{m}, \bar{\boldsymbol{m}}\}$ denote the null tetrad constructed from the spin basis, as described in Section 3.1.9. The NP convention for the directional derivatives along the directions given by the null tetrad is

$$
\begin{aligned}
D &\equiv l^a \nabla_a = o^A \bar{o}^{A'} \nabla_{AA'} = \nabla_{\mathbf{00'}}, \\
\Delta &\equiv n^a \nabla_a = \iota^A \bar{\iota}^{A'} \nabla_{AA'} = \nabla_{\mathbf{11'}}, \\
\delta &\equiv m^a \nabla_a = o^A \bar{\iota}^{A'} \nabla_{AA'} = \nabla_{\mathbf{01'}}, \\
\bar{\delta} &\equiv \bar{m}^a \nabla_a = \iota^A \bar{o}^{A'} \nabla_{AA'} = \nabla_{\mathbf{10'}}.
\end{aligned}
$$

The spin coefficients

In what follows, it is assumed that the connection $\boldsymbol{\nabla}$ is Levi-Civita so that $\nabla_{AA'}\epsilon_{BC} = 0$. The NP convention for the spin coefficients of $\boldsymbol{\nabla}$ is given by:

$$
\begin{aligned}
\epsilon &= \Gamma_{\mathbf{00'}}{}^{\mathbf{0}}{}_{\mathbf{0}} = -\Gamma_{\mathbf{00'}}{}^{\mathbf{1}}{}_{\mathbf{1}} = \Gamma_{\mathbf{00'10}}, \\
\alpha &= \Gamma_{\mathbf{10'}}{}^{\mathbf{0}}{}_{\mathbf{0}} = -\Gamma_{\mathbf{10'}}{}^{\mathbf{1}}{}_{\mathbf{1}} = \Gamma_{\mathbf{10'10}}, \\
\beta &= \Gamma_{\mathbf{01'}}{}^{\mathbf{0}}{}_{\mathbf{0}} = -\Gamma_{\mathbf{01'}}{}^{\mathbf{1}}{}_{\mathbf{1}} = \Gamma_{\mathbf{01'10}}, \\
\gamma &= \Gamma_{\mathbf{11'}}{}^{\mathbf{0}}{}_{\mathbf{0}} = -\Gamma_{\mathbf{11'}}{}^{\mathbf{1}}{}_{\mathbf{1}} = \Gamma_{\mathbf{11'10}},
\end{aligned}
$$

$$
\begin{aligned}
\pi &= \Gamma_{\mathbf{00'}}{}^{\mathbf{0}}{}_{\mathbf{1}} = \Gamma_{\mathbf{00'11}}, & \kappa &= -\Gamma_{\mathbf{00'}}{}^{\mathbf{1}}{}_{\mathbf{0}} = \Gamma_{\mathbf{00'00}}, \\
\lambda &= \Gamma_{\mathbf{10'}}{}^{\mathbf{0}}{}_{\mathbf{1}} = \Gamma_{\mathbf{10'11}}, & \rho &= -\Gamma_{\mathbf{10'}}{}^{\mathbf{1}}{}_{\mathbf{0}} = \Gamma_{\mathbf{10'00}}, \\
\mu &= \Gamma_{\mathbf{01'}}{}^{\mathbf{0}}{}_{\mathbf{1}} = \Gamma_{\mathbf{01'11}}, & \sigma &= -\Gamma_{\mathbf{01'}}{}^{\mathbf{1}}{}_{\mathbf{0}} = \Gamma_{\mathbf{01'00}}, \\
\nu &= \Gamma_{\mathbf{11'}}{}^{\mathbf{0}}{}_{\mathbf{1}} = \Gamma_{\mathbf{11'11}}, & \tau &= -\Gamma_{\mathbf{11'}}{}^{\mathbf{1}}{}_{\mathbf{0}} = \Gamma_{\mathbf{11'00}}.
\end{aligned}
$$

The above spin coefficients can be expressed entirely in terms of the directional derivatives D, Δ, δ, $\bar{\delta}$ applied to the null frame vectors or, alternatively, applied to the spin basis $\{\boldsymbol{o}, \boldsymbol{\iota}\}$. See O'Donnell (2003) and Stewart (1991) for details on this. Explicit expressions of the spin coefficients in terms of *curls* (antisymmetrised derivatives) have been worked out in Cocke (1989).

The Ricci and Weyl tensors

The NP conventions to denote the components of the Weyl spinor Ψ_{ABCD} with respect to $\{o, \iota\}$ are:

$$\Psi_0 \equiv \Psi_{ABCD}o^A o^B o^C o^D, \quad \Psi_1 \equiv \Psi_{ABCD}o^A o^B o^C \iota^D, \quad \Psi_2 \equiv \Psi_{ABCD}o^A o^B \iota^C \iota^D,$$
$$\Psi_3 \equiv \Psi_{ABCD}o^A \iota^B \iota^C \iota^D, \quad \Psi_4 \equiv \Psi_{ABCD}\iota^A \iota^B \iota^C \iota^D.$$

The conventions for the components of the trace-free Ricci spinor $\Phi_{AA'BB'}$ are:

$$\Phi_{00} \equiv \Phi_{AA'BB'}o^A o^B \bar{o}^{A'} \bar{o}^{B'}, \qquad \Phi_{01} \equiv \Phi_{AA'BB'}o^A o^B \bar{o}^{A'} \bar{\iota}^{B'},$$
$$\Phi_{02} \equiv \Phi_{AA'BB'}o^A o^B \bar{\iota}^{A'} \bar{\iota}^{B'}, \qquad \Phi_{10} \equiv \Phi_{AA'BB'}o^A \iota^B \bar{o}^{A'} \bar{o}^{B'},$$
$$\Phi_{11} \equiv \Phi_{AA'BB'}o^A \iota^B \bar{o}^{A'} \bar{\iota}^{B'}, \qquad \Phi_{12} \equiv \Phi_{AA'BB'}o^A \iota^B \bar{\iota}^{A'} \bar{\iota}^{B'},$$
$$\Phi_{20} \equiv \Phi_{AA'BB'}\iota^A \iota^B \bar{o}^{A'} \bar{o}^{B'}, \qquad \Phi_{21} \equiv \Phi_{AA'BB'}\iota^A \iota^B \bar{o}^{A'} \bar{\iota}^{B'},$$
$$\Phi_{22} \equiv \Phi_{AA'BB'}\iota^A \iota^B \bar{\iota}^{A'} \bar{\iota}^{B'}.$$

Notice that in both lists of definitions the value of the index denotes the number of contractions with the spinor ι.

The NP formalism makes use of the symbol Λ to denote a multiple of the trace of the Ricci tensor. The relation to the Ricci scalar is

$$R = -24\Lambda.$$

The Newman-Penrose field equations

Newman and Penrose (1962) provided explicit expressions of the Ricci and the Bianchi identities in terms of their notation for the spin connection coefficients and the components of Ψ_{ABCD} and $\Phi_{AA'BB'}$. These equations are collectively called the *Newman-Penrose field equations.* Explicit expressions are available in O'Donnell (2003), Penrose and Rindler (1986) and Stewart (1991). Besides the NP field equations, the formalism consists also of explicit expressions for the commutators of the directional derivatives D, Δ, δ, $\bar{\delta}$. Expressions for the source-free Maxwell equations are available in the literature as well; see, for example, the appendix in Stewart (1991).

4
Space spinors

This chapter discusses a framework for spinors in which a further structure is introduced – a so-called *Hermitian inner product*. The resulting formalism will be referred to as ***space spinors*** or $SU(2,\mathbb{C})$-***spinors***. The space spinor formalism can be used to describe the geometry of three-dimensional Riemannian manifolds and, more generally, foliations of spacetime. Moreover, it can also be used to provide a description of the hyperplanes associated to a congruence of timelike curves.

The notion of space spinors was first introduced in Sommers (1980); see also Sen (1981). It provides a systematic approach to the construction of evolution equations which can be regarded as a spinorial version of the $1+3$ formalism for tensors. Space spinors are used in several other areas of relativity such as quantum gravity (see e.g. Ashtekar (1991)), the construction of quasi-local notions of energy (see e.g. Szabados (2009)) and global aspects of the geometry of 3-manifolds (see e.g. Bäckdahl and Valiente Kroon (2010a); Beig and Szabados (1997); Tod (1984)).

4.1 Hermitian inner products and 2-spinors

Let $(\mathcal{M}, \boldsymbol{g})$ denote a four-dimensional Lorentzian manifold. As in Chapter 3, it is assumed that at each point $p \in \mathcal{M}$ one has a two-dimensional simplectic vector space $\mathfrak{S}|_p(\mathcal{M})$ as given by Definition 3.1. One has the following definition:

Definition 4.1 (***Hermitian inner product***) *A* ***Hermitian inner product*** *on a simplectic two-dimensional vector space* $\mathfrak{S}$ *is a function* $\langle\langle\cdot,\cdot\rangle\rangle : \mathfrak{S}\times\mathfrak{S} \rightarrow \mathbb{C}$ *which is:*

(i) ***Hermitian****; that is, given* $\boldsymbol{\xi}$, $\boldsymbol{\eta} \in \mathfrak{S}$

$$\langle\langle \boldsymbol{\xi}, \boldsymbol{\eta} \rangle\rangle = \overline{\langle\langle \boldsymbol{\eta}, \boldsymbol{\xi} \rangle\rangle}$$

(ii) ***linear in the second entry;*** *that is, given* $\boldsymbol{\xi}$, $\boldsymbol{\eta}$, $\boldsymbol{\zeta} \in \mathfrak{S}$, $z \in \mathbb{C}$

$$\langle\langle \boldsymbol{\xi}, \boldsymbol{\eta} + z\boldsymbol{\zeta} \rangle\rangle = \langle\langle \boldsymbol{\xi}, \boldsymbol{\eta} \rangle\rangle + z\langle\langle \boldsymbol{\xi}, \boldsymbol{\zeta} \rangle\rangle$$

(iii) ***positive definite;*** *that is, given* $\boldsymbol{\xi} \in \mathfrak{S}$

$$\langle\langle \boldsymbol{\xi}, \boldsymbol{\xi} \rangle\rangle \geq 0$$

and $\langle\langle \boldsymbol{\xi}, \boldsymbol{\xi} \rangle\rangle = 0$ *if and only if* $\boldsymbol{\xi} = 0$.

From conditions (i) and (ii) it follows that a Hermitian inner product is *antilinear* in the first entry; that is, given $\boldsymbol{\xi}$, $\boldsymbol{\eta}$, $\boldsymbol{\zeta} \in \mathfrak{S}$, $z \in \mathbb{C}$, one has

$$\langle\langle \boldsymbol{\xi} + z\boldsymbol{\zeta}, \boldsymbol{\eta} \rangle\rangle = \langle\langle \boldsymbol{\xi}, \boldsymbol{\eta} \rangle\rangle + \bar{z}\langle\langle \boldsymbol{\zeta}, \boldsymbol{\eta} \rangle\rangle.$$

4.1.1 Hermitian conjugation

In what follows, given a spacetime $(\mathcal{M}, \boldsymbol{g})$, assume that for each point $p \in \mathcal{M}$, the vector space $\mathfrak{S}|_p(\mathcal{M})$ is endowed with a Hermitian inner product which changes smoothly from point to point. The Hermitian inner product can be expressed in terms of a Hermitian spinor $\varpi_{AA'} \in \mathfrak{S}_{AA'}(\mathcal{M})$ such that

$$\langle\langle \boldsymbol{\xi}, \boldsymbol{\eta} \rangle\rangle = \varpi_{AA'}\bar{\xi}^{A'}\eta^{A}. \tag{4.1}$$

It can be verified that the right-hand side of the above expression satisfies conditions (i) and (ii) of Definition 4.1. Given a spinor basis $\{\epsilon_{\boldsymbol{A}}{}^{A}\}$, the components of $\varpi_{AA'}$ with respect to the basis are given by $\varpi_{\boldsymbol{AA'}} \equiv \varpi_{AA'}\epsilon_{\boldsymbol{A}}{}^{A}\bar{\epsilon}_{\boldsymbol{A'}}{}^{A'}$. The components $\varpi_{\boldsymbol{AA'}}$ can be thought of as the entries of a (2×2) matrix $(\varpi_{\boldsymbol{AA'}})$. The *positivity condition* (iii) of Definition 4.1 requires the above matrix to be diagonalisable and to have positive eigenvalues. Thus, it is natural to consider a (not necessarily normalised) basis $\{\epsilon_{\boldsymbol{A}}{}^{A}\}$ for which $(\varpi_{\boldsymbol{AA'}})$ takes the diagonal form

$$(\varpi_{\boldsymbol{AA'}}) = \begin{pmatrix} \varpi_{\boldsymbol{00'}} & 0 \\ 0 & \varpi_{\boldsymbol{11'}} \end{pmatrix}.$$

The scaling of the basis $\{\epsilon_{\boldsymbol{A}}{}^{A}\}$ can be fixed, without loss of generality, so that $(\varpi_{\boldsymbol{AA'}})$ is the identity matrix. *In the rest of the book, whenever a Hermitian inner product is discussed, it will be assumed that a spin basis* $\{\epsilon_{\boldsymbol{A}}{}^{A}\}$ *has been chosen so that*

$$(\varpi_{\boldsymbol{AA'}}) = \begin{pmatrix} 1 & 0 \\ 0 & 1 \end{pmatrix}. \tag{4.2}$$

A direct consequence of the above normalisation condition is that one can write

$$\varpi_{AA'} = o_A\bar{o}_{A'} + \iota_A\bar{\iota}_{A'} = \epsilon^{\boldsymbol{1}}{}_A\bar{\epsilon}^{\boldsymbol{1'}}{}_{A'} + \epsilon^{\boldsymbol{0}}{}_A\bar{\epsilon}^{\boldsymbol{0'}}{}_{A'}, \tag{4.3a}$$

$$\varpi_A{}^{A'} = o_A\bar{o}^{A'} + \iota_A\bar{\iota}^{A'} = \epsilon^{\mathbf{1}}{}_A\bar{\epsilon}_{\mathbf{0}'}{}^{A'} - \epsilon^{\mathbf{0}}{}_A\bar{\epsilon}_{\mathbf{1}}{}^{A'}, \tag{4.3b}$$

$$\varpi^{AA'} = o^A\bar{o}^{A'} + \iota^A\bar{\iota}^{A'} = \epsilon_{\mathbf{0}}{}^A\bar{\epsilon}_{\mathbf{0}'}{}^{A'} + \epsilon_{\mathbf{1}}{}^A\bar{\epsilon}_{\mathbf{1}'}{}^{A'}. \tag{4.3c}$$

From these expressions it follows that

$$\varpi_{AA'}\varpi^{BA'} = o_A\iota^B - \iota_A o^B = \epsilon^{\mathbf{1}}{}_A\epsilon_{\mathbf{1}}{}^B + \epsilon^{\mathbf{0}}{}_A\epsilon_{\mathbf{0}}{}^B.$$

Thus,

$$\varpi_{AA'}\varpi^{A'B} = \delta_A{}^B. \tag{4.4}$$

Notice, in particular, that $\varpi_{AA'}\varpi^{AA'} = 2$.

The spinor $\varpi_{AA'}$ induces an ***operation of Hermitian conjugation*** $^+$: $\mathfrak{S}^\bullet(\mathcal{M}) \to \mathfrak{S}^\bullet(\mathcal{M})$. Given $\mu_A \in \mathfrak{S}(\mathcal{M})$, we define its Hermitian conjugate μ^+_A via

$$\mu^+_A \equiv \varpi_A{}^{A'}\bar{\mu}_{A'}. \tag{4.5}$$

It follows then that one can write

$$\langle\langle \boldsymbol{\xi}, \boldsymbol{\eta} \rangle\rangle = -\xi^+{}_A\eta^A = \eta_A\xi^{+A}.$$

Observe that as a consequence of the *see-saw rule*, Equations (3.4a) and (3.4b), one has $\mu^{+A} \equiv -\varpi^A{}_{A'}\bar{\mu}^{A'}$. The Hermitian conjugation is extended to higher valence spinors by requiring

$$(\boldsymbol{\mu\lambda})^+ = \boldsymbol{\mu}^+\boldsymbol{\lambda}^+$$

for $\boldsymbol{\mu}$, $\boldsymbol{\lambda} \in \mathfrak{S}^\bullet(\mathcal{M})$. It is a consequence of the normalisation condition (4.2) that

$$\mu^{++}_{A_1\cdots A_k} = (-1)^k\mu_{A_1\ldots A_k}.$$

Furthermore, $\mu^{+A}\mu_A = 0$ if and only if $\mu_A = 0$, the latter as a result of condition (iii) of Definition 3.1. Using the representation of $\varpi_A{}^{A'}$ given by (4.3b) one finds that

$$o^+_A = \iota_A, \qquad \iota^+_A = -o_A, \tag{4.6a}$$

$$o^{+A} = \iota^A, \qquad \iota^{+A} = -o^A. \tag{4.6b}$$

Hence, the normalisation leading to (4.3a)–(4.3c) is equivalent to the normalisation condition

$$o_A o^{+A} = 1,$$

for the spinor $\boldsymbol{o}$. Notice also that, as a consequence of the previous discussion, a non-zero spinor and its Hermitian conjugate are linearly independent. Finally, a calculation using the expression of ϵ_{AB} in terms of o_A and ι_A, yields

$$\epsilon^+{}_{AB} = \varpi_A{}^{A'}\varpi_B{}^{B'}\epsilon_{A'B'} = \epsilon_{AB}.$$

Remark. In the rest of the book, when working with spinor structures endowed with a Hermitian product, it will always be assumed that a spin basis satisfying relations (4.6a) and (4.6b) has been chosen.

4.2 The space spinor formalism

A consequence of the existence of an operation of Hermitian conjugation is that, given a spinor $\boldsymbol{\xi} \in \mathfrak{S}^{\bullet}(\mathcal{M})$, its complex conjugate $\bar{\boldsymbol{\xi}}$ and its Hermitian conjugate $\boldsymbol{\xi}^{+}$ contain the same information. This observation allows the introduction of a ***spinorial formalism*** based entirely on spinors with unprimed indices by contracting all the primed indices in the spinors with $\varpi_A{}^{A'}$. Given $\xi_{A_1\cdots A_p A'_1\cdots A'_q}$, we define its ***space spinor counterpart*** $\xi_{A_1\cdots A_p B_1\cdots B_q}$ as

$$\xi_{A_1\cdots A_p B_1\cdots B_q} \equiv \varpi_{B_1}{}^{A'_1}\cdots\varpi_{B_q}{}^{A'_q}\xi_{A_1\cdots A_p A'_1\cdots A'_q}. \tag{4.7}$$

The above expression can be inverted by recalling the normalisation condition (4.4) to yield

$$\xi_{A_1\cdots A_p A'_1\cdots A'_q} = (-1)^q \varpi^{B_1}{}_{A'_1}\cdots\varpi^{B_q}{}_{A'_q}\xi_{A_1\cdots A_p B_1\cdots B_q}.$$

Thus, the information contained in a spinor with primed indices and its space spinor counterpart is equivalent.

4.2.1 The Hermitian product and three-dimensional vectors

The operation of Hermitian conjugation gives rise to a notion of *reality* for spinors. More precisely, a spinor $\mu_{A_1B_1\cdots A_kB_k}$ with an *even number of indices* will be said to be ***real*** if

$$\mu^{+}_{A_1B_1\cdots A_kB_k} = (-1)^k \mu_{A_1B_1\cdots A_kB_k},$$

and ***imaginary*** if

$$\mu^{+}_{A_1B_1\cdots A_kB_k} = (-1)^{k+1} \mu_{A_1B_1\cdots A_kB_k}.$$

Consider now a *symmetric* valence-2 spinor $v^{AB} \in \mathfrak{S}^{\bullet}(\mathcal{M})$. Given a space spinor basis $\{\boldsymbol{o}, \boldsymbol{\iota}\}$ such that $\boldsymbol{\iota} \equiv \boldsymbol{o}^{+}$, one can write

$$v^{AB} = a o^A o^B + b \iota^A \iota^B + c o^{(A}\iota^{B)}, \tag{4.8}$$

with $a,\ b,\ c \in \mathbb{C}$. The Hermitian conjugate of v^{AB} is given by

$$v^{+AB} = \bar{a}\iota^A\iota^B + \bar{b} o^A o^B - \bar{c}\iota^{(A}o^{B)}.$$

If v^{AB} is *real*, that is, $v^{+AB} = -v^{AB}$, then $\bar{b} = -a$ and $\bar{c} = c$. Thus, *the real spinor v^{AB} has only three real components so that it describes a three-dimensional vector v^i.* This argument can be extended to higher valence *real* space spinors so that

$$\xi_{A_1B_1\cdots A_kB_k}{}^{C_1D_1\cdots C_mD_m} = \xi_{(A_1B_1)\cdots(A_kB_k)}{}^{(C_1D_1)\cdots(C_mD_m)},$$

if real, can be regarded as the space spinor counterpart of a three-dimensional tensor $\xi_{i_1\cdots i_k}{}^{j_1\cdots j_m}$ – every pair of symmetric spinor indices is associated to a spatial tensor index. One can summarise the previous discussion in the following:

Lemma 4.1 (***the distribution associated to a Hermitian product***) *A Hermitian spinor* $\varpi_{AA'}$ *on* $\mathfrak{S}^\bullet(\mathcal{M})$ *induces a three-dimensional distribution* Π *on* $\mathcal{M}$.

Observation. The distribution Π may not possess integrable manifolds.

The consequences of Lemma 4.1 can be further elaborated by considering the spinorial counterpart of the projector $h_a{}^b$ associated to the distribution Π. To this end let

$$h_{AA'}{}^{BB'} \equiv \delta_A{}^B\delta_{A'}{}^{B'} - \frac{1}{2}\varpi_{AA'}\varpi^{BB'}.$$

It can be readily verified that

$$h_{AA'}{}^{BB'}h_{BB'}{}^{CC'} = h_{AA'}{}^{CC'}, \qquad h_{AA'}{}^{BB'}\varpi_{BB'} = 0.$$

Now, given $v_{AA'} \in \mathfrak{S}^\bullet(\mathcal{M})$ denote by v_{AB} its space spinor counterpart. A calculation then shows that

$$v_{(AB)} = \varpi_B{}^{A'}h_{AA'}{}^{CC'}v_{CC'} = \varpi_{(A}{}^{A'}v_{B)A'}.$$

Thus, *the spinor* $h_{AA'}{}^{BB'}$ *is the projector associated to the distribution induced by* $\varpi_{AA'}$. The space spinor version of $h_{AA'}{}^{BB'}$ is given by $h_{ABCD} \equiv \varpi_B{}^{A'}\varpi_D{}^{C'}$ $h_{AA'CC'}$. It can be readily verified that $h_{ABCD} = h_{(AB)(CD)}$. Using the Jacobi identity (3.5) one can show that

$$h_{ABCD} \equiv -\epsilon_{A(C}\epsilon_{D)B}. \tag{4.9}$$

It can also be verified that

$$h_{ABPQ}h^{PQCD} = h_{AB}{}^{CD} \equiv \epsilon_A{}^{(C}\epsilon_B{}^{D)}, \qquad h_{ABCD}h^{ABCD} = h_{PQ}{}^{PQ} = 3.$$

In addition, one has

$$h^+_{ABCD} = h_{ABCD},$$

so that h_{ABCD} is a *real space spinor*. Moreover, given $v^{AB} = v^{(AB)}$ and $u_{AB} = u_{(AB)}$ one has that

$$v^{AB}h_{ABCD} = v_{CD}, \qquad u_{AB}h^{ABCD} = u^{CD}, \qquad u_{AB}h_{CD}{}^{AB} = u_{CD}.$$

Finally, it is observed that if the spinor v^{AB} is real, then using the decomposition of Equation (4.8) one readily finds that

$$h_{ABCD}v^{AB}v^{CD} = v_{AB}v^{AB} = 2ab - \frac{1}{2}c^2 = -2|b|^2 - \frac{1}{2}c^2 \leq 0.$$

Thus, given $p \in \mathcal{M}$, *the spinor* h_{ABCD} *gives rise to a negative definite inner product on* $\Pi|_p \subset T|_p(\mathcal{M})$. Accordingly, the vectors in Π are spatial with respect to the metric $\boldsymbol{g}$. As Π may not possess integrable submanifolds, h_{ABCD} is not necessarily the spinorial counterpart of a (negative definite) three-dimensional Riemannian metric of a spacelike submanifold of $\mathcal{M}$.

4.2.2 Spatial Infeld-van der Waerden symbols

The relation between space spinors and three-dimensional vectors can be formalised by means of suitable *soldering objects.* At each point $p \in \mathcal{M}$ consider a $\boldsymbol{g}$-orthonormal basis $\{\boldsymbol{e}_i\}$ of $\Pi|_p$ and let $\{\boldsymbol{\omega}^i\}$ denote the associated cobasis. One has then that $\boldsymbol{g}(\boldsymbol{e}_i, \boldsymbol{e}_j) = -\delta_{ij}$. In addition, let $\{\boldsymbol{\epsilon}_A{}^A\}$ be a spin basis satisfying $\epsilon_1^{+A} = \epsilon_0{}^A$; compare Equation (4.6b). ***Spatial Infeld-van der Waerden symbols*** can be defined from the spacetime Infeld-van der Waerden symbols $\sigma_{\boldsymbol{a}}{}^{\boldsymbol{AA'}}$ and $\sigma^{\boldsymbol{a}}{}_{\boldsymbol{AA'}}$ through the relations

$$\sigma_{\boldsymbol{i}}{}^{\boldsymbol{AB}} \equiv -\varpi^{(\boldsymbol{A}}{}_{\boldsymbol{A'}}\sigma_{\boldsymbol{i}}{}^{\boldsymbol{B})\boldsymbol{A'}}, \qquad \sigma^{\boldsymbol{i}}{}_{\boldsymbol{AB}} \equiv \varpi_{(\boldsymbol{A}}{}^{\boldsymbol{A'}}\sigma^{\boldsymbol{i}}{}_{\boldsymbol{B})\boldsymbol{A'}}.$$

$$h_{\boldsymbol{ABCD}} = -\sigma^{\boldsymbol{i}}{}_{\boldsymbol{AB}}\sigma^{\boldsymbol{j}}{}_{\boldsymbol{CD}}\delta_{\boldsymbol{ij}}, \qquad \sigma_{\boldsymbol{i}}{}^{\boldsymbol{AB}}\sigma^{\boldsymbol{j}}{}_{\boldsymbol{AB}} = \delta_{\boldsymbol{i}}{}^{\boldsymbol{j}}. \tag{4.10}$$

The explicit expressions for the spatial Infeld-van der Waerden symbols are given by

$$\sigma_{\boldsymbol{1}}{}^{\boldsymbol{AB}} = \frac{1}{\sqrt{2}}\begin{pmatrix} -1 & 0 \\ 0 & 1 \end{pmatrix}, \quad \sigma_{\boldsymbol{2}}{}^{\boldsymbol{AB}} = \frac{1}{\sqrt{2}}\begin{pmatrix} -\mathrm{i} & 0 \\ 0 & -\mathrm{i} \end{pmatrix}, \quad \sigma_{\boldsymbol{3}}{}^{\boldsymbol{AB}} = \frac{1}{\sqrt{2}}\begin{pmatrix} 0 & 1 \\ 1 & 0 \end{pmatrix},$$

and

$$\sigma^{\boldsymbol{1}}{}_{\boldsymbol{AB}} = \frac{1}{\sqrt{2}}\begin{pmatrix} -1 & 0 \\ 0 & 1 \end{pmatrix}, \quad \sigma^{\boldsymbol{2}}{}_{\boldsymbol{AB}} = \frac{1}{\sqrt{2}}\begin{pmatrix} \mathrm{i} & 0 \\ 0 & \mathrm{i} \end{pmatrix}, \quad \sigma^{\boldsymbol{3}}{}_{\boldsymbol{AB}} = \frac{1}{\sqrt{2}}\begin{pmatrix} 0 & 1 \\ 1 & 0 \end{pmatrix}.$$

Given the above, the components of a three-dimensional vector $\boldsymbol{v}$ and a three-dimensional covector $\boldsymbol{\xi}$ can be put in correspondence with symmetric valence-2 real spinors via the formulae

$$v^{\boldsymbol{i}} \mapsto v^{\boldsymbol{AB}} = v^{\boldsymbol{i}}\sigma_{\boldsymbol{i}}{}^{\boldsymbol{AB}}, \qquad \xi_{\boldsymbol{i}} \mapsto \xi_{\boldsymbol{AB}} = \xi_{\boldsymbol{i}}\sigma^{\boldsymbol{i}}{}_{\boldsymbol{AB}},$$

or more explicitly

$$(v^1, v^2, v^3) \mapsto \frac{1}{\sqrt{2}}\begin{pmatrix} -v^1 - \mathrm{i}v^2 & v^3 \\ v^3 & v^1 - \mathrm{i}v^2 \end{pmatrix}, \tag{4.11a}$$

$$(\xi_1, \xi_2, \xi_3) \mapsto \frac{1}{\sqrt{2}}\begin{pmatrix} -\xi_1 + \mathrm{i}\xi_2 & \xi_3 \\ \xi_3 & \xi_1 + \mathrm{i}\xi_2 \end{pmatrix}. \tag{4.11b}$$

It can be verified that

$$\langle \boldsymbol{\xi}, \boldsymbol{v} \rangle = \xi_{\boldsymbol{i}}v^{\boldsymbol{i}} = \xi_{\boldsymbol{AB}}v^{\boldsymbol{AB}} = -\xi_1 v^1 - \xi_2 v^2 - \xi_3 v^3.$$

The above correspondence between space spinors and three-dimensional vectors can be readily extended to higher rank tensors. For example, given the components $T_{ij}{}^{k}$ of a tensor $T_{ij}{}^{k}$ one has the correspondence

$$T_{ij}{}^{k} \mapsto T_{ABCD}{}^{EF} \equiv \sigma^{i}{}_{AB}\sigma^{j}{}_{CD}\sigma_{k}{}^{EF}T_{ij}{}^{k}.$$

Finally, it is observed that $\sigma^{i}{}_{AB}$ and $\sigma_{i}{}^{AB}$ can be used to define an alternative set of basis and cobasis $\{e_{AB}\}$, $\{\omega^{AB}\}$ for $\Pi|_p$ through the relations

$$e_{AB} \equiv \sigma^{i}{}_{AB}e_i, \qquad \omega^{AB} \equiv \sigma_{i}{}^{AB}\omega^{i}. \tag{4.12}$$

In terms of these definitions the symbols $\sigma^{i}{}_{AB}$ are the components of the frame $\{e_{AB}\}$ with respect to itself. It can be verified that

$$\langle \omega^{AB}, e_{CD}\rangle = h_{CD}{}^{AB}, \qquad g(e_{AB}, e_{CD}) = h_{ABCD}.$$

4.2.3 Changes of basis and $SU(2,\mathbb{C})$ transformations

To characterise the class of transformations preserving the structure of the Hermitian inner product it is convenient to consider a change of spin basis

$$\tilde{\epsilon}_{A}{}^{A} = O_{A}{}^{P}\epsilon_{P}{}^{A}, \qquad \tilde{\epsilon}^{A}{}_{A} = O^{A}{}_{P}\epsilon^{P}{}_{A},$$

where $(O_{A}{}^{P})$ and $(O^{A}{}_{P})$ are $SL(2,\mathbb{C})$ matrices such that

$$O_{A}{}^{P}O^{B}{}_{P} = \delta_{A}{}^{B}.$$

It follows that

$$\begin{aligned}\tilde{\varpi}_{AA'} &= \varpi_{AA'}\tilde{\epsilon}_{A}{}^{A}\bar{\tilde{\epsilon}}_{A'}{}^{A'}\\ &= \varpi_{AA'}\left(O_{A}{}^{P}\epsilon_{P}{}^{A}\right)\left(\bar{O}_{A'}{}^{P'}\bar{\epsilon}_{P'}{}^{A'}\right) = \varpi_{PP'}O_{A}{}^{P}\bar{O}_{A'}{}^{P'}.\end{aligned}$$

Hence, if one requires $O_{A}{}^{P}$ to be such that both $\varpi_{AA'}$ and $\tilde{\varpi}_{AA'}$ are the identity matrix – compare Equation (4.2) – then $O_{A}{}^{P}$ and $\bar{O}_{A'}{}^{P'}$ have to be inverses of each other; that is, the transformation described by the matrix $(O_{A}{}^{B})$ is an $SU(2,\mathbb{C})$ ***transformation***. This property explains the alternative name of $SU(2,\mathbb{C})$ spinors used to describe spinorial structures endowed with a Hermitian product; see, for example, Ashtekar (1991). It is a direct consequence of the previous discussion that the notions of real and imaginary space spinors as discussed in Section 4.2.1 are *invariant under* $SU(2,\mathbb{C})$ *transformations.*

It can be readily verified that $SU(2,\mathbb{C})$ *transformations are related to three-dimensional rotations*, that is, $O(3)$ ***transformations***. As $SU(2,\mathbb{C})$ transformations are a special case of $SL(2,\mathbb{C})$ transformations, it follows that $\epsilon_{AB} = O_{A}{}^{C}O_{B}{}^{D}\epsilon_{CD}$; that is, $\tilde{\epsilon}_{AB} = \epsilon_{AB}$. From the latter one has that

$$h_{ABCD} = O_{A}{}^{E}O_{B}{}^{F}O_{C}{}^{G}O_{D}{}^{H}h_{EFGH}.$$

Contracting the above expression with spatial Infeld-van der Waerden symbols one obtains

$$\delta_{ij} = O_i{}^k O_j{}^l \delta_{kl},$$

where the matrix $(O_i{}^k)$ with elements given by

$$O_i{}^k \equiv \sigma_i{}^{AB} \sigma^k{}_{EF} O_A{}^E O_B{}^F$$

is an $O(3)$-transformation, that is, a three-dimensional real matrix preserving the identity matrix.

4.2.4 Spinors on three-dimensional manifolds

In this section it is assumed that the distribution Π associated to the Hermitian spinor $\varpi_{AA'}$ has an integral manifold $\mathcal{S}$. It follows that $\mathcal{S}$ is a spacelike hypersurface of the spacetime manifold $\mathcal{M}$, and the restriction of Π to $\mathcal{S}$ coincides with the tangent bundle $T(\mathcal{S})$. Vectors and covectors in $T(\mathcal{S})$ are associated to symmetric valence-2 real spinors.

Under the assumptions of the previous paragraph, consistently with the discussion of Section 4.2.1, one has that the spinor $h_{ABCD} = -\epsilon_{A(C}\epsilon_{D)B}$ is the spinorial counterpart of the three-dimensional (negative definite) Riemannian metric $\boldsymbol{h}$ induced on $\mathcal{S}$ by $\boldsymbol{g}$. One can write

$$\boldsymbol{h} = -\delta_{ij}\boldsymbol{\omega}^i \otimes \boldsymbol{\omega}^j = h_{ABCD}\boldsymbol{\omega}^{AB} \otimes \boldsymbol{\omega}^{CD}$$

with the coframe $\{\boldsymbol{\omega}^{AB}\}$ defined as in Equation (4.12). In a natural manner, the spinor $\varpi_{AA'}$ can be identified with the normal to $\mathcal{S}$.

An alternative point of view

The discussion of spinors on three-dimensional manifolds outlined in the previous paragraphs assumes that $\mathcal{S}$ is a spacelike hypersurface of a spacetime $(\mathcal{M}, \boldsymbol{g})$. A more intrinsic perspective can be obtained by *postulating* the existence of a spinorial structure over $\mathcal{S}$, to be denoted by $\mathfrak{S}(\mathcal{S})$, endowed with an operation of Hermitian conjugation $^+ : \mathfrak{S}(\mathcal{S}) \to \mathfrak{S}(\mathcal{S})$ satisfying the properties discussed in Section 4.1.1. This point of view leads one to consider conditions ensuring the existence of this ***space spinor structure***. An example of a sufficient condition is that the vacuum Einstein constraint equations (see Chapter 11) can be solved on $\mathcal{S}$. If this is the case, the three-dimensional manifold can be regarded as a spacelike hypersurface of a spacetime $(\mathcal{M}, \boldsymbol{g})$; see Chapter 14. The spacetime $(\mathcal{M}, \boldsymbol{g})$ is globally hyperbolic and, thus, admits a spinor structure; see Proposition 3.4. The $\boldsymbol{g}$-normal to $\mathcal{S}$ in $\mathcal{M}$ induces the required operation of Hermitian conjugation.

In what follows, the notational conventions of Section 3.1.4 are adopted, and one writes $\mathfrak{S}^\bullet(\mathcal{S})$, $\mathfrak{S}_A(\mathcal{S})$, $\mathfrak{S}^A(\mathcal{S}), \ldots$ to denote the various bundles associated to $\mathfrak{S}(\mathcal{S})$.

Totally symmetric spinors

Spinors provide a simple representation of the operation of taking the *symmetric trace-free part of a three-dimensional tensor.*

Proposition 4.1 (***space spinor representation of trace-free three-dimensional tensors***) *Let $T_{A_1B_1\cdots A_pB_p}$ denote the spinorial counterpart of a three-dimensional real tensor $T_{i_1\cdots i_p}$. One has that*

$$T_{A_1B_1\cdots A_pB_p} = T_{(A_1B_1\cdots A_pB_p)} \qquad \textit{if and only if} \qquad T_{i_1\cdots i_p} = T_{\{i_1\cdots i_p\}}.$$

Proof Any possible contraction of $T_{(A_1B_1\cdots A_pB_p)}$ with h^{ABCD} must vanish so that $T_{i_1\cdots i_p}$ must be trace free. Conversely, if $T_{i_1\cdots i_p} = T_{\{i_1\cdots i_p\}}$, then one has that

$$h^{A_1B_1A_2B_2}T_{A_1B_1A_2B_2\cdots A_pB_p} = T^{PQ}{}_{PQ\cdots A_pB_p} = 0.$$

Using the decomposition (3.8) together with the symmetries of $T_{A_1B_1A_2B_2\cdots A_pB_p}$ in the indices ${}_{A_1B_1A_2B_2}$ one concludes that

$$T_{A_1B_1A_2B_2\cdots A_pB_p} = T_{(A_1B_1A_2B_2)\cdots A_pB_p}.$$

Now, considering the contraction of the pair ${}_{A_2B_2}$ with pairs outside the symmetrisation bracket and repeating the previous argument as many times as necessary one concludes that $T_{A_1B_1A_2B_2\cdots A_pB_p}$ must be completely symmetric. □

4.2.5 Timelike congruences and Hermitian products

Assume now that the spacetime $(\mathcal{M}, \boldsymbol{g})$ has some *privileged future directed timelike vector* $\boldsymbol{\tau}$ with parameter τ.[1] The vector $\boldsymbol{\tau}$ does not need to be hypersurface orthogonal. Let $\tau^{AA'}$ denote the spinorial counterpart of $\boldsymbol{\tau}$ and consider the normalisation $\boldsymbol{g}(\boldsymbol{\tau},\boldsymbol{\tau}) = 2$. As discussed in Section 2.7.1, $\boldsymbol{\tau}$ defines a distribution on $\mathcal{M}$. Let $\mathcal{S}_\tau$ denote the hyperplanes generated by $\boldsymbol{\tau}$; as $\boldsymbol{\tau}$ is not hypersurface orthogonal, the hyperplanes are not, in general, the tangent bundles to the leaves of a foliation of $\mathcal{M}$. The timelike spinor $\tau_{AA'}$ induces a Hermitian product $\langle\langle \boldsymbol{\xi}, \boldsymbol{\eta}\rangle\rangle = \tau_{AA'}\bar{\xi}^{A'}\eta^A$ for ξ^A, $\eta^A \in \mathfrak{S}(\mathcal{S})$. Indeed, as $\tau^{AA'}$ is the spinorial counterpart of a spacetime vector, it is a Hermitian spinor, so that

$$\overline{\tau_{AA'}\bar{\xi}^{A'}\eta^A} = \tau_{AA'}\bar{\eta}^{A'}\xi^A.$$

Furthermore, given that $\tau^{AA'}$ is timelike future directed and $\xi^A\bar{\xi}^{A'}$ describes a future-directed null vector, it follows that $\tau_{AA'}\xi^A\bar{\xi}^{A'} \geq 0$. Thus, formula (4.1)

[1] It is possible to construct a "space spinor" formalism adapted to spacelike congruences with tangent vector $\rho^{AA'}$; see e.g. Szabados (1994). This requires adapting some of the formulae given in the preceding sections. In particular, the associated Hermitian product needs to be negative definite. Moreover, the analogue of Equation (4.13) is given by $\rho_{AA'}\rho^{BA'} = \epsilon^B{}_A$ so that $\rho_{AA'}\rho^{AA'} = -2$.

and the various subsequent expressions in Section 4.1 can be used for the choice $\varpi_{AA'} = \tau_{AA'}$.

From the discussion in Section 4.1.1 it follows that there exists a spin basis $\{\epsilon_{\boldsymbol{A}}{}^{A}\}$ such that

$$\tau^{AA'} = \epsilon_{\boldsymbol{0}}{}^{A}\epsilon_{\boldsymbol{0'}}{}^{A'} + \epsilon_{\boldsymbol{1}}{}^{A}\epsilon_{\boldsymbol{1'}}{}^{A'}.$$

In particular, one has that

$$\tau_{AA'}\tau^{BA'} = \epsilon_{A}{}^{B}. \tag{4.13}$$

Space spinor split of general spacetime spinors

The tensorial counterpart of a spinor $\mu_{A_1A'_1\cdots A_kA'_k}$ can be expanded in terms of the spatial frame $\{\boldsymbol{e_{AB}}\}$ if and only if it is *spatial* with respect to $\boldsymbol{\tau}$, that is, if the k conditions

$$\tau^{A_1A'_1}\mu_{A_1A'_1\cdots A_kA'_k} = 0, \qquad \cdots \qquad \tau^{A_kA'_k}\mu_{A_1A'_1\cdots A_kA'_k} = 0$$

hold. In this case, its space spinor counterpart is given by

$$\mu_{A_1B_1\cdots A_kB_k} = \tau_{B_1}{}^{A'_1}\cdots\tau_{B_k}{}^{A'_k}\mu_{A_1A'_1\cdots A_kA'_k} = \mu_{(A_1B_1)\cdots(A_kB_k)}.$$

To deal with the spinorial counterparts of tensors which are not spatial in the sense described above, one makes use of the ***projector***

$$h^{BB'}{}_{AA'} \equiv \epsilon_{A}{}^{B}\epsilon_{A'}{}^{B'} - \frac{1}{2}\tau_{AA'}\tau^{BB'},$$

which takes a spinor $\xi_{A_1A'_1\cdots A_kA'_k}$ onto the *spatial spinor*

$$\xi_{A_1A'_1\cdots A_kA'_k}h^{A_1A'_1}{}_{B_1B'_1}\cdots h^{A_kA'_k}{}_{B_kB'_k}.$$

The space spinor version of the above spatial spinor is obtained by contracting the primed indices with $\tau_{A}{}^{A'}$ as in formula (4.7). In particular, this procedure applied to the projector $h_{AA'BB'}$ yields h_{ABCD}. The non-spatial components of $\xi_{A_1A'_1\cdots A_kA'_k}$ can be obtained by a full contraction of a primed-unprimed pair of indices with $\tau^{AA'}$.

An alternative way of looking at the projection procedure described in the previous paragraphs is the following: given the spinorial counterpart $\xi_{A_1A'_1\cdots A_1A'_k}$ of a (in principle non-spatial) tensor, define

$$\xi_{A_1B_1\cdots A_kB_k} \equiv \tau_{B_1}{}^{A'_1}\cdots\tau_{B_k}{}^{A'_k}\xi_{A_1A'_1\cdots A_1A'_k}.$$

Then $\xi_{(A_1B_1)\cdots(A_kB_k)}$ encodes the spatial part of $\xi_{A_1A'_1\cdots A_1A'_k}$, while $\xi_{P_1}{}^{P_1}\cdots{}_{P_k}{}^{P_k}$ corresponds to its pure time component. Mixed time-spatial components have the form $\xi_{P}{}^{P}{}_{(A_2B_2)\cdots(A_kB_k)}$, and so on.

As a particular example of the previous discussion one has that for a Hermitian spinor $v_{AA'} \in \mathfrak{S}^\bullet(\mathcal{M})$ it holds that

$$v_{AA'} = \frac{1}{2}\tau_{AA'}v - \tau^Q{}_{A'}v_{(QA)}$$

where $v \equiv v_{PP'}\tau^{PP'}$ and $v_{AB} \equiv \tau_B{}^{A'}v_{AA'}$. Observing that $v = v_Q{}^Q$ one can write, alternatively, that

$$v_{AB} = \frac{1}{2}\epsilon_{AB}v + v_{(AB)}.$$

The $1+3$ *split of frame and the metric*

Given a $\boldsymbol{g}$-orthonormal frame $\{\boldsymbol{e}_{AA'}\}$ and its coframe $\{\boldsymbol{\omega}^{AA'}\}$, the discussion of the previous paragraphs implies that they can be written as

$$\boldsymbol{e}_{AA'} = \frac{1}{2}\tau_{AA'}\boldsymbol{e} - \tau^{\boldsymbol{B}}{}_{A'}\boldsymbol{e}_{AB},$$

$$\boldsymbol{\omega}^{AA'} = \frac{1}{2}\tau^{AA'}\boldsymbol{\omega} + \tau_C{}^{A'}\boldsymbol{\omega}^{CA},$$

where the various vectors and covectors in the decomposition are given by

$$\boldsymbol{e} \equiv \tau^{PP'}\boldsymbol{e}_{PP'}, \qquad \boldsymbol{e}_{AB} \equiv \tau_{(A}{}^{P'}\boldsymbol{e}_{B)P'},$$
$$\boldsymbol{\omega} \equiv \tau_{PP'}\boldsymbol{\omega}^{PP'}, \qquad \boldsymbol{\omega}^{AB} \equiv -\tau^{(A}{}_{P'}\boldsymbol{\omega}^{B)P'}.$$

Now, recalling that $\langle \boldsymbol{\omega}^{AA'}, \boldsymbol{e}_{BB'}\rangle = \epsilon_B{}^A\epsilon_{B'}{}^{A'}$, one obtains that

$$\langle \boldsymbol{\omega}, \boldsymbol{e}\rangle = 2, \qquad \langle \boldsymbol{\omega}, \boldsymbol{e}_{AB}\rangle = 0,$$
$$\langle \boldsymbol{\omega}^{AB}, \boldsymbol{e}\rangle = 0, \qquad \langle \boldsymbol{\omega}^{AB}, \boldsymbol{e}_{CD}\rangle = h^{AB}{}_{CD}.$$

Using the above pairings together with expression (3.29) one obtains the following $1+3$ *split of* $\boldsymbol{g}$:

$$\boldsymbol{g} = \frac{1}{2}\boldsymbol{\omega}\otimes\boldsymbol{\omega} + h_{ABCD}\boldsymbol{\omega}^{AB}\otimes\boldsymbol{\omega}^{CD}. \tag{4.14}$$

In particular, one has that

$$h_{ABCD} \equiv \boldsymbol{g}(\boldsymbol{e}_{AB}, \boldsymbol{e}_{CD}) = -\epsilon_{A(C}\epsilon_{D)B}.$$

If $\boldsymbol{\tau}$ is hypersurface orthogonal, then the vectors $\{\boldsymbol{e}_{AB}\}$ and the covectors $\{\boldsymbol{\omega}^{AB}\}$ are intrinsic to the hypersurfaces $\mathcal{S}_{\boldsymbol{\tau}}$ orthogonal to $\boldsymbol{\tau}$; thus, they can be regarded as belonging to $T(\mathcal{S}_{\boldsymbol{\tau}})$ and $T^*(\mathcal{S}_{\boldsymbol{\tau}})$, respectively. In addition,

$$\boldsymbol{h} \equiv h_{ABCD}\boldsymbol{\omega}^{AB}\otimes\boldsymbol{\omega}^{CD}$$

corresponds to the (negative definite, Riemannian) metric induced by $\boldsymbol{g}$ on $\mathcal{S}_{\boldsymbol{\tau}}$. Let ϵ_{ijk} denote the volume form of the three-dimensional metric $\boldsymbol{h}$, and let

ϵ_{ABCDEF} be its spinorial counterpart. Using the antisymmetry properties of ϵ_{ABCDEF} it can be expressed in terms of ϵ_{AB} as

$$\epsilon_{ABCDEF} = \frac{\mathrm{i}}{\sqrt{2}}(\epsilon_{AC}\epsilon_{BE}\epsilon_{DF} + \epsilon_{BD}\epsilon_{AF}\epsilon_{CE}). \tag{4.15}$$

Furthermore, it can be checked that

$$\epsilon_{ABCDEF}\epsilon^{ABCDEF} = -6.$$

Alternatively, Equation (4.15) can be obtained from Equation (3.25), by suitable contactions with $\tau^{AA'}$. More precisely, one has that

$$\epsilon_{CDEFGH} = \frac{1}{\sqrt{2}}\tau^{AA'}\tau_D{}^{C'}\tau_F{}^{E'}\tau_H{}^{G'}\epsilon_{AA'CC'EE'GG'}.$$

4.3 Calculus of space spinors

This section discusses the notion of covariant derivative in the context of the space spinor formalism. For simplicity of the presentation, it is assumed that one has a situation as described in Section 4.2.5 where the spinor $\varpi_{AA'}$ is given by the spinorial counterpart $\tau_{AA'}$ of the tangent vector $\boldsymbol{\tau}$ to a timelike congruence in $(\mathcal{M}, \boldsymbol{g})$. Moreover, it is also assumed that $\nabla_{AA'}$ is the spinorial counterpart of the Levi-Civita connection of the metric $\boldsymbol{g}$.

4.3.1 The Sen connection

The spinor $\tau^{AA'}$ can be used to obtain a *space spinor version* of the spacetime spinorial covariant derivative $\nabla_{AA'}$. More precisely, one can define

$$\nabla_{AB} \equiv \tau_B{}^{A'}\nabla_{AA'}.$$

The latter, in turn, can be written in terms of its irreducible components as

$$\nabla_{AB} = \frac{1}{2}\epsilon_{AB}\mathcal{P} + \mathcal{D}_{AB}, \tag{4.16}$$

where

$$\mathcal{P} \equiv \tau^{AA'}\nabla_{AA'}, \qquad \mathcal{D}_{AB} \equiv \tau_{(B}{}^{A'}\nabla_{A)A'}.$$

The operator $\mathcal{P}$ is the ***directional derivative*** of the connection $\boldsymbol{\nabla}$ in the direction of $\boldsymbol{\tau}$. The differential operator $\mathcal{D}_{AB}$ is the so-called ***Sen connection*** of $\boldsymbol{\nabla}$ relative to the vector field $\boldsymbol{\tau}$. In view of these definitions one can also write

$$\nabla_{AA'} = \frac{1}{2}\tau_{AA'}\mathcal{P} - \tau_{A'}{}^Q\mathcal{D}_{AQ}.$$

The timelike vector $\boldsymbol{\tau}$ is completely arbitrary; in particular, *it is not assumed to be hypersurface orthogonal.* This has several consequences; most notably, the Sen connection has, in general, a non-vanishing torsion which can be expressed in terms of the covariant derivative of $\tau^{AA'}$. Furthermore, even in the case when $\boldsymbol{\tau}$ is hypersurface orthogonal, $\mathcal{D}_{AB}$ does not coincide with the Levi-Civita connection $\boldsymbol{D}$ of the intrinsic 3-metric of the hypersurfaces $\mathcal{S}_{\boldsymbol{\tau}}$ orthogonal to $\boldsymbol{\tau}$. Finally, it is pointed out that $\mathcal{D}_{AB}$ *is not a real differential operator* in the sense that $\mathcal{D}^{+}_{AB} \neq -\mathcal{D}_{AB}$.

The derivative of $\tau^{AA'}$

For future use, it is convenient to define

$$\chi_{ABCD} \equiv \frac{1}{\sqrt{2}} \tau_D{}^{C'} \nabla_{AB} \tau_{CC'}. \tag{4.17}$$

Using the split (4.16) one obtains the decomposition

$$\chi_{ABCD} = \frac{1}{2}\epsilon_{AB}\chi_{CD} + \chi_{(AB)CD}$$

where

$$\chi_{AB} \equiv \frac{1}{\sqrt{2}} \tau_B{}^{A'} \mathcal{P} \tau_{AA'}, \qquad \chi_{(AB)CD} \equiv \frac{1}{\sqrt{2}} \tau_D{}^{C'} \mathcal{D}_{AB} \tau_{CC'}.$$

It can be verified that the above spinors satisfy the following symmetry and reality properties:

$$\chi_{AB} = \chi_{(AB)} = -\chi^{+}_{AB}, \qquad \chi_{(AB)CD} = \chi_{(AB)(CD)} = \chi^{+}_{(AB)CD}.$$

The spinor χ_{AB} corresponds to the *acceleration vector* of $\boldsymbol{\tau}$, while $\chi_{(AB)CD}$ is related to the *Weingarten tensor* of the distribution defined by $\boldsymbol{\tau}$. It can be checked that the distribution is integrable if and only if $\chi^{Q}{}_{(BC)Q} = 0$. In this case $\boldsymbol{\tau}$ is hypersurface orthogonal, and χ_{ABCD} corresponds to the extrinsic curvature of the orthogonal hypersurfaces $\mathcal{S}_{\boldsymbol{\tau}}$.

The hypersurface orthogonal case

If $\boldsymbol{\tau}$ is hypersurface orthogonal, given a spinor μ_C, the covariant derivative D_{AB} defined by

$$D_{AB}\mu_C \equiv \mathcal{D}_{AB}\mu_C + \frac{1}{\sqrt{2}} \chi_{(AB)C}{}^{Q} \mu_Q \tag{4.18}$$

can be verified to be torsion-free. As $\mathcal{D}_{AB}\epsilon_{CD} = 0$ and using that $\chi_{ABCD} = \chi_{AB(CD)}$, one concludes that $D_{AB}\epsilon_{CD} = 0$. Thus, D_{AB} *is metric and must coincide with the (spinorial counterpart of the) Levi-Civita connection of the leaves of the foliation defined by* $\boldsymbol{\tau}$. It can be further verified from Equation (4.18) that

$$\left(D_{AB}\mu_C\right)^{+} = -D_{AB}\mu^{+}_C,$$

so that D_{AB} is a *real* differential operator in the sense of Section 4.2.1.

Remark. The derivative D_{AB} as defined in Equation (4.18) provides an explicit example of the notion of space spinor covariant derivative to be introduced in Section 4.3.3.

4.3.2 Space spinor split of the spacetime connection coefficients

Following the notation of Section 3.2.2, let $\Gamma_{AA'CD}$ denote the spin connection coefficients of a Levi-Civita connection $\nabla_{AA'}$ with respect to some spin basis $\{\epsilon_A{}^A\}$. Its space spinor counterpart Γ_{ABCD} is defined by

$$\Gamma_{ABCD} \equiv \tau_B{}^{A'}\Gamma_{AA'CD}.$$

The spin coefficients Γ_{ABCD} satisfy no specific reality conditions. However, sometimes it is convenient to have a split of Γ_{ABCD} into real and imaginary parts. One has that

$$\begin{aligned}\tau_B{}^{A'}(\nabla_{AA'}\tau_{CC'})\tau_D{}^{C'} &= -\tau_B{}^{A'}\Gamma_{AA'}{}^Q{}_C\tau_{QC'}\tau_D{}^{C'} - \tau_B{}^{A'}\bar{\Gamma}_{A'A}{}^{Q'}{}_{C'}\tau_{CQ'}\tau_D{}^{C'}\\ &= -\Gamma_{ABCD} + \tau_B{}^{A'}\tau_C{}^{Q'}\tau_{A'}{}^E\tau_{Q'}{}^F\tau_{C'}{}^G\Gamma^+_{EAFG}\tau_D{}^{C'}\\ &= -\Gamma_{ABCD} - \delta_B{}^E\delta_C{}^F\delta_D{}^G\Gamma^+_{EAFG}\\ &= -\Gamma_{ABCD} - \Gamma^+_{BACD},\end{aligned}$$

where it has been used that $e_{AA'}(\tau_{CC'}) = 0$, the identity

$$\bar{\Gamma}_{A'AB'C'} = -\tau_{A'}{}^E\tau_{B'}{}^F\tau_{C'}{}^G\Gamma^+_{EAFG},$$

and the identity (4.13). Hence, it follows that χ_{ABCD} corresponds, essentially, to the *real part* of Γ_{ABCD}; that is,

$$\chi_{ABCD} = -\frac{1}{\sqrt{2}}(\Gamma_{ABCD} + \Gamma^+_{ABCD}).$$

The reality of the above expression follows from $\Gamma^{++}_{ABCD} = \Gamma_{ABCD}$. The *imaginary part* of Γ_{ABCD} is given by

$$\xi_{ABCD} = \frac{1}{\sqrt{2}}(\Gamma_{ABCD} - \Gamma^+_{ABCD}).$$

Inverting the definitions of χ_{ABCD} and ξ_{ABCD} it follows then that

$$\begin{aligned}\Gamma_{ABCD} &= \frac{1}{\sqrt{2}}(\xi_{ABCD} - \chi_{ABCD}),\\ &= \frac{1}{\sqrt{2}}(\xi_{ABCD} - \chi_{(AB)CD}) - \frac{1}{2\sqrt{2}}\epsilon_{AB}\chi_{CD}.\end{aligned}$$

Observe the symmetry conditions

$$\chi_{ABCD} = \chi_{AB(CD)}, \qquad \xi_{ABCD} = \xi_{(AB)(CD)}.$$

4.3.3 Intrinsic derivatives

When working with a three-dimensional Riemannian $(\mathcal{S}, \boldsymbol{h})$ it is convenient to make use of an intrinsic notion of covariant derivative, a so-called ***space spinor covariant derivative*** D_{AB}, compatible with operation of Hermitian conjugation, which is the spinorial counterpart of the Levi-Civita connection $\boldsymbol{D}$ of the metric $\boldsymbol{h}$. One regards D_{AB} as a map

$$D_{AB} : \mathfrak{S}^{C\cdots D}{}_{E\cdots F}(\mathcal{S}) \to \mathfrak{S}^{C\cdots D}{}_{ABE\cdots F}(\mathcal{S}).$$

The properties of the operator D_{AB} have to be consistent with those of the operator defined in Equation (4.18). It is required to satisfy:

(i) ***Symmetry.*** Given $\zeta^{C\cdots D}{}_{E\cdots F} \in \mathfrak{S}^{\bullet}(\mathcal{S})$ one has

$$D_{AB}\zeta^{C\cdots D}{}_{E\cdots F} = D_{(AB)}\zeta^{C\cdots D}{}_{E\cdots F}.$$

(ii) ***Linearity.*** Given $\zeta^{C\cdots D}{}_{E\cdots F}$, $\eta^{C\cdots D}{}_{E\cdots F} \in \mathfrak{S}^{\bullet}(\mathcal{S})$ one has

$$D_{AB}(\zeta^{C\cdots D}{}_{E\cdots F} + \eta^{C\cdots D}{}_{E\cdots F}) = D_{AB}\zeta^{C\cdots D}{}_{E\cdots F} + D_{AB}\eta^{C\cdots D}{}_{E\cdots F}.$$

(iii) ***Leibnitz rule.*** Given $\zeta^{C\cdots D}{}_{E\cdots F}$, $\xi^{G\cdots H}{}_{P\cdots Q} \in \mathfrak{S}^{\bullet}(\mathcal{S})$ one has

$$\begin{aligned} D_{AB}(\zeta^{C\cdots D}{}_{E\cdots F}\xi^{G\cdots H}{}_{P\cdots Q}) = {} & \xi^{G\cdots H}{}_{P\cdots Q} D_{AB}\zeta^{C\cdots D}{}_{E\cdots F} \\ & + \zeta^{C\cdots D}{}_{E\cdots F} D_{AB}\zeta^{G\cdots H}{}_{P\cdots Q}. \end{aligned}$$

(iv) ***Reality.*** Given $\zeta^{C\cdots D}{}_{E\cdots F} \in \mathfrak{S}^{\bullet}(\mathcal{S})$ one has

$$(D_{AB}\zeta^{C\cdots D}{}_{E\cdots F})^{+} = -D_{AB}\zeta^{+C\cdots D}{}_{E\cdots F}$$

(v) ***Action on scalars.*** Given a scalar $\phi \in \mathfrak{X}(\mathcal{S})$, then $D_{AB}\phi$ is the spinorial counterpart of $D_i\phi$.

(vi) ***Representation of derivations.*** Given a derivation $\mathcal{D}$, there exists a spinor $\xi^{AB} \in \mathfrak{S}^{\bullet}(\mathcal{S})$ such that

$$\mathcal{D}\zeta^{C\cdots D}{}_{E\cdots F} = \xi^{PQ} D_{PQ}\zeta^{C\cdots D}{}_{E\cdots F}$$

for all $\zeta^{C\cdots D}{}_{E\cdots F} \in \mathfrak{S}^{C\cdots D}{}_{E\cdots F}$.

(vii) ***Compatibility with the ϵ-spinor.*** The operator D_{AB} satisfies $D_{AB}\epsilon_{CD} = 0$ so that, in addition, $D_{AB}h_{CDEF} = 0$.

(viii) ***No torsion.*** For $\phi \in \mathfrak{X}(\mathcal{S})$ one has that $D_{AB}D_{CD}\phi = D_{CD}D_{AB}\phi$.

The space spinor spin coefficients

Let $\{e_i\}$ and $\{\omega^i\}$ denote, respectively, an $\boldsymbol{h}$-orthonormal basis and cobasis on $\mathcal{S}$. One defines the ***spatial connection coefficients*** $\gamma_i{}^k{}_j$ via the equation

$$D_i e_j = \gamma_i{}^k{}_j e_k. \tag{4.19}$$

In what follows, it will be assumed that the connection $\boldsymbol{D}$ has spinorial counterpart D_{AB} satisfying conditions (i)–(viii) of the previous section. Let $\{e_{AB}\}$ and $\{\omega^{AB}\}$ denote, respectively, the vector basis and cobasis obtained from $\{e_i\}$ and $\{\omega^i\}$ through the correspondences in (4.12) and let D_{AB} denote the associated covariant directional derivative.

The spinorial counterpart of the spatial connection coefficients $\gamma_{AB}{}^{CD}{}_{EF}$ can be obtained by contraction of $\gamma_i{}^k{}_j$ with the spatial Infeld-van der Waerden symbols so that the reality condition

$$\gamma^{+}_{AB}{}^{CD}{}_{EF} = -\gamma_{AB}{}^{CD}{}_{EF} \tag{4.20}$$

holds. Now, defining the ***space spinor directional covariant derivative*** $D_{AB} \equiv \sigma^i{}_{AB} D_i$, the spinorial counterpart of (4.19) can be written as

$$D_{AB} e_{EF} = \gamma_{AB}{}^{CD}{}_{EF} e_{CD}.$$

Hence, one has

$$\gamma_{AB}{}^{CD}{}_{EF} = \langle \omega^{CD}, D_{AB} e_{EF} \rangle,$$

so that $\gamma_{AB}{}^{CD}{}_{EF}$ has the symmetries

$$\gamma_{AB}{}^{CD}{}_{EF} = \gamma_{(AB)}{}^{(CD)}{}_{(EF)}.$$

Now, as $D_{AB} h_{CDEF} = 0$, it follows then from

$$D_{AB} h_{CDEF} = e_{AB}(h_{CDEF}) - \gamma_{AB}{}^{PQ}{}_{CD} h_{PQEF} - \gamma_{AB}{}^{PQ}{}_{EF} h_{CDPQ},$$

that

$$\gamma_{ABCDEF} = -\gamma_{ABEFCD}.$$

This antisymmetry can be exploited to obtain the decomposition

$$\begin{aligned}\gamma_{AB}{}^{CD}{}_{EF} &= \frac{1}{2}\gamma_{AB}{}^{PD}{}_{PF}\delta_E{}^C + \frac{1}{2}\gamma_{AB}{}^{CP}{}_{EP}\delta_F{}^D \\ &= \gamma_{AB}{}^D{}_F \delta_E{}^C + \gamma_{AB}{}^C{}_E \delta_F{}^D,\end{aligned}$$

where the ***space spinor spin coefficients***, $\gamma_{AB}{}^D{}_F = \gamma_{(AB)}{}^D{}_F$, have been defined by

$$\gamma_{AB}{}^D{}_F \equiv \frac{1}{2}\gamma_{AB}{}^{PD}{}_{PF}. \tag{4.21}$$

Observing the reality condition (4.20) and that $\epsilon^{+}_{AB} = \epsilon_{AB}$, it follows that

$$\gamma^{+}_{AB}{}^C{}_D = -\gamma_{AB}{}^C{}_D;$$

that is, the space spinor connection coefficients are imaginary. A computation similar to the one performed in Section 3.2.2 to express the spacetime spin coefficients in terms of derivatives of the spin basis shows that

$$\gamma_{\boldsymbol{AB}}{}^{\boldsymbol{C}}{}_{\boldsymbol{D}} = \epsilon^{\boldsymbol{C}}{}_{Q} D_{\boldsymbol{AB}} \epsilon_{\boldsymbol{D}}{}^{Q} = -\epsilon_{\boldsymbol{D}}{}^{Q} D_{\boldsymbol{AB}} \epsilon^{\boldsymbol{C}}{}_{Q}.$$

From these expressions, it can be shown that given spinors κ_A and μ^A with components $\kappa_{\boldsymbol{A}}$ and $\mu^{\boldsymbol{A}}$ with respect to the space spinor basis $\{\epsilon_{\boldsymbol{A}}{}^{A}\}$, one has

$$\begin{aligned} D_{\boldsymbol{AB}} \kappa_{\boldsymbol{C}} &= e_{\boldsymbol{AB}}(\kappa_{\boldsymbol{C}}) - \gamma_{\boldsymbol{AB}}{}^{\boldsymbol{Q}}{}_{\boldsymbol{C}} \kappa_{\boldsymbol{Q}}, \\ D_{\boldsymbol{AB}} \mu^{\boldsymbol{C}} &= e_{\boldsymbol{AB}}(\mu^{\boldsymbol{C}}) + \gamma_{\boldsymbol{AB}}{}^{\boldsymbol{C}}{}_{\boldsymbol{Q}} \mu^{\boldsymbol{Q}}, \end{aligned}$$

where $D_{\boldsymbol{AB}} \kappa_{\boldsymbol{C}} \equiv \epsilon_{\boldsymbol{C}}{}^{Q} D_{\boldsymbol{AB}} \kappa_{Q}$ and $D_{\boldsymbol{AB}} \mu^{\boldsymbol{C}} \equiv \epsilon^{\boldsymbol{C}}{}_{Q} D_{\boldsymbol{AB}} \mu^{Q}$.

The three-dimensional curvature spinors

As $\boldsymbol{D}$ is being assumed to be the Levi-Civita connection of a three-dimensional negative definite metric $\boldsymbol{h}$, it follows that the spinorial counterpart $r_{ABCDEFGH}$ of the Riemann tensor r_{ijkl} of $\boldsymbol{D}$ satisfies the antisymmetry property

$$r_{ABCDEFGH} = -r_{CDABEFGH}.$$

Hence, one has the decomposition

$$r_{ABCDEFGH} = -r_{ACEFGH}\epsilon_{BD} - r_{BDEFGH}\epsilon_{AC}, \tag{4.22}$$

with

$$r_{ACEFGH} \equiv \frac{1}{2} r_{AQC}{}^{Q}{}_{EFGH}, \qquad r_{ACEFGH} = r_{(AC)EFGH}.$$

Now, as $r_{ABCDEF} = -r_{ABEFCD}$ one has further that

$$r_{ABCDEF} = r_{ABCE}\epsilon_{DF} + r_{ABDF}\epsilon_{CE},$$

with

$$r_{ABCE} = \frac{1}{2} r_{ABCQE}{}^{Q}, \qquad r_{ABCE} = r_{AB(CE)}.$$

As a consequence of the symmetry $r_{ABCDEFGH} = r_{EFGHABCD}$, the spinor r_{ABCD} inherits the symmetry $r_{ABCD} = r_{CDAB}$. Taking into account all these symmetries in the general decomposition for a general valence-4 spinor, Equation (3.8), one concludes that

$$r_{ABCD} = r_{(ABCD)} + \frac{1}{3} r_{PQ}{}^{PQ} h_{ABCD}.$$

In what follows let s_{ABCD} and r denote, respectively, the ***spinorial counterpart of the trace-free Ricci tensor*** and the ***Ricci scalar*** of the connection $\boldsymbol{D}$. One has that

$$s_{ABCD} = r_{(ABCD)}, \qquad r = -4 r_{PQ}{}^{PQ}.$$

Hence, one finds that r_{ABCDEF} can be written as

$$r_{ABCDEF} = \left(\frac{1}{2}s_{ABCE} - \frac{1}{12}rh_{ABCE}\right)\epsilon_{DF} + \left(\frac{1}{2}s_{ABDF} - \frac{1}{12}rh_{ABDF}\right)\epsilon_{CE}. \tag{4.23}$$

Using an argument similar to the one employed in Section 3.2.3 one finds that the commutator of the covariant derivative D_{AB} satisfies

$$(D_{AB}D_{CD} - D_{CD}D_{AB})\mu^E = r^E{}_{FABCD}\mu^F.$$

Finally, for completeness it is noticed that the three-dimensional second Bianchi identity takes, in the present context, the form

$$D^{PQ}s_{PQAB} = \frac{1}{6}D_{AB}r.$$

This last expression can be obtained from multiplying by ϵ^{ijk} the tensorial Bianchi identity

$$D_i r_{jklm} + D_j r_{kilm} + D_k r_{ijlm} = 0,$$

and considering its spinorial counterpart using Equations (4.22) and (4.23).

4.4 Further reading

The notions of space spinor and space spinor splits were originally introduced in Sommers (1980); see also Sen (1981). A monograph on space spinors is Torres del Castillo (2003). An alternative discussion, having applications in quantum gravity in mind, is given in an appendix of Ashtekar (1991). The space spinor formalism was first used in Friedrich (1988, 1991) to analyse the conformal field equations. Further developments can be found in Friedrich (1995, 1998c), and a slightly different perspective on these ideas is given in Frauendiener (1998a).

5
Conformal geometry

Conformal geometry is concerned with the properties of *angle-preserving* geometric transformations. Conformal geometry, as a branch of differential geometry, has a long story going back to the work of Cotton, Schouten and Weyl; see, for example, Cotton (1899), Schouten (1921) and Weyl (1918, 1968). It remains an active area of research; compare the monograph by Fefferman and Graham (2012).

The approach to the use of conformal methods in general relativity followed in this book goes back to the seminal work by R. Penrose in the 1960s; see Penrose (1963, 1964). Penrose's ideas allowed to reformulate, in a geometric manner, the study of the asymptotic behaviour of the gravitational field. Since then, conformal methods have provided a valuable tool for the analysis of global aspects of the Einstein field equations and their solutions. Conformal methods have also been useful in the construction of exact solutions to the Einstein field equations; see Stephani et al. (2003).

This chapter provides an introduction to the notions of conformal geometry to be used in the later parts of this book. The organisation of this chapter is geared towards applications.

5.1 Basic concepts of conformal geometry

This section discusses the basic notions of conformal geometry that will be used throughout this book.

5.1.1 Conformal rescalings and transformations

The key notion in conformal geometry is that of a ***conformal rescaling***. In what follows, let $\tilde{\boldsymbol{g}}$ and $\boldsymbol{g}$ denote two metrics on a manifold $\tilde{\mathcal{M}}$. The metrics

$\tilde{\boldsymbol{g}}$ and $\boldsymbol{g}$ are said to be ***conformally related*** (or simply ***conformal***) to each other if there exists a positive $\Xi \in \mathfrak{X}(\tilde{\mathcal{M}})$ such that

$$\boldsymbol{g} = \Xi^2 \tilde{\boldsymbol{g}}. \tag{5.1}$$

The scalar Ξ is called the ***conformal factor***. Throughout this book, the symbol Ξ will be used to denote a generic conformal factor on a four-dimensional manifold.

The conformal rescaling in Equation (5.1) gives rise to an equivalence relation among the set of metrics over $\tilde{\mathcal{M}}$. The ***conformal class*** of a metric $\tilde{\boldsymbol{g}}$, to be denoted by $[\tilde{\boldsymbol{g}}]$, is the collection of metrics conformally related to $\tilde{\boldsymbol{g}}$. A conformal class is also called a ***conformal structure***. From Equation (5.1) it follows that the contravariant metrics $\tilde{\boldsymbol{g}}^\sharp$ and $\boldsymbol{g}^\sharp$ are related by

$$\boldsymbol{g}^\sharp = \Xi^{-2} \tilde{\boldsymbol{g}}^\sharp;$$

that is, $g^{ab} = \Xi^{-2} \tilde{g}^{ab}$, so as to ensure that $\tilde{g}_{ab}\tilde{g}^{bc} = \delta_a{}^c$ and $g_{ab}g^{bc} = \delta_a{}^c$.

Closely related to the notion of conformally related metrics is the concept of conformal transformations. To discuss this idea, let $\tilde{\mathcal{M}}$ and $\mathcal{M}$ denote two manifolds with metrics $\tilde{\boldsymbol{g}}$ and $\boldsymbol{g}$, respectively. A ***conformal transformation*** (also called ***conformorphism***) is a diffeomorphism $\varphi : \tilde{\mathcal{M}} \rightarrow \mathcal{M}$ such that the pull-back of $\boldsymbol{g}$ is conformal to $\tilde{\boldsymbol{g}}$. That is, one has that

$$\varphi^* \boldsymbol{g} = \Xi^2 \tilde{\boldsymbol{g}}. \tag{5.2}$$

Notice that as φ is a diffeomorphism, then $(\varphi^*)^{-1}$ is well defined and the last expression could have been written, alternatively, as $\boldsymbol{g} = (\varphi^*)^{-1} \left(\Xi^2 \tilde{\boldsymbol{g}}\right)$.

A special case of the previous discussion occurs when $\tilde{\boldsymbol{g}}$ is a flat metric – in the Lorentzian four-dimensional case the Minkowski metric $\boldsymbol{\eta}$ and in the Riemannian three-dimensional case the Euclidean metric $\boldsymbol{\delta}$. In these cases one then says that $\boldsymbol{g}$ is ***conformally flat***. Determining whether a given conformal class contains the flat metric is a classical problem in conformal geometry; see Section 5.2.3.

The conformal group

As before, let $[\tilde{\boldsymbol{g}}]$ denote the conformal class of a Lorentzian metric $\tilde{\boldsymbol{g}}$ on a manifold $\tilde{\mathcal{M}}$. Consider a frame $\{\tilde{\boldsymbol{e}}_{\boldsymbol{a}}\}$ which is orthonormal with respect to $\tilde{\boldsymbol{g}}$. If $\{\tilde{\boldsymbol{\omega}}^{\boldsymbol{a}}\}$ denotes the associated coframe, one has that

$$\tilde{\boldsymbol{g}} = \eta_{\boldsymbol{ab}} \tilde{\boldsymbol{\omega}}^{\boldsymbol{a}} \otimes \tilde{\boldsymbol{\omega}}^{\boldsymbol{b}}, \qquad \text{that is,} \qquad \tilde{\boldsymbol{g}}(\tilde{\boldsymbol{e}}_{\boldsymbol{a}}, \tilde{\boldsymbol{e}}_{\boldsymbol{a}}) = \eta_{\boldsymbol{ab}}.$$

In order to investigate the type of transformations of $\{\tilde{\boldsymbol{\omega}}_{\boldsymbol{a}}\}$ which lead to another metric $\boldsymbol{g} \in [\tilde{\boldsymbol{g}}]$, write

$$\tilde{\boldsymbol{\omega}}^{\boldsymbol{a}} = K^{\boldsymbol{a}}{}_{\boldsymbol{c}} \boldsymbol{\omega}^{\boldsymbol{c}},$$

with $(K^a{}_c)$ denoting some transformation matrix and $\{\boldsymbol{\omega}_a\}$ another orthonormal frame. The condition on the matrix $(K^a{}_c)$ so that it leads to another member of the conformal class, say, $\boldsymbol{g} = \Xi^2\tilde{\boldsymbol{g}}$, is then given by

$$\tilde{\boldsymbol{g}} = \eta_{ab}K^a{}_c K^b{}_d \boldsymbol{\omega}^c \otimes \boldsymbol{\omega}^d = \Xi^{-2}\eta_{cd}\boldsymbol{\omega}^c \otimes \boldsymbol{\omega}^d.$$

The latter expression suggests writing

$$K^a{}_b = \Xi^{-1}\Lambda^a{}_b,$$

where $(\Lambda^a{}_b)$ is a Lorentz transformation; that is, $\Lambda^a{}_c\Lambda^b{}_d\eta_{ab} = \eta_{cd}$. The ***group of (four-dimensional) Lorentz transformations*** will be denoted by $O(1,3)$. It follows that at a point $p \in \tilde{\mathcal{M}}$ the group of transformations taking a $\tilde{\boldsymbol{g}}$-orthonormal frame to a frame which is orthonormal with respect to another metric in the conformal class $[\boldsymbol{g}]$, the so-called ***conformal group*** $CO(1,3)$, is given by $CO(1,3) = \mathbb{R}^+ \times O(1,3)$. The previous discussion can be adapted to the case of three-dimensional Riemannian metrics. In that case, the conformal group, denoted by $CO(3)$, is given by $CO(3) = \mathbb{R}^+ \times O(3)$, where $O(3)$ denotes the ***group of three-dimensional orthogonal transformations (rotations)***.

5.1.2 Conformal extensions and conformal compactifications

If a smooth mapping $\varphi : \tilde{\mathcal{M}} \to \mathcal{M}$ satisfying condition (5.2) is injective but not surjective (i.e. $\varphi(\tilde{\mathcal{M}}) \subsetneq \mathcal{M}$), then one says that $\mathcal{M}$ is a ***conformal extension*** of $\tilde{\mathcal{M}}$. An important type of conformal extensions are the so-called conformal compactifications. A ***conformal compactification*** of a manifold $\tilde{\mathcal{M}}$ with metric $\tilde{\boldsymbol{g}}$ is a conformal transformation $\varphi : \tilde{\mathcal{M}} \to \mathcal{U}$ where $\mathcal{U}$ is a relatively compact (i.e. the closure of $\mathcal{U}$ is compact), connected, open set of a manifold $\mathcal{M}$ such that

$$\boldsymbol{g} = (\varphi^*)^{-1}(\Xi^2\tilde{\boldsymbol{g}}) \quad \text{in } \mathcal{U},$$

with a conformal factor Ξ such that:

(i) $\Xi > 0$ in $\mathcal{U}$.
(ii) $\Xi = 0$ on $\partial\mathcal{U}$, the boundary of the open set $\mathcal{U}$. The set $\partial\mathcal{U}$ is called the ***conformal boundary*** of $\tilde{\mathcal{M}}$.

Examples of conformal extensions will be discussed in Chapter 6.

5.2 Conformal transformation formulae

The discussion of Section 2.4.4 can be applied to obtain the transformation formulae relating the curvature tensors of the Levi-Civita connections $\tilde{\boldsymbol{\nabla}}$ and $\boldsymbol{\nabla}$ of two metrics $\tilde{\boldsymbol{g}}$ and $\boldsymbol{g}$ related to each other by Equation (5.1).

5.2.1 Transformation formulae for the connection

As a first step, one needs to find the specific form of the transition tensor $Q_a{}^c{}_b$ – see Equation (2.13). The first observation is that as the connections $\tilde{\nabla}$ and ∇ are torsion free, it follows from Equation (2.15) that the transition tensor is symmetric; that is, one has that

$$Q_a{}^c{}_b = Q_{(a}{}^c{}_{b)}.$$

Using formula (2.14) one has that

$$\nabla_a g_{bc} - \tilde{\nabla}_a g_{bc} = -Q_a{}^d{}_b g_{dc} - Q_a{}^d{}_c g_{bd}.$$

From $\nabla_a g_{bc} = 0$ and $\tilde{\nabla}_a g_{bc} = \tilde{\nabla}_a(\Xi^2 \tilde{g}_{bc}) = 2\Xi\tilde{\nabla}_a \Xi \tilde{g}_{bc}$ (as $\tilde{\nabla}_a \tilde{g}_{bc} = 0$) one finds that

$$2(\Xi^{-1}\nabla_a \Xi) g_{bc} = Q_a{}^d{}_b g_{dc} + Q_a{}^d{}_c g_{bd}.$$

Two further companion equations can be obtained from the latter by permuting cyclically the indices ${}_{abc}$. Adding two of them and subtracting the third one, one can solve for $Q_a{}^c{}_b$ to find

$$Q_a{}^c{}_b = \Xi^{-1}(\nabla_a \Xi \delta_b{}^c + \nabla_b \Xi \delta_a{}^c - \nabla_d \Xi g^{dc} g_{ab}).$$

This last expression can be rewritten in a more concise form as

$$Q_a{}^c{}_b = S_{ab}{}^{cd}(\Xi^{-1}\nabla_d \Xi), \tag{5.3}$$

where

$$S_{ab}{}^{cd} \equiv \delta_a{}^c \delta_b{}^d + \delta_a{}^d \delta_b{}^c - g_{ab} g^{cd}.$$

To simplify the presentation of the various transformation formulae, let

$$\Upsilon_a \equiv \Xi^{-1}\nabla_a \Xi, \qquad \Upsilon_a{}^c{}_b \equiv S_{ab}{}^{cd}\Upsilon_d.$$

Hence, one can write schematically that

$$\boldsymbol{\nabla} - \tilde{\boldsymbol{\nabla}} = \boldsymbol{S}(\boldsymbol{\Upsilon}), \tag{5.4}$$

and Equation (5.3) yields $Q_a{}^c{}_b = \Upsilon_a{}^c{}_b$. The tensor $\boldsymbol{S}$ appeared in Section 2.5.2 in the decomposition of the Riemann tensor; see Equation (2.21b). Using Equation (5.1) one finds that

$$\delta_a{}^c \delta_b{}^d + \delta_a{}^d \delta_b{}^c - g_{ab} g^{cd} = \delta_a{}^c \delta_b{}^d + \delta_a{}^d \delta_b{}^c - \tilde{g}_{ab} \tilde{g}^{cd}.$$

Hence, the tensor $\boldsymbol{S}$ is independent of the representative of the conformal class; that is, it is an invariant of $[\tilde{\boldsymbol{g}}]$.

5.2.2 Transformation formulae for the curvature

Combining the results of Section 2.4.4 with the expression for the transition tensor of Equation (5.3) one obtains a transformation rule for the Riemann tensor:

$$R^c{}_{dab} - \tilde{R}^c{}_{dab} = 2(\nabla_{[a}\Upsilon_{b]}{}^c{}_d + \Upsilon_{[a}{}^c{}_{|e|}\Upsilon_{b]}{}^e{}_d). \tag{5.5}$$

Some of the transformation formulae for the various concomitants of the Riemann tensor are dimension dependent; thus, they are analysed separately.

The 4-dimensional case

In the *four-dimensional case* one has that the Ricci and Schouten tensors and Ricci scalar of the connections $\tilde{\boldsymbol{\nabla}}$ and $\boldsymbol{\nabla}$ are related to each other, respectively, by the expressions

$$R_{ab} - \tilde{R}_{ab} = -\frac{2}{\Xi}\nabla_a\nabla_b\Xi - g_{ab}g^{cd}\left(\frac{1}{\Xi}\nabla_c\nabla_d\Xi - \frac{3}{\Xi^2}\nabla_c\Xi\nabla_d\Xi\right), \tag{5.6a}$$

$$L_{ab} - \tilde{L}_{ab} = -\frac{1}{\Xi}\nabla_a\nabla_b\Xi + \frac{1}{2\Xi^2}\nabla_c\Xi\nabla^c\Xi\, g_{ab}, \tag{5.6b}$$

$$R - \frac{1}{\Xi^2}\tilde{R} = -\frac{6}{\Xi}\nabla_c\nabla^c\Xi + \frac{12}{\Xi^2}\nabla_c\Xi\nabla^c\Xi. \tag{5.6c}$$

Using the tensor $S_{ab}{}^{cd}$, one can rewrite the transformation formula for the Schouten tensor, Equation (5.6b), in the alternative form

$$L_{ab} - \tilde{L}_{ab} = \nabla_a\Upsilon_b + \frac{1}{2}S_{ab}{}^{cd}\Upsilon_c\Upsilon_d. \tag{5.7}$$

By letting $\vartheta \equiv \Xi^{-1}$, the transformation rule for the Ricci tensor can be rewritten as

$$6\nabla_a\nabla^a\vartheta - R\vartheta = -\tilde{R}\vartheta^3.$$

Using the irreducible decomposition of the Riemann tensor, Equation (2.21b), as a definition for the Weyl tensor, together with Equations (5.5) and (5.6b), one finds that

$$C^c{}_{dab} = \tilde{C}^c{}_{dab}.$$

In other words, *the Weyl tensor is an invariant of the conformal class* $[\tilde{\boldsymbol{g}}]$. Using this invariance and the transformation law for the connection, a calculation leads to the important identity

$$\nabla_a(\Xi^{-1}C^a{}_{bcd}) = \Xi^{-1}\tilde{\nabla}_a C^a{}_{bcd}. \tag{5.8}$$

A further tensor which will play a role in the present treatment of conformal geometry is the so-called ***Cotton tensor*** of $\tilde{\boldsymbol{\nabla}}$. This tensor is defined as

$$\tilde{Y}_{abc} \equiv \tilde{\nabla}_a\tilde{L}_{bc} - \tilde{\nabla}_b\tilde{L}_{ac}.$$

Notice that by construction $\tilde{Y}_{abc} = \tilde{Y}_{[ab]c}$. The Cotton tensor is closely related to the Weyl tensor. To see this, consider the second Bianchi identity

$$\tilde{\nabla}_{[e}\tilde{R}^{a}{}_{|b|cd]} = 0, \tag{5.9}$$

satisfied by the Riemann tensor of the metric $\tilde{\boldsymbol{g}}$; see Section 2.4.3. Now, as seen in Section 2.5.2, for a Levi-Civita connection, the Riemann tensor $\tilde{R}^{a}{}_{bcd}$ can be decomposed in terms of the Weyl tensor $C^{a}{}_{bcd}$ and the Schouten tensor $\tilde{L}_{ab}$ as

$$\tilde{R}^{a}{}_{bcd} = C^{a}{}_{bcd} + 2(\tilde{g}^{a}{}_{[c}\tilde{L}_{d]b} - \tilde{g}_{b[c}\tilde{L}_{d]}{}^{a}). \tag{5.10}$$

Substituting the latter into Equation (5.9) one obtains

$$2(\tilde{g}_{b[c}\tilde{\nabla}_{e}\tilde{L}_{d]}{}^{a} - \tilde{g}^{a}{}_{[c}\tilde{\nabla}_{e}\tilde{L}_{d]b}) = \tilde{\nabla}_{[e}C^{a}{}_{|b|cd]}.$$

Contracting the indices ^{a} and $_{e}$ one obtains

$$\tilde{\nabla}_{c}\tilde{L}_{db} - \tilde{\nabla}_{d}\tilde{L}_{cb} = \tilde{\nabla}_{a}C^{a}{}_{bcd}. \tag{5.11}$$

That is,

$$\tilde{Y}_{cdb} = \tilde{\nabla}_{a}C^{a}{}_{bcd}. \tag{5.12}$$

In particular, one sees that if $C^{a}{}_{bcd} = 0$, then $\tilde{Y}_{cdb} = 0$. Moreover, as a consequence of the first Bianchi identity for the Weyl tensor, $\tilde{Y}_{[abc]} = 0$. The Riemann tensor $R^{a}{}_{bcd}$ of the connection $\boldsymbol{\nabla}$ satisfies equations analogous to (5.9) and (5.10). It follows by the same computation described above that

$$\nabla_{c}L_{db} - \nabla_{d}L_{cb} = \nabla_{a}C^{a}{}_{bcd}. \tag{5.13}$$

Alternatively, defining the Cotton tensor of $\boldsymbol{\nabla}$, $Y_{cdb} \equiv \nabla_{c}L_{db} - \nabla_{d}L_{cb}$, one can write

$$Y_{cdb} = \nabla_{a}C^{a}{}_{bcd}. \tag{5.14}$$

Combining Equations (5.8), (5.12) and (5.14) one finds that the transformation rule for the Cotton tensor is given by:

$$Y_{cdb} - \tilde{Y}_{cdb} = \Upsilon_{a}C^{a}{}_{bcd}. \tag{5.15}$$

The three-dimensional case

In the case of a three-dimensional manifold, let $\boldsymbol{h} = \Omega^2\tilde{\boldsymbol{h}}$ – throughout, the symbol Ω will be used to denote a generic conformal factor on a manifold of dimension three. One has that

$$r_{ij} - \tilde{r}_{ij} = -\frac{1}{\Omega}D_{i}D_{j}\Omega - h_{ij}h^{kl}\left(\frac{1}{\Omega}D_{k}D_{l}\Omega - \frac{2}{\Omega^2}D_{k}\Omega D_{l}\Omega\right), \tag{5.16a}$$

$$l_{ij} - \tilde{l}_{ij} = -\frac{1}{\Omega}D_{i}D_{j}\Omega + \frac{1}{2\Omega^2}D_{k}\Omega D^{k}\Omega\, h_{ij}, \tag{5.16b}$$

$$r - \frac{1}{\Omega^2}\tilde{r} = -\frac{4}{\Omega}D_{i}D^{i}\Omega + \frac{6}{\Omega^2}D_{i}\Omega D^{i}\Omega. \tag{5.16c}$$

where D_i denotes the Levi-Civita covariant derivative of the metric $\boldsymbol{h}$, and r_{ij}, l_{ij}, r correspond to its Ricci and Schouten tensors and its Ricci scalar, respectively. The transformation law of the Schouten tensor is of particular interest. *Comparing Equations (5.6b) and* (5.16b) *one sees that although the definition of the Schouten tensor is dimension dependent, its transformation formula is not.*

Letting $\vartheta \equiv \Omega^{-1/2}$, the transformation law for the Ricci scalar can be recast as

$$8D_iD^i\vartheta - r\vartheta = -\tilde{r}\vartheta^5. \tag{5.17}$$

This expression plays an important role in the discussion of the Einstein constraint equations; see Chapter 11.

Given the three-dimensional Schouten tensor $\tilde{l}_{ij}$, its associated Cotton tensor $\tilde{y}_{ijk}$ is given by

$$\tilde{y}_{ijk} \equiv \tilde{D}_i\tilde{l}_{jk} - \tilde{D}_j\tilde{l}_{ik}. \tag{5.18}$$

Using the transformation rule (5.16b), a computation shows that

$$y_{ijk} = \tilde{y}_{ijk}.$$

That is, in three dimensions the Cotton tensor is conformally invariant. Sometimes it is more convenient to work with its Hodge dual, the so-called ***Cotton-York tensor***, given by

$$\tilde{y}_{ij} = -\frac{1}{2}\tilde{y}_{klj}\epsilon_i{}^{kl}.$$

It can be readily verified that

$$y_{ij} = y_{ji}, \qquad y_i{}^i = 0, \qquad D^i y_{ij} = 0.$$

Moreover, the Cotton-York tensor satisfies the transformation rule

$$y_{ij} = \Omega^{-1}\tilde{y}_{ij}. \tag{5.19}$$

5.2.3 Characterising conformal flatness

Given a conformal class $[\boldsymbol{g}]$ on a manifold $\mathcal{M}$, an important question is whether the flat metric belongs to it, so that $\boldsymbol{g}$ is conformally flat. Conformally flat metrics are a source of geometric intuition in general relativity as they have a simpler curvature tensor depending on the Schouten tensor only. Conformal flatness is characterised by the following classical result:

Theorem 5.1 (***Weyl-Schouten theorem***)

(i) Let $(\mathcal{M}, \boldsymbol{g})$ be a manifold with metric of dimension $n \geq 3$. The metric $\boldsymbol{g}$ is conformally flat if and only if the Cotton tensor of $\boldsymbol{g}$ vanishes.

(ii) Let $(\mathcal{M}, \boldsymbol{g})$ be a manifold with metric of dimension $n \geq 4$. The metric $\boldsymbol{g}$ is conformally flat if and only if the Weyl tensor of $\boldsymbol{g}$ vanishes.

Proof A direct computation shows that if a metric is conformally flat, then both its Cotton and Weyl tensors vanish; this proves the *if* part.

In order to prove the *only if* part, one uses the fact that if the Weyl tensor vanishes then, for dimensions $n \geq 4$, the Cotton tensor vanishes; compare Equation (5.14). In view of Equation (5.15) one concludes that the vanishing of the Cotton tensor holds for any metric in the conformal class. From this point onwards, the proofs for the various dimensions are similar. For simplicity, only the four-dimensional case is considered.

Given a metric $\boldsymbol{g}$ in the conformal class, one needs to find a conformal factor Ξ such that $\boldsymbol{g} = \Xi^2 \boldsymbol{\eta}$ where $\boldsymbol{\eta}$ is the flat Minkowski metric. Motivated by the transformation law for the Schouten tensor, Equation (5.6b), consider the equation

$$\nabla_a \alpha_b + \alpha_a \alpha_b - \frac{1}{2} \alpha_c \alpha^c g_{ab} = -L_{ab}. \tag{5.20}$$

The latter can be read as an *overdetermined* partial differential equation for the covector α_a. Given a solution to Equation (5.20), an antisymmetrisation yields that $\nabla_{[a}\alpha_{b]} = 0$ so that α_a is a *closed covector*. Thus, locally α_a is exact and can be written as $\alpha_a = \nabla_a(\ln \Xi) = \Xi^{-1}\nabla_a \Xi$ for some function Ξ. Comparing Equation (5.20) with (5.6b) one concludes that the Schouten tensor of the metric $\tilde{\boldsymbol{g}} = \Xi^{-2}\boldsymbol{g}$ must vanish. As $C^a{}_{bcd} = 0$, the whole Riemann tensor of $\tilde{\boldsymbol{g}}$ must vanish. Consequently, one concludes that $\tilde{\boldsymbol{g}} = \boldsymbol{\eta}$.

Hence, to conclude the proof one needs to show that Equation (5.20) admits a solution under the assumption that $C^a{}_{bcd} = 0$ and $Y_{abc} = 0$. Applying ∇_c to Equation (5.20), antisymmetrising on ${}_{ca}$ and finally using the commutator of covariant derivatives, one finds the *integrability condition*

$$R^d{}_{bca}\alpha_d + 2\alpha_{[a}\nabla_{c]}\alpha_b + 2\alpha_e \nabla_{[c}\alpha^e g_{a]b} = 0. \tag{5.21}$$

Now, as $C^a{}_{bcd} = 0$ one has that

$$R^a{}_{bcd} = 2(\delta^a{}_{[c}L_{d]b} - g_{b[c}L_{d]}{}^a).$$

Using the latter expression for the Riemann tensor together with Equation (5.20), one finds that the integrability condition (5.21) is automatically satisfied. A general version of the Frobenius theorem ensures the existence of a solution α_b to Equation (5.20); see, for example, Choquet-Bruhat et al. (1982) or Spivak (1970). □

5.3 Weyl connections

As in the previous sections, let $\tilde{\boldsymbol{\nabla}}$ denote the Levi-Civita connection of a metric $\tilde{\boldsymbol{g}}$ on $\tilde{\mathcal{M}}$. Some of the applications of conformal geometry to be considered in this book give rise to connections which are not necessarily the Levi-Civita connection of a metric but, nevertheless, respect the conformal class. A ***Weyl connection*** is a torsion-free connection $\hat{\boldsymbol{\nabla}}$ such that

$$\hat{\nabla}_a \tilde{g}_{bc} = -2\,\tilde{f}_a \tilde{g}_{bc}, \tag{5.22}$$

for some arbitrary covector $\tilde{f}_a$.

The transition tensor $Q_a{}^c{}_b$ relating the connections $\tilde{\boldsymbol{\nabla}}$ and $\hat{\boldsymbol{\nabla}}$ can be obtained using an argument similar to the one employed in Section 5.2.1 to compute the transition tensor of a conformal rescaling. One finds that

$$Q_a{}^c{}_b = S_{ab}{}^{cd} \tilde{f}_d.$$

Schematically one writes

$$\hat{\boldsymbol{\nabla}} - \tilde{\boldsymbol{\nabla}} = \boldsymbol{S}(\tilde{\boldsymbol{f}}).$$

If the covector $\tilde{\boldsymbol{f}}$ is exact, so that on suitable open sets it can be written in the form $\boldsymbol{f} = -\Xi^{-1}\mathbf{d}\Xi$ with some smooth function $\Xi > 0$, then the Weyl connection $\hat{\boldsymbol{\nabla}}$ is, in fact, the Levi-Civita connection of the metric $\boldsymbol{g} = \Xi^2\tilde{\boldsymbol{g}}$.

The condition $\hat{\nabla}_a \delta_b{}^c = 0$ satisfied by a generic connection together with the relation $\delta_b{}^c = \tilde{g}_{bd}\tilde{g}^{dc}$ and the defining property of a Weyl connection, Equation (5.22), show that

$$\hat{\nabla}_a \tilde{g}^{bc} = 2\,\tilde{f}_a \tilde{g}^{bc}.$$

Using the above expressions one readily obtains that

$$\begin{aligned}\hat{\nabla}_e S_{ab}{}^{cd} &= \hat{\nabla}_e(\delta_a{}^c \delta_b{}^d + \delta_b{}^c \delta_a{}^d - \tilde{g}_{ab}\tilde{g}^{cd}) \\ &= -\hat{\nabla}_e(\tilde{g}_{ab}\tilde{g}^{cd}) = -\hat{\nabla}_e \tilde{g}_{ab}\tilde{g}^{cd} - \tilde{g}_{ab}\hat{\nabla}_e \tilde{g}^{cd} = 0.\end{aligned}$$

In what follows, let $\hat{R}^a{}_{bcd}$ denote the Riemann tensor of the Weyl connection $\hat{\boldsymbol{\nabla}}$. This tensor possesses the basic symmetry $\hat{R}^a{}_{bcd} = -\hat{R}^a{}_{bdc}$. As the connection $\hat{\boldsymbol{\nabla}}$ has vanishing torsion, it follows that $\hat{R}^a{}_{bcd}$ satisfies the *first* and *second Bianchi identities* in the form:

$$\hat{R}^a{}_{[bcd]} = 0, \tag{5.23a}$$

$$\hat{\nabla}_{[e}\hat{R}^a{}_{|b|cd]} = 0. \tag{5.23b}$$

5.3.1 Weyl propagation

To investigate the relation between Weyl connections and the conformal class $[\tilde{\boldsymbol{g}}]$, consider a curve γ with parameter $\mathrm{s} \in I \subseteq \mathbb{R}$ on $(\tilde{\mathcal{M}}, \tilde{\boldsymbol{g}})$ with tangent $\dot{\boldsymbol{x}} \in T(\tilde{\mathcal{M}})$. A vector $\boldsymbol{u} \in T(\tilde{\mathcal{M}})$ is said to be ***Weyl propagated*** along γ if it is parallely propagated along γ with respect to a Weyl connection $\hat{\boldsymbol{\nabla}}$; that is, $\boldsymbol{u}$ satisfies the equation

$$\hat{\boldsymbol{\nabla}}_{\dot{\boldsymbol{x}}}\boldsymbol{u} = 0.$$

Writing the latter in terms of the Levi-Civita connection $\tilde{\nabla}$ one has that

$$\begin{aligned}\dot{x}^a\tilde{\nabla}_a u^b &= -\dot{x}^a S_{ac}{}^{be}u^c\tilde{f}_e,\\ &= \tilde{g}_{cd}u^c\dot{x}^d\tilde{f}^b - u^c\tilde{f}_c\dot{x}^b - \dot{x}^c\tilde{f}_c u^b.\end{aligned}$$

In index-free notation one has that

$$\tilde{\nabla}_{\dot{x}}\boldsymbol{u} = \tilde{\boldsymbol{g}}(\boldsymbol{u},\dot{\boldsymbol{x}})\tilde{\boldsymbol{f}}^\sharp - \langle\tilde{\boldsymbol{f}},\boldsymbol{u}\rangle\dot{\boldsymbol{x}} - \langle\tilde{\boldsymbol{f}},\dot{\boldsymbol{x}}\rangle\boldsymbol{u}.$$

Let $\{\boldsymbol{e_a}\}$ denote an arbitrary frame which is Weyl propagated along γ so that $\hat{\nabla}_{\dot{x}}\boldsymbol{e_a}=0$. Letting $\tilde{g}_{\boldsymbol{ab}} \equiv \tilde{\boldsymbol{g}}(\boldsymbol{e_a},\boldsymbol{e_b})$, a computation then shows that

$$\hat{\nabla}_{\dot{x}}\tilde{g}_{\boldsymbol{ab}} = \hat{\nabla}_{\dot{x}}(\tilde{\boldsymbol{g}}(\boldsymbol{e_a},\boldsymbol{e_b})) = -2\langle\tilde{\boldsymbol{f}},\dot{\boldsymbol{x}}\rangle\tilde{g}_{\boldsymbol{ab}}. \tag{5.24}$$

Consequently, one obtains $\hat{\nabla}_{\dot{x}}(\ln\tilde{g}_{\boldsymbol{ab}}) = -2\langle\tilde{\boldsymbol{f}},\dot{\boldsymbol{x}}\rangle$. The latter equation can be solved to give

$$\tilde{g}_{\boldsymbol{ab}}(\eta) = \tilde{g}_{\boldsymbol{ab}}(\eta_\star)\exp\left(-2\int_{\mathrm{s}_\star}^{\mathrm{s}}\langle\tilde{\boldsymbol{f}},\dot{\boldsymbol{x}}\rangle\mathrm{d}s'\right)$$

along the curve $x(\mathrm{s})$. *Thus, one finds that Weyl connections respect the conformal class in the sense that parallel propagation of a metric using a Weyl connection leads to a metric in the same conformal class.* Notice also that Equation (5.24) allows one to conclude that if the frame is *orthogonal* at some point along the curve, then it is orthogonal elsewhere on γ – the normalisation, however, is lost.

5.3.2 Transformation formulae for the curvature

The transformation formulae between the curvature tensors of the Levi-Civita connection $\tilde{\nabla}$ and the Weyl connection $\hat{\nabla}$ follow directly from the general discussion of Section 2.4.4.

In what follows let $\tilde{f}_a{}^c{}_b \equiv S_{ab}{}^{cd}\tilde{f}_d$. If $\hat{R}^a{}_{bcd}$ denotes the Riemann tensor of $\hat{\nabla}$, then one has that

$$\hat{R}^a{}_{bcd} - R^a{}_{bcd} = 2(\tilde{\nabla}_{[c}\tilde{f}_{d]}{}^a{}_b + \tilde{f}_e{}^a{}_{[c}\tilde{f}_{d]}{}^e{}_b), \tag{5.25a}$$

$$\begin{aligned}&= 2(\delta^a{}_{[c}\tilde{\nabla}_{d]}\tilde{f}_b + \tilde{\nabla}_{[c}\tilde{f}^a\tilde{g}_{d]b} - \delta^a{}_b\tilde{\nabla}_{[c}\tilde{f}_{d]}\\ &\quad -\delta^a{}_{[c}\tilde{f}_{d]}\tilde{f}_b + \tilde{g}_{b[c}\tilde{f}_{d]}\tilde{f}^a + \delta^a{}_{[c}\tilde{g}_{d]b}\tilde{f}_e\tilde{f}^e).\end{aligned} \tag{5.25b}$$

Note that the above transformation law involves both the symmetric and antisymmetric parts of the covariant derivative $\tilde{\nabla}_a\tilde{f}_b$.

A transformation formula for the Ricci tensor $\hat{R}_{bd} \equiv \hat{R}^a{}_{bad}$ can be obtained directly from Equation (5.25b):

$$\hat{R}_{cd} - \tilde{R}_{cd} = -3\tilde{\nabla}_d\tilde{f}_c + \tilde{\nabla}_c\tilde{f}_d + 2\,\tilde{f}_c\tilde{f}_d - \tilde{g}_{cd}\left(\tilde{\nabla}_e\tilde{f}^e + 2\,\tilde{f}_e\tilde{f}^e\right). \tag{5.26}$$

Now, as there is no canonical metric to lower or raise indices in expressions involving a Weyl connection, it is conventional to choose a representative of the conformal class, say, $\tilde{\boldsymbol{g}}$, and use it to compute traces. In this spirit one defines the Ricci scalar of the Weyl connection via $\hat{R} \equiv \tilde{g}^{ab}\hat{R}_{ab}$. It can then be directly computed that

$$\hat{R} - \tilde{R} = -6\tilde{\nabla}_a \tilde{f}^a - 6\,\tilde{f}_a\tilde{f}^a. \tag{5.27}$$

Combining the transformation formula for the Riemann tensor, Equation (5.25a), with the irreducible decomposition of the Riemann tensor $\tilde{R}^a{}_{bcd}$ given by Equation (2.21b), one can find an analogous decomposition for the Riemann tensor $\hat{R}^a{}_{bcd}$ of $\hat{\boldsymbol{\nabla}}$:

$$\begin{aligned} \hat{R}^c{}_{dab} &= C^c{}_{dab} + 2S_{d[a}{}^{ce}\hat{L}_{b]e}, \\ &= C^c{}_{dab} + 2(\tilde{g}^c{}_{[a}\hat{L}_{b]d} - \delta^c{}_d\hat{L}_{[ab]} - \tilde{g}_{d[a}\hat{L}_{b]}{}^c), \end{aligned} \tag{5.28a}$$

where

$$\hat{L}_{ab} = \frac{1}{2}\left(\hat{R}_{(ab)} - \frac{1}{2}\hat{R}_{[ab]} - \frac{1}{6}\tilde{g}_{ab}\hat{R}\right)$$

is the ***Schouten tensor of the Weyl connection*** $\hat{\boldsymbol{\nabla}}$. *This definition is independent of the choice of the representative of the conformal class.* Making use of the transformation laws for the Ricci tensor and scalar, Equations (5.26) and (5.27), one finds that

$$\tilde{L}_{ab} - \hat{L}_{ab} = \tilde{\nabla}_a\tilde{f}_b - \tilde{f}_a\tilde{f}_b + \frac{1}{2}\tilde{g}_{ab}\tilde{f}^c\tilde{f}_c, \tag{5.29a}$$

$$= \tilde{\nabla}_a\tilde{f}_b - \frac{1}{2}S_{ab}{}^{cd}\tilde{f}_c\tilde{f}_d, \tag{5.29b}$$

$$= \hat{\nabla}_a\tilde{f}_b + \frac{1}{2}S_{ab}{}^{cd}\tilde{f}_c\tilde{f}_d. \tag{5.29c}$$

Finally, it is observed that letting $\hat{R}_{abcd} \equiv \tilde{g}_{ae}\hat{R}^e{}_{bcd}$, it follows from the discussion in the previous paragraphs that

$$\hat{R}_{abcd} = \hat{R}_{[ab]cd} + 2\tilde{g}_{ab}\hat{\nabla}_{[c}f_{d]}, \tag{5.30a}$$

$$= \hat{R}_{[ab]cd} - 2\tilde{g}_{ab}\hat{L}_{[cd]}. \tag{5.30b}$$

These formulae show in an explicit way how the usual symmetries of the curvature tensor are obstructed by the covector defining a Weyl connection.

5.4 Spinorial expressions

This section discusses the spinorial counterparts of the tensorial expressions obtained in the previous sections of this chapter.

5.4.1 Conformal rescalings

As in previous sections, let $\tilde{g}$ and g denote two metrics on $\tilde{\mathcal{M}}$ related to each other by the conformal rescaling (5.1). Following the discussion of Chapter 3, the spinorial counterparts of $\tilde{g}$ and g are given by

$$\tilde{g}_{AA'BB'} = \tilde{\epsilon}_{AB}\tilde{\epsilon}_{A'B'}, \qquad g_{AA'BB'} = \epsilon_{AB}\epsilon_{A'B'};$$

compare Equation (3.15). Hence, it is natural to consider the transformation laws

$$\epsilon_{AB} = \Xi\tilde{\epsilon}_{AB}, \qquad \epsilon^{AB} = \Xi^{-1}\tilde{\epsilon}^{AB},$$
$$\epsilon_{A'B'} = \Xi\tilde{\epsilon}_{A'B'}, \qquad \epsilon^{A'B'} = \Xi^{-1}\tilde{\epsilon}^{A'B'}.$$

Let $\{\tilde{o}_A, \tilde{\iota}_A\}$ and $\{o_A, \iota_A\}$ denote two spin bases satisfying, respectively, the conditions

$$\tilde{\epsilon}_{AB} = \tilde{o}_A\tilde{\iota}_B - \tilde{\iota}_A\tilde{o}_B, \qquad \epsilon_{AB} = o_A\iota_B - \iota_A o_B.$$

There are several possible transformation rules between the two spin bases which are consistent with the above equations and with the rescaling (5.1). Namely, one has:

$$o_A = \tilde{o}_A, \quad \iota_A = \Xi\tilde{\iota}_A, \quad o^A = \Xi^{-1}\tilde{o}^A, \quad \iota^A = \tilde{\iota}^A, \tag{5.31a}$$

$$o_A = \Xi\tilde{o}_A, \quad \iota_A = \tilde{\iota}_A, \quad o^A = \tilde{o}^A, \quad \iota^A = \Xi^{-1}\tilde{\iota}^A, \tag{5.31b}$$

$$o_A = \Xi^{1/2}\tilde{o}_A, \quad \iota_A = \Xi^{1/2}\tilde{\iota}_A, \quad o^A = \Xi^{-1/2}\tilde{o}^A, \quad \iota^A = \Xi^{-1/2}\tilde{\iota}^A. \tag{5.31c}$$

The choice of the most convenient transformation rule depends on the nature of the application at hand; see, for example, Chapter 10.

Transformation rules for the connection and curvature

In what follows let $\Upsilon_{AA'} \equiv \Xi^{-1}\nabla_{AA'}\Xi$ denote the spinorial counterpart of the covector Υ_a. Let also $\Upsilon_a{}^c{}_b \equiv S_{ab}{}^{cd}\Upsilon_d$. Its spinorial counterpart is given by

$$\Upsilon_{AA'}{}^{CC'}{}_{BB'} = \delta_A{}^C\delta_{A'}{}^{C'}\Upsilon_{BB'} + \delta_B{}^C\delta_{B'}{}^{C'}\Upsilon_{AA'} - \epsilon_{AB}\epsilon_{A'B'}\Upsilon^{CC'}.$$

By rewriting

$$\delta_A{}^C\delta_{A'}{}^{C'}\Upsilon_{BB'} + \delta_B{}^C\delta_{B'}{}^{C'}\Upsilon_{AA'} = \delta_A{}^C\epsilon_{A'}{}^{C'}\epsilon_{B'}{}^{D'}\Upsilon_{BD'} + \delta_B{}^C\epsilon_{B'}{}^{C'}\epsilon_{A'}{}^{D'}\Upsilon_{AD'},$$

and using the Jacobi identity (3.5), one finds that

$$\Upsilon_{AA'}{}^{CC'}{}_{BB'} = \Upsilon_{AA'}{}^C{}_B\delta_{B'}{}^{C'} + \bar{\Upsilon}_{A'A}{}^{C'}{}_{B'}\delta_B{}^C,$$

where

$$\Upsilon_{AA'}{}^C{}_B \equiv \delta_A{}^C\Upsilon_{BA'}.$$

The reduced coefficient $\Upsilon_{AA'}{}^{C}{}_{B}$ can be used to obtain the transformation laws relating the covariant derivatives of spinors. In particular, one has for arbitrary spinors κ_A, $\mu_{A'}$, ξ^A and $\eta^{A'}$ that

$$\begin{aligned}
\tilde{\nabla}_{AA'}\kappa_B &= \nabla_{AA'}\kappa_B + \Upsilon_{BA'}\kappa_A,\\
\tilde{\nabla}_{AA'}\mu_{B'} &= \nabla_{AA'}\mu_{B'} + \Upsilon_{AB'}\mu_{A'},\\
\tilde{\nabla}_{AA'}\xi^B &= \nabla_{AA'}\xi^B - \delta_A{}^B\Upsilon_{CA'}\xi^C,\\
\tilde{\nabla}_{AA'}\eta^{B'} &= \nabla_{AA'}\eta^{B'} - \delta_{A'}{}^{B'}\Upsilon_{AC'}\eta^{C'}.
\end{aligned}$$

These expressions can be extended, in a direct way, to higher valence spinors. For the curvature spinors, it can be verified that

$$\begin{aligned}
\tilde{\Psi}_{ABCD} &= \Psi_{ABCD},\\
\tilde{\Phi}_{AA'BB'} &= \Phi_{AA'BB'} + \Xi^{-1}\nabla_{A(A'}\nabla_{B')B}\Xi.
\end{aligned}$$

5.4.2 Weyl connections

In what follows, let $\hat{\nabla}_{AA'}$ denote the spinorial counterpart of the Weyl connection $\hat{\boldsymbol{\nabla}}$ defined by Equation (5.22). To determine expressions for $\hat{\nabla}_{AA'}\tilde{\epsilon}_{BC}$ and $\hat{\nabla}_{AA'}\tilde{\epsilon}^{BC}$ one notices that the spinorial version of Equation (5.22) is

$$\hat{\nabla}_{AA'}(\tilde{\epsilon}_{BC}\tilde{\epsilon}_{B'C'}) = -2\,\tilde{f}_{AA'}\tilde{\epsilon}_{BC}\tilde{\epsilon}_{B'C'},$$

so that

$$\tilde{\epsilon}_{B'C'}\hat{\nabla}_{AA'}\tilde{\epsilon}_{BC} + \tilde{\epsilon}_{BC}\hat{\nabla}_{AA'}\tilde{\epsilon}_{B'C'} = -2\,\tilde{f}_{AA'}\tilde{\epsilon}_{BC}\tilde{\epsilon}_{B'C'}.$$

The latter is satisfied if one sets

$$\hat{\nabla}_{AA'}\tilde{\epsilon}_{BC} = -\tilde{f}_{AA'}\tilde{\epsilon}_{BC}.$$

From this expression and using that $\hat{\nabla}_{AA'}\delta_B{}^C = 0$, one can readily compute $\hat{\nabla}_{AA'}\tilde{\epsilon}^{BC}$. One finds that

$$\hat{\nabla}_{AA'}\tilde{\epsilon}^{BC} = \tilde{f}_{AA'}\tilde{\epsilon}^{BC}.$$

Decomposition of the spin connection coefficients of a Weyl connection

Let $\{\tilde{\epsilon}_{\boldsymbol{A}}{}^{A}\}$ denote a spin basis with respect to $\tilde{\epsilon}_{AB}$. Following the general discussion on spin connection coefficients of Section 3.2.2 – compare Equation (3.33) – the spinorial counterparts of the connection coefficients $\tilde{\Gamma}_{\boldsymbol{a}}{}^{\boldsymbol{b}}{}_{\boldsymbol{c}}$ and $\hat{\Gamma}_{\boldsymbol{a}}{}^{\boldsymbol{b}}{}_{\boldsymbol{c}}$ admit the decompositions

$$\begin{aligned}
\tilde{\Gamma}_{\boldsymbol{AA'}}{}^{\boldsymbol{BB'}}{}_{\boldsymbol{CC'}} &= \tilde{\Gamma}_{\boldsymbol{AA'}}{}^{\boldsymbol{B}}{}_{\boldsymbol{C}}\delta_{\boldsymbol{C'}}{}^{\boldsymbol{B'}} + \bar{\tilde{\Gamma}}_{\boldsymbol{AA'}}{}^{\boldsymbol{B'}}{}_{\boldsymbol{C'}}\delta_{\boldsymbol{C}}{}^{\boldsymbol{B}},\\
\hat{\Gamma}_{\boldsymbol{AA'}}{}^{\boldsymbol{BB'}}{}_{\boldsymbol{CC'}} &= \hat{\Gamma}_{\boldsymbol{AA'}}{}^{\boldsymbol{B}}{}_{\boldsymbol{C}}\delta_{\boldsymbol{C'}}{}^{\boldsymbol{B'}} + \bar{\hat{\Gamma}}_{\boldsymbol{AA'}}{}^{\boldsymbol{B'}}{}_{\boldsymbol{C'}}\delta_{\boldsymbol{C}}{}^{\boldsymbol{B}}.
\end{aligned}$$

The spinorial counterpart of the equation

$$\hat{\Gamma}_a{}^b{}_c = \tilde{\Gamma}_a{}^b{}_c + \delta_a{}^b \tilde{f}_c + \delta_c{}^b \tilde{f}_a - \eta_{ac}\tilde{f}^b$$

is given by

$$\hat{\Gamma}_{AA'}{}^{BB'}{}_{CC'} = \Gamma_{AA'}{}^{BB'}{}_{CC'} + \delta_A{}^B \delta_{A'}{}^{B'} \tilde{f}_{CC'} + \delta_C{}^B \delta_{C'}{}^{B'} \tilde{f}_{AA'} - \epsilon_{AC}\epsilon_{A'C'}\tilde{f}^{BB'}.$$

Now, by rewriting

$$\delta_A{}^B \delta_{A'}{}^{B'} \tilde{f}_{CC'} = \delta_A{}^B \epsilon_{A'}{}^{B'} \epsilon_{C'}{}^{D'} \tilde{f}_{CD'}, \quad \delta_C{}^B \delta_{C'}{}^{B'} \tilde{f}_{AA'} = \delta_C{}^B \epsilon_{C'}{}^{B'} \epsilon_{A'}{}^{D'} \tilde{f}_{AD'}$$

and using the Jacobi identity (3.5), one finds that

$$\delta_A{}^B \delta_{A'}{}^{B'} \tilde{f}_{CC'} + \delta_C{}^B \delta_{C'}{}^{B'} \tilde{f}_{AA'} - \epsilon_{AC}\epsilon_{A'C'}\tilde{f}^{BB'} = \delta_A{}^B \delta_{C'}{}^{B'} \tilde{f}_{CA'} + \delta_C{}^B \delta_{A'}{}^{B'} \tilde{f}_{AC'}.$$

Hence,

$$\hat{\Gamma}_{AA'}{}^{BB'}{}_{CC'} = (\tilde{\Gamma}_{AA'}{}^B{}_C + \delta_A{}^B \tilde{f}_{CA'})\delta_{C'}{}^{B'} + (\bar{\tilde{\Gamma}}_{AA'}{}^{B'}{}_{C'} + \delta_{A'}{}^{B'} \tilde{f}_{AC'})\delta_C{}^B,$$

so that

$$\hat{\Gamma}_{AA'}{}^B{}_C = \tilde{\Gamma}_{AA'}{}^B{}_C + \delta_A{}^B \tilde{f}_{CA'}. \tag{5.32}$$

In particular, as $\tilde{\Gamma}_{AA'BC} = \tilde{\Gamma}_{AA'(BC)}$, it follows that

$$\hat{\Gamma}_{AA'}{}^Q{}_Q = \tilde{f}_{AA'}.$$

Decomposition of the curvature tensors

The discussion of the decomposition of the spinorial counterpart of a general Riemann tensor given in Section 3.2.3 can be applied to the case of a Weyl connection. In particular, if $\hat{R}^{AA'}{}_{BB'CC'DD'}$ denotes the spinorial counterpart of the Riemann tensor of a Weyl connection $\hat{\nabla}$, one has that Equation (3.35) gives, in the present context, the decomposition

$$\hat{R}_{AA'BB'CC'DD'} = \epsilon_{A'B'}\hat{R}_{ABCC'DD'} + \epsilon_{AB}\bar{\hat{R}}_{A'B'CC'DD'},$$

where

$$\begin{aligned}\hat{R}_{ABCC'DD'} &\equiv \hat{R}_{(AB)CC'DD'} + \frac{1}{2}\epsilon_{AB}(\hat{\nabla}_{CC'}\tilde{f}_{DD'} - \hat{\nabla}_{DD'}\tilde{f}_{CC'}), \\ &= \hat{R}_{(AB)CC'DD'} - \frac{1}{2}\epsilon_{AB}(\hat{L}_{CC'DD'} - \hat{L}_{DD'CC'}),\end{aligned}$$

and $\hat{L}_{AA'BB'}$ denotes the spinorial counterpart of the Schouten tensor of $\hat{\boldsymbol{\nabla}}$. A more detailed expression is given by

$$\hat{R}_{ABCC'DD'} = -\Psi_{ABCD}\epsilon_{C'D'} + \hat{L}_{BC'DD'}\epsilon_{AC} - \hat{L}_{BD'CC'}\epsilon_{AD}. \tag{5.33}$$

The spinorial counterpart $\hat{L}_{AA'BB'}$ of the Schouten tensor admits, in turn, the decomposition

$$\hat{L}_{AA'BB'} = \Phi_{AA'BB'} - \frac{1}{24}R\epsilon_{AB}\epsilon_{A'B'} + \Phi_{AB}\epsilon_{A'B'} + \bar{\Phi}_{A'B'}\epsilon_{AB}$$

where $\Phi_{AA'BB'}$ represents the trace-free part of $\frac{1}{2}\hat{R}_{(ab)}$, while Φ_{AB} describes the antisymmetric tensor $\frac{1}{4}\hat{R}_{[ab]}$.

5.5 Conformal geodesics

This section discusses a class of invariants of the conformal structure of a spacetime $(\tilde{\mathcal{M}}, \tilde{\boldsymbol{g}})$. To motivate the discussion let $x(\mathrm{s})$, $\mathrm{s} \in I \subset \mathbb{R}$, denote a curve on $\tilde{\mathcal{M}}$ with tangent given by $\boldsymbol{x}' \equiv \mathrm{d}x/\mathrm{ds}$. The curve $x(\mathrm{s})$ is a geodesic if it satisfies the equation $\tilde{\nabla}_{\boldsymbol{x}'}\boldsymbol{x}' = 0$. The transformation rule of the covariant derivative $\tilde{\boldsymbol{\nabla}}$ under the conformal rescaling (5.1) implies, in turn, the equation

$$\nabla_{\boldsymbol{x}'}\boldsymbol{x}' = 2\langle \boldsymbol{\Upsilon}, \boldsymbol{x}'\rangle\boldsymbol{x}' - \boldsymbol{g}(\boldsymbol{x}', \boldsymbol{x}')\boldsymbol{\Upsilon}^\sharp. \tag{5.34}$$

Let $\tau = \tau(\mathrm{s})$ denote a new parameter. Writing $\dot{\boldsymbol{x}} \equiv \mathrm{d}\boldsymbol{x}/\mathrm{d}\tau$ and $\tau' \equiv \mathrm{d}\tau/\mathrm{ds}$, the chain rule yields $\boldsymbol{x}' = \tau'\dot{\boldsymbol{x}}$, so that Equation (5.34) implies

$$\tau'^2\nabla_{\dot{\boldsymbol{x}}}\dot{\boldsymbol{x}} = (2\langle \boldsymbol{\Upsilon}, \dot{\boldsymbol{x}}\rangle\tau'^2 - \tau'')\dot{\boldsymbol{x}} - \tau'^2\boldsymbol{g}(\dot{\boldsymbol{x}}, \dot{\boldsymbol{x}})\boldsymbol{\Upsilon}^\sharp.$$

This last expression suggests choosing the parameter τ so that it satisfies the condition

$$\tau'' = 2\langle \boldsymbol{\Upsilon}, \dot{\boldsymbol{x}}\rangle\tau'^2.$$

As $\boldsymbol{\Upsilon}$ is known along the curve, this equation can be read as a second-order ordinary differential equation for τ. Thus, it can always be solved locally so that

$$\tau'^2\nabla_{\dot{\boldsymbol{x}}}\dot{\boldsymbol{x}} = -\tau'^2\boldsymbol{g}(\boldsymbol{x}', \boldsymbol{x}')\boldsymbol{\Upsilon}^\sharp.$$

It follows that only when the curve $x(\mathrm{s})$ is null (i.e. $\boldsymbol{g}(\boldsymbol{x}', \boldsymbol{x}') = 0$) is it possible to reparametrise so that $x(\mathrm{s})$ is a geodesic. *Hence, timelike or spacelike geodesics are not, in general, conformal invariants.*

5.5.1 Basic definitions

A ***conformal geodesic*** on a spacetime $(\tilde{\mathcal{M}}, \tilde{\boldsymbol{g}})$ is a pair $(x(\tau), \boldsymbol{\beta}(\tau))$ consisting of a curve $x(\tau)$ on $\tilde{\mathcal{M}}$, $\tau \in I \subset \mathbb{R}$, with tangent $\dot{\boldsymbol{x}}(\tau)$ and a covector $\boldsymbol{\beta}(\tau)$ along $x(\tau)$ satisfying the equations

$$\tilde{\nabla}_{\dot{\boldsymbol{x}}}\dot{\boldsymbol{x}} = -2\langle \boldsymbol{\beta}, \dot{\boldsymbol{x}}\rangle \dot{\boldsymbol{x}} + \tilde{\boldsymbol{g}}(\dot{\boldsymbol{x}}, \dot{\boldsymbol{x}})\boldsymbol{\beta}^{\sharp}, \tag{5.35a}$$

$$\tilde{\nabla}_{\dot{\boldsymbol{x}}}\boldsymbol{\beta} = \langle \boldsymbol{\beta}, \dot{\boldsymbol{x}}\rangle \boldsymbol{\beta} - \frac{1}{2}\tilde{\boldsymbol{g}}^{\sharp}(\boldsymbol{\beta}, \boldsymbol{\beta})\dot{\boldsymbol{x}}^{\flat} + \tilde{\boldsymbol{L}}(\dot{\boldsymbol{x}}, \cdot), \tag{5.35b}$$

where $\tilde{\boldsymbol{L}}$ denotes the *Schouten tensor* of the Levi-Civita connection $\tilde{\boldsymbol{\nabla}}$. Associated to a conformal geodesic, it is convenient to consider a frame $\{\boldsymbol{e}_a\}$ which is *Weyl propagated* along $x(\tau)$ so that

$$\tilde{\nabla}_{\dot{\boldsymbol{x}}}\boldsymbol{e}_a = -\langle \boldsymbol{\beta}, \boldsymbol{e}_a\rangle \dot{\boldsymbol{x}} - \langle \boldsymbol{\beta}, \dot{\boldsymbol{x}}\rangle \boldsymbol{e}_a + \tilde{\boldsymbol{g}}(\boldsymbol{e}_a, \dot{\boldsymbol{x}})\boldsymbol{\beta}^{\sharp}. \tag{5.36}$$

Initial data for the conformal geodesic Equations (5.35a) and (5.35b) consist of an initial position, an initial direction for the curve and an initial value for the covector:

$$x_\star \in \tilde{\mathcal{M}}, \quad \dot{\boldsymbol{x}}_\star \in T|_{x_\star}(\tilde{\mathcal{M}}), \quad \boldsymbol{\beta}_\star \in T^*|_{x_\star}(\tilde{\mathcal{M}}). \tag{5.37}$$

Piccard's theorem – see, for example, Hartman (1987) – ensures the existence of a unique conformal geodesic $(x(\tau), \boldsymbol{\beta}(\tau))$ near $x_\star$ satisfying for given $\tau_\star \in \mathbb{R}$

$$x(\tau_\star) \equiv x_\star, \quad \dot{\boldsymbol{x}}(\tau_\star) \equiv \dot{\boldsymbol{x}}_\star, \quad \boldsymbol{\beta}(\tau_\star) \equiv \boldsymbol{\beta}_\star.$$

A direct computation using Equations (5.35a) and (5.35b) yields the relations

$$\tilde{\nabla}_{\dot{\boldsymbol{x}}}\left(\tilde{\boldsymbol{g}}(\dot{\boldsymbol{x}}, \dot{\boldsymbol{x}})\right) = -2\langle \boldsymbol{\beta}, \dot{\boldsymbol{x}}\rangle \tilde{\boldsymbol{g}}(\dot{\boldsymbol{x}}, \dot{\boldsymbol{x}}), \tag{5.38a}$$

$$\tilde{\nabla}_{\dot{\boldsymbol{x}}}\langle \boldsymbol{\beta}, \dot{\boldsymbol{x}}\rangle = -\langle \boldsymbol{\beta}, \dot{\boldsymbol{x}}\rangle^2 + \frac{1}{2}\tilde{\boldsymbol{g}}(\dot{\boldsymbol{x}}, \dot{\boldsymbol{x}})\tilde{\boldsymbol{g}}^{\sharp}(\boldsymbol{\beta}, \boldsymbol{\beta}) + \tilde{\boldsymbol{L}}(\dot{\boldsymbol{x}}, \dot{\boldsymbol{x}}), \tag{5.38b}$$

$$\tilde{\nabla}_{\dot{\boldsymbol{x}}}\left(\tilde{\boldsymbol{g}}^{\sharp}(\boldsymbol{\beta}, \boldsymbol{\beta})\right) = \langle \boldsymbol{\beta}, \dot{\boldsymbol{x}}\rangle \tilde{\boldsymbol{g}}^{\sharp}(\boldsymbol{\beta}, \boldsymbol{\beta}) + 2\tilde{\boldsymbol{L}}(\dot{\boldsymbol{x}}, \boldsymbol{\beta}^{\sharp}). \tag{5.38c}$$

In particular, from Equation (5.38a) it follows that if $\tilde{\boldsymbol{g}}(\dot{\boldsymbol{x}}, \dot{\boldsymbol{x}}) = 0$ at some point along the conformal geodesic, one has that $\tilde{\boldsymbol{g}}(\dot{\boldsymbol{x}}, \dot{\boldsymbol{x}}) = 0$ everywhere else. This null conformal geodesic can, in turn, be reparametrised so that it coincides with a null geodesic of $\tilde{\boldsymbol{g}}$.

Expressions in abstract index notation

For later use, it is observed that the conformal geodesic equations can be written in abstract index notation using the tensor $S_{ab}{}^{cd}$ as

$$\dot{x}^c\tilde{\nabla}_c\dot{x}^a = -S_{ef}{}^{ac}\dot{x}^e\dot{x}^f\beta_c,$$

$$\dot{x}^c\tilde{\nabla}_c\beta_a = \frac{1}{2}S_{ca}{}^{ef}\beta_e\beta_f\dot{x}^c + \tilde{L}_{ca}\dot{x}^c,$$

$$\dot{x}^c\tilde{\nabla}_c e_{\boldsymbol{a}}{}^a = -S_{cd}{}^{af}e_{\boldsymbol{a}}{}^d\dot{x}^c\beta_f.$$

5.5.2 Conformal geodesics and changes of connection

The motivation behind the notion of conformal geodesics is not directly apparent from the defining Equations (5.35a) and (5.35b). Their relevance becomes apparent only once one considers their transformation rules under conformal rescalings and transitions to Weyl connections.

As in the previous section, let $(x(\tau), \boldsymbol{\beta}(\tau))$ denote a solution to the conformal geodesic Equations (5.35a) and (5.35b) on a spacetime $(\tilde{\mathcal{M}}, \tilde{\boldsymbol{g}})$. Given $\check{\boldsymbol{f}} \in T^*(\tilde{\mathcal{M}})$ one can define a Weyl connection $\check{\boldsymbol{\nabla}}$ via the relation

$$\check{\boldsymbol{\nabla}} \equiv \tilde{\boldsymbol{\nabla}} + \boldsymbol{S}(\check{\boldsymbol{f}}). \tag{5.39}$$

A computation using Equations (5.35a) and (5.35b) shows that $(x(\tau), \check{\boldsymbol{\beta}}(\tau))$ with

$$\check{\boldsymbol{\beta}}(\tau) \equiv \boldsymbol{\beta}(\tau) - \check{\boldsymbol{f}}(\tau) \tag{5.40}$$

is a solution to the $\check{\boldsymbol{\nabla}}$-conformal geodesic equations:

$$\check{\nabla}_{\dot{\boldsymbol{x}}}\dot{\boldsymbol{x}} = -2\langle\check{\boldsymbol{\beta}}, \dot{\boldsymbol{x}}\rangle\dot{\boldsymbol{x}} + \tilde{\boldsymbol{g}}(\dot{\boldsymbol{x}}, \dot{\boldsymbol{x}})\check{\boldsymbol{\beta}}^\sharp,$$
$$\check{\nabla}_{\dot{\boldsymbol{x}}}\check{\boldsymbol{\beta}} = \langle\check{\boldsymbol{\beta}}, \dot{\boldsymbol{x}}\rangle\check{\boldsymbol{\beta}} - \frac{1}{2}\tilde{\boldsymbol{g}}^\sharp(\check{\boldsymbol{\beta}}, \check{\boldsymbol{\beta}})\dot{\boldsymbol{x}}^\flat + \check{\boldsymbol{L}}(\dot{\boldsymbol{x}}, \cdot),$$

where $\check{\boldsymbol{L}}$ denotes the Schouten tensor of the Weyl connection $\check{\boldsymbol{\nabla}}$. The latter is given by

$$\check{L}_{ab} = \tilde{L}_{ab} - \check{\nabla}_a \check{f}_b - \frac{1}{2} S_{ab}{}^{cd} \check{f}_c \check{f}_d.$$

Thus, one concludes that conformal geodesics are invariants of $[\tilde{\boldsymbol{g}}]$. Notice, in particular, that one could have chosen $\check{\boldsymbol{f}} = -\Xi^{-1}\mathbf{d}\Xi$ for some positive $\Xi \in \mathfrak{X}(\tilde{\mathcal{M}})$ so that the change of connections given by Equation (5.39) corresponds, in fact, to a conformal rescaling of $\tilde{\boldsymbol{g}}$.

Now, choosing $\check{\boldsymbol{f}}(\tau) = \boldsymbol{\beta}(\tau)$ one has that $\check{\boldsymbol{\beta}}(\tau) = 0$, so that the $\check{\boldsymbol{\nabla}}$-conformal geodesic equations reduce to:

$$\check{\nabla}_{\dot{\boldsymbol{x}}}\dot{\boldsymbol{x}} = 0, \qquad \check{\boldsymbol{L}}(\dot{\boldsymbol{x}}, \cdot) = 0. \tag{5.41}$$

Moreover, the frame propagation Equation (5.36) yields

$$\check{\nabla}_{\dot{\boldsymbol{x}}}\boldsymbol{e}_{\boldsymbol{a}} = 0.$$

Hence, given a congruence of conformal geodesics on $(\tilde{\mathcal{M}}, \tilde{\boldsymbol{g}})$, there exists a Weyl connection $\check{\boldsymbol{\nabla}}$ on $[\tilde{\boldsymbol{g}}]$ with respect to which the curves $x(\tau)$ are (affine) geodesics and the frame $\{\boldsymbol{e}_{\boldsymbol{a}}\}$ is parallely propagated. This observation justifies the name *conformal geodesics* given to a solution to Equations (5.35a) and (5.35b). *Thus, conformal geodesics not only are an invariant of the conformal structure, but also single out a particular Weyl connection on the conformal class* $[\boldsymbol{g}]$.

5.5.3 Reparametrisations

Given two solutions to the conformal geodesic Equations (5.35a) and (5.35b), $(x(\tau), \boldsymbol{\beta}(\tau))$ and $(\bar{x}(\bar{\tau}), \bar{\boldsymbol{\beta}}(\bar{\tau}))$, it is natural to ask under which conditions $x(\tau)$ and $\bar{x}(\bar{\tau})$ coincide locally (as sets of points) so that $\tau = \tau(\bar{\tau})$ and $x(\tau(\bar{\tau})) = \bar{x}(\bar{\tau})$. Let $\dot{\boldsymbol{x}} \equiv \mathrm{d}x/\mathrm{d}\tau$ and $\bar{\boldsymbol{x}}' \equiv \mathrm{d}\bar{x}/\mathrm{d}\bar{\tau}$ denote the corresponding tangent vectors and assume that $\tilde{\boldsymbol{g}}(\dot{\boldsymbol{x}}, \dot{\boldsymbol{x}}) \neq 0$ and $\tilde{\boldsymbol{g}}(\bar{\boldsymbol{x}}', \bar{\boldsymbol{x}}') \neq 0$. By definition, the tangent vector $\boldsymbol{x}'$ satisfies

$$\tilde{\nabla}_{\bar{\boldsymbol{x}}'} \bar{\boldsymbol{x}}' = -2\langle \bar{\boldsymbol{\beta}}, \bar{\boldsymbol{x}}' \rangle \bar{\boldsymbol{x}}' + \tilde{\boldsymbol{g}}(\bar{\boldsymbol{x}}', \bar{\boldsymbol{x}}') \bar{\boldsymbol{\beta}}^\sharp, \tag{5.42a}$$

$$\tilde{\nabla}_{\bar{\boldsymbol{x}}'} \bar{\boldsymbol{\beta}} = \langle \bar{\boldsymbol{\beta}}, \bar{\boldsymbol{x}}' \rangle \bar{\boldsymbol{\beta}} - \frac{1}{2} \tilde{\boldsymbol{g}}^\sharp(\bar{\boldsymbol{\beta}}, \bar{\boldsymbol{\beta}}) \bar{\boldsymbol{x}}'^\flat + \tilde{\boldsymbol{L}}(\bar{\boldsymbol{x}}', \cdot). \tag{5.42b}$$

Now, letting $\tau' \equiv \mathrm{d}\tau/\mathrm{d}\bar{\tau}$ one has that

$$\bar{\boldsymbol{x}}' = \tau' \dot{\boldsymbol{x}}, \qquad \tilde{\nabla}_{\bar{\boldsymbol{x}}'} \bar{\boldsymbol{x}}' = \tau'' \dot{\boldsymbol{x}} + \tau'^2 \tilde{\nabla}_{\dot{\boldsymbol{x}}} \dot{\boldsymbol{x}}.$$

Substituting the latter into Equation (5.42a) and using (5.35a) to eliminate $\tilde{\nabla}_{\dot{\boldsymbol{x}}} \dot{\boldsymbol{x}}$ one obtains

$$\tau'' \dot{\boldsymbol{x}} + 2\tau'^2 \langle \bar{\boldsymbol{\beta}} - \boldsymbol{\beta}, \dot{\boldsymbol{x}} \rangle \dot{\boldsymbol{x}} + \tau'^2 \tilde{\boldsymbol{g}}(\dot{\boldsymbol{x}}, \dot{\boldsymbol{x}})(\boldsymbol{\beta}^\sharp - \bar{\boldsymbol{\beta}}^\sharp) = 0. \tag{5.43}$$

It follows from this last equation that the difference $\bar{\boldsymbol{\beta}}^\sharp - \boldsymbol{\beta}^\sharp$ has components only along $\dot{\boldsymbol{x}}$. Hence, one can write

$$\bar{\boldsymbol{\beta}} - \boldsymbol{\beta} = \alpha \dot{\boldsymbol{x}}^\flat, \tag{5.44}$$

for some scalar α. Substituting into Equation (5.43) one obtains the differential equation

$$\tau'' + \alpha \tau'^2 \tilde{\boldsymbol{g}}(\dot{\boldsymbol{x}}, \dot{\boldsymbol{x}}) = 0. \tag{5.45}$$

Combining Equations (5.35a), (5.35b), (5.42b) and (5.44) one obtains

$$\dot{\alpha} = 2\langle \boldsymbol{\beta}, \dot{\boldsymbol{x}} \rangle \alpha + \frac{1}{2} \tilde{\boldsymbol{g}}(\dot{\boldsymbol{x}}, \dot{\boldsymbol{x}}) \alpha^2. \tag{5.46}$$

Equations (5.44), (5.45) *and* (5.46) *encode the requirement that the curves* $x(\tau)$ *and* $\bar{x}(\bar{\tau})$ *coincide as sets.* Using Equation (5.38a) together with Equation (5.46) one finds that

$$\tilde{\nabla}_{\dot{\boldsymbol{x}}} \left(\alpha \tilde{\boldsymbol{g}}(\dot{\boldsymbol{x}}, \dot{\boldsymbol{x}}) \right) = \frac{1}{2} \left(\alpha \tilde{\boldsymbol{g}}(\dot{\boldsymbol{x}}, \dot{\boldsymbol{x}}) \right)^2.$$

This last equation can be solved to give

$$\alpha \tilde{\boldsymbol{g}}(\dot{\boldsymbol{x}}, \dot{\boldsymbol{x}}) = \frac{2\alpha_\star \tilde{\boldsymbol{g}}(\dot{\boldsymbol{x}}_\star, \dot{\boldsymbol{x}}_\star)}{1 - \alpha_\star \tilde{\boldsymbol{g}}(\dot{\boldsymbol{x}}_\star, \dot{\boldsymbol{x}}_\star)(\tau - \tau_\star)},$$

where $\alpha_\star \equiv \alpha(\tau_\star)$, $\dot{\boldsymbol{x}}_\star \equiv \dot{\boldsymbol{x}}(\tau_\star)$ and $\tau_\star$ denotes some fiducial value of the parameter τ. Using Equations (5.44) and (5.45) one finally finds that:

$$\bar{\boldsymbol{x}}' = \frac{4\varkappa}{1 + 2\varkappa\alpha_\star \tilde{\boldsymbol{g}}(\dot{\boldsymbol{x}}_\star, \dot{\boldsymbol{x}}_\star)(\tau - \tau_\star)}\dot{\boldsymbol{x}}, \tag{5.47a}$$

$$\bar{\boldsymbol{\beta}} = \boldsymbol{\beta} + \frac{2\alpha_\star \tilde{\boldsymbol{g}}(\dot{\boldsymbol{x}}_\star, \dot{\boldsymbol{x}}_\star)}{(1 - \alpha_\star \tilde{\boldsymbol{g}}(\dot{\boldsymbol{x}}_\star, \dot{\boldsymbol{x}}_\star)(\tau - \tau_\star))\,\tilde{\boldsymbol{g}}(\dot{\boldsymbol{x}}, \dot{\boldsymbol{x}})}\dot{\boldsymbol{x}}^\flat, \tag{5.47b}$$

$$\tau = \tau_\star + \frac{4\varkappa(\bar{\tau} - \bar{\tau}_\star)}{1 + 2\varkappa\alpha_\star \tilde{\boldsymbol{g}}(\dot{\boldsymbol{x}}, \dot{\boldsymbol{x}})(\bar{\tau} - \bar{\tau}_\star)}, \tag{5.47c}$$

with $\varkappa$ a non-zero real constant. One can summarise the previous discussion in the following lemma:

Lemma 5.1 (***admissible reparametrisations of conformal geodesics***) *The admissible reparametrisations taking (non-null) conformal geodesics into (non-null) conformal geodesics are given by fractional transformations of the form*

$$\tau \mapsto \frac{a\tau + b}{c\tau + d}, \tag{5.48}$$

with $a, b, c, d \in \mathbb{R}$.

If $\alpha_\star = 0$, then Equation (5.47c) shows that the reparametrisation reduces to an affine parameter transformation. Notice also, that with a suitable choice of constants, it is always possible to choose a parametrisation such that $\tau \to \infty$ for a given value of $\bar{\tau}$. This property of conformal geodesics is in stark contrast to the behaviour of *standard* geodesics.

A final remark concerning the reparametrisation of conformal curves follows from evaluating Equations (5.47a) and (5.47b) at $\tau_\star$. One finds that $\bar{\boldsymbol{x}}'_\star = 4\varkappa\dot{\boldsymbol{x}}_\star$ and $\bar{\boldsymbol{\beta}}_\star = \bar{\boldsymbol{\beta}}_\star + \alpha_\star \dot{\boldsymbol{x}}^\flat_\star$. Consequently, the transformations of initial data given by

$$\dot{\boldsymbol{x}}_\star \mapsto 4\varkappa\dot{\boldsymbol{x}}_\star, \qquad \boldsymbol{\beta}_\star \mapsto \boldsymbol{\beta}_\star + \alpha_\star \dot{\boldsymbol{x}}^\flat_\star, \tag{5.49}$$

preserve the set of points covered by the conformal geodesics. *From the discussion in the previous paragraphs it follows that the transformation of initial data* (5.49) *implies a reparametrisation of the resulting curves.*

5.5.4 Geodesics as conformal geodesics

It is of natural interest to investigate the relation between conformal geodesics and metric geodesics. For a null conformal geodesic this relation can be readily established. If $(\bar{x}(\bar{\tau}), \bar{\boldsymbol{\beta}}(\bar{\tau}))$ denotes a null conformal geodesic, it follows readily from Equation (5.42a) that

$$\tilde{\nabla}_{\bar{\boldsymbol{x}}'}\bar{\boldsymbol{x}}' = -2\langle \boldsymbol{\beta}, \bar{\boldsymbol{x}}' \rangle \bar{\boldsymbol{x}}'.$$

Thus, using an argument similar to the one discussed at the beginning of Section 5.5, one finds that *null conformal geodesics are, up to a reparametrisation, null geodesics.*

The situation for non-null conformal geodesics is more complicated and requires restrictions of the Schouten tensor of the spacetime. One has the following result (see Friedrich and Schmidt (1987)):

Lemma 5.2 (***standard geodesics as conformal geodesics***) *Any non-null $\tilde{\boldsymbol{g}}$-geodesic in an Einstein spacetime $(\tilde{\mathcal{M}}, \tilde{\boldsymbol{g}})$ is, up to a reparametrisation, a non-null conformal geodesic.*

Proof Let $x(\tau)$ denote a solution to the metric geodesic equation $\tilde{\nabla}_{\dot{\boldsymbol{x}}}\dot{\boldsymbol{x}} = 0$. Consider a reparametrisation of the curve of the form $\tau = \tau(\bar{\tau})$. The analysis in Section 5.5.3 suggests completing $x(\bar{\tau})$ to a conformal geodesic using an ansatz of the form $\bar{\boldsymbol{\beta}} = \alpha(\bar{\tau})\dot{\boldsymbol{x}}^\flat$. Writing, as in the previous section, $\bar{\boldsymbol{x}}' = \tau'\dot{\boldsymbol{x}}$, Equation (5.42a) readily leads to the condition

$$\tau'' + \alpha\tau'^2\tilde{\boldsymbol{g}}(\dot{\boldsymbol{x}}, \dot{\boldsymbol{x}}) = 0,$$

where it is noticed that $\tilde{\boldsymbol{g}}(\dot{\boldsymbol{x}}, \dot{\boldsymbol{x}})$ is constant along the curve as it is a $\tilde{\boldsymbol{g}}$-geodesic. To obtain an equation for α one substitutes the ansatz for $\bar{\boldsymbol{\beta}}$ into (5.42b) and notices that $\tilde{\nabla}_{\dot{\boldsymbol{x}}}\bar{\boldsymbol{\beta}} = \alpha'\dot{\boldsymbol{x}}^\flat$ so that

$$\alpha'\dot{\boldsymbol{x}}^\flat = \frac{1}{2}\alpha^2\tau'\tilde{\boldsymbol{g}}(\dot{\boldsymbol{x}}, \dot{\boldsymbol{x}})\dot{\boldsymbol{x}}^\flat + \tau'\tilde{\boldsymbol{L}}(\dot{\boldsymbol{x}}, \cdot).$$

The solvability of this equation depends on the available information about $\tilde{\boldsymbol{L}}$. In the case of an Einstein space one has that $\tilde{\boldsymbol{L}}(\dot{\boldsymbol{x}}, \cdot) = \frac{1}{6}\lambda\dot{\boldsymbol{x}}^\flat$ so that one obtains

$$\alpha' = \frac{1}{2}\alpha^2\tau'\tilde{\boldsymbol{g}}(\dot{\boldsymbol{x}}, \dot{\boldsymbol{x}}) + \frac{1}{6}\lambda\tau',$$

which can always be solved – at least locally. □

A partial converse of Lemma 5.2 is given by:

Lemma 5.3 (***conformal geodesics as metric geodesics***) *Let $(\tilde{\mathcal{M}}, \tilde{\boldsymbol{g}})$ be a Einstein spacetime and let $\boldsymbol{g} = \Xi^2\tilde{\boldsymbol{g}}$ be a further metric on $\tilde{\mathcal{M}}$. A conformal geodesic $(\bar{x}(\bar{\tau}), \bar{\boldsymbol{\beta}}(\bar{\tau}))$ with respect to the metric $\boldsymbol{g}$ is, up to a reparametrisation, a $\tilde{\boldsymbol{g}}$-geodesic if there exists a function $\alpha(\bar{\tau})$ such that*

$$\bar{\boldsymbol{\beta}} = -\boldsymbol{\Upsilon} + \alpha\bar{\boldsymbol{x}}'^\flat.$$

Proof The geodesic equation $\tilde{\nabla}_{\dot{\boldsymbol{x}}}\dot{\boldsymbol{x}} = 0$ implies, under the conformal rescaling $\boldsymbol{g} = \Xi^2\tilde{\boldsymbol{g}}$, the equation

$$\nabla_{\dot{\boldsymbol{x}}}\dot{\boldsymbol{x}} = 2\langle\boldsymbol{\Upsilon}, \dot{\boldsymbol{x}}\rangle\dot{\boldsymbol{x}} - \boldsymbol{g}(\dot{\boldsymbol{x}}, \dot{\boldsymbol{x}})\boldsymbol{\Upsilon}^\sharp. \tag{5.50}$$

It follows from the analysis of Section 5.5.3 that, to reparametrise the conformal geodesic equations for the metric $\boldsymbol{g}$ to yield Equation (5.50), one needs to have a parameter α such that $\bar{\boldsymbol{\beta}} = -\boldsymbol{\Upsilon} + \alpha \bar{\boldsymbol{x}}'^{\flat}$. □

5.5.5 Conformal factors associated to congruences of conformal geodesics

In what follows, for simplicity it will be assumed that the spacetime $(\tilde{\mathcal{M}}, \tilde{\boldsymbol{g}})$ can be covered by a *non-intersecting* congruence of conformal geodesics. The congruence of conformal geodesics can be used to single out a metric $\boldsymbol{g} \in [\tilde{\boldsymbol{g}}]$ by means of a conformal factor Θ such that

$$\boldsymbol{g}(\dot{\boldsymbol{x}}, \dot{\boldsymbol{x}}) = 1, \qquad \boldsymbol{g} = \Theta^2 \tilde{\boldsymbol{g}}. \tag{5.51}$$

That is, the tangent vector field of the congruence of conformal geodesics is $\boldsymbol{g}$-normalised – *accordingly, the parameter τ of the geodesics corresponds to the $\boldsymbol{g}$-proper time.* It follows by applying $\tilde{\nabla}_{\dot{\boldsymbol{x}}}$ to the first equation in (5.51) and using the conformal geodesic Equation (5.35a) that

$$\dot{\Theta} = \langle \boldsymbol{\beta}, \dot{\boldsymbol{x}} \rangle \Theta, \tag{5.52}$$

where $\dot{\Theta} \equiv \tilde{\nabla}_{\dot{\boldsymbol{x}}} \Theta$. Thus, by prescribing $\Theta_\star \equiv \Theta(\tau_\star)$ at some fiduciary value $\tau_\star \in \mathbb{R}$ along the conformal geodesic one finds that the value of Θ is fully determined by Equation (5.52). If the initial value $\Theta_\star$ is chosen to vary smoothly along the curves on the congruence, one readily obtains a conformal factor for the whole of the spacetime. It is important to remark that this conformal factor depends on the particular congruence of conformal geodesics; a different choice of congruence would lead to a different Θ and, hence, to a different conformal metric $\boldsymbol{g}$. *Thus, if the congruence of conformal geodesics is specified by a prescription of initial data of the form given in* (5.37) *on an initial hypersurface $\mathcal{S}$, then $\boldsymbol{g}$ is determined in an implicit way by the initial data for the congruence and by $\Theta_\star$.* In the remainder of this section it will be shown that for metrics $\tilde{\boldsymbol{g}}$ satisfying the vacuum Einstein equations this correspondence can be made explicit.

A direct consequence of Equations (5.38a) and (5.52) is that

$$\tilde{\nabla}_{\dot{\boldsymbol{x}}} \left(\boldsymbol{g}(\dot{\boldsymbol{x}}, \dot{\boldsymbol{x}}) \right) = 0.$$

Hence, one sees that a conformal geodesic that is, respectively, timelike, null or spacelike at a given point in $\tilde{\mathcal{M}}$ preserves its causal character throughout the whole curve. Further computations using the conformal geodesic Equations (5.35a) and (5.35b) and the relations (5.38a)–(5.38c) and (5.52) show that

$$\ddot{\Theta} = \frac{1}{2} \Theta \tilde{\boldsymbol{g}}(\dot{\boldsymbol{x}}, \dot{\boldsymbol{x}}) \tilde{\boldsymbol{g}}^\sharp(\boldsymbol{\beta}, \boldsymbol{\beta}) + \Theta \tilde{\boldsymbol{L}}(\dot{\boldsymbol{x}}, \dot{\boldsymbol{x}}), \tag{5.53a}$$

$$\dddot{\Theta} = \left(\tilde{\nabla}_{\dot{\boldsymbol{x}}} (\tilde{\boldsymbol{L}}(\dot{\boldsymbol{x}}, \dot{\boldsymbol{x}})) + \tilde{\boldsymbol{L}}(\dot{\boldsymbol{x}}, \boldsymbol{\beta}^\sharp) \tilde{\boldsymbol{g}}(\dot{\boldsymbol{x}}, \dot{\boldsymbol{x}}) + \langle \boldsymbol{\beta}, \dot{\boldsymbol{x}} \rangle \tilde{\boldsymbol{L}}(\dot{\boldsymbol{x}}, \dot{\boldsymbol{x}}) \right) \Theta. \tag{5.53b}$$

Moreover, if $\{e_a\}$ denotes a g-orthonormal frame, that is, $g(e_a, e_b) = \eta_{ab}$, propagated according to Equation (5.36) with $e_0 = \dot{x}$, one readily finds that

$$\tilde{\nabla}_{\dot{x}}(\Theta\langle\beta, e_a\rangle) = \Theta\tilde{L}(\dot{x}, e_a) + \frac{1}{2}\Theta\tilde{g}^\sharp(\beta,\beta)\tilde{g}(\dot{x}, e_a). \tag{5.54}$$

Notice that for the frame $\{e_a\}$ one has, in addition, that $\tilde{\nabla}_{\dot{x}}(g(e_a, e_b)) = 0$. The expressions discussed in the previous paragraph lead to the following result first proven in Friedrich (1995):

Proposition 5.1 (***the canonical conformal factor associated to a conformal geodesic***) *Let $(\tilde{\mathcal{M}}, \tilde{g})$ denote an Einstein spacetime. Suppose that $(x(\tau), \beta(\tau))$ is a solution to the conformal geodesic equations (5.35a) and* (5.35b) *and that $\{e_a\}$ is a g-orthonormal frame propagated along the curve according to Equation (5.36). If $g = \Theta^2\tilde{g}$ is such that $g(\dot{x}, \dot{x}) = 1$, then the conformal factor Θ satisfies*

$$\Theta(\tau) = \Theta_\star + \dot{\Theta}_\star(\tau - \tau_\star) + \frac{1}{2}\ddot{\Theta}_\star(\tau - \tau_\star)^2, \tag{5.55}$$

where the coefficients $\Theta_\star \equiv \Theta(\tau_\star)$, $\dot{\Theta}_\star \equiv \dot{\Theta}(\tau_\star)$ and $\ddot{\Theta}_\star \equiv \ddot{\Theta}(\tau_\star)$ are constant along the conformal geodesic and are subject to the constraints

$$\dot{\Theta}_\star = \langle\beta_\star, \dot{x}_\star\rangle\Theta_\star, \qquad \Theta_\star\ddot{\Theta}_\star = \frac{1}{2}\tilde{g}^\sharp(\beta_\star, \beta_\star) + \frac{1}{6}\lambda. \tag{5.56}$$

Furthermore, along each conformal geodesic

$$\Theta\beta_0 = \dot{\Theta}, \qquad \Theta\beta_i = \Theta_\star\beta_{i\star}, \tag{5.57}$$

where $\beta_a \equiv \langle\beta, e_a\rangle$.

Proof For an Einstein spacetime the Schouten tensor is given by $\tilde{L} = \frac{1}{6}\lambda\tilde{g}$. Substituting this expression into Equation (5.53b), one finds that $\dddot{\Theta} = 0$ so that Equation (5.55) follows. The constraints (5.56) follow from Equations (5.52) and (5.53a). Finally, the relations in (5.57) follow from (5.52) and (5.54). □

5.5.6 The $\tilde{g}$-adapted equations

As a consequence of the normalisation condition (5.51), the parameter τ is the g-proper time of the curve $x(\tau)$. In some computations it is more convenient to consider a parametrisation in terms of a $\tilde{g}$-proper time $\tilde{\tau}$. To this end, consider the parameter transformation $\tilde{\tau} = \tilde{\tau}(\tau)$ given by

$$\frac{d\tau}{d\tilde{\tau}} = \Theta, \qquad \text{so that} \qquad \tilde{\tau} = \tilde{\tau}_\star + \int_{\tau_\star}^{\tau} \frac{ds}{\Theta(s)}, \tag{5.58}$$

with inverse $\tau = \tau(\tilde{\tau})$. In what follows, write $\tilde{x}(\tilde{\tau}) \equiv x(\tau(\tilde{\tau}))$. It can then be verified that

$$\tilde{x}' \equiv \frac{d\tilde{x}}{d\tilde{\tau}} = \frac{d\tau}{d\tilde{\tau}}\frac{dx}{d\tau} = \Theta\dot{x}, \tag{5.59}$$

so that $\tilde{\boldsymbol{g}}(\tilde{\boldsymbol{x}}', \tilde{\boldsymbol{x}}') = 1$. Hence, $\tilde{\tau}$ is, indeed, the $\tilde{\boldsymbol{g}}$-proper time of the curve $\tilde{x}$. Now, consider, consistent with Equation (5.47b), the split

$$\boldsymbol{\beta} = \tilde{\boldsymbol{\beta}} + \varpi \dot{\boldsymbol{x}}^\flat, \qquad \varpi \equiv \frac{\langle \boldsymbol{\beta}, \dot{\boldsymbol{x}} \rangle}{\tilde{\boldsymbol{g}}(\dot{\boldsymbol{x}}, \dot{\boldsymbol{x}})}, \tag{5.60}$$

where the covector $\tilde{\boldsymbol{\beta}}$ satisfies

$$\langle \tilde{\boldsymbol{\beta}}, \dot{\boldsymbol{x}} \rangle = 0, \qquad \boldsymbol{g}^\sharp(\boldsymbol{\beta}, \boldsymbol{\beta}) = \langle \boldsymbol{\beta}, \dot{\boldsymbol{x}} \rangle^2 + \boldsymbol{g}^\sharp(\tilde{\boldsymbol{\beta}}, \tilde{\boldsymbol{\beta}}). \tag{5.61}$$

It can be readily verified that

$$\tilde{\boldsymbol{g}}(\dot{\boldsymbol{x}}, \dot{\boldsymbol{x}}) = \Theta^{-2}, \qquad \langle \boldsymbol{\beta}, \dot{\boldsymbol{x}} \rangle = \Theta^{-1}\dot{\Theta}, \qquad \varpi = \Theta\dot{\Theta}. \tag{5.62}$$

Using the split (5.60) in Equations (5.35a) and (5.35b) and taking into account the relations in (5.59), (5.61) and (5.62), one obtains the following ***$\tilde{g}$-adapted equations for the conformal geodesics***:

$$\tilde{\nabla}_{\tilde{\boldsymbol{x}}'} \tilde{\boldsymbol{x}}' = \tilde{\boldsymbol{\beta}}^\sharp, \tag{5.63a}$$

$$\tilde{\nabla}_{\tilde{\boldsymbol{x}}'} \tilde{\boldsymbol{\beta}} = \beta^2 \tilde{\boldsymbol{x}}'^\flat + \tilde{\boldsymbol{L}}(\tilde{\boldsymbol{x}}', \cdot) - \tilde{\boldsymbol{L}}(\tilde{\boldsymbol{x}}', \tilde{\boldsymbol{x}}')\tilde{\boldsymbol{x}}'^\flat, \tag{5.63b}$$

with $\beta^2 \equiv -\tilde{\boldsymbol{g}}^\sharp(\tilde{\boldsymbol{\beta}}, \tilde{\boldsymbol{\beta}})$ – observe that as a consequence of (5.61) the covector $\tilde{\boldsymbol{\beta}}$ is spacelike, and, thus, the definition of β^2 makes sense. The Weyl propagation Equation (5.36) can also be cast in a $\tilde{\boldsymbol{g}}$-adapted form. A calculation shows that

$$\tilde{\nabla}_{\tilde{\boldsymbol{x}}'}(\Theta \boldsymbol{e}_{\boldsymbol{a}}) = -\langle \tilde{\boldsymbol{\beta}}, \Theta \boldsymbol{e}_{\boldsymbol{a}} \rangle \tilde{\boldsymbol{x}}'.$$

Equation (5.63a) provides a clear-cut interpretation of the covector $\tilde{\boldsymbol{\beta}}$ – it corresponds to the *physical acceleration* of the conformal curve. Recalling that $\tilde{\boldsymbol{g}} = \Theta^2 \boldsymbol{g}$ and using (5.61) together with Equation (5.57) of Proposition 5.1 one finds that

$$\beta^2 = -\tilde{\boldsymbol{g}}^\sharp(\tilde{\boldsymbol{\beta}}, \tilde{\boldsymbol{\beta}}) = -\Theta^2 \boldsymbol{g}^\sharp(\tilde{\boldsymbol{\beta}}, \tilde{\boldsymbol{\beta}}) = \Theta^2 \delta^{\boldsymbol{ij}} \beta_{\boldsymbol{i}} \beta_{\boldsymbol{j}} = \Theta_\star^2 \delta^{\boldsymbol{ij}} \beta_{\boldsymbol{i}\star} \beta_{\boldsymbol{j}\star}. \tag{5.64}$$

That is, β^2 is a constant along the conformal geodesic. Using Equation (5.63a) to eliminate $\tilde{\boldsymbol{\beta}}$ in Equation (5.63b), one obtains a third-order differential equation for the curve $\tilde{x}(\tilde{\tau})$:

$$\tilde{\nabla}_{\tilde{\boldsymbol{x}}'} \tilde{\nabla}_{\tilde{\boldsymbol{x}}'} \tilde{\boldsymbol{x}}' = \beta^2 \tilde{\boldsymbol{x}}' + \tilde{\boldsymbol{L}}^\sharp(\tilde{\boldsymbol{x}}', \cdot) - \tilde{\boldsymbol{L}}(\tilde{\boldsymbol{x}}', \tilde{\boldsymbol{x}}')\tilde{\boldsymbol{x}}'. \tag{5.65}$$

A computation making use of the expressions derived in this section shows that

$$\tilde{\nabla}_{\tilde{\boldsymbol{x}}'}\big(\tilde{\boldsymbol{g}}(\tilde{\boldsymbol{\beta}}, \tilde{\boldsymbol{\beta}})\big) = 2\tilde{\boldsymbol{L}}(\tilde{\boldsymbol{x}}', \tilde{\boldsymbol{\beta}}).$$

Consequently, unless $(\tilde{\mathcal{M}}, \tilde{\boldsymbol{g}})$ is an Einstein spacetime the acceleration of the curve cannot be constant. This is related to an open question concerning the behaviour of conformal geodesics discussed in Tod (2012): if a conformal geodesic γ enters every neighbourhood of a point p, does γ necessarily pass through p with a finite limiting velocity and acceleration? This potential pathological behaviour is known as ***spiralling***; see Figure 5.1. This does not happen for

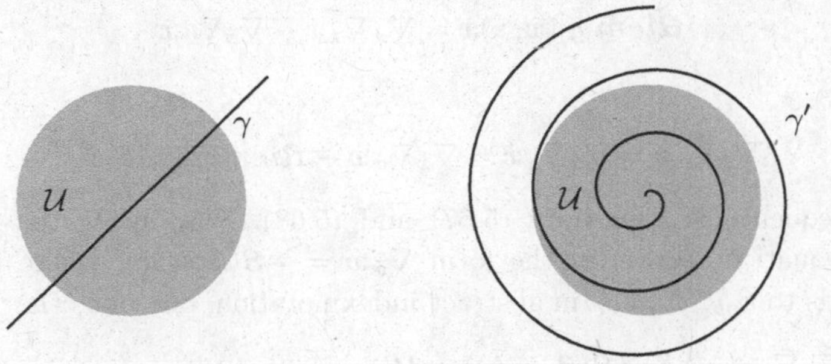

Figure 5.1 Spiralling of conformal geodesics: (left) a standard geodesic γ entering a geodesically convex ball $\mathcal{U}$ must leave it in finite proper time; (right) by contrast, a conformal geodesic γ' may not leave $\mathcal{U}$ and spiral towards a point.

standard geodesics, for if a geodesic enters a *geodesically convex ball*, then it must leave it too; see Section 11.6.2 for a discussion of the notion of geodesically convex ball. Using Piccard's existence theorem for ordinary differential equations – see, for example, Hartman (1987) – on Equation (5.65), it follows that spiralling can occur only if either $\tilde{\boldsymbol{\beta}}$ or $\tilde{\boldsymbol{x}}'$ diverge.

5.5.7 The conformal geodesic deviation equations

An important issue arising in applications involving congruences of conformal geodesics is that of deciding whether the congruence develops caustics, that is, points where it becomes singular. To address this one needs to consider the *conformal geodesic deviation equations* for the congruence. The deviation of these equations is analogous to the one leading to the geodesic deviation equation for standard geodesics; see Section 2.4.5.

In what follows let

$$(x_\eta(\tau), \boldsymbol{\beta}_\eta(\tau)) \equiv (x(\tau,\eta), \boldsymbol{\beta}(\tau,\eta))$$

denote a family of conformal geodesics depending smoothly on a parameter $\eta \in \mathbb{R}$. Following the notation used in previous sections for fixed η, let $\dot{\boldsymbol{x}}$ denote the tangent vector to the curves of the congruence. The ***deviation vector*** and ***deviation covector*** are defined, respectively, by

$$\boldsymbol{z} \equiv \partial_\eta x, \qquad \boldsymbol{\zeta} \equiv \tilde{\nabla}_{\boldsymbol{z}} \boldsymbol{\beta}. \tag{5.66}$$

A short computation shows that

$$[\dot{\boldsymbol{x}}, \boldsymbol{z}] = \tilde{\nabla}_{\dot{\boldsymbol{x}}} \boldsymbol{z} - \tilde{\nabla}_{\boldsymbol{z}} \dot{\boldsymbol{x}} = 0, \tag{5.67}$$

so that $\boldsymbol{z}$ is a well-defined *deviation vector*; compare Section 2.4.5. Moreover, making use of the definition of the Riemann tensor given by Equation (2.9), one has that

$$\boldsymbol{Riem}[\tilde{g}](\dot{\boldsymbol{x}}, \boldsymbol{z})\dot{\boldsymbol{x}} = \tilde{\nabla}_{\dot{\boldsymbol{x}}}\tilde{\nabla}_{\boldsymbol{z}}\dot{\boldsymbol{x}} - \tilde{\nabla}_{\boldsymbol{z}}\tilde{\nabla}_{\dot{\boldsymbol{x}}}\dot{\boldsymbol{x}}. \tag{5.68}$$

Hence,

$$\tilde{\nabla}_{\dot{\boldsymbol{x}}}\tilde{\nabla}_{\dot{\boldsymbol{x}}}\boldsymbol{z} = \tilde{\nabla}_{\dot{\boldsymbol{x}}}\tilde{\nabla}_{\boldsymbol{z}}\dot{\boldsymbol{x}} = \tilde{\nabla}_{\boldsymbol{z}}\tilde{\nabla}_{\dot{\boldsymbol{x}}}\dot{\boldsymbol{x}} + \boldsymbol{Riem}[\tilde{g}](\dot{\boldsymbol{x}}, \boldsymbol{z})\dot{\boldsymbol{x}},$$

as a consequence of Equations (5.67) and (5.68). Now, using the conformal geodesic equation (5.35a) in the form $\tilde{\nabla}_{\dot{\boldsymbol{x}}}\dot{\boldsymbol{x}} = -\boldsymbol{S}(\boldsymbol{\beta}; \dot{\boldsymbol{x}}, \dot{\boldsymbol{x}})$, where $\boldsymbol{S}(\boldsymbol{\beta}; \dot{\boldsymbol{x}}, \dot{\boldsymbol{x}})$ corresponds to $S_{ab}{}^{cd}\dot{x}^a\dot{x}^b\beta_c$ in abstract index notation, one finds that

$$\begin{aligned}\tilde{\nabla}_{\dot{\boldsymbol{x}}}\tilde{\nabla}_{\dot{\boldsymbol{x}}}\boldsymbol{z} &= -\tilde{\nabla}_{\boldsymbol{z}}(\boldsymbol{S}(\boldsymbol{\beta}; \dot{\boldsymbol{x}}, \dot{\boldsymbol{x}})) + \boldsymbol{Riem}[\tilde{g}](\dot{\boldsymbol{x}}, \boldsymbol{z})\dot{\boldsymbol{x}} \\ &= -\boldsymbol{S}(\tilde{\nabla}_{\boldsymbol{z}}\boldsymbol{\beta}; \dot{\boldsymbol{x}}, \dot{\boldsymbol{x}}) - 2\boldsymbol{S}(\boldsymbol{\beta}; \tilde{\nabla}_{\boldsymbol{z}}\dot{\boldsymbol{x}}, \dot{\boldsymbol{x}}) + \boldsymbol{Riem}[\tilde{g}](\dot{\boldsymbol{x}}, \boldsymbol{z})\dot{\boldsymbol{x}}. \end{aligned} \tag{5.69}$$

A similar computation shows that

$$\begin{aligned}\tilde{\nabla}_{\dot{\boldsymbol{x}}}\boldsymbol{\zeta} &= \tilde{\nabla}_{\dot{\boldsymbol{x}}}\tilde{\nabla}_{\boldsymbol{z}}\boldsymbol{\beta} = \tilde{\nabla}_{\boldsymbol{z}}\tilde{\nabla}_{\dot{\boldsymbol{x}}}\boldsymbol{\beta} - \boldsymbol{\beta}\cdot\boldsymbol{Riem}[\tilde{g}](\dot{\boldsymbol{x}}, \boldsymbol{z}) \\ &= \frac{1}{2}\tilde{\nabla}_{\boldsymbol{z}}(\boldsymbol{\beta}\cdot\boldsymbol{S}(\boldsymbol{\beta}; \dot{\boldsymbol{x}}, \cdot)) + \tilde{\nabla}_{\boldsymbol{z}}(\tilde{\boldsymbol{L}}(\dot{\boldsymbol{x}}, \cdot)) - \boldsymbol{\beta}\cdot\boldsymbol{Riem}[\tilde{g}](\dot{\boldsymbol{x}}, \boldsymbol{z}) \\ &= -\boldsymbol{\beta}\cdot\boldsymbol{Riem}[\tilde{g}](\dot{\boldsymbol{x}}, \boldsymbol{z}) + \tilde{\nabla}_{\boldsymbol{z}}\tilde{\boldsymbol{L}}(\dot{\boldsymbol{x}}, \cdot) + \tilde{\boldsymbol{L}}(\tilde{\nabla}_{\boldsymbol{z}}\dot{\boldsymbol{x}}, \cdot) + \frac{1}{2}\tilde{\nabla}_{\boldsymbol{z}}\boldsymbol{\beta}\cdot\boldsymbol{S}(\boldsymbol{\beta}; \dot{\boldsymbol{x}}, \cdot) \\ &\quad + \frac{1}{2}\boldsymbol{\beta}\cdot\boldsymbol{S}(\tilde{\nabla}_{\boldsymbol{z}}\boldsymbol{\beta}; \dot{\boldsymbol{x}}, \cdot) + \frac{1}{2}\boldsymbol{\beta}\cdot\boldsymbol{S}(\boldsymbol{\beta}; \tilde{\nabla}_{\dot{\boldsymbol{x}}}\boldsymbol{z}, \cdot)\end{aligned} \tag{5.70}$$

where, in the third line, Equation (5.35b) in the form

$$\tilde{\nabla}_{\dot{\boldsymbol{x}}}\boldsymbol{\beta} = \frac{1}{2}\boldsymbol{\beta}\cdot\boldsymbol{S}(\boldsymbol{\beta}; \dot{\boldsymbol{x}}, \cdot) + \tilde{\boldsymbol{L}}(\dot{\boldsymbol{x}}, \cdot)$$

has been used. Finally, taking into account the definitions in (5.66) in Equations (5.69) and (5.70), one obtains the ***conformal geodesic deviation equations***:

$$\tilde{\nabla}_{\dot{\boldsymbol{x}}}\tilde{\nabla}_{\dot{\boldsymbol{x}}}\boldsymbol{z} = \boldsymbol{Riem}[\tilde{g}](\dot{\boldsymbol{x}}, \boldsymbol{z})\dot{\boldsymbol{x}} - \boldsymbol{S}(\boldsymbol{\zeta}; \dot{\boldsymbol{x}}, \dot{\boldsymbol{x}}) - 2\boldsymbol{S}(\boldsymbol{\beta}; \dot{\boldsymbol{x}}, \tilde{\nabla}_{\dot{\boldsymbol{x}}}\boldsymbol{z}), \tag{5.71a}$$

$$\begin{aligned}\tilde{\nabla}_{\dot{\boldsymbol{x}}}\boldsymbol{\zeta} &= -\boldsymbol{\beta}\cdot\boldsymbol{Riem}[\tilde{g}](\dot{\boldsymbol{x}}, \boldsymbol{z}) + \tilde{\nabla}_{\boldsymbol{z}}\tilde{\boldsymbol{L}}(\dot{\boldsymbol{x}}, \cdot) + \tilde{\boldsymbol{L}}(\tilde{\nabla}_{\boldsymbol{z}}\dot{\boldsymbol{x}}, \cdot) + \frac{1}{2}\boldsymbol{\zeta}\cdot\boldsymbol{S}(\boldsymbol{\beta}; \dot{\boldsymbol{x}}, \cdot) \\ &\quad + \frac{1}{2}\boldsymbol{\beta}\cdot\boldsymbol{S}(\boldsymbol{\zeta}; \dot{\boldsymbol{x}}, \cdot) + \frac{1}{2}\boldsymbol{\beta}\cdot\boldsymbol{S}(\boldsymbol{\beta}; \tilde{\nabla}_{\dot{\boldsymbol{x}}}\boldsymbol{z}, \cdot),\end{aligned} \tag{5.71b}$$

where

$$\boldsymbol{S}(\boldsymbol{\beta}; \boldsymbol{u}, \boldsymbol{v}) \equiv \langle\boldsymbol{\beta}, \boldsymbol{u}\rangle\boldsymbol{v} + \langle\boldsymbol{\beta}, \boldsymbol{v}\rangle\boldsymbol{u} - \tilde{\boldsymbol{g}}(\boldsymbol{u}, \boldsymbol{v})\boldsymbol{\beta}^\sharp,$$

$$\boldsymbol{\alpha}\cdot\boldsymbol{S}(\boldsymbol{\beta}; \boldsymbol{u}, \cdot) \equiv \langle\boldsymbol{\alpha}, \boldsymbol{u}\rangle\boldsymbol{\beta} + \langle\boldsymbol{\beta}, \boldsymbol{u}\rangle\boldsymbol{\alpha} - \tilde{\boldsymbol{g}}^\sharp(\boldsymbol{\alpha}, \boldsymbol{\beta})\boldsymbol{u}^\flat,$$

for $\boldsymbol{u}, \boldsymbol{v} \in T(\tilde{\mathcal{M}})$ and $\boldsymbol{\alpha} \in T^*(\tilde{\mathcal{M}})$. In standard abstract index notation $\boldsymbol{S}(\boldsymbol{\beta}; \boldsymbol{u}, \boldsymbol{v})$ corresponds to the expression $S_{ab}{}^{cd}u^a v^b \beta_c$, while $\boldsymbol{\alpha}\cdot\boldsymbol{S}(\boldsymbol{\beta}; \boldsymbol{u}, \cdot)$, to $S_{ab}{}^{cd}u^a\beta_c\alpha_d$.

A ***caustic*** in a conformal geodesic is a point along the curve for which $\boldsymbol{z} = 0$. Caustics of conformal geodesics are more complicated than caustics of metric geodesics since, for a given tangent vector, there exists a three-parameter family of conformal geodesics with the same tangent vector. Moreover, the analysis of Equation (5.71a) requires the simultaneous consideration of the evolution equation of the deviation covector $\boldsymbol{\zeta}$, Equation (5.71b). This feature can be useful in applications: Equation (5.71a) has two extra terms, $-\boldsymbol{S}(\boldsymbol{\zeta};\dot{\boldsymbol{x}},\dot{\boldsymbol{x}})$ and $-2\boldsymbol{S}(\boldsymbol{\beta};\dot{\boldsymbol{x}},\tilde{\nabla}_{\dot{\boldsymbol{x}}}\boldsymbol{z})$, not appearing in the standard geodesic deviation equation; under suitable circumstances these terms may be used to counteract the natural tendency of the curvature to develop caustics.

The $\tilde{\mathbf{g}}$-adapted conformal geodesic deviation equations

Following the strategy discussed in Section 5.5.6, one can rewrite the conformal geodesic deviation equations in a way adapted to the metric $\tilde{\boldsymbol{g}}$. To this end define the ***$\tilde{\boldsymbol{g}}$-adapted deviation vector and covector***

$$\tilde{\boldsymbol{z}} \equiv \partial_\lambda \tilde{\boldsymbol{x}}, \qquad \tilde{\boldsymbol{\zeta}} \equiv \tilde{\nabla}_{\tilde{\boldsymbol{z}}}\tilde{\boldsymbol{\beta}}.$$

Now, observing that $[\tilde{\boldsymbol{x}}',\tilde{\boldsymbol{z}}] = 0$, a computation taking into account the $\tilde{\boldsymbol{g}}$-adapted conformal geodesic Equations (5.63a) and (5.63b) and the commutator of covariant derivatives leads to the following ***$\tilde{\boldsymbol{g}}$-adapted conformal geodesic deviation equations***:

$$\tilde{\nabla}_{\tilde{\boldsymbol{x}}'}\tilde{\nabla}_{\tilde{\boldsymbol{x}}'}\tilde{\boldsymbol{z}} = \boldsymbol{Riem}[\tilde{\boldsymbol{g}}](\tilde{\boldsymbol{x}}',\tilde{\boldsymbol{z}})\tilde{\boldsymbol{x}}' + \tilde{\boldsymbol{\zeta}}^\sharp, \tag{5.72a}$$

$$\tilde{\nabla}_{\tilde{\boldsymbol{x}}'}\tilde{\boldsymbol{\zeta}} = -\tilde{\boldsymbol{\beta}}\cdot\boldsymbol{Riem}[\tilde{\boldsymbol{g}}](\tilde{\boldsymbol{x}}',\tilde{\boldsymbol{z}}) + (\tilde{\nabla}_{\tilde{\boldsymbol{z}}}\beta^2)\tilde{\boldsymbol{x}}'^\flat + \beta^2\tilde{\nabla}_{\tilde{\boldsymbol{x}}'}\tilde{\boldsymbol{z}}^\flat. \tag{5.72b}$$

A computation exploiting the fact that the connection $\tilde{\boldsymbol{\nabla}}$ is assumed to be torsion free gives

$$\tilde{\nabla}_{\tilde{\boldsymbol{x}}'}\tilde{\nabla}_{\tilde{\boldsymbol{z}}}\beta^2 = \tilde{\nabla}_{\tilde{\boldsymbol{z}}}\tilde{\nabla}_{\tilde{\boldsymbol{x}}'}\beta^2 = 0,$$

where the last equality follows from the fact that β^2 is constant along a given conformal geodesic; see Equation (5.64). Hence, the components of the terms with $\tilde{\boldsymbol{x}}'^\flat$ and $\tilde{\nabla}_{\tilde{\boldsymbol{x}}'}\tilde{\boldsymbol{z}}^\flat$ in Equation (5.72b) are constant and can be evaluated at some fiducial time.

5.6 Further reading

Basic references for applications of conformal geometry in general relativity are Penrose and Rindler (1984, 1986) and Stewart (1991). A discussion of the properties of the Weyl and Cotton tensor can be found in García et al. (2004).

The first systematic treatments of conformal geodesics in the context of general relativity can be found in Schmidt (1986) and Friedrich and Schmidt (1987). A discussion of Weyl connections making use of the more general language of fibre bundles is given in Friedrich (1995); a brief presentation of the subject in the spirit of this chapter can be found in Friedrich (2002). A discussion of the properties of conformal geodesics in the context of general relativity can be found in Friedrich (2003a); a more technical discussion can be found in the earlier reference Friedrich (1995). Properties of conformal geodesics have been explored from a different perspective in Tod (2012).

The results of Proposition 5.1 strongly depend on the hypothesis that $(\tilde{\mathcal{M}}, \tilde{\boldsymbol{g}})$ is an Einstein space – in other words, $\tilde{\boldsymbol{g}}$ satisfies the vacuum Einstein equations. To get around this restriction, a more general class of curves has been introduced in Lübbe and Valiente Kroon (2012). These curves are a suitable generalisation of the conformal geodesics which allow the recovery of the conclusions of Proposition 5.1 for general spacetimes and, thus, provide a systematic way of identifying the conformal boundary of non-vacuum spacetimes. A discussion of the associated deviation equations with explicit expressions for the case of warped-product spacetimes is given in Lübbe and Valiente Kroon (2013a).

A detailed mathematical theory of conformal connections can be found in Ogiue (1967) and Kobayashi (1995). A more recent monograph on the subject is Fefferman and Graham (2012). Conformal geometry is naturally related to twistor theory; a discussion of this and related topics such as *tractors* can be found in Eastwood (1996).

The reader interested in surveys on research in conformal geometry is referred to Kulkarni and Pinkall (1988), Chang et al. (2007) and Branson et al. (2004) as suitable entry points to the literature in the subject.

Part II

General relativity and conformal geometry

6
Conformal extensions of exact solutions

Exact solutions to the Einstein field equations are the prime source of geometric and physical intuition in general relativity. This chapter revisits some of the classical exact solutions of general relativity (the Minkowski, de Sitter, anti-de Sitter and Schwarzschild spacetimes) from the point of view of conformal geometry. In addition, a general discussion of the construction of Penrose diagrams of static spherically symmetric spacetimes is provided. Most of the material in this chapter can be considered as *classic* – complementary discussions can be found in, for example, Hawking and Ellis (1973) and Griffiths and Podolský (2009). In view of the applications in the later parts of the book, particular emphasis is given to the construction of explicit congruences of conformal geodesics in the exact solutions.

6.1 Preliminaries

6.1.1 Spherical symmetry

In what follows, let $SO(3)$ denote the group of homogeneous linear transformations of $\mathbb{R}^3$ onto itself which preserve the Euclidean length of vectors and the orientation of the space. A spacetime $(\mathcal{M}, \boldsymbol{g})$ is said to be ***spherically symmetric*** if the group $SO(3)$ acts by isometry on $(\mathcal{M}, \boldsymbol{g})$ with simply connected, complete, spacelike two-dimensional orbits; see, for example, Ehlers (1973). Two points $p, q \in SO(3)$ are said to be in the same orbit if there is an element of the group $SO(3)$ taking p to q. Given a spherically symmetric spacetime it is natural to introduce the ***quotient manifold*** $\mathcal{Q} \equiv \mathcal{M}/SO(3)$, that is, the manifold obtained from $\mathcal{M}$ by identifying points on the same orbit. The manifold $\mathcal{Q}$ inherits from $(\mathcal{M}, \boldsymbol{g})$ a two-dimensional Lorentzian metric γ, the ***quotient metric***. Let Γ denote the subset of $\mathcal{Q}$ corresponding to the fixed points of the action of $SO(3)$. If Γ is non-empty, then it can be shown that it is a connected timelike boundary of $\mathcal{Q}$ – the ***centre of symmetry***. A spherically spacetime can have none, one or two centres; see Künzle (1967).

Given a spherically symmetric spacetime $(\mathcal{M}, \boldsymbol{g})$, there exists a function $\varrho : \mathcal{Q} \to \mathbb{R}$ such that the spacetime metric $\boldsymbol{g}$ can be written in the ***warped product form***

$$\boldsymbol{g} = \boldsymbol{\gamma} + \varrho^2 \boldsymbol{\sigma}, \tag{6.1}$$

where $\boldsymbol{\sigma}$ is the ***standard metric*** of $\mathbb{S}^2$ given, in the usual spherical coordinates (θ, φ), by

$$\boldsymbol{\sigma} = \mathbf{d}\theta \otimes \mathbf{d}\theta + \sin^2\theta \mathbf{d}\varphi \otimes \mathbf{d}\varphi.$$

The function ϱ is not necessarily an areal coordinate.

6.1.2 The 3-sphere

The unit ***3-sphere*** $\mathbb{S}^3$ is the three-dimensional submanifold of $\mathbb{R}^4$ defined by

$$\mathbb{S}^3 \equiv \{(w, x, y, z) \in \mathbb{R}^4 \,|\, w^2 + x^2 + y^2 + z^2 = 1\}.$$

The ***standard Euclidean metric*** in $\mathbb{R}^4$ induces, in a natural way, a 3-metric $\boldsymbol{\hbar}$ on $\mathbb{S}^3$, the ***standard metric*** of $\mathbb{S}^3$. The metric $\boldsymbol{\hbar}$ is best expressed using ***spherical coordinates*** (ψ, θ, φ) such that

$$w = \cos\psi, \quad x = \sin\psi\cos\theta, \quad y = \sin\psi\sin\theta\cos\varphi, \quad z = \sin\psi\sin\theta\sin\varphi,$$

taking the range $0 \leq \psi \leq \pi$, $0 \leq \theta \leq \pi$ and $0 \leq \varphi < 2\pi$. For simplicity of presentation, in what follows the degeneracy of the spherical coordinate system (ψ, θ, φ) will be ignored as it can be dealt with by introducing further coordinate charts. In terms of these coordinates one has

$$\boldsymbol{\hbar} = \mathbf{d}\psi \otimes \mathbf{d}\psi + \sin^2\psi \boldsymbol{\sigma}.$$

Conventionally, the point given by $\psi = 0$ will be called the ***north pole***, while the one with $\psi = \pi$ will be called the ***south pole***.

At every point $p \in \mathbb{S}^3$ the restriction of the coordinates (w, x, y, z) can be used to construct suitable *local coordinates*. For example, if $w(p) > 0$, then the coordinates $(x^\alpha) = (x, y, z)$ constitute a well-defined system of local coordinates on the northern hemisphere of $\mathbb{S}^3$.

A frame on $\mathbb{S}^3$

A direct computation shows that the vector fields on $T(\mathbb{R}^4)$

$$\mathbf{c}_1 \equiv w\frac{\boldsymbol{\partial}}{\boldsymbol{\partial} z} - z\frac{\boldsymbol{\partial}}{\boldsymbol{\partial} w} + x\frac{\boldsymbol{\partial}}{\boldsymbol{\partial} y} - y\frac{\boldsymbol{\partial}}{\boldsymbol{\partial} x}, \tag{6.2a}$$

$$\mathbf{c}_2 \equiv w\frac{\boldsymbol{\partial}}{\boldsymbol{\partial} y} - y\frac{\boldsymbol{\partial}}{\boldsymbol{\partial} w} + z\frac{\boldsymbol{\partial}}{\boldsymbol{\partial} x} - x\frac{\boldsymbol{\partial}}{\boldsymbol{\partial} z}, \tag{6.2b}$$

$$\mathbf{c}_3 \equiv w\frac{\boldsymbol{\partial}}{\boldsymbol{\partial} x} - x\frac{\boldsymbol{\partial}}{\boldsymbol{\partial} w} + y\frac{\boldsymbol{\partial}}{\boldsymbol{\partial} z} - z\frac{\boldsymbol{\partial}}{\boldsymbol{\partial} y}, \tag{6.2c}$$

are linearly independent and tangent to $\mathbb{S}^3$; hence, they can be regarded as globally defined vectors on $T(\mathbb{S}^3)$. This point of view will be used systematically in this book. As $\boldsymbol{c}_i \neq 0$ on $\mathbb{S}^3$ one has, in fact, a globally defined frame on the 3-sphere. Moreover, it can be shown that

$$\boldsymbol{\hbar}(\boldsymbol{c}_i, \boldsymbol{c}_j) = \delta_{ij}.$$

Accordingly, the vectors $\{\boldsymbol{c}_i\}$ are $\boldsymbol{\hbar}$-orthogonal. A direct calculation shows that

$$[\boldsymbol{c}_1, \boldsymbol{c}_2] = 2\boldsymbol{c}_3, \qquad [\boldsymbol{c}_2, \boldsymbol{c}_3] = 2\boldsymbol{c}_1, \qquad [\boldsymbol{c}_3, \boldsymbol{c}_1] = 2\boldsymbol{c}_2.$$

The above expressions can be more concisely written as

$$[\boldsymbol{c}_i, \boldsymbol{c}_j] = 2\epsilon_{ij}{}^{k}\boldsymbol{c}_k,$$

where ϵ_{ijk} denotes the components of the volume form in $\mathbb{R}^3$. In particular, one has that $\epsilon_{123} = 1$. The above commutators can be combined with the Cartan structure equations – see Equations (2.41) and (2.42) – to compute the connection coefficients $\gamma_i{}^j{}_k$ with respect to the frame $\{\boldsymbol{c}_i\}$. One obtains the concise expression

$$\gamma_i{}^j{}_k = -\epsilon_i{}^j{}_k.$$

The compactification of $\mathbb{R}^3$ into $\mathbb{S}^3$

An important example of conformal compactification is the so-called ***point compactification*** of the Euclidean space $\mathbb{R}^3$ into $\mathbb{S}^3$. Let

$$\boldsymbol{\delta} = \mathbf{d}r \otimes \mathbf{d}r + r^2\mathbf{d}\theta \otimes \mathbf{d}\theta + r^2 \sin^2\theta \mathbf{d}\varphi \otimes \mathbf{d}\varphi$$

denote the standard (negative definite) three-dimensional Euclidean metric in spherical coordinates with $0 \leq r < \infty$, $0 \leq \theta \leq \pi$ and $0 \leq \varphi < 2\pi$.

An explicit computation shows that the Cotton tensor – see Equation (5.18) – of the metrics $\boldsymbol{\delta}$ and $\boldsymbol{\hbar}$ vanish so that they must be conformally related; compare Theorem 5.1. In order to make this correspondence explicit, write

$$\boldsymbol{\hbar} = \omega^2\boldsymbol{\delta} \tag{6.3}$$

where ω is a conformal factor to be determined. Expressing the radial coordinate as $r = r(\psi)$, one finds from Equation (6.3) the conditions

$$\omega^2 r'^2 = 1, \qquad r^2\omega^2 = \sin^2\psi, \tag{6.4}$$

where $'$ denotes the derivative with respect to ψ. A solution to the equations in (6.4) is given by

$$\omega = \frac{2}{\alpha}\sin^2\frac{\psi}{2}, \qquad r(\psi) = \alpha\cot\frac{\psi}{2}, \tag{6.5}$$

where α is a real constant. Notice that $r \to \infty$ as $\psi \to 0$. Thus, the transformation given by (6.5) is a compactification of $\mathbb{R}^3$ sending the north pole of $\mathbb{S}^3$ to the point at infinity in $\mathbb{R}^3$, while the south pole of $\mathbb{S}^3$ is sent to the origin of $\mathbb{R}^3$. An alternative solution to equations (6.4), sending the south pole to the point at infinity and the north pole to the origin, is given by

$$\omega = \frac{2}{\alpha}\cos^2\frac{\psi}{2}, \qquad r(\psi) = \alpha\tan\frac{\psi}{2}, \tag{6.6}$$

as can be verified by an explicit computation.

6.1.3 The Einstein static universe

The ***Einstein static universe*** – sometimes also called the ***Einstein cosmos*** or ***Einstein cylinder*** – is the spacetime $(\mathcal{M}_{\mathscr{E}}, \boldsymbol{g}_{\mathscr{E}})$ given by

$$\mathcal{M}_{\mathscr{E}} \equiv \mathbb{R}\times\mathbb{S}^3, \qquad \boldsymbol{g}_{\mathscr{E}} \equiv \mathbf{d}T\otimes\mathbf{d}T - \boldsymbol{\hbar}. \tag{6.7}$$

It can be readily verified that $\boldsymbol{\partial}_T$ is a timelike Killing vector of $\boldsymbol{g}_{\mathscr{E}}$ so that the solution is indeed static. Moreover, as $(\mathbb{S}^3, \boldsymbol{\sigma})$ is a homogeneous and isotropic Riemannian manifold, it follows that $(\mathcal{M}_{\mathscr{E}}, \boldsymbol{g}_{\mathscr{E}})$ is *spatially homogeneous and isotropic.*

A computation shows that

$$\boldsymbol{Weyl}[\boldsymbol{g}_{\mathscr{E}}] = 0, \qquad R[\boldsymbol{g}_{\mathscr{E}}] = -6, \tag{6.8a}$$

$$\boldsymbol{Schouten}[\boldsymbol{g}_{\mathscr{E}}] = \frac{1}{2}(\mathbf{d}T\otimes\mathbf{d}T + \boldsymbol{\hbar}). \tag{6.8b}$$

Hence, one sees that $(\mathcal{M}_{\mathscr{E}}, \boldsymbol{g}_{\mathscr{E}})$ is conformally flat. A discussion of the properties of the Einstein static universe as a solution to the Einstein field equations with a perfect fluid matter source can be found in, for example, Griffiths and Podolský (2009) and Hawking and Ellis (1973).

Finally, it is observed that the Einstein static universe is spherically symmetric. Comparing the metric $\boldsymbol{g}_{\mathscr{E}}$ in (6.7) with the warped product metric (6.1) it is natural to set

$$\boldsymbol{\gamma}_{\mathscr{E}} \equiv \mathbf{d}T\otimes\mathbf{d}T - \mathbf{d}\psi\otimes\mathbf{d}\psi, \qquad \varrho_{\mathscr{E}} \equiv \sin\psi,$$

so that (T,ψ) can be used as coordinates of the quotient manifold $\mathcal{Q}_{\mathscr{E}} \equiv (\mathbb{R}\times\mathbb{S}^3)/SO(3) \approx \mathbb{R}\times[0,\pi]$.

A class of conformal geodesics in the Einstein static universe

In what follows, consider the congruence of curves on $(\mathcal{M}_{\mathscr{E}}, \boldsymbol{g}_{\mathscr{E}})$ given by

$$x(\tau) = (\tau, x_\star), \qquad \tau\in\mathbb{R}, \tag{6.9}$$

with $x_\star \in \mathbb{S}^3$ fixed. Varying $x_\star$ over $\mathbb{S}^3$ one obtains a non-intersecting timelike congruence covering the whole of $\mathcal{M}_{\mathscr{E}}$. It can be verified that the curves (6.9) are geodesics for $\boldsymbol{g}_{\mathscr{E}}$ with proper time τ and tangent vector $\dot{\boldsymbol{x}} = \boldsymbol{\partial}_T$.

The curves (6.9) can be recast as conformal geodesics. To see this, one follows the argument of Lemma 5.2 and introduces a parameter $\bar{\tau}$ such that $\tau = \tau(\bar{\tau})$ and makes use of the ansatz

$$\bar{\boldsymbol{\beta}} \equiv \alpha(\bar{\tau})\dot{\boldsymbol{x}}^\flat = \alpha(\bar{\tau})\mathbf{d}T.$$

Substituting the above expression into the conformal geodesic Equations (5.42a) and (5.42b) and taking into account formula (6.8b) for the Schouten tensor of the Einstein universe one finds the equations

$$\tau'' + \alpha\tau'^2 = 0, \qquad \alpha' = \frac{1}{2}\tau'(\alpha^2+1),$$

with $'$ denoting differentiation with respect to $\bar{\tau}$. A solution to the above equations is given by

$$\tau = 2\arctan\frac{\bar{\tau}}{2}, \qquad \alpha = \frac{\bar{\tau}}{2}. \tag{6.10}$$

Now, one has that $\langle\bar{\boldsymbol{\beta}}, \boldsymbol{x}'\rangle = \alpha\tau'$ so that the conformal factor $\bar{\Theta}$ satisfying the condition $\bar{\Theta}^2\boldsymbol{g}_{\mathscr{E}}(\boldsymbol{x}',\boldsymbol{x}') = 1$ obeys the equation $\bar{\Theta}' = \langle\bar{\boldsymbol{\beta}},\boldsymbol{x}'\rangle\bar{\Theta}$ with initial condition $\bar{\Theta}_\star = 1$. The differential equation for $\bar{\Theta}$ can be solved to give

$$\bar{\Theta} = 1 + \frac{1}{4}\bar{\tau}^2. \tag{6.11}$$

It can be verified that

$$\bar{\boldsymbol{\beta}} = \frac{1}{2}\bar{\tau}\mathbf{d}T = \bar{\Theta}^{-1}\mathbf{d}\bar{\Theta}.$$

Using the conformal factor $\bar{\Theta}$ one obtains a conformal representation of the Einstein universe with metric $\bar{\boldsymbol{g}}_{\mathscr{E}} \equiv \bar{\Theta}^2\boldsymbol{g}_{\mathscr{E}}$ so that

$$\bar{\boldsymbol{g}}_{\mathscr{E}} = \mathbf{d}\bar{\tau}\otimes\mathbf{d}\bar{\tau} - \left(1+\frac{1}{4}\bar{\tau}^2\right)^2\boldsymbol{\hbar},$$

where the parameter $\bar{\tau}$ has been introduced as the new time coordinate. This conformal representation of the Einstein cylinder will be known as the ***expanding Einstein cylinder***. Notice that $\boldsymbol{x}' = \bar{\Theta}^{-1}\dot{\boldsymbol{x}}$ so that $\bar{\boldsymbol{g}}_{\mathscr{E}}(\boldsymbol{x}',\boldsymbol{x}') = 1$. It can be readily verified that the congruence is integrable and that the curves are orthogonal to the surfaces of constant $\bar{\tau}$.

6.2 The Minkowski spacetime

The Minkowski solution $(\tilde{\mathcal{M}},\tilde{\boldsymbol{\eta}})$ is the spacetime given by $\tilde{\mathcal{M}} = \mathbb{R}^4$ and

$$\tilde{\boldsymbol{\eta}} = \eta_{\mu\nu}\mathbf{d}x^\mu\otimes\mathbf{d}x^\nu, \tag{6.12}$$

where $(x^\mu) = (t,x,y,z)$ and $\eta_{\mu\nu} \equiv \mathrm{diag}(1,-1,-1,-1)$. Alternatively, using spherical coordinates one can write

$$\tilde{\boldsymbol{\eta}} = \mathbf{d}t\otimes\mathbf{d}t - r^2\mathbf{d}r\otimes\mathbf{d}r - r^2\boldsymbol{\sigma}. \tag{6.13}$$

Using the expression of the Minkowski metric in Cartesian coordinates one readily sees that $\boldsymbol{Riem}[\tilde{\boldsymbol{\eta}}] = 0$ so that, in particular, $\boldsymbol{Ric}[\tilde{\boldsymbol{\eta}}] = 0$; that is, $\tilde{\boldsymbol{\eta}}$ is a solution to the vacuum Einstein field equations with vanishing cosmological constant. Moreover, one has that $\boldsymbol{Weyl}[\tilde{\boldsymbol{\eta}}] = 0$ so that $\tilde{\boldsymbol{\eta}}$ is conformally flat and, thus, conformal to the metric of the Einstein cylinder. This relation is analysed in the next section.

6.2.1 The compactification into the Einstein cylinder

A standard procedure for the construction of conformal extensions of Lorentzian manifolds is to make use of pairs of so-called *null coordinates*. In the present case a convenient choice is given by

$$u \equiv t - r, \qquad v \equiv t + r. \tag{6.14}$$

Conventionally, the coordinate is called a ***retarded time***, while v is an ***advanced time***. It can be readily verified that $\tilde{\boldsymbol{\eta}}^\sharp(\mathbf{d}u, \mathbf{d}u) = \tilde{\boldsymbol{\eta}}^\sharp(\mathbf{d}v, \mathbf{d}v) = 0$. It follows that

$$\tilde{\boldsymbol{\eta}} = \frac{1}{2}(\mathbf{d}u \otimes \mathbf{d}v + \mathbf{d}v \otimes \mathbf{d}u) - \frac{1}{4}(v-u)^2\boldsymbol{\sigma}.$$

In order to have $r \geq 0$ one has the restriction $u \leq v$. The present analysis is mainly concerned with the behaviour at infinity; thus, it is natural to introduce a further transformation of coordinates:

$$u \equiv \tan U, \quad v \equiv \tan V, \quad U,\ V \in (-\tfrac{1}{2}\pi, \tfrac{1}{2}\pi), \quad U \leq V.$$

From the relations

$$\mathbf{d}u = \frac{1}{\cos^2 U}\mathbf{d}U = (1+u^2)\mathbf{d}U, \quad \mathbf{d}v = \frac{1}{\cos^2 V}\mathbf{d}V = (1+v^2)\mathbf{d}V,$$

and the identity

$$v - u = \tan V - \tan U = \frac{\sin(V-U)}{\cos U \cos V},$$

one obtains

$$\tilde{\boldsymbol{\eta}} = \frac{1}{4\cos^2 U \cos^2 V}\left(2(\mathbf{d}U \otimes \mathbf{d}V + \mathbf{d}V \otimes \mathbf{d}U) - \sin^2(U-V)\boldsymbol{\sigma}\right).$$

This last expression suggests defining the *unphysical metric* $\boldsymbol{\eta} \equiv \Xi_{\mathscr{M}}^2 \tilde{\boldsymbol{\eta}}$ where

$$\Xi_{\mathscr{M}} \equiv 2\cos U \cos V, \tag{6.15}$$

so that

$$\boldsymbol{\eta} = 2(\mathbf{d}U \otimes \mathbf{d}V + \mathbf{d}V \otimes \mathbf{d}U) - \sin^2(U-V)\boldsymbol{\sigma}.$$

The conformal factor $\Xi_{\mathscr{M}}$ vanishes whenever $U = \pm\frac{1}{2}\pi$ or $V = \pm\frac{1}{2}\pi$. In order to investigate the situation in more detail one introduces the final change of coordinates

$$\psi \equiv V - U, \quad T \equiv V + U.$$

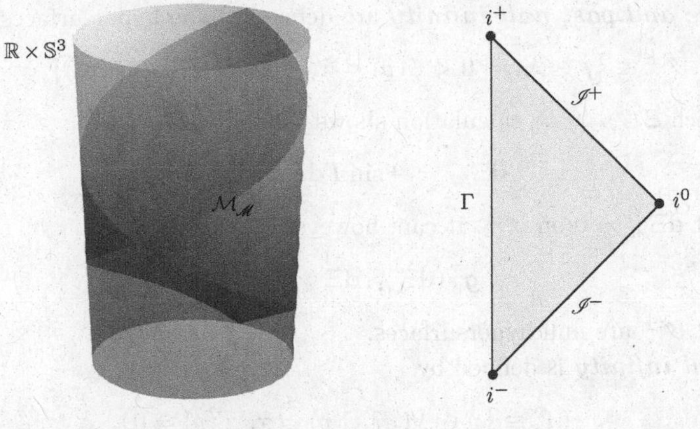

Figure 6.1 Conformal extension of the Minkowski spacetime. Left, conformal embedding of the Minkowski spacetime in the Einstein cylinder: the shaded region corresponds to the set $\mathcal{M}_{\mathscr{M}}$ of equation (6.21). Right, Penrose diagram of the Minkowski spacetime: the line Γ corresponds to the axis of symmetry, the points i^0, i^+ and i^- are spatial infinity, future timelike infinity and past timelike infinity, respectively. Finally, $\mathscr{I}^+$ and $\mathscr{I}^-$ are future and past null infinity; see main text for further details.

Using standard trigonometric identities, one can rewrite the conformal factor (6.15) in terms of the coordinates T and ψ to obtain

$$\Xi_{\mathscr{M}} = \cos T + \cos\psi. \tag{6.16}$$

Thus, one ends up with the metric

$$\boldsymbol{\eta} = \mathbf{d}T\otimes\mathbf{d}T - \mathbf{d}\psi\otimes\mathbf{d}\psi - \sin^2\psi\boldsymbol{\sigma}. \tag{6.17}$$

Thus, one has that $\boldsymbol{\eta} = \boldsymbol{g}_{\mathscr{E}}$. Consequently, *the rescaling procedure described in the previous paragraphs compactifies the Minkowski spacetime into a region of the Einstein cylinder*; see Figure 6.1, left panel. The *standard coordinates* (t, r) on the Minkowski spacetime are related to the (T, ψ) coordinates on the Einstein cylinder via the formulae:

$$t = \frac{\sin T}{\cos T + \cos\psi}, \qquad r = \frac{\sin\psi}{\cos T + \cos\psi}. \tag{6.18}$$

It follows from the previous discussion that the Minkowski spacetime $(\mathbb{R}^4, \tilde{\boldsymbol{\eta}})$ is conformal to the domain

$$\tilde{\mathcal{M}}_{\mathscr{M}} \equiv \big\{p\in\mathcal{M}_{\mathscr{E}} \mid 0\leq\psi(p)<\pi,\ \ \psi(p)-\pi<T(p)<\pi-\psi(p)\big\},$$

on which $\Xi_{\mathscr{M}} > 0$. In addition to $\tilde{\mathcal{M}}_{\mathscr{M}}$, it is convenient to single out a number of subsets of $\mathcal{M}_{\mathscr{E}}$ playing a special role in the discussion of the asymptotic behaviour of the Minkowski spacetime:

(a) ***Future and past null infinity*** are defined as the hypersurfaces

$$\mathscr{I}^{\pm} \equiv \{p \in \mathcal{M}_{\mathscr{E}} \mid 0 < \psi(p) < \pi,\ T(p) = \pm(\pi - \psi(p))\}, \tag{6.19}$$

on which $\Xi_{\mathscr{M}} = 0$. A calculation shows that

$$\mathbf{d}\Xi_{\mathscr{M}} = -\sin T\mathbf{d}T - \sin\psi\mathbf{d}\psi, \tag{6.20}$$

so that $\mathbf{d}\Xi_{\mathscr{M}} \neq 0$ on $\mathscr{I}^{\pm}$. It can, however, be verified that

$$\boldsymbol{g}_{\mathscr{E}}(\mathbf{d}\Xi_{\mathscr{M}}, \mathbf{d}\Xi_{\mathscr{M}})|_{\mathscr{I}^{\pm}} = 0,$$

so that $\mathscr{I}^{\pm}$ are null hypersurfaces.

(b) ***Spatial infinity*** is defined by

$$i^0 \equiv \{p \in \mathcal{M}_{\mathscr{E}} \mid \psi(p) = \pi,\ T(p) = 0\}.$$

Inspection of expression (6.7) for the metric $\boldsymbol{g}_{\mathscr{E}}$ shows that the radius of the 2-sphere defined by $T = 0$ and $\psi = 0$ vanishes. Accordingly, i^0 consists of a single point. Evaluating the differential (6.20) at i^0 one finds that

$$\mathbf{d}\Xi_{\mathscr{M}}|_{i^0} = 0, \qquad \boldsymbol{Hess}\,\Xi_{\mathscr{M}}|_{i^0} = -\boldsymbol{g}_{\mathscr{E}}|_{i^0}.$$

(c) ***Future and past timelike infinity*** is defined as

$$i^{\pm} \equiv \{p \in \mathcal{M}_{\mathscr{E}} \mid \psi(p) = 0,\ T(p) = \pm\pi\}.$$

Again, from the metric (6.7) it follows that the 2-spheres defined by $T = \pm\pi$ and $\psi = 0$ have vanishing radius so that both i^+ and i^- correspond to points. Using (6.20) one finds that

$$\mathbf{d}\Xi_{\mathscr{M}}|_{i^{\pm}} = 0, \qquad \boldsymbol{Hess}\,\Xi_{\mathscr{M}}|_{i^{\pm}} = \boldsymbol{g}_{\mathscr{E}}|_{i^{\pm}}.$$

The motivation for the above definitions follows from the analysis of geodesics; see below. It is important to point out that by convention i^0, $i^{\pm} \notin \mathscr{I}^{\pm}$. Finally, it is convenient to define the manifold with boundary

$$\mathcal{M}_{\mathscr{M}} = \tilde{\mathcal{M}}_{\mathscr{M}} \cup \mathscr{I}^+ \cup \mathscr{I}^- \cup i^+ \cup i^- \cup i^0, \tag{6.21}$$

which will be called the ***conformally extended Minkowski manifold***.

The Penrose diagram of the Minkowski spacetime

The spherical symmetry of the Minkowski spacetime can be exploited to provide a diagrammatic representation of the global structure of the spacetime known as a ***Penrose (or Penrose-Carter) diagram***. It follows from the discussion in Section 6.1.1 that the action of the group $SO(3)$ on $\mathcal{M}_{\mathscr{M}}$ gives rise to the quotient manifold with coordinates (T, ψ) given by

$$\mathcal{Q}_{\mathscr{M}} \equiv \{p \in \mathcal{Q}_{\mathscr{E}} \mid 0 \leq \psi(p) \leq \pi,\ \ \psi(p) - \pi \leq T(p) \leq \pi - \psi(p)\}.$$

In what follows, in a slight abuse of notation, the projections of $\mathscr{I}^{\pm}$, $i^{\pm}$, $i^0 \subset \mathcal{M}_{\mathscr{M}}$ on the quotient manifold $\mathcal{Q}_{\mathscr{M}}$ will be denoted, again, by the same symbols.

Clearly, $\mathscr{I}^+$, $i^\pm$, $i^0 \subset \partial\mathcal{Q}_{\mathscr{M}}$; however, $\partial\mathcal{Q}_{\mathscr{M}}$ has a further component consisting of the centre of symmetry

$$\Gamma \equiv \{p \in \mathcal{Q}_{\mathscr{M}} \mid \psi(p) = 0, \quad -\pi < T(p) < \pi\}.$$

As the conformal metric $\boldsymbol{\eta}$ is the standard one on the Einstein cylinder, it follows from the discussion in Section 6.1.3 that the quotient metric inherited by $\boldsymbol{\eta}$ on $\mathcal{Q}_{\mathscr{M}}$ is the two-dimensional Minkowski metric

$$\boldsymbol{\gamma}_{\mathscr{M}} = \mathbf{d}T \otimes \mathbf{d}T - \mathbf{d}\psi \otimes \mathbf{d}\psi.$$

Given the above, the Penrose diagram of the Minkowski spacetime is simply the depiction of $\mathcal{Q}_{\mathscr{M}}$ as a subset of $\mathbb{R}^2$ as shown in Figure 6.1, right panel. A discussion of the construction of Penrose diagrams for more general spacetimes is given in Section 6.5.2.

Analysis of the behaviour of geodesics

Intuition on the various features of the construction described in the previous paragraphs can be obtained by analysing the behaviour of various types of physical metric geodesics. To this end one notices the following formulae that can be verified using the coordinate transformations taking the original Minkowski metric of Equation (6.13) into the metric (6.17):

$$\sin T = \frac{2t}{\sqrt{(1+(t-r)^2)\,(1+(t+r)^2)}}, \tag{6.22a}$$

$$\cos T = \frac{1-t^2+r^2}{\sqrt{(1+(t-r)^2)\,(1+(t+r)^2)}}, \tag{6.22b}$$

$$\cos\psi = \frac{1+t^2-r^2}{\sqrt{(1+(t-r)^2)\,(1+(t+r)^2)}}. \tag{6.22c}$$

(a) **Spacelike geodesics.** Radial spacelike geodesics in the Minkowski spacetime can be described using the radial coordinate r as a parameter. It follows then that the time coordinate of the curves is given by

$$t = ar + t_\star, \qquad a^2 < 1, \qquad t_\star \in \mathbb{R}.$$

For $r \to \infty$ it follows from (6.22a)–(6.22c) that

$$\sin T \to 0, \qquad \cos T \to 1, \qquad \cos\psi \to -1.$$

Hence one concludes that $T \to 0$ and $\psi \to \pi$ as the curve escapes to infinity. *Thus, in the unphysical picture, spacelike radial geodesics finish at the same point, spatial infinity i^0, independently of the value of a.*

(b) **Timelike geodesics.** For concreteness, consider the family of geodesics described by

$$t = ar + t_\star, \qquad |a| > 1.$$

It can be verified that as $r \to \infty$ one has the limits

$$\sin T \to 0, \qquad \cos T \to -1, \qquad \cos\psi \to 1.$$

Depending on whether $\sin T$ approaches 0 from the right or the left, the latter limits correspond to either $T \to \pi$ and $\psi \to 0$ or $T \to -\pi$ and $\psi \to 0$. *Thus, the timelike geodesics start and finish, respectively, at i^- and i^+.*

(c) **Null geodesics.** Consider, for example, the family of *outgoing* null geodesics described by the condition $u = u_\star$, where $u_\star$ is a constant and u is the null coordinate defined in (6.14). Now, taking the limit $v \to \infty$ one finds that

$$\sin T \to \frac{1}{\sqrt{1+u_\star^2}}, \qquad \cos T \to \frac{u_\star}{\sqrt{1+u_\star^2}}, \qquad \cos\psi \to -\frac{u_\star}{\sqrt{1+u_\star^2}}.$$

Thus, in the limit one has that $T = \pi - \psi$, corresponding to future null infinity $\mathscr{I}^+$. Similarly, for incoming geodesics described by the condition $v = v_\star$, $v_\star$ a constant and with v as defined in (6.14), one finds that the limit points lie on the line $T - \psi = \pi$, corresponding to past null infinity $\mathscr{I}^-$. *Summarising, incoming null geodesics start at $\mathscr{I}^-$ while outgoing null geodesics end at $\mathscr{I}^+$.*

6.2.2 Compactifications adapted to spatial infinity

The discussion of the structure of spatial infinity is better carried out in an alternative conformal representation. Intuitively, the region of spacetime associated with the spatial infinity of the Minkowski spacetime $(\mathbb{R}^4, \tilde{\boldsymbol{\eta}})$ is contained in the domain $\tilde{\mathcal{D}} \equiv \{p \in \mathbb{R}^4 \mid \eta_{\mu\nu}x^\mu(p)x^\nu(p) < 0\}$, the ***complement of the light cone through the origin***, where (x^μ) denote the standard Cartesian coordinates. Now, consider the coordinate inversion defined by

$$y^\mu = -\frac{x^\mu}{X^2}, \qquad x^\mu = -\frac{y^\mu}{Y^2},$$

where $X^2 \equiv \eta_{\mu\nu}x^\mu x^\nu$ and $Y^2 \equiv \eta_{\mu\nu}y^\mu y^\nu$. This coordinate transformation maps $\tilde{\mathcal{D}}$ onto itself. Moreover, a computation yields

$$\mathbf{d}y^\mu = -\frac{1}{X^2}\left(\delta^\mu{}_\lambda - \frac{2}{X^2}x^\mu \eta_{\lambda\nu}x^\nu\right)\mathbf{d}x^\lambda,$$

so that

$$\eta_{\mu\nu}\mathbf{d}y^\mu \otimes \mathbf{d}y^\nu = X^{-4}\eta_{\mu\nu}\mathbf{d}x^\mu \otimes \mathbf{d}x^\nu.$$

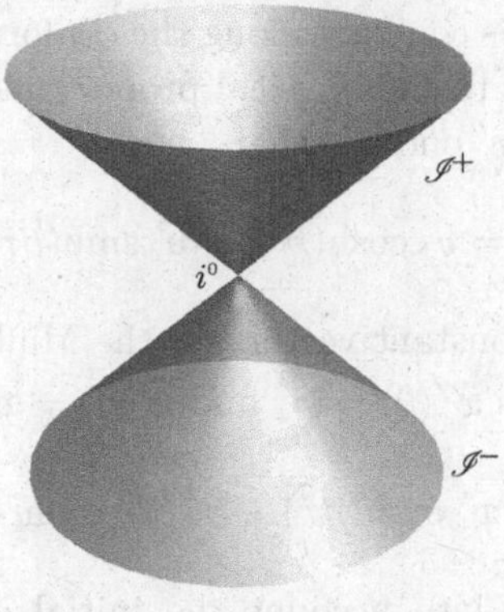

Figure 6.2 The Minkowski spacetime close to null and spatial infinity: the conformal boundary corresponds to the surface of the cones, while the interior of the spacetime corresponds to the exterior of the light cones; see main text for further details.

This computation suggests introducing the conformal factor $\Xi = 1/X^2$. Hence, one concludes that

$$\boldsymbol{\eta} = \Xi^2 \tilde{\boldsymbol{\eta}} = \Xi^2 \eta_{\mu\nu} \mathbf{d}x^\mu \otimes \mathbf{d}x^\nu.$$

Hence, *one has a conformal representation of the Minkowski spacetime which is also flat.* An inspection shows that the boundary $\partial\tilde{\mathcal{D}}$ decomposes into the sets

$$\begin{aligned} \mathscr{I}^+ &= \{p \in \mathbb{R}^4 \mid y^0(p) > 0,\ \eta_{\mu\nu} y^\mu(p) y^\nu(p) = 0\}, \\ \mathscr{I}^- &= \{p \in \mathbb{R}^4 \mid y^0(p) < 0,\ \eta_{\mu\nu} y^\mu(p) y^\nu(p) = 0\}, \\ i^0 &= \{p \in \mathbb{R}^4 \mid (y^\mu(p)) = (0,0,0,0)\}; \end{aligned}$$

see Figure 6.2. An analysis similar to the one carried out in Section 6.2.1 shows that these sets admit the interpretation of future null infinity, past null infinity and spatial infinity, respectively. More precisely, $\mathscr{I}^+$ ($\mathscr{I}^-$) can be thought of as being generated by the end points of future (past) directed null geodesics while all spatial geodesics eventually run into the point i^0. The null hypersurfaces $\mathscr{I}^+$ form the null cone through the point i^0. Defining the manifold with boundary $\mathcal{D} \equiv \tilde{\mathcal{D}} \cup \partial\tilde{\mathcal{D}}$ one observes that the conformal metric $\boldsymbol{\eta}$ extends smoothly through the boundary.

6.2.3 Conformal geodesics in the Minkowski spacetime

Conformal geodesics in the Minkowski spacetime can be computed using the version of the equations adapted to the physical metric; see Section 5.5.6. Using standard Cartesian coordinates so that all the Christoffel symbols vanish, the third-order Equation (5.65) reduces to

$$\tilde{\boldsymbol{x}}''' = \beta^2 \tilde{\boldsymbol{x}}', \qquad \beta^2 \equiv -\tilde{\boldsymbol{\eta}}^\sharp(\tilde{\boldsymbol{\beta}}, \tilde{\boldsymbol{\beta}}), \tag{6.23}$$

where it is recalled that β^2 is constant along the conformal geodesic and $'$ denotes differentiation with respect to the physical proper time $\tilde{\tau}$. Regarding (6.23) as a second-order equation for $\tilde{\boldsymbol{x}}'$ one has that

$$\tilde{\boldsymbol{x}}' = \boldsymbol{v}_1 \cosh(\beta\tilde{\tau}) + \boldsymbol{v}_2 \sinh(\beta\tilde{\tau}),$$

where $\boldsymbol{v}_1$ and $\boldsymbol{v}_2$ are two constant vectors on the Minkowski spacetime. Making use of the initial conditions $\tilde{\boldsymbol{x}}''(0) = \tilde{\boldsymbol{\beta}}^\sharp_\star$ and $\tilde{\boldsymbol{x}}'(0) = \boldsymbol{x}'_\star$ one finds that

$$\tilde{\boldsymbol{x}}' = \boldsymbol{x}'_\star \cosh(\beta\tilde{\tau}) + \beta^{-1}\tilde{\boldsymbol{\beta}}^\sharp_\star \sinh(\beta\tilde{\tau}).$$

A final integration taking into account the initial condition $\tilde{x}(0) = x_\star$ and Equation (5.63a) yields

$$\begin{aligned}\tilde{x}(\tilde{\tau}) &= x_\star + \beta^{-1}\boldsymbol{x}'_\star \sinh(\beta\tilde{\tau}) + \beta^{-2}\tilde{\boldsymbol{\beta}}^\sharp_\star \cosh(\beta\tilde{\tau}) - \beta^2\tilde{\boldsymbol{\beta}}^\sharp_\star,\\ \tilde{\boldsymbol{\beta}}(\tilde{\tau}) &= \beta\boldsymbol{x}'^\flat_\star \sinh(\beta\tilde{\tau}) + \tilde{\boldsymbol{\beta}}_\star \cosh(\beta\tilde{\tau}),\end{aligned}$$

where in the first expression, in an abuse of notation, the vectors $\boldsymbol{x}'_\star$ and $\tilde{\boldsymbol{\beta}}^\sharp_\star$ are understood as describing points in $\mathbb{R}^4$. To rewrite this general solution in terms of the unphysical proper time τ one makes use of formula (5.58). A computation yields the formulae:

$$x(\tau) = x_\star + \Theta_\star\Theta^{-1}(\tau)\left(\dot{\boldsymbol{x}}_\star\tau + \frac{1}{2}\tilde{\boldsymbol{\eta}}(\dot{\boldsymbol{x}}_\star, \dot{\boldsymbol{x}}_\star)\boldsymbol{\beta}^\sharp_\star\tau^2\right), \tag{6.24a}$$

$$\boldsymbol{\beta}(\tau) = (1 + \tau\langle\boldsymbol{\beta}_\star, \dot{\boldsymbol{x}}_\star\rangle)\boldsymbol{\beta}_\star - \frac{1}{2}\tilde{\boldsymbol{\eta}}^\sharp(\boldsymbol{\beta}_\star, \boldsymbol{\beta}_\star)\dot{\boldsymbol{x}}_\star\tau, \tag{6.24b}$$

where

$$\Theta(\tau) = \Theta_\star\left(1 + \langle\boldsymbol{\beta}_\star, \dot{\boldsymbol{x}}_\star\rangle\tau + \frac{1}{4}\tilde{\boldsymbol{\eta}}(\dot{\boldsymbol{x}}_\star, \dot{\boldsymbol{x}}_\star)\tilde{\boldsymbol{\eta}}^\sharp(\boldsymbol{\beta}_\star, \boldsymbol{\beta}_\star)\tau^2\right).$$

Conformal geodesics which satisfy $\tilde{\boldsymbol{\eta}}(\dot{\boldsymbol{x}}, \dot{\boldsymbol{x}}) = 0$ at some point coincide, following the discussion of Section 5.5.4, with null geodesics. Those with $\boldsymbol{\beta}_\star = 0$ are standard geodesics of the Minkowski spacetime. Now, if $\dot{\boldsymbol{x}}_\star$ is spacelike or timelike one can assume, without loss of generality, that $\langle\boldsymbol{\beta}_\star, \dot{\boldsymbol{x}}_\star\rangle = 0$ – following the discussion of Section 5.5.3 this can always be achieved through a reparametrisation of the form given by Equation (5.48) of Lemma 5.1. If $\dot{\boldsymbol{x}}_\star$ and $\boldsymbol{\beta}^\sharp_\star$ generate a timelike 2-surface and $\dot{\boldsymbol{x}}_\star$ is timelike, then the conformal geodesic is a hyperbola in the plane tangent to that 2-surface. An example of such type of curve is given by the expression

$$x(\tau) = \left(\frac{4\tau}{4 - a^2\tau^2}, \frac{1}{a} + \frac{2a\tau^2}{4 - a^2\tau^2}, 0, 0\right), \qquad |\tau| \leq \frac{2}{a} \tag{6.25}$$

where $a^{-2} \equiv -\tilde{\boldsymbol{\eta}}(\dot{\boldsymbol{x}}, \dot{\boldsymbol{x}})$. Examples of these conformal geodesics are depicted in Figure 6.3.

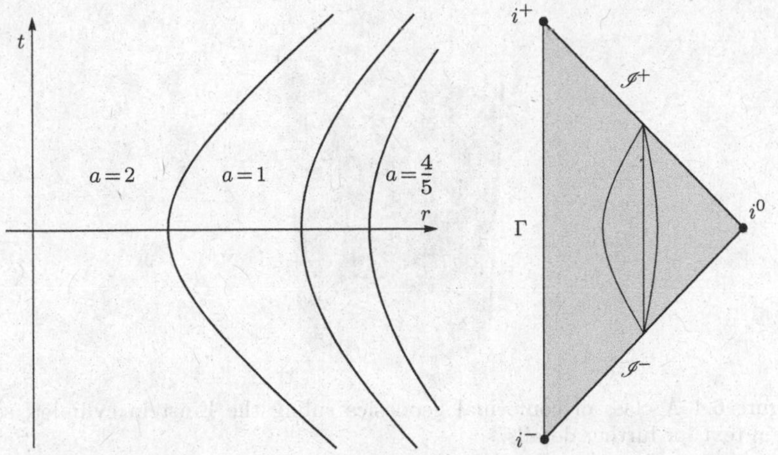

Figure 6.3 Examples of conformal geodesics in the Minkowski spacetime: left, plot in (t, r)-coordinates of the curve (6.25) for the parameter choices $a = \frac{4}{5}$, 1, 2; right, location of the curves in the Penrose diagram of the Minkowski spacetime. Notice that the curves intersect the conformal boundary at the same points. The diagram is quantitatively correct.

A special class of conformal geodesics

As a consequence of the transformation properties of the conformal geodesic equations, the family of curves on the Einstein cylinder given by Equation (6.9) defines a congruence of conformal geodesics on the Minkowski spacetime.

Recalling that $\bar{\boldsymbol{g}}_{\mathscr{E}} \equiv \bar{\Theta}^2 \boldsymbol{g}_{\mathscr{E}}$ and $\boldsymbol{g}_{\mathscr{E}} \equiv \Xi^2_{\mathscr{M}} \tilde{\boldsymbol{\eta}}$ where $\bar{\Theta}$ and $\Xi_{\mathscr{M}}$ are the conformal factors given, respectively, by Equations (6.11) and (6.16), one has that $\bar{\boldsymbol{g}}_{\mathscr{E}} = \Theta^2_{\mathscr{M}} \tilde{\boldsymbol{\eta}}$ with $\Theta_{\mathscr{M}} \equiv \Xi_{\mathscr{M}} \bar{\Theta}$. Along the conformal curves one has that

$$\Xi_{\mathscr{M}} = \cos\tau + \cos\psi = \left(\frac{4 - \bar{\tau}^2}{4 + \bar{\tau}^2} \right) + \cos\psi,$$

where the second equality is obtained from the reparametrisation formula (6.10) and standard trigonometric identities. Moreover, one finds that along the conformal geodesics

$$\Theta_{\mathscr{M}} = 2\cos^2\frac{\psi}{2} \left(1 - \frac{1}{4}\tan^2\frac{\psi}{2}\bar{\tau}^2 \right).$$

To obtain the covector $\boldsymbol{\beta}_{\mathscr{M}}$ associated to the solution of the $\tilde{\boldsymbol{\eta}}$-conformal geodesic equations let

$$\boldsymbol{\Upsilon}_{\mathscr{M}} \equiv \Xi^{-1}_{\mathscr{M}} \mathbf{d}\Xi_{\mathscr{M}} = -\frac{\sin\tau \mathbf{d}T + \sin\psi \mathbf{d}\psi}{\cos\tau + \cos\psi}.$$

Recalling that $\tau = 2\arctan\frac{1}{2}\bar{\tau}$ it follows, using standard trigonometric identities, that

Figure 6.4 A class of conformal geodesics ruling the Einstein cylinder; see main text for further details.

$$\boldsymbol{\Upsilon}_{\mathscr{M}} = -\Theta_{\mathscr{M}}^{-1}\left(\bar{\tau}\mathbf{d}T + \left(1+\frac{1}{4}\bar{\tau}^2\right)\sin\psi\mathbf{d}\psi\right).$$

Finally, defining $\boldsymbol{\beta}_{\mathscr{M}} \equiv \bar{\boldsymbol{\beta}}+\boldsymbol{\Upsilon}_{\mathscr{M}}$, one concludes from the transformation formulae for the solutions of the conformal geodesic equations that the pair $(x(\bar{\tau}), \boldsymbol{\beta}_{\mathscr{M}}(\bar{\tau}))$ with

$$x(\bar{\tau}) = \left(2\arctan\frac{\bar{\tau}}{2}, x_\star\right),$$
$$\boldsymbol{\beta}_{\mathscr{M}}(\bar{\tau}) = \left(\arctan\frac{\bar{\tau}}{2} - \Theta_{\mathscr{M}}^{-1}\bar{\tau}\right)\mathbf{d}T - \Theta_{\mathscr{M}}^{-1}\left(1+\frac{1}{4}\bar{\tau}^2\right)\sin\psi\mathbf{d}\psi,$$

is a solution to the $\tilde{\boldsymbol{\eta}}$-conformal geodesic equations. Notice, in particular, that

$$\boldsymbol{\beta}_{\mathscr{M}}(0) = -\frac{\sin\psi}{1+\cos\psi}\mathbf{d}\psi.$$

A depiction of the above class of conformal geodesics is given in Figure 6.4.

6.2.4 Hyperboloids in the Minkowski spacetime

An important class of spacelike hypersurfaces in the Minkowski spacetime is given by the ***standard hyperboloids***

$$\mathcal{H}_k = \{p \in \mathbb{R}^4 \,|\, t^2(p) - r^2(p) = k\}, \qquad k > 0. \tag{6.26}$$

A direct computation reveals that the unit normal vector to these hypersurfaces is given by

$$\boldsymbol{\nu}^\sharp = \frac{1}{\sqrt{k}}(t\boldsymbol{\partial}_t + r\boldsymbol{\partial}_r).$$

Using this expression one can verify that the extrinsic curvature of the hyperboloids is *pure trace*, that is, proportional to the intrinsic metric of $\mathcal{H}_k$. The mean curvature (i.e. the trace of the extrinsic curvature) is given by

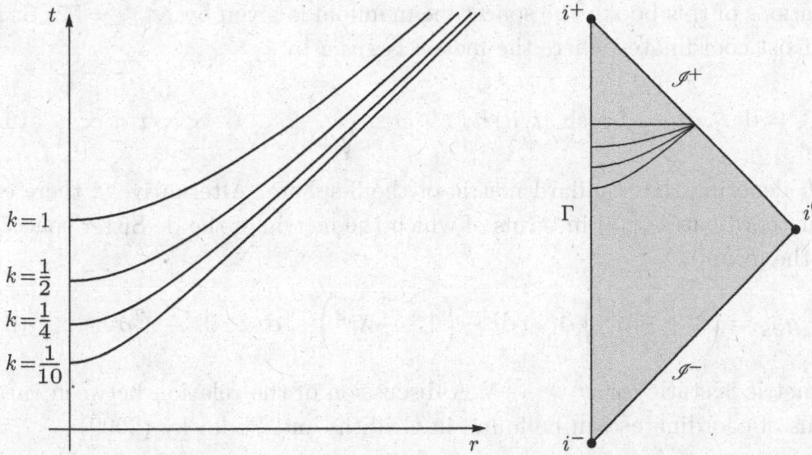

Figure 6.5 Examples of hyperboloids in the Minkowski spacetime: left, the standard hyperboloids $\mathcal{H}_k$ with $k = 1, \frac{1}{2}, \frac{1}{4}, \frac{1}{10}$ in (t, r)-coordinates (see Equation (6.26)); right, location of the hyperboloids in the Penrose diagram. The diagram is quantitatively correct.

$$\tilde{K} = \frac{3}{\sqrt{k}}.$$

That is, the standard hyperboloids are surfaces of *constant mean curvature.* Making use of the coordinates (U, V) introduced in Section 6.2.1, the defining equation for the hyperboloids can be rewritten as $\tan U \tan V = k$. The hyperboloids intersect null infinity whenever $V = \frac{1}{2}\pi$. It follows that, in this case, $U = 0$. Thus, the hyperboloids $\mathcal{H}_k$ intersect null infinity at the same points independent of the value of k. The hypersurfaces differ from each other by the angle α at which they intersect null infinity. One can compute that

$$\tan \alpha = -\frac{\mathrm{d}U}{\mathrm{d}V} = k.$$

In particular, if $k^2 = 1$, one has that $\alpha = \frac{1}{4}\pi$. This particular hyperboloid corresponds to a horizontal line $T = \frac{1}{2}\pi$ in the Penrose diagram of the Minkowski spacetime; see Figure 6.5.

A more general class of hyperboloids can be obtained by translating the standard hyperboloids (6.26). To this end, one considers the defining equation $k = (t - t_\star)^2 - r^2$ for fixed k and $t_\star$. Varying $t_\star$ one obtains a family of ***translated hyperboloids*** $\mathcal{H}_{t_\star,k}$. The intersection of the $\mathcal{H}_{t_\star,k}$ now depends on the value of $t_\star$: if $V = \frac{1}{2}\pi$, it follows that $U = \arctan t_\star$.

6.3 The de Sitter spacetime

The ***de Sitter spacetime*** $(\tilde{\mathcal{M}}_{dS}, \tilde{\boldsymbol{g}}_{dS})$ is the solution to the vacuum Einstein field equations $\boldsymbol{Ric}[\tilde{\boldsymbol{g}}] = \lambda \tilde{\boldsymbol{g}}$ with negative constant Ricci scalar, in the signature conventions of this book. The spacetime manifold is given by $\tilde{\mathcal{M}}_{dS} = \mathbb{R}\times\mathbb{S}^3$ and there exist coordinates where the metric is given by

$$\tilde{\boldsymbol{g}}_{dS} = \mathbf{d}t\otimes\mathbf{d}t - a^2\cosh^2(t/a)\,\boldsymbol{\hbar}, \qquad a \equiv \sqrt{\frac{3}{|\lambda|}}, \qquad -\infty < t < \infty, \tag{6.27}$$

with $\boldsymbol{\hbar}$ denoting the standard metric of the 3-sphere. Alternatively, there exist further coordinates $(\bar{t},\bar{r})$ in terms of which the metric of the de Sitter spacetime takes the form

$$\tilde{\boldsymbol{g}}_{dS} = \left(1+\frac{1}{3}\lambda\bar{r}^2\right)\mathbf{d}\bar{t}\otimes\mathbf{d}\bar{t} - \left(1+\frac{1}{3}\lambda\bar{r}^2\right)^{-1}\mathbf{d}\bar{r}\otimes\mathbf{d}\bar{r} - \bar{r}^2\boldsymbol{\sigma}. \tag{6.28}$$

This metric is static for $\bar{r}^2 > -\frac{1}{3}\lambda$. A discussion of the relation between various systems of coordinates can be found in Griffiths and Podolský (2009).

To construct a conformal extension of the de Sitter spacetime it is convenient to introduce a new coordinate T via the condition

$$\mathbf{d}t = a\cosh(t/a)\,\mathbf{d}T.$$

Fixing the constant of integration by requiring that $T=0$ if $t=0$ one obtains

$$T = 2a\arctan e^t - \frac{1}{2}a\pi, \qquad t = \ln\tan\left(\frac{T}{2a}+\frac{1}{4}\pi\right),$$

or, equivalently $\tan(T/2)=\tanh(t/2a)$. Using standard trigonometric identities the latter can be recast as

$$\cos T = \frac{1}{\cosh(t/a)}.$$

Thus, one concludes that

$$\tilde{\boldsymbol{g}}_{dS} = a^2\cosh^2 t\,(\mathbf{d}T\otimes\mathbf{d}T - \boldsymbol{\hbar}).$$

The latter expression suggests introducing the conformal factor

$$\Xi_{dS} = \frac{1}{a\cosh(t/a)} = \frac{1}{a}\cos T, \tag{6.29}$$

so that the conformal metric $\Xi^2_{dS}\tilde{\boldsymbol{g}}$ is, again, that of the Einstein cylinder. It follows that the locus of points for which $\Xi_{dS}=0$ corresponds to $T=\pm\frac{1}{2}\pi$; notice that $T\to\pm\frac{1}{2}\pi$ as $t\to\pm\infty$. In view of the latter, one defines ***future and past conformal infinity***, respectively, as

$$\mathscr{I}^{\pm}_{dS} \equiv \left\{ p\in\mathcal{M}_{\mathscr{E}} \;\middle|\; T(p) = \pm\frac{\pi}{2}\right\}. \tag{6.30}$$

The terminology for these sets will be justified in the next section. From the previous discussion it follows that the de Sitter spacetime is conformal to the domain

$$\tilde{\mathcal{M}}_{dS} \equiv \left\{ p\in\mathcal{M}_{\mathscr{E}} \;\middle|\; -\frac{\pi}{2} < T(p) < \frac{\pi}{2}\right\}$$

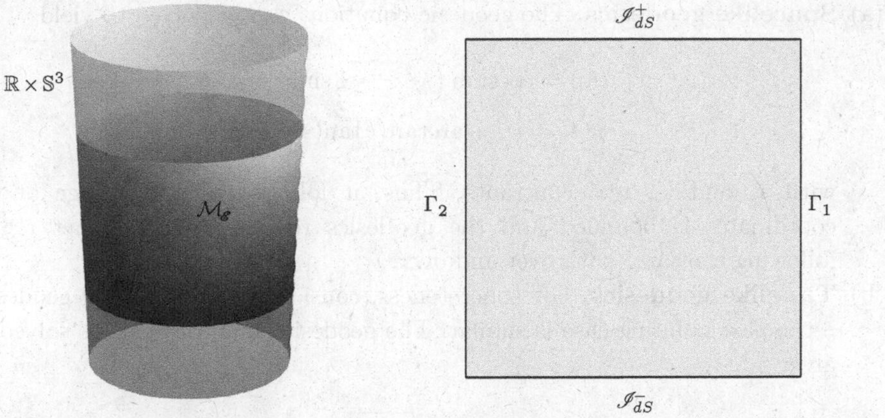

Figure 6.6 Conformal extension of the de Sitter spacetime. Left, conformal embedding of the de Sitter spacetime in the Einstein cylinder: the shaded region corresponds to the set $\mathcal{M}_{dS}$ of Equation (6.31). Right, Penrose diagram of the de Sitter spacetime: the lines Γ_1 and Γ_2 correspond to the axes of symmetry, while $\mathscr{I}^+_{dS}$ and $\mathscr{I}^-_{dS}$ denote, respectively, future and past conformal infinity; see main text for further details.

of the Einstein cylinder. Moreover, one sets

$$\mathcal{M}_{dS} \equiv \tilde{\mathcal{M}}_{dS} \cup \mathscr{I}^+_{dS} \cup \mathscr{I}^-_{dS}. \tag{6.31}$$

To construct the Penrose diagram of the de Sitter spacetime one considers the quotient domain $\mathcal{Q}_{dS} \equiv \mathcal{M}_{dS}/SO(3)$. The boundary $\partial\mathcal{Q}_{dS}$ of the quotient manifold consists of the projection of the conformal boundary (to be denoted again by $\mathscr{I}^{\pm}_{dS}$) and the two centres of symmetry Γ_1 and Γ_2 given, respectively, by the conditions $\psi = 0$ and $\psi = \pi$. A depiction of the Penrose diagram of the de Sitter spacetime is given in Figure 6.6, right panel.

6.3.1 Behaviour of geodesics

As in the case of the Minkowski spacetime, intuition about the conformal representation of the de Sitter spacetime can be obtained through the analysis of the behaviour of geodesics. For simplicity set $\lambda = -3$ so that $a = 1$ in the line element (6.27). The geodesic equations can be found to be

$$t'^2 - \cosh^2 t\, \psi'^2 = \epsilon, \qquad \psi' = \frac{\ell}{\cosh^2 t},$$

where ℓ is a constant, ϵ takes the values 1, 0 or -1 depending on whether one considers timelike, null or spacelike geodesics and $'$ denotes differentiation with respect to an affine parameter s. As in the case of the Minkowski spacetime one distinguishes the following three cases:

(a) **Spacelike geodesics.** The geodesic equations can be solved to yield

$$t(\mathrm{s}) = \operatorname{arcsinh}\left(\sqrt{\ell^2-1}\sin(\mathrm{s}-\mathrm{s}_\star)\right),$$
$$\psi(\mathrm{s}) = \psi_\star + \arctan\big(\ell\tan(\mathrm{s}-\mathrm{s}_\star)\big),$$

with $\mathrm{s}_\star$ and $\psi_\star$ real constants. Thus, it follows that the range of the coordinates is bounded and the geodesics remain in a compact region following the same path over and over.

(b) **Timelike geodesics.** For concreteness, consider future-pointing geodesics – the past pointing case is similar. The geodesic equations can be solved to give:

$$t(\mathrm{s}) = \operatorname{arcsinh}\left(\sqrt{\ell^2+1}\sinh(\mathrm{s}-\mathrm{s}_\star)\right),$$
$$\psi(\mathrm{s}) = \psi_\star + \arctan\big(\ell\tanh(\mathrm{s}-\mathrm{s}_\star)\big).$$

It follows that for $\mathrm{s}\to\infty$ one has $t\to\infty$. Thus, one obtains the limit points

$$T = \frac{\pi}{2}, \qquad \psi = \psi_\star + \arctan\ell.$$

The geodesics approach a definite point on the spacelike hypersurface $\mathscr{I}^+_{dS}$ defined in (6.30).

(c) **Null geodesics.** In this case, the geodesic equations give the solution

$$t(\mathrm{s}) = \operatorname{arcsinh}\mathrm{s},$$
$$\psi(\mathrm{s}) = \psi_\star + \arctan(\ell\mathrm{s}).$$

From these expressions it follows, on the one hand, that if $\mathrm{s}\to\infty$, then

$$t\to\infty, \quad T\to\frac{\pi}{2}, \quad \psi\to\psi_\star+\frac{\pi}{2},$$

while on the other, if $\mathrm{s}\to-\infty$, then

$$t\to-\infty, \quad T\to-\frac{\pi}{2}, \quad \psi\to\psi_\star-\frac{\pi}{2}.$$

Hence, the null geodesics start and finish, respectively, at $\mathscr{I}^-_{dS}$ and $\mathscr{I}^+_{dS}$, in antipodal points on $\mathbb{S}^3$.

6.3.2 Conformal geodesics in the de Sitter spacetime

As a consequence of the conformal invariance of conformal geodesics, the curves in the Einstein universe discussed in Section 6.1.3 are also conformal geodesics of the de Sitter spacetime.

Making use of the relations $\boldsymbol{g}_{\mathscr{E}} = \Xi^2_{dS}\tilde{\boldsymbol{g}}_{dS}$ and $\bar{\boldsymbol{g}}_{\mathscr{E}} = \bar{\Theta}^2\boldsymbol{g}_{\mathscr{E}}$ where $\bar{\Theta}$ and Ξ_{dS} are the conformal factors given, respectively, by Equations (6.11) and (6.29), one finds that $\bar{\boldsymbol{g}}_{\mathscr{E}} = \Theta_{dS}^2 \tilde{\boldsymbol{g}}_{dS}$ with $\Theta_{dS} \equiv \bar{\Theta}\Xi_{dS}$. A calculation using the first of the equations in (6.10) shows that

$$\Xi_{dS} = a\cos\tau = a\left(\frac{4-\bar{\tau}^2}{4+\bar{\tau}^2}\right),$$

where in a slight abuse of notation *the coordinate T has been replaced by the parameter of the curves* τ. Hence, one finds that

$$\Theta_{dS} = a\left(1 - \frac{1}{4}\bar{\tau}^2\right),$$

so that Θ_{dS} vanishes at $\bar{\tau} = \pm 2$. To construct the covector associated to the congruence of conformal geodesics consider $\boldsymbol{\Upsilon}_{dS} \equiv \Xi_{dS}{}^{-1}\mathbf{d}\Xi_{dS}$. A calculation then shows that

$$\boldsymbol{\Upsilon}_{dS} = -\frac{16\bar{\tau}}{16-\bar{\tau}^4}\mathbf{d}\bar{\tau} = -\frac{4\bar{\tau}}{4-\bar{\tau}^2}\mathbf{d}\tau.$$

Letting $\boldsymbol{\beta}_{dS} \equiv \bar{\boldsymbol{\beta}} + \boldsymbol{\Upsilon}_{dS}$, it follows from the transformation laws for conformal geodesics in Section 5.5.2 that the pair $(x(\bar{\tau}), \boldsymbol{\beta}_{ds}(\bar{\tau}))$, with

$$x(\bar{\tau}) = \left(2\arctan\frac{\bar{\tau}}{2}, x_\star\right), \qquad \boldsymbol{\beta}_{dS}(\bar{\tau}) = -\left(\frac{2\bar{\tau}}{4-\bar{\tau}^2}\right)\mathbf{d}\bar{\tau}, \tag{6.32}$$

is a solution to the $\tilde{\boldsymbol{g}}_{dS}$-conformal geodesic equations with parameter $\bar{\tau}$. Notice, in particular, that at the Cauchy surface given by $\bar{\tau} = 0$ one has that $\boldsymbol{\beta}_{dS}(0) = 0$.

Following the discussion from the previous paragraph, the surface given by the condition $\bar{\tau} = -2$ represents past null infinity $\mathscr{I}_{dS}^-$. In some applications, one needs to prescribe initial data for the congruence of conformal geodesics at $\mathscr{I}_{dS}^-$. In this case, it is convenient to introduce the further reparametrisation $\hat{\tau} = \bar{\tau} + 2$ so that

$$\Theta_{dS} = \hat{\tau} - \frac{1}{4}\hat{\tau}^2, \qquad \boldsymbol{\beta}_{dS} = -\left(\frac{2\hat{\tau}-4}{4\hat{\tau}+\hat{\tau}^2}\right)\mathbf{d}\hat{\tau}.$$

6.4 The anti-de Sitter spacetime

The ***anti-de Sitter spacetime*** is given by the manifold $\tilde{\mathcal{M}}_{adS} \approx \mathbb{R}^4$ equipped with the metric

$$\tilde{\boldsymbol{g}}_{adS} = \cosh^2 r\,\mathbf{d}t \otimes \mathbf{d}t - a^2\left(\mathbf{d}r \otimes \mathbf{d}r + \sinh^2 r\,\boldsymbol{\sigma}\right), \qquad a \equiv \sqrt{\frac{3}{\lambda}},$$

with $t \in \mathbb{R}$ and $r \in (0, \infty)$. Strictly speaking, this spacetime is the so-called ***universal covering space of the anti-de Sitter spacetime*** – the *classical* anti-de Sitter spacetime has a periodic time coordinate and, thus, closed timelike curves. As in the case of the de Sitter spacetime, it is possible to introduce coordinates $(\bar{t}, \bar{r})$ in terms of which the metric takes the form

$$\tilde{\boldsymbol{g}}_{adS} = \left(1 + \frac{1}{3}\lambda\bar{r}^2\right)\mathbf{d}\bar{t} \otimes \mathbf{d}\bar{t} - \left(1 + \frac{1}{3}\lambda\bar{r}^2\right)^{-1}\mathbf{d}\bar{r} \otimes \mathbf{d}\bar{r} - \bar{r}^2\boldsymbol{\sigma}. \tag{6.33}$$

As (in the signature conventions used in this book) $\lambda > 0$, there are no horizons in the anti-de Sitter spacetime; see, for example, Griffiths and Podolský (2009).

To obtain a conformal representation of this spacetime, it is convenient to consider a new radial coordinate via the condition

$$\mathbf{d}r = \cosh r \mathbf{d}\psi,$$

so that $\psi = 2\arctan e^r - \frac{1}{2}\pi$. This condition is equivalent to $\tan\psi = \sinh r$. Hence, one has that $\psi \in [0, \frac{1}{2}\pi]$. Setting $T = t/a$, a calculation then shows that

$$\tilde{\boldsymbol{g}}_{adS} = a^2 \cosh^2 r \left(\mathbf{d}T \otimes \mathbf{d}T - \boldsymbol{\hbar}\right).$$

The latter suggests introducing the conformal factor

$$\Xi_{adS} = \frac{a}{\cosh r} = a\cos\psi. \tag{6.34}$$

Thus, the conformal metric $\Xi_{adS}^2 \tilde{\boldsymbol{g}}$ is, again, that of the Einstein cylinder. From the previous discussion it follows that the anti-de Sitter spacetime is conformal to the domain

$$\tilde{\mathcal{M}}_{adS} = \left\{ p \in \mathcal{M}_{\mathscr{E}} \;\middle|\; 0 \leq \psi(p) < \frac{\pi}{2} \right\},$$

of the Einstein cylinder. Notice, however, that in contrast to the conformal representation of the de Sitter spacetime, the conformal anti-de Sitter spacetime does not cover the whole spatial sections of the cylinder. In particular, one has that the conformal factor Ξ_{adS} vanishes at $\psi = \frac{1}{2}\pi$; that is, the conformal boundary is, in this case, a timelike hypersurface. Following standard usage, define the ***conformal infinity of the anti-de Sitter spacetime*** as

$$\mathscr{I}_{adS} \equiv \left\{ p \in \mathcal{M}_{\mathscr{E}} \;\middle|\; \psi(p) = \frac{\pi}{2} \right\}.$$

One also defines

$$\mathcal{M}_{adS} \equiv \tilde{\mathcal{M}}_{adS} \cup \mathscr{I}_{adS}. \tag{6.35}$$

The Penrose diagram for the anti-de Sitter spacetime is constructed by considering the quotient domain $\mathcal{Q}_{adS} \equiv \mathcal{M}_{adS}/SO(3)$ with boundary $\partial\mathcal{Q}_{adS} = \mathscr{I}_{adS} \cup \Gamma$ where $\mathscr{I}_{adS}$ denotes the projection of null infinity onto $\mathcal{Q}_{\mathscr{E}}$ and Γ denotes the centre of symmetry given by the condition $\psi = 0$. A depiction of the Penrose diagram is given in Figure 6.7.

6.4.1 Geodesics in the anti-de Sitter spacetime

In what follows, for simplicity assume that $\lambda = 3$ so that $a = 1$. Radial geodesics in the anti-de Sitter spacetime are described by the equations

$$\cosh^2 r\, T'^2 - r'^2 = \epsilon, \qquad T' = \frac{\ell}{\cosh^2 r}.$$

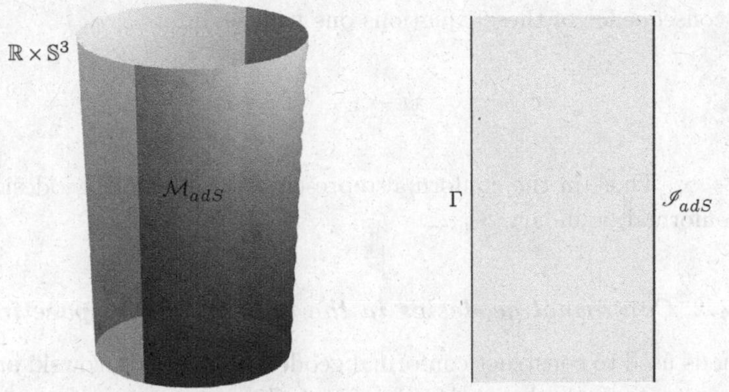

Figure 6.7 Conformal extension of the anti-de Sitter spacetime. Left, conformal embedding of the anti-de Sitter spacetime in the Einstein cylinder: the shaded region corresponds to the set $\mathcal{M}_{adS}$ of Equation (6.35). Right, Penrose diagram of the anti-de Sitter spacetime: the line Γ corresponds to the axis of symmetry, while $\mathscr{I}_{adS}$ denotes conformal infinity; see main text for further details.

These equations can be obtained from those of the de Sitter spacetime by the replacements $t \mapsto r$, $\psi \mapsto T$ and $\epsilon \mapsto -\epsilon$; see Section 6.3.1. Again, one has three cases to consider:

(a) **Spacelike geodesics.** In this case, the geodesic equations can be solved to give

$$r(\mathrm{s}) = \operatorname{arcsinh}\left(\sqrt{\ell^2+1}\sinh(\mathrm{s}-\mathrm{s}_\star)\right),$$
$$T(\mathrm{s}) = T_\star + \arctan(\ell\tanh(\mathrm{s}-\mathrm{s}_\star)),$$

with $\mathrm{s}_\star$ and $T_\star$ real constants. Thus, for $\mathrm{s}\to\infty$ one obtains the limits

$$r\to\infty, \quad \psi\to\frac{\pi}{2}, \quad T\to T_\star+\arctan\ell.$$

As a consequence, in the conformal representation, radial spacelike geodesics approach the conformal boundary $\mathscr{I}_{adS}$.

(b) **Timelike geodesics.** For simplicity only future-oriented geodesics are considered. The solution to the geodesic equations is then given by

$$r(\mathrm{s}) = \operatorname{arcsinh}\left(\sqrt{\ell^2-1}\sin(\mathrm{s}-\mathrm{s}_\star)\right),$$
$$T(\mathrm{s}) = T_\star + \arctan(\ell\tan(\mathrm{s}-\mathrm{s}_\star)).$$

Accordingly, the coordinate r is periodic while τ grows unbounded – the limit points of these curves are not in the Einstein cylinder.

(c) **Null geodesics.** In this case, the solution to the geodesic equations is

$$r(\mathrm{s}) = \operatorname{arcsinh}\mathrm{s},$$
$$T(\mathrm{s}) = T_\star + \arctan\mathrm{s}.$$

As a consequence of these equations one has the limits

$$r \to \infty, \quad \psi \to \frac{\pi}{2}, \quad T \to T_\star + \frac{\pi}{2},$$

as s $\to \infty$. Thus, in the conformal representation, the null geodesics end at the conformal boundary $\mathscr{I}_{adS}$.

6.4.2 Conformal geodesics in the anti-de Sitter spacetime

The methods used to construct conformal geodesics in the Minkowski and the de Sitter spacetimes can also be used in the anti-de Sitter spacetime. For conciseness, the discussion is restricted to the class of conformal geodesics arising from the curves (6.9) in the Einstein universe.

Using that $\boldsymbol{g}_{\mathscr{E}} = \Xi^2_{adS}\tilde{\boldsymbol{g}}_{adS}$ and that $\bar{\boldsymbol{g}}_{\mathscr{E}} = \bar{\Theta}^2\boldsymbol{g}_{\mathscr{E}}$ where $\bar{\Theta}$ and Ξ_{adS} are the conformal factors given by Equations (6.11) and (6.34), one finds that $\bar{\boldsymbol{g}}_{\mathscr{E}} = \Theta^2_{adS}\tilde{\boldsymbol{g}}_{adS}$, where

$$\Theta_{adS} \equiv \bar{\Theta}\Xi_{adS} = a\cos\psi\left(1 + \frac{1}{4}\bar{\tau}^2\right).$$

Letting

$$\boldsymbol{\Upsilon}_{adS} \equiv \Xi^{-1}_{adS}\mathbf{d}\Xi_{adS} = -\tan\psi\mathbf{d}\psi,$$

one finds that the associated covector is given by

$$\begin{aligned}\boldsymbol{\beta}_{adS} &\equiv \bar{\boldsymbol{\beta}} + \boldsymbol{\Upsilon}_{adS} = \frac{1}{2}\bar{\tau}\mathbf{d}T - \tan\psi\mathbf{d}\psi \\ &= \frac{2\bar{\tau}}{4+\bar{\tau}^2}\mathbf{d}\bar{\tau} - \tan\psi\mathbf{d}\psi.\end{aligned}$$

The expression for the actual curve is, as in the case of the de Sitter spacetime, given by

$$x(\bar{\tau}) = \left(2\text{arctan}\frac{\bar{\tau}}{2}, x_\star\right). \tag{6.36}$$

An important property of this non-intersecting congruence of conformal geodesics is that *curves that for some value of the parameter $\bar{\tau}$ are at the conformal boundary $\mathscr{I}_{adS}$ remain in it for all values of $\bar{\tau}$; this observation follows from the fact that the curve given by Equation* (6.36) *is constant in the spatial directions.*

Remark. As $\text{arctan}\frac{1}{2}\bar{\tau} \to \frac{1}{2}\pi$ as $\bar{\tau} \to \infty$ and $\tau = 2\text{arctan}\frac{1}{2}\bar{\tau}$, the parameter $\bar{\tau}$ does not cover the whole Einstein cylinder and only exhausts the *slab* $[-\pi, \pi]\times\mathbb{S}^3$. In order to continue the conformal geodesic to other portions of the anti-de Sitter spacetime one has to introduce a reparametrisation of the curve by means of a fractional transformation as discussed in Lemma 5.1.

6.5 Conformal extensions of static and stationary black hole spacetimes

A natural extension of the discussion of the previous sections is the analysis of the conformal structure of spacetimes describing black holes. The more complicated topology of these spacetimes and the presence of singularities and horizons make this analysis a much more challenging endeavour. In fact, several aspects of the conformal structure of static and stationary black holes are open research questions.

6.5.1 The Schwarzschild spacetime

The Schwarzschild spacetime, being static and spherically symmetric, is the simplest type of black hole spacetime. The *Birkhoff theorem* states that any spherically symmetric solution to the vacuum Einstein field equations with vanishing cosmological constant is, in fact, isometric to the Schwarzschild spacetime; see, for example, Misner et al. (1973). Moreover, the black hole uniqueness theorems show that the Schwarzschild spacetime is the only static black hole spacetime; see, for example, Chruściel et al. (2012b) for an entry point to the extensive literature on this topic.

The ***Schwarzschild metric*** is given in standard (t, r) coordinates by the line element

$$\tilde{\boldsymbol{g}}_{\mathscr{S}} = \left(1 - \frac{2m}{r}\right) \mathbf{d}t \otimes \mathbf{d}t - \left(1 - \frac{2m}{r}\right)^{-1} \mathbf{d}r \otimes \mathbf{d}r - r^2 \boldsymbol{\sigma}, \tag{6.37}$$

with m the so-called ***mass parameter***. The reader interested in a discussion of the various aspects of the Schwarzschild spacetime is referred to, for example, Griffiths and Podolský (2009) and Hawking and Ellis (1973).

To obtain a conformal extension of the Schwarzschild spacetime it is convenient to make use of coordinates adapted to the light-cone structure of the spacetime. Accordingly, one introduces the ***advanced*** and ***retarded null Eddington-Finkelstein coordinates***

$$u \equiv t - r - 2m \log |r - 2m|, \qquad v \equiv t + r + 2m \log |r - 2m|,$$

so that the line element (6.37) transforms into

$$\tilde{\boldsymbol{g}}_{\mathscr{S}} = \frac{1}{2}\left(1 - \frac{2m}{r}\right) (\mathbf{d}u \otimes \mathbf{d}v + \mathbf{d}v \otimes \mathbf{d}u) - r^2 \boldsymbol{\sigma},$$

where the relation between r and the coordinates (u, v) is given implicitly by the condition

$$r + 2m \log |r - 2m| = \frac{1}{2}(v - u).$$

The singular behaviour of the metric at $r = 2m$ is then removed by means of a reparametrisation of the null coordinates. Namely, one sets

$$U \equiv -4me^{-u/4m}, \qquad V \equiv 4me^{v/4m},$$

so that one obtains

$$\tilde{\boldsymbol{g}}_{\mathscr{S}} = \frac{m}{r}e^{-r/2m}(\mathbf{d}U \otimes \mathbf{d}V + \mathbf{d}V \otimes \mathbf{d}U) - r^2\boldsymbol{\sigma}, \tag{6.38}$$

where r is now given implicitly by the condition

$$UV = -8m(r - 2m)e^{r/2m}.$$

The horizon is then given by the condition $UV = 0$ while the singularity corresponds to $UV = 16m^2$. The line element in Equation (6.38) is the so-called ***Kruskal-Székeres form*** of the Schwarzschild spacetime. It provides the maximal analytic extension of the Schwarzschild metric (6.37). Inspection of the admissible range of coordinates in Equation (6.38) shows that the resulting maximal manifold has the topology of $\mathbb{R} \times \mathbb{R} \times \mathbb{S}^2$.

To compactify the Kruskal-Székeres form of the Schwarzschild metric one introduces a further coordinate transformation:

$$\bar{U} \equiv \arctan\left(\frac{U}{4m}\right), \qquad \bar{V} \equiv \arctan\left(\frac{V}{4m}\right)$$

where

$$-\frac{1}{2}\pi < \bar{U} < \frac{1}{2}\pi, \qquad -\frac{1}{2}\pi < \bar{V} < \frac{1}{2}\pi, \qquad -\frac{1}{2}\pi < \bar{U} + \bar{V} < \frac{1}{2}\pi.$$

It follows then that

$$\mathbf{d}U = 4m \sec^2 \bar{U} \mathbf{d}\bar{U}, \qquad \mathbf{d}V = 4m \sec^2 \bar{V} \mathbf{d}\bar{V},$$

so that the line element (6.38) transforms into

$$\tilde{\boldsymbol{g}}_{\mathscr{S}} = \sec^2 \bar{U} \sec^2 \bar{V} \left(\frac{16m^3}{r}e^{-r/2m}(\mathbf{d}\bar{U} \otimes \mathbf{d}\bar{V} + \mathbf{d}\bar{V} \otimes \mathbf{d}\bar{U}) - r^2 \cos^2 \bar{U} \cos^2 \bar{V}\boldsymbol{\sigma}\right).$$

It is, therefore, natural to consider a conformal factor of the form

$$\Xi_{\mathscr{S}} = \cos \bar{U} \cos \bar{V},$$

so that $\boldsymbol{g}_{\mathscr{S}} = \Xi^2_{\mathscr{S}}\tilde{\boldsymbol{g}}_{\mathscr{S}}$ is given by

$$\boldsymbol{g}_{\mathscr{S}} = \frac{16m^3}{r}e^{-r/2m}(\mathbf{d}\bar{U} \otimes \mathbf{d}\bar{V} + \mathbf{d}\bar{V} \otimes \mathbf{d}\bar{U}) - r^2 \cos^2 \bar{U} \cos^2 \bar{V}\boldsymbol{\sigma},$$

where $r = r(\bar{U}, \bar{V})$. This conformal metric is singular at $r = 0$ (the singularity). In order to discuss the structure of the conformal boundary of the Schwarzschild spacetime, it is convenient to introduce the coordinates

$$T \equiv \bar{V} + \bar{U}, \qquad \psi \equiv \bar{V} - \bar{U},$$

so that $T \in [-\pi, \pi]$, $\psi \in [-\pi, \pi]$. One sees that the maximal analytic extension of the Schwarzschild spacetime is conformal to the interior of the domain $\mathcal{M}_{\mathscr{S}} \subset (-\pi, \pi) \times [-\pi, \pi] \times \mathbb{S}^2$ with boundary given by

$$\partial \mathcal{M}_{\mathscr{S}} = \mathscr{I}_1^+ \cup \mathscr{I}_2^+ \cup \mathscr{I}_1^- \cup \mathscr{I}_2^- \cup i_1^0 \cup i_2^0 \cup i_1^+ \cup i_2^+ \cup i_1^- \cup i_2^-,$$

where by analogy with the analysis of the conformal boundary of the Minkowski spacetime one defines the various components of ***null infinity*** as

$$\mathscr{I}_1^+ \equiv \left\{ \bar{V} = \frac{1}{2}\pi \right\}, \qquad \mathscr{I}_2^+ \equiv \left\{ \bar{U} = \frac{1}{2}\pi \right\},$$
$$\mathscr{I}_1^- \equiv \left\{ \bar{U} = -\frac{1}{2}\pi \right\}, \qquad \mathscr{I}_2^- \equiv \left\{ \bar{V} = -\frac{1}{2}\pi \right\},$$

and the two components of ***spatial infinity*** as

$$i_1^0 \equiv \{T = 0,\ \psi = \pi\}, \qquad i_2^0 \equiv \{T = 0,\ \psi = -\pi\}.$$

Finally, the ***timelike infinities*** are given by

$$i_1^\pm \equiv \left\{ T = \pm\pi,\ \psi = \frac{1}{2}\pi \right\}, \qquad i_2^\pm \equiv \left\{ T = \pm\pi,\ \psi = -\frac{1}{2}\pi \right\}.$$

An analysis of the geodesics on the Schwarzschild spacetime justifies the name given to the various components of $\partial\mathcal{M}_{\mathscr{S}}$. Observe that the singularities at $r = 0$ are not included as part of the boundary $\partial\mathcal{M}_{\mathscr{S}}$. In this representation, the spatial infinities i_1^0 and i_2^0 can be seen to correspond to two points on the conformal manifold. Further properties of the conformal structure of the Schwarzschild spacetime – in particular, the nature of i^0 – will be analysed in the context of the *conformal Einstein field equations* in Chapter 20. Finally, the Penrose diagram of the Schwarzschild spacetime can be readily obtained by considering the quotient manifold $\mathcal{Q}_{\mathscr{S}} = \mathcal{M}_{\mathscr{S}}/SO(3)$; the resulting diagram is given in Figure 6.8.

Conformal geodesics in the Schwarzschild spacetime

A detailed analysis of a class of conformal geodesics in this spacetime can be found in Friedrich (2003a) where it is shown that the Schwarzschild spacetime can be completely covered by a (non-singular) congruence of conformal geodesics. This congruence is adapted to the spherical symmetry of the spacetime.

6.5.2 Conformal extensions of other static, spherically symmetric spacetimes

The procedure to construct a conformal extension of the Schwarzschild spacetime discussed in Section 6.5.1 can be generalised to include a wide class of *static, spherically symmetric* spacetimes. In this section, an adaptation of a general procedure given on Walker (1970) is discussed. This discussion illuminates the conformal diagram of a number of spacetimes.

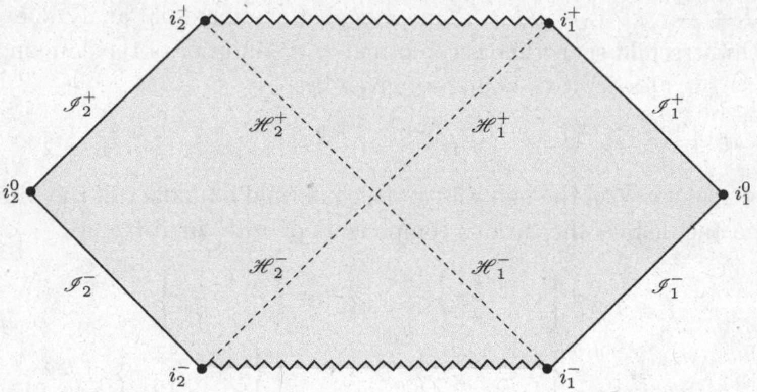

Figure 6.8 Penrose diagram of the Schwarzschild spacetime. The null hypersurfaces $\mathscr{I}_1^\pm$ and $\mathscr{I}_2^\pm$ correspond to the four different components of null infinity, while the points i_1^0, i_2^0 and $i_1^\pm$, $i_2^\pm$ denote, respectively, the various locations of spatial and timelike infinities. The serrated lines denote the singularities, and $\mathscr{H}_1^\pm$ and $\mathscr{H}_2^\pm$ correspond to the various components of the horizon; see the main text for further details.

In what follows, let $(\tilde{\mathcal{M}}, \tilde{\boldsymbol{g}})$ denote a spherically symmetric spacetime endowed with a further Killing vector $\boldsymbol{\partial}_t$. The following considerations will be independent of the matter content of the spacetime; hence, the spacetime is not assumed to be a vacuum. Attention will be restricted to spacetimes in which it is possible to find coordinates (t, r) such that the metric $\tilde{\boldsymbol{g}}$ takes the form

$$\tilde{\boldsymbol{g}} = F(r)\mathbf{d}t \otimes \mathbf{d}t - F(r)^{-1}\mathbf{d}r \otimes \mathbf{d}r - r^2\boldsymbol{\sigma}.$$

The coordinate r is an areal coordinate; that is, the area of a 2-sphere described by the conditions $t =$ constant, $r =$ constant is $4\pi r^2$. The function $F(r)$ is the norm of the Killing vector $\boldsymbol{\partial}_t$. When $F(r) > 0$ the Killing vector $\boldsymbol{\partial}_t$ is timelike, and, thus, the metric $\tilde{\boldsymbol{g}}$ is ***static***.

To simplify the presentation, the subsequent analysis will make use of the quotient manifold $\tilde{\mathcal{Q}} = \tilde{\mathcal{M}}/SO(3)$. The two-dimensional quotient metric $\tilde{\boldsymbol{\gamma}}$ induced by $\tilde{\boldsymbol{g}}$ on $\tilde{\mathcal{Q}}$ is given by

$$\tilde{\boldsymbol{\gamma}} = F(r)\mathbf{d}t \otimes \mathbf{d}t - F(r)^{-1}\mathbf{d}r \otimes \mathbf{d}r.$$

The Levi-Civita connection associated to the Lorentzian metric $\tilde{\boldsymbol{\gamma}}$ will be denoted by $\tilde{\mathcal{D}}$. Let $\dot{\boldsymbol{x}} = (\dot{t}, \dot{r})$ denote the tangent vector to an affinely parametrised geodesic in $\tilde{\mathcal{Q}}$; here, and in what follows, a dot ($\dot{}$) denotes differentiation with respect to an affine parameter. The geodesic equation $\tilde{\mathcal{D}}_{\dot{\boldsymbol{x}}}\dot{\boldsymbol{x}} = 0$ can be integrated once to yield

$$\dot{t} = \varkappa F, \qquad \dot{r} = \sqrt{\varkappa^2 - \epsilon F},$$

where $\varkappa$ is a constant and $\epsilon \equiv \tilde{\boldsymbol{\gamma}}(\dot{\boldsymbol{x}}, \dot{\boldsymbol{x}})$. As $\tilde{\boldsymbol{\gamma}}$ is a two-dimensional metric, the only invariant of the curvature of $\tilde{\boldsymbol{\gamma}}$ is the Ricci scalar $R[\tilde{\boldsymbol{\gamma}}] = F''$, where $'$ denotes differentiation with respect to r. In the remainder of this section it will be shown that if F and F'' are finite for all $r \in \mathbb{R}$, then every geodesic in $\tilde{\mathcal{Q}}$ can be extended until it is complete. If, on the other hand, F or F'' become unbounded for some value $r_{\not{\iota}}$, then only those geodesics along which $r = r_{\not{\iota}}$ within a finite affine distance from some point in $\tilde{\mathcal{Q}}$ are incomplete and inextendible; hence, $\tilde{\mathcal{Q}}$ and also $\tilde{\mathcal{M}}$ are singular. The extensions obtained from the following considerations are *maximal*.

Elementary blocks. In what follows, assume that F has a finite number of zeros, to be denoted by a_i, $i = 1, \ldots, n$ with $a_1 < \cdots < a_n$. If F approaches a constant finite value as $r \to \infty$, so that $R[\tilde{\gamma}] = F'' \to 0$, then one can redefine coordinates so that $\lim_{r\to\infty} F = \pm 1$. In this case $\tilde{\mathcal{Q}}$ is *asymptotically flat* and a null conformal boundary similar to that of the Minkowski spacetime can be constructed; an analogous discussion can be made in the case $r \to -\infty$. A different possible asymptotic behaviour occurs when F becomes unbounded as $r \to \infty$. Using the de Sitter and the anti-de Sitter metrics in static form as given by Equations (6.28) and (6.33) one sees that the behaviour $F \to -\infty$ and $F \to \infty$ as $r \to \infty$ corresponds, respectively, to *de Sitter-like* and *anti-de Sitter-like* asymptotic regions. These regions can be compactified to obtain conformal boundaries similar to those of the de Sitter and anti-de Sitter spacetimes, that is, given, respectively, by spacelike and timelike hypersurfaces.

When F vanishes, the orbits of the timelike Killing vector become null; that is, one has a ***Killing horizon***. This suggests dividing $\tilde{\mathcal{Q}}$ into $n + 1$ regions (***blocks***). Each of these regions is bounded by two of the Killing horizons, by a Killing horizon and conformal infinity, or by a Killing horizon and a singular line at $r = r_{\not{\iota}}$ for $r_{\not{\iota}}$ fixed. The maximal extension of $\tilde{\mathcal{Q}}$ is found by *gluing* together elementary blocks along their boundaries (***seams***). In what follows, for a ***non-singular seam*** it will be understood one where $F = 0$ and F'' is finite, while a ***singular seam*** will be one where F or F'' (or both) are unbounded. Blocks can be glued together only along non-singular seams across which F'' is *smooth*.

In each region

$$\tilde{\mathcal{Q}}_i \equiv \{(t, r) \in \tilde{\mathcal{Q}} \mid t \in \mathbb{R},\ r \in [a_i, a_{i+1}]\},$$

fix some value $r_i \in (a_i, a_{i+1})$ of r and define *null coordinates* via

$$u_i \equiv t - \int_{r_i}^{r} F^{-1}(\mathrm{s})\mathrm{ds}, \qquad v_i \equiv t + \int_{r_i}^{r} F^{-1}(\mathrm{s})\mathrm{ds}.$$

In terms of these new coordinates the metric $\tilde{\gamma}$ takes the form

$$\tilde{\gamma} = \frac{1}{2} F(r)(\mathbf{d}u_i \otimes \mathbf{d}v_i + \mathbf{d}v_i \otimes \mathbf{d}u_i).$$

This form of the metric is smooth for u_i, $v_i \in \mathbb{R}$ if $F(r)$ is smooth. The coordinate r will be regarded as a function of (u_i, v_i) given, implicitly, as the solution to the equations

$$\frac{1}{2}(v_i - u_i) = \int_{r_i}^{r} F^{-1}(\mathrm{s})\mathrm{ds}, \qquad \frac{1}{2}(u_i + v_i) = t. \tag{6.39}$$

The construction can be extended to singular blocks by setting $r_i = r_{\not{i}}$.

In the non-singular case, the integrals

$$\int_{a_i}^{r} F^{-1}(\mathrm{s})\mathrm{ds}, \qquad \int_{r_i}^{a_{i+1}} F^{-1}(\mathrm{s})\mathrm{ds},$$

are divergent as the points $r = a_i$, a_{i+1} are poles of the integrand. From this observation together with the formulae in (6.39), assuming $F > 0$ in $\tilde{\mathcal{Q}}_i$, one deduces the limits:

(a) If $r \to a_{i+1}$ and v_i is finite, then $u_i \to -\infty$ and $t \to -\infty$.
(b) If $r \to a_{i+1}$ and u_i is finite, then $v_i \to +\infty$ and $t \to +\infty$.
(c) If $r \to a_i$ and v_i is finite, then $u_i \to +\infty$ and $t \to +\infty$.
(d) If $r \to a_i$ and u_i is finite, then $v_i \to -\infty$ and $t \to -\infty$.

The setting described by the above limits is depicted in Figure 6.9, left panel. The coordinates (u_i, v_i) can be compactified via

$$U_i \equiv \arctan u_i, \qquad V_i \equiv \arctan v_i,$$

with U_i, $V_i \in [-\frac{1}{2}\pi, \frac{1}{2}\pi]$ so that $\tilde{\gamma}$ can be rewritten as

$$\tilde{\gamma} = \frac{1}{2}F(r)\sec^2 U_i \sec^2 V_i(\mathbf{d}U_i \otimes \mathbf{d}V_i + \mathbf{d}V_i \otimes \mathbf{d}U_i). \tag{6.40}$$

In what follows, for simplicity, given a regular block $\tilde{\mathcal{Q}}_i$ with coordinate $r \in [a_i, a_{i+1}]$, it is assumed that the zeros of $F(r)$ are such that

$$F(r)\sec^2 U_i \sec^2 V_i < \infty \quad \text{as} \quad U_i \to \pm\frac{1}{2}\pi \quad \text{or} \quad V_i \to \pm\frac{1}{2}\pi.$$

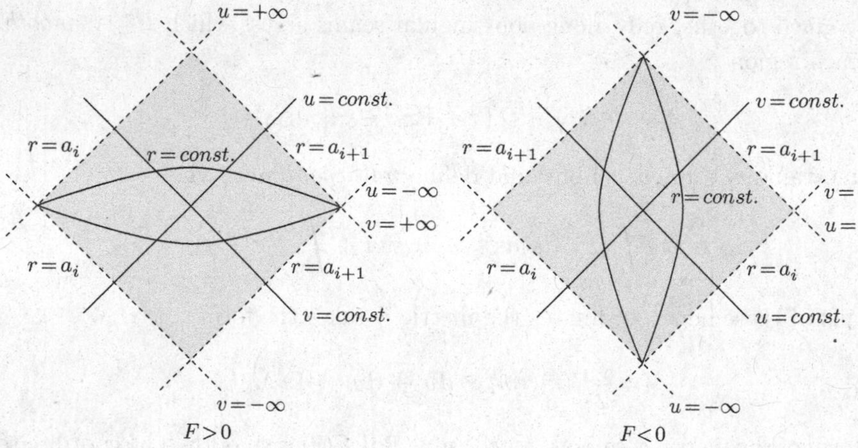

Figure 6.9 Coordinates in a regular block without asymptotic regions; see main text for details.

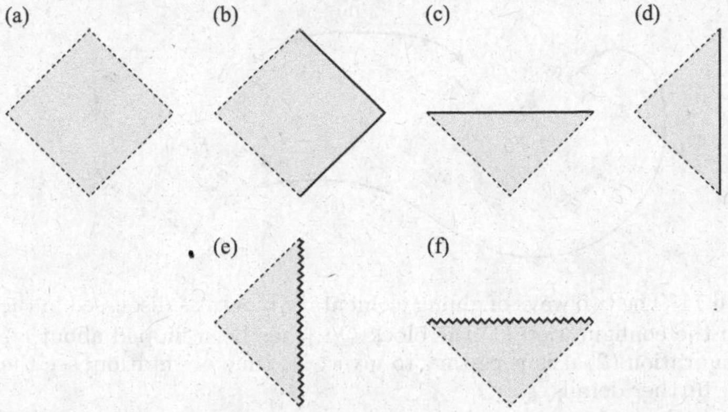

Figure 6.10 Elementary blocks for the construction of Penrose diagrams: (a) non-singular block without asymptotic regions; (b) non-singular block with a null $\mathscr{I}$; (c) non-singular block with a spacelike $\mathscr{I}$; (d) non-singular block with a timelike $\mathscr{I}$; (e) block with a timelike singularity; (f) block with a spacelike singularity.

In the case of blocks $\tilde{\mathcal{Q}}_i$ with $r \in [a_i, \pm\infty)$ corresponding to asymptotic regions, the metric (6.40) allows us to read out a conformal factor. The particular form of the conformal factor depends on the particular nature of the asymptotic end:

$$\Xi_i = \begin{cases} \cos U_i \cos V_i & \text{if } F(r) \to 1 \text{ as } r \to \infty \\ \dfrac{\cos U_i \cos V_i}{\sqrt{F(r)}} & \text{if } F(r) \to \infty \text{ as } r \to \infty. \end{cases}$$

The resulting construction can be depicted in a conformal diagram.

The discussion of the case $F < 0$ in $\tilde{\mathcal{Q}}_i$ is analogous. In this case the orbits of the Killing vector $\boldsymbol{\partial}_t$ are spacelike, and the hypersurfaces of constant r are timelike. The behaviour of the various coordinates in a regular elementary block is summarised in Figure 6.9, right panel. A depiction of the various elementary blocks is given in Figure 6.10.

Flipping of blocks. The convention used in drawing the diagrams in Figure 6.9 is that the coordinate r increases from left to right (if $F > 0$) and from bottom to top (if $F < 0$). As this is a mere convention, it is possible to flip the blocks about r_i. This operation effectively interchanges the roles of u and v. In addition, as the metric $\tilde{\gamma}$ is independent of the coordinate t, one has the discrete symmetry $t \mapsto -t$ which allows further flipping of blocks with respect to the surfaces of constant t – vertically if $F > 0$ and horizontally if $F < 0$.

Gluing blocks. All geodesics such that $r = a_i$ for some finite value of the affine parameter are incomplete. These geodesics can be extended by gluing blocks along non-singular seams. The convention followed in gluing the blocks is that the time coordinate in each block, t if $F(r) > 0$ and r if $F(r) < 0$, changes

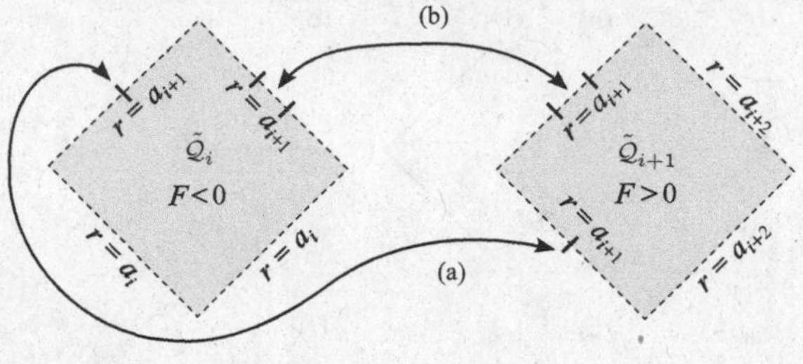

Figure 6.11 The two ways of gluing elementary blocks as discussed in the main text. In the configuration (1) the block $\tilde{\mathcal{Q}}_{i+1}$ has been flipped about r_i, while in configuration (2) it is necessary to invert its time orientation; see the main text for further details.

vertically. Consider, for example, the gluing of a block $\tilde{\mathcal{Q}}_{i+1}$ with $F(r) > 0$ and a block $\tilde{\mathcal{Q}}_i$ with $F(r) < 0$. By suitably flipping the blocks, they can be glued together in the two ways shown in Figure 6.11. Under the assumption that F'' is smooth at $r = a_{i+1}$ (and hence also the curvature), the *gluing of blocks* is equivalent to showing that the null hypersurfaces are continuous along the seams, that is, that there is a coordinate system covering both blocks in a neighbourhood of $r = a_{i+1}$ in a smooth fashion. This construction is implemented through ***Eddington-Finkelstein type coordinates.***

As a first example consider configuration (1) of Figure 6.11 where the block $\tilde{\mathcal{Q}}_i$ is glued to a block $\tilde{\mathcal{Q}}_{i+1}$ which has been flipped about r_i. Direct inspection reveals that while advanced null coordinates exhaust at the gluing seam (i.e. they become infinite), a null retarded coordinate extends to the two blocks $\tilde{\mathcal{Q}}_i$ and $\tilde{\mathcal{Q}}_{i+1}$. Accordingly, one sets

$$\mathbf{d}u = \mathbf{d}t - F^{-1}(r)\mathbf{d}r,$$

so that

$$\tilde{\gamma} = F(r)\mathbf{d}u \otimes \mathbf{d}u + (\mathbf{d}u \otimes \mathbf{d}r + \mathbf{d}r \otimes \mathbf{d}u). \tag{6.41}$$

Now, allowing $r \in [a_i, a_{i+2}]$ one finds that the coordinates (u, r) cover both blocks in configuration (1) of Figure 6.11 – the resulting combined block is shown in configuration (1) of Figure 6.12. In particular, the coordinate u is finite at $r = a_{i+1}$ and the metric (6.41) is smooth for $r \in (a_i, a_{i+2})$.

In order to perform the gluing in configuration (2) of Figure 6.11, one needs to flip the block $\tilde{\mathcal{Q}}_{i+1}$ about t. Direct inspection shows that for this configuration retarded null coordinates exhaust at the gluing seam. Accordingly, one introduces advanced null coordinates

$$\mathbf{d}v = \mathbf{d}t + F^{-1}(r)\mathbf{d}r,$$

so as to obtain

$$\tilde{\gamma} = F(r)\mathbf{d}v \otimes \mathbf{d}v - (\mathbf{d}v \otimes \mathbf{d}r + \mathbf{d}r \otimes \mathbf{d}v). \tag{6.42}$$

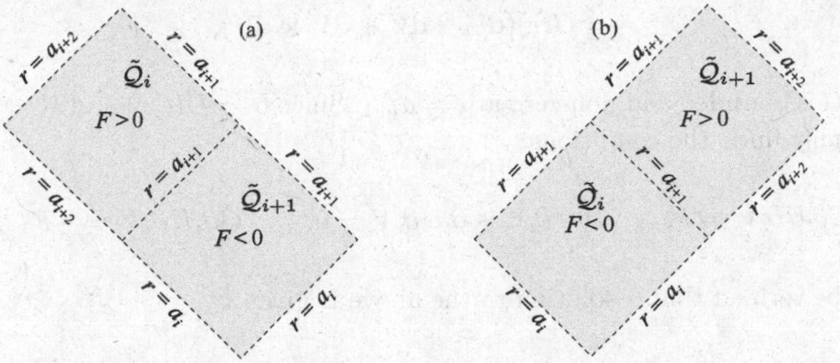

Figure 6.12 The two composite blocks obtained from the the gluing procedures in Figure 6.11.

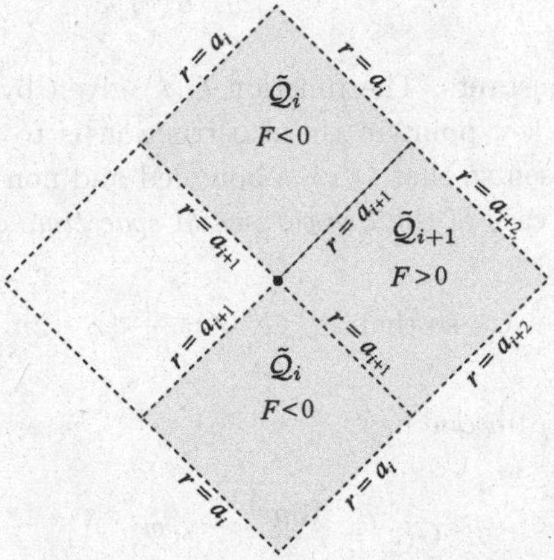

Figure 6.13 Gluing applied to three blocks simultaneously. A fourth block can be glued using the Kruskal construction – see the main text for further details.

The coordinate v is finite at $r = a_{i+1}$. Thus, the pair (v, r) covers the composite block. One can verify that the metric (6.42) is smooth for $r \in (a_i, a_{i+2})$; the resulting combined block is shown in configuration (2) of Figure 6.12.

Gluing à la Kruskal. The gluings discussed in the last paragraphs can be performed simultaneously; the resulting configuration is shown in Figure 6.13. In addition, one could also glue a further region $\tilde{\mathcal{Q}}_{i+1}$ obtained from $\tilde{\mathcal{Q}}_{i+1}$ by applying the reflection in the time coordinate. Notice, however, that the point p is not covered by either of the coordinates (u, r) and (v, r). Depending on the particular form of $F(r)$ it may be possible to obtain a single coordinate patch for the four blocks; this is the case, for example, in the Schwarzschild spacetime. The procedure to do this makes use of a generalisation of the ***Kruskal coordinates***; see Figure 6.13. The general strategy is to find coordinates (U, V) such that the metric takes the form

$$\tilde{\gamma} = G(r)(\mathbf{d}U \otimes \mathbf{d}V + \mathbf{d}V \otimes \mathbf{d}U),$$

where G is bounded and non-zero at $r = a_{i+1}$. Since $U = U(t,r)$ and $V = V(t,r)$ one readily finds the conditions

$$G\partial_t U \partial_t V = F, \qquad \partial_r U \partial_t V + \partial_t U \partial_r V = 0, \qquad G\partial_r U \partial_r V = -F^{-1}.$$

It can be verified that a solution to the above is given by

$$U(t,r) = a \exp\left(bt + b\int \frac{\mathrm{d}r}{F(r)} \right), \qquad V(t,r) = \frac{1}{a}\exp\left(-bt + b\int \frac{\mathrm{d}r}{F(r)} \right),$$
$$G(r) = \frac{F(r)}{b^2}\exp\left(-2b\int \frac{\mathrm{d}r}{F(r)} \right),$$

where a and b are constants. The function $G(r)$ given by the above formulae can be singular. The key point in the construction is to analyse whether the constant b can be chosen so that $G(r)$ is bounded and non-zero at $r = a_{i+1}$. As an illustration, in the case of the *Schwarzschild spacetime* one has that

$$F(r) = 1 - \frac{2m}{r} \qquad \text{so that} \qquad G(r) = \frac{1}{b^2 r}(r-2m)^{1-4mb}e^{-2br}.$$

Hence, choosing $b = 1/4m$ one has

$$G(r) = \frac{16m^2}{r}e^{-r/2m},$$

which is bounded and non-zero at $r = 2m$, the location of the horizon. By contrast, in the case of the *extremal Reissner-Nordström spacetime* – see Equation (6.43) – one has

$$F(r) = \left(1 - \frac{m}{r}\right)^2 \quad \text{so that} \quad G(r) = \frac{1}{b^2}(r-m)^{2-4mb}\exp\left(-2br + \frac{6m^2 b}{r-m} \right).$$

In this case one cannot find a value of b which makes $G(r)$ finite and non-zero at $r = m$. In particular, the choice $b = 1/2m$ yields

$$G(r) = \frac{1}{4m^2}\exp\left(-\frac{r}{m} + \frac{3m}{r-m} \right),$$

which is singular at $r = m$. More generally, *if $F(r)$ is a rational function, it can be shown that the value b can be chosen so that $G(r)$ is finite and non-zero at $r = a_{i+1}$ if a_{i+1} is a non-repeated zero of $F(r)$*; see Walker (1970) for the details of the proof.

Some examples

The procedure described in the previous paragraphs can be employed to construct the Penrose diagrams and conformal compactifications of a number of well-known spherically symmetric spacetimes. In particular, one has the following (details can be found in the given references):

The non-extremal Reissner-Nordström spacetime. This is the solution to the Einstein-Maxwell field equations given by the metric

$$\tilde{g} = \left(1 - \frac{2m}{r} + \frac{q^2}{r^2}\right) \mathbf{d}t \otimes \mathbf{d}t - \left(1 - \frac{2m}{r} + \frac{q^2}{r^2}\right)^{-1} \mathbf{d}r \otimes \mathbf{d}r - r^2 \boldsymbol{\sigma},$$

with $q^2 < m^2$. In this case $F(r) = 1 - 2m/r + q^2/r^2$ has two zeros. One can identify three elementary blocks: an asymptotically flat region, a standard regular block and a block with a timelike singularity. In this case Kruskal's construction can be employed to glue four blocks simultaneously. The resulting Penrose diagram is given in Figure 6.14; see Carter (1973).

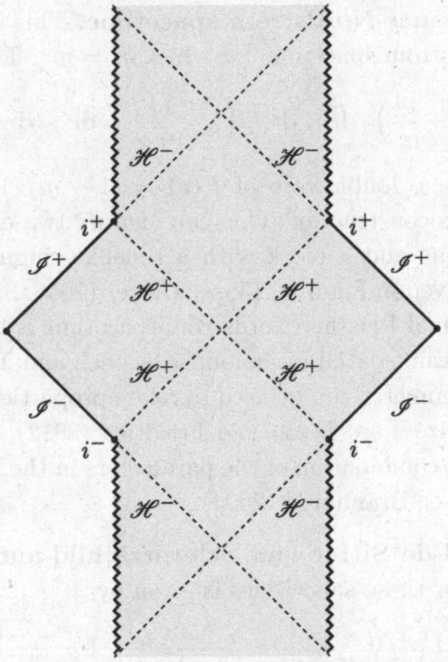

Figure 6.14 Penrose diagram of the Reissner-Nordström spacetime in the non-extremal ($q^2 < m^2$) case. The points i^0 correspond to the various spatial infinities, the points $i^\pm$ to future and past timelike infinity, respectively, and the lines $\mathscr{I}^\pm$ to the various components of null infinity. The dashed lines $\mathscr{H}^\pm$ correspond to the various horizons. Finally, the serrated lines denote the singularities.

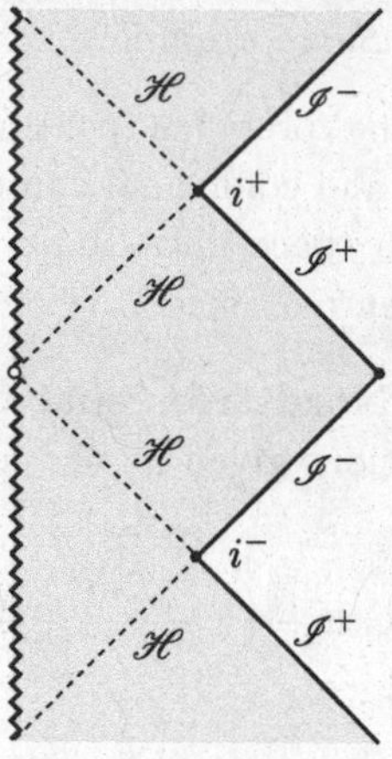

Figure 6.15 Penrose diagram of the Reissner-Nordström spacetime in the extremal case ($q^2 = m^2$). The point i^0 corresponds to spatial infinity, the points $i^{\pm}$ to future and past timelike infinity, respectively, and the lines $\mathscr{I}^{\pm}$ to the various components of null infinity. The dashed lines labeled by $\mathscr{H}$ correspond to the various horizons. Finally, the serrated lines denote the singularities.

The extremal Reissner-Nordström spacetime. This is the particular case of the Reissner-Nordström spacetime for which $q^2 = m^2$. The metric is given by

$$\tilde{\boldsymbol{g}} = \left(1 - \frac{m}{r}\right)^2 \mathbf{d}t \otimes \mathbf{d}t - \left(1 - \frac{m}{r}\right)^{-2} \mathbf{d}r \otimes \mathbf{d}r - r^2\boldsymbol{\sigma}. \tag{6.43}$$

In this case, one has a double zero of $F(r) = (1 - m/r)^2$. Thus, one cannot make use of Kruskal's construction. One can identify two elementary blocks: an asymptotically flat one and a block with a timelike singularity. The resulting Penrose diagram is given in Figure 6.15; see Carter (1966a, 1971). An interesting property of the extremal Reissner-Nordström spacetime is that it is conformally invariant under a certain spatial inversion; see Couch and Torrence (1984). This discrete conformal symmetry can be used to relate properties of null infinity with properties of the horizon; see Bizon and Friedrich (2012). A similar symmetry exists for a particular combination of the parameters in the Reissner-Nordström-de Sitter spacetime; see Brännlund (2004).

The Schwarzschild-de Sitter and Schwarzschild-anti de Sitter spacetimes. The metric for these spacetimes is given by

$$\tilde{\boldsymbol{g}} = \left(1 - \frac{2m}{r} + \frac{1}{3}\lambda r^2\right) \mathbf{d}t \otimes \mathbf{d}t - \left(1 - \frac{2m}{r} + \frac{1}{3}\lambda r^2\right)^{-1} \mathbf{d}r \otimes \mathbf{d}r - r^2\boldsymbol{\sigma},$$

where it is assumed that $m > 0$. If $\lambda > 0$ (the *anti-de Sitter* case), then it can be verified that $F(r) = 1 - 2m/r - \lambda r^2/3$ has only one real root corresponding to a *black hole-type horizon.* The resulting diagram is given in Figure 6.16. If $\lambda < 0$ (the *de Sitter* case) and $0 < -9\lambda m^2 < 1$, then $F(r)$ can be shown to have two

Figure 6.16 Penrose diagram of the Schwarzschild-anti de Sitter spacetime. The vertical lines $\mathscr{I}$ denote the two components of conformal infinity, while the dashed lines labeled by $\mathscr{H}^{\pm}$ denote the various components of the horizon. The serrated lines denote the singularities.

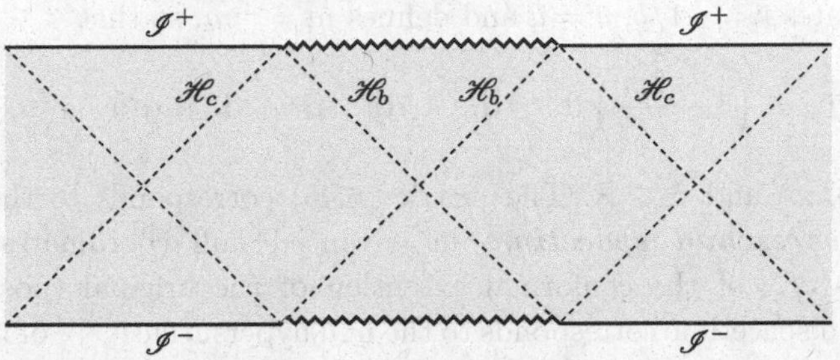

Figure 6.17 Penrose diagram of the Schwarzschild-de Sitter spacetime. The horizontal lines labeled by $\mathscr{I}^{\pm}$ correspond to the various components of conformal infinity. The dashed lines $\mathscr{H}_c$ and $\mathscr{H}_b$ denote, respectively, the cosmological and black hole horizons. The serrated lines indicate the location of the singularities.

positive real roots corresponding, respectively, to a *black hole-type horizon* and a *cosmological type horizon*. The resulting blocks can be arranged in a periodic diagram as given in Figure 6.17; see, for example, Griffiths and Podolský (2009). In this case it is also possible to make topological identifications; see Beig and Heinzle (2005). The cases $-9\lambda m^2 = 1$ and $-9\lambda m^2 > 1$ correspond, respectively, to the so-called *extremal* and *hyperextremal* cases.

Other examples of spacetimes amenable to the general construction described in this section are the Nariai solution, the Reissner-Nordström-de Sitter and the Reissner-Nordström-anti de Sitter solutions; see, for example, Brill and Hayward (1994) for a detailed discussion.

6.5.3 Extending across the conformal boundary

In Schmidt and Walker (1983) it has been observed that the conformal representations of some spacetimes can be extended across the conformal boundary. In the case of the Schwarzschild solution this is best seen by considering the metric written in terms of a *retarded null coordinate*:

$$\tilde{\boldsymbol{g}}_{\mathscr{S}} = \left(1 - \frac{2m}{r}\right) \mathbf{d}u \otimes \mathbf{d}u + (\mathbf{d}u \otimes \mathbf{d}r + \mathbf{d}r \otimes \mathbf{d}u) - r^2 \boldsymbol{\sigma}, \tag{6.44}$$

where, in particular, $u \in \mathbb{R}$. Defining $\varrho \equiv 1/r$, a calculation yields that

$$\varrho^2 \tilde{\boldsymbol{g}}_{\mathscr{S}} = \varrho^2 (1 - 2m\varrho)\, \mathbf{d}u \otimes \mathbf{d}u - (\mathbf{d}u \otimes \mathbf{d}\varrho + \mathbf{d}\varrho \otimes \mathbf{u}) - \boldsymbol{\sigma}. \tag{6.45}$$

In this representation, future null infinity $\mathscr{I}^+$ is given by the condition $\varrho = 0$. The key observation is that the metric (6.45) can be analytically extended by allowing ϱ to take negative values. To identify the spacetime on the other side of $\mathscr{I}^+$ one undoes the conformal rescaling to obtain

$$\bar{\boldsymbol{g}}_{\mathscr{S}} = (1 - 2m\varrho)\, \mathbf{d}u \otimes \mathbf{d}u - \frac{1}{\varrho^2}(\mathbf{d}u \otimes \mathbf{d}\varrho + \mathbf{d}\varrho \otimes \mathbf{d}u) - \frac{1}{\varrho^2}\boldsymbol{\sigma},$$

where $\varrho \in (-\infty, 0)$. To bring the metric to a more familiar form one introduces new coordinates $\bar{r} = -1/\varrho$, $\bar{v} = u$ and defines $\bar{m} = -m$, so that

$$\bar{\boldsymbol{g}}_{\mathscr{S}} = \left(1 - \frac{2\bar{m}}{\bar{r}}\right) \mathbf{d}\bar{v} \otimes \mathbf{d}\bar{v} - (\mathbf{d}\bar{v} \otimes \mathbf{d}\bar{r} + \mathbf{d}r \otimes \mathbf{d}\bar{v}) - \bar{r}^2 \sigma, \tag{6.46}$$

with $\bar{r} \in (0, \infty)$ and $\bar{v} \in \mathbb{R}$. The metric (6.46) corresponds to the ***negative mass Schwarzschild spacetime*** in advanced null coordinates. The null hypersurface $\mathscr{I}^+$ of the conformal extension of the original (positive mass) Schwarzschild spacetime corresponds to the null hypersurface $\mathscr{I}^-$ of the negative mass Schwarzschild spacetime. It is important to point out that the spacetimes described by the metrics (6.44) and (6.46) are causally disconnected; however, at the level of the conformal structure, they are an extension of each other. This situation is depicted, at the level of Penrose diagrams, in Figure 6.18.

The ideas described in the previous paragraphs can be used to construct so-called ***maximal conformal extensions of the Schwarzschild spacetime.*** Further details can be found in Schmidt and Walker (1983). The construction described in this section can be adapted to other spacetimes, for example, the Reissner-Nordström solution.

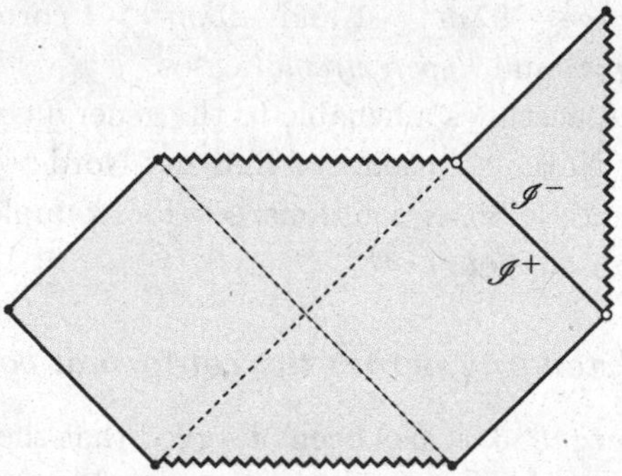

Figure 6.18 Extending the conformal structure of the Schwarzschild spacetime through null infinity. The future null infinity of the positive mass Schwarzschild spacetime is identified with the past null infinity of the negative mass Schwarzschild spacetime, see main text for further details. The points denoted by white dots are excluded from the discussion.

6.6 Further reading

The discussion presented in this chapter has been restricted to the analysis of the conformal structure of static, spherically symmetric spacetimes. Some aspects of this discussion can be adapted to the analysis of other exact solutions like the Kerr and Kerr-Newman spacetimes; this is discussed in, for example, Carter (1973), Hawking and Ellis (1973) and Griffiths and Podolský (2009). Another class of spacetimes amenable to an explicit discussion of its conformal structure is that of the Friedman-Lemaître-Robertson-Walker (FLRW) cosmological models; see again Hawking and Ellis (1973) and Griffiths and Podolský (2009).

An alternative discussion of Penrose diagrams of spherically symmetric spacetimes which allows for dynamic configurations can be found in appendix C of Dafermos and Rodnianski (2005). The idea of a Penrose diagram can be adapted to the analysis of suitable two-dimensional timelike totally geodesic hypersurfaces of non-spherically symmetric spacetimes. This idea has been particularly fruitful in the case of the Kerr and Kerr-Newman spacetime; see, for example, Carter (1966b, 1968, 1973), Hawking and Ellis (1973) and Griffiths and Podolský (2009). This strategy, has been adapted to a variety of situations in Chruściel et al. (2012a).

7

Asymptotic simplicity

The analysis of the conformal structure of exact solutions carried out in Chapter 6 exhibited a number of common features among the various spacetimes considered. The most conspicuous one is that they all admit a smooth conformal extension which attaches a boundary to the spacetime. This *conformal boundary* represents *points at infinity.* It is natural to ask whether this property is shared by a larger class of spacetimes. This question leads to the notion of *asymptotic simplicity.* In formulating this notion one tries to strike a delicate balance: the definition should be strong enough so that it excludes clearly *pathological* situations, but at the same time it should leave enough room to include interesting spacetimes that go beyond the obvious explicit examples. The original definition of asymptotic simplicity is due to Penrose (1963, 1964, 1965). This definition has had a lasting influence on the field of mathematical relativity, in general, and in the applications of conformal methods to the analysis of global properties of spacetimes, in particular.

7.1 Basic definitions

The following definition of asymptotic simplicity is adapted from Hawking and Ellis (1973):

Definition 7.1 (***asymptotically simple spacetimes***) *A spacetime* $(\tilde{\mathcal{M}}, \tilde{\boldsymbol{g}})$ *is said to be* ***asymptotically simple*** *if there exists a smooth, oriented, time-oriented, causal*[1] *spacetime* $(\mathcal{M}, \boldsymbol{g})$ *and a smooth function* Ξ *on* $\mathcal{M}$ *such that:*

(i) $\mathcal{M}$ *is a manifold with boundary* $\mathscr{I} \equiv \partial\mathcal{M}$.
(ii) $\Xi > 0$ *on* $\mathcal{M} \setminus \mathscr{I}$, *and* $\Xi = 0$, $\mathbf{d}\Xi \neq 0$ *on* $\mathscr{I}$.

[1] A ***causal spacetime*** is one in which there exist no closed timelike or null (i.e. causal) curves; see also Chapter 14.

(iii) There exists an embedding $\varphi : \tilde{\mathcal{M}} \to \mathcal{M}$ *such that* $\varphi(\tilde{\mathcal{M}}) = \mathcal{M} \setminus \mathscr{I}$ *and*

$$\varphi^* \boldsymbol{g} = \Xi^2 \tilde{\boldsymbol{g}}.$$

(iv) Each null geodesic of $(\tilde{\mathcal{M}}, \tilde{\boldsymbol{g}})$ *acquires two distinct endpoints on* $\mathscr{I}$.

The spacetime $(\tilde{\mathcal{M}}, \tilde{\boldsymbol{g}})$ is called the ***physical spacetime***, while $(\mathcal{M}, \boldsymbol{g})$ is known as the ***unphysical spacetime***. The boundary $\mathscr{I}$ is generally known as ***conformal infinity***. In the cases where $\mathscr{I}$ corresponds to a null hypersurface one calls it ***null infinity***. More informally, $\mathscr{I}$ is also called *scri*[2] – a shortened version of *script I*. In a slight abuse of notation, one usually identifies $\tilde{\mathcal{M}}$ and $\mathcal{M} \setminus \mathscr{I}$ so that one writes $\boldsymbol{g} = \Xi^2 \tilde{\boldsymbol{g}}$; see, for example, the examples discussed in Chapter 6. In what follows phrases like "***at infinity***" are to be understood as meaning *in a suitable neighbourhood* of $\mathscr{I}$ in $\mathcal{M}$.

Definition 7.1 allows for a non-vanishing matter content. Spacetimes for which, in addition, one has that $R_{ab} = 0$ in a neighbourhood of $\mathscr{I}$ in $\varphi^{-1}(\mathcal{M})$ are sometimes called ***asymptotically empty and simple***.

Remarks

(a) **Restriction on the conformal class.** Definition 7.1 imposes restrictions only on the conformal class of the admissible spacetimes $(\tilde{\mathcal{M}}, \tilde{\boldsymbol{g}})$. It does not single out any specific conformal representation; that is, it does not provide a *canonical* unphysical spacetime $(\mathcal{M}, \boldsymbol{g})$.

(b) **Conformal infinity is a hypersurface.** The boundary $\mathscr{I}$ as introduced in point (i) in Definition 7.1 is a well-defined three-dimensional hypersurface of $\mathcal{M}$ with normal given by $\mathbf{d}\Xi$. In particular, sets where $\mathbf{d}\Xi = 0$ – such as spatial infinity i^0 and the timelike infinities $i^{\pm}$ of the Minkowski spacetime – are excluded from $\mathscr{I}$. Points of this type, if present, will still be regarded as belonging to the conformal boundary but will be treated separately.

(c) **Conformal infinity is at infinity.** Points (ii) and (iii) of Definition 7.1 ensure that the boundary $\mathscr{I}$ shares the key properties of the null infinity of the Minkowski, de Sitter and anti-de Sitter spacetimes. To see that this is the case one needs to analyse the behaviour of null geodesics. The transformation behaviour of null geodesics under conformal rescalings has already been discussed in Section 5.5. In what follows, let $\tilde{\mathrm{s}}$ and s denote, respectively, $\tilde{\boldsymbol{g}}$-affine and $\boldsymbol{g}$-affine parameters of a null geodesic $\gamma \subset \tilde{\mathcal{M}}$. It follows then that $\tilde{\mathrm{s}}$ and s are related to each other by the equation

$$\frac{\mathrm{d}\tilde{\mathrm{s}}}{\mathrm{d}\mathrm{s}} = \frac{1}{\Xi^2}.$$

Without loss of generality, one can choose the unphysical affine parameter s to vanish at $\mathscr{I}$; that is, $\Xi = O(\mathrm{s}^{\alpha})$ along the null geodesic with $\alpha > 0$.

[2] Remarkably, the word *scri* is pronounced in the same way as the Polish word *scraj* meaning *boundary*.

Now, as $\mathbf{d}\Xi \neq 0$ at $\mathscr{I}$, one concludes that, in fact, $\alpha \geq 1$. Hence, $\tilde{s} \to \infty$ as $\Xi \to 0$ – that is, from the *physical point of view (as measured by the affine parameter* $\tilde{s}$*) the null geodesic never reaches the conformal boundary* $\mathscr{I}$. Thus, the conformal boundary lies at infinity from the perspective of the physical metric $\tilde{\boldsymbol{g}}$.

(d) **Smoothness of the conformal extension and decay.** As will be discussed in Chapter 10 the *smoothness assumption* in Definition 7.1 imposes a sharp decay behaviour on the gravitational field at infinity – in particular, it leads to what is known as the ***peeling behaviour*** of the Weyl tensor. There are variations of Definition 7.1 in which the smoothness requirement is relaxed to admit conformal extensions of class C^k for some suitable positive integer k; see Penrose and Rindler (1986). The physical relevance of these weaker regularity conditions is a delicate technical point which cannot be satisfactorily assessed by just looking at specific examples. These weaker regularity conditions lead to a different asymptotic behaviour of the gravitational field.

(e) **Matter and causal nature of null infinity.** As already mentioned, Definition 7.1 allows for the presence of matter in the spacetime. If the energy-momentum tensor of the matter models has a suitable decay at infinity, then the causal nature of $\mathscr{I}$ is fixed by the sign of the cosmological constant λ: it is spacelike if $\lambda < 0$, null if $\lambda = 0$ and timelike if $\lambda > 0$; see Theorem 10.1.

(f) **The completeness requirement.** Point (iv) in Definition 7.1 is a completeness condition which, in particular, excludes spacetimes such as the Schwarzschild solution in which there exist null geodesics which do not reach $\mathscr{I}$ – not only those falling into the black hole region, but also those lying in the ***photon sphere*** at $r = 3m$; see Wald (1984).

(g) **Regular solutions which are not asymptotically simple.** That a spacetime is smooth and geodesically complete is not a guarantee that it admits a smooth conformal extension. An example of this is given by the so-called *Nariai spacetime* described by

$$\tilde{\mathcal{M}} = \mathbb{R} \times (\mathbb{S}^1 \times \mathbb{S}^2), \qquad \tilde{\boldsymbol{g}} = \mathbf{d}t \otimes \mathbf{d}t - \cosh^2 t \mathbf{d}\psi \otimes \mathbf{d}\psi - \boldsymbol{\sigma}, \tag{7.1}$$

which is a solution to the vacuum Einstein field equations with $\lambda = -1$. In addition to being geodesically complete, the Nariai spacetime is also *globally hyperbolic*; see Section 14.1. Remarkably, *the Nariai spacetime does not even admit a patch of a conformal boundary.* To see this, assume one has a conformal extension with the required properties. The standard conformal transformation laws imply that

$$\tilde{C}_{abcd}\tilde{C}^{abcd} = \Xi^4 C_{abcd} C^{abcd}.$$

Thus, if the solution admits a smooth conformal extension, then $\tilde{C}_{abcd}\tilde{C}^{abcd} = 0$. On the other hand, a direct computation with the line element (7.1) shows that $\tilde{C}_{abcd}\tilde{C}^{abcd} = \text{constant} \neq 0$. This is a contradiction, and accordingly there cannot exist a piece of conformal boundary which is C^2. This argument is adapted from Friedrich (2015a); an alternative topological argument has been given in Beyer (2009a).

In order to consider spacetimes for which the completeness condition (iv) in Definition 7.1 does not hold, one introduces the further notion of ***weakly asymptotically simple spacetimes***, that is, spacetimes whose *asymptotic region* is diffeomorphic to that of an asymptotically simple spacetime. More precisely, one has the following:

Definition 7.2 (***weakly asymptotically simple spacetimes***) *A spacetime* $(\tilde{\mathcal{M}}, \tilde{\boldsymbol{g}})$ *is said to be* ***weakly asymptotically simple*** *if there exists an asymptotically simple spacetime* $(\tilde{\mathcal{M}}', \tilde{\boldsymbol{g}}')$ *and a neighbourhood* $\mathcal{U}'$ *of* $\mathscr{I}' \equiv \partial\mathcal{M}'$ *such that* $\varphi^{-1}(\mathcal{U}') \cap \tilde{\mathcal{M}}'$ *is isometric to an open subspace* $\tilde{\mathcal{U}}$ *of* $\tilde{\mathcal{M}}$.

Basic examples of asymptotically simple spacetimes have been given in Chapter 6. Notoriously, all the given examples are time independent. More generally, it can be shown that stationary solutions to the vacuum equations $R_{ab} = 0$ with a suitable behaviour at infinity are at least weakly asymptotically simple; see Damour and Schmidt (1990) and Dain (2001b). Thus, it is natural to ask whether there are *dynamic* solutions to the Einstein field equations. At the level of exact solutions, the closest examples are given by the spacetimes known as ***boost-rotation symmetric spacetimes*** – see, for example, Bičák and Schmidt (1989), Bičák (2000) and Griffiths and Podolský (2009) – and in particular the so-called ***C-metric*** – see Ashtekar and Dray (1981). All these exact solutions contain some pathologies (e.g. naked singularities, piercing of null infinity) which prevent them from being true examples of asymptotically simple spacetimes.

A detailed discussion of the properties of asymptotic simple spacetimes requires the conformal Einstein field equations and is deferred to Chapter 10.

7.2 Other related definitions

The definition of asymptotic simplicity makes neither reference to nor restricts the behaviour of the conformal spacetime $(\mathcal{M}, \boldsymbol{g})$ at spatial infinity. Several authors have introduced more refined definitions of asymptotic simplicity in which further requirements on the behaviour of the gravitational field at null infinity are prescribed, as in, for example, Persides (1979), or at spatial infinity, as in the concept of ***asymptotically empty and flat spacetime at null and spatial infinity*** of Ashtekar and Hansen (1978); see also Persides (1980). Similar ideas have been pursued by a number of authors in an attempt to analyse the structure of timelike infinity; see, for example, Persides (1982a,b), Porrill (1982), Cutler (1989) and Herberthson and Ludwig (1994).

The aim of the definitions mentioned in the previous paragraph and similar other proposals for the analysis of the asymptotic structure of spacetime is to identify a *minimal* number of assumptions on the asymptotic structure which, in turn, can be used to develop a formalism to construct physical and geometrical notions of interest. A critique to this approach is that, a priori, they do not provide any information on the genericity of the assumptions or on the size of the class of spacetimes they contain. Moreover, it is not clear how these spacetimes can be constructed. As pointed out in Geroch (1976), pages 3–4:

> Conditions too strong will have the effect of eliminating solutions which would seem clearly to represent isolated systems; conditions too weak may have the effect of admitting too many solutions, or what is worse, may result in a structure which is so weak that potentially useful aspects of the asymptotic behaviour of one's fields are lost in a sea of bad behaviour ··· There are no correct or incorrect definitions, only more or less useful ones.

The point of view pursued in this book is that rather than making assumptions on the nature of spatial, null or timelike infinity, the structures of the conformal boundary of a spacetime should arise as a result of the evolution of some initial data set for the Einstein field equations.

7.3 Penrose's proposal

Asymptotically empty and simple spacetimes (i.e. spacetimes with a vanishing cosmological constant and matter suitably decaying at infinity) play an important role in the approach to the analysis of ***isolated systems*** in general relativity put forward by Penrose (1963, 1964). The notion of an isolated system is a convenient idealisation of astrophysical systems where the effects of the cosmological expansion are ignored. This notion allows one to define concepts of clear physical interest such as the total energy of a system or its mass loss due to gravitational radiation. Intuitively, an isolated system should behave asymptotically like the Minkowski spacetime. Penrose, based on earlier work by Bondi et al. (1962) and Sachs (1962b), takes this idea further; see also Friedrich (2002, 2004):

Penrose's proposal. *Far fields of isolated gravitating systems behave like those of asymptotically simple spacetimes in the sense that they can be smoothly extended to null infinity after a suitable conformal rescaling.*

In other words, if a spacetime $(\tilde{\mathcal{M}}, \tilde{g})$ describes an isolated system, then it should be weakly asymptotically simple.

As pointed out in Remark (d) earlier in the chapter, the requirement of smoothness results in a very definite decay behaviour of the gravitational field at infinity. Whether this behaviour is actually realised in solutions to the Einstein equations, and if so to what extent, is a delicate question which will be analysed in later chapters of this book.

7.4 Further reading

The literature on asymptotic simplicity and other definitions of asymptotic flatness is dauntingly vast and is best accessed through reviews. There are a good number of references covering various periods and aspects of the topic. Penrose (1964, 1967) gives an overview of the early ideas and results on asymptotic simplicity; Geroch (1976) provides a good discussion on the physical motivation of the study of isolated systems in general relativity, the notion of conserved quantities and asymptotic symmetries; Schmidt (1978), Newman and Tod (1980) and Ashtekar (1980, 1984) provide alternative discussions of these topics and Friedrich (1992, 1998a, 1999) provides reviews of the notion of asymptotic simplicity from the point of view of the conformal Einstein field equations and the construction of global solutions. More recent reviews of the topic can be found in Frauendiener (2004), Ashtekar (2014) and Friedrich (2015a).

8
The conformal Einstein field equations

To use conformal rescalings to analyse the global existence of asymptotically simple spacetimes one requires a suitable *conformal representation* of the Einstein field equations. The naive direct approach to this problem is to make use of the transformation law of the Ricci tensor. However, this leads to equations which are *singular* at the conformal boundary, so that the standard theory of partial differential equations (PDEs) cannot be applied. Remarkably, by introducing new variables, it is possible to obtain a system of equations for various *conformal fields* which is regular even at the conformal boundary and whose solutions imply, in turn, solutions to the Einstein field equations – this construction was first done in Friedrich (1981b). These equations are known as the *conformal Einstein field equations.*

This chapter provides derivations of two versions of the conformal field equations introduced by Friedrich: the so-called *standard conformal Einstein field equations* written in terms of the Levi-Civita connection of a conformally rescaled (unphysical) spacetime, and the *extended conformal field equations* which are given in terms of a Weyl connection. These two versions of the conformal equations can be expressed in tensorial, frame or spinorial form. The presentation in this chapter allows for the presence of general classes of matter models. It also provides a discussion of some basic properties of the equations, in particular, their conformal covariance and their relation to the Einstein field equations.

8.1 A singular equation for the conformal metric

Assume one has two spacetimes $(\tilde{\mathcal{M}}, \tilde{\boldsymbol{g}})$ and $(\mathcal{M}, \boldsymbol{g})$ which are related to each other by means of a conformal transformation given by

$$g_{ab} = \Xi^2 \tilde{g}_{ab}. \tag{8.1}$$

Following the conventions of Section 7.1, $(\tilde{\mathcal{M}}, \tilde{\boldsymbol{g}})$ is called the ***physical spacetime***, while $(\mathcal{M}, \boldsymbol{g})$ is known as the ***unphysical spacetime***. For simplicity, the discussion in this section is restricted to the case $\tilde{R}_{ab} = 0$.

From the discussion in Chapter 5, the conformal rescaling (8.1) implies the transformation law

$$R_{ab} = \tilde{R}_{ab} - 2\Xi^{-1}\nabla_a\nabla_b\Xi - g_{ab}g^{cd}\left(\Xi^{-1}\nabla_c\nabla_d\Xi - 3\Xi^{-2}\nabla_c\Xi\nabla_d\Xi\right) \tag{8.2}$$

for the Ricci tensor. Combining this expression with the vacuum Einstein field equations one obtains the following *conformal vacuum Einstein field equation*:

$$R_{ab} - \frac{1}{2}Rg_{ab} = -2\Xi^{-1}(\nabla_a\nabla_b\Xi - \nabla_c\nabla^c\Xi g_{ab}) - 3\Xi^{-2}\nabla_c\Xi\nabla^c\Xi g_{ab}. \tag{8.3}$$

The latter equation can be interpreted as an Einstein field equation for the unphysical metric $\boldsymbol{g}$ with an *unphysical matter* with energy-momentum tensor T_{ab} given by

$$T_{ab} \equiv -2\Xi^{-1}\left(\nabla_a\nabla_b\Xi - \nabla_c\nabla^c\Xi g_{ab}\right) - 3\Xi^{-2}\nabla_c\Xi\nabla^c\Xi g_{ab}.$$

Equation (8.3) contains factors of Ξ^{-1} which become singular at $\Xi = 0$. Following the discussion of Chapter 7, such points correspond to the conformal boundary of the spacetime – a region of the unphysical spacetime $(\mathcal{M}, \boldsymbol{g})$ for which one would like to be able to make analytic statements. This is not possible for Equation (8.3) as the standard theory of PDEs assumes equations which are *formally regular*. It is important to observe that multiplying Equation (8.3) by Ξ^2 does not improve the state of affairs as one has then an equation whose *principal part* (i.e. the terms containing the higher order derivatives) vanishes at $\Xi = 0$.

8.2 The metric regular conformal field equations

In what follows, it will be shown that by introducing new variables and reinterpreting old ones, it is possible to obtain a set of equations which is regular even at the conformal boundary. Under suitable conditions, a solution of this system implies a solution to the *physical* Einstein field equations.

The analysis of this section assumes a general matter content of the spacetime so that

$$\tilde{R}_{ab} - \frac{1}{2}\tilde{R}\tilde{g}_{ab} + \lambda\tilde{g}_{ab} = \tilde{T}_{ab} \tag{8.4}$$

and

$$\tilde{\nabla}^a\tilde{T}_{ab} = 0.$$

From the above it follows directly that

$$\tilde{R} = 4\lambda - \tilde{T}, \tag{8.5a}$$

$$\tilde{L}_{ab} \equiv \frac{1}{2}\tilde{R}_{ab} + \frac{1}{12}\tilde{R}\tilde{g}_{ab} = \frac{1}{2}\tilde{T}_{ab} + \frac{1}{6}(\lambda - \tilde{T})\tilde{g}_{ab}, \tag{8.5b}$$

where $\tilde{T} \equiv \tilde{g}^{ab}\tilde{T}_{ab}$ and $\tilde{L}_{ab}$ denotes the physical Schouten tensor.

8.2.1 The regularisation of the transformation law for the Schouten tensor

The starting point of the construction is the *singular* transformation law for the Ricci tensor given by Equation (8.2). In practice, it is more convenient to work with the Schouten tensor than with the Ricci tensor. The analogue of Equation (8.2) for the Schouten tensor is given by

$$L_{ab} = \tilde{L}_{ab} - \Xi^{-1}\nabla_a\nabla_b\Xi + \frac{1}{2}\Xi^{-2}\nabla_c\Xi\nabla^c\Xi\, g_{ab}. \tag{8.6}$$

Formally, the most singular term in this equation is $\frac{1}{2}\Xi^{-2}\nabla_c\Xi\nabla^c\Xi$. From the transformation law

$$R = \Xi^{-2}\tilde{R} - 6\Xi^{-1}\nabla_c\nabla^c\Xi + 12\Xi^{-2}\nabla_c\Xi\nabla^c\Xi, \tag{8.7}$$

it follows that

$$\Xi^{-2}\nabla_c\Xi\nabla^c\Xi = \frac{1}{12}\left(R - \Xi^{-2}\tilde{R}\right) + \frac{1}{2}\Xi^{-1}\nabla_c\nabla^c\Xi. \tag{8.8}$$

The right-hand side of the last expression contains the singular term $-\frac{1}{12}\Xi^{-2}\tilde{R}$. Yet substituting Equation (8.8) into (8.6), some cancellations occur. Making use of Equation (8.5b) one obtains

$$L_{ab} = \frac{1}{2}\tilde{T}_{ab} + \frac{1}{6}(\lambda - \tilde{T})\tilde{g}_{ab} - \Xi^{-1}\nabla_a\nabla_b\Xi + \frac{1}{24}\left(R - \Xi^{-2}\tilde{R}\right) g_{ab} + \frac{1}{4}\Xi^{-1}\nabla^c\nabla_c\Xi g_{ab}.$$

Now, defining the ***Friedrich scalar***

$$s \equiv \frac{1}{4}\nabla^c\nabla_c\Xi + \frac{1}{24}R\Xi, \tag{8.9}$$

and writing $\Xi^{-2}\tilde{R}g_{ab} = \tilde{R}\tilde{g}_{ab}$, one obtains

$$\begin{aligned} L_{ab} &= \frac{1}{2}\tilde{T}_{ab} + \left(\frac{1}{6}\lambda - \frac{1}{6}\tilde{T} - \frac{1}{24}\tilde{R}\right)\tilde{g}_{ab} - \Xi^{-1}\nabla_a\nabla_b\Xi + \Xi^{-1}sg_{ab}, \\ &= \frac{1}{2}\tilde{T}_{ab} - \frac{1}{8}\tilde{T}\tilde{g}_{ab} - \Xi^{-1}\nabla_a\nabla_b\Xi + \Xi^{-1}sg_{ab}, \end{aligned} \tag{8.10}$$

where in the last expression, Equation (8.5a) has been used. The last expression brings about the question of the transformation law for the energy-momentum tensor $\tilde{T}_{ab}$ upon the conformal rescaling $\boldsymbol{g} = \Xi^2\tilde{\boldsymbol{g}}$. As $\tilde{T}_{ab}$ *is not a geometric object* derived from the metric $\tilde{\boldsymbol{g}}$ and concomitants thereof, one is free to choose the transformation law which best suits the analysis. As will be further elaborated in Chapter 9, a convenient choice is to define the ***unphysical energy-momentum tensor*** T_{ab} as

$$T_{ab} \equiv \Xi^{-2}\tilde{T}_{ab}.$$

It follows then that

$$\frac{1}{2}\tilde{T}_{ab} - \frac{1}{8}\tilde{T}\tilde{g}_{ab} = \Xi^2\left(\frac{1}{2}T_{ab} - \frac{1}{8}Tg_{ab}\right) = \frac{1}{2}\Xi^2 T_{\{ab\}}, \tag{8.11}$$

where $T \equiv g^{ab}T_{ab}$ so that $\tilde{T} = \Xi^4 T$ and $T_{\{ab\}}$ denotes the $\boldsymbol{g}$-trace-free part of T_{ab}. Substituting Equation (8.11) into Equation (8.10) one obtains

$$L_{ab} = \frac{1}{2}\Xi^2 T_{\{ab\}} - \Xi^{-1}\nabla_a\nabla_b\Xi + \Xi^{-1}s g_{ab}. \tag{8.12}$$

This last equation still contains formally singular terms. To get around this problem, one reads it not as determining the components of the conformal metric $\boldsymbol{g}$ contained in L_{ab}, but as conditions on the second covariant derivative of the conformal factor Ξ. Adopting this point of view, and multiplying Equation (8.12) by Ξ one obtains

$$\nabla_a\nabla_b\Xi = -\Xi L_{ab} + s g_{ab} + \frac{1}{2}\Xi^3 T_{\{ab\}}. \tag{8.13}$$

Equation (8.13) promotes the fields s and L_{ab} to the level of unknowns for which suitable equations need to be constructed. This will be done in the following sections.

8.2.2 The equation for s

In order to construct an equation for s, one applies ∇_c to Equation (8.13) and obtains

$$\begin{aligned}\nabla_c\nabla_a\nabla_b\Xi = &-\nabla_c\Xi L_{ab} - \Xi\nabla_c L_{ab} + \nabla_c s g_{ab} \\ &+ \frac{3}{2}\Xi^2\nabla_c\Xi T_{\{ab\}} + \frac{1}{2}\Xi^3\nabla_c T_{\{ab\}}.\end{aligned} \tag{8.14}$$

By commuting covariant derivatives, the right-hand side of this equation can be rewritten as

$$\nabla_c\nabla_a\nabla_b\Xi = \nabla_a\nabla_c\nabla_b\Xi - R^d{}_{bca}\nabla_d\Xi.$$

Hence, contracting the indices $_b$ and $_c$ one finds that Equation (8.14) implies

$$\begin{aligned}\nabla_a(\nabla^c\nabla_c\Xi) + R_{ca}\nabla^c\Xi = &-L_{ca}\nabla^c\Xi - \Xi\nabla^c L_{ac} + \nabla_a s \\ &+ \frac{3}{2}\Xi^2\nabla^c\Xi T_{\{ac\}} + \frac{1}{2}\Xi^3\nabla^c T_{\{ac\}}.\end{aligned} \tag{8.15}$$

Now, the definition of the field s, Equation (8.9), implies that

$$\nabla_a(\nabla^c\nabla_c\Xi) = 4\nabla_a s - \frac{1}{6}\Xi\nabla_a R - \frac{1}{6}R\nabla_a\Xi.$$

Using this expression in (8.15) and observing that

$$R_{ab} = 2L_{ab} + \frac{1}{6}R g_{ab}, \tag{8.16}$$

one obtains

$$3\nabla_a s - \frac{1}{6}\Xi\nabla_a R = -3L_{ac}\nabla^c\Xi - \Xi\nabla^c L_{ac} + \frac{3}{2}\Xi^2\nabla^c\Xi T_{\{ac\}} + \frac{1}{2}\Xi^3\nabla^c T_{\{ac\}}.$$

Now, let $G_{ab} \equiv R_{ab} - \frac{1}{2}Rg_{ab}$ denote the Einstein tensor of the metric $\boldsymbol{g}$. One has that $\nabla^a G_{ab} = 0$. This last equation can be rewritten in terms of the Schouten tensor as

$$\nabla^c L_{ca} - \frac{1}{6}\nabla_a R = 0. \tag{8.17}$$

Making use of this last expression one obtains

$$\nabla_a s = -L_{ac}\nabla^c\Xi + \frac{1}{2}\Xi^2\nabla^c\Xi T_{\{ac\}} + \frac{1}{6}\Xi^3\nabla^c T_{\{ca\}}. \tag{8.18}$$

This is a suitable equation for s.

8.2.3 The equations for the curvature

Equation (8.13) brings the unphysical Schouten tensor L_{ab} into play. Thus, one needs to obtain an equation which can be regarded as a differential condition on L_{ab}. The natural place to look for such an equation is the second Bianchi identity; see Section 2.4.3. In Section 5.2.2, it has been shown that the second Bianchi identity together with the decomposition of the Riemann tensor in terms of the Weyl and Schouten tensors lead to the expressions

$$\tilde{\nabla}_c\tilde{L}_{db} - \tilde{\nabla}_d\tilde{L}_{cb} = \tilde{\nabla}_a C^a{}_{bcd},$$
$$\nabla_c L_{db} - \nabla_d L_{cb} = \nabla_a C^a{}_{bcd};$$

compare Equations (5.11) and (5.13). As it stands, the second of the above equations is not a satisfactory differential condition for L_{ab} as it contains, in its right-hand side, the divergence of the Weyl tensor. One needs to find an expression for the latter in terms of undifferentiated fields. Observe that the right-hand side of this equation can be expressed in terms of the physical energy-momentum tensor $\tilde{T}_{ab}$ using formula (8.5b). This will not be done at this point. Instead, it is more convenient to expresses it in terms of the *physical Cotton* tensor $\tilde{Y}_{cdb} \equiv \tilde{\nabla}_c\tilde{L}_{db} - \tilde{\nabla}_d\tilde{L}_{cb}$, so that

$$\tilde{\nabla}_a C^a{}_{bcd} = \tilde{Y}_{cdb}. \tag{8.19}$$

Now, one would like to express the divergence $\tilde{\nabla}_a C^a{}_{bcd}$ in terms of an expression involving the covariant derivative $\boldsymbol{\nabla}$. For this, one makes use of the identity

$$\nabla_a(\Xi^{-1}C^a{}_{bcd}) = \Xi^{-1}\tilde{\nabla}_a C^a{}_{bcd};$$

see Equation (5.8). Making use of the latter in Equation (8.19) one obtains

$$\nabla_a(\Xi^{-1}C^a{}_{bcd}) = \Xi^{-1}\tilde{Y}_{cdb}.$$

This equation seems to lead to a dead end because of the Ξ^{-1} terms appearing on both sides, and which do not cancel out. However, defining the ***rescaled Weyl tensor***

$$d^a{}_{bcd} \equiv \Xi^{-1}C^a{}_{bcd}, \tag{8.20}$$

and the ***rescaled Cotton tensor***

$$T_{cdb} \equiv \Xi^{-1}\tilde{Y}_{cdb}, \tag{8.21}$$

one obtains the formally *regular equation*

$$\nabla_a d^a{}_{bcd} = T_{cdb}. \tag{8.22}$$

This last equation suggests that the Weyl tensor $C^a{}_{bcd}$ be replaced by the rescaled Weyl tensor $d^a{}_{bcd}$ in the construction of a regular set of conformal field equations. In Chapter 10 it will be seen that the definitions of $d^a{}_{bcd}$ and T_{cdb} are justified in the sense that under suitable assumptions *the tensors $d^a{}_{bcd}$ and T_{cdb} are regular at the points where $\Xi = 0$*; see, in particular, Theorem 10.3.

One is now in the position of returning to the analysis of the equation for the Schouten tensor. Writing $C^a{}_{bcd}$ in terms of $d^a{}_{bcd}$ one obtains

$$\begin{aligned}\nabla_c L_{db} - \nabla_d L_{cb} &= \nabla_a(\Xi d^a{}_{bcd}) \\ &= \nabla_a \Xi d^a{}_{bcd} + \Xi \nabla_a d^a{}_{bcd}.\end{aligned}$$

Finally, using Equation (8.22) in the last term yields

$$\nabla_c L_{db} - \nabla_d L_{cb} = \nabla_a \Xi d^a{}_{bcd} + \Xi T_{cdb}, \tag{8.23}$$

which, again, is formally regular if $\Xi = 0$.

8.2.4 The regularised transformation rule for the Ricci scalar

To relate solutions of the conformal field equations to solutions of the Einstein field equations, one also needs to consider a regularised version of the transformation rule for the Ricci scalar, Equation (8.7). Multiplying this transformation law by Ξ^2 and rearranging the various terms one obtains

$$\tilde{R} = \Xi^2 R + 6\Xi \nabla_c \nabla^c \Xi - 12 \nabla_c \Xi \nabla^c \Xi.$$

Finally, using Equations (8.5a) and (8.9) one concludes that

$$\lambda = 6\Xi s - 3\nabla_c \Xi \nabla^c \Xi + \frac{1}{4}\Xi^4 T. \tag{8.24}$$

To understand the role of this equation it is useful to compute the derivative of its right-hand side. One has that

$$\begin{aligned}\nabla_a \Big(6\Xi s - 3\nabla_c \Xi \nabla^c \Xi + \frac{1}{4}\Xi^4 T \Big) \\ &= 6\nabla_a \Xi s + 6\Xi \nabla_a s - 6\nabla_a \nabla_c \Xi \nabla^c \Xi + \Xi^3 \nabla_a \Xi T + \frac{1}{4}\Xi^4 \nabla_a T \\ &= \Xi^4 (\nabla^c T_{ca} + \Xi^{-1} \nabla_a \Xi T),\end{aligned}$$

where in the second equality Equations (8.13) and (8.18) have been used to remove, respectively, the terms $\nabla_a \nabla_c \Xi$ and $\nabla_a \Xi$.

As will be further discussed in Chapter 9, the tensors $\tilde{T}_{ab}$ and T_{ab} satisfy the relation

$$g^{bc}\nabla_b T_{ca} = \Xi^{-4}\tilde{g}^{bc}(\tilde{\nabla}_b\tilde{T}_{ca} - \Xi^{-1}\tilde{\nabla}_a\Xi\tilde{T}_{bc}).$$

Hence, if $\tilde{\nabla}^b\tilde{T}_{ba} = 0$ it follows that

$$\nabla^c T_{ca} + \Xi^{-1}\nabla_a\Xi T = 0. \tag{8.25}$$

This last relation implies that

$$\nabla_a\left(6\Xi s - 3\nabla_c\Xi\nabla^c\Xi + \frac{1}{4}\Xi^4 T\right) = 0.$$

One has the following result:

Lemma 8.1 (***propagation of the cosmological constant***) *If Equations* (8.13), (8.18) *and* (8.25) *are satisfied on* $\mathcal{M}$ *and, in addition, Equation (8.24) holds at a point* $p \in \mathcal{M}$*, then Equation (8.24) is also satisfied on* $\mathcal{M}$.

Thus, Equation (8.24) plays the role of a *constraint* which is preserved, upon evolution, by virtue of the other conformal field equations.

8.2.5 Properties of the metric conformal field equations

The discussion of the previous sections is summarised in the following list of equations:

$$\nabla_a\nabla_b\Xi = -\Xi L_{ab} + s g_{ab} + \frac{1}{2}\Xi^3 T_{\{ab\}}, \tag{8.26a}$$

$$\nabla_a s = -L_{ac}\nabla^c\Xi + \frac{1}{2}\Xi^2\nabla^c\Xi T_{\{ac\}} + \frac{1}{6}\Xi^3\nabla^c T_{\{ca\}}, \tag{8.26b}$$

$$\nabla_c L_{db} - \nabla_d L_{cb} = \nabla_a\Xi d^a{}_{bcd} + \Xi T_{cdb}, \tag{8.26c}$$

$$\nabla_a d^a{}_{bcd} = T_{cdb}, \tag{8.26d}$$

$$6\Xi s - 3\nabla_c\Xi\nabla^c\Xi + \frac{1}{4}\Xi^4 T = \lambda. \tag{8.26e}$$

These are known as the (regular) ***metric conformal Einstein field equations***. *Equations* (8.26a)–(8.26e) *should be read as differential conditions for the fields* Ξ, s, L_{ab}, $d^a{}_{bcd}$. As already mentioned, Equation (8.26e) plays the role of a constraint. At the points where $\Xi \neq 0$, these equations are complemented by the physical conservation equation $\tilde{\nabla}^a\tilde{T}_{ab} = 0$ expressed in terms of conformal quantities:

$$\nabla^c T_{ca} + \Xi^{-1}\nabla_a\Xi T = 0. \tag{8.27}$$

Observe that in contrast to Equations (8.26a)–(8.26e), *Equation (8.27) is not formally regular at* $\Xi = 0$. This equation will be analysed in more detail in Chapter 9.

In what follows, for a ***solution to the metric conformal Einstein field equations*** it will be understood that a collection of fields

$$(g_{ab}, \Xi, s, L_{ab}, d^a{}_{bcd}, T_{ab})$$

satisfies Equations (8.26a)–(8.26e) and (8.27).

Remark. The discussion so far has not considered an equation for the components of the metric g_{ab}. To obtain the required condition assume that the Schouten tensor L_{ab} is determined through Equation (8.26c) and consider the relation (8.16) expressed in terms of some local coordinates (x^μ):

$$R_{\mu\nu} = 2L_{\mu\nu} + \frac{1}{6}Rg_{\mu\nu}.$$

Recalling that the components $R_{\mu\nu}$ can be expressed in terms of second-order derivatives of the components of the metric, one can read the previous expression as a differential condition for $g_{\mu\nu}$. To cast this equation in the form of some recognisable type of PDE one needs to make a particular choice of coordinates; see the discussions in Section 13.5.1 and in the Appendix of Chapter 13 on the *reduced Einstein field equations.*

The conformal vacuum Einstein field equations

An important particular case of Equations (8.26a)–(8.26e) occurs when $\tilde{T}_{ab} = 0$ on the whole of $\tilde{\mathcal{M}}$. Then one also has that $T_{abc} = 0$, and the conformal field equations reduce to:

$$\nabla_a\nabla_b\Xi = -\Xi L_{ab} + sg_{ab}, \tag{8.28a}$$
$$\nabla_a s = -L_{ac}\nabla^c\Xi, \tag{8.28b}$$
$$\nabla_c L_{db} - \nabla_d L_{cb} = \nabla_a\Xi d^a{}_{bcd}, \tag{8.28c}$$
$$\nabla_a d^a{}_{bcd} = 0, \tag{8.28d}$$
$$6\Xi s - 3\nabla_c\Xi\nabla^c\Xi = \lambda. \tag{8.28e}$$

Equations (8.28a)–(8.28e) are known as the ***conformal vacuum Einstein field equations.***

The metric conformal field equations and the Einstein field equations

Any solution to the Einstein field equations satisfies Equations (8.26a)–(8.26e) for *any* (smooth) choice of conformal factor Ξ. The converse of this observation is given in the following result.

Proposition 8.1 (***solutions of the conformal Einstein field equations as solutions to the Einstein field equations***) *Let*

$$(g_{ab}, \Xi, s, L_{ab}, d^a{}_{bcd}, T_{ab})$$

denote a solution to Equations (8.26a)–(8.26d) *and* (8.27) *such that* $\Xi \neq 0$ *on an open set* $\mathcal{U} \subset \mathcal{M}$. *If, in addition, Equation (8.26e) is satisfied at a point* $p \in \mathcal{U}$, *then the metric* $\tilde{g}_{ab} = \Xi^{-2} g_{ab}$ *is a solution to the Einstein field Equations (8.4) on* $\mathcal{U}$.

Proof It will be first shown that the Schouten tensor $\tilde{L}_{ab}$ of the metric $\tilde{g}_{ab} = \Xi^{-2} g_{ab}$ satisfies Equation (8.5b). Notice that the metric $\tilde{g}_{ab}$ is well defined on $\mathcal{U}$ as $\Xi \neq 0$. The transformation law for the Schouten tensor under conformal rescalings gives

$$\tilde{L}_{ab} = L_{ab} + \Xi^{-1}\nabla_a\nabla_b\Xi - \frac{1}{2}\Xi^{-2}\nabla_c\Xi\nabla^c\Xi g_{ab}.$$

Using Equations (8.26a) and (8.26b) the latter simplifies to

$$\tilde{L}_{ab} = \frac{1}{2}\tilde{T}_{ab} + \frac{1}{6}(\lambda - \tilde{T})\tilde{g}_{ab},$$

as required. In order to conclude that the Einstein field equations hold, one also needs to compute the Ricci scalar of the metric $\tilde{g}_{ab}$. As a consequence of Lemma 8.1 one has that Equation (8.26e) holds on the whole of $\mathcal{U}$. From the latter, again using (8.26a) and (8.26b) and recalling that $\tilde{T} = \Xi^4 T$, it follows that $\tilde{R} = 4\lambda - \tilde{T}$. Combining the obtained expressions for $\tilde{L}_{ab}$ and $\tilde{R}$ one readily concludes that (8.4) is indeed satisfied. □

Conformal freedom and conformal gauge

Consider a solution $(g_{ab}, \Xi, s, L_{ab}, d^a{}_{bcd}, T_{ab})$ to the metric conformal field Equations (8.26a)–(8.26e) and (8.27). As a consequence of Proposition 8.1, one has that $\tilde{\boldsymbol{g}} = \Xi^{-2}\boldsymbol{g}$ and $\tilde{T}_{ab} = \Xi^2 T_{ab}$, give rise to a solution of the Einstein field equations as long as $\Xi \neq 0$. Consider now another conformal factor $\acute{\Xi}$. From $\acute{\Xi}$, together with the physical fields $\tilde{g}_{ab}$ and $\tilde{T}_{ab}$, one can construct, by direct computation using the definitions of Sections 8.2.1–8.2.3, a collection of conformal fields $(\acute{g}_{ab}, \acute{\Xi}, \acute{s}, \acute{L}_{ab}, \acute{d}^a{}_{bcd}, \acute{T}_{ab})$. In particular, one has that $\acute{g}_{ab} = \acute{\Xi}^2\tilde{g}_{ab}$ and $\acute{T}_{ab} = \acute{\Xi}^{-2}T_{ab}$. These fields constitute, in turn, a solution to the metric conformal field equations. That is, they satisfy

$$\acute{\nabla}_a\acute{\nabla}_b\acute{\Xi} = -\acute{\Xi}\acute{L}_{ab} + \acute{s}\acute{g}_{ab} + \frac{1}{2}\acute{\Xi}^3\acute{T}_{\{ab\}},$$
$$\acute{\nabla}_a\acute{s} = -\acute{L}_{ac}\acute{\nabla}^c\acute{\Xi} + \frac{1}{2}\acute{\Xi}^2\acute{\nabla}^c\acute{\Xi}\acute{T}_{\{ac\}} + \frac{1}{6}\acute{\Xi}^3\acute{\nabla}^c\acute{T}_{\{ca\}},$$
$$\acute{\nabla}_c\acute{L}_{db} - \acute{\nabla}_d\acute{L}_{cb} = \acute{\nabla}_a\Xi\acute{d}^a{}_{bcd} + \acute{\Xi}\acute{T}_{cdb},$$
$$\acute{\nabla}_a\acute{d}^a{}_{bcd} = \acute{T}_{cdb},$$
$$6\acute{\Xi}\acute{s} - 3\acute{\nabla}_c\acute{\Xi}\acute{\nabla}^c\acute{\Xi} + \frac{1}{4}\acute{\Xi}^4\acute{T} = \lambda,$$
$$\acute{\nabla}^c\acute{T}_{ca} + \acute{\Xi}^{-1}\acute{\nabla}_a\acute{\Xi}\acute{T} = 0.$$

The unphysical metrics $\boldsymbol{g}$ and $\acute{\boldsymbol{g}}$ are conformally related to each other: one has that $\acute{\boldsymbol{g}} = \kappa^2 \boldsymbol{g}$ with $\kappa \equiv \acute{\Xi}\Xi^{-1}$, $\Xi \neq 0$. Using the transformation formulae of Chapter 5, one can express the solution $(\acute{g}_{ab}, \acute{\Xi}, \acute{s}, \acute{L}_{ab}, \acute{d}^a{}_{bcd}, \acute{T}_{ab})$ in terms of $(g_{ab}, \Xi, s, L_{ab}, d^a{}_{bcd}, T_{ab})$ and κ. One has that

$$\acute{\Xi} = \kappa\Xi, \qquad \acute{g}_{ab} = \kappa^2 g_{ab}, \tag{8.29a}$$

$$\acute{s} = \kappa^{-1}s + \kappa^{-2}\nabla_c\kappa\nabla^c\Xi + \frac{1}{2}\kappa^{-3}\Xi\nabla_c\kappa\nabla^c\kappa, \tag{8.29b}$$

$$\acute{L}_{ab} = L_{ab} - \kappa^{-1}\nabla_a\nabla_b\kappa + \frac{1}{2}\kappa^{-2}\nabla_c\kappa\nabla^c\kappa\, g_{ab}, \tag{8.29c}$$

$$\acute{d}^a{}_{bcd} = \kappa d^a{}_{bcd}, \tag{8.29d}$$

$$\acute{T}_{ab} = \kappa^{-2}T_{ab}. \tag{8.29e}$$

The two sets of solutions to the metric conformal field equations are said to be ***conformally related***.

From the discussion of the previous paragraphs it follows that there exists an infinite number of solutions to the metric conformal field equations giving rise to the same solution of the Einstein field equations. This is a manifestation of the ***conformal invariance*** of the equations. This conformal invariance is tied to a ***conformal freedom (or gauge)*** which, in turn, manifests itself in the properties of the unphysical metric $\boldsymbol{g}$. This conformal freedom has to be fixed in some way if one is to apply the theory of PDEs to the metric conformal field equations.

The issue of the conformal gauge discussed in the previous paragraph is closely related to the Ricci scalar R of the unphysical metric $\boldsymbol{g}$. The scalar R does not explicitly appear in the conformal Equations (8.26a)–(8.26e) and (8.27). Hence, it is not determined by the equations. Of course, given a solution $(g_{ab}, \Xi, s, L_{ab}, d^a{}_{bcd}, T_{ab})$, one can readily compute R. In general, conformally related solutions to the metric conformal field equations will give rise to different Ricci scalars. In order to understand better the connection between the conformal gauge and the Ricci scalar, consider a metric $\acute{\boldsymbol{g}}$ conformally related to $\boldsymbol{g}$ via $\acute{\boldsymbol{g}} = \kappa^2\boldsymbol{g}$. The transformation law for the Ricci scalar under conformal transformations implies that

$$6\nabla^a\nabla_a\kappa - R\kappa = -\acute{R}\kappa^3, \tag{8.30}$$

from where $\acute{R}$ can be determined. Alternatively, if $\acute{R}$ is an arbitrary scalar on $\mathcal{M}$, then Equation (8.30) can be read as a linear wave equation for κ. Given suitable initial data for this equation, it can be solved locally. The solution κ gives, in turn, the metric $\acute{\boldsymbol{g}} = \kappa^2\boldsymbol{g}$. From this point of view, the scalar field $\acute{R}$ plays the role of a ***conformal gauge source function***. In particular, one could choose $\acute{R} = 0$. As will be seen in later chapters of this book, this choice, despite its simplicity, is not necessarily the best one.

8.3 Frame and spinorial formulation of the conformal field equations

8.3.1 The frame formulation

This section provides a discussion of a frame formulation of the conformal Einstein field equations. This version of the field equations is more flexible than the metric one.

General definitions, frame fields

In what follows, consider a set of frame fields $\{e_a\}$, $a = 0, \ldots, 3$ which is orthonormal with respect to the metric g. Frames of this type will be said to be ***g*-orthonormal**. One has that

$$g(e_a, e_b) = \eta_{ab} = \mathrm{diag}(1, -1, -1, -1).$$

Following the conventions of Chapter 2, let $\Gamma_a{}^c{}_b = \langle \omega^c, \nabla_a e_b \rangle$ denote the connection coefficients of the connection ∇. As a consequence of the metric compatibility of ∇ one has that

$$\Gamma_a{}^d{}_b \eta_{dc} + \Gamma_a{}^d{}_c \eta_{bd} = 0.$$

The components, $\Sigma_a{}^c{}_b$ of the torsion of ∇ are given by the relation

$$\Sigma_a{}^c{}_b e_c = [e_a, e_b] - (\Gamma_a{}^c{}_b - \Gamma_b{}^c{}_a) e_c.$$

In the case of ∇ one naturally has that $\Sigma_a{}^c{}_b = 0$.

The geometric and the algebraic curvature

The discussion of the conformal field equations in terms of a frame formalism requires the expression of the components $R^c{}_{dab}$ of the Riemann tensor $R^c{}_{dab}$ with respect to the frame $\{e_a\}$; see Equation (2.31). Let $P^c{}_{dab}$ denote the right-hand side of equation (2.31), namely,

$$\begin{aligned} P^c{}_{dab} &\equiv e_a(\Gamma_b{}^c{}_d) - e_b(\Gamma_a{}^c{}_d) \\ &\quad + \Gamma_f{}^c{}_d(\Gamma_b{}^f{}_a - \Gamma_a{}^f{}_b) + \Gamma_b{}^f{}_d \Gamma_a{}^c{}_f - \Gamma_a{}^f{}_d \Gamma_b{}^c{}_f. \end{aligned}$$

In what follows, $P^c{}_{dab}$ will be known as the ***geometric curvature***. To complete the discussion one also needs to consider the decomposition of the Riemann tensor in terms of the Weyl tensor $C^c{}_{dab}$ and the Schouten tensor L_{ab}; see Equation (2.21b). The frame version of the decomposition is given by

$$R^c{}_{dab} = C^c{}_{dab} + 2 S_{d[a}{}^{ce} L_{b]e},$$

where, consistent with the general conventions of Chapter 2, $C^c{}_{dab}$ and L_{ab} denote, respectively, the components of the tensors $C^c{}_{dab}$ and L_{ab} with respect to $\{e_a\}$. The components $\rho^c{}_{dab}$ of the ***algebraic curvature*** are given by

$$\rho^c{}_{dab} \equiv \Xi d^c{}_{dab} + 2 S_{d[a}{}^{ce} L_{b]e}.$$

In the above definition it has been used that $C^{c}{}_{dab} = \Xi d^{c}{}_{dab}$. The geometric and algebraic curvature serve as useful shorthands of expressions which will be repeatedly used. Observe, in particular, that the equation $P^{c}{}_{dab} = \rho^{c}{}_{dab}$ encodes the idea that the fields $C^{c}{}_{dab}$ and L_{ab} correspond to the components of the Weyl and Schouten tensor of the connection defined by $\Gamma_{a}{}^{b}{}_{c}$.

The frame zero quantities and the frame conformal field equations

The frame version of the conformal field Equations (8.26a)–(8.26e) and (8.27) are readily obtained by contraction with the frame $\{e_a\}$ and the coframe $\{\omega^a\}$. One obtains

$$\nabla_a\nabla_b\Xi = -\Xi L_{ab} + s\eta_{ab} + \frac{1}{2}\Xi^3 T_{\{ab\}},$$
$$\nabla_a s = -L_{ac}\nabla^c\Xi + \frac{1}{2}\Xi^2\nabla^c\Xi T_{\{ac\}} + \frac{1}{6}\Xi^3\nabla^c T_{\{ca\}},$$
$$\nabla_c L_{db} - \nabla_d L_{cb} = \nabla_a\Xi d^{a}{}_{bcd} + \Xi T_{cdb},$$
$$\nabla_a d^{a}{}_{bcd} = T_{cdb},$$
$$6\Xi s - 3\nabla_c\Xi\nabla^c\Xi + \frac{1}{4}\Xi^4 T = \lambda,$$

and

$$\nabla^c T_{ca} + \Xi^{-1}\nabla_a\Xi T = 0,$$

where the directional derivative ∇_a acts on components of tensorial fields according to the rules in (2.28). The above frame conformal field equations will be complemented by the *structure equations*

$$\Sigma_{a}{}^{c}{}_{b}e_c = 0,$$
$$P^{c}{}_{dab} = \rho^{c}{}_{dab},$$

which express that for the connection $\boldsymbol{\nabla}$, its torsion must vanish and its geometric and algebraic curvature must coincide.

For convenience of the subsequent discussion one introduces a set of ***zero quantities***:

$$\Sigma_{ab} \equiv \Sigma_{a}{}^{c}{}_{b}e_c, \tag{8.31a}$$
$$\Xi^{c}{}_{dab} \equiv P^{c}{}_{dab} - \rho^{c}{}_{dab}, \tag{8.31b}$$
$$Z_{ab} \equiv \nabla_a\nabla_b\Xi + \Xi L_{ab} - s\eta_{ab} - \frac{1}{2}\Xi^3 T_{\{ab\}}, \tag{8.31c}$$
$$Z_a \equiv \nabla_a s + L_{ac}\nabla^c\Xi - \frac{1}{2}\Xi^2\nabla^c\Xi T_{\{ac\}} - \frac{1}{6}\Xi^3\nabla^c T_{\{ca\}}, \tag{8.31d}$$
$$\Delta_{cdb} \equiv \nabla_c L_{db} - \nabla_d L_{cb} - \nabla_a\Xi d^{a}{}_{bcd} - \Xi T_{cdb}, \tag{8.31e}$$
$$\Lambda_{bcd} \equiv \nabla_a d^{a}{}_{bcd} - T_{cdb}, \tag{8.31f}$$
$$Z \equiv 6\Xi s - 3\nabla_c\Xi\nabla^c\Xi + \frac{1}{4}\Xi^4 T - \lambda, \tag{8.31g}$$
$$M_a \equiv \nabla^c T_{ca} + \Xi^{-1}\nabla_a\Xi T. \tag{8.31h}$$

In terms of the above zero quantities, the ***frame version of the conformal field equations*** can be compactly written as

$$\Sigma_{ab} = 0, \qquad \Xi^{c}{}_{dab} = 0, \qquad Z_{ab} = 0, \qquad Z_{a} = 0, \tag{8.32a}$$

$$\Delta_{cdb} = 0, \qquad \Lambda_{bcd} = 0, \qquad Z = 0, \qquad M_{a} = 0. \tag{8.32b}$$

Accordingly, a ***solution to the frame conformal Einstein field equations*** is a collection $(e_a, \Gamma_a{}^b{}_c, \Xi, s, L_{ab}, d^a{}_{bcd}, T_{ab})$ satisfying Equations (8.32a) and (8.32b). The equations associated to the zero quantities Σ_{ab} and $\Xi^c{}_{dab}$ provide differential conditions for the components of the frame vectors $\{e_a\}$ and for the connection coefficients $\Gamma_a{}^b{}_c$. The role of the equations associated to the zero quantities Z_{ab}, Z_a, Δ_{cdb}, Λ_{bcd}, Z and M_a is similar to that of their metric counterparts in Section 8.2.

By considering a frame version of the conformal field equations, one introduces a further gauge freedom into the system. This gauge freedom corresponds to the Lorentz transformations preserving the $\boldsymbol{g}$-orthonormality of the frame vectors $\{e_a\}$. In this case one speaks of a ***frame gauge freedom***. As in the case of the conformal freedom discussed in Section 8.2.5, this freedom needs to be fixed in order to be able to apply the methods of the theory of PDEs. These issues will be discussed further in Chapter 13.

The frame conformal field equations and the Einstein field equations

As in the case of the metric conformal field equations, a solution to the frame conformal field equations implies, under suitable conditions, a solution to the Einstein field equations; see Proposition 8.1. More precisely, one has:

Proposition 8.2 (***solutions to the frame conformal field equations as solutions to the Einstein field equations***) *Let*

$$(e_a, \Gamma_a{}^b{}_c, \Xi, s, L_{ab}, d^a{}_{bcd}, T_{ab})$$

denote a solution to the frame conformal field Equations (8.32a) *and* (8.32b) *with* $\Gamma_a{}^c{}_b$ *satisfying the metric compatibility condition*

$$\Gamma_a{}^d{}_b \eta_{dc} + \Gamma_a{}^d{}_c \eta_{bd} = 0,$$

and such that

$$\Xi \neq 0, \quad \det(\eta^{ab} e_a \otimes e_b) \neq 0,$$

on an open set $\mathcal{U} \subset \mathcal{M}$. *Then the metric* $\tilde{g} = \Xi^{-2} \eta_{ab} \omega^a \otimes \omega^b$, *where* $\{\omega^a\}$ *is the dual frame to* $\{e_a\}$, *is a solution to the Einstein field Equations (8.4) on* $\mathcal{U}$.

Proof As a consequence of the metric compatibility assumption and $\Sigma_{ab} = 0$, the coefficients $\Gamma_a{}^c{}_b$ can be interpreted as the connection coefficients of a Levi-Civita connection with respect to the frame $\{e_a\}$. By the uniqueness of the Levi-Civita connection, $\boldsymbol{g} = \eta_{ab}\boldsymbol{\omega}^a \otimes \boldsymbol{\omega}^b$ is the metric associated to this connection. Notice that by assumption $\boldsymbol{g}$ is well defined on $\mathcal{U}$. Furthermore, because of $\Xi^c{}_{dab} = 0$ and exploiting the uniqueness of the decomposition of the Riemann tensor in terms of the Weyl and the Schouten tensors, it follows that L_{ab} are the components, with respect to the frame $\{\boldsymbol{e}_a\}$, of the Schouten tensor of the metric $\boldsymbol{g}$. From here, following arguments analogous to those used in the proof of Proposition 8.1 one concludes that $\tilde{\boldsymbol{g}} = \Xi^{-2}\eta_{ab}\boldsymbol{\omega}^a \otimes \boldsymbol{\omega}^b$ and $\tilde{T}_{ab} = \Xi^2 \omega^a{}_a \omega^b{}_b T_{ab}$ give a solution to the Einstein field Equations (8.4) on $\mathcal{U}$. □

8.3.2 Spinorial formulation of the conformal field equations

The frame conformal field equations lead, in a natural way, to a spinorial formulation. This formulation of the equations reveals in a more clear fashion the inherent algebraic structure of the equations and provides a systematic procedure for the construction of evolution equations. The formulation discussed in this section is not an abstract spinor formulation, but rather a *frame spinor formulation.*

General remarks concerning the spinorial formulation

Following the discussion in Section 3.1.13, the $\boldsymbol{g}$-orthonormal frame $\{\boldsymbol{e}_a\}$ gives rise to a frame $\{\boldsymbol{e}_{AA'}\}$ such that $\boldsymbol{g}(\boldsymbol{e}_{AA'}, \boldsymbol{e}_{BB'}) = \epsilon_{AB}\epsilon_{A'B'}$; that is, $\{\boldsymbol{e}_{AA'}\}$ is a null tetrad. In what follows, let

$$\Sigma_{AA'}{}^{CC'}{}_{BB'}, \quad P^{CC'}{}_{DD'AA'BB'}, \quad \rho^{CC'}{}_{DD'AA'BB'}, \quad T_{AA'BB'},$$
$$L_{AA'BB'}, \quad d^{AA'}{}_{BB'CC'DD'}, \quad T_{AA'BB'CC'},$$

denote, respectively, the spinorial counterparts of the fields

$$\Sigma_a{}^c{}_b, \quad P^c{}_{dab}, \quad \rho^c{}_{dab}, \quad T_{ab}, \quad L_{ab}, \quad d^a{}_{bcd}, \quad T_{abc}.$$

The spinorial counterpart of the geometric curvature, $P^{CC'}{}_{DD'AA'BB'}$, is expressed in terms of the spinorial connection coefficients $\Gamma_{AA'}{}^{BB'}{}_{CC'}$. These, in turn, can be expressed in terms of the *reduced spin connection coefficients* $\Gamma_{AA'}{}^B{}_C$; see formula (3.33). As the connection $\boldsymbol{\nabla}$ is metric, it follows that $\Gamma_{AA'BC} = \Gamma_{AA'(BC)}$; compare Section 3.2.2. By analogy to the split of the spinorial counterpart of the curvature tensor – Equation (3.35) – one can split the geometric curvature as

$$P^{CC'}{}_{DD'AA'BB'} = P^C{}_{DAA'BB'}\delta_{D'}{}^{C'} + \bar{P}^{C'}{}_{D'AA'BB'}\delta_D{}^C.$$

In what follows, the discussion will make use only of the **reduced spinorial geometric curvature**

$$\begin{aligned}
P^C{}_{DAA'BB'} \equiv\ & \boldsymbol{e}_{AA'}(\Gamma_{BB'}{}^C{}_D) - \boldsymbol{e}_{BB'}(\Gamma_{AA'}{}^C{}_D) \\
& - \Gamma_{FB'}{}^C{}_D\Gamma_{AA'}{}^F{}_B - \Gamma_{BF'}{}^C{}_D\bar{\Gamma}_{AA'}{}^{F'}{}_{B'} + \Gamma_{FA'}{}^C{}_D\Gamma_{BB'}{}^F{}_A \\
& + \Gamma_{AF'}{}^C{}_D\bar{\Gamma}_{BB'}{}^{F'}{}_{A'} + \Gamma_{AA'}{}^C{}_F\Gamma_{BB'}{}^F{}_D - \Gamma_{BB'}{}^C{}_F\Gamma_{AA'}{}^F{}_D.
\end{aligned}$$

The spinorial algebraic curvature has a similar split. Its information is encoded in the field

$$\rho^C{}_{DAA'BB'} \equiv -\Psi^C{}_{DAB}\epsilon_{A'B'} + L_{DB'AA'}\delta_B{}^C - L_{DA'BB'}\delta_A{}^C,$$

where it is recalled that Ψ_{ABCD} is the Weyl spinor; see Equation (3.43). One then introduces the totally symmetric ***rescaled Weyl spinor*** ϕ_{ABCD} as

$$\phi_{ABCD} \equiv \Xi^{-1}\Psi_{ABCD}.$$

Consistent with Equation (3.43), ϕ_{ABCD} is related to the spinorial counterpart of $d^a{}_{bcd}$ via

$$d_{AA'BBCC'DD'} = -\phi_{ABCD}\epsilon_{A'B'}\epsilon_{C'D'} - \bar{\phi}_{A'B'C'D'}\epsilon_{AB}\epsilon_{CD}. \tag{8.33}$$

Hence, the reduced spinorial algebraic curvature can be written as

$$\rho^C{}_{DAA'BB'} \equiv -\Xi\phi^C{}_{DAB}\epsilon_{A'B'} + L_{DB'AA'}\delta_B{}^C - L_{DA'BB'}\delta_A{}^C.$$

The spinorial counterpart of $T_{\{ab\}}$, the symmetric trace-free part of T_{ab}, is given by $T_{(AB)(A'B')}$; compare Equation (3.12). Finally, exploiting the antisymmetry $T_{cdb} = -T_{dcb}$ of the rescaled Cotton tensor, one has the split

$$T_{CC'DD'BB'} = T_{CDBB'}\epsilon_{C'D'} + \bar{T}_{C'D'BB'}\epsilon_{CD}, \tag{8.34}$$

where $T_{CDBB'} \equiv \frac{1}{2}T_{CQ'D}{}^{Q'}{}_{BB'}$. Observe that $T_{CDBB'} = T_{(CD)BB'}$.

The spinorial zero quantities

The spinorial counterparts of the frame conformal Einstein field equations are obtained by suitable contraction with the Infeld-van der Waerden symbols. Simpler expressions are obtained if one takes into account the remarks made in the previous subsection. It is convenient to introduce the following ***spinorial zero quantities***:

$$\Sigma_{AA'BB'} \equiv [e_{AA'}, e_{BB'}] - (\Gamma_{AA'}{}^{CC'}{}_{BB'} - \Gamma_{BB'}{}^{CC'}{}_{AA'})e_{CC'}, \tag{8.35a}$$

$$\Xi^C{}_{DAA'BB'} \equiv P^C{}_{DAA'BB'} - \rho^C{}_{DAA'BB'}, \tag{8.35b}$$

$$Z_{AA'BB'} \equiv \nabla_{AA'}\nabla_{BB'}\Xi + \Xi L_{AA'BB'} - s\epsilon_{AB}\epsilon_{A'B'} - \frac{1}{2}\Xi^3 T_{(AB)(A'B')}, \tag{8.35c}$$

$$Z_{AA'} \equiv \nabla_{AA'}s + L_{AA'CC'}\nabla^{CC'}\Xi - \frac{1}{2}\Xi^2\nabla^{CC'}\Xi T_{(AC)(A'C')} - \frac{1}{6}\Xi^3\nabla^{CC'}T_{(AC)(A'C')}, \tag{8.35d}$$

$$\Delta_{CC'DD'BB'} \equiv \nabla_{CC'}L_{DD'BB'} - \nabla_{DD'}L_{CC'BB'} \tag{8.35e}$$

$$- \nabla_{AA'}\Xi d^{AA'}{}_{BB'CC'DD'} - \Xi T_{CC'DD'BB'}, \tag{8.35f}$$

$$\Lambda_{BB'CC'DD'} \equiv \nabla_{AA'} d^{AA'}{}_{BB'CC'DD'} - T_{CC'DD'BB'}, \tag{8.35g}$$

$$Z \equiv 6\Xi s - 3\nabla_{CC'}\nabla^{CC'} + \frac{1}{4}\Xi^4 T - \lambda, \tag{8.35h}$$

$$M_{AA'} \equiv \nabla^{CC'} T_{CC'AA'} + \Xi^{-1}\nabla_{AA'}\Xi T. \tag{8.35i}$$

Hence, the spinorial conformal Einstein field equations are given in terms of the above zero quantities as

$$\Sigma_{AA'BB'} = 0, \quad \Xi^{C}{}_{DAA'BB'} = 0, \quad Z_{AA'BB'} = 0, \quad Z_{AA'} = 0, \tag{8.36a}$$

$$\Delta_{CC'DD'BB'} = 0, \quad \Lambda_{BB'CC'DD'} = 0, \quad Z = 0, \quad M_{AA'} = 0. \tag{8.36b}$$

A reduced set of zero quantities can be obtained by explicitly making use of the antisymmetry of several of the spinorial zero quantities. In particular, it is noticed that as $\Delta_{CC'DD'BB'} = -\Delta_{DD'CC'BB'}$ and $\Lambda_{BB'CC'DD'} = -\Lambda_{BB'DD'CC'}$ one can write

$$\Delta_{CC'DD'BB'} = \Delta_{CDBB'}\epsilon_{C'D'} + \bar{\Delta}_{C'D'BB'}\epsilon_{CD},$$

$$\Lambda_{BB'CC'DD'} = \Lambda_{BB'CD}\epsilon_{C'D'} + \bar{\Lambda}_{B'BC'D'}\epsilon_{CD},$$

where

$$\Delta_{CDBB'} \equiv \frac{1}{2}\Delta_{CQ'D}{}^{Q'}{}_{BB'}, \qquad \Lambda_{BB'CD} \equiv \frac{1}{2}\Lambda_{BB'CQ'D}{}^{Q'}.$$

A direct computation using the splits (8.33) and (8.34) yields

$$\Delta_{CDBB'} = \nabla_{(C}{}^{Q'} L_{D)Q'BB'} + \nabla^{Q}{}_{B'}\Xi\phi_{CDBQ} + \Xi T_{CDBB'}, \tag{8.37a}$$

$$\Lambda_{BB'CD} = \nabla^{Q}{}_{B'}\phi_{BCDQ} + T_{CDBB'}. \tag{8.37b}$$

Thus, an equivalent spinorial formulation of the conformal field equations is given by

$$\Sigma_{AA'BB'} = 0, \quad \Xi^{C}{}_{DAA'BB'} = 0, \quad Z_{AA'BB'} = 0, \quad Z_{AA'} = 0 \tag{8.38a}$$

$$\Delta_{CDBB'} = 0, \quad \Lambda_{BB'CD} = 0, \quad Z = 0, \quad M_{AA'} = 0. \tag{8.38b}$$

The antisymmetry of the zero quantities $\Sigma_{AA'BB'}$ and $\Xi^{C}{}_{DAA'BB'}$ can also be exploited to obtain reduced zero quantities $\Sigma_{AB} = \Sigma_{(AB)}$ and $\Xi^{C}{}_{DAB} = \Xi^{C}{}_{D(AB)}$. This strategy will not be pursued further.

The spinorial conformal field equations and the Einstein field equations

As a consequence of the equivalence between spinorial and frame expressions discussed in Section 3.1.9, it follows that each of the two spinorial formulations of the conformal field Equations (8.36a) and (8.36b) or (8.38a) and (8.38b) is equivalent to the frame conformal field Equations (8.32a) and (8.32b). Thus, an analogue of Proposition 8.2 holds for the spinorial conformal field equations with the metric

$$\tilde{g} = \Xi^{-2}\epsilon_{AB}\epsilon_{A'B'}\omega^{AA'} \otimes \omega^{BB'},$$

yielding the required solution to the Einstein field equations. In this last expression $\{\omega^{AA'}\}$ denotes the duals of the frame $\{e_{AA'}\}$.

8.3.3 Conformal freedom in the frame and spinorial conformal field equations

The transformation laws for the various conformal fields under a conformal gauge change follow from the tensorial version given in (8.29a)–(8.29e). As before, assume that one has two metrics $\boldsymbol{g}$ and $\acute{\boldsymbol{g}}$ such that $\boldsymbol{g} = \kappa^2 \acute{\boldsymbol{g}}$. Consider now a $\boldsymbol{g}$-orthonormal frame $\{\boldsymbol{e_a}\}$ with associated coframe $\{\boldsymbol{\omega^a}\}$. From

$$\boldsymbol{g}(\boldsymbol{e_a}, \boldsymbol{e_b}) = \kappa^2 \acute{\boldsymbol{g}}(\boldsymbol{e_a}, \boldsymbol{e_b}) = \eta_{\boldsymbol{ab}},$$

it follows that $\{\acute{\boldsymbol{e}}_{\boldsymbol{a}}\}$ and $\{\acute{\boldsymbol{\omega}}^{\boldsymbol{a}}\}$, with

$$\acute{\boldsymbol{e}}_{\boldsymbol{a}} \equiv \kappa \boldsymbol{e_a}, \qquad \acute{\boldsymbol{\omega}}^{\boldsymbol{a}} \equiv \kappa^{-1} \boldsymbol{\omega^a},$$

are a $\acute{\boldsymbol{g}}$-orthonormal frame and a $\acute{\boldsymbol{g}}$-orthonormal coframe, respectively. As a consequence, the tensorial transformation formulae (8.29a)–(8.29e) may pick up factors of κ depending on whether they are contracted with $\boldsymbol{e_a}$ or $\acute{\boldsymbol{e}}_{\boldsymbol{a}}$. For, example

$$\begin{aligned} \acute{d}^{\boldsymbol{a}}{}_{\boldsymbol{bcd}} &\equiv \acute{\omega}^{\boldsymbol{a}}{}_a \acute{e}_{\boldsymbol{b}}{}^b \acute{e}_{\boldsymbol{c}}{}^c \acute{e}_{\boldsymbol{d}}{}^d \acute{d}^a{}_{bcd} \\ &= \kappa^3 \omega^{\boldsymbol{a}}{}_a e_{\boldsymbol{b}}{}^b e_{\boldsymbol{c}}{}^c e_{\boldsymbol{d}}{}^d d^a{}_{bcd} = \kappa^3 d^{\boldsymbol{a}}{}_{\boldsymbol{bcd}}. \end{aligned}$$

Similar considerations lead to

$$\begin{aligned} \acute{L}_{\boldsymbol{ab}} &= \kappa^2 L_{\boldsymbol{ab}} + \kappa^2 \nabla_{\boldsymbol{a}}(\kappa^{-1}\nabla_{\boldsymbol{b}}\kappa) - \frac{1}{2} S_{\boldsymbol{ab}}{}^{\boldsymbol{cd}} \nabla_{\boldsymbol{c}}\kappa \nabla_{\boldsymbol{d}}\kappa, \\ \acute{T}_{\boldsymbol{ab}} &= \kappa^2 T_{\boldsymbol{ab}}, \end{aligned}$$

where

$$\begin{aligned} S_{\boldsymbol{ab}}{}^{\boldsymbol{cd}} &\equiv e_{\boldsymbol{a}}{}^a e_{\boldsymbol{b}}{}^b \omega^{\boldsymbol{c}}{}_c \omega^{\boldsymbol{d}}{}_d S_{ab}{}^{cd} \\ &\equiv \delta_{\boldsymbol{a}}{}^{\boldsymbol{c}} \delta_{\boldsymbol{b}}{}^{\boldsymbol{d}} + \delta_{\boldsymbol{a}}{}^{\boldsymbol{d}} \delta_{\boldsymbol{b}}{}^{\boldsymbol{c}} - \eta_{\boldsymbol{ab}} \eta^{\boldsymbol{cd}}. \end{aligned}$$

The spinorial counterparts of the conformal fields obey similar transformations. If $\{\epsilon_{\boldsymbol{A}}{}^A\}$ and $\{\acute{\epsilon}_{\boldsymbol{A}}{}^A\}$ denote the spin dyads associated, respectively, to the frame vectors $\{\boldsymbol{e_a}\}$ and $\{\acute{\boldsymbol{e}}_{\boldsymbol{a}}\}$, then

$$\acute{\epsilon}_{\boldsymbol{A}}{}^A = \kappa \epsilon_{\boldsymbol{A}}{}^A.$$

As a consequence one has, for example, that

$$\acute{\Psi}_{\boldsymbol{ABCD}} = \kappa^3 \Psi_{\boldsymbol{ABCD}}.$$

8.4 The extended conformal Einstein field equations

The conformal Einstein field equations discussed in the previous sections are expressed in terms of the Levi-Civita connection of the unphysical metric $\boldsymbol{g}$. This section provides a more general version of the equations by rewriting them in terms of a *Weyl connection*. The resulting system of equations is known as the ***extended conformal Einstein field equations***. The use of Weyl connections introduces a further freedom in the equations. This freedom can be exploited to incorporate *conformally privileged gauges*. The idea of reexpressing the vacuum conformal field equations in terms of a Weyl connection was first introduced in Friedrich (1995). Further discussions can be found in Friedrich (1998c, 2002, 2004). The extension of these ideas to the matter case has been given in Lübbe and Valiente Kroon (2012, 2013b).

In what follows, for ease of presentation, the discussion in this section is restricted to the vacuum case.

Basic setting

As in the previous sections of this chapter, let $\boldsymbol{g}$ denote an *unphysical* Lorentzian metric related to a *physical* metric $\tilde{\boldsymbol{g}}$ via $\boldsymbol{g} = \Xi^2\tilde{\boldsymbol{g}}$. The metric $\tilde{\boldsymbol{g}}$ is assumed to satisfy the vacuum Einstein field equations. Let $\boldsymbol{\nabla}$ and $\tilde{\boldsymbol{\nabla}}$ denote, respectively, the Levi-Civita connections of the metrics $\boldsymbol{g}$ and $\tilde{\boldsymbol{g}}$.

In what follows, consider a Weyl connection $\hat{\boldsymbol{\nabla}}$ defined via

$$\hat{\boldsymbol{\nabla}} - \boldsymbol{\nabla} = \boldsymbol{S}(\boldsymbol{f}), \tag{8.39}$$

where $\boldsymbol{f}$ is a smooth covector. As

$$\boldsymbol{\nabla} - \tilde{\boldsymbol{\nabla}} = \boldsymbol{S}(\Xi^{-1}\mathbf{d}\Xi),$$

it follows that $\hat{\boldsymbol{\nabla}} - \tilde{\boldsymbol{\nabla}} = \boldsymbol{S}(\boldsymbol{f} + \Xi^{-1}\mathbf{d}\Xi)$. Hence, defining

$$\boldsymbol{\beta} \equiv \boldsymbol{f} + \Xi^{-1}\mathbf{d}\Xi,$$

one has that

$$\hat{\boldsymbol{\nabla}} - \tilde{\boldsymbol{\nabla}} = \boldsymbol{S}(\boldsymbol{\beta}).$$

It is convenient to define

$$\boldsymbol{d} \equiv \Xi\boldsymbol{f} + \mathbf{d}\Xi, \tag{8.40}$$

so that $\boldsymbol{d} = \Xi\boldsymbol{\beta}$.

As the Weyl connection $\hat{\boldsymbol{\nabla}}$ is torsion free, it follows that its Riemann curvature tensor $\hat{R}^c{}_{dab}$ can be decomposed in terms of its Schouten tensor $\hat{L}_{ab}$ and the Weyl tensor of the conformal class of $C^c{}_{dab}$; see Equation (5.28a). Using the latter and recalling the definition of the rescaled Weyl tensor $d^c{}_{dab}$, Equation (8.20), one obtains the equation

$$\hat{R}^c{}_{dab} = 2S_{d[a}{}^{ce}\hat{L}_{b]e} + \Xi d^c{}_{dab}.$$

Consistent with the discussion in Section 5.3.2, Equations (5.29a)–(5.29c), the Schouten tensors of the connections $\tilde{\boldsymbol{\nabla}}$, $\boldsymbol{\nabla}$ and $\hat{\boldsymbol{\nabla}}$ are related to each other via

$$\hat{L}_{ab} - L_{ab} = \nabla_a f_b - \frac{1}{2} S_{ab}{}^{cd} f_c f_d,$$
$$\hat{L}_{ab} - \tilde{L}_{ab} = \hat{\nabla}_a \beta_b + \frac{1}{2} S_{ab}{}^{cd} \beta_c \beta_d,$$
$$L_{ab} - \tilde{L}_{ab} = \nabla_a(\Xi^{-1}\nabla_b \Xi) + \frac{1}{2}\Xi^{-2} S_{ab}{}^{cd} \nabla_c \Xi \nabla_d \Xi.$$

Taking into account the above expressions and recalling that $\hat{\boldsymbol{\nabla}} \boldsymbol{S} = 0$ one has that

$$\begin{aligned}
\hat{\nabla}_c \hat{L}_{db} - \hat{\nabla}_d \hat{L}_{cb} &= (\hat{\nabla}_c L_{db} - S_{cd}{}^{ef} f_e L_{fb} - S_{cb}{}^{ef} f_e L_{df}) \\
&\quad -(\hat{\nabla}_d L_{cb} - S_{dc}{}^{ef} f_e L_{fb} - S_{db}{}^{ef} f_e L_{cf}) \\
&= \hat{\nabla}_c \hat{L}_{db} - \hat{\nabla}_d \hat{L}_{cb} + (\hat{\nabla}_c \hat{\nabla}_d - \hat{\nabla}_d \hat{\nabla}_c) f_b \\
&\quad + S_{db}{}^{ef} f_e (\hat{\nabla}_c f_f - L_{cf}) - S_{cb}{}^{ef} f_e (\hat{\nabla}_d f_f - L_{df}).
\end{aligned}$$

A further computation using Equation (5.29a) and the definition of the tensor $\boldsymbol{S}$ yields

$$\begin{aligned}
S_{db}{}^{ef} f_e (\hat{\nabla}_c f_f - L_{cf}) - S_{cb}{}^{ef} f_e (\hat{\nabla}_d f_f - L_{df}) &= (S_{cb}{}^{ef} \hat{L}_{df} - S_{db}{}^{ef} \hat{L}_{cf}) f_e \\
&= 2 S_{b[c}{}^{ef} \hat{L}_{d]f} f_e.
\end{aligned}$$

Hence, recalling the split (5.28a) of the Riemann tensor one obtains

$$\hat{\nabla}_c \hat{L}_{db} - \hat{\nabla}_d \hat{L}_{cb} = \nabla_c L_{db} - \nabla_d L_{cb} + f_a C^a{}_{bcd}.$$

Thus, the Weyl connection version of the vacuum Cotton equation is given by

$$\hat{\nabla}_a \hat{L}_{bc} - \hat{\nabla}_b \hat{L}_{ac} = d_e d^e{}_{cab}.$$

Now, for the Bianchi Equation (8.22) one has that

$$\begin{aligned}
\hat{\nabla}_a d^a{}_{bcd} &= \nabla_a d^a{}_{bcd} - S_{ah}{}^{fa} f_f d^h{}_{bcd} + S_{ab}{}^{fh} f_f d^a{}_{hcd} \\
&\quad + S_{ac}{}^{fh} f_f d^a{}_{bhd} + S_{ad}{}^{fh} f_f d^a{}_{bch} \\
&= \nabla_a d^a{}_{bcd} - f_a d^a{}_{dcb} + f_a d^a{}_{cdb} \\
&= \nabla_a d^a{}_{bcd} - f_a d^a{}_{bcd},
\end{aligned}$$

where in the last line it has been used that $d^a{}_{bcd}$ satisfies the first Bianchi identity $d^a{}_{bcd} + d^a{}_{cdb} + d^a{}_{dbc} = 0$. Hence, Equation (8.22) expressed in terms of the Weyl connection $\hat{\boldsymbol{\nabla}}$ takes the form:

$$\hat{\nabla}_a d^a{}_{bcd} = f_a d^a{}_{bcd}.$$

As a summary of this section one has the two equations:

$$\hat{\nabla}_a \hat{L}_{bc} - \hat{\nabla}_b \hat{L}_{ac} = d_d d^d{}_{cab}, \tag{8.41a}$$
$$\hat{\nabla}^d d_{dcab} = f_d d^d{}_{cab}. \tag{8.41b}$$

These two equations will be regarded as the core of the *extended conformal Einstein field equations*. They provide differential conditions on the Schouten tensor of the Weyl connection and the rescaled Weyl tensor.

8.4.1 The frame version of the extended conformal field equations

Equations (8.41a) and (8.41b) need to be supplemented with equations which provide information about the metric $\boldsymbol{g}$ associated to the conformal factor Ξ and which also allow to determine the covector $\boldsymbol{f}$ giving rise to the Weyl connection $\hat{\boldsymbol{\nabla}}$. The most convenient way of doing this is to make use of a frame formalism.

As in Section 8.3.1 let $\{\boldsymbol{e_a}\}$, $\boldsymbol{a} = \boldsymbol{0}, \ldots, \boldsymbol{3}$ denote a frame field which is $\boldsymbol{g}$-orthonormal so that $\boldsymbol{g}(\boldsymbol{e_a}, \boldsymbol{e_b}) = \eta_{\boldsymbol{ab}}$. As $\boldsymbol{\nabla}$ is the Levi-Civita connection of $\boldsymbol{g}$, its connection coefficients, $\Gamma_{\boldsymbol{a}}{}^{\boldsymbol{c}}{}_{\boldsymbol{b}} = \langle \boldsymbol{\omega^c}, \nabla_{\boldsymbol{a}} \boldsymbol{e_b} \rangle$, satisfy the metric compatibility condition of Equation (2.29).

Let now $\hat{\boldsymbol{\nabla}}$ denote the Weyl connection constructed from the Levi-Civita connection $\boldsymbol{\nabla}$ and the covector $\boldsymbol{f}$ using Equation (8.39). If $\hat{\Gamma}_{\boldsymbol{a}}{}^{\boldsymbol{c}}{}_{\boldsymbol{b}} = \langle \boldsymbol{\omega^c}, \hat{\nabla}_{\boldsymbol{a}} \boldsymbol{e_b} \rangle$ denotes the connection coefficients of $\hat{\boldsymbol{\nabla}}$ with respect to the frame $\{\boldsymbol{e_a}\}$, one has that

$$
\begin{aligned}
\hat{\Gamma}_{\boldsymbol{a}}{}^{\boldsymbol{c}}{}_{\boldsymbol{b}} &= \Gamma_{\boldsymbol{a}}{}^{\boldsymbol{c}}{}_{\boldsymbol{b}} + S_{\boldsymbol{ab}}{}^{\boldsymbol{cd}} f_{\boldsymbol{d}}, && (8.42a)\\
&= \Gamma_{\boldsymbol{a}}{}^{\boldsymbol{c}}{}_{\boldsymbol{b}} + \delta_{\boldsymbol{a}}{}^{\boldsymbol{c}} f_{\boldsymbol{b}} + \delta_{\boldsymbol{b}}{}^{\boldsymbol{c}} f_{\boldsymbol{a}} - \eta_{\boldsymbol{ab}} \eta^{\boldsymbol{cd}} f_{\boldsymbol{d}}. && (8.42b)
\end{aligned}
$$

In particular, one has that

$$
f_{\boldsymbol{a}} = \frac{1}{4} \hat{\Gamma}_{\boldsymbol{a}}{}^{\boldsymbol{b}}{}_{\boldsymbol{b}}, \qquad (8.43)
$$

as $\Gamma_{\boldsymbol{a}}{}^{\boldsymbol{b}}{}_{\boldsymbol{b}} = 0$ in the case of a metric connection.

Let $\hat{\Sigma}_{\boldsymbol{a}}{}^{\boldsymbol{c}}{}_{\boldsymbol{b}}$ denote the torsion of the connection $\hat{\boldsymbol{\nabla}}$. Using the transformation formula for the torsion under change of connections, Equation (2.15), together with Equation (8.42a), one obtains

$$
\hat{\Sigma}_{\boldsymbol{a}}{}^{\boldsymbol{c}}{}_{\boldsymbol{b}} - \Sigma_{\boldsymbol{a}}{}^{\boldsymbol{c}}{}_{\boldsymbol{b}} = -2 S_{[\boldsymbol{ab}]}{}^{\boldsymbol{cd}} f_{\boldsymbol{d}} = 0.
$$

Thus,

$$
\hat{\Sigma}_{\boldsymbol{a}}{}^{\boldsymbol{c}}{}_{\boldsymbol{b}} = 0,
$$

as $\Sigma_{\boldsymbol{a}}{}^{\boldsymbol{c}}{}_{\boldsymbol{b}} = 0$. As in Section 8.3.1 it is convenient to distinguish between the *geometric curvature* $\hat{P}^{\boldsymbol{c}}{}_{\boldsymbol{dab}}$ – that is, the expression for the components of the Riemann tensor of the connection $\hat{\boldsymbol{\nabla}}$ in terms of the connection coefficients $\hat{\Gamma}_{\boldsymbol{a}}{}^{\boldsymbol{c}}{}_{\boldsymbol{b}}$ – and the *algebraic curvature* $\hat{\rho}^{\boldsymbol{c}}{}_{\boldsymbol{dab}}$ – that is, the expression of the Riemann tensor in terms of the Schouten and Weyl tensors. These are given by

$$
\begin{aligned}
\hat{P}^{\boldsymbol{c}}{}_{\boldsymbol{dab}} &\equiv \boldsymbol{e_a}(\hat{\Gamma}_{\boldsymbol{b}}{}^{\boldsymbol{c}}{}_{\boldsymbol{d}}) - \boldsymbol{e_b}(\hat{\Gamma}_{\boldsymbol{a}}{}^{\boldsymbol{c}}{}_{\boldsymbol{d}}) \\
&\quad + \hat{\Gamma}_{\boldsymbol{f}}{}^{\boldsymbol{c}}{}_{\boldsymbol{d}}(\hat{\Gamma}_{\boldsymbol{b}}{}^{\boldsymbol{f}}{}_{\boldsymbol{a}} - \hat{\Gamma}_{\boldsymbol{a}}{}^{\boldsymbol{f}}{}_{\boldsymbol{b}}) + \hat{\Gamma}_{\boldsymbol{b}}{}^{\boldsymbol{f}}{}_{\boldsymbol{d}} \hat{\Gamma}_{\boldsymbol{a}}{}^{\boldsymbol{c}}{}_{\boldsymbol{f}} - \hat{\Gamma}_{\boldsymbol{a}}{}^{\boldsymbol{f}}{}_{\boldsymbol{d}} \hat{\Gamma}_{\boldsymbol{b}}{}^{\boldsymbol{c}}{}_{\boldsymbol{f}}, \\
\hat{\rho}^{\boldsymbol{c}}{}_{\boldsymbol{dab}} &\equiv \Xi d^{\boldsymbol{c}}{}_{\boldsymbol{dab}} + 2 S_{\boldsymbol{d}[\boldsymbol{a}}{}^{\boldsymbol{ce}} \hat{L}_{\boldsymbol{b}]\boldsymbol{e}}.
\end{aligned}
$$

In analogy to the discussion of Section 8.3.1, it is convenient to introduce a set of *geometric zero quantities* associated to the various equations. In the present case let:

$$\hat{\Sigma}_{ab} \equiv [e_a, e_b] - (\hat{\Gamma}_a{}^c{}_b - \hat{\Gamma}_b{}^c{}_a)e_c, \tag{8.44a}$$

$$\hat{\Xi}^c{}_{dab} \equiv \hat{P}^c{}_{dab} - \rho^c{}_{dab}, \tag{8.44b}$$

$$\hat{\Delta}_{cdb} \equiv \hat{\nabla}_c \hat{L}_{db} - \hat{\nabla}_d \hat{L}_{cb} - d_a d^a{}_{bcd}, \tag{8.44c}$$

$$\hat{\Lambda}_{bcd} \equiv \hat{\nabla}_a d^a{}_{bcd} - f_a d^a{}_{bcd}. \tag{8.44d}$$

Now, taking into account Equation (8.43) one has that

$$\begin{aligned}
\hat{\Xi}^c{}_{cab} &= e_a(\hat{\Gamma}_b{}^c{}_c) - e_b(\hat{\Gamma}_a{}^c{}_c) + \hat{\Gamma}_d{}^c{}_c(\hat{\Gamma}_b{}^d{}_a - \hat{\Gamma}_a{}^d{}_b) - 2S_{c[a}{}^{ce}\hat{L}_{b]e}, \\
&= 4\left(e_a(f_b) - e_b(f_a) - f_d(\hat{\Gamma}_b{}^d{}_a - \hat{\Gamma}_a{}^d{}_b) - \hat{L}_{ba} + \hat{L}_{ab}\right), \\
&= 4\left(\hat{\nabla}_a f_b - \hat{\nabla}_b f_a - \hat{L}_{ba} + \hat{L}_{ab}\right).
\end{aligned}$$

In view of the latter, it is convenient to define

$$\hat{\Xi}_{ab} \equiv \frac{1}{4}\hat{\Xi}^c{}_{cab} = \hat{\nabla}_a f_b - \hat{\nabla}_b f_a - \hat{L}_{ab} + \hat{L}_{ba}. \tag{8.45}$$

In terms of the zero quantities discussed in the previous paragraphs, one defines the *extended conformal vacuum Einstein field equations* as the conditions

$$\hat{\Sigma}_{ab} = 0, \qquad \hat{\Xi}^c{}_{dab} = 0, \qquad \hat{\Delta}_{cdb} = 0, \qquad \hat{\Lambda}_{bcd} = 0. \tag{8.46}$$

These equations yield differential conditions, respectively, for the components of the frame $\{e_a\}$, the spin coefficients $\hat{\Gamma}_a{}^c{}_b$ (including the components f_a of the covector $\boldsymbol{f}$), the components of the Schouten tensor $\hat{L}_{ab}$ and the components of the rescaled Weyl tensor $d^a{}_{bcd}$. In contrast to the standard conformal field Equations (8.32a) and (8.32b), there are no equations which can be regarded as differential conditions on the conformal factor Ξ and the components d_a of the covector $\boldsymbol{d}$. As will be seen in Chapter 13, these objects will be fixed through gauge conditions.

In order to relate the extended conformal field Equations (8.46) to the Einstein field equations, one introduces further zero quantities:

$$\delta_a \equiv d_a - \Xi f_a - \hat{\nabla}_a \Xi, \tag{8.47a}$$

$$\gamma_{ab} \equiv \hat{L}_{ab} - \hat{\nabla}_a \beta_b - \frac{1}{2} S_{ab}{}^{cd} \beta_c \beta_d + \frac{1}{6}\lambda \Xi^{-2} \eta_{ab}, \tag{8.47b}$$

$$\varsigma_{ab} \equiv \hat{L}_{[ab]} - \hat{\nabla}_{[a} f_{b]}. \tag{8.47c}$$

The associated equations

$$\delta_a = 0, \qquad \gamma_{ab} = 0, \qquad \varsigma_{ab} = 0, \tag{8.48}$$

will be treated as constraints. The first equation expresses the relation between the covectors $\boldsymbol{d}$, $\boldsymbol{f}$ and the conformal factor Ξ. The second equation encodes the relation between the components of the Schouten tensor of the Weyl connection $\hat{L}_{ab}$ and the physical Schouten tensor via the Einstein field equations – this constraint is the analogue of the standard conformal equation $Z_{ab} = 0$. The role of the equations in (8.48) is similar to that of the equation $Z = 0$ of the standard conformal field equations.

In the particular case when $f_a = 0$ it follows from (8.40) that $d_a = \nabla_a \Xi$. Hence, one has that $\hat{\nabla} = \nabla$. Under these circumstances the extended conformal field Equations (8.46) reduce to

$$\Sigma_{ab} = 0, \qquad \Xi^c{}_{dab} = 0, \qquad \Delta_{cdb} = 0, \qquad \Lambda_{bcd} = 0,$$

where the zero quantities Σ_{ab}, $\Xi^c{}_{dab}$, Δ_{cdb} and Λ_{bcd} are as defined in Section 8.3.1.

The conformal covariance of the equations

As in the case of the standard conformal field equations, the extended conformal field equations discussed in the previous section are conformally covariant. To make this statement more precise, consider a spacetime $(\tilde{\mathcal{M}}, \tilde{\boldsymbol{g}})$ and two metrics $\boldsymbol{g}$ and $\acute{\boldsymbol{g}}$ conformally related to $\tilde{\boldsymbol{g}}$ via

$$\boldsymbol{g} = \Xi^2 \tilde{\boldsymbol{g}}, \qquad \acute{\boldsymbol{g}} = \acute{\Xi}^2 \tilde{\boldsymbol{g}},$$

so that $\boldsymbol{g} = \kappa^2 \acute{\boldsymbol{g}}$ with $\kappa \equiv \Xi \acute{\Xi}^{-1}$. Let $\boldsymbol{\nabla}$, $\acute{\boldsymbol{\nabla}}$ denote, respectively, the Levi-Civita connections of the metrics $\boldsymbol{g}$ and $\acute{\boldsymbol{g}}$. One has that

$$\boldsymbol{\nabla} - \tilde{\boldsymbol{\nabla}} = \boldsymbol{S}(\Xi^{-1}\mathbf{d}\Xi), \qquad \acute{\boldsymbol{\nabla}} - \tilde{\boldsymbol{\nabla}} = \boldsymbol{S}(\acute{\Xi}^{-1}\mathbf{d}\acute{\Xi}), \tag{8.49}$$

and, furthermore,

$$\boldsymbol{\nabla} - \acute{\boldsymbol{\nabla}} = \boldsymbol{S}(\kappa^{-1}\mathbf{d}\kappa).$$

In addition, consider the covectors $\boldsymbol{f}$ and $\acute{\boldsymbol{f}}$ and define by means of these two the Weyl connections $\hat{\boldsymbol{\nabla}}$ and $\check{\boldsymbol{\nabla}}$ via

$$\hat{\boldsymbol{\nabla}} - \boldsymbol{\nabla} = \boldsymbol{S}(\boldsymbol{f}), \qquad \check{\boldsymbol{\nabla}} - \acute{\boldsymbol{\nabla}} = \boldsymbol{S}(\acute{\boldsymbol{f}}). \tag{8.50}$$

Combining Equations (8.49) and (8.50) one finds that the relation between the physical connection $\tilde{\boldsymbol{\nabla}}$ and the Weyl connections $\hat{\boldsymbol{\nabla}}$ and $\check{\boldsymbol{\nabla}}$ is given by

$$\hat{\boldsymbol{\nabla}} - \tilde{\boldsymbol{\nabla}} = \boldsymbol{S}(\boldsymbol{\beta}), \qquad \check{\boldsymbol{\nabla}} - \tilde{\boldsymbol{\nabla}} = \boldsymbol{S}(\acute{\boldsymbol{\beta}}),$$

where

$$\boldsymbol{\beta} \equiv \boldsymbol{f} + \Xi^{-1}\mathbf{d}\Xi, \qquad \acute{\boldsymbol{\beta}} \equiv \acute{\boldsymbol{f}} + \acute{\Xi}^{-1}\mathbf{d}\acute{\Xi}.$$

Combining these expressions one finds that

$$\begin{aligned}\hat{\boldsymbol{\nabla}} - \check{\boldsymbol{\nabla}} &= \boldsymbol{S}(\boldsymbol{\beta} - \acute{\boldsymbol{\beta}}) \\ &= \boldsymbol{S}(\boldsymbol{k} + \kappa^{-1}\mathbf{d}\kappa),\end{aligned}$$

with $\boldsymbol{k} \equiv \boldsymbol{f} - \acute{\boldsymbol{f}}$. Hence, letting $\boldsymbol{d} = \Xi\boldsymbol{\beta}$ and $\acute{\boldsymbol{d}} = \acute{\Xi}\acute{\boldsymbol{\beta}}$, one concludes that

$$\boldsymbol{d} = \kappa^{-1}\acute{\boldsymbol{d}} + \kappa\acute{\Xi}\boldsymbol{k} + \acute{\Xi}\mathbf{d}\kappa. \tag{8.51}$$

Assume now that the fields $(\boldsymbol{e}_{\boldsymbol{a}}, \hat{\Gamma}_{\boldsymbol{a}}{}^{\boldsymbol{b}}{}_{\boldsymbol{c}}, \hat{L}_{\boldsymbol{ab}}, d^{\boldsymbol{a}}{}_{\boldsymbol{bcd}}, \Xi, d_{\boldsymbol{a}})$ constitute a solution to the extended conformal field Equations (8.46). Then, proceeding in analogy to the discussion in Section 8.3.3, one finds that the fields $(\acute{\boldsymbol{e}}_{\boldsymbol{a}}, \check{\Gamma}_{\boldsymbol{a}}{}^{\boldsymbol{b}}{}_{\boldsymbol{c}}, \check{L}_{\boldsymbol{ab}}, \acute{d}^{\boldsymbol{a}}{}_{\boldsymbol{bcd}}, \acute{\Xi}, \acute{d}_{\boldsymbol{a}})$ with

$$
\begin{aligned}
\acute{\boldsymbol{e}}_{\boldsymbol{a}} &= \kappa \boldsymbol{e}_{\boldsymbol{a}} \\
\check{\Gamma}_{\boldsymbol{a}}{}^{\boldsymbol{b}}{}_{\boldsymbol{c}} &= \kappa \hat{\Gamma}_{\boldsymbol{a}}{}^{\boldsymbol{b}}{}_{\boldsymbol{c}} + \delta_{\boldsymbol{c}}{}^{\boldsymbol{b}} \hat{\nabla}_{\boldsymbol{a}} \kappa - \kappa S_{\boldsymbol{ac}}{}^{\boldsymbol{bd}} (k_{\boldsymbol{d}} + \kappa^{-1} \hat{\nabla}_{\boldsymbol{d}} \kappa), \\
\check{L}_{\boldsymbol{ab}} &= \kappa^2 \hat{L}_{\boldsymbol{ab}} - \kappa^2 \hat{\nabla}_{\boldsymbol{a}} (k_{\boldsymbol{b}} + \kappa^{-1} \hat{\nabla}_{\boldsymbol{b}} \kappa) \\
&\quad - \frac{1}{2} \kappa^2 S_{\boldsymbol{ab}}{}^{\boldsymbol{cd}} (k_{\boldsymbol{c}} + \kappa^{-1} \hat{\nabla}_{\boldsymbol{c}} \kappa)(k_{\boldsymbol{d}} + \kappa^{-1} \hat{\nabla}_{\boldsymbol{d}} \kappa), \\
\acute{d}^{\boldsymbol{a}}{}_{\boldsymbol{bcd}} &= \kappa^3 d^{\boldsymbol{a}}{}_{\boldsymbol{bcd}}, \\
\acute{\Xi} &= \kappa^{-1} \Xi, \\
\acute{d}_{\boldsymbol{a}} &= \kappa d_{\boldsymbol{a}} - \Xi \hat{\nabla}_{\boldsymbol{a}} \kappa - \kappa \Xi k_{\boldsymbol{a}},
\end{aligned}
$$

are also a solution of the extended conformal equations. Observe that the $\hat{\nabla}$-quantities are components with respect to the frame $\{\boldsymbol{e}_{\boldsymbol{a}}\}$ which is $\boldsymbol{g}$-orthonormal, while the $\check{\nabla}$-quantities are components on the $\{\acute{\boldsymbol{e}}_{\boldsymbol{a}}\}$ frame which is $\acute{\boldsymbol{g}}$-orthonormal.

8.4.2 The spinorial version of the extended conformal field equations

The frame formulation of the extended conformal field equations discussed in the previous subsection leads directly to its spinorial counterpart. The strategy is analogous to the one adopted in Section 8.3.2.

The spinorial counterpart of the $\boldsymbol{g}$-orthogonal frame $\{\boldsymbol{e}_{\boldsymbol{a}}\}$ is given by the null tetrad $\{\boldsymbol{e}_{\boldsymbol{AA'}}\}$ satisfying $\boldsymbol{g}(\boldsymbol{e}_{\boldsymbol{AA'}}, \boldsymbol{e}_{\boldsymbol{BB'}}) = \epsilon_{\boldsymbol{AB}} \epsilon_{\boldsymbol{A'B'}}$. Furthermore, let $\hat{\nabla}_{\boldsymbol{AA'}} \equiv \boldsymbol{e}_{\boldsymbol{AA'}}{}^{a} \hat{\nabla}_a$. Similarly, let

$$
\hat{\Sigma}_{\boldsymbol{AA'}}{}^{\boldsymbol{CC'}}{}_{\boldsymbol{BB'}}, \qquad \hat{P}^{\boldsymbol{CC'}}{}_{\boldsymbol{DD'AA'BB'}}, \qquad \hat{\rho}^{\boldsymbol{CC'}}{}_{\boldsymbol{DD'AA'BB'}}, \\
\hat{L}_{\boldsymbol{AA'BB'}}, \qquad d_{\boldsymbol{AA'}}, \qquad f_{\boldsymbol{AA'}},
$$

denote, respectively, the spinorial counterparts of the fields

$$
\Sigma_{\boldsymbol{a}}{}^{\boldsymbol{c}}{}_{\boldsymbol{b}}, \qquad P^{\boldsymbol{c}}{}_{\boldsymbol{dab}}, \qquad \rho^{\boldsymbol{c}}{}_{\boldsymbol{dab}}, \qquad \hat{L}_{\boldsymbol{ab}}, \qquad d_{\boldsymbol{a}}, \qquad f_{\boldsymbol{a}}.
$$

The spinorial counterpart of the geometric curvature, $\hat{P}^{\boldsymbol{CC'}}{}_{\boldsymbol{DD'AA'BB'}}$, is given in terms of the spinorial connection coefficients $\hat{\Gamma}_{\boldsymbol{AA'}}{}^{\boldsymbol{BB'}}{}_{\boldsymbol{CC'}}$. These, in turn, can be expressed in terms of the *reduced* spin connection coefficients $\hat{\Gamma}_{\boldsymbol{AA'}}{}^{\boldsymbol{B}}{}_{\boldsymbol{C}}$ by

$$
\hat{\Gamma}_{\boldsymbol{AA'}}{}^{\boldsymbol{BB'}}{}_{\boldsymbol{CC'}} = \hat{\Gamma}_{\boldsymbol{AA'}}{}^{\boldsymbol{B}}{}_{\boldsymbol{C}} \delta_{\boldsymbol{C'}}{}^{\boldsymbol{B'}} + \bar{\hat{\Gamma}}_{\boldsymbol{A'A}}{}^{\boldsymbol{B'}}{}_{\boldsymbol{C'}} \delta_{\boldsymbol{C}}{}^{\boldsymbol{B}}, \tag{8.52}
$$

consistent with formula (3.33). The reduced Weyl spin connection coefficients are related to the unphysical spin connection coefficients via

$$
\hat{\Gamma}_{\boldsymbol{AA'}}{}^{\boldsymbol{B}}{}_{\boldsymbol{C}} = \Gamma_{\boldsymbol{AA'}}{}^{\boldsymbol{B}}{}_{\boldsymbol{C}} + \delta_{\boldsymbol{A}}{}^{\boldsymbol{B}} f_{\boldsymbol{CA'}}, \qquad \hat{\Gamma}_{\boldsymbol{AA'}}{}^{\boldsymbol{Q}}{}_{\boldsymbol{Q}} = f_{\boldsymbol{AA'}};
$$

see Equation (5.32).

The geometric and algebraic Weyl curvature admit, respectively, the splits

$$\hat{P}^{CC'}{}_{DD'AA'BB'} = \hat{P}^{C}{}_{DAA'BB'}\delta_{D'}{}^{C'} + \bar{\hat{P}}^{C'}{}_{D'AA'BB'}\delta_{D}{}^{C},$$
$$\hat{\rho}^{CC'}{}_{DD'AA'BB'} = \hat{\rho}^{C}{}_{DAA'BB'}\delta_{D'}{}^{C'} + \bar{\hat{\rho}}^{C'}{}_{D'AA'BB'}\delta_{D}{}^{C}.$$

The formula giving the reduced geometric curvature $\hat{P}^{C}{}_{DAA'BB'}$ in terms of the reduced spin connection coefficients is identical to that for a Levi-Civita connection. Namely, one has that

$$\begin{aligned}\hat{P}^{C}{}_{DAA'BB'} \equiv\ & e_{AA'}(\hat{\Gamma}_{BB'}{}^{C}{}_{D}) - e_{BB'}(\hat{\Gamma}_{AA'}{}^{C}{}_{D}) \\ & - \hat{\Gamma}_{FB'}{}^{C}{}_{D}\hat{\Gamma}_{AA'}{}^{F}{}_{B} - \hat{\Gamma}_{BF'}{}^{C}{}_{D}\bar{\hat{\Gamma}}_{AA'}{}^{F'}{}_{B'} + \hat{\Gamma}_{FA'}{}^{C}{}_{D}\hat{\Gamma}_{BB'}{}^{F}{}_{A} \\ & + \hat{\Gamma}_{AF'}{}^{C}{}_{D}\bar{\hat{\Gamma}}_{BB'}{}^{F'}{}_{A'} + \hat{\Gamma}_{AA'}{}^{C}{}_{E}\hat{\Gamma}_{BB'}{}^{E}{}_{D} - \hat{\Gamma}_{BB'}{}^{C}{}_{E}\hat{\Gamma}_{AA'}{}^{E}{}_{D}.\end{aligned}$$

In particular, it can be verified that

$$\begin{aligned}\hat{P}^{Q}{}_{QAA'BB'} &= e_{AA'}(f_{BB'}) - e_{BB'}(f_{AA'}) + \hat{\Gamma}_{AA'}{}^{Q}{}_{B}f_{QB'} + \bar{\hat{\Gamma}}_{AA'}{}^{Q'}{}_{B'}f_{BQ'} \\ &\quad - \hat{\Gamma}_{BB'}{}^{Q}{}_{A}f_{QA'} - \bar{\hat{\Gamma}}_{BB'}{}^{Q'}{}_{A'}f_{AQ'} \\ &= \hat{\nabla}_{AA'}f_{BB'} - \hat{\nabla}_{BB'}f_{AA'}.\end{aligned}$$

Hence, one can write

$$\hat{P}_{ABCC'DD'} = \hat{P}_{(AB)CC'DD'} + \frac{1}{2}\epsilon_{AB}(\hat{\nabla}_{CC'}f_{DD'} - \hat{\nabla}_{DD'}f_{CC'}).$$

The reduced algebraic curvature spinor satisfies a similar expression. Namely, one has that

$$\hat{\rho}_{ABCC'DD'} = \hat{\rho}_{(AB)CC'DD'} - \frac{1}{2}\epsilon_{AB}(\hat{L}_{CC'DD'} - \hat{L}_{DD'CC'}),$$

with

$$\hat{\rho}_{(AB)CC'DD'} = -\Xi\phi_{ABCD}\epsilon_{C'D'} + \hat{L}_{BC'DD'}\epsilon_{AC} - \hat{L}_{BD'CC'}\epsilon_{AD};$$

compare Equation (5.33).

The objects discussed in the previous paragraphs can be used, in turn, to define the zero quantities:

$$\hat{\Sigma}_{AA'BB'} \equiv [e_{AA'}, e_{BB'}] - (\hat{\Gamma}_{AA'}{}^{CC'}{}_{BB'} - \hat{\Gamma}_{BB'}{}^{CC'}{}_{AA'})e_{CC'}, \tag{8.53a}$$
$$\hat{\Xi}^{C}{}_{DAA'BB'} \equiv \hat{P}^{C}{}_{DAA'BB'} - \hat{\rho}^{C}{}_{DAA'BB'}, \tag{8.53b}$$
$$\hat{\Delta}_{CC'DD'BB'} \equiv \hat{\nabla}_{CC'}\hat{L}_{DD'BB'} - \hat{\nabla}_{DD'}\hat{L}_{CC'BB'} \tag{8.53c}$$
$$- d_{AA'}d^{AA'}{}_{BB'CC'DD'}, \tag{8.53d}$$
$$\hat{\Lambda}_{BB'CC'DD'} \equiv \hat{\nabla}_{AA'}d^{AA'}{}_{BB'CC'DD'}$$
$$- f_{AA'}d^{AA'}{}_{BB'CC'DD'} \tag{8.53e}$$

One can exploit the symmetries of some of the above zero quantities to obtain reduced zero quantities. In particular, one can write

$$\hat{\Lambda}_{BB'CC'DD'} = \hat{\Lambda}_{BB'CD}\epsilon_{C'D'} + \bar{\hat{\Lambda}}_{B'BC'D'}\epsilon_{CD},$$

with

$$\hat{\Lambda}_{BB'CD} \equiv \frac{1}{2}\hat{\Lambda}_{BB'CQ'D}{}^{Q'}.$$

A similar idea can be applied to $\hat{\Sigma}_{AA'BB'}$, $\hat{\Xi}^{C}{}_{DAA'BB'}$ and $\hat{\Delta}_{CC'DD'BB'}$. This idea will not be pursued here.

The spinorial version of the extended conformal Einstein field equations is expressed in terms of the above zero quantities as:

$$\hat{\Sigma}_{AA'BB'} = 0, \quad \hat{\Xi}^{C}{}_{DAA'BB'} = 0, \quad \hat{\Delta}_{CC'DD'BB'} = 0, \tag{8.54a}$$

$$\hat{\Lambda}_{BB'CD} = 0. \tag{8.54b}$$

Finally, let $\delta_{AA'}$, $\varsigma_{AA'BB'}$ and $\gamma_{AA'BB'}$ denote the spinorial counterparts of the zero quantities δ_{a} and γ_{ab}. One then requires that

$$\delta_{AA'} = 0, \qquad \varsigma_{AA'BB'} = 0, \qquad \gamma_{AA'BB'} = 0. \tag{8.55}$$

8.4.3 The extended conformal Einstein field equations and the Einstein field equations

As in the case of the other versions of the conformal field equations discussed in this chapter, it is important to analyse the precise relation between the extended conformal field equations and the (physical) Einstein field equations. One has the following:

Proposition 8.3 (***solutions to the extended conformal field equations as solutions to the Einstein field equations***) *Let*

$$(e_{a}, \hat{\Gamma}_{a}{}^{b}{}_{c}, \hat{L}_{ab}, d^{a}{}_{bcd})$$

denote a solution to the extended conformal field Equations (8.46) for some choice of the conformal gauge fields (Ξ, d_{a}) satisfying the supplementary Equations (8.48). Furthermore, suppose

$$\Xi \neq 0, \quad \det(\eta^{ab} e_{a} \otimes e_{b}) \neq 0,$$

on an open subset $\mathcal{U} \subset \mathcal{M}$. Then the metric $\tilde{g} = \Xi^{-2}\eta_{ab}\omega^{a} \otimes \omega^{b}$, where $\{\omega^{a}\}$ is the dual frame to $\{e_{a}\}$, is a solution to the Einstein field equations (8.4) on $\mathcal{U}$.

Proof As a consequence of the conformal equation $\hat{\Sigma}_{ab} = 0$, the fields $\hat{\Gamma}_{a}{}^{b}{}_{c}$ can be interpreted as the connection coefficients, with respect to the frame field $\{e_{a}\}$, of a torsion-free connection $\hat{\nabla}$. In order to show that $\hat{\nabla}$ is a Weyl connection, one needs to compute $\hat{\nabla}_a\eta_{bc}$. This is best done using spinors. As $e_{AA'}(\epsilon_{BC}) = 0$, one has that

$$\begin{aligned}\hat{\nabla}_{AA'}\epsilon_{BC} &= -\hat{\Gamma}_{AA'}{}^{Q}{}_{B}\epsilon_{QC} - \hat{\Gamma}_{AA'}{}^{Q}{}_{C}\epsilon_{BQ} = -\hat{\Gamma}_{AA'CB} + \hat{\Gamma}_{AA'BC} \\ &= -\hat{\Gamma}_{AA'}{}^{Q}{}_{Q}\epsilon_{BC} = -f_{AA'}\epsilon_{BC}.\end{aligned}$$

Recalling that $\epsilon_{AB}\epsilon_{A'B'}$ is the spinorial counterpart of η_{ab} and observing the split (8.52) one concludes that $\hat{\nabla}_a\eta_{bc} = -2f_a\eta_{bc}$; that is, $\hat{\boldsymbol{\nabla}}$ is a Weyl connection. Now, as $\hat{\Xi}^c{}_{dab} = 0$, the fields $\hat{L}_{ab}$ and $\Xi d^c{}_{dab}$ obtained as a solution to the extended conformal field equations correspond to, respectively, the Schouten tensor and the Weyl tensor of the connection $\hat{\boldsymbol{\nabla}}$, as a consequence of the uniqueness of the decomposition in terms of irreducible components.

Given the Weyl connection $\hat{\boldsymbol{\nabla}}$, one can define a new connection $\boldsymbol{\nabla}$ via $\boldsymbol{\nabla} \equiv \hat{\boldsymbol{\nabla}} - \boldsymbol{S}(\boldsymbol{f})$. By construction, this connection is metric. The Schouten tensor of $\boldsymbol{\nabla}$ is then given by

$$L_{ab} = \hat{L}_{ab} - \nabla_a f_b + \frac{1}{2}S_{ab}{}^{cd}f_c f_d.$$

As $\hat{\Xi}^c{}_{dab} = 0$, it follows that $\hat{\Xi}_{ab}$ as defined by Equation (8.45) also vanishes. As $\hat{\boldsymbol{\nabla}}$ is torsion free, so is $\boldsymbol{\nabla}$. Hence, one concludes that $\boldsymbol{\nabla}$ must be the Levi-Civita connection of the metric $\boldsymbol{g} \equiv \eta_{ab}\boldsymbol{\omega}^a \otimes \boldsymbol{\omega}^b$. The latter expression is a well-defined Lorentzian metric on $\mathcal{U}$ as its determinant is, by hypothesis, non-vanishing.

Finally, one defines a *physical connection* $\tilde{\boldsymbol{\nabla}}$ via $\tilde{\boldsymbol{\nabla}} \equiv \boldsymbol{\nabla} - \boldsymbol{S}(\Xi^{-1}\mathbf{d}\Xi)$. As $\delta_a = 0$ it follows that $d_a = f_a + \hat{\nabla}_a\Xi$ so that $\tilde{\boldsymbol{\nabla}}$ is the Levi-Civita connection of the metric $\tilde{\boldsymbol{g}} \equiv \Xi^{-2}\eta_{ab}\boldsymbol{\omega}^a \otimes \boldsymbol{\omega}^b$. The latter is well defined as long as $\Xi \neq 0$. The Schouten tensor of $\tilde{\boldsymbol{\nabla}}$ is given by

$$\tilde{L}_{ab} = L_{ab} - \nabla_a(\Xi^{-1}\nabla_b\Xi) - \Xi^{-2}S_{ab}{}^{cd}\nabla_c\Xi\nabla_d\Xi.$$

As a consequence of $\delta_a = 0$ and $\gamma_{ab} = 0$ one concludes that

$$\tilde{L}_{ab} = \frac{1}{6}\lambda\Xi^{-2}\eta_{ab}.$$

Thus, $\tilde{\boldsymbol{g}}$ is a solution to the vacuum Einstein field equations on $\mathcal{U}$. □

Remark. Given the equivalence between the frame and spinorial versions of the extended conformal field equations, the latter result also provides the connection between the spinorial extended conformal field equations and the Einstein field equations.

8.5 Further reading

The standard conformal Einstein field equations were first introduced in Friedrich (1981a,b, 1982). General aspects of the Cauchy problem of the conformal equations were first discussed in Friedrich (1983); see also Friedrich (1984). A systematic discussion of gauge issues and hyperbolic reductions of the equations has been given in Friedrich (1985). A discussion of the conformal field equations with trace-free matter was first given in Friedrich (1991). The extended conformal field equations were introduced in Friedrich (1995). Reviews discussing various aspects of the conformal field equations can be found in Friedrich (2002, 2004). The extended conformal field equations with (trace-free) matter were first discussed in Lübbe and Valiente Kroon (2012).

The conformal field equations can be related to other geometrical objects (*twistors*); see Frauendiener and Sparling (2000). The extended conformal field equations can be set in a more geometrical framework involving the language of differential forms. A discussion of this has been given in Friedrich (1995). Certain applications of the conformal equations require the use of a *lift* of the equations to a suitable fibre bundle. A discussion of this type of procedure can be found in Friedrich (1986b, 1998c).

A different approach to the construction of regular conformal field equations based on the *Fefferman-Graham obstructions* – see, for example, Graham and Hirachi (2005) – has been elaborated in Anderson (2005a) and Anderson and Chruściel (2005). This approach gives rise to suitable field equations for an arbitrary number of odd space dimensions and has been used to prove (global and semi-global) existence and stability results of higher dimensional asymptotically simple spacetimes.

In the context of conformal geometry, given a metric $\boldsymbol{g}$, it is natural to ask whether there exists a further metric in the conformal class $[\boldsymbol{g}]$ which is an Einstein space. This problem has been addressed in, for example, Baston and Mason (1987), Kozameh et al. (1985) and Mason (1986).

9
Matter models

This chapter provides a discussion of various matter models amenable to a treatment by means of conformal techniques. These matter models can be used as matter sources for the conformal Einstein field equations discussed in Chapter 8. The matter models to be considered are the electromagnetic field, radiation perfect fluids and the conformally invariant scalar field. These matter models share the property of having an energy-momentum tensor which is *trace free*. This property leads to simple transformation laws for the equations satisfied by the matter models. Moreover, the *unphysical equations* obtained by means of these transformations are regular at points where the conformal factor vanishes.

9.1 General properties of the conformal treatment of matter models

The fundamental object in the description of a matter model in general relativity is its ***energy-momentum tensor*** $\tilde{T}_{ab}$. The equations describing the model are then given by

$$\tilde{\nabla}^a \tilde{T}_{ab} = 0. \tag{9.1}$$

The energy-momentum tensor is related to the curvature of the spacetime via the Einstein field equations; see Equation (8.4). Despite this connection, a conformal transformation $\boldsymbol{g} = \Xi^2 \tilde{\boldsymbol{g}}$ does not directly imply a transformation rule for the ***physical energy-momentum tensor*** $\tilde{T}_{ab}$. Nevertheless, it is convenient to define an ***unphysical energy-momentum tensor*** T_{ab} when rewriting Equation (9.1) in terms of geometric quantities derived from the rescaled metric $\boldsymbol{g}$.

9.1.1 The unphysical energy-momentum tensor

There is considerable freedom in a possible definition of T_{ab}. Guiding principles are simplicity in both the definition and the resulting form of the unphysical version of Equation (9.1). Arguably, the simplest definition of the unphysical energy-momentum tensor is one which is homogeneous with respect to the conformal factor Ξ. Accordingly, set

$$T_{ab} = \Xi^{-2}\tilde{T}_{ab}. \tag{9.2}$$

It follows then that

$$g^{ab}\nabla_a T_{bc} = \Xi^{-4}\tilde{g}^{ab}\tilde{\nabla}_a\tilde{T}_{bc} - \Xi^{-5}\tilde{\nabla}_c\Xi\tilde{g}^{ab}\tilde{T}_{ab}. \tag{9.3}$$

Hence, Equation (9.1) *implies the equation* $\nabla^a T_{ab} = 0$ *only if*

$$\tilde{T} = 0, \qquad \tilde{T} \equiv \tilde{g}^{ab}\tilde{T}_{ab}.$$

This observation justifies definition (9.2) and the importance given in this chapter to trace-free matter models. As a result of the homogeneous nature of the transformation law in Equation (9.2), T_{ab} is trace free if $\tilde{T}_{ab}$ is trace free.

In the case of matter models with $\tilde{T}_{ab} \neq 0$, define $T \equiv g^{ab}T_{ab}$, so that $T = \Xi^{-4}\tilde{T}$. It follows from Equations (9.1) and (9.3) that

$$\nabla^a T_{ab} = \Xi^{-1}\nabla_b\Xi T.$$

This is an equation which is formally singular at the points where $\Xi = 0$. Dealing with this singularity is the essential problem faced in the analysis of general matter models by means of conformal methods.

9.1.2 The rescaled Cotton tensor

As discussed in Chapter 8, the matter fields couple to the conformal Einstein field equations through the ***rescaled Cotton tensor*** T_{abc}; compare Equations (8.22) and (8.23). Recall that the physical Cotton tensor is given by $T_{abc} = \Xi^{-1}\tilde{Y}_{abc}$ where

$$\tilde{Y}_{abc} = \tilde{\nabla}_a\tilde{L}_{bc} - \tilde{\nabla}_b\tilde{L}_{ac};$$

compare Equation (8.21). One readily finds that

$$T_{abc} = \frac{1}{2}\Xi^{-1}\left(\tilde{\nabla}_a\tilde{T}_{bc} - \tilde{\nabla}_b\tilde{T}_{ac}\right) - \frac{1}{6}\Xi^{-1}\left(\tilde{\nabla}_a\tilde{T}\tilde{g}_{bc} - \tilde{\nabla}_b\tilde{T}\tilde{g}_{ac}\right),$$

where it has been used that the physical Schouten tensor $\tilde{L}_{ab}$ is related to the physical energy-momentum tensor via Equation (8.5b). In what follows, attention will be restricted to the trace-free matter case so that

$$T_{abc} = \frac{1}{2}\Xi^{-1}\left(\tilde{\nabla}_a\tilde{T}_{bc} - \tilde{\nabla}_b\tilde{T}_{ac}\right).$$

The latter can be reexpressed in terms of the unphysical connection $\boldsymbol{\nabla}$ and the unphysical energy-momentum tensor T_{ab}. A computation using the methods of Chapter 5 yields

$$T_{abc} = \Xi\nabla_{[a}T_{b]c} + \nabla_{[a}\Xi T_{b]c} + g_{c[a}T_{b]e}\nabla^e\Xi. \tag{9.4}$$

From the above expression it follows that T_{abc} is regular whenever $\Xi = 0$ if T_{ab} is smooth at the conformal boundary.

Equation (9.4) can be expressed in terms of derivatives of the (conformal) matter fields. This feature complicates the construction of suitable conformal evolution equations as it introduces further derivatives of the fields into the principal part of the equations. This difficulty can be overcome by introducing evolution equations for the *derivatives of the matter fields* which cannot be eliminated with the equation $\nabla^a T_{ab} = 0$. This analysis depends on the specific properties of the matter model under consideration.

9.2 The Maxwell field

The ***electromagnetic*** or ***Maxwell field*** is the prototype of a relativistic matter model that can be treated by means of conformal methods. The Maxwell field is described by an antisymmetric tensor $\tilde{F}_{ab}$ – the ***Faraday tensor***. In terms of the latter, the source-free Maxwell equations are given by

$$\tilde{\nabla}^a \tilde{F}_{ab} = 0, \tag{9.5a}$$

$$\tilde{\nabla}_{[a} \tilde{F}_{bc]} = 0. \tag{9.5b}$$

Multiplying Equation (9.5b) by the volume form $\tilde{\epsilon}^{dabc}$, one obtains the alternative expression

$$\tilde{\nabla}^a \tilde{F}^*_{ab} = 0, \tag{9.6}$$

where $\tilde{F}^*_{ab} \equiv -\frac{1}{2}\tilde{\epsilon}_{ab}{}^{cd}\tilde{F}_{cd}$ denotes the ***dual Faraday tensor***. Now, introducing the ***self-dual Faraday tensor***

$$\tilde{\mathcal{F}}_{ab} \equiv \tilde{F}_{ab} + \mathrm{i}\tilde{F}^*_{ab},$$

it follows from (9.5a) and (9.6) that

$$\tilde{\nabla}^a \tilde{\mathcal{F}}_{ab} = 0. \tag{9.7}$$

This last equation contains the same information as Equations (9.5a) and (9.5b). The energy-momentum tensor of the electromagnetic field is quadratic in the Faraday tensor. It is given by

$$\tilde{T}_{ab} = \tilde{F}_{ac}\tilde{F}_b{}^c - \frac{1}{4}\tilde{g}_{ab}\tilde{F}_{cd}\tilde{F}^{cd}.$$

It can be readily verified that $\tilde{T} = 0$. Making use of the dual $\tilde{F}^*_{ab}$ one obtains the alternative expressions

$$\tilde{T}_{ab} = \frac{1}{2}\left(\tilde{F}_{ac}\tilde{F}_b{}^c + \tilde{F}^*_{ac}\tilde{F}^{*c}_b\right) \tag{9.8a}$$

$$= \frac{1}{2}\tilde{\mathcal{F}}_{ac}\bar{\tilde{\mathcal{F}}}_b{}^c. \tag{9.8b}$$

It can be readily verified that the Maxwell Equations (9.5a) and (9.5b) imply that $\tilde{\nabla}^a\tilde{T}_{ab} = 0$.

Conformal transformation properties

The source-free Maxwell Equations (9.5a), (9.5b) and (9.6) are *conformally invariant.* In order to see this, assume that $(\mathcal{M}, \boldsymbol{g})$ is a conformal extension of a spacetime $(\tilde{\mathcal{M}}, \tilde{\boldsymbol{g}})$ with $\boldsymbol{g} = \Xi^2 \tilde{\boldsymbol{g}}$, and define the ***conformal (unphysical) Faraday tensor*** via

$$F_{ab} \equiv \tilde{F}_{ab}. \tag{9.9}$$

It follows from this definition that $F^*_{ab} = \tilde{F}^*_{ab}$. Moreover, using the transformation laws between the connections $\tilde{\boldsymbol{\nabla}}$ and $\boldsymbol{\nabla}$ one finds that Equations (9.5a), (9.5b) and (9.6) imply

$$\nabla^a F_{ab} = 0, \qquad \nabla_{[a} F_{bc]} = 0, \qquad \nabla^a F^*_{ab} = 0, \tag{9.10}$$

which shows the conformal invariance of the equations. Let $\hat{\boldsymbol{\nabla}}$ be a Weyl connection defined via $\hat{\boldsymbol{\nabla}} - \boldsymbol{\nabla} = \boldsymbol{S}(\boldsymbol{f})$ with $\boldsymbol{f}$ a covector. A further computation yields

$$\hat{\nabla}^a F_{ab} = 0, \qquad \hat{\nabla}_{[a} F_{bc]} = 0, \qquad \hat{\nabla}^a F^*_{ab} = 0.$$

Consistent with the transformation law (9.2) for the energy-momentum tensor one finds that

$$T_{ab} = F_{ac} F_b{}^c - \frac{1}{4} g_{ab} F_{cd} F^{cd} = \frac{1}{2} \left(F_{ac} F_b{}^c + F^*_{ac} F^{*c}_b \right).$$

Substituting the last expression in Equation (9.4) for the rescaled Cotton tensor one obtains

$$\begin{aligned} 2T_{abc} = {} & \nabla_{[a} F_{b]d} F_c{}^d + F_{d[a} \nabla_{b]} F_c{}^d + \nabla_{[a} F^*_{b]d} F^{*d}_c + F^*_{d[a} \nabla_{b]} F^{*d}_c \\ & + \nabla_{[a} \Xi F_{b]d} F_c{}^d + \nabla_{[a} \Xi F^*_{b]d} F^{*d}_c \\ & + g_{c[a} F_{b]e} F_d{}^e \nabla^d \Xi + g_{c[a} F^*_{b]e} F^{*e}_d \nabla^d \Xi. \end{aligned}$$

A direct inspection shows that the first four terms of the right-hand side contain derivatives of the Faraday tensor which cannot be eliminated using the (conformal) Maxwell Equations (9.10). Thus, it is necessary to consider equations for the derivatives of F_{ab}. A suitable equation can be obtained from the commutator of covariant derivatives applied to F_{ab}. More precisely, one has that

$$\nabla_a \nabla_b F_{cd} - \nabla_b \nabla_a F_{cd} = -R^e{}_{cab} F_{ed} - R^e{}_{dab} F_{ce}.$$

In view of this equation one introduces the ***auxiliary field*** $F_{abc} \equiv \nabla_a F_{bc}$ so that

$$\nabla_a F_{bcd} - \nabla_b F_{acd} = -R^e{}_{cab} F_{ed} - R^e{}_{dab} F_{ce}. \tag{9.11}$$

By construction one has that

$$F_{abc} = F_{a[bc]}, \qquad F_{[abc]} = 0, \qquad F^a{}_{ac} = 0.$$

9.2.1 The spinorial form of the Maxwell equations

The spinorial treatment of the Maxwell field is a direct consequence of the decomposition of spinors in irreducible components; see Section 3.1.6. The spinorial formulation of the Maxwell equations offers a number of computational advantages and makes more evident the similarities between the gravitational and electromagnetic fields.

In what follows, let $\tilde{F}_{AA'BB'}$ denote the ***spinorial counterpart of the Faraday tensor*** $\tilde{F}_{ab}$. By exploiting the antisymmetry of $\tilde{F}_{ab}$, it follows from Equation (3.13) that there exists a symmetric spinor $\tilde{\phi}_{AB}$, the ***Maxwell spinor***, such that

$$\tilde{F}_{AA'BB'} = \tilde{\phi}_{AB}\tilde{\epsilon}_{A'B'} + \bar{\tilde{\phi}}_{A'B'}\tilde{\epsilon}_{AB}, \qquad \tilde{\phi}_{AB} = \frac{1}{2}\tilde{F}_{AQ'B}{}^{Q'}. \tag{9.12}$$

Using the decomposition (9.12) it follows that

$$\tilde{\mathcal{F}}_{AA'BB'} = 2\tilde{\phi}_{AB}\tilde{\epsilon}_{A'B'},$$

where $\tilde{\mathcal{F}}_{AA'BB'}$ is the spinorial counterpart of the self-dual Faraday tensor $\tilde{\mathcal{F}}_{ab}$. Taking into account Equation (9.7) one obtains

$$\tilde{\nabla}^{A}{}_{A'}\tilde{\phi}_{AB} = 0. \tag{9.13}$$

This last equation is known as the ***spinorial Maxwell equation***. A further computation using Equation (9.8b) shows that the spinorial counterpart of the energy-momentum tensor takes the simple form

$$\tilde{T}_{AA'BB'} = \tilde{\phi}_{AB}\bar{\tilde{\phi}}_{A'B'}.$$

Behaviour under conformal rescalings

The definition of the unphysical (conformal) Faraday tensor given in Equation (9.9) suggests introducing the ***unphysical Maxwell spinor*** ϕ_{AB} as

$$\phi_{AB} \equiv \Xi^{-1}\tilde{\phi}_{AB}. \tag{9.14}$$

The factor Ξ^{-1} in the above definition is necessary to compensate for the factor Ξ picked up by the spinor $\tilde{\epsilon}_{AB}$ in Equation (9.12). It follows from Equation (9.13) and the transformation law of the connection upon conformal rescalings $\boldsymbol{g} = \Xi^2\tilde{\boldsymbol{g}}$ given in Section 5.4 that

$$\nabla^{Q}{}_{A'}\phi_{BQ} = 0. \tag{9.15}$$

That is, the transformation rule (9.14) makes the spinorial Maxwell Equation (9.13) conformally invariant – this result was to be expected in view of the equations in (9.10). One readily sees the similarities between Equation (9.15) and the spinorial Bianchi identity $\nabla^{AA'}\phi_{ABCD} = 0$. Consistent with Equation (9.2) one finds that the spinorial counterpart of the unphysical energy-momentum tensor is given by

$$T_{AA'BB'} = \phi_{AB}\bar{\phi}_{A'B'}.$$

Equation (9.15) can be expressed in terms of a Weyl connection $\hat{\boldsymbol{\nabla}} = \boldsymbol{\nabla} + \boldsymbol{S}(\boldsymbol{f})$ as

$$\hat{\nabla}^Q{}_{A'}\phi_{BQ} = f^Q{}_{A'}\phi_{BQ}.$$

In order to write a spinorial version of Equation (9.11) it is observed that the action of the commutator of covariant derivatives on the spinor ϕ_{AB} is given by

$$\nabla_{AA'}\nabla_{BB'}\phi_{CD} - \nabla_{BB'}\nabla_{AA'}\phi_{CD} = -\phi_{QD}R^Q{}_{CAA'BB'} - \phi_{CQ}R^Q{}_{DAA'BB'}.$$

Letting $\psi_{AA'BC} \equiv \nabla_{AA'}\phi_{BC}$, one obtains

$$\nabla_{AA'}\psi_{BB'CD} - \nabla_{BB'}\psi_{AA'CD} = -2\phi_{Q(C}R^Q{}_{D)AA'BB'}.$$

By construction, the auxiliary spinor $\psi_{AA'BC}$ possesses the symmetries

$$\psi_{AA'BC} = \psi_{AA'CB}, \qquad \psi^Q{}_{A'BQ} = 0.$$

9.3 The scalar field

A scalar field $\tilde{\phi}$ satisfying the wave equation

$$\tilde{\nabla}^a\tilde{\nabla}_a\tilde{\phi} = 0 \tag{9.16}$$

is a convenient matter model to couple to the Einstein field equations. It provides a way of incorporating dynamical degrees of freedom in spherically symmetric configurations; see, for example, Choptuik (1993) and Gundlach and Martín-García (2007). This idea has been exploited in a number of analyses of ***cosmic censorship*** and the formation of black holes through gravitational collapse; see, for example, Christodoulou (1986) and Dafermos (2003, 2005).

Unfortunately, as a direct computation shows, Equation (9.16) does not have good conformal transformation properties. This difficulty can be fixed by considering a modified version – the so-called ***conformally invariant scalar field equation***

$$\tilde{\nabla}^a\tilde{\nabla}_a\tilde{\phi} - \frac{1}{6}\tilde{R}\tilde{\phi} = 0, \tag{9.17}$$

where $\tilde{R}$ denotes the Ricci scalar of the physical spacetime metric $\tilde{\boldsymbol{g}}$. Letting, as usual, $\boldsymbol{g} = \Xi^2\tilde{\boldsymbol{g}}$ and defining the ***unphysical (conformal) scalar*** ϕ as

$$\phi \equiv \Xi^{-1}\tilde{\phi},$$

one finds, after a calculation using the transformation rule for the Ricci scalar Equation (5.6c), that

$$\nabla^a\nabla_a\phi - \frac{1}{6}R\phi = 0, \tag{9.18}$$

where R denotes the Ricci scalar of $\boldsymbol{g}$. An energy-momentum tensor for Equation (9.17) is given by

$$\tilde{T}_{ab} = \tilde{\nabla}_a\tilde{\phi}\tilde{\nabla}_b\tilde{\phi} - \frac{1}{4}\tilde{g}_{ab}\tilde{\nabla}_c\tilde{\phi}\tilde{\nabla}^c\tilde{\phi} - \frac{1}{2}\tilde{\phi}\tilde{\nabla}_a\tilde{\nabla}_b\tilde{\phi} + \frac{1}{2}\tilde{\phi}^2\tilde{L}_{ab}. \tag{9.19}$$

A peculiarity of the above expression is the presence of the curvature terms $\tilde{L}_{ab}$ in the right-hand side of the energy-momentum tensor. Using the Einstein field Equations (8.4), the Schouten tensor can be reexpressed in terms of the energy-momentum tensor, so that Equation (9.19) takes the form

$$\tilde{T}_{ab} = \left(1-\frac{1}{4}\tilde{\phi}^2\right)^{-1}\left(\tilde{\nabla}_a\tilde{\phi}\tilde{\nabla}_b\tilde{\phi} - \frac{1}{4}\tilde{g}_{ab}\tilde{\nabla}_c\tilde{\phi}\tilde{\nabla}^c\tilde{\phi} - \frac{1}{2}\tilde{\phi}\tilde{\nabla}_a\tilde{\nabla}_b\tilde{\phi} + \frac{1}{12}(\lambda - \tilde{T})\tilde{\phi}^2\tilde{g}_{ab}\right).$$

Taking the trace of Equation (9.19) one finds that

$$\tilde{T} \equiv \tilde{g}^{ab}\tilde{T}_{ab} = -\frac{1}{2}\tilde{\phi}\left(\tilde{\nabla}^a\tilde{\nabla}_a\tilde{\phi} - \frac{1}{6}\tilde{R}\tilde{\phi}^2\right).$$

Thus, the energy-momentum tensor of Equation (9.19) is trace free if and only if the conformally invariant wave Equation (9.17) is satisfied. A lengthier computation using the commutator of covariant derivatives and the Bianchi identity in the form $\tilde{\nabla}^a\tilde{L}_{ab} = \frac{1}{6}\tilde{\nabla}_b\tilde{R}$ shows that

$$\tilde{\nabla}^a\tilde{T}_{ab} = \tilde{\nabla}_b\tilde{\phi}\left(\tilde{\nabla}^c\tilde{\nabla}_c\tilde{\phi} - \frac{1}{6}\tilde{R}\tilde{\phi}\right) - \frac{1}{2}\tilde{\phi}\tilde{\nabla}_b\left(\tilde{\nabla}^c\tilde{\nabla}_c\tilde{\phi} - \frac{1}{6}\tilde{R}\tilde{\phi}\right).$$

One concludes that $\tilde{T}_{ab}$ is divergence free if and only if Equation (9.17) holds. Finally, using the transformation law for the Schouten tensor under conformal rescalings, Equation (5.6b), one finds that

$$T_{ab} = \nabla_a\phi\nabla_b\phi - \frac{1}{4}g_{ab}\nabla_c\phi\nabla^c\phi - \frac{1}{2}\phi\nabla_a\nabla_b\phi + \frac{1}{2}\phi^2 L_{ab},$$

so that $T_{ab} = \Xi^{-2}\tilde{T}_{ab}$. It follows from the previous discussion that

$$\nabla^a T_{ab} = 0, \qquad g^{ab}T_{ab} = 0.$$

Spinorial description

The straightforward spinorial counterpart of Equation (9.19) is given by

$$\begin{aligned}\tilde{T}_{AA'BB'} = {}& \tilde{\nabla}_{AA'}\tilde{\phi}\tilde{\nabla}_{BB'}\tilde{\phi} - \frac{1}{4}\tilde{\epsilon}_{AB}\tilde{\epsilon}_{A'B'}\tilde{\nabla}_{PP'}\tilde{\phi}\tilde{\nabla}^{PP'}\tilde{\phi} \\ & - \frac{1}{2}\tilde{\phi}\tilde{\nabla}_{AA'}\tilde{\nabla}_{BB'}\tilde{\phi} + \frac{1}{2}\tilde{\phi}^2\tilde{L}_{AA'BB'}.\end{aligned}$$

Applying the decomposition formula (3.12) to $\tilde{\nabla}_{AA'}\tilde{\phi}\tilde{\nabla}_{BB'}\tilde{\phi}$ and $\tilde{\nabla}_{AA'}\tilde{\nabla}_{BB'}\tilde{\phi}$ one finds that

$$\tilde{\nabla}_{AA'}\tilde{\phi}\tilde{\nabla}_{BB'}\tilde{\phi} = \tilde{\nabla}_{A(A'}\tilde{\phi}\tilde{\nabla}_{B')B}\tilde{\phi} + \frac{1}{4}\tilde{\epsilon}_{AB}\tilde{\epsilon}_{A'B'}\tilde{\nabla}_{PP'}\tilde{\phi}\tilde{\nabla}^{PP'}\tilde{\phi},$$

$$\tilde{\nabla}_{AA'}\tilde{\nabla}_{BB'}\tilde{\phi} = \tilde{\nabla}_{A(A'}\tilde{\nabla}_{B')B}\tilde{\phi} + \frac{1}{4}\tilde{\epsilon}_{AB}\tilde{\epsilon}_{A'B'}\tilde{\nabla}_{PP'}\tilde{\nabla}^{PP'}\tilde{\phi},$$

where it has been used that

$$\tilde{\nabla}_{P(A'}\tilde{\phi}\tilde{\nabla}^{P}{}_{B')}\tilde{\phi} = 0, \qquad \tilde{\nabla}_{P(A'}\tilde{\nabla}^{P}{}_{B')}\tilde{\phi} = 0.$$

The above formulae, together with the wave equation (9.17), imply the following alternative spinorial expression for the energy-momentum tensor:

$$\tilde{T}_{AA'BB'} = \tilde{\nabla}_{A(A'}\tilde{\phi}\tilde{\nabla}_{B')B}\tilde{\phi} - \frac{1}{2}\tilde{\phi}\tilde{\nabla}_{A(A'}\tilde{\nabla}_{B')B}\tilde{\phi} + \frac{1}{2}\tilde{\phi}^2\tilde{\Phi}_{AA'BB'}$$

where $\tilde{\Phi}_{AA'BB'}$ is the (physical) trace-free Ricci spinor. The unphysical spacetime version of the above equation follows directly by removing the $\tilde{}$ of the various fields.

9.3.1 Equations for the derivatives of the scalar field

As in the case of the electromagnetic field, the coupling of the conformally invariant scalar field to the conformal field equations through the rescaled Cotton tensor T_{abc} involves derivatives of ϕ. Indeed, a calculation exploiting the fact that $\nabla_{[a}\nabla_{b]}\phi = 0$ shows that

$$\begin{aligned}\nabla_{[a}T_{b]c} = \frac{3}{2}\nabla_{[b}\phi\nabla_{a]}\nabla_c\phi - \frac{1}{2}g_{c[b}\nabla_{a]}\nabla_e\phi\nabla^e\phi + \phi\nabla_{[a}\phi L_{b]c}\\ - \frac{1}{2}\phi\nabla_{[a}\nabla_{b]}\nabla_c\phi + \frac{1}{2}\phi^2\nabla_{[a}L_{b]c}.\end{aligned}$$

The terms in the second line of the preceding equation can be rewritten using the commutator

$$\nabla_{[a}\nabla_{b]}\nabla_c\phi = -\frac{1}{2}R^e{}_{cab}\nabla_e\phi$$

and the Cotton Equation (8.23). Putting everything together in Equation (9.4) and rearranging one obtains

$$\begin{aligned}\left(1 - \frac{1}{4}\phi^2\Xi^2\right)T_{abc} = \frac{3}{2}\Xi\nabla_{[b}\phi\nabla_{a]}\nabla_c\phi - \frac{1}{2}\Xi g_{c[b}\nabla_{a]}\nabla_e\phi\nabla^e\phi + \phi\nabla_{[a}\phi L_{b]c}\\ + \frac{1}{4}\Xi\phi R^e{}_{cab}\nabla_e\phi + \frac{1}{4}\Xi\phi^2\nabla_e\Xi R^e{}_{cab} + \nabla_{[a}\Xi T_{b]c} + g_{c[a}T_{b]e}\nabla^e\Xi.\end{aligned}$$

The above expression contains first and second derivatives of ϕ which cannot be eliminated using the wave Equation (9.18). Accordingly, field equations for these derivatives need to be constructed.

In what follows, let $\phi_a \equiv \nabla_a\phi$, $\phi_{ab} \equiv \nabla_a\nabla_b\phi$. As $\boldsymbol{\nabla}$ is torsion free one has that $\phi_{[ab]} = 0$ and one can write

$$\phi_{ab} = \phi_{\{ab\}} + \frac{1}{4}g_{ab}\phi_e{}^e = \phi_{\{ab\}} + \frac{1}{24}g_{ab}R\phi, \tag{9.20}$$

where in the second equality one has used Equation (9.18) in the form $\phi_e{}^e = \frac{1}{6}R\phi$. Regarding ϕ_a and ϕ_{ab} as further field unknowns one obtains the *field equations*

$$\nabla_a\phi - \phi_a = 0, \qquad \nabla_a\phi_b - \phi_{\{ab\}} - \frac{1}{24}g_{ab}R\phi = 0, \qquad \nabla^e\phi_e - \frac{1}{6}R\phi = 0.$$

To obtain equations for ϕ_{ab} one considers the commutator of covariant derivatives applied to $\nabla_c\phi$ in the form

$$\nabla_{[a}\phi_{b]c} = -\frac{1}{2}R^d{}_{cab}\phi_d.$$

Letting $\psi_{ab} \equiv \phi_{\{ab\}}$ and using the decomposition (9.20) one obtains

$$\nabla_{[a}\psi_{b]c} - \frac{1}{24}\left(Rg_{c[a}\phi_{b]} + \phi g_{c[b}\nabla_{a]}R\right) = -\frac{1}{2}R^d{}_{cab}\phi_d.$$

Finally, an equation for the trace term $\phi_e{}^e$ is obtained by differentiating Equation (9.18) so that

$$\nabla_a\phi_e{}^e - \frac{1}{6}\left(\phi_a R + \phi\nabla_a R\right) = 0.$$

9.3.2 Relation to other wave equations

Solutions to the conformally invariant wave Equation (9.17) on a spacetime $(\tilde{\mathcal{M}}, \tilde{\boldsymbol{g}})$ are related to solutions of the standard wave equation on a conformally related spacetime $(\grave{\mathcal{M}}, \grave{\boldsymbol{g}})$ through a transformation first discussed in Bekenstein (1974): the scalar field $\tilde{\phi}$ can be used to define a metric $\hat{\boldsymbol{g}}$ conformally related to $\tilde{\boldsymbol{g}}$ via

$$\grave{\boldsymbol{g}} = \grave{\Xi}^2\tilde{\boldsymbol{g}}, \qquad \grave{\Xi} \equiv 1 - \frac{1}{4}\tilde{\phi}^2.$$

It follows from a direct computation that the scalar field

$$\grave{\phi} \equiv \sqrt{6}\arctan\frac{1}{2}\tilde{\phi}$$

is a solution of the equation

$$\grave{\nabla}_a\grave{\nabla}^a\grave{\phi} = 0.$$

As noticed in Bičák et al. (2010), this observation can be turned into a procedure to construct solutions to the Einstein-scalar field equations out of vacuum static solutions; see also Buchdahl (1959).

9.4 Perfect fluids

Perfect fluids constitute an important class of matter models for the Einstein field equations. In the cosmological context, perfect fluids are used to describe the matter content of the universe at a suitably large scale; see, for example, Ellis et al. (2012). Given a spacetime $(\tilde{\mathcal{M}}, \tilde{\boldsymbol{g}})$, the ***energy-momentum tensor of a perfect fluid*** with ***4-velocity*** $\tilde{u}^a$, ***pressure*** $\tilde{p}$ and ***density*** $\tilde{\varrho}$ is given by

$$\tilde{T}_{ab} = (\tilde{\varrho} + \tilde{p})\tilde{u}_a\tilde{u}_b - \tilde{p}\tilde{g}_{ab}, \tag{9.21}$$

with $\tilde{u}^a$ satisfying the normalisation condition $\tilde{u}_a\tilde{u}^a = 1$. The equations of motion for the fields $\tilde{u}^a$, $\tilde{\varrho}$ and $\tilde{p}$ are given by $\tilde{\nabla}^a\tilde{T}_{ab} = 0$. This last equation gives four equations for six unknowns. The normalisation of $\tilde{u}^a$ can be used to eliminate one of the components of the 4-velocity (usually the time component). To close the system a *phenomenological constitutive relation* linking the pressure and the density must be prescribed. A standard assumption made on perfect fluids is to have the density and the pressure related to each other by means of a ***barotropic equation of state*** $\tilde{p} = f(\tilde{\varrho})$ with f a smooth function of the density $\tilde{\varrho}$. From Equation (9.21) it follows that $\tilde{T} = \tilde{\varrho} - 3\tilde{p}$. Thus, the energy-momentum tensor of a perfect fluid is trace free if and only if

$$\tilde{p} = \frac{1}{3}\tilde{\varrho}. \tag{9.22}$$

This constitutive relation is known as the ***equation of state of radiation***. *In what follows, the discussion will be restricted to perfect fluids satisfying the equation of state* (9.22). To discuss the perfect fluid in the conformally rescaled spacetime $(\mathcal{M}, \boldsymbol{g})$ with $\boldsymbol{g} = \Xi^2\tilde{\boldsymbol{g}}$ it is convenient to consider the following unphysical conformal fields

$$u_a \equiv \Xi\tilde{u}_a, \qquad \varrho \equiv \Xi^{-4}\tilde{\varrho}, \qquad p \equiv \Xi^{-4}\tilde{p}.$$

The above definitions are consistent with the transformation law for the energy-momentum tensor of Equation (9.2). Moreover, it follows that $p = \frac{1}{3}\varrho$, so that the unphysical energy-momentum tensor takes the form

$$T_{ab} = \frac{4}{3}\varrho u_a u_b - \frac{1}{3}\varrho g_{ab} \quad \text{with} \quad \nabla^a T_{ab} = 0. \tag{9.23}$$

Moreover, one has that $u_a u^a = 1$, so that differentiating along u^a one finds that

$$u^a\nabla_a(u_b u^b) = 0.$$

From this expression it follows that if $u_a u^a = 1$ at some point in a flow line, then $u_a u^a = 1$ everywhere along the flow line. From Equation (9.23) it readily follows that

$$\frac{4}{3}(u_a u^c\nabla_c\varrho + \varrho u_a\nabla_c u^c + \varrho u^c\nabla_c u_a) - \frac{1}{3}\nabla_a\varrho = 0.$$

Contracting this equation, respectively, with u^a and $g_{ab} - u_a u_b$ one obtains

$$u^a\nabla_a\varrho + \frac{4}{3}\varrho\nabla_a u^a = 0,$$
$$\frac{4}{3}\varrho u^c\nabla_c u_a + \frac{1}{3}u_a u^c\nabla_c\varrho - \frac{1}{3}\nabla_a\varrho = 0.$$

These equations are the conformal versions of the ***equation of energy conservation*** and the ***equations of motion***; see, for example, Choquet-Bruhat (2008). A discussion on how to use these equations to construct suitable evolution equations for the fields ϱ and the spatial components u^i of the fluid 4-velocity can be found in the same reference.

9.5 Further reading

A further matter model amenable to a treatment by means of conformal methods is the Yang-Mills field. The Yang-Mills equations can be regarded as a suitable generalisation of the Maxwell equations; see, for example, Frankel (2003) for a discussion. The conformal field equations with matter source given by a Yang-Mills field of arbitrary gauge group have been discussed in Friedrich (1991). The discussion of the Maxwell field presented in this chapter is adapted from that reference. A treatment of the conformal Einstein-Maxwell system by means of Weyl connections is given in Lübbe and Valiente Kroon (2012).

The discussion of the conformal field equations coupled to the conformally invariant wave equation was first given in Hübner (1995). An alternative approach to the analysis of the conformal Einstein field equations with a scalar field can be found in Bičák et al. (2010). In Friedrich (2015b) it has been shown that the Einstein–massive scalar field system has good conformal properties if the mass of the scalar field and the cosmological constant satisfy the relation $3m^2 = -2\lambda$.

Finally, the conformal Einstein-Euler equations have been analysed in Lübbe and Valiente Kroon (2013b) and used to prove the future non-linear stability of perturbations of Friedman-Lemaître-Robinson-Walker cosmological models with a radiation fluid. Analyses of the Einstein-Euler system not making use of conformal methods can be found in Rodnianski and Speck (2013) and Speck (2012).

The purpose of this chapter has been to present a discussion of matter models with properties which make them suitable sources for the conformal field equations. However, conformal methods have also been used for other types of constructions. As an example, one has Bičák and Krtouš (2001, 2002) where the conformal invariance of the Maxwell equations has been exploited to construct the analogue of the Born solution (describing the motion of a pair of uniformly accelerated charges) in the de Sitter spacetime.

10

Asymptotics

This chapter discusses some basic consequences of the notion of *asymptotic simplicity* introduced in Chapter 7. As already mentioned, the main motivation behind this definition is to provide a characterisation of a broad class of spacetimes in which *universal structures* can be identified. Once this has been done, the idea is to use these structures to define in a rigorous manner concepts of physical interest.

The characterisation of the gravitational field through the analysis of its asymptotic behaviour has a long tradition dating back to the early works by Bondi et al. (1962), Sachs (1962b) and Newman and Penrose (1962). These studies culminated in the identification of gravitational radiation as a real physical phenomenon. The developments of this *classical theory* have been treated extensively in the literature; see, for example, Geroch (1976), Penrose and Rindler (1986), Stewart (1991) and Frauendiener (2004). The readers interested in the historic development of this idea are referred to Kennefick (2007).

Despite the important insights provided by the classical theory of asymptotics of general relativity, this approach has the weakness of being, to some extent, *formal.* More precisely, it relies on a number of assumptions about the nature of solutions to the Einstein field equations – say, for example, the regularity of the conformal boundary – which are hard to verify for a *suitably large class of spacetimes.* This point is key: the theory of asymptotics of the gravitational field comes fully into life when combined with the (conformal) field equations and methods of the theory of partial differential equations. This remark does not disown the fundamental insights into the behaviour of the gravitational field that formal asymptotic analyses have produced, but rather insists on the need to carry the subject further.

Arguably, the most important consequence of asymptotic simplicity is the set of results collectively known as ***peeling*** – that is, a detailed description of the asymptotic behaviour of the gravitational field expressed in terms of the components of the Weyl tensor. The peeling behaviour is the main subject of this chapter. The basic assumptions behind the peeling results are the main subject of Chapter 20. Complementary to the discussion of the peeling behaviour, this chapter contains a detailed discussion of a gauge prescription for the analysis of the structure of the gravitational field at the conformal boundary of Minkowski-like spacetimes, the so-called ***NP gauge***. The chapter concludes with a brief overview of other aspects of the theory of the asymptotics of the gravitational field which are sligthly outside the main focus of this book: the Bondi mass, the BMS group and the so-called Newman-Penrose constants.

10.1 Basic set up: general structure of the conformal boundary

In what follows let $(\tilde{\mathcal{M}}, \tilde{\boldsymbol{g}})$ be an *asymptotically simple spacetime* in the sense of Definition 7.1 and let $(\mathcal{M}, \boldsymbol{g}, \Xi)$ denote an associated conformal extension. As in Section 7.1, let $\mathscr{I}$ denote part of the conformal boundary characterised by the requirements

$$\Xi = 0, \qquad \mathbf{d}\Xi \neq 0. \tag{10.1}$$

Much of the analysis of the present chapter is based on the evaluation of the various conformal field equations at $\mathscr{I}$. In what follows, *the notation* $\simeq$ *will be used to indicate that a certain equality holds at* $\mathscr{I}$. In terms of this notation, the conditions in (10.1) can be rewritten as

$$\Xi \simeq 0, \qquad \mathbf{d}\Xi \not\simeq 0.$$

The basic observation concerning the set $\mathscr{I}$ is that its causal nature is determined by the sign of the cosmological constant λ. This result follows from a direct inspection of the conformal Einstein field equations; see, for example, Equations (8.26a)–(8.26e) in Section 8.2.5. One has that:

Theorem 10.1 (***causal nature of the conformal boundary***) *Suppose that the Friedrich scalar s is finite at $\mathscr{I}$ and that $T = o(\Xi^{-4})$. Then $\mathscr{I}$ is a null, spacelike or timelike hypersurface, respectively, depending on whether $\lambda = 0$, $\lambda < 0$ or $\lambda > 0$.*

Proof The normal to the hypersurface $\mathscr{I}$ is given by $\nabla_a \Xi$. From Equation (8.24) one directly has that

$$\nabla_a \Xi \nabla^a \Xi \simeq -\frac{1}{3}\lambda, \tag{10.2}$$

as by hypothesis $\Xi^4 T \to 0$ if $\Xi \to 0$ and s is finite at $\mathscr{I}$. □

A discussion of the ***order symbols*** o and O used in the previous and other results of this chapter can be found in the Appendix to Chapter 11.

Remark. Spacetimes with $\lambda = 0$ will be said to be ***Minkowski-like***, those with $\lambda < 0$ ***de Sitter-like*** and those with $\lambda > 0$ ***anti-de Sitter-like***.

The regularity of s at $\mathscr{I}$ can be rephrased in terms of a sufficiently rapid decay of the physical energy-momentum tensor $\tilde{T}_{ab}$. Using the conformal field Equation (8.13) it follows that a sufficient condition for $\nabla_a \nabla_b \Xi$ and s to be finite at $\mathscr{I}$ is that $T_{\{ab\}} = o(\Xi^{-3})$. In this case one concludes that

$$\nabla_a \nabla_b \Xi \simeq s g_{ab}. \tag{10.3}$$

It follows from the transformation formulae of the energy-momentum tensor, Equation (9.2), that if $T_{\{ab\}} = o(\Xi^{-3})$, then, in fact, $\tilde{T}_{\{ab\}} = O(\Xi^3)$; see also the discussion in Stewart (1991). If, in addition, one has that R is finite at $\mathscr{I}$, then expression (10.3) reduces to

$$\nabla_a \nabla_b \Xi \simeq \frac{1}{4} \nabla^c \nabla_c \Xi g_{ab}.$$

The spinorial version of the above expression is

$$\nabla_{A(A'} \nabla_{B')B} \Xi \simeq 0. \tag{10.4}$$

The latter is usually known as the ***asymptotic Einstein condition***; see, for example, Penrose and Rindler (1986).

10.1.1 Topology of the conformal boundary

As will be seen in Chapter 15, there exists considerable freedom in the specification of the topology of de Sitter-like spacetimes. By contrast, the case of a vanishing cosmological constant is much more restrictive:

Theorem 10.2 (***topology of $\mathscr{I}$ for asymptotically Minkowskian spacetimes***) *Let $(\tilde{\mathcal{M}}, \tilde{\boldsymbol{g}})$ denote an asymptotically simple spacetime with $\lambda = 0$ and let $(\mathcal{M}, \boldsymbol{g}, \Xi)$ denote a conformal extension thereof. Then $\mathscr{I}$ consists of two disjoint components $\mathscr{I}^-$ and $\mathscr{I}^+$, each one having the topology of $\mathbb{R} \times \mathbb{S}^2$.*

A discussion of the proof of the above theorem goes beyond the scope of this book. The interested reader is referred to Newman (1989) for a proof and for a discussion on pitfalls in earlier arguments in Penrose (1965) and Geroch (1971b, 1976); see also Hawking and Ellis (1973). Remarkably, this result depends on the satisfactory resolution of the so-called *Poincaré conjecture*; see, for example, Gowers (2008) for an introduction to this (now solved) classical problem in mathematics. Vacuum spacetimes with a vanishing cosmological constant and a conformal infinity with sections which are toroidal, that is, having the topology of $\mathbb{R} \times \mathbb{S} \times \mathbb{S}$ have been considered in the literature; see Schmidt (1996). Note that as a consequence of Theorem 10.2 these spacetimes must exhibit some type of pathology – and, in particular, they cannot be asymptotically simple.

The behaviour of points in the conformal extension of an asymptotically simple spacetime for which both $\Xi = 0$ and $\mathbf{d}\Xi = 0$ will be analysed from various perspectives in Chapters 16, 18 and 20.

10.1.2 Further properties of the case $\lambda = 0$

In this section let $\lambda = 0$ throughout so that the asymptotically simple spacetime $(\tilde{\mathcal{M}}, \tilde{\boldsymbol{g}})$ has a null conformal boundary. For ease of the exposition, attention is restricted to the vacuum case.

As a consequence of Theorem 10.1 the physical spacetime manifold $\tilde{\mathcal{M}}$ must lie either towards the past or the future of $\mathscr{I}$ – intuitively, this assertion seems natural; however, a detailed argument requires the ideas of the discussion on Lorentzian causality in Chapter 14. Consistent with the discussion of conformal extensions of exact solutions in Chapter 6, $\mathscr{I}^+$ (i.e. ***future null infinity***) will denote the set on which null geodesics attain a future endpoint while $\mathscr{I}^-$ (i.e. ***past null infinity***) corresponds to the set of past endpoints of null geodesics. A null hypersurface has the property of being generated by null geodesics; that is, each $p \in \mathscr{I}^\pm$ lies on exactly one null geodesic which is everywhere tangent to $\mathscr{I}^\pm$. Accordingly, each of $\mathscr{I}^+$ and $\mathscr{I}^-$ can be regarded as the union of these ***generators*** (or ***rays***). Complementary to the latter is the notion of a ***cut of null infinity***, that is, a two-dimensional surface $\mathscr{C}$ which intersects each generator exactly once. As a result of Theorem 10.2 one has that $\mathscr{C} \approx \mathbb{S}^2$.

The subsequent discussion will, for simplicity, be restricted to $\mathscr{I}^+$ – an analogous discussion follows, *mutatis mutandis*, for $\mathscr{I}^-$. By definition, the normal to $\mathscr{I}^+$ is given by $\mathbf{d}\Xi$. As $\boldsymbol{g}^\sharp(\mathbf{d}\Xi, \mathbf{d}\Xi) \simeq 0$, it follows that the vector $\boldsymbol{N} \equiv -\boldsymbol{g}^\sharp(\mathbf{d}\Xi, \cdot)$ satisfies $\langle \mathbf{d}\Xi, \boldsymbol{N}\rangle = 0$ and, thus, is tangent to $\mathscr{I}^+$ – and, in particular, to its null generators.

As $\mathscr{I}^+$ is a hypersurface of $\mathcal{M}$, there exists an embedding $\varphi : \mathscr{I}^+ \to \mathcal{M}$. Let $\boldsymbol{q} \equiv \varphi^*\boldsymbol{g}$ denote the metric induced on $\mathscr{I}^+$ by $\boldsymbol{g}$. *The metric $\boldsymbol{q}$ is degenerate.* To see this, write $\mathbf{d}\Xi$ in coordinates adapted to $\mathscr{I}^+$; it follows that $\varphi^*(\mathbf{d}\Xi) = 0$ so that $\varphi^*(\boldsymbol{N}^\flat) = 0$. Thus, from $\boldsymbol{N}^\flat = \boldsymbol{g}(\boldsymbol{N}, \cdot)$ one concludes that $\boldsymbol{q}(\boldsymbol{N}, \cdot) = 0$ as claimed – observe that as $\boldsymbol{N}$ is tangent to $\mathscr{I}^+$, it follows that it has a well-defined pull-back.

To analyse the behaviour of the metric $\boldsymbol{q}$ along the generators of $\mathscr{I}^+$ consider the Lie derivative $\pounds_{\boldsymbol{N}}\boldsymbol{q}$. To compute it start from

$$\begin{aligned}\pounds_{\boldsymbol{N}} g_{ab} &= N^c\nabla_c g_{ab} + \nabla^c N_a g_{cb} + \nabla^c N_b g_{ac}\\ &= \nabla_b N_a + \nabla_a N_b = 2\nabla_a N_b,\end{aligned}$$

as $\nabla_a\nabla_b\Xi = -\nabla_a N_b = -\nabla_b N_a$. Hence, using Equation (10.3) it follows that

$$\pounds_{\boldsymbol{N}}\boldsymbol{q} = -s\boldsymbol{q}. \tag{10.5}$$

The trace-free part of $\pounds_{\boldsymbol{N}}\boldsymbol{q}$ is called the ***shear tensor*** ς of the congruence of generators of $\mathscr{I}^+$ – it describes the tendency of a sphere of points in the congruence to be deformed into an ellipsoid with the same volume. As Equation (10.5) is pure trace, it follows that $\varsigma = 0$. Thus, *the congruence of generators of $\mathscr{I}^+$ is shear free.* This result is a consequence of the conformal field equations via Equation (10.3) so that from the conformal invariance of the equations it follows that the shear-freeness of the congruence of generators is a property independent of the particular choice of conformal factor.

The conformal gauge freedom inherent in the construction of a conformal extension can be exploited to gain further insight into the structure of null infinity. Given a conformal extension $(\mathcal{M}, \boldsymbol{g}, \Xi)$ consider $\vartheta > 0$ and define a conformally related metric $\boldsymbol{g}'$ via $\boldsymbol{g}' = \vartheta^2 \boldsymbol{g}$. The transformation rule of the Friedrich scalar s – see Equation (8.29b) – yields that

$$s' \simeq \vartheta^{-1} s - \vartheta^{-2} N^c \nabla_c \vartheta.$$

Thus, if initially $s \neq 0$, one can always find a further conformal representation $(\mathcal{M}, \boldsymbol{g}', \Xi')$ for which $s' = 0$ if one imposes the condition

$$N^c \nabla_c \vartheta = \vartheta s. \tag{10.6}$$

Notice that the above equation can be rewritten as $\pounds_{\boldsymbol{N}} \vartheta = \vartheta s$, and, accordingly, it can be read as an ordinary differential equation along the generators of null infinity. It is important to observe that once condition (10.6) has been imposed, one is still left with the freedom of specifying a further rescaling $\boldsymbol{g}'' = \varkappa^2 \boldsymbol{g}'$ such that $\pounds_{\boldsymbol{N}'} \varkappa = 0$.

The conformal gauge implied by condition (10.6) yields, together with Equation (10.5), that

$$\pounds_{\boldsymbol{N}'} \boldsymbol{q}' = 0; \tag{10.7}$$

that is, the intrinsic metric of $\mathscr{I}^+$ is *Lie dragged* along the generators of null infinity. Each of the cuts $\mathscr{C}$ of null infinity inherits from the metric $\boldsymbol{q}$ on $\mathscr{I}^+$ a metric $\boldsymbol{k}$ which is *non-degenerate*. As a consequence of Equation (10.7), if one considers any other cut $\mathscr{C}'$, one obtains the same induced metric $\boldsymbol{k}$. Now, any metric on a two-dimensional surface which is topologically $\mathbb{S}^2$ is conformal to the standard metric of $\mathbb{S}^2$, $\boldsymbol{\sigma}$ – this fact is a consequence of the so-called ***Riemann mapping theorem***; see, for example, Krantz (2006), chapter 4. Hence, one can write $\boldsymbol{k} = \theta^2 \boldsymbol{\sigma}$ for some conformal factor $\theta > 0$ on $\mathbb{S}^2$. Under a further conformal gauge transformation $\boldsymbol{g}'' = \varkappa^2 \boldsymbol{g}'$ such that $\pounds_{\boldsymbol{N}'} \varkappa = 0$ (see the previous paragraph), one can then always assume that the gauge has been chosen so that $\boldsymbol{k} = \boldsymbol{\sigma}$. Under these circumstances the conformal gauge freedom is reduced to a function $\varkappa$ such that $\varkappa \simeq 1$.

10.2 Peeling properties

One of the most important results of the theory of asymptotics of the gravitational field is the so-called ***Peeling theorem*** – a precise prescription of the decay of the Weyl tensor of an asymptotically simple spacetime. The Peeling theorem is based on the important observation that the Weyl tensor of an asymptotically simple spacetime must vanish on $\mathscr{I}$. As will be seen in the following, this observation follows in a quite straight forward manner if $\lambda \neq 0$. A more subtle argument is required if $\lambda = 0$.

In what follows, let Ψ_{ABCD} denote the Weyl spinor, and recall that $\Psi_{ABCD} = \Xi\phi_{ABCD}$. The subsequent analysis is best carried out with the spinorial conformal Einstein field equations expressed with respect to a spin dyad $\{\epsilon_{\boldsymbol{A}}{}^{A}\}$; see Section 8.3.2. *In this formulation of the field equations the fields are scalars. Hence, they can readily be evaluated at the conformal boundary without the need of pull-backs.* One has the following:

Theorem 10.3 (***vanishing of the Weyl tensor at*** $\mathscr{I}$) *Assume that* $\Psi_{\boldsymbol{ABCD}}$ *is smooth at* $\mathscr{I}$*. If* $\lambda \neq 0$ *and the physical Cotton tensor satisfies* $\tilde{Y}_{\boldsymbol{abc}} = o(\Xi^{-1})$ *at* $\mathscr{I}$*, then* $\Psi_{\boldsymbol{ABCD}} = 0$ *at* $\mathscr{I}$*. If* $\lambda = 0$*, the same conclusion follows if* $\tilde{Y}_{\boldsymbol{abc}} = o(\Xi^{-1})$ *and* $\nabla_{\boldsymbol{d}}\tilde{Y}_{\boldsymbol{abc}} = o(\Xi^{-1})$.

Proof (case $\lambda \neq 0$*)* The starting point of the analysis is the Bianchi equation

$$\nabla^{\boldsymbol{Q}}{}_{\boldsymbol{A}'}\phi_{\boldsymbol{ABCQ}} + T_{\boldsymbol{BCAA}'} = 0;$$

compare the spinorial conformal Einstein Equation (8.37b). Now, recalling that $\phi_{\boldsymbol{ABCD}} = \Xi^{-1}\Psi_{\boldsymbol{ABCD}}$ and $T_{\boldsymbol{BCAA}'} = \Xi^{-1}\tilde{Y}_{\boldsymbol{BCAA}'}$ it follows that

$$\nabla^{\boldsymbol{Q}}{}_{\boldsymbol{A}'}\Xi\Psi_{\boldsymbol{ABCQ}} - \Xi\nabla^{\boldsymbol{Q}}{}_{\boldsymbol{A}'}\Psi_{\boldsymbol{ABCQ}} = \Xi\tilde{Y}_{\boldsymbol{BCAA}'}. \tag{10.8}$$

Hence, using $\tilde{Y}_{\boldsymbol{abc}} = o(\Xi^{-1})$ one finds that $\nabla^{\boldsymbol{Q}}{}_{\boldsymbol{A}'}\Xi\,\Psi_{\boldsymbol{ABCQ}} \simeq 0$. Contracting with $\nabla_{\boldsymbol{DA}'}\Xi$ one obtains

$$\nabla_{\boldsymbol{DA}'}\Xi\nabla_{\boldsymbol{Q}}{}^{\boldsymbol{A}'}\Xi\,\Psi^{\boldsymbol{Q}}{}_{\boldsymbol{ABC}} \simeq 0. \tag{10.9}$$

Now, using Equation (10.2) one has

$$\nabla_{\boldsymbol{DA}'}\Xi\nabla_{\boldsymbol{Q}}{}^{\boldsymbol{A}'}\Xi = \frac{1}{2}\nabla_{\boldsymbol{PP}'}\Xi\nabla^{\boldsymbol{PP}'}\Xi\,\epsilon_{\boldsymbol{DQ}} \simeq -\frac{3}{2}\lambda\,\epsilon_{\boldsymbol{DQ}}.$$

Substituting the latter in (10.9) one finds that $\lambda\Psi_{\boldsymbol{ABCD}}\simeq 0$. Hence, $\Psi_{\boldsymbol{ABCD}} = 0$ on $\mathscr{I}$. □

Proof (case $\lambda = 0$*)* Again, one has that

$$\nabla^{\boldsymbol{AA}'}\Xi\,\Psi_{\boldsymbol{ABCD}} \simeq 0. \tag{10.10}$$

In this case, however, $\nabla^{\boldsymbol{AA}'}\Xi$ is the spinorial counterpart of a null vector. Hence, there exists a spinor $\iota^{\boldsymbol{A}}$ such that

$$\nabla^{\boldsymbol{AA}'}\Xi = \iota^{\boldsymbol{A}}\bar{\iota}^{\boldsymbol{A}'}. \tag{10.11}$$

It follows from Equation (10.10) that there exists a scalar field ψ such that

$$\Psi_{\boldsymbol{ABCD}} \simeq \psi\iota_{\boldsymbol{A}}\iota_{\boldsymbol{B}}\iota_{\boldsymbol{C}}\iota_{\boldsymbol{D}}. \tag{10.12}$$

In order to extract further information consider Equation (10.8) – which is also valid in the case $\lambda = 0$ – and apply $\nabla_{EE'}$ to both sides. The assumptions on $\tilde{Y}_{CDBB'}$ imply that

$$\nabla_{EE'}\nabla^{Q}{}_{B'}\Xi\Psi_{ABCQ} + \nabla^{Q}{}_{B'}\Xi\nabla_{EE'}\Psi_{ABCQ} - \nabla_{EE'}\Xi\nabla^{Q}{}_{B'}\Psi_{ABCQ} \simeq 0.$$

Symmetrising on $_{E'}$ and $_{B'}$, and using the asymptotic Einstein condition (10.4) one concludes that

$$\nabla^{Q}{}_{(B'}\Xi\nabla_{E')E}\Psi_{ABCQ} - \nabla_{E(E'}\Xi\nabla^{Q}{}_{B')}\Psi_{ABCQ} \simeq 0. \tag{10.13}$$

Now, using identity (3.6) to interchange the indices $_{E}$ and $_{Q}$ one obtains

$$\nabla^{Q}{}_{(B'}\Xi\nabla_{E')Q}\Psi_{ABCE} - \nabla_{Q(E'}\Xi\nabla^{Q}{}_{B')}\Psi_{ABCE}$$
$$- \epsilon_{EQ}\epsilon^{ST}\left(\nabla^{Q}{}_{(B'}\Xi\nabla_{E')S}\Psi_{ABCT} + \nabla_{S(E'}\Xi\nabla^{Q}{}_{B')}\Psi_{ABCT}\right) \simeq 0,$$

which in view of Equation (10.13) reduces to

$$\nabla^{Q}{}_{(B'}\Xi\nabla_{E')Q}\Psi_{ABCE} \simeq 0.$$

Using the decomposition (10.11) in this last equation one obtains

$$\iota^{Q}\bar{\iota}_{(B'}\nabla_{E')Q}\Psi_{ABCE} \simeq 0.$$

Contracting the latter with $\bar{\iota}^{B'}$ and observing that $\bar{\iota}_{E'} \neq 0$, one concludes that

$$\iota^{Q}\bar{\iota}^{B'}\nabla_{QB'}\Psi_{ABCE} \simeq 0. \tag{10.14}$$

Thus,

$$\iota^{Q}\nabla_{QE'}\Psi_{ABCE} \simeq \alpha\bar{\iota}_{E'}\zeta_{ABCE}$$

for some scalar α and a spinor $\zeta_{ABCE} \neq 0$. Substituting back into (10.14) one concludes that $\alpha = 0$ so that one has

$$\iota^{Q}\nabla_{QE'}\Psi_{ABCE} \simeq 0. \tag{10.15}$$

In order to bring this last result into a more convenient form one completes the spinor ι^{A} to a spin basis $\{\epsilon_{A}{}^{A}\} = \{o^{A}, \iota^{A}\}$ with $o_{A}\iota^{A} = 1$ so that $\iota^{A} = \delta_{1}{}^{A}$ and $o^{A} = \delta_{0}{}^{A}$. Thus, contracting Equation (10.15) with $\bar{o}^{E'}$ and substituting (10.12) into Equation (10.15) one obtains

$$\iota^{Q}\bar{o}^{E'}\nabla_{QE'}(\psi\iota_{A}\iota_{B}\iota_{C}\iota_{E}) = \nabla_{10'}(\psi\iota_{A}\iota_{B}\iota_{C}\iota_{E}) \simeq 0. \tag{10.16}$$

The above expression is to be regarded as a differential equation for ψ over the cuts of $\mathscr{I}^{+}$. To conclude the argument one makes use of the ***formalism of the*** $\eth$ ***and*** $\bar{\eth}$ ***operators*** as discussed in the Appendix to this chapter. Accordingly, in what follows it is assumed that one has a conformal representation for which the cuts are metric unit spheres $\mathbb{S}^2$. Contracting (10.16) with $o^{A}o^{B}o^{C}o^{E}$ one obtains

$$\bar{\eth}\psi \simeq 0.$$

Now, from $\psi = \Psi_{ABCD}o^A o^B o^C o^D$ it follows that ψ has spin-weight 2. Hence, using Lemma 10.1 in the Appendix to this chapter it follows that $\psi \simeq 0$ and thus Ψ_{ABCD} vanishes at $\mathscr{I}$. □

Remark. The above result strongly depends on the fact that for an asymptotically simple spacetime with $\lambda = 0$ one has that $\mathscr{I} \approx \mathbb{R} \times \mathbb{S}^2$. For the spacetimes with *toroidal* null infinities considered in Schmidt (1996), the crucial Lemma 10.1 does not hold – see Frauendiener and Szabados (2001) – and the desired conclusion cannot be obtained.

A more detailed description

To obtain a more detailed description of the *peeling behaviour*, it is necessary to introduce further structure. In what follows, consider a null geodesic γ in $(\mathcal{M}, \boldsymbol{g})$ reaching $\mathscr{I}$ at a point p and let $\tilde{\gamma}$ denote the corresponding null geodesic on $(\tilde{\mathcal{M}}, \tilde{\boldsymbol{g}})$. At a point $q \in \tilde{\gamma}$ one can choose a spin dyad $\{\tilde{\boldsymbol{o}}, \tilde{\boldsymbol{\iota}}\}$ such that the tangent to $\tilde{\gamma}$ is given by the vector $\tilde{\boldsymbol{l}}$ with spinorial counterpart $\tilde{l}^{AA'} = \tilde{o}^A \bar{\tilde{o}}^{A'}$. The spin dyad can be naturally propagated along $\tilde{\gamma}$ by requiring

$$\tilde{D}\tilde{o}^A = 0, \qquad \tilde{D}\tilde{\iota}^A = 0, \tag{10.17}$$

where $\tilde{D} \equiv \tilde{l}^a \tilde{\nabla}_a = \tilde{o}^A \bar{\tilde{o}}^{A'} \tilde{\nabla}_{AA'}$ in standard Newman-Penrose (NP) notation. Now, let $\tilde{r}$ denote an affine parameter along $\tilde{\gamma}$. It follows that $\tilde{D} = \mathrm{d}/\mathrm{d}\tilde{r}$. In order to rewrite the above expressions in terms of quantities defined on the unphysical spacetime $(\mathcal{M}, \boldsymbol{g})$ it is convenient to consider the transformation

$$o_A = \tilde{o}_A, \qquad o^A = \Xi^{-1}\tilde{o}^A, \qquad \iota_A = \Xi\tilde{\iota}_A, \qquad \iota^A = \tilde{\iota}^A; \tag{10.18}$$

compare Equations (5.31a)–(5.31c) in Chapter 5. Using the transformation laws under conformal transformations for the covariant derivatives it follows from (10.17) that

$$Do^A = 0, \qquad D\iota^A = \left(\Xi^{-1}\bar{\delta}\Xi\right)o^A,$$

where $\bar{\delta} \equiv \bar{m}^a \nabla_a = \iota^A \bar{o}^{A'} \nabla_{AA'}$. The second of the above expressions is potentially singular at $\mathscr{I}$ – observe, however, that as $\Xi \simeq 0$, it follows that $\bar{\delta}\Xi \simeq 0$ as $\bar{\boldsymbol{m}}$ is intrinsic to $\mathscr{I}$. Thus, the spin dyad $\{\boldsymbol{o}, \boldsymbol{\iota}\}$ is well defined and regular at $\mathscr{I}$. Now, from $Do^A = 0$ it follows that the null geodesic γ is affinely parametrised. Let r denote a possible affine parameter. Its origin and scaling can be chosen so that

$$r = 0, \qquad \text{and} \qquad D\Xi = \frac{\mathrm{d}\Xi}{\mathrm{d}r} = -1 \qquad \text{at} \quad p \in \mathscr{I}.$$

From Remark (c) in Section 7.1 it follows that

$$\frac{\mathrm{d}\tilde{r}}{\mathrm{d}r} = \frac{1}{\Xi^2}$$

where $\tilde{r}$ is an affine parameter in the physical spacetime $(\tilde{\mathcal{M}}, \tilde{\boldsymbol{g}})$. Hence, one concludes that

$$\tilde{r} = O(\Xi^{-1}) \qquad \text{near } \mathscr{I}. \tag{10.19}$$

Making use of the above relations one obtains the following, more detailed, version of the peeling behaviour:

Theorem 10.4 (***Peeling theorem***) *Let $(\tilde{\mathcal{M}}, \tilde{\boldsymbol{g}})$ denote an asymptotically simple spacetime with $\lambda = 0$ for which the hypotheses of Theorem 10.3 hold. Moreover, let*

$$\tilde{\psi}_0 \equiv \Psi_{ABCD}\tilde{o}^A\tilde{o}^B\tilde{o}^C\tilde{o}^D, \quad \tilde{\psi}_1 \equiv \Psi_{ABCD}\tilde{\iota}^A\tilde{o}^B\tilde{o}^C\tilde{o}^D, \quad \tilde{\psi}_2 \equiv \Psi_{ABCD}\tilde{\iota}^A\tilde{\iota}^B\tilde{o}^C\tilde{o}^D,$$
$$\tilde{\psi}_3 \equiv \Psi_{ABCD}\tilde{\iota}^A\tilde{\iota}^B\tilde{\iota}^C\tilde{o}^D, \qquad \tilde{\psi}_4 \equiv \Psi_{ABCD}\tilde{\iota}^A\tilde{\iota}^B\tilde{\iota}^C\tilde{\iota}^D,$$

then

$$\tilde{\psi}_0 = O(\tilde{r}^{-5}), \qquad \tilde{\psi}_1 = O(\tilde{r}^{-4}), \qquad \tilde{\psi}_2 = O(\tilde{r}^{-3})$$
$$\tilde{\psi}_3 = O(\tilde{r}^{-2}), \qquad \tilde{\psi}_4 = O(\tilde{r}^{-1}).$$

Proof Let

$$\psi_0 \equiv \Psi_{ABCD}o^Ao^Bo^Co^D, \quad \ldots \quad \psi_4 \equiv \Psi_{ABCD}\iota^A\iota^B\iota^C\iota^D.$$

It follows from Theorem 10.3 that $\psi_k = O(\Xi)$. Now, using the transformation rules (10.18) one has that

$$\tilde{\psi}_k = \Xi^{4-k}\psi_k.$$

Thus, recalling (10.19), one finds the desired result. □

Combining the definitions of the fields $\tilde{\psi}_k$ with the corresponding decays given by Theorem 10.4 one obtains a detailed expression for the asymptotic behaviour of the Weyl spinor. It can be written schematically as

$$\Psi_{ABCD} = \frac{[N]_{ABCD}}{\tilde{r}} + \frac{[III]_{ABCD}}{\tilde{r}^2} + \frac{[II]_{ABCD}}{\tilde{r}^3} + \frac{[I]_{ABCD}}{\tilde{r}^4} + O(\tilde{r}^{-5}), \tag{10.20}$$

where $[N]_{ABCD}$, $[III]_{ABCD}$, $[II]_{ABCD}$ and $[I]_{ABCD}$ represent, respectively, totally symmetric spinors of ***Petrov type*** N, III, II and I; for a concise discussion of the ***Petrov classification of the Weyl tensor*** using spinors, see Stewart (1991). For Petrov type N Weyl tensors the spinor Ψ_{ABCD} has four repeated principal null directions. They are associated to gravitational plane waves. Similarly, a spacetime with a Weyl spinor of Petrov type III has three repeated principal null directions; one of Petrov type II has two principal directions, while one of Petrov type I is algebraically general. The observation that a repeated principal null direction is lost at each order in the expansion (10.20) justifies the name of *peeling* in analogy to the peeling of a fruit; see Figure 10.1.

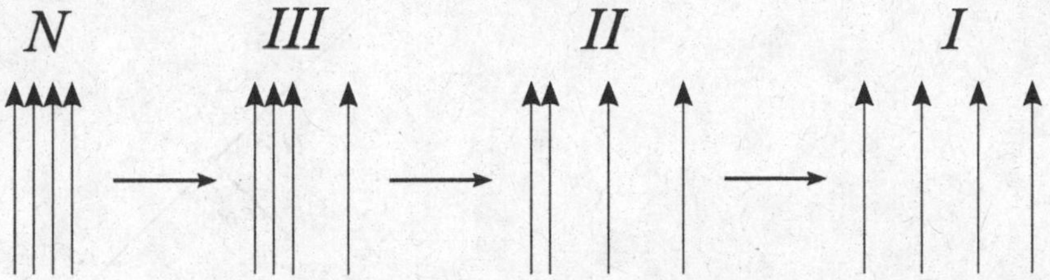

Figure 10.1 Schematic representation of the Peeling theorem: the leading behaviour of the Weyl tensor corresponds to that of a plane wave (Petrov type N). More general behaviour is observed as one looks into higher order terms.

Remark. The key assumption in the derivation of the peeling behaviour is the smoothness of the Weyl tensor at $\mathscr{I}^+$. A careful inspection of the arguments in the previous sections shows that the smoothness requirement can be relaxed and that the conclusions of Theorems 10.3 and 10.4 can be recovered if it is assumed that Ψ_{ABCD} is of class C^{k_*} at $\mathscr{I}^+$ for some positive integer k_*. A determination of a *sharp value* of k_* will not be pursued here. One of the challenges in the construction of spacetimes satisfying the peeling behaviour or, more generally, spacetimes which are asymptotically simple is to ensure that their Weyl tensor has the required regularity at the conformal boundary. The latter will be a recurrent idea in the remainder of this book. The analysis of the *non-linear stability of the Minkowski spacetime* in Christodoulou and Klainerman (1993) renders a Weyl tensor with a limited regularity at $\mathscr{I}^+$ for which only a *partial peeling behaviour* of the form

$$\tilde{\psi}_0 = O(\tilde{r}^{-1}), \qquad \tilde{\psi}_1 = O(\tilde{r}^{-2}), \qquad \tilde{\psi}_2 = O(\tilde{r}^{-3}),$$
$$\tilde{\psi}_3 = O(\tilde{r}^{-7/2}), \qquad \tilde{\psi}_4 = O(\tilde{r}^{-7/2}),$$

can be recovered; see, for example, Friedrich (1992) for a discussion.

10.3 The Newman-Penrose gauge

The analysis leading to the Peeling theorem shows the advantages of using a gauge which is adapted to the geometry of null infinity. In this section this idea is further elaborated. The resulting ***Newman-Penrose gauge*** allows one to obtain further insights into the properties of asymptotically simple spacetimes.

10.3.1 The construction of the gauge

As in the previous section let $(\mathcal{M}, \boldsymbol{g}, \Xi)$ denote a conformal extension of an asymptotically simple spacetime $(\tilde{\mathcal{M}}, \tilde{\boldsymbol{g}})$ with $\lambda = 0$. For conciseness, the subsequent discussion will be restricted to future null infinity $\mathscr{I}^+$. An analogous discussion can be readily adapted for $\mathscr{I}^-$.

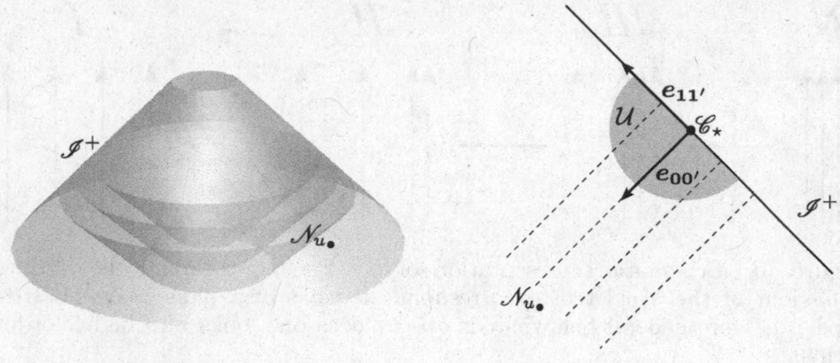

Figure 10.2 Schematic representation of the setting for the construction of the NP gauge. The NP gauge is based on a fiduciary cut $\mathscr{C}_\star$ on $\mathscr{I}^+$ and is valid in a neighbourhood $\mathcal{U}$ of the conformal boundary. The vector $e_{11'}$ is tangent to the generators of null infinity, while $e_{00'}$ generates the outgoing null hypersurfaces $\mathscr{N}_{u_\bullet}$. See the main text for further details.

In what follows, let $\{\boldsymbol{e}_{\boldsymbol{AA'}}\}$ be a frame satisfying $\boldsymbol{g}(\boldsymbol{e}_{\boldsymbol{AA'}}, \boldsymbol{e}_{\boldsymbol{BB'}}) = \epsilon_{\boldsymbol{AB}}\epsilon_{\boldsymbol{A'B'}}$ defined in a neighbourhood $\mathcal{U}$ of $\mathscr{I}^+$. The frame will be said to be ***adapted to*** $\mathscr{I}^+$ if – see Figure 10.2:

(i) The vector $\boldsymbol{e}_{\boldsymbol{11'}}$ is tangent to $\mathscr{I}^+$ and is parallely propagated along its generators; that is, one has

$$\nabla_{\boldsymbol{11'}}\boldsymbol{e}_{\boldsymbol{11'}} \simeq 0.$$

(ii) On $\mathcal{U}$ there exists a function u (a ***retarded time***) which can be regarded as an affine parameter of the generators of $\mathscr{I}^+$ such that $\boldsymbol{e}_{\boldsymbol{11'}}(u) \simeq 1$. The retarded time is constant on null hypersurfaces transverse to $\mathscr{I}^+$ and satisfies $\boldsymbol{e}_{\boldsymbol{00}} = \boldsymbol{g}^\sharp(\mathbf{d}u, \cdot)$. It follows that $\boldsymbol{e}_{\boldsymbol{00}}$ is tangent to the hypersurfaces

$$\mathscr{N}_{u_\bullet} = \{p \in \mathcal{U} \,|\, u(p) = u_\bullet\},$$

where $u_\bullet$ is a constant. Moreover, $\boldsymbol{e}_{\boldsymbol{00'}}$ is tangent to the null generators of $\mathscr{N}_{u_\bullet}$.

(iii) The fields $\{\boldsymbol{e}_{\boldsymbol{AA'}}\}$ are tangent to the cuts $\mathscr{C}_{u_\bullet} \equiv \mathscr{N}_{u_\bullet} \cap \mathscr{I}^+$ and parallely propagated along the direction of $\boldsymbol{e}_{\boldsymbol{00'}}$. That is, one has

$$\nabla_{\boldsymbol{00'}}\boldsymbol{e}_{\boldsymbol{AA'}} = 0 \qquad \text{on} \qquad \mathscr{N}_{u_\bullet}.$$

Using the definition of the spin-connection coefficients it follows from the above requirements that

$$\Gamma_{\boldsymbol{10'11}} \simeq 0, \qquad \Gamma_{\boldsymbol{11'11}} \simeq 0, \tag{10.21a}$$

$$\Gamma_{\boldsymbol{10'00}} = \bar{\Gamma}_{\boldsymbol{1'00'0'}}, \qquad \Gamma_{\boldsymbol{11'00}} = \bar{\Gamma}_{\boldsymbol{1'00'1'}} + \Gamma_{\boldsymbol{01'01}} \qquad \text{on } \mathcal{U}, \tag{10.21b}$$

$$\Gamma_{\boldsymbol{00'AB}} = 0 \qquad \text{on } \mathcal{U}. \tag{10.21c}$$

The condition $\Gamma_{10'11} \simeq 0$ is, in fact, another way of expressing the fact that the congruence of null generators of $\mathscr{I}^+$ is shear free. This can be seen by evaluating the conformal field Equation (8.35c)

$$\nabla_{AA'}\nabla_{BB'}\Xi = -\Xi L_{AA'BB'} + s\epsilon_{AB}\epsilon_{A'B'} \tag{10.22}$$

at $\mathscr{I}^+$ for ${}_{AA'}\,{}_{BB'} = {}_{10'10'}$. It follows that $\Gamma_{10'11}e_{00'}(\Xi) \simeq 0$, but $e_{00'}(\Xi) \not\simeq 0$ so that one concludes $\Gamma_{10'11} \simeq 0$ as claimed.

Remark. The discussion of the previous sections shows that an adapted frame can always be obtained in a neighbourhood $\mathcal{U}$ of $\mathscr{I}^+$. The key observation is that $\boldsymbol{N} = \boldsymbol{g}^\sharp(\mathrm{d}\Xi, \cdot)$ is tangent to the null generators of $\mathscr{I}^+$ so that one can set $\boldsymbol{e}_{11'}$ proportional to $\boldsymbol{N}$. A suitable choice of affine parameter for $\boldsymbol{N}$ renders the retarded time u and hence the frame vector $\boldsymbol{e}_{00'}$. The rest of the frame is then naturally completed by looking at a basis on the tangent bundle of the cuts $\mathscr{C}_{u_\bullet}$.

Following the ideas of Section 10.1.2, the gauge can be further specialised by considering a suitable conformal rescaling. Accordingly, consider

$$\boldsymbol{g} \mapsto \boldsymbol{g}' = \vartheta^2\boldsymbol{g}, \qquad \Xi \mapsto \Xi' = \vartheta\Xi. \tag{10.23}$$

The above rescaling will be used to obtain an *improved adapted frame* $\{\boldsymbol{e}'_{\boldsymbol{AA'}}\}$. For an arbitrary conformal factor $\vartheta > 0$ and an arbitrary function $\varkappa > 0$ which is constant along the generators of $\mathscr{I}^+$ set

$$\boldsymbol{e}'_{11'} \simeq \vartheta^{-2}\varkappa\boldsymbol{e}_{11'}; \tag{10.24}$$

compare the discussion in Section 10.1.2. In addition, define a further parameter $u' = u'(u)$ such that $\mathrm{d}u'/\mathrm{d}u = \varkappa^{-1}\vartheta^2$. Integrating along the generators of null infinity one finds that

$$u' = \frac{1}{\varkappa}\int_{u_\star}^{u} \vartheta^2(\mathrm{s})\mathrm{ds} + u'_\star.$$

The real constants $u_\star$ and $u'_\star$ are fixed so that they identify a certain *fiduciary cut* $\mathscr{C}_\star \equiv \mathscr{C}_{u_\star}$. In what follows, for convenience, *the symbol* $\overset{\star}{\simeq}$ *is used to denote equality at* $\mathscr{C}_\star$. It can be verified that $\boldsymbol{e}'_{11'}$ is parallely propagated and that $\boldsymbol{e}'_{11'}(u') = 1$. The transformation rule (10.24) is supplemented at $\mathscr{C}_\star$ by

$$\boldsymbol{e}'_{00'} \overset{\star}{\simeq} \varkappa^{-1}\boldsymbol{e}_{00'}, \qquad \boldsymbol{e}'_{01'} \overset{\star}{\simeq} \vartheta^{-1}\boldsymbol{e}_{01'}. \tag{10.25}$$

It can be verified that $\boldsymbol{g}'(\boldsymbol{e}'_{\boldsymbol{AA'}}, \boldsymbol{e}'_{\boldsymbol{BB'}}) = \epsilon_{\boldsymbol{AB}}\epsilon_{\boldsymbol{A'B'}}$ on $\mathscr{C}_\star$.

As seen in Section 10.1.2, $\mathscr{C}_\star \approx \mathbb{S}^2$ so that the metric $\boldsymbol{k}_\star$ induced by $\boldsymbol{g}'$ on $\mathscr{C}_\star$ is conformal to the standard metric $\boldsymbol{\sigma}$ of $\mathbb{S}^2$. Accordingly, the conformal factor ϑ can be chosen on $\mathscr{C}_\star$ so that $\boldsymbol{k}_\star \overset{\star}{\simeq} \boldsymbol{\sigma}$. A calculation using the transformation laws of Chapter 5 shows that the rescaling (10.23) and the conditions (10.24) and (10.25) imply on $\mathscr{C}_\star$

$$\Gamma'_{10'00} = \varkappa^{-1}\big(\Gamma_{10'00} - \vartheta^{-1}\boldsymbol{e}_{00'}(\vartheta)\big), \tag{10.26a}$$

$$\Gamma'_{01'11} = \varkappa\vartheta^{-2}\big(\Gamma_{01'11} + \vartheta^{-1}\boldsymbol{e}_{11'}(\vartheta)\big). \tag{10.26b}$$

Hence, by a suitable choice of $\mathbf{d}\vartheta$ and $\varkappa$ it is possible to ensure that

$$\Gamma'_{\mathbf{10'00}} \overset{\star}{\simeq} 0, \quad \Gamma'_{\mathbf{01'11}} \overset{\star}{\simeq} 0, \qquad e'_{\mathbf{00'}}(\Xi') \overset{\star}{\simeq} \text{constant} \neq 0. \tag{10.27}$$

A convenient way of prescribing the conformal factor ϑ off $\mathscr{C}_\star$ follows from the transformation law for the trace-free part of the Ricci tensor Φ_{ab} under the rescaling (10.23):

$$\Phi'_{ab} - \Phi_{ab} = -2\vartheta^{-1}\Big((\nabla_a\nabla_b\vartheta - 2\vartheta^{-1}\nabla_a\vartheta\nabla_b\vartheta) - \frac{1}{4}g_{ab}(\nabla_c\nabla^c)\vartheta - 2\vartheta^{-1}\nabla_c\vartheta\nabla^c\vartheta)\Big);$$

see Equation (5.6a). Transvecting this last equation with $\boldsymbol{e}_{\mathbf{11'}} \otimes \boldsymbol{e}_{\mathbf{11'}}$ it follows that if ϑ satisfies the equation

$$\boldsymbol{e}_{\mathbf{11'}}(\boldsymbol{e}_{\mathbf{11'}}(\vartheta)) - 2\vartheta^{-1}\big(\boldsymbol{e}_{\mathbf{11'}}(\vartheta)\big)^2 \simeq \vartheta\Phi_{22}, \tag{10.28}$$

then $\Phi'_{22} \simeq 0$. By means of the substitution $z = \vartheta^{-1}$, Equation (10.28) can be read as a second-order linear ordinary differential equation for ϑ^{-1} along the generators of $\mathscr{I}^+$. Thus, this equation can always be solved, at least in a neighbourhood of $\mathscr{C}_\star$ on $\mathscr{I}^+$ to ensure that

$$\Phi'_{22} \simeq 0. \tag{10.29}$$

This last construction also fixes the value of $e'_{\mathbf{01'}}(\vartheta)$ on $\mathscr{I}^+$.

The initial data for Equation (10.28) on the fiduciary cut $\mathscr{C}_\star$ is chosen so that $\boldsymbol{e}_{\mathbf{11'}}(\vartheta) \overset{\star}{\simeq} -\Gamma_{\mathbf{01'11}}$ consistent with Equation (10.27); compare Equation (10.26b).

Now, taking into account Equations (10.21a) and (10.29), one has that the Ricci identity – compare the conformal field Equation (8.35b) of Chapter 8 – gives for the values $\boldsymbol{AA'} = \mathbf{11'}$, $\boldsymbol{BB'} = \mathbf{01'}$ and $\boldsymbol{CD} = \mathbf{11}$ that

$$\boldsymbol{e}_{\mathbf{11'}}(\Gamma'_{\mathbf{01'11}}) + \big(\Gamma'_{\mathbf{01'11}}\big)^2 + \Gamma'_{\mathbf{01'11}}\bar{\Gamma}'_{\mathbf{1'10'1'}} = 0.$$

The latter equation can be interpreted as a homogeneous differential equation along the generators of $\mathscr{I}^+$ for the reduced spin connection coefficient $\Gamma'_{\mathbf{01'11}}$. As a consequence of the initial condition (10.27) on $\mathscr{C}_\star$, it follows that $\Gamma'_{\mathbf{01'11}} \simeq 0$.

The construction described in the previous paragraphs provides a specification of the conformal factor ϑ and of the function $\varkappa$ which fixes the frame vector $\boldsymbol{e}'_{\mathbf{11'}}$ completely on $\mathscr{I}^+$. Notice, however, that the vectors $\boldsymbol{e}'_{\mathbf{01'}}$ and $\boldsymbol{e}'_{\mathbf{10'}}$ (tangent to the cuts $\mathscr{C}_{u_\bullet}$) are determined up to a rotation of the form

$$\boldsymbol{e}'_{\mathbf{01'}} \mapsto e^{\mathrm{i}c}\boldsymbol{e}'_{\mathbf{01'}}, \qquad \boldsymbol{e}'_{\mathbf{10'}} \mapsto e^{-\mathrm{i}c}\boldsymbol{e}'_{\mathbf{10'}}, \tag{10.30}$$

with c a *real phase* on $\mathscr{I}^+$. A rotation on $T(\mathscr{C}_{u_\bullet})$ can be exploited to obtain additional simplifications in the spin connection coefficients. A calculation using the definition of the spin connection coefficients and taking into account that $\nabla_{\mathbf{11'}}\boldsymbol{e}'_{\mathbf{11}} \simeq 0$ gives that

$$\Gamma'_{\mathbf{11'01}} \simeq -\frac{1}{2}\langle \boldsymbol{\omega}'^{\mathbf{10}}, \nabla'_{\mathbf{11'}}\boldsymbol{e}'_{\mathbf{10'}}\rangle.$$

Under the rotation (10.30) the above relation transforms as

$$\Gamma'_{11'01} \mapsto \frac{\mathrm{i}}{2}e'_{11'}(c) - \frac{1}{2}\Gamma'_{11'01}, \qquad \text{on } \mathscr{I}^+.$$

Thus, given a particular choice of vectors $e'_{01'}$ and $e'_{10'}$ on $\mathscr{I}^+$, by solving the equation

$$e'_{11'}(c) \simeq -\frac{\mathrm{i}}{2}\Gamma'_{11'01}, \qquad \text{with } c \stackrel{\star}{\simeq} 0,$$

along the generators of $\mathscr{I}^+$, it is always possible to rotate the basis according to (10.30) so as to ensure that $\Gamma'_{11'01} \simeq 0$. In the following, it will be assumed that $e'_{01'}$ and $e'_{10'}$ have been chosen so that the latter is the case.

The choice of vectors $e'_{01'}$ and $e'_{10'}$ has some further consequences. Evaluating the primed version of Equation (10.22) at $\mathscr{I}^+$ for ${}_{AA'} = {}_{01'}$ and ${}_{BB'} = {}_{10'}$ one finds that $\nabla'_{01'}\nabla'_{10'}\Xi' \simeq -s'$. Now, as $\Xi' = 0$ on $\mathscr{I}^+$ and e'_{01} is tangent to $\mathscr{I}^+$, it follows from $\nabla'_{01'}\nabla'_{10'}\Xi' = \nabla'_{01'}e'_{10'}(\Xi') \simeq 0$ that $s' \simeq 0$ and that

$$\nabla'_{AA'}\nabla'_{BB'}\Xi' \simeq 0. \tag{10.31}$$

This last expression can be regarded as a strengthened version of the asymptotic Einstein condition (10.4). In particular, for ${}_{AA'} = {}_{11'}$ and ${}_{BB'} = {}_{00'}$ Equation (10.31) implies that $\nabla'_{11'}(e_{00'}(\Xi')) \simeq 0$ so that $e_{00'}(\Xi')$ is constant along the generators of $\mathscr{I}^+$. Moreover, setting ${}_{AA'} = {}_{00'}$ and ${}_{BB'} = {}_{01'}$ and using that $e_{01'}(\Xi') \simeq 0$ one finds that

$$\Gamma'_{01'}{}^{Q}{}_{0}e'_{Q0'}(\Xi') + \bar{\Gamma}'_{1'0}{}^{Q'}{}_{0'}e'_{0Q'}(\Xi') \simeq 0.$$

Expanding and using, again, that $e_{01'}(\Xi') \simeq 0$ and recalling (10.21b) one finds that $\Gamma'_{11'00}e'_{00'}(\Xi') \simeq 0$. However, $e'_{00'}(\Xi') \not\simeq 0$ so that one concludes that $\Gamma'_{11'00} \simeq 0$.

To conclude, it is observed that although Equation (10.28) fixed the derivative $e'_{11'}(\vartheta)$ along $\mathscr{I}^+$, the derivative $e'_{00'}(\vartheta)$ still remains free. A convenient way of fixing $e'_{00'}(\vartheta)$ can be obtained from the transformation law for the Ricci scalar – see Equation (5.6c) – which, in the present context, takes the form

$$R[g'] = \vartheta^{-2}R[g] + 12\vartheta^{-2}\nabla'_a\vartheta\nabla'^a\vartheta - 6\vartheta^{-1}\nabla'_a\nabla'^a\vartheta.$$

A natural requirement is to set $R[g'] = 0$ on $\mathscr{I}^+$ so that along the generators of $\mathscr{I}^+$ one obtains the equation

$$e'_{11'}(e'_{00'}(\vartheta)) - 2\vartheta^{-1}e'_{11'}(\vartheta)e'_{00'}(\vartheta) \simeq F', \tag{10.32}$$

where

$$F' \equiv \mathrm{Re}\bigg(e'_{01'}(e'_{10'}(\vartheta)) - 2\Gamma'_{01'01}e'_{10}(\vartheta) - 2\vartheta^{-1}e'_{01'}(\vartheta)e'_{10}(\vartheta) + \frac{1}{12}\vartheta^{-1}R[g]\bigg).$$

Equation (10.32) can be regarded as a linear differential equation for $e'_{00'}(\vartheta)$ along the generators of $\mathscr{I}^+$ with a non-homogeneous term F' which consists of quantities which are already known along $\mathscr{I}^+$. Equation (10.32) is supplemented by the condition $e'_{\mathbf{00'}}(\vartheta) = \varkappa^{-1}\vartheta\Gamma_{\mathbf{10'00}} \overset{\star}{\simeq} 0$ consistent with Equation (10.27). It follows that

$$R[\boldsymbol{g}'] \simeq 0. \tag{10.33}$$

Using the Ricci identity, Equation (8.35b), taking into account the conformal gauge condition (10.33) and the conditions on the spin connection coefficients, gives for the values $_{\boldsymbol{AA'}} = {}_{\mathbf{11'}}$, $_{\boldsymbol{BB'}} = {}_{\mathbf{10'}}$ and $_{\boldsymbol{CD}} = {}_{\mathbf{00}}$ a *homogeneous ordinary differential equation* for $\Gamma'_{\mathbf{10'00}}$ along the generators of $\mathscr{I}^+$. Observing the initial condition (10.27) the latter implies that $\Gamma'_{\mathbf{10'00}} \simeq 0$. Finally, a further use of the Ricci identities gives $\Phi'_{12} = \Phi_{21} \simeq 0$.

The construction of the previous paragraphs is rounded up with the introduction of adapted coordinates. On the fiduciary cut $\mathscr{C}_\star \approx \mathbb{S}^2$ one chooses some coordinates $\theta = (\theta^{\mathcal{A}})$ $\mathcal{A} = 2, 3$ and extends them along $\mathscr{I}^+$ by requiring them to be constant along the null generators. On the hypersurfaces $\mathcal{N}_{u'}$ transverse to $\mathscr{I}^+$ it is natural to identify an affine parameter r' of the null generators of these hypersurfaces in such a way that $e'_{\mathbf{00'}}(r') = 1$ and $r' \simeq 0$. The coordinates $\theta = (\theta^{\mathcal{A}})$ are propagated off $\mathscr{I}^+$ in such a way that they are constant along the generators of $\mathcal{N}_{u'}$. As a result of this construction one obtains ***Bondi coordinates*** $x = (u', r', \theta^{\mathcal{A}})$ in the neighbourhood $\mathcal{U}$ of $\mathscr{I}^+$.

Summary of the construction

The lengthy construction in this section can be summarised in the following proposition (for ease of presentation the $'$ in the objects associated to the *improved adapted frame* has been dropped from the expressions):

Proposition 10.1 (***the NP gauge at*** $\mathscr{I}^+$) *Let* $(\tilde{\mathcal{M}}, \tilde{\boldsymbol{g}})$ *denote an asymptotically simple spacetime. Locally, it is always possible to find a conformal extension* $(\mathcal{M}, \boldsymbol{g}, \Xi)$ *for which*

$$R[\boldsymbol{g}] \simeq 0$$

and an adapted frame $\{\boldsymbol{e}_{\boldsymbol{AA'}}\}$ *such that the associated spin connection coefficients* $\Gamma_{\boldsymbol{AA'BC}}$ *satisfy*

$$\Gamma_{\mathbf{00'}\boldsymbol{BC}} \simeq 0, \qquad \Gamma_{\mathbf{11'}\boldsymbol{BC}} \simeq 0,$$
$$\Gamma_{\mathbf{01'11}} \simeq 0, \qquad \Gamma_{\mathbf{10'00}} \simeq 0, \qquad \Gamma_{\mathbf{10'11}} \simeq 0$$
$$\bar{\Gamma}_{\mathbf{1'00'1'}} + \Gamma_{\mathbf{01'01}} \simeq 0.$$

In addition, one has that

$$\Phi_{12} \simeq 0, \qquad \Phi_{22} \simeq 0,$$

and $\boldsymbol{e}_{\mathbf{00'}}(\Xi)$ *is constant on* $\mathscr{I}^+$.

A quick inspection reveals that in the gauge associated to Proposition 10.1 the only non-zero spin connection coefficients on $\mathscr{I}^+$ are given by $\Gamma_{\mathbf{01'00}}$, $\Gamma_{\mathbf{00'01}}$ and $\Gamma_{\mathbf{10'01}}$ which in standard NP notation correspond, respectively, to σ, α, β. On $\mathscr{I}^+$ the connection coefficients α and β satisfy $\alpha + \bar{\beta} \simeq 0$ and describe, essentially, *the connection of the intrinsic metric of the cuts of* $\mathscr{I}^+$; that is, the connection of the standard metric of $\mathbb{S}^2$, $\boldsymbol{\sigma}$. The remaining spin connection coefficient, $\sigma = \Gamma_{\mathbf{01'00}}$, encodes the (non-trivial) dynamical degrees of freedom in the set up. Its relation with the notion of ***gravitational radiation*** will be briefly explored in the next subsection.

10.3.2 The radiation field and the news function

To explore the relation between the spin connection coefficient σ and the notion of gravitational radiation it is convenient to expand the Ricci, Cotton and Bianchi identities – that is, the conformal field Equations (8.35b), (8.37a) and (8.37b) – in terms of the gauge given by Proposition 10.1. An inspection of the components of the Ricci identity not used in the derivation of the NP gauge, taking into account that $\Psi_{ABCD} \simeq 0$, provides the relations

$$\Phi_{00} \simeq -\sigma\bar{\sigma}, \qquad \Phi_{01} \simeq -\bar{\eth}\sigma, \qquad \Phi_{02} \simeq -\dot{\sigma},$$

where $\dot{}$ denotes differentiation with respect to the retarded time u. In addition, one also finds

$$\Phi_{11} \simeq \delta\alpha - \bar{\delta}\beta + 4\alpha\beta.$$

As α and β describe the Levi-Civita connection of the standard metric of $\mathbb{S}^2$, it can be readily verified that Φ_{11} corresponds, essentially, to the curvature of $\mathbb{S}^2$ – recall that in two-dimensional manifolds the curvature is encoded in the Ricci scalar.

The relation between σ and the components of the rescaled Weyl tensor can be established by inspection of the Bianchi identity (8.37b) at $\mathscr{I}^+$. Choosing, for convenience Ξ so that $\boldsymbol{e}_{\mathbf{00'}}(\Xi) \simeq -1$, one finds that

$$\phi_4 \simeq -\ddot{\bar{\sigma}}, \qquad \phi_3 \simeq -\eth\dot{\bar{\sigma}}.$$

Moreover, one also obtains the *constraint*

$$\phi_2 + \sigma\dot{\bar{\sigma}} + \eth^2\bar{\sigma} \simeq \bar{\phi}_2 + \bar{\sigma}\dot{\sigma} + \bar{\eth}^2\sigma.$$

In view of the *Peeling theorem*, Theorem 10.4, the component ϕ_4 describes the leading term of the gravitational field – the so-called ***radiation field*** or ***outgoing field***. In particular, if $\dot{\bar{\sigma}}$ is constant along $\mathscr{I}^+$ one has that $\phi_4 \simeq 0$, $\phi_3 \simeq 0$, and one interprets this situation as describing the absence of gravitational radiation – that is why $\dot{\bar{\sigma}}$ is sometimes called the ***news function***. The component ϕ_2 is interpreted as describing the *Coulomb part* of the gravitational field while ϕ_1 and ϕ_0 are associated with ***incoming radiation***; see Szekeres (1965).

10.4 Other aspects of asymptotics

The present chapter provides a minimalistic account of the theory of asymptotics of the gravitational field. A detailed account would go beyond the scope of this book. It is, nevertheless, of interest to briefly highlight certain topics.

10.4.1 The Bondi mass

The analysis of the asymptotics of the gravitational field allows one to describe in a rigorous manner the loss of energy of an isolated system due to gravitational radiation. This physical process is described in terms of the so-called Bondi mass; see Trautman (1958), Bondi et al. (1962), Sachs (1962b) and also Penrose (1965). In terms of the notation introduced in this chapter, the ***Bondi mass*** $m_{\mathscr{B}}$ over a cut $\mathscr{C}$ of $\mathscr{I}^+$ is given by the surface integral

$$m_{\mathscr{B}} \equiv -\frac{1}{2}\int_{\mathscr{C}} (\phi_2 + \sigma\dot{\bar{\sigma}})\mathrm{d}S.$$

A concise deduction of the above expression can be found in Stewart (1991). Moreover, it can be shown that under suitable assumptions $m_{\mathscr{B}} \geq 0$; see Ludvigsen and Vickers (1981, 1982). A further calculation renders that

$$\dot{m}_{\mathscr{B}} = -\frac{1}{2}\int_{\mathscr{C}} |\dot{\sigma}|^2 \mathrm{d}S \leq 0.$$

The above inequality is called the ***Bondi mass-loss*** formula and encodes the loss of mass of an isolated system due to the energy that is carried away by (outgoing) gravitational radiation.

10.4.2 The Bondi-Metzner-Sachs group

As already mentioned, one of the central objectives of the theory of asymptotics of the gravitational field is to identify universal structures in a wide class of spacetimes and, in turn, use these to extract physical insight into the behaviour of isolated systems in general relativity. An example of this type of universal structures is given by the so-called ***Bondi-Metzner-Sachs (BMS) group***; see Sachs (1962a), Bondi et al. (1962)and Newman and Penrose (1966).

In what follows let $(u, r, \theta^{\mathcal{A}})$ denote a Bondi coordinate system defined in a neighbourhood of the future null infinity $\mathscr{I}^+$ of an asymptotically simple spacetime. The BMS group is defined by the following transformations on the u and $\theta = (\theta^{\mathcal{A}})$ coordinates:

$$u' = K(\theta)\big(u - \alpha(\theta)\big), \tag{10.34a}$$
$$\theta'^{\mathcal{A}} = \theta'^{\mathcal{A}}(\theta^2, \theta^3), \tag{10.34b}$$

where the map $(\theta^{\mathcal{A}}) \mapsto (\theta'^{\mathcal{A}})$ is a conformal transformation of $\mathbb{S}^2$ onto itself, and $K(\theta)$ is the associated conformal factor so that

$$\boldsymbol{\sigma}' = K^2\boldsymbol{\sigma},$$

and where $\alpha(\theta)$ is an arbitrary smooth real function on $\mathbb{S}^2$. The particular BMS transformations for which $\theta'^{\mathcal{A}} = \theta^{\mathcal{A}}$ are called ***supertranslations***. Under a supertranslation, the system of null hypersurfaces $\mathscr{N}_{u_\bullet}$ with $u_\bullet$ constant is transformed into a different system $\mathscr{N}_{u'_\bullet}$. Expanding the function $\alpha(\theta)$ in terms of spherical harmonics Y_{lm} – see the Appendix to this chapter – one finds that

$$\alpha(\theta) = \sum_{l=0}^{\infty} \sum_{m=-l}^{l} a_{lm} Y_{lm},$$

with $a_{lm} \in \mathbb{C}$. Thus, the supertranslations are an *infinite-dimensional subgroup* of the BMS group. The particular (four-dimensional) case for which $a_{lm} = 0$ for $l > 2$ is called the ***translations subgroup***.

Generic asymptotically simple spacetimes do not possess Killing vectors – in the conformal picture Killing vectors of the physical spacetime correspond to *conformal Killing vectors*. The BMS group arises from a notion of ***asymptotic symmetries*** which ensures the existence of non-trivial solutions for generic spacetimes, that is, a diffeomorphism $\varphi : \mathscr{I}^+ \to \mathscr{I}^+$ satisfying the conditions

$$\varphi^* \boldsymbol{q} = \vartheta^2 \boldsymbol{q}, \qquad \varphi_* \boldsymbol{N} = \vartheta^{-1} \boldsymbol{N}, \tag{10.35}$$

for some function $\vartheta > 0$ and where the tensor fields $\boldsymbol{q}$ and $\boldsymbol{N}$ are as given in Section 10.1.2. It can be verified that the BMS transformations (10.34a) and (10.34b) satisfy the conditions in (10.35) with $K = \vartheta$. A particular type of asymptotic symmetries corresponds to those generated by an ***asymptotic Killing vector***, that is, a field $\boldsymbol{\xi}$ on $\mathscr{I}^+$ satisfying the conditions

$$\pounds_{\boldsymbol{\xi}} \boldsymbol{q} = 2\vartheta \boldsymbol{q}, \qquad \pounds_{\boldsymbol{\xi}} \boldsymbol{N} = -\vartheta \boldsymbol{N}.$$

Given an asymptotically simple spacetime $(\tilde{\mathcal{M}}, \tilde{\boldsymbol{g}})$ endowed with a Killing vector $\tilde{\boldsymbol{\xi}}$, let $(\mathcal{M}, \boldsymbol{g}, \Xi)$ denote a conformal extension thereof. Given that $0 = \pounds_{\tilde{\boldsymbol{\xi}}} \tilde{\boldsymbol{g}} = \pounds_{\tilde{\boldsymbol{\xi}}} (\Xi^{-2} \boldsymbol{g})$, it follows that

$$\pounds_{\tilde{\boldsymbol{\xi}}} \boldsymbol{g} = 2(\Xi^{-1} \tilde{\boldsymbol{\xi}}(\Xi))\, \boldsymbol{g}, \tag{10.36}$$

for $\Xi \neq 0$, so that $\tilde{\boldsymbol{\xi}}$ is a conformal Killing vector of $\boldsymbol{g}$ on $\tilde{\mathcal{M}}$. Since this vector is determined by the smooth metric $\boldsymbol{g}$, it extends smoothly to $\mathscr{I}^+$ as a vector $\boldsymbol{\xi}$. Now, the left-hand side of Equation (10.36) extends smoothly to $\mathscr{I}^+$, and, therefore, the right-hand side does so too. It follows that

$$\boldsymbol{\xi}(\Xi) = \alpha' \Xi, \tag{10.37}$$

with α' a smooth function such that $\alpha = O(\Xi^0)$ so that $\boldsymbol{\xi}$ is tangent to $\mathscr{I}^+$. From Equation (10.36) one concludes that

$$\pounds_{\boldsymbol{\xi}} \boldsymbol{q} = 2\alpha' \boldsymbol{q}, \qquad \pounds_{\boldsymbol{\xi}} \boldsymbol{N} = -\alpha' \boldsymbol{N}.$$

Accordingly, *any Killing vector of $(\tilde{\mathcal{M}}, \tilde{\boldsymbol{g}})$ admits a unique extension to a vector on $\mathscr{I}^+$ and which defines an asymptotic Killing vector.* The maximum number of linearly independent Killing vectors in a four-dimensional manifold is 10. Accordingly, Killing vectors can give rise, at most, to 10 asymptotic Killing vectors. By means of a direct calculation, it is possible to show that the function α' in Equation (10.37) and the function α appearing in (10.34a) are the same. Thus, the BMS transformations (10.34a) and (10.34b) are asymptotic symmetries. In particular, the translations subgroup can be put in correspondence with the asymptotic Killing vectors arising from translations in the Minkowski spacetime.

For further details on the structure and properties of the BMS group, see, for example, Penrose and Rindler (1986) and Schmidt et al. (1975). A discussion of the properties of Killing vectors in asymptotically simple spacetimes can be found in Ashtekar and Xanthopoulos (1978) and Ashtekar and Schmidt (1980).

10.4.3 Newman-Penrose constants

In Newman and Penrose (1965) – see also Newman and Penrose (1968) and Penrose and Rindler (1986) – it has been shown that in an asymptotically simple spacetime $(\tilde{\mathcal{M}}, \tilde{\boldsymbol{g}})$ there exists a set of ten quantities defined as integrals over cuts of null infinity which are *absolutely conserved* in the sense that their value is independent of the particular cut $\mathscr{C}$ on which they are evaluated – the so-called ***Newman-Penrose constants***. In terms of the adapted frame $\{\boldsymbol{e}_{\boldsymbol{AA'}}\}$ of Proposition 10.1 these constants are given by

$$G_m \equiv \int_{\mathscr{C}} {}_2\bar{Y}_{2m}\boldsymbol{e}_{\boldsymbol{00}}(\phi_0)\mathrm{d}S, \qquad m = -2, \ldots, 2,$$

where ${}_2Y_{2m}$ is a ***spin-weighted spherical harmonic***; see the Appendix to this chapter. A discussion of the relation between the above expression and the original formula of Newman and Penrose can be found in Friedrich and Kánnár (2000a).

There exists no general consensus about the physical meaning or interpretation of the Newman-Penrose constants. An explicit computation for stationary spacetimes shows that they are of the form

$$(\text{mass}) \times (\text{quadrupole}) - (\text{dipole})^2;$$

see, for example, Bäckdahl (2009). Evaluations of the Newman-Penrose spacetimes for dynamic spacetimes can be found in Friedrich and Schmidt (1987) and Friedrich and Kánnár (2000a). In particular, in the former reference it is shown that for spacetimes possessing a conformal extension which includes the points i^+ and i^- the Newman-Penrose constants correspond, essentially, to the value of the rescaled Weyl spinor $\phi_{\boldsymbol{ABCD}}$ at those points. Electrovacuum asymptotically simple spacetimes have a suitable generalisation of these absolutely conserved constants; see Exton et al. (1969).

10.5 Further reading

An excellent introduction to the theory of asymptotics of the gravitational field is given in Stewart (1991) where the subject is called "asymptopia". A related account can be found in Penrose and Rindler (1986). A convenient entry point to the extensive literature on the subject can be found in the review of Frauendiener (2004). A detailed discussion of the ideas and general philosophy behind the treatment of the asymptotics of the gravitational field by means of conformal methods can be found in Geroch (1976). Accounts similar in spirit to the latter can be found in Ashtekar (1980, 1987). A slightly different perspective on the subject can be found in Friedrich (1992); see also Friedrich (1998a, 1999). A recent review on the subject of asymptotics is given in Ashtekar (2014).

Appendix: spin-weighted functions

Let $\{\boldsymbol{o}, \boldsymbol{\iota}\}$ denote a spinorial dyad defined on a spacetime $(\mathcal{M}, \boldsymbol{g})$ and let $\{\boldsymbol{l}, \boldsymbol{n}, \boldsymbol{m}, \bar{\boldsymbol{m}}\}$ denote the associated null tetrad. As discussed in Section 3.1.10, the null vectors $\boldsymbol{m}$ and $\bar{\boldsymbol{m}}$ span a spacelike subspace of $T(\mathcal{M})$ which is orthogonal to both $\boldsymbol{l}$ and $\boldsymbol{n}$. Of particular interest is the case when this subspace corresponds to the tangent bundle of a compact two-dimensional submanifold $\mathscr{C}$ of $\mathcal{M}$. *In the following it is assumed that this is the case.* From the expression

$$\boldsymbol{g} = \boldsymbol{l} \otimes \boldsymbol{n} + \boldsymbol{n} \otimes \boldsymbol{l} - \boldsymbol{m} \otimes \bar{\boldsymbol{m}} - \bar{\boldsymbol{m}} \otimes \boldsymbol{m}$$

of the metric $\boldsymbol{g}$ in terms of the null tetrad, it follows that the intrinsic metric $\boldsymbol{\sigma}$ induced by $\boldsymbol{g}$ on $\mathscr{C}$ is given by

$$\boldsymbol{\sigma} = -\boldsymbol{m} \otimes \bar{\boldsymbol{m}} - \bar{\boldsymbol{m}} \otimes \boldsymbol{m}.$$

There is a certain *gauge freedom* in the above expression since ***spin-boosts*** of the form

$$\boldsymbol{o} \mapsto e^{\frac{1}{2}\mathrm{i}c}\boldsymbol{o}, \qquad \boldsymbol{\iota} \mapsto e^{-\frac{1}{2}\mathrm{i}c}\boldsymbol{\iota}, \tag{10.38}$$

with arbitrary $c \in \mathbb{R}$ which imply the transition

$$\boldsymbol{m} \mapsto e^{\mathrm{i}c}\boldsymbol{m}, \qquad \bar{\boldsymbol{m}} \mapsto e^{-\mathrm{i}c}\bar{\boldsymbol{m}},$$

leave the metric $\boldsymbol{\sigma}$ unchanged.

Given a spinor $\eta_{A_1\cdots A_n A'_1\cdots A'_m}$ of valence $n+m$, it is natural to consider the behaviour of its components with respect to the dyad $\{\boldsymbol{o}, \boldsymbol{\iota}\}$ under the spin boost (10.38). For example, given p, q, r, $t \in \mathbb{N}$ such that $p+q=n$, $r+t=m$, the scalar

$$\eta \equiv \eta_{A_1\cdots A_p B_1\cdots B_q A'_1\cdots A'_r B'_1\cdots B'_t} o^{A_1}\cdots o^{A_p}\iota^{B_1}\cdots\iota^{B_q}\bar{o}^{A'_1}\cdots\bar{o}^{A'_r}\bar{\iota}^{B'_1}\cdots\bar{\iota}^{B'_t} \tag{10.39}$$

has a transformation given by

$$\eta \mapsto e^{\frac{1}{2}\mathrm{i}(p+t-q-r)\vartheta}\eta.$$

One says, then, that η has ***spin weight*** $s = \frac{1}{2}(p+t-q-r)$. The spin weight of all the possible components of $\eta_{A_1\cdots A_n A'_1\cdots A'_m}$ lies in the range $-m-n \leq s \leq m+n$.

In what follows, we adopt the standard Newman-Penrose conventions to denote the directional covariant derivatives with respect to $\boldsymbol{m}$ and $\bar{\boldsymbol{m}}$ and let $\delta \equiv m^a\nabla_a$, $\bar{\delta} \equiv \bar{m}^a\nabla_a$. Generically, the directional derivatives δ and $\bar{\delta}$ acting on a spin-weighted scalar do not give rise to scalars with a well-defined spin weight. To amend this deficiency it is convenient to define operators $\eth$ and $\bar{\eth}$ which, acting on scalars with a given spin weight, give rise to new scalars with a well-defined spin weight. Given the spin-weighted scalar η of Equation (10.39), the action of $\eth$ and $\bar{\eth}$ is defined to be

$$\begin{aligned}\eth\eta \equiv\, & o^{A_1}\cdots o^{A_p}\iota^{B_1}\cdots\iota^{B_q}\bar{o}^{A'_1}\cdots\bar{o}^{A'_r}\bar{\iota}^{B'_1}\cdots\bar{\iota}^{B'_t}\\ &\times\delta(\eta\,\iota_{A_1}\cdots\iota_{A_p}o_{B_1}\cdots o_{B_q}\bar{\iota}_{A'_1}\cdots\bar{\iota}_{A'_r}\bar{o}_{B'_1}\cdots\bar{o}_{B'_t}), \qquad (10.40a)\\ \bar{\eth}\eta \equiv\, & o^{A_1}\cdots o^{A_p}\iota^{B_1}\cdots\iota^{B_q}\bar{o}^{A'_1}\cdots\bar{o}^{A'_r}\bar{\iota}^{B'_1}\cdots\bar{\iota}^{B'_t}\\ &\times\bar{\delta}(\eta\,\iota_{A_1}\cdots\iota_{A_p}o_{B_1}\cdots o_{B_q}\bar{\iota}_{A'_1}\cdots\bar{\iota}_{A'_r}\bar{o}_{B'_1}\cdots\bar{o}_{B'_t}). \qquad (10.40b)\end{aligned}$$

The operators $\eth$ and $\bar{\eth}$ are complex conjugates of each other in the sense that $\overline{\eth\eta} = \bar{\eth}\bar{\eta}$. If the scalar η has spin weight s, one can verify that $\eth\eta$ and $\bar{\eth}\eta$ have, respectively, spin weights $s+1$ and $s-1$. Furthermore, $\eth\eta$ and $\bar{\eth}\eta$ satisfy the Leibnitz rule. In order to obtain alternative expressions for $\eth\eta$ and $\bar{\eth}\eta$ let

$$\alpha \equiv o^A\bar{\delta}\iota_A = \iota^A\bar{\delta}o_A, \qquad \beta \equiv o^A\delta\iota_A = \iota^A\delta o_A,$$

consistent with standard Newman-Penrose notation. Expanding (10.40a) and (10.40b) and using the above definitions one obtains

$$\begin{aligned}\eth\eta &= (-1)^{p+r}\left(\delta\eta + ((q-p)\beta + (t-r)\bar{\alpha})\,\eta\right),\\ \bar{\eth}\eta &= (-1)^{p+r}\left(\bar{\delta}\eta + ((q-p)\bar{\beta} + (t-r)\alpha)\,\eta\right).\end{aligned}$$

A computation with the above expressions shows that

$$(\bar{\eth}\eth - \eth\bar{\eth})\eta = s\eta.$$

The above expressions are convenient for the discussion of ***spin-weighted harmonics***. In terms of standard spherical harmonics Y_{lm}, these are given by

$${}_0Y_{lm} \equiv Y_{lm},$$

and for $s \neq 0$

$${}_sY_{lm} \equiv \begin{cases} (-1)^s\sqrt{\dfrac{2^s(l-s)!}{(l+s)!}}\,\eth^s Y_{lm} & 0 < s \leq l\\ \sqrt{\dfrac{(l+s)!}{2^s(l-s)!}}\,\bar{\eth}^{-s}Y_{lm} & -l \leq s < 0\\ 0 & \text{otherwise;}\end{cases}$$

see, for example, Stewart (1991) for further discussion.

Of special relevance for Theorem 10.3 is the following result:

Lemma 10.1 *Assume $\mathscr{C}$ to be diffeomorphic to $\mathbb{S}^2$ and let η denote a smooth scalar on $\mathscr{C}$ having spin weight s. If $\eth\eta = 0$ and $s < 0$, then $\eta = 0$. Similarly, if $\bar{\eth}\eta = 0$ and $s > 0$, then $\eta = 0$.*

Proofs of this result can be found in Penrose and Rindler (1984) and Stewart (1991). Remarkably, *the result depends on the topology (genus) of $\mathscr{C}$*; see Frauendiener and Szabados (2001). For example, the above result is not valid for surfaces diffeomorphic to the 2-torus $\mathbb{S} \times \mathbb{S}$. It is of interest to point out that Lemma 10.1 is equivalent to the statement that there exist no non-zero symmetric trace-free, divergence-free, rank 2 tensor fields on $\mathbb{S}^2$; see Beig (1985) and Frauendiener and Szabados (2001).

Part III

Methods of the theory of partial differential equations

11
The conformal constraint equations

This chapter analyses the intrinsic equations implied by the conformal Einstein field equations on non-null hypersurfaces. These equations are known as the ***conformal constraint equations***. They play an essential role in the construction of initial data sets for the conformal field equations and in the identification of boundary conditions. Not surprisingly, these conformal constraint equations are closely related to the standard ***Einstein constraint equations*** – consequently, this chapter starts by considering the properties of the latter.

The solvability and behaviour of solutions to the conformal constraint equations is closely related to the nature of the underlying three-dimensional manifold on which the equations are imposed. As a consequence, this chapter also provides a discussion of general properties of asymptotically Euclidean and asymptotically hyperboloidal 3-manifolds from a conformal point of view. The systematic analysis of the constraint equations relies on methods of elliptic partial differential equations. Hence, this chapter provides a discussion of some of the basic notions of this theory.

An important aspect of the conformal constraint equations – the so-called *propagation of the constraints* – is discussed in Chapter 13. The analysis of the constraint equations on null hypersurfaces is treated in Chapter 18.

11.1 General setting and basic formulae

Let $(\tilde{\mathcal{M}}, \tilde{\boldsymbol{g}})$ denote a spacetime satisfying the Einstein field equations. In what follows, it will be assumed that $(\tilde{\mathcal{M}}, \tilde{\boldsymbol{g}})$ can be conformally extended to an unphysical spacetime $(\mathcal{M}, \boldsymbol{g})$. Accordingly, there exists an embedding $\phi : \tilde{\mathcal{M}} \rightarrow \mathcal{M}$ and a conformal factor Ξ such that $\phi^*\boldsymbol{g} = \Xi^2\tilde{\boldsymbol{g}}$. Now, let $\tilde{\mathcal{S}}$ denote a three-dimensional submanifold of $\tilde{\mathcal{M}}$ and let $\varphi : \tilde{\mathcal{S}} \rightarrow \tilde{\mathcal{M}}$ denote the associated embedding. As the composition $\phi \circ \varphi : \tilde{\mathcal{S}} \rightarrow \mathcal{M}$ is also an embedding, the three-dimensional manifold $\tilde{\mathcal{S}}$ can be regarded, in turn, as a submanifold of $\mathcal{M}$. As discussed in Section 2.7.3, the spacetime metric $\tilde{\boldsymbol{g}}$ induces a metric $\tilde{\boldsymbol{h}}$ on $\tilde{\mathcal{S}}$ via

$\tilde{\boldsymbol{h}} = \varphi^*\tilde{\boldsymbol{g}}$. Similarly, regarding $\tilde{\mathcal{S}}$ as a hypersurface on $\mathcal{M}$, the unphysical metric $\boldsymbol{g}$ also induces a metric $\boldsymbol{h}$ via the pull-back $\boldsymbol{h} = (\phi \circ \varphi)^*\boldsymbol{g}$. A calculation shows that

$$\boldsymbol{h} = (\phi \circ \varphi)^*\boldsymbol{g} = (\varphi^* \circ \phi^*)\boldsymbol{g} = \varphi^*(\Xi^2|_{\tilde{\mathcal{S}}}\tilde{\boldsymbol{g}}) = \Omega^2\varphi^*\tilde{\boldsymbol{g}}$$

where $\Omega \equiv \Xi^2|_{\tilde{\mathcal{S}}}$ is the restriction of Ξ to the hypersurface $\tilde{\mathcal{S}}$. Following the conventions of previous chapters, $\boldsymbol{h} = \Omega^2\varphi^*\tilde{\boldsymbol{g}}$ will often be written as

$$\boldsymbol{h} = \Omega^2\tilde{\boldsymbol{g}}.$$

Now, let $\tilde{\boldsymbol{\nu}}$ and $\boldsymbol{\nu}$ denote, respectively, the $\tilde{\boldsymbol{g}}$-unit and $\boldsymbol{g}$-unit normals of $\tilde{\mathcal{S}}$ and define

$$\epsilon \equiv \tilde{\boldsymbol{g}}(\tilde{\boldsymbol{\nu}}, \tilde{\boldsymbol{\nu}}) = \boldsymbol{g}(\boldsymbol{\nu}, \boldsymbol{\nu}).$$

In accordance with the signature convention $(+ - --)$, the hypersurface $\tilde{\mathcal{S}}$ is spacelike if $\epsilon = 1$ and timelike if $\epsilon = -1$. It follows that

$$\boldsymbol{\nu} = \Xi\tilde{\boldsymbol{\nu}}, \qquad \boldsymbol{\nu}^\sharp = \Xi^{-1}\tilde{\boldsymbol{\nu}}^\sharp$$

or, using index notation, $\nu_a = \Xi\tilde{\nu}_a$ and $\nu^a = \Xi^{-1}\tilde{\nu}^a$. In what follows, the indices of objects in $\tilde{\mathcal{M}}$ are raised/lowered using the metric $\tilde{\boldsymbol{g}}$, while the indices of objects on $\mathcal{M}$ are moved using $\boldsymbol{g}$.

11.1.1 The transformation formulae for the extrinsic curvature

Having discussed the relation between the 3-metrics and the unit normals to $\tilde{\mathcal{S}}$, one is in the position to consider the relation between the extrinsic curvatures $\tilde{\boldsymbol{K}}$ and $\boldsymbol{K}$. Given *spatial vectors* $\boldsymbol{u}$, $\boldsymbol{v} \in T(\tilde{\mathcal{S}})$ – so that $\langle\tilde{\boldsymbol{\nu}}, \boldsymbol{u}\rangle = \langle\tilde{\boldsymbol{\nu}}, \boldsymbol{v}\rangle = 0$ – one has that

$$\tilde{\boldsymbol{K}}(\boldsymbol{u}, \boldsymbol{v}) = \langle\tilde{\boldsymbol{\nabla}}_{\boldsymbol{u}}\tilde{\boldsymbol{\nu}}, \boldsymbol{v}\rangle, \qquad \boldsymbol{K}(\boldsymbol{u}, \boldsymbol{v}) = \langle\boldsymbol{\nabla}_{\boldsymbol{u}}\boldsymbol{\nu}, \boldsymbol{v}\rangle;$$

see Equation (2.43). Recalling that $\boldsymbol{\nabla} - \tilde{\boldsymbol{\nabla}} = \boldsymbol{S}(\boldsymbol{\Upsilon})$ one readily has that

$$\boldsymbol{\nabla}_{\boldsymbol{u}}\boldsymbol{\nu} = \tilde{\boldsymbol{\nabla}}_{\boldsymbol{u}}\boldsymbol{\nu} - \boldsymbol{S}(\boldsymbol{\Upsilon}, \boldsymbol{\nu}; \boldsymbol{u});$$

the minus sign arises from the fact that $\boldsymbol{\nu}$ is a covector. In abstract index notation $\boldsymbol{S}(\boldsymbol{\Upsilon}, \boldsymbol{\nu}; \boldsymbol{u})$ is given by $S_{ab}{}^{cd}\Upsilon_c\nu_d u^b$ from where a short calculation gives that

$$\begin{aligned} S_{ab}{}^{cd}\Upsilon_c\nu_d u^b &= S_{ab}{}^{cd}\tilde{\nabla}_c\Xi\tilde{\nu}_d u^b, \\ &= \tilde{\nabla}_b\Xi\tilde{\nu}_a u^b - \tilde{g}_{ab}u^b\tilde{g}^{cd}\tilde{\nabla}_c\Xi\tilde{\nu}_d \\ &= (u^c\tilde{\nabla}_c\Xi)\tilde{\nu}_a - \Xi\Sigma\tilde{g}_{ac}u^c, \end{aligned}$$

where

$$\Sigma \equiv g^{ab}\nabla_a\Xi\nu_b = \boldsymbol{g}^\sharp(\mathbf{d}\Xi, \boldsymbol{\nu}) = \Xi^{-1}\tilde{\boldsymbol{g}}^\sharp(\mathbf{d}\Xi, \tilde{\boldsymbol{\nu}})$$

is the derivative of Ξ in the direction of the $\boldsymbol{g}$-unit normal to $\tilde{\mathcal{S}}$. Accordingly, one has that

$$\boldsymbol{S}(\boldsymbol{\Upsilon},\boldsymbol{\nu};\boldsymbol{u}) = \boldsymbol{u}(\Xi)\tilde{\boldsymbol{\nu}} - \Xi\Sigma\tilde{\boldsymbol{g}}(\boldsymbol{u},\cdot),$$

from where, recalling that $\boldsymbol{u}$, $\boldsymbol{v} \in T(\tilde{\mathcal{S}})$ so that $\tilde{\boldsymbol{g}}(\boldsymbol{u},\boldsymbol{v}) = \tilde{\boldsymbol{h}}(\boldsymbol{u},\boldsymbol{v})$, it follows that

$$\begin{aligned}\boldsymbol{K}(\boldsymbol{u},\boldsymbol{v}) &= \langle \tilde{\boldsymbol{\nabla}}_{\boldsymbol{u}}\boldsymbol{\nu},\boldsymbol{v}\rangle - \boldsymbol{u}(\Xi)\langle\tilde{\boldsymbol{\nu}},\boldsymbol{v}\rangle + \Omega\Sigma\langle\tilde{\boldsymbol{g}}(\boldsymbol{u},\cdot),\boldsymbol{v}\rangle \\ &= \Omega\langle\tilde{\boldsymbol{\nabla}}_{\boldsymbol{u}}\tilde{\boldsymbol{\nu}},\boldsymbol{v}\rangle + \Omega\Sigma\tilde{\boldsymbol{h}}(\boldsymbol{u},\boldsymbol{v}) \\ &= \Omega\left(\tilde{\boldsymbol{K}}(\boldsymbol{u},\boldsymbol{v}) + \Sigma\tilde{\boldsymbol{h}}(\boldsymbol{u},\boldsymbol{v})\right),\end{aligned}$$

where to pass from the first to the second line it has been used that $\langle\tilde{\boldsymbol{\nu}},\boldsymbol{v}\rangle = 0$ as $T(\tilde{\mathcal{S}})$.

Summarising, the calculations in the previous paragraphs show that

$$h_{ij} = \Omega^2\tilde{h}_{ij}, \tag{11.1a}$$

$$K_{ij} = \Omega\big(\tilde{K}_{ij} + \Sigma\tilde{h}_{ij}\big). \tag{11.1b}$$

These are the basic transformation formulae for the remainder of this chapter. Taking the trace of the transformation formula for the extrinsic curvature, Equation (11.1b), it follows that

$$\Omega K = \tilde{K} + 3\Sigma,$$

where $\tilde{K} \equiv \tilde{h}^{ij}\tilde{K}_{ij}$ and $K \equiv h^{ij}K_{ij}$ – these scalars are sometimes called, respectively, the physical and unphysical ***mean curvature*** of $\tilde{\mathcal{S}}$. The scalars Σ, K admit a geometric interpretation: if $\Sigma = K = 0$, then, necessarily, $\tilde{K} = 0$ and the hypersurface $\tilde{\mathcal{S}}$ is ***maximal*** in $\tilde{\mathcal{M}}$ with respect to both the metrics $\tilde{\boldsymbol{g}}$ and $\boldsymbol{g}$ – that is, it encloses a maximum volume for a given area.

11.1.2 Decompositions in electric and magnetic parts

A key ingredient in the analysis of the conformal constraint equations is the ***decomposition in electric and magnetic parts*** of tensors with antisymmetric pairs of indices. Let $\mathcal{S}$ denote a hypersurface on a spacetime $(\mathcal{M},\boldsymbol{g})$, and let $\boldsymbol{\nu}$ denote the unit normal to the hypersurface. The ***projector*** to $\mathcal{S}$ is the tensor $h_a{}^b$ given by

$$h_a{}^b \equiv \delta_a{}^b - \epsilon\nu_a\nu^b.$$

It follows that

$$h_a{}^b\nu_b = 0, \qquad h_a{}^b h_b{}^c = h_a{}^c.$$

Furthermore, using the properties of the spacetime volume form ϵ_{abcd} – see Section 2.5.3 – one can deduce that

$$h_a{}^{[c}h_b{}^{d]} = -\frac{1}{2}\epsilon_{abe}\epsilon^{cde}, \tag{11.2}$$

where $\epsilon_{abe} \equiv \epsilon_{fabe}\nu^f$ is the ***three-dimensional volume form***.

Now, let F_{ab} denote an antisymmetric tensor of rank 2 and let $F^*_{ab} \equiv -\frac{1}{2}\epsilon_{ab}{}^{cd}F_{cd}$ denote its Hodge dual. Its ***electric*** and ***magnetic parts*** are defined, respectively, to be

$$F_a \equiv F_{cb}\nu^b h_a{}^c, \qquad F^*_a \equiv F^*_{cb}\nu^b h_a{}^c.$$

It can be verified that

$$F_a\nu^a = F^*_a\nu^a = 0, \qquad h_a{}^b F_b = F_a, \qquad h_a{}^b F^*_b = F^*_a,$$

so that the electric and magnetic parts are said to be ***spatial*** tensors. Together, F_a and F^*_a encode the same information as the original tensor F_{ab}. In order to see this, one writes

$$\begin{aligned} F_{ab} &= F_{cd}\delta_a{}^c\delta_b{}^d = F_{cd}(h_a{}^c + \epsilon\nu_a\nu^c)(h_b{}^d + \epsilon\nu_b\nu^d) \\ &= F_{cd}h_a{}^c h_b{}^d + \epsilon F_{cd}h_a{}^c\nu_b\nu^d + \epsilon F_{cd}h_b{}^d\nu_a\nu^c \\ &= 2\epsilon F_{[a}\nu_{b]} + F_{cd}h_a{}^c h_b{}^d. \end{aligned} \tag{11.3}$$

The term $F_{cd}h_a{}^c h_b{}^d$ is, in turn, manipulated using the identity (11.2) as follows:

$$\begin{aligned} F_{cd}h_a{}^c h_b{}^d = F_{cd}h_a{}^{[c}h_b{}^{d]} &= -\frac{1}{2}F_{cd}\epsilon^{fcde}\nu_f\epsilon_{abe} \\ &= F^*_{ef}\nu^f\epsilon_{ab}{}^e = F^*_e\epsilon_{ab}{}^e. \end{aligned} \tag{11.4}$$

Thus, combining Equations (11.3) and (11.4), one concludes that

$$F_{ab} = 2\epsilon F_{[a}\nu_{b]} + F^*_e\epsilon^e{}_{ab}.$$

The decomposition in electric and magnetic parts can be extended to tensors W_{abcd} with the same symmetries as the Weyl tensor; such tensors are sometimes known as ***Weyl candidates***. By analogy to the rank-2 case one defines the $\boldsymbol{\nu}$-electric and $\boldsymbol{\nu}$-magnetic parts of W_{abcd} to be

$$W_{ac} \equiv W_{ebfd}\nu^b\nu^d h_a{}^e h_c{}^f, \qquad W^*_{ac} \equiv W^*_{ebfd}\nu^b\nu^d h_a{}^e h_c{}^f,$$

with $W^*_{abcd} \equiv -\frac{1}{2}\epsilon_{cd}{}^{ef}W_{abef}$ denoting the *right Hodge dual* of W_{abcd}. In the subsequent discussion it is convenient to consider

$$W_{abc} \equiv W_{efgh}\nu^f h_a{}^e h_b{}^g h_c{}^h.$$

It can be verified that

$$W^*_{ab} = -\frac{1}{2}W_{acd}\epsilon_b{}^{cd}.$$

As in the rank-2 case, the tensors W_{ab} and W^*_{ab} (or, alternatively, W_{ab} and W_{abc}) encode the same information as W_{abcd}. The argument to show this equivalence is similar to that of the rank-2 case:

$$\begin{aligned}
W_{abcd} &= W_{efgh}\delta_a{}^e\delta_b{}^f\delta_c{}^g\delta_d{}^h \\
&= W_{efgh}(h_a{}^e + \epsilon\nu_a\nu^e)(h_b{}^f + \epsilon\nu_b\nu^f)(h_c{}^g + \epsilon\nu_c\nu^g)(h_d{}^h + \epsilon\nu_d\nu^h) \\
&= W_{efgh}h_a{}^e h_b{}^f h_c{}^g h_d{}^h + \epsilon W_{cab}\nu_d - \epsilon W_{deb}h_a{}^e\nu_c + \epsilon W_{acd}\nu_b \\
&\quad + W_{ac}\nu_b\nu_d - W_{ad}\nu_b\nu_c - \epsilon W_{bcd}\nu_a - W_{bc}\nu_a\nu_d + W_{bd}\nu_a\nu_c. \qquad (11.5)
\end{aligned}$$

From the definition of the magnetic part W^*_{ab} it follows that

$$W_{abc} = \epsilon^e{}_{bc}W^*_{ae}. \qquad (11.6)$$

Moreover, using that

$$\epsilon_{abc}\epsilon^{def} = -6\delta_a{}^{[d}\delta_b{}^e\delta_c{}^{f]}, \qquad (11.7)$$

it follows that

$$\begin{aligned}
W_{efgh}h_a{}^e h_b{}^f h_c{}^g h_d{}^h &= \frac{1}{4}W_{efgh}\epsilon^{efz}\epsilon_{abz}\epsilon^{ghx}\epsilon_{cdx} \\
&= {}^*W^*_{rzsx}\nu^r\nu^s\epsilon_{ab}{}^z\epsilon_{cd}{}^x = -W_{zx}\epsilon_{ab}{}^z\epsilon_{cd}{}^x \\
&= W_{ca}h_{bd} + W_{db}h_{ac} - W_{cb}h_{ad} - W_{da}h_{bc}. \qquad (11.8)
\end{aligned}$$

Combining Equations (11.5), (11.6) and (11.8) one obtains the desired decomposition of W_{abcd} in terms of W_{ab} and W^*_{ab}:

$$W_{abcd} = 2\epsilon(l_{b[c}W_{d]a} - l_{a[c}W_{d]b}) - 2(\nu_{[c}W^*_{d]e}\epsilon^e{}_{ab} + \nu_{[a}W^*_{b]e}\epsilon^e{}_{cd}), \qquad (11.9)$$

where $l_{ab} \equiv h_{ab} - \epsilon\nu_a\nu_b$. A similar computation renders

$$W^*_{abcd} = 2\nu_{[a}W_{b]e}\epsilon^e{}_{cd} - 4W_{e[a}\epsilon_{b]}{}^e{}_{[c}\nu_{d]} - 4\nu_{[a}W^*_{b][c}\nu_{d]} - W^*_{ef}\epsilon^e{}_{ab}\epsilon^f{}_{cd}. \qquad (11.10)$$

Expressions in terms of an adapted frame

The decomposition discussed in the previous paragraphs acquires a particularly simple form when supplemented with a frame $\{\boldsymbol{e_a}\}$ adapted to the hypersurface $\mathcal{S}$. For such a frame, the projection of a particular index with respect to the normal corresponds to replacement of the corresponding frame index with $_\perp$ while the spatial part of a tensor is given by the replacement of the spacetime frame indices $_{\boldsymbol{a}}, _{\boldsymbol{b}}, _{\boldsymbol{c}}, \ldots$ with the spatial frame indices $_{\boldsymbol{i}}, _{\boldsymbol{j}}, _{\boldsymbol{k}}, \ldots$ In particular, the three-dimensional volume form satisfies $\epsilon_{\boldsymbol{ijk}} = \epsilon_{\perp\boldsymbol{ijk}}$, and the electric and magnetic parts of the antisymmetric tensor F_{ab} are represented, respectively, by

$$F_{\boldsymbol{i}} = F_{\boldsymbol{i}\perp}, \qquad F^*_{\boldsymbol{i}} = F^*_{\boldsymbol{i}\perp}.$$

In the case of the Weyl candidate W_{abcd} one has that the tensors W_{ab}, W^*_{ab} and W_{abc} correspond to

$$W_{ij} = W_{i\perp j\perp}, \qquad W^*_{ij} = W^*_{i\perp j\perp}, \qquad W_{ijk} = W_{i\perp jk}.$$

11.2 Basic notions of elliptic equations

Elliptic differential operators arise naturally in the study of the constraint equations of general relativity on spacelike hypersurfaces. In view of this, some basic properties of elliptic operators on Riemannian manifolds are briefly discussed.

Let $(\mathcal{S}, \boldsymbol{h})$ denote a Riemannian three-dimensional manifold with $\boldsymbol{h}$ a negative definite metric. A ***linear differential operator*** of order M over $\mathcal{S}$ is a map between tensor bundles

$$\mathbf{L} : \mathfrak{T}_{i_1\cdots i_S}(\mathcal{S}) \to \mathfrak{T}_{k_1\cdots k_N}(\mathcal{S}), \qquad S, N \in \mathbb{N},$$

of the form

$$(\mathbf{L}v)_{k_1\cdots k_N} \equiv \sum_{r=0}^{M} a^{j_1\cdots j_r i_1\cdots i_S}{}_{k_1\cdots k_N} D_{j_1}\cdots D_{j_r} v_{i_1\cdots i_S}, \tag{11.11}$$

for a smooth $v_{i_1\cdots i_S} \in \mathfrak{T}_{i_1\cdots i_S}(\mathcal{S})$ and where the coefficients $a^{j_1\cdots j_r i_1\cdots i_S}{}_{k_1\cdots k_N}$ are smooth functions over $\mathcal{S}$. The ***principal part*** of $\mathbf{L}$ consists of the terms in Equation (11.11) with the highest order derivatives, that is,

$$a^{j_1\cdots j_M i_1\cdots i_S}{}_{k_1\cdots k_N} D_{j_1}\cdots D_{j_M} v_{i_1\cdots i_S}.$$

Closely related to the principal part is the ***symbol of*** $\mathbf{L}$, $\boldsymbol{\sigma}_{\mathbf{L}}(\boldsymbol{\xi})$, defined pointwise on $\mathcal{S}$, for $\boldsymbol{\xi} \in T^*|_p(\mathcal{S})$ as the linear map

$$\boldsymbol{\sigma}_{\mathbf{L}}(\boldsymbol{\xi}) : T_{i_1\cdots i_S}|_p(\mathcal{S}) \to T_{k_1\cdots k_N}|_p(\mathcal{S}),$$

given by

$$(\boldsymbol{\sigma}_{\mathbf{L}}(\boldsymbol{\xi})v)_{k_1\cdots k_N} \equiv a^{j_1\cdots j_M i_1\cdots i_S}{}_{k_1\cdots k_N} \xi_{j_1}\cdots \xi_{j_M} v_{i_1\cdots i_S}.$$

Observe that the symbol is obtained by the formal replacement of the derivatives $D_i \mapsto \xi_i$ in the principal part of the operator. The symbol $\boldsymbol{\sigma}_{\mathbf{L}}(\boldsymbol{\xi})$ determines the nature of the differential operator. In particular, $\mathbf{L}$ is said to be ***underdetermined elliptic*** at $p \in \mathcal{S}$ if $\boldsymbol{\sigma}_{\mathbf{L}}(\boldsymbol{\xi})$ is *surjective* for all $\boldsymbol{\xi} \neq 0$; $\mathbf{L}$ is ***overdetermined elliptic*** at $p \in \mathcal{S}$ if $\boldsymbol{\sigma}_{\mathbf{L}}(\boldsymbol{\xi})$ is *injective.* Finally, $\mathbf{L}$ is ***elliptic*** if $\boldsymbol{\sigma}_{\mathbf{L}}(\boldsymbol{\xi})$ is *bijective*, that is, if it is injective and surjective. If the coefficients $a^{j_1\cdots j_r i_1\cdots i_S}{}_{k_1\cdots k_N}$ in the operator (11.11) depend not only on the point on $\mathcal{S}$ but also on the derivatives $D_{j_1}\cdots D_{j_l}$, $l < r$, then $\mathbf{L}$ is said to be ***quasilinear***. The definitions of (underdetermined, overdetermined) elliptic differential operators extend in a natural way to the quasilinear case.

The paradigmatic example of an elliptic operator is the ***Laplace operator*** of the metric $\boldsymbol{h}$:

$$\Delta_{\boldsymbol{h}}\phi \equiv h^{ij}D_iD_j\phi, \qquad \phi \in \mathfrak{X}(\mathcal{S}).$$

In this case the operator is equal to its principal part. Moreover, its symbol is given by $h^{ij}\xi_i\xi_j < 0$ for $\xi_i \neq 0$ (as a consequence of negative-definiteness), from where it follows that the symbol is a bijection and, hence, $\Delta_{\boldsymbol{h}}$ is an elliptic operator. Particular examples of overdetermined and underdetermined elliptic operators are discussed in Section 11.3.3.

Associated to the differential operator $\mathbf{L}$ in (11.11) one has its ***formal adjoint*** $\mathbf{L}^*$ given by

$$(\mathbf{L}^*u)^{i_1\cdots i_S} \equiv \sum_{r=0}^{M}(-1)^r D_{j_1}\cdots D_{j_r}(a^{j_1\cdots j_r i_1\cdots i_S}{}_{k_1\cdots k_N}u^{k_1\cdots k_N}),$$

for smooth $u^{k_1\cdots k_N} \in \mathfrak{T}^{k_1\cdots k_N}(\mathcal{S})$. The above expression comes from the identity between ***inner products***

$$\int_{\mathcal{S}}(\mathbf{L}v)_{k_1\cdots k_N}u^{k_1\cdots k_N}\mathbf{d}\mu_{\boldsymbol{h}} = \int_{\mathcal{S}} v_{i_1\cdots i_S}(\mathbf{L}^*u)^{i_1\cdots i_S}\mathbf{d}\mu_{\boldsymbol{h}}, \tag{11.12}$$

which is obtained by repeated integration by parts. In the previous expression, $\mathbf{d}\mu_{\boldsymbol{h}}$ denotes the volume element of $\boldsymbol{h}$. For simplicity, in the identity (11.12) it is assumed that $\mathcal{S}$ is a compact manifold so that the integrals are well defined. Important for the subsequent discussion is the fact (verifiable using the definitions given in the previous paragraphs) that $\mathbf{L}$ is an underdetermined elliptic operator if and only if $\mathbf{L}^*$ is overdetermined elliptic. Moreover, if $\mathbf{L}$ is underdetermined elliptic, then $\mathbf{L}\circ\mathbf{L}^*$ is elliptic.

The interested reader is referred to appendix II in Choquet-Bruhat (2008) for further details on the theory of elliptic equations. An alternative summary can be found in the appendix of Besse (2008).

11.3 The Hamiltonian and momentum constraints

Before proceeding to analyse the conformal constraint equations, it is convenient to discuss the intrinsic equations implied by the Einstein field equations

$$\tilde{R}_{ab} - \frac{1}{2}\tilde{R}\tilde{g}_{ab} + \lambda\tilde{g}_{ab} = \tilde{T}_{ab}$$

on a non-null hypersurface of a spacetime $(\tilde{\mathcal{M}}, \tilde{\boldsymbol{g}})$ – the so-called Einstein constraint equations.

11.3.1 Derivation of the Einstein constraint equations

Starting from the *Gauss-Codazzi identity*, Equation (2.47) and contracting with $\tilde{h}^{ik}$ one obtains

$$\begin{aligned}\tilde{r}_{jl} + \tilde{K}\tilde{K}_{jl} - \tilde{K}^{k}{}_{j}\tilde{K}_{kl} &= \tilde{h}^{ik}\tilde{R}_{ijkl}\\ &= \eta^{ab}\tilde{R}_{ajbl} - \epsilon\tilde{R}_{\perp j\perp l}\\ &= \tilde{R}_{jl} - \epsilon\tilde{R}_{\perp j\perp l}.\end{aligned}$$

Contracting this last equation with $\tilde{h}^{jl}$ one finally obtains

$$\begin{aligned}\tilde{r} + \tilde{K}^2 - \tilde{K}_{jl}\tilde{K}^{jl} &= \tilde{h}^{jl}\tilde{R}_{jl} - \epsilon\tilde{h}^{jl}\tilde{R}_{\perp j\perp l}\\ &= \eta^{ab}\tilde{R}_{ab} - \epsilon\tilde{R}_{\perp\perp} - \epsilon\eta^{ab}\tilde{R}_{\perp a\perp b}\\ &= \tilde{R} - 2\epsilon\tilde{R}_{\perp\perp}.\end{aligned}$$

Similarly, starting from the *Codazzi-Mainardi identity, Equation* (2.48), and contracting with $\tilde{h}^{ij}$ one has that

$$\begin{aligned}\tilde{D}^{j}\tilde{K}_{kj} - \tilde{D}_{k}\tilde{K} &= \tilde{h}^{ij}\tilde{R}_{i\perp jk}\\ &= \eta^{ij}\tilde{R}_{i\perp jk} = \tilde{R}_{\perp k},\end{aligned}$$

where to pass from the first to the second line one uses that $\tilde{R}_{\perp\perp jk} = 0$.

Using the Einstein field equations in the frame component form

$$\tilde{R}_{ab} - \frac{1}{2}\eta_{ab}\tilde{R} + \lambda\eta_{ab} = \tilde{T}_{ab}$$

one obtains the so-called ***Einstein constraint equations***

$$\tilde{r} + \tilde{K}^2 - \tilde{K}_{jl}\tilde{K}^{jl} = 2(\lambda - \epsilon\tilde{\varrho}), \tag{11.13a}$$

$$\tilde{D}^{j}\tilde{K}_{kj} - \tilde{D}_{k}\tilde{K} = \tilde{j}_{k}, \tag{11.13b}$$

where

$$\tilde{\varrho} \equiv \tilde{T}_{\perp\perp}, \qquad \tilde{j}_{k} \equiv \tilde{T}_{\perp k}$$

are, respectively, the ***energy density*** and the components of the ***energy flux vector*** of the energy-momentum tensor in the direction of $\tilde{\boldsymbol{\nu}}$. Equations (11.13a) and (11.13b) are known, respectively, as the ***Hamiltonian constraint*** and the ***momentum constraint***. The tensorial version of Equations (11.13a) and (11.13b) is given by

$$\tilde{r} + \tilde{K}^2 - \tilde{K}_{jl}\tilde{K}^{jl} = 2(\lambda - \epsilon\tilde{\varrho}), \qquad \tilde{D}^{j}\tilde{K}_{kj} - \tilde{D}_{k}\tilde{K} = \tilde{j}_{k}. \tag{11.14}$$

Finally, it is observed that in index-free notation the constraint equations can be written as

$$r[\tilde{\boldsymbol{h}}] + (\mathbf{tr}_{\tilde{\boldsymbol{h}}}\tilde{\boldsymbol{K}})^2 - |\tilde{\boldsymbol{K}}|^2_{\tilde{\boldsymbol{h}}} = 2(\lambda - \epsilon\tilde{\varrho}), \qquad \mathbf{div}_{\tilde{\boldsymbol{h}}}\tilde{\boldsymbol{K}} - \mathbf{grad}\,\mathbf{tr}_{\tilde{\boldsymbol{h}}}\tilde{\boldsymbol{K}} = \tilde{\boldsymbol{j}}.$$

In what follows, a collection $(\tilde{\mathcal{S}}, \tilde{\boldsymbol{h}}, \tilde{\boldsymbol{K}}, \tilde{\varrho}, \tilde{\boldsymbol{j}})$ such that the negative definite metric $\tilde{\boldsymbol{h}}$ and the symmetric rank-2 tensor $\tilde{\boldsymbol{K}}$ satisfy the Einstein constraints (11.14) with $\epsilon = 1$ on the three-dimensional manifold $\tilde{\mathcal{S}}$ will be known as an ***initial data set*** for the Einstein field equations. If $\tilde{\varrho} = 0$ and $\tilde{\boldsymbol{j}} = 0$, one speaks of a ***vacuum initial data set***.

An important class of initial data sets is that for which $\tilde{\boldsymbol{K}} = 0$ and $\tilde{\boldsymbol{j}} = 0$, so that one is left only with the Hamiltonian constraint in the form

$$r[\tilde{\boldsymbol{h}}] = 2(\lambda - \tilde{\rho}).$$

Such an initial data set is called ***time reflection symmetric*** (or ***time symmetric*** for short); it follows from the properties of the *Einstein reduced equations* that for this type of initial data one has $\partial_t h_{\alpha\beta} = 0$ on the initial hypersurface $\tilde{\mathcal{S}}$ so that the resulting solution to the Einstein field equations is invariant under the replacement $t \mapsto -t$.

11.3.2 The conformal Hamiltonian and momentum constraint equations

Regarding, as in Section 11.1, the three-dimensional manifold $\tilde{\mathcal{S}}$ as a hypersurface on both $(\tilde{\mathcal{M}}, \tilde{\boldsymbol{g}})$ and $(\mathcal{M}, \boldsymbol{g})$, it follows from a computation using the transformation rules (11.1a) and (11.1b) together with the transformation rules for the Ricci scalar, Equation (5.16c), that Equation (11.14) can be reexpressed in terms of unphysical quantities as:

$$2\Omega D_i D^i \Omega - 3 D_i \Omega D^i \Omega + \frac{1}{2}\Omega^2 r - 3\epsilon\Sigma^2 \\ + \frac{1}{2}\Omega^2 (K^2 - K_{ij}K^{ij}) + 2\epsilon\Omega\Sigma K = \lambda - \epsilon\Omega^4 \varrho, \qquad (11.15a)$$

$$\Omega^3 D^i (\Omega^{-2} K_{ik}) - \Omega (D_k K - 2\Omega^{-1} D_k \Sigma) = \Omega^3 j_k, \qquad (11.15b)$$

where

$$\varrho \equiv \Omega^{-4} \tilde{\varrho}, \qquad j_k \equiv \Omega^{-3} \tilde{j}_k, \qquad (11.16)$$

denote, respectively, the ***unphysical energy density*** and the ***flux vector***.

11.3.3 The Hamiltonian and momentum constraint as an elliptic system

The Einstein constraint Equations (11.14) on a spacelike manifold $\tilde{\mathcal{S}}$ (i.e. $\epsilon = 1$) have been studied extensively in the literature; see, for example, Bartnik and Isenberg (2004) for a review of the topic and see also Choquet-Bruhat (2008), chapter 7, and Choquet-Bruhat and York (1980). In this section an adaptation of the so-called ***conformal method*** of Licnerowicz, Choquet-Bruhat and York to analyse the conformal Hamiltonian and momentum constraints (11.15a) and (11.15b) will be discussed; see, for example, York (1971, 1972). This approach works directly on a compact *unphysical manifold* $\mathcal{S}$ which is a conformal extension of the physical manifold $\tilde{\mathcal{S}}$. The key idea in this analysis is to show that these constraint equations imply an elliptic system of equations for suitable conformal fields. Proceeding in this way, one also obtains an insight into the nature of the *freely specifiable data* in the Einstein constraints. The use of a compact manifold $\mathcal{S}$ simplifies some of the technical aspects of the analysis. This approach to the Einstein constraint equations has been advocated in Friedrich (1988, 1998c, 2004, 2013), Dain and Friedrich (2001) and Beig and O'Murchadha (1991, 1994).

Following the discussion of the previous paragraph, let $(\mathcal{S}, \boldsymbol{h})$ denote a compact Riemannian manifold with $\boldsymbol{h}$ negative definite and set $\epsilon = 1$ so that $\mathcal{S}$ can be regarded as a spacelike hypersurface of an unphysical spacetime $(\mathcal{M}, \boldsymbol{g})$. In what follows, for simplicity, it is assumed that the matter fields ϱ and $\boldsymbol{j}$ are known on $\mathcal{S}$.

The first step to transform Equations (11.15a) and (11.15b) into an elliptic system is given by the transformation law of the three-dimensional Ricci scalar, Equation (5.17), which suggests introducing a conformal factor ϑ satisfying $\Omega = \vartheta^{-2}$. By substituting this definition into Equation (11.15a) one finds that

$$\Delta_{\boldsymbol{h}}\vartheta - \frac{1}{8}r[\boldsymbol{h}]\vartheta = \frac{1}{8}(K_{ij}K^{ij} - K^2)\vartheta + \frac{1}{4}(\vartheta^{-3}\varrho - \vartheta^5\lambda) + \frac{3}{4}\Sigma^2\vartheta^5 - \frac{1}{2}\vartheta^3\Sigma K, \tag{11.17}$$

where, as before, $\Delta_{\boldsymbol{h}} \equiv h^{ij}D_iD_j$ and the notation $r[\boldsymbol{h}]$ has been used to make explicit the dependence of the Ricci scalar on the metric $\boldsymbol{h}$. Following the standard use in the literature, this equation will be known as the ***Licnerowicz equation***. If the fields $\boldsymbol{h}$ (and hence $r[\boldsymbol{h}]$), K_{ij}, K, ϱ and Σ are known, this last equation can be read as a *non-linear elliptic equation* determining ϑ. For future use, it is convenient to define the ***Yamabe operator*** $\mathbf{L}_{\boldsymbol{h}} : \mathfrak{X}(\mathcal{S}) \to \mathfrak{X}(\mathcal{S})$ as

$$\mathbf{L}_{\boldsymbol{h}}\vartheta \equiv \Delta_{\boldsymbol{h}}\vartheta - \frac{1}{8}r[\boldsymbol{h}]\vartheta, \tag{11.18}$$

so that Equation (11.17) can be rewritten as

$$\mathbf{L}_{\boldsymbol{h}}\vartheta = \frac{1}{8}(K_{ij}K^{ij} - K^2)\vartheta + \frac{1}{4}(\varrho - \lambda)\vartheta^{-3} + \frac{1}{2}\Sigma\vartheta^3\left(K - \frac{1}{6}\vartheta^2\Sigma\right).$$

The Yamabe operator has nice conformal transformation properties; see Equation (11.23) below.

Equation (11.15b) suggests that the extrinsic curvature K_{ij} should be split into a trace-free part multiplied by a power of the conformal factor and a pure trace part. In this spirit one writes

$$K_{ij} = \vartheta^{-4}\psi_{ij} + \frac{1}{3}Kh_{ij}, \qquad h^{ij}\psi_{ij} = 0,$$

which, substituted into (11.15b), yields

$$D^i\psi_{ij} = \frac{2}{3}\vartheta^6 D_j(\vartheta^{-2}K) - 2\vartheta^{-6}D_j\Sigma + j_j.$$

In view of the latter, it is convenient to reintroduce the physical trace $\tilde{K} = \Omega K = \vartheta^{-2}K$ so that one obtains

$$D^i\psi_{ij} = \frac{2}{3}\vartheta^6 D_j\tilde{K} - 2\vartheta^{-4}D_j\Sigma + j_j. \tag{11.19}$$

This last equation is to be read as an equation for the trace-free tensor ψ_{ij}. If $\tilde{K}$ is a constant and $\Sigma = 0$, then Equations (11.17) and (11.19) decouple.

Following the discussion of Section 11.2 it can be verified that the principal part of Equation (11.19) is *underdetermined elliptic*. To transform Equation (11.19) into an elliptic equation one makes use of a so-called ***York splitting***; see York (1973). One considers an ansatz for ψ_{ij} of the form

$$\psi_{ij} = D_i\varsigma_j + D_j\varsigma_i - \frac{2}{3}h_{ij}D_k\varsigma^k + \psi'_{ij}, \tag{11.20}$$

where ς_i is some covector on $\mathcal{S}$ and ψ'_{ij} is a freely specifiable symmetric and trace-free tensor. The operator $(\boldsymbol{\mathcal{L}}_{\boldsymbol{h}}\varsigma)_i$ defined by

$$(\boldsymbol{\mathcal{L}}_{\boldsymbol{h}}\varsigma)_i \equiv D_i\varsigma_j + D_j\varsigma_i - \frac{2}{3}h_{ij}D_k\varsigma^k,$$

is called the ***conformal Killing operator***. It can be verified to be the formal adjoint of the divergence operator acting on symmetric trace-free tensors. Substituting the ansatz (11.20) into Equation (11.19) one obtains

$$\Delta_{\boldsymbol{h}}\varsigma_j + D^iD_j\varsigma_i - \frac{2}{3}D_jD_k\varsigma^k = \frac{2}{3}\vartheta^6 D_j\tilde{K} - 2\vartheta^{-4}D_j\Sigma + j_j - D^i\psi'_{ij}. \tag{11.21}$$

The symbol of this equation can be seen to be

$$(\boldsymbol{\sigma}_{\mathbf{div}\circ\boldsymbol{\mathcal{L}}}(\boldsymbol{\xi})\boldsymbol{\varsigma})_j = \xi^i\xi_i\varsigma_j + \xi^i\xi_j\varsigma_i - \frac{2}{3}\xi_j\xi^k\varsigma_k.$$

Contracting with ξ^j one immediately finds that

$$(\boldsymbol{\sigma}_{\mathbf{div}\circ\boldsymbol{\mathcal{L}}}(\boldsymbol{\xi})\boldsymbol{\varsigma})_j\xi^j = (\xi_i\xi^i)(\varsigma_k\varsigma^k) + \frac{1}{3}(\varsigma_i\xi^i)^2 > 0 \quad \text{for} \quad \xi_i, \varsigma_j \neq 0.$$

Thus, it follows that (11.21) is a linear elliptic equation for the covector ς_i. The freely specifiable data for this equation is the symmetric trace-free tensor ψ'_{ij}. As in the case of Equation (11.19) it decouples from the Licnerowicz Equation (11.17) if $\tilde{K}$ is constant and $\Sigma = 0$. The analysis of the coupled system (11.17)–(11.19) is much more challenging; see, for example, Holst et al. (2008a,b).

Gauge freedom

The conformal method described in the previous paragraphs has a ***conformal gauge freedom***. More precisely, if ϕ is a positive function on $\mathcal{S}$, then a direct computation shows that the transitions

$$h_{ij} \mapsto \phi^4 h_{ij}, \quad \psi_{ij} \mapsto \phi^{-2}\psi_{ij}, \quad \Omega \mapsto \phi^2\Omega, \quad K_{ij} \mapsto \phi^2 K_{ij}, \tag{11.22a}$$

$$\Sigma \mapsto \phi^2\Sigma, \quad \varrho \mapsto \phi^{-8}\varrho, \quad j_i \mapsto \phi^{-6} j_i, \tag{11.22b}$$

yield another solution to the conformal constraint Equations (11.15a) and (11.15b) with the same physical data $(\tilde{\boldsymbol{h}}, \tilde{\boldsymbol{K}})$. This gauge freedom can be exploited to simplify certain specific computations. In particular, letting $\boldsymbol{h}' = \phi^4\boldsymbol{h}$, a calculation using the transformation laws for conformal transformations shows that

$$\phi^{-5}\big(\Delta_{\boldsymbol{h}} - \tfrac{1}{8}r[\boldsymbol{h}]\big)\vartheta = \big(\Delta_{\boldsymbol{h}'} - \tfrac{1}{8}r[\boldsymbol{h}']\big)(\phi^{-1}\vartheta), \tag{11.23}$$

that is,

$$\phi^{-5}\mathbf{L}_{\boldsymbol{h}}[\vartheta] = \mathbf{L}_{\boldsymbol{h}'}(\phi^{-1}\vartheta).$$

11.3.4 The Yamabe problem

A classic question of Differential Geometry is the so-called ***Yamabe problem*** which, given a compact three-dimensional Riemannian manifold $(\mathcal{S}, \boldsymbol{h})$, asks whether it is possible to conformally rescale the (smooth) metric $\boldsymbol{h}$ to a metric with constant Ricci scalar; see Yamabe (1960). This problem requires finding a positive conformal factor ω and a constant $r_\bullet$ satisfying the equation

$$\Delta_{\boldsymbol{h}}\omega = \frac{1}{8}\big(r[\boldsymbol{h}]\omega - r_\bullet\omega^5\big), \tag{11.24}$$

which follows from the transformation equation for the three-dimensional Ricci scalar Equation (5.17). The Yamabe problem has been solved in the affirmative; see Trudinger (1968), Aubin (1976) and Schoen (1984). In particular, one has the following (e.g. Lee and Parker (1987); O'Murchadha (1988)):

Theorem 11.1 (***resolution of the Yamabe problem***) *Let $\boldsymbol{h}$ be a smooth Riemannian metric on a compact manifold $\mathcal{S}$. There exists a smooth, positive definite function ω on $\mathcal{S}$ such that $r[\omega^4\boldsymbol{h}]$ is constant.*

Theorem 11.1 allows the classification of Riemannian metrics according to whether they can be rescaled to a metric with constant Ricci scalar which is positive, negative or zero – a given metric $\boldsymbol{h}$ cannot be rescaled to two different metrics with constant curvature of different signs. Thus, the resulting ***Yamabe classes*** are conformal invariants. As will be seen in Section 11.5, this observation plays a role in the construction of initial data sets on compact manifolds. Remarkably, the analogous *Yamabe problem on non-compact manifolds* turns out not to be true as shown by a number of counterexamples; see, for example, Zhiren (1988).

11.4 The conformal constraint equations

Having analysed the standard Einstein constraint equations, focus is now on the constraint equations implied by the conformal Einstein field equations. These equations can be regarded as an extension of the conformal Hamiltonian and momentum constraints (11.15a) and (11.15b).

11.4.1 The derivation of the equations

In this section the frame version of the conformal Einstein field equations, Equations (8.32a) and (8.32b), are considered. By making use of an orthonormal frame adapted to the geometry of the hypersurface under consideration, as described in Section 2.7.3, the split of the equations follows almost directly.

In what follows, let $(\mathcal{M}, \boldsymbol{g})$ denote an *unphysical spacetime* and let $\mathcal{S}$ denote a hypersurface thereof. As in Section 11.1.1, let Σ denote the covariant derivative in the direction of the $\boldsymbol{g}$-unit normal. The evaluation of a spacetime frame index in the direction of the unit normal (i.e. the values $\boldsymbol{0}$ or $\boldsymbol{3}$ depending on the causal character of $\mathcal{S}$) will be indicated by the symbol $_\perp$.

The constraints implied by Z_{ab}. Given

$$Z_{\boldsymbol{ab}} \equiv \nabla_{\boldsymbol{a}}\nabla_{\boldsymbol{b}}\Xi + \Xi L_{\boldsymbol{ab}} - s\eta_{\boldsymbol{ab}} - \frac{1}{2}\Xi^3 T_{\{\boldsymbol{ab}\}}, \tag{11.25}$$

the information of the conformal equation $Z_{\boldsymbol{ab}} = 0$ which is intrinsic to the hypersurface $\mathcal{S}$ is encoded in the components

$$Z_{\boldsymbol{ij}} = 0, \qquad Z_{\perp\boldsymbol{i}} = 0. \tag{11.26}$$

In order to obtain explicit intrinsic expressions for these equations it is observed that

$$\nabla_{\boldsymbol{a}}\nabla_{\boldsymbol{b}}\Xi \equiv \boldsymbol{e_a}{}^a \boldsymbol{e_b}{}^b \nabla_a\nabla_b\Xi = \boldsymbol{e_a}(\boldsymbol{e_b}(\Xi)) - \Gamma_{\boldsymbol{a}}{}^{\boldsymbol{c}}{}_{\boldsymbol{b}}\boldsymbol{e_c}(\Xi).$$

Hence, in particular, one has that

$$\begin{aligned}\nabla_{\boldsymbol{i}}\nabla_{\boldsymbol{j}}\Xi &= \boldsymbol{e_i}(\boldsymbol{e_j}(\Xi)) - \Gamma_{\boldsymbol{i}}{}^{\boldsymbol{c}}{}_{\boldsymbol{j}}\boldsymbol{e_c}(\Xi)\\ &= \boldsymbol{e_i}(\boldsymbol{e_j}(\Xi)) - \Gamma_{\boldsymbol{i}}{}^{\boldsymbol{k}}{}_{\boldsymbol{j}}\boldsymbol{e_k}(\Xi) - \Gamma_{\boldsymbol{i}}{}^{\perp}{}_{\boldsymbol{j}}\boldsymbol{e}_\perp(\Xi)\\ &= \boldsymbol{e_i}(D_{\boldsymbol{j}}\Xi) - \gamma_{\boldsymbol{i}}{}^{\boldsymbol{k}}{}_{\boldsymbol{j}} D_{\boldsymbol{k}}\Xi + \epsilon K_{\boldsymbol{ij}}\Sigma\\ &= D_{\boldsymbol{i}}D_{\boldsymbol{j}}\Xi + \epsilon K_{\boldsymbol{ij}}\Sigma,\end{aligned} \tag{11.27}$$

where, in the last term of the last line, Equation (2.45) for the extrinsic curvature has been used. A similar computation shows that

$$
\begin{aligned}
\nabla_i \nabla_\perp \Xi &= e_i(e_\perp(\Xi)) - \Gamma_i{}^c{}_\perp e_c(\Xi) \\
&= e_i(\Sigma) - \Gamma_i{}^k{}_\perp e_k(\Xi) - \Gamma_i{}^\perp{}_\perp \Sigma \\
&= D_i \Sigma - \epsilon K_i{}^k D_k \Xi,
\end{aligned} \tag{11.28}
$$

where, in the third line, it has been used that $\Gamma_a{}^\perp{}_\perp = 0$ as a consequence of the metricity of the connection and the fact that $K_i{}^k = \Gamma_i{}^k{}_\perp$.

Substituting the above expressions into Equation (11.26) and taking into account definition (11.25) one obtains the constraint equations

$$
\begin{aligned}
D_i D_j \Omega &= -\epsilon K_{ij} \Sigma - \Omega L_{ij} + s h_{ij} + \frac{1}{2}\Omega\left(T_{ij} - \frac{1}{4} T h_{ij}\right), \\
D_j \Sigma &= K_j{}^k D_k \Omega - \Omega L_j + \frac{1}{2}\Omega^3 j_j,
\end{aligned}
$$

where

$$
L_i \equiv L_{i\perp}, \qquad \Omega \equiv \Xi|_{\mathcal{S}}.
$$

The constraints implied by Z_a. Given

$$
Z_a \equiv \nabla_a s + L_{ac} \nabla^c \Xi - \frac{1}{2}\Xi^2 \nabla^c \Xi T_{\{ac\}} - \frac{1}{6}\Xi^3 \nabla^c T_{\{ca\}}, \tag{11.29}
$$

the intrinsic information of the equation $Z_a = 0$ is encoded in the components

$$
Z_i = 0. \tag{11.30}
$$

Now, the spatial components of the term $L_{ab}\nabla^b \Omega$ in Equation (11.29) can be expanded as

$$
\begin{aligned}
L_{ib} \nabla^b \Omega &= L_{ib} \eta^{ba} \nabla_a \Omega \\
&= L_{i\perp} \eta^{\perp\perp} \nabla_\perp \Omega + L_{ik} \eta^{kl} \nabla_l \Omega \\
&= \epsilon L_i \Sigma + L_{ik} D^k \Omega.
\end{aligned}
$$

By similar arguments one concludes that

$$
\begin{aligned}
\nabla^c \Xi T_{\{ic\}} &= \epsilon \Sigma j_i - T_{ik} D^k \Omega - \frac{1}{4} D_i T, \\
\nabla^c T_{\{ic\}} &= \epsilon \nabla_\perp j_i + D^k T_{ki} - \frac{1}{4} D_i T.
\end{aligned}
$$

It is important to observe in $\nabla^c T_{\{ic\}}$ the presence of the term $\nabla_\perp j_i$ which requires further information about the matter model in order to be cast in a form intrinsic to the hypersurface $\mathcal{S}$. In the case of trace-free matter, one has that $\nabla^c T_{\{ic\}} = 0$, so that no further considerations are required.

From the discussion in the previous paragraphs it follows that Equation (11.30) can be reexpressed as

$$
\begin{aligned}
D_i s = & -\epsilon L_i \Sigma - L_{ik} D^k \Omega + \frac{1}{2}\Omega^2 \big(\epsilon \Sigma j_i - T_{ik} D^k \Omega - \frac{1}{4} D_i T\big) \\
& + \frac{1}{6}\Omega^3 \big(\epsilon \nabla_\perp j_i + D^k T_{ki} - \frac{1}{4} D_i T\big).
\end{aligned}
$$

The constraints implied by Δ_{cdb}. Given

$$
\Delta_{cdb} \equiv \nabla_c L_{db} - \nabla_d L_{cb} - \nabla_a \Xi d^a{}_{bcd} - \Xi T_{cdb}, \tag{11.31}
$$

the information intrinsic to the hypersurface $\mathcal{S}$ of the conformal equation $\Delta_{cdb} = 0$ is encoded in the components

$$
\Delta_{ijk} = 0, \qquad \Delta_{ij\perp} = 0. \tag{11.32}
$$

A calculation similar to that leading to Equations (11.27) and (11.28) yields

$$
\begin{aligned}
\nabla_i L_{jk} &= D_i L_{jk} + \epsilon K_{ik} L_j, \\
\nabla_i L_j &= D_i L_j + K_i{}^k L_{kj}.
\end{aligned}
$$

Given the components d_{abcd} of the rescaled Weyl tensor with respect to the adapted frame $\{e_a\}$, it is convenient to define

$$
d_{ij} \equiv d_{i\perp j\perp}, \qquad d_{ijk} \equiv d_{i\perp jk}.
$$

Following the discussion of Section 11.1.2, d_{ij} corresponds to the components of the electric part of the rescaled Weyl tensor, while d_{ijk} encodes the information of the magnetic part. It can be verified that

$$
d_{ij} = d_{ji}, \qquad d^i{}_i = 0, \qquad d_{ijk} = -d_{ikj}, \qquad d_{[ijk]} = 0, \tag{11.33a}
$$

$$
d_{ijkl} = 2(h_{i[k} d_{l]j} + h_{j[l} d_{k]i}). \tag{11.33b}
$$

It follows from the latter expressions, together with (11.31), that the constraints (11.32) can be reexpressed as

$$
\begin{aligned}
D_i L_{jk} - D_j L_{ik} &= -\epsilon \Sigma d_{ijk} + D^l \Omega d_{lkij} - \epsilon (K_{ik} L_j - K_{jk} L_i) + \Omega T_{ijk}, \\
D_i L_j - D_j L_i &= D^l \Omega d_{lij} + K_i{}^k L_{jk} - K_j{}^k L_{ik} + \Omega J_{ij},
\end{aligned}
$$

where

$$
J_{jk} \equiv T_{jk\perp}.
$$

The constraints implied by Λ_{bcd}. Given

$$
\Lambda_{bcd} \equiv \nabla_a d^a{}_{bcd} - T_{cdb},
$$

as a consequence of the decomposition of the Weyl tensor in electric and magnetic parts, it follows that the information of the equation $\Lambda_{bcd} = 0$ which is intrinsic to the hypersurface $\mathcal{S}$ is contained in the components

$$\Lambda_{\perp ij} = 0, \qquad \Lambda_{\perp j \perp} = 0. \tag{11.34}$$

Observing that

$$\nabla^a d_{aijk} = \eta^{ab}\big(e_a(d_{bijk}) - \Gamma_a{}^c{}_b d_{cijk} - \Gamma_a{}^c{}_i d_{bcjk} - \Gamma_a{}^c{}_j d_{bick} - \Gamma_a{}^c{}_k d_{bijc}\big),$$

one concludes, by arguments similar to those used to obtain Equations (11.27) and (11.28), that

$$\nabla_a d^a{}_{\perp jk} = D^i d_{ijk} + \epsilon(K^i{}_k d_{ji} - K^i{}_j d_{ki}),$$
$$\nabla_a d^a{}_{\perp j \perp} = D^i d_{ij} - K^{ik} d_{ijk}.$$

It follows from the previous discussion that the constraint Equations (11.34) can be reexpressed as

$$D^i d_{ijk} = \epsilon(K^i{}_j d_{ki} - K^i{}_k d_{ji}) + J_{jk},$$
$$D^i d_{ij} = K^{ik} d_{ijk} + J_j,$$

where

$$J_{jk} \equiv T_{jk\perp}, \qquad J_j \equiv T_{j\perp\perp}.$$

The explicit form of J_{jk} and J_j depends on the matter model under consideration. In the case of the electromagnetic field, they can be expressed in terms of the electric and magnetic parts of the Faraday tensor and their spatial derivatives.

The constraint $\mathrm{Z} = 0$**.** Recall that

$$Z \equiv 6\Xi s - 3\nabla_c \Xi \nabla^c \Xi + \frac{1}{4}\Xi^4 T - \lambda.$$

As discussed in Section 8.2.4 the equation $Z = 0$ is, in fact, a constraint equation whose propagation is ensured by the other conformal field equations; see Lemma 8.1. Following the procedure employed in the decomposition of the other conformal equations, it can be expressed in terms of quantities intrinsic to the hypersurface $\mathcal{S}$ as

$$\lambda = 6\Omega s - 3\epsilon\Sigma^2 - 3D_k\Omega D^k\Omega + \frac{1}{4}\Omega^4 T.$$

11.4.2 The Gauss-Codazzi and Codazzi-Mainardi equations in terms of conformal fields

The intrinsic equations discussed in the previous section are supplemented by the Gauss-Codazzi and Codazzi-Mainardi equations, Equations (2.47) and (2.48)

$$R_{ijkl} = r_{ijkl} + K_{ik}K_{jl} - K_{il}K_{jk},$$
$$R_{i\perp jk} = D_j K_{ki} - D_k K_{ji},$$

expressed in terms of conformal fields. As a consequence of the decomposition of the four-dimensional Riemann tensor R_{abcd} in terms of the Weyl and Schouten tensor, Equation (2.21b), one has that

$$R_{abcd} = \Xi d_{abcd} + \eta_{ac}L_{db} - \eta_{ad}L_{cb} + L_{ac}\eta_{db} - L_{ad}\eta_{cb},$$

while the three-dimensional Riemann tensor r_{ijkl} can be expressed in terms of the three-dimensional Schouten tensor l_{ij} as

$$r_{ijkl} = h_{ik}l_{lj} - h_{il}l_{kj} + h_{jl}l_{ki} - h_{jk}l_{li}, \qquad l_{ij} \equiv r_{ij} - \frac{1}{4}rh_{ij};$$

see Equation (2.40). A direct calculation using the above expressions yields the two additional constraint equations

$$D_j K_{ki} - D_k K_{ji} = \Omega d_{ijk} + h_{ij}L_k - h_{ik}L_j,$$
$$l_{ij} = \Omega d_{ij} + L_{ij} - K_k{}^k\Big(K_{ij} - \frac{1}{4}Kh_{ij}\Big) + K_{ki}K_j{}^k - \frac{1}{4}K_{kl}K^{kl}h_{ij}.$$

These equations provide the link between the spatial curvature tensor l_{ij} and the spacetime curvature as described by d_{ab}, d_{abc}, L_{ab} and L_a.

11.4.3 Summary of the equations and basic properties of the conformal constraint equations

As a summary of the discussion of the previous sections, the conformal constraint equations are collected:

$$D_iD_j\Omega = -\epsilon\Sigma K_{ij} - \Omega L_{ij} + sh_{ij} + \frac{1}{2}\Omega^3\Big(T_{ij} - \frac{1}{4}Th_{ij}\Big), \tag{11.35a}$$

$$D_i\Sigma = K_i{}^k D_k\Omega - \Omega L_i + \frac{1}{2}\Omega^3 j_i, \tag{11.35b}$$

$$\begin{aligned} D_i s = {} & -\epsilon L_i\Sigma - L_{ik}D^k\Omega + \frac{1}{2}\Omega^2\Big(\epsilon\Sigma j_i - T_{ik}D^k\Omega - \frac{1}{4}D_iT\Big) \\ & + \frac{1}{6}\Omega^3\Big(\epsilon\nabla_\perp j_i + D^kT_{ki} - \frac{1}{4}D_iT\Big), \end{aligned} \tag{11.35c}$$

$$\begin{aligned} D_iL_{jk} - D_jL_{ik} = {} & -\epsilon\Sigma d_{kij} + D^l\Omega d_{lkij} \\ & - \epsilon(K_{ik}L_j - K_{jk}L_i) + \Omega T_{ijk}, \end{aligned} \tag{11.35d}$$

$$D_iL_j - D_jL_i = D^l\Omega d_{lij} + K_i{}^kL_{jk} - K_j{}^kL_{ik} + \Omega T_{ij\perp}, \tag{11.35e}$$

$$D^k d_{kij} = \epsilon(K^k{}_i d_{jk} - K^k{}_j d_{ik}) + J_{ij}, \tag{11.35f}$$

$$D^i d_{ij} = K^{ik} d_{ijk} + J_j, \tag{11.35g}$$

$$\lambda = 6\Omega s - 3\epsilon\Sigma^2 - 3D_k\Omega D^k\Omega + \frac{1}{4}\Omega^4 T, \tag{11.35h}$$

$$D_j K_{ki} - D_k K_{ji} = \Omega d_{ijk} + h_{ij} L_k - h_{ik} L_j, \tag{11.35i}$$

$$l_{ij} = \Omega d_{ij} + L_{ij} - K(K_{ij} - \frac{1}{4}K h_{ij}) + K_{ki}K_j{}^k - \frac{1}{4}K_{kl}K^{kl} h_{ij}. \tag{11.35j}$$

Using the identity (11.7) and recalling that $d^*_{ij} = -\frac{1}{2} d_{ikl}\epsilon_j{}^{kl}$, Equations (11.35f) and (11.35g) can be rewritten in the alternative form

$$D^i d^*_{ij} = -\epsilon\epsilon_j{}^{kl} K^i{}_k d_{li} - \frac{1}{2}\epsilon_j{}^{kl} J_{kl}, \tag{11.36a}$$

$$D^i d_{ij} = \epsilon^l{}_{jk} K^{ik} d^*_{il} + J_j. \tag{11.36b}$$

The conformal constraint Equations (11.35a)–(11.35j) are not independent since integrability conditions have been used in their derivation. A list of various relations between the vacuum constraint equations can be found in Friedrich (1983). In particular, it can be shown that

$$D_i\left(6\Omega s - 3\epsilon\Sigma^2 - 3D_k\Omega D^k\Omega + \frac{1}{4}\Omega^4 T\right) = 0,$$

consistent with the fact that the left-hand side of Equation (11.35h) equals the cosmological constant λ.

For future reference, it is observed that from Equation (11.35j) it follows that

$$r_{ij} = \Omega d_{ij} + L_{ij} + L_k{}^k h_{ij} - K K_{ij} + K_{ik}K^k{}_j, \tag{11.37a}$$

$$r = 4L_k{}^k - K^2 + K_{ij}K^{ij}. \tag{11.37b}$$

The vacuum version of the conformal constraint equations is obtained by setting the *matter fields* T_{ij}, T, j_i, T_{ijk}, J_i, J_{ij} equal to zero. In the derivation of the conformal constraint Equations (11.35a)–(11.35j), it has been assumed that the connection $\boldsymbol{D}$ is the Levi-Civita connection of the intrinsic metric $\boldsymbol{h}$. Thus, by analogy to the full conformal field equations one also has the relations

$$\sigma_i{}^k{}_j = 0, \qquad \Pi^k{}_{lij} = \pi^k{}_{lij}, \tag{11.38}$$

where $\sigma_i{}^k{}_j$, $\Pi^k{}_{lij}$ and $\pi^k{}_{lij}$ denote, respectively, the components of the ***torsion***, the ***geometric curvature*** and the ***algebraic curvature*** of the connection $\boldsymbol{D}$. Explicitly, one has that

$$\sigma_i{}^k{}_j \boldsymbol{e}_k \equiv [\boldsymbol{e}_i, \boldsymbol{e}_j] - (\gamma_i{}^k{}_j - \gamma_j{}^k{}_i)\boldsymbol{e}_k,$$

$$\Pi^k{}_{lij} \equiv \boldsymbol{e}_i(\gamma_j{}^k{}_l) - \boldsymbol{e}_j(\gamma_i{}^k{}_l) + \gamma_m{}^k{}_l(\gamma_j{}^m{}_i - \gamma_i{}^m{}_j) + \gamma_j{}^m{}_l\gamma_i{}^k{}_m - \gamma_i{}^m{}_l\gamma_j{}^k{}_m,$$

$$\pi_{klij} \equiv h_{ik}l_{lj} - h_{il}l_{kj} + h_{jl}l_{ki} - h_{jk}l_{li}.$$

Given a collection of matter fields on $\mathcal{S}$,

$$\mathbf{m}_\star \equiv (T_{ij}, T, \varrho, j_i, \nabla_\perp j_i, J_i, J_{ij}),$$

by a ***solution to the conformal constraint equations*** on $\mathcal{S}$ it will be understood a collection

$$\mathbf{u}_\star \equiv (\Omega, \Sigma, s, \boldsymbol{e}_i, \gamma_i{}^{\boldsymbol{k}}{}_j, K_{ij}, L_{ij}, L_i, d_{ij}, d_{ijk})$$

satisfying Equations (11.35a)–(11.35j) together with the supplementary conditions (11.38).

The relation between the conformal constraint Equations (11.35a)–(11.35j) and the conformal Hamiltonian and momentum constraints (11.15a)–(11.15b) is summarised in the following lemma.

Lemma 11.1 (***relation between the solutions to the Einstein constraints and the conformal constraints***) *A solution to the conformal constraints* (11.35a)–(11.35j) *for a collection* $\mathbf{m}_\star$ *of matter fields implies a solution to the conformal Hamiltonian and momentum constraints* (11.15a) *and* (11.15b). *Conversely, a solution of* (11.15a) *and* (11.15b) *together with a collection of matter fields* $\mathbf{m}_\star$ *gives rise to a solution to* (11.35a)–(11.35j) *on the points of* $\mathcal{S}$ *for which* $\Omega \neq 0$.

Proof Using Equations (11.35a) and (11.35h) to eliminate $L_{\boldsymbol{k}}{}^{\boldsymbol{k}}$ one readily obtains the conformal Hamiltonian Equation (11.15a). Similarly, starting from Equation (11.35i) and using Equation (11.35b) to eliminate L_i one obtains the conformal momentum constraint (11.15b). Thus, any solution to the conformal constraints (11.35a)–(11.35j) implies a solution to the conformal Hamiltonian and momentum constraints, Equations (11.15a) and (11.15b).

Assume now one has a collection $(\Omega, \boldsymbol{h}, \boldsymbol{K}, \Sigma, \varrho, j_i)$ satisfying Equations (11.15a) and (11.15b) together with a collection $(T_{ij}, T, J_i, J_{ij}, \nabla_\perp j_i)$ consistent with the matter fields ϱ and j_i. Let now $\{\boldsymbol{e}_i\}$ denote an $\boldsymbol{h}$-orthonormal frame. Using this frame one can compute the components l_{ij} and K_{ij} of the three-dimensional Schouten tensor and of the extrinsic curvature. If $\Omega \neq 0$, one can use the conformal constraint (11.35h) to compute the field s. Next, one makes use of Equations (11.35a) and (11.35b) to compute L_{ij} and L_i. A computation using the commutator of the covariant derivative D_i shows that Equation (11.35c) is automatically satisfied. Once the components L_{ij} and L_i are known, one can use Equations (11.35i) and (11.35j), respectively, to compute d_{ijk} and d_{ij} – it can be verified that the resulting fields are trace free. A final computation using the three-dimensional Bianchi identity in the form

$$D^i r_{ij} = \frac{1}{4} D_j r,$$

together with the irreducible decomposition of the three-dimensional Riemann tensor r_{ijkl}, the decomposition of d_{ijkl} into the electric and magnetic parts and the commutator of D_i shows that Equations (11.35d)–(11.35g) are also automatically satisfied. Thus, the fields obtained constitute the required solution to the conformal constraint equations; see Friedrich (1983) for further details. □

Remark. In order to make assertions about the behaviour of solutions to the conformal constraint equations at points where $\Omega = 0$, the equations need to be supplemented with boundary conditions. Several different classes of boundary conditions on three-dimensional manifolds will be considered: ***compact manifolds***, ***asymptotically Euclidean manifolds*** and ***hyperboloidal manifolds.***

11.4.4 The conformal constraints at the conformal boundary

By construction, the conformal constraint equations can be evaluated in a regular manner at a non-null hypersurface belonging to the conformal boundary of spacetime. By definition such a hypersurface satisfies the conditions

$$\Omega = 0, \qquad \mathbf{d}\Omega \neq 0.$$

Following the convention introduced in Chapter 6 this hypersurface will be denoted by $\mathscr{I}$. The null case will be discussed in Chapter 18.

The defining properties of the hypersurface $\mathscr{I}$ lead to a number of simplifications in the conformal constraint equations. In particular, $\mathbf{d}\Omega$ is normal to $\mathscr{I}$ so that, in terms of a tetrad adapted to the hypersurface, one has $D_i\Omega = 0$. Assuming that the matter fields T_{ij}, T and T_{ijk} are smooth at $\mathscr{I}$ one finds that on the hypersurface the conformal constraints (11.35a)–(11.35j) imply the equations

$$s h_{ij} \simeq \epsilon \Sigma K_{ij}, \tag{11.39a}$$
$$D_i\Sigma \simeq 0, \tag{11.39b}$$
$$D_i s \simeq -\epsilon L_i \Sigma, \tag{11.39c}$$
$$D_i L_{jk} - D_j L_{ik} \simeq -\epsilon \Sigma d_{ijk} - \epsilon (K_{ik} L_j - K_{jk} L_i), \tag{11.39d}$$
$$D_i L_j - D_j L_i \simeq K_i{}^k L_{jk} - K_j{}^k L_{ik}, \tag{11.39e}$$
$$D^k d_{kij} \simeq \epsilon \big(K^k{}_i d_{jk} - K^k{}_j d_{ik}\big) + J_{ij}, \tag{11.39f}$$
$$D^i d_{ij} \simeq K^{ik} d_{ijk} + J_j, \tag{11.39g}$$
$$\lambda \simeq -3\epsilon \Sigma^2, \tag{11.39h}$$
$$D_j K_{ki} - D_k K_{ji} \simeq h_{ij} L_k - h_{ik} L_j, \tag{11.39i}$$
$$l_{ij} \simeq L_{ij} - K\left(K_{ij} - \frac{1}{4} K h_{ij}\right) + K_{ki} K_j{}^k - \frac{1}{4} K_{kl} K^{kl} h_{ij}, \tag{11.39j}$$

where $\simeq$ denotes equality at the conformal boundary. From Equations (11.39b) and (11.39h) it follows that Σ is a constant on $\mathscr{I}$ with a value given by

$\Sigma = \sqrt{-\epsilon\lambda/3}$ – observe that if $\epsilon = 1$, then $\lambda < 0$, and if $\epsilon = -1$, then $\lambda > 0$, for the previous expression to make sense. Moreover, from Equation (11.39a) the extrinsic curvature of $\mathscr{I}$ is proportional to the intrinsic metric.

A procedure for constructing solutions to Equations (11.39a)–(11.39j) in the vacuum case (so that $J_{jk} = 0$, $J_j = 0$) has been given in Friedrich (1986a, 1995). The fundamental idea is to identify the function s and the 3-metric $\boldsymbol{h}$ on $\mathscr{I}$ as freely specifiable data. Instead of working directly with s, it is more convenient to use a smooth function $\varkappa \in \mathfrak{X}(\mathcal{S})$ such that

$$s \simeq \Sigma\varkappa. \tag{11.40}$$

It follows directly from Equations (11.39a), (11.39c) and (11.39j) that

$$K_{ij} \simeq \epsilon\varkappa h_{ij}, \qquad L_i \simeq -\epsilon D_i\varkappa, \qquad L_{ij} \simeq l_{ij} + \frac{1}{2}\varkappa^2 h_{ij}. \tag{11.41}$$

Substituting these expressions into Equation (11.39d) one obtains, after some simplification, that

$$d_{ijk} \simeq -\epsilon\Sigma^{-1} y_{ijk} \tag{11.42}$$

where $y_{ijk} \equiv D_i l_{jk} - D_j l_{ik}$ denote the components of the Cotton tensor of the metric $\boldsymbol{h}$; see Section 5.2.2. Alternatively, one can write

$$d^*_{ij} \simeq -\epsilon\Sigma^{-1} y_{ij},$$

with $y_{ij} \equiv -\frac{1}{2} y_{klj}\epsilon_i{}^{kl}$ the components of the ***Bach tensor***. It can be verified that the integrability conditions (11.39e), (11.39f) and (11.39i) are automatically satisfied by (11.41) and (11.42). Finally, by substituting into Equation (11.39g) one obtains that

$$D^i d_{ij} \simeq 0.$$

This is the only differential condition that has to be solved in this procedure. This can be done by means of a York splitting so as to obtain an elliptic equation for the components of a covector.

The discussion of the previous paragraph is summarised in the following:

Proposition 11.1 (***solutions to the conformal constraint equations at the conformal boundary***) *Given a three-dimensional metric* $\boldsymbol{h}$, *an* $\boldsymbol{h}$*-divergence-free and trace-free field* d_{ij} *and a smooth function* $\varkappa$, *the fields* s, K_{ij}, L_i, L_{ij}, d_{ijk} *as given by Equations* (11.40), (11.41) *and* (11.42) *constitute a solution to the vacuum conformal constraint equations with* $\Omega = 0$.

As will be seen in later chapters, a solution to Equations (11.39a)–(11.39j) constitutes, in the case of $\epsilon = 1$ (i.e. $\mathscr{I}$ spacelike), initial data at, say, past null infinity for de Sitter-like spacetimes. In the case $\epsilon = -1$ (i.e. $\mathscr{I}$ timelike), the solution gives boundary data for an anti-de Sitter-like spacetime.

Remark. The procedure indicated in the previous paragraphs can be extended to the matter case if the field J_i is known.

Exploiting the conformal freedom

The conformal freedom inherent to the conformal field equations can be employed to express the solution to the conformal constraint equations at the conformal boundary in an even simpler form. Recall the discussion in Section 8.2.5 on the transformation properties of the various fields appearing in the conformal field equations. In particular, it follows from Equation (8.29b) that, under a rescaling of the form $\boldsymbol{g}' = \vartheta^2 \boldsymbol{g}$ which implies a rescaling

$$\boldsymbol{h}' \simeq \vartheta^2 \boldsymbol{h}$$

of the intrinsic metric of $\mathscr{I}$, the field s on $\mathscr{I}$ transforms as

$$s' \simeq \left(\vartheta^{-1} s + \vartheta^{-2} \nabla^a \vartheta \nabla_a \Xi\right).$$

In particular, *it is always possible to choose* ϑ *at* $\mathscr{I}$ *so that locally*

$$s' \simeq 0.$$

Accordingly, in this particular conformal gauge one has that Equation (11.40) implies $\varkappa' = 0$ and, moreover,

$$K'_{ij} \simeq 0, \qquad L'_i \simeq 0, \qquad L'_{ij} \simeq l_{ij}.$$

11.5 The constraints on compact manifolds

An important class of initial data sets for the Einstein field equations involves physical 3-manifolds $\tilde{\mathcal{S}}$ which are compact. This type of initial data set is of relevance in the discussion of cosmological models. In particular, in the vacuum case with negative cosmological constant one expects these initial data sets to give rise to de Sitter-like spacetimes. Initial data sets on compact manifolds have been studied extensively in the literature, and there is a good understanding of the required conditions on the free data in order to ensure existence of solutions to the Einstein constraint equations; see, for example, Isenberg (1995).

For this type of initial data one can set, without loss of generality, $\Omega = 1$ and $\Sigma = 0$ and let $\mathcal{S} = \tilde{\mathcal{S}}$. For simplicity of the presentation, in the remainder of this section the discussion is restricted to the vacuum case. Furthermore, it is assumed that the physical mean curvature $\tilde{K}$ is constant so that Equations (11.17) and (11.21) decouple from each other. The fundamental tool in the analysis of the solvability of the constraint equations is given by the ***maximum principle*** for the Laplacian of a Riemannian metric. A convenient formulation of this result is given by (see Isenberg (1995)):

Proposition 11.2 (***maximum principle for compact manifolds***) *Let $(\mathcal{S}, \boldsymbol{h})$ denote a Riemannian manifold with $\mathcal{S}$ compact. Given a smooth $\psi \in \mathfrak{X}(\mathcal{S})$ such that $\Delta_{\boldsymbol{h}}\psi$ has the same sign on the whole of $\mathcal{S}$, then ψ must be a constant.*

As a consequence of the above principle, the equation

$$\Delta_{\boldsymbol{h}}\psi = F(x, \psi),$$

with $\psi > 0$ has no solution if $F(x, \psi)$ does not change sign on $\mathcal{S}$ except for the case where $F(x, \psi) = 0$. Using this observation it is easy to see that certain combinations of free data cannot give rise to solutions of the constraint equations. As an example, consider time-symmetric data (i.e. $\tilde{\boldsymbol{K}} = 0$) with vanishing cosmological constant on a compact manifold $\mathcal{S}$. As a consequence of the conformal gauge freedom given in Equations (11.22a) and (11.22b) and of the *Yamabe theorem*, Theorem 11.1, one can assume that $r[\boldsymbol{h}]$ is a negative constant on $\mathcal{S}$ – such a metric is said to be of ***positive Yamabe class***. An example of this situation is $\mathbb{S}^3$ with its standard metric. One is then left with a Licnerowicz equation of the form

$$\Delta_{\boldsymbol{h}}\vartheta = \frac{1}{8}r[\boldsymbol{h}]\vartheta.$$

If ϑ is required to be positive everywhere on $\mathcal{S}$, it follows that $\Delta_{\boldsymbol{h}}\vartheta < 0$ everywhere so that no positive solution can exist since, as a consequence of the maximum principle, ϑ must be a constant so that $\Delta_{\boldsymbol{h}}\vartheta = 0$ which is a contradiction. To get around this situation one can consider initial data with a negative (i.e. de Sitter-like) cosmological constant. Keeping the time symmetry of the initial data and the condition $r[\boldsymbol{h}] < 0$, one obtains the Licnerowicz equation

$$\Delta_{\boldsymbol{h}}\vartheta = \frac{1}{8}r[\boldsymbol{h}]\vartheta - \frac{1}{4}\lambda\vartheta^3. \tag{11.43}$$

The right-hand side of this equation has no definite sign for positive ϑ, so there is no obstruction to the existence of solutions. In any case, a further argument (not discussed here) is required to show that Equation (11.43) does indeed have a solution.

The methods in Isenberg (1995) allow one to prove the following proposition:

Proposition 11.3 (***solvability of the Einstein constraints with cosmological constant on a compact manifold***) *Let $(\mathcal{S}, \boldsymbol{h})$ be a Riemannian manifold with $\mathcal{S} \approx \mathbb{S}^3$ and $\boldsymbol{h}$ conformal to a metric with constant negative Ricci scalar (positive Yamabe class). Then the vacuum Einstein constraints with de Sitter-like cosmological constant have a solution for an arbitrary choice of the seed metric $\boldsymbol{h}$, trace-free tensor ψ'_{ij} and constant physical mean curvature $\tilde{K}$.*

The initial data sets given by this proposition will be used to construct de Sitter-like spacetimes in Chapter 15.

11.6 Asymptotically Euclidean manifolds

Spacetimes with $\lambda = 0$ can be thought of as describing *isolated systems* for which the effects of cosmological expansion are neglected. An important class of these spacetimes consists of those solutions to the Einstein field equations which are asymptotically simple in the sense of Definition 7.1, that is, asymptotically simple and empty. Proposition 14.3 shows that these spacetimes are globally hyperbolic, suggesting a systematic procedure for their construction through suitable initial data prescribed on a Cauchy hypersurface.

In order to develop intuition, it is convenient to look at the *Minkowski spacetime* $(\mathbb{R}^4, \tilde{\boldsymbol{\eta}})$. A foliation of this spacetime is given by the hypersurfaces of constant time t. These hypersurfaces are Riemannian manifolds of the form $(\mathbb{R}^3, -\boldsymbol{\delta})$. It can be verified that these hypersurfaces are extrinsically flat; that is, their extrinsic curvature $\tilde{\boldsymbol{K}} = 0$ vanishes. Of course, these are not the only possible types of Cauchy hypersurfaces in this spacetime.

As a second example, consider the Schwarzschild spacetime. In terms of the so-called Schwarzschild ***isotropic radial coordinate***

$$\bar{r} \equiv \frac{1}{2}\left(r - m + \sqrt{r(r-2m)}\right),$$

the line element of the spacetime can be rewritten as

$$\tilde{\boldsymbol{g}}_{\mathscr{S}} = \left(\frac{1 - m/2\bar{r}}{1 + m/2\bar{r}}\right)^2 \mathbf{d}t \otimes \mathbf{d}t - \left(1 + \frac{m}{2\bar{r}}\right)^4 (\mathbf{d}\bar{r} \otimes \mathbf{d}\bar{r} + \bar{r}^2 \boldsymbol{\sigma}).$$

An example of a Cauchy hypersurface for this spacetime is given by the $t = 0$ hypersurface. One can verify that the intrinsic metric and the extrinsic curvature of this hypersurface are given, respectively, by

$$\tilde{\boldsymbol{h}}_{\mathscr{S}} = -\left(1 + \frac{m}{2\bar{r}}\right)^4 \boldsymbol{\delta}, \qquad \tilde{\boldsymbol{K}}_{\mathscr{S}} = 0. \tag{11.44}$$

The most general form of the above initial data set is obtained by performing a translation of the radial coordinate to obtain

$$\tilde{\boldsymbol{h}}_{\mathscr{S}} = -\left(1 + \frac{m}{2|y - y_0|}\right)^4 \boldsymbol{\delta}, \tag{11.45}$$

with $|y - y_0|^2 \equiv (y^1 - y_0^1)^2 + (y^2 - y_0^2)^2 + (y^3 - y_0^3)^2$ where $(y^\alpha) = (y^1, y^2, y^3)$ are standard Cartesian coordinates and $(y_0^\alpha) \in \mathbb{R}^3$ arbitrary. Observe that the metric $\tilde{\boldsymbol{h}}_{\mathscr{S}}$ is, in fact, conformally flat and that $\tilde{\boldsymbol{h}}_{\mathscr{S}} \to -\boldsymbol{\delta}$ as $\bar{r} \to \infty$. Moreover, one has that

$$\left(1 + \frac{m}{2\bar{r}}\right)^4 = 1 + \frac{2m}{\bar{r}} + O\left(\frac{1}{\bar{r}^2}\right). \tag{11.46}$$

To understand the behaviour as $\bar{r} \to 0$, it is observed that under the coordinate inversion $\breve{r} \equiv m^2/4\bar{r}$ one has that

$$\tilde{\boldsymbol{h}}_{\mathscr{S}} = -\left(1 + \frac{m}{2\breve{r}}\right)^4 (\mathbf{d}\breve{r} \otimes \mathbf{d}\breve{r} + \breve{r}^2 \boldsymbol{\sigma}).$$

Figure 11.1 Embedding diagram of the Einstein-Rosen bridge in the standard time-symmetric Schwarzschild hypersurface. The diagram is obtained as the surface of revolution of the curve $z = \pm\ln(1+\sqrt{r^2-1})$; see Morris and Thorne (1988) for more details.

Thus, the behaviour of the metric $\tilde{\boldsymbol{h}}$ is identical for both $\bar{r}, \breve{r} \to \infty$. There is a discrete reflexion symmetry with respect to the two-dimensional surface $\{r = m/2\}$. Thus, the topology of the hypersurface is $\mathcal{S} \approx \mathbb{R} \times \mathbb{S}^2$. One says that $\mathcal{S}$ has a ***non-trivial topology*** with two asymptotically flat regions (see next section) joined by a so-called ***Einstein-Rosen bridge***. A representation of this is given in Figure 11.1.

An example of an initial data set with non-vanishing extrinsic curvature is given by the family of conformally flat initial data sets for the Schwarzschild spacetime with extrinsic curvature given by

$$\tilde{K}^{\alpha\beta} = \frac{A}{|y|^3}(3y^\alpha y^\beta + |y|^2\delta^{\alpha\beta}),$$

where $|y|^2 = \delta_{\alpha\beta}y^\alpha y^\beta$ and (y^α) are, again, standard Cartesian coordinates; see Beig and O'Murchadha (1998), Estabrook et al. (1973) and Reinhart (1973). It can be verified that $\tilde{K} = \tilde{h}^{\alpha\beta}\tilde{K}_{\alpha\beta} = 0$ as $h_{\alpha\beta}y^\alpha y^\beta = -|y|^2$. This hypersurface has the nontrivial topology of $\mathbb{R}\times\mathbb{S}^2$. However, in contrast to the time-symmetric case, the conformal factor ϑ cannot be written in a closed form. Nevertheless, the leading terms of its asymptotic expansion are the same as in Equation (11.46) with $|y|$ playing the role of the radial coordinate $\bar{r}$.

11.6.1 Definition in terms of physical fields

The hypersurfaces discussed in the previous paragraphs are examples of ***asymptotically Euclidean manifolds***. Given a three-dimensional manifold $\tilde{\mathcal{S}}$, an ***asymptotic end*** is a subset $\tilde{\mathcal{E}} \subset \tilde{\mathcal{S}}$ which is diffeomorphic to the complement of a closed ball on $\mathbb{R}^3$; that is,

$$\tilde{\mathcal{E}} \approx \left\{(y^\alpha) \in \mathbb{R}^3 \mid |y| > r_0\right\},$$

where r_0 is some positive real number and $|y|^2 \equiv \delta_{\alpha\beta}y^\alpha y^\beta$. By identifying $\tilde{\mathcal{E}}$ with the complement of a ball, the triple $\underline{y} = (y^\alpha)$ can be used as coordinates on the asymptotic end – so-called ***asymptotically Cartesian coordinates***. The hypersurfaces of the Minkowski and Schwarzschild spacetimes have, respectively, one and two asymptotic ends. More generally, a three-dimensional manifold $\tilde{\mathcal{S}}$ is said to have N ***asymptotically flat ends*** if there exists a compact subset of $\tilde{\mathcal{S}}$ such that its complement is the union of disjoint subsets $\tilde{\mathcal{E}}_k$, $k = 1, 2, \ldots, N$, each of which is an asymptotic end. In terms of the above concepts one can now introduce the key definition of this section:

Definition 11.1 (***asymptotically Euclidean manifolds***) *An initial data set for the vacuum Einstein field equations* $(\tilde{\mathcal{S}}, \tilde{\boldsymbol{h}}, \tilde{\boldsymbol{K}})$ *is said to be* ***asymptotically Euclidean*** *if* $\tilde{\mathcal{S}}$ *is a three-dimensional manifold with* N *asymptotically flat ends* $\tilde{\mathcal{E}}_k$, $k = 1, \ldots, N$ *such that on each* $\tilde{\mathcal{E}}_k$ *the 3-metric and the extrinsic curvature satisfy, in terms of asymptotically Cartesian coordinates on the end, the asymptotic behaviour*

$$\tilde{h}_{\alpha\beta} = -\left(1 + \frac{2m_k}{|y|}\right)\delta_{\alpha\beta} + O_2\left(\frac{1}{|y|^2}\right), \tag{11.47a}$$

$$\tilde{K}_{\alpha\beta} = O_1\left(\frac{1}{|y|^2}\right), \tag{11.47b}$$

where m_k, $k = 1, \ldots, N$ *are constants.*

The notation O_1 and O_2 in Equations (11.47a) and (11.47b) is explained in the Appendix to this chapter. More general notions of asymptotic flatness for three-dimensional manifolds have been considered in the literature; see, for example, Chaljub-Simon (1982), Chaljub-Simon and Choquet-Bruhat (1980), Choquet-Bruhat and York (1980) and Christodoulou and O'Murchadha (1981). Their precise formulation require the use of the notion of weighted Sobolev spaces; see, for example, appendix I of Choquet-Bruhat (2008) and Bartnik (1986). These definitions are tailored for the analysis of the elliptic equations arising from the constraint equations.

The asymptotic conditions in Definition 11.1 ensure the finiteness of the ***ADM-linear momentum*** and ***ADM-angular momentum***[1] of each asymptotic end. These asymptotic quantities are given, respectively, by the surface integrals

$$P_\alpha \equiv \frac{1}{8\pi}\lim_{r\to\infty}\int_{\mathcal{S}_r}(\tilde{K}_{\alpha\beta} - \tilde{K}\tilde{h}_{\alpha\beta})n^\beta \mathrm{d}S_{\tilde{h}},$$

$$J_\alpha \equiv \frac{1}{8\pi}\lim_{r\to\infty}\int_{\mathcal{S}_r}\tilde{\epsilon}_{\alpha\beta\gamma}y^\beta(\tilde{K}^{\gamma\delta} - \tilde{K}\tilde{h}^{\gamma\delta})n_\delta \mathrm{d}S_{\tilde{h}}$$

with

$$\mathcal{S}_r \equiv \{(y^\alpha) \in \tilde{\mathcal{S}} \mid |y| = r\},$$

[1] ADM stands for Arnowitt-Deser-Miser, pioneers of the *Hamiltonian* formulation of general relativity; see Arnowitt et al. (1962) and Arnowitt et al. (2008) for a republication of this classical review.

n^α its outward pointing normal and $\mathrm{d}S_{\tilde{\boldsymbol{h}}}$ the surface element induced by $\tilde{\boldsymbol{h}}$ on S_r. The constants m_k in Definition 11.1 correspond to the ***ADM mass*** of each of the asymptotic ends. They are also computable as surface integrals of the *sphere at infinity* via the expression

$$m = -\frac{1}{16\pi}\lim_{r\to\infty}\int_{S_r}\tilde{h}^{\alpha\beta}(\partial_\alpha\tilde{h}_{\beta\gamma} - \partial_\gamma\tilde{h}_{\alpha\beta})n^\gamma \mathrm{d}S_{\tilde{\boldsymbol{h}}}.$$

Strictly speaking, m is the *time component* of a 4-vector, the ***ADM 4-momentum***, whose *spatial components* are given by P^α; thus, it is more accurately described as an *energy*.

Definition 11.1 can be extended to initial data sets with matter. In these situations, decay conditions for the matter sources which are compatible with the decay for $\tilde{\boldsymbol{h}}$ and $\tilde{\boldsymbol{K}}$ are given by (11.47a) and (11.47b). Direct inspection of the constraint Equation (11.14) suggests that

$$\tilde{\varrho} = O\left(\frac{1}{|y|^3}\right), \qquad \tilde{j}_\alpha = O\left(\frac{1}{|y|^3}\right).$$

These conditions can be refined via a more careful analysis of the constraint equations.

It is possible to have an initial data set with several asymptotic ends, some of which are not asymptotically Euclidean. The simplest example is given by the extremal Reissner-Nordström spacetime; see Equation (6.43). The intrinsic metric of the hypersurface $t = 0$ is given, in terms of the extremal Reissner-Nordström ***isotropic radial coordinate*** $\bar{r} = r - m$, by

$$\tilde{\boldsymbol{h}} = -\left(1 + \frac{m}{\bar{r}}\right)^2 \boldsymbol{\delta}. \tag{11.48}$$

Clearly

$$\left(1 + \frac{m}{\bar{r}}\right)^2 = 1 + \frac{2m}{\bar{r}} + O\left(\frac{1}{\bar{r}^2}\right) \quad \text{as } \bar{r} \to \infty.$$

Thus, for large $\bar{r}$, the extremal Reissner-Nordström 3-metric (11.48) has an asymptotically Euclidean end. To discuss the behaviour as $\bar{r} \to 0$, consider the new radial coordinate $\check{r} = -\ln \bar{r}$, so that $\check{r} \to \infty$ as $\bar{r} \to 0$. It follows that in terms of this coordinate the metric (11.48) can be rewritten as

$$\tilde{\boldsymbol{h}} = -(m + e^{-\check{r}})(\mathbf{d}\check{r} \otimes \mathbf{d}\check{r} + \boldsymbol{\sigma}).$$

This metric approaches a constant multiple of the standard metric of the cylinder $\mathbb{R}^+ \times \mathbb{S}^2$ as $\check{r} \to \infty$. Accordingly, one speaks of a ***cylindrical asymptotic end***. A similar type of asymptotic behaviour can be found, for example, in hypersurfaces of the extremal Kerr spacetime; see, for example, Dain and Gabach-Clement (2011).

11.6.2 Definition using conformal rescalings

The notion of asymptotically Euclidean manifolds can be strengthened by requiring the physical hypersurface $\tilde{\mathcal{S}}$ to have a conformal extension which is a ***point compactification***. This approach provides a more geometrical setting for the discussion of the asymptotic behaviour of the various fields, that is, independent of the use of particular *asymptotically Cartesian coordinates.* This point of view was first introduced by Geroch (1972b).

Definition 11.2 (***asymptotically Euclidean and regular manifolds***) *A three-dimensional Riemannian manifold $(\tilde{\mathcal{S}}, \tilde{\boldsymbol{h}})$ will be said to be* ***asymptotically Euclidean and regular*** *if there exists a three-dimensional, orientable, compact manifold $(\mathcal{S}, \boldsymbol{h})$ with points $i_k \in \mathcal{S}$, $k = 1, \ldots, N$ with N some integer, a diffeomorphism $\varphi : \mathcal{S} \setminus \{i_1, \ldots, i_N\} \to \tilde{\mathcal{S}}$ and a function $\Omega \in C^2$ such that:*

(i) $\Omega(i_k) = 0$, $\mathbf{d}\Omega(i_k) = 0$, $\mathbf{Hess}\,\Omega(i_k) = -2\boldsymbol{h}(i_k)$.
(ii) $\Omega > 0$ *on* $\mathcal{S} \setminus \{i_1, \ldots, i_N\}$.
(iii) $\boldsymbol{h} = \Omega^2 \varphi^* \tilde{\boldsymbol{h}}$ *on* $\mathcal{S} \setminus \{i_1, \ldots, i_N\}$ *with* $\boldsymbol{h} \in C^2(\mathcal{S}) \cap C^\infty(\mathcal{S} \setminus \{i_1, \ldots, i_N\})$.

More generally, a function $\Lambda^{1/2}$ such that Λ satisfies conditions (i) in the above definition is called an ***asymptotic distance function***. The function Λ does not need to be defined globally on $\mathcal{S}$.

When no confusion arises, condition (iii) in Definition 11.2 will simply be written as $\boldsymbol{h} = \Omega^2 \tilde{\boldsymbol{h}}$ so that $\mathcal{S} \setminus \{i_1, \ldots, i_N\}$ and $\tilde{\mathcal{S}}$ are identified. As will be seen in the following, for asymptotically Euclidean and regular manifolds, suitable neighbourhoods of the points i_k – the ***points at infinity*** – are mapped to the asymptotic ends of $\tilde{\mathcal{S}}$. Thus, one can use local differential geometry to discuss the asymptotic properties of the initial data set $(\tilde{\mathcal{S}}, \tilde{\boldsymbol{h}})$. The question of the differentiability of Ω and $\boldsymbol{h}$ at $i_1, \ldots, i_N$ will be addressed later in this subsection. Definition 11.2 is purely Riemannian; that is, it makes no reference to the extrinsic curvature. The behaviour of the extrinsic curvature at the points at infinity will be discussed in the subsequent paragraphs.

There is some ***conformal gauge freedom*** in Definition 11.2. A replacement of the form

$$\boldsymbol{h} \mapsto \phi^4 \boldsymbol{h}, \qquad \Omega \mapsto \phi^2 \Omega, \tag{11.49}$$

with $\phi(i_k) = 1$ gives rise to the same physical metric $\tilde{\boldsymbol{h}} = \Omega^{-4}\boldsymbol{h}$ and preserves the boundary conditions in point (i) of the definition. This gauge freedom can be used to select conformal metrics with special properties. For example, given a particular point at infinity i, and choosing ϕ such that

$$\Delta_{\boldsymbol{h}}\phi - \frac{1}{8} r[\boldsymbol{h}]\phi = 0 \qquad \text{on } \mathcal{B}_\varepsilon(i), \tag{11.50}$$

with $\mathcal{B}_\varepsilon(i)$ the ball of radius ε centred at i for some $\varepsilon > 0$, it follows from Equation (11.23) that

$$r[\boldsymbol{h}'] = 0 \qquad \text{on } \mathcal{B}_\varepsilon(i).$$

A general property of elliptic equations with smooth coefficients is that they can always be solved locally; see, for example, Besse (2008) and Garabedian (1986). Thus, the requirement (11.50) can always be satisfied. In other words, the conformal metric $\boldsymbol{h}$ can always be chosen so that it vanishes in a neighbourhood of one of the points at infinity. In general, this statement is not true globally.

Normal coordinates around i

The consequences of Definition 11.2 are better analysed by means of ***normal coordinates***. Consider the set of $\boldsymbol{h}$-geodesics $\gamma_{\boldsymbol{v}} \subset \mathcal{S}$ starting at a particular point at infinity i (i.e. $\gamma_{\boldsymbol{v}}(0) = i$) with initial velocity $\boldsymbol{v} \in T|_i(\mathcal{S})$. Moreover, let $\mathcal{T}$ denote the subset of $T|_i(\mathcal{S})$ defined by

$$\mathcal{T} \equiv \big\{\boldsymbol{v} \in T|_i(\mathcal{S}) \,\big|\, \gamma_{\boldsymbol{v}} \text{ is defined on an interval containing } [0,1]\big\}.$$

On the set $\mathcal{T}$ one can define the ***exponential map at*** i, $\exp_i : \mathcal{T} \to \mathcal{S}$, through the condition $\exp_i(\boldsymbol{v}) = \gamma_{\boldsymbol{v}}(1)$; that is, the exponential map sends the vector $\boldsymbol{v}$ to the point at a unit parameter distance along the unique geodesic through i with initial velocity $\boldsymbol{v}$. It can be shown that *there exists a neighbourhood* $\mathcal{Q} \subset T|_i(\mathcal{S})$ *of the vector* $\mathbf{0}$ *such that the exponential map at i gives a diffeomorphism onto a neighbourhood* $\mathcal{U} \subset \mathcal{S}$ *of i*; see, for example, O'Neill (1983) for a proof. If $\boldsymbol{v} \in \mathcal{Q}$ implies that $\lambda\boldsymbol{v} \in \mathcal{Q}$ for all $\lambda \in [0,1]$, then one says that $\mathcal{Q}$ is ***star shaped*** and $\mathcal{U} = \exp_i(\mathcal{Q})$ is called a ***normal neighbourhood of*** i. In particular, if $\mathcal{U} = \mathcal{B}_\varepsilon(i)$, the open ball of radius $\varepsilon > 0$ with respect to $\boldsymbol{h}$, one has a ***geodesic ball***.

In what follows, assume one has a normal neighbourhood $\mathcal{U}$ around i and that one is given an orthonormal basis $\{\boldsymbol{e_i}\}$ for $T|_i(\mathcal{S})$. Given $p \in \mathcal{U}$ and $\boldsymbol{v}$ such that $p = \exp(\boldsymbol{v})$, then writing $\boldsymbol{v} = x^{\boldsymbol{i}}\boldsymbol{e_i}$ one can use the components $\underline{x} = (x^{\boldsymbol{i}}) \in \mathbb{R}^3$ as coordinates for the point p – these are the ***normal coordinates*** determined by the basis $\{\boldsymbol{e_i}\}$. For consistency, the normal coordinates will be written as (x^α) rather than $(x^{\boldsymbol{i}})$. In terms of normal coordinates a geodesic through the origin has the form $x(\mathrm{s}) = (\mathrm{s}\,x^\alpha)$ where s is an affine parameter. As $\dot{\boldsymbol{x}} = (x^\alpha)$ and $\ddot{\boldsymbol{x}} = 0$, it follows from the geodesic equation that $\gamma_\beta{}^\alpha{}_\gamma(i)x^\beta x^\gamma = 0$ with $\gamma_\beta{}^\alpha{}_\gamma$ being the Christoffel symbols of the metric $\boldsymbol{h}$. As this has to hold for any geodesic on $\mathcal{U}$, one concludes that $\gamma_\beta{}^\alpha{}_\gamma(i) = 0$. It also follows that $\partial_\alpha h_{\beta\gamma} = 0$, so that one can write

$$h_{\alpha\beta} = -\delta_{\alpha\beta} + O(|x|^2) \qquad \text{close to } i, \tag{11.51}$$

where $|x|^2 \equiv \delta_{\alpha\beta}x^\alpha x^\beta$. Moreover, from the above construction it follows that

$$x^\alpha h_{\alpha\beta} = -\delta_{\alpha\beta}x^\alpha. \tag{11.52}$$

For future use it is observed that, in terms of normal coordinates and sufficiently close to i, one has

$$\mathrm{d}\mu_{\boldsymbol{h}} = |x|^2 \mathrm{d}\sigma + O(|x|^3), \tag{11.53}$$

where $\mathrm{d}\mu_{\boldsymbol{h}}$ is the volume element of the metric $\boldsymbol{h}$ and $\mathrm{d}\sigma$ denotes the area element of the unit 2-sphere $\mathbb{S}^2$.

For later use, it is convenient to define the ***(square of the) geodesic distance*** $\Gamma^2 \equiv |x|^2$. One has that Γ^2 is a smooth function of the normal coordinates. It can be verified that

$$h^{\alpha\beta} D_\alpha \Gamma D_\beta \Gamma = -4\Gamma \tag{11.54}$$

and that

$$\Gamma(i) = 0, \qquad D_\alpha \Gamma(i) = 0, \qquad D_\alpha D_\beta \Gamma(i) = -2h_{\alpha\beta}, \qquad D_\alpha D_\beta D_\gamma \Gamma(i) = 0.$$

Hence, Γ satisfies the boundary conditions (i) in Definition 11.2 so it is an asymptotic distance function. Observe that, in general, Γ is not defined globally on $\mathcal{S}$.

Remark. The results obtained using normal coordinates can be strengthened by exploiting the conformal freedom in (11.49). In particular, a conformal factor can always be found such that the Riemann curvature tensor of the resulting rescaled metric vanishes at i. In order to see this, given the metric $\boldsymbol{h}$, let $\Omega' \equiv e^f$ with $f \in \mathfrak{X}(\mathcal{S})$ such that

$$f = \frac{1}{2} x^\alpha x^\beta l_{\alpha\beta}(i) \quad \text{on } \mathcal{B}_\varepsilon(i),$$

where $l_{\alpha\beta}(i)$ denotes the components with respect to the normal coordinates of the three-dimensional Schouten tensor at i. A calculation shows that

$$\Omega'(i) = 1, \qquad D_\alpha \Omega'(i) = 0, \qquad D_\alpha D_\beta \Omega'(i) = l_{\alpha\beta}(i).$$

Hence, using the conformal transformation formula for the Schouten tensor (5.16b) one finds that the metric $\boldsymbol{h}' \equiv \Omega'^2 \boldsymbol{h}$ satisfies $l'_{\alpha\beta}(i) = 0$. As in dimension 3 the Riemann tensor is completely determined by the Schouten tensor one concludes that $r'_{\alpha\beta\gamma\delta}(i) = 0$ as claimed. The metric $\boldsymbol{h}'$ satisfies the *improved* expansion

$$h'_{\alpha\beta} = -\delta_{\alpha\beta} + O(|x|^3) \qquad \text{close to } i;$$

compare with (11.51).

The construction described in the previous paragraph is not the only possible way of exploiting the conformal gauge freedom. Depending on the particular analysis, other choices may be more convenient – for example, the *conformal normal gauge* introduced in Friedrich and Schmidt (1987) and Friedrich (1998c) or the *central harmonic gauge* used in Friedrich (2013).

Asymptotically Euclidean data versus asymptotically Euclidean and regular data

It is useful to compare the two definitions of asymptotic flatness presented in this Section: Definitions 11.1 and 11.2. Condition (i) in Definition 11.2 restricts the form of the conformal factor Ω in a neighbourhood $\mathcal{B}_a(i_k)$ of a given point at infinity i_k. More precisely, one has that

$$\Omega = |x|^2 f(\underline{x}) \quad \text{near } i_k, \tag{11.55}$$

where f is continuous with $f(0) = 1$. Given the normal coordinates (x^α) on $\mathcal{B}_a(i_k)$, one can introduce inversion coordinates $y^\alpha \equiv x^\alpha/|x|^2$ so that

$$\tilde{h}_{\alpha\beta} = \Omega^{-2} h_{\alpha\beta} = -\delta_{\alpha\beta} + O(|y|^{-1}) \quad \text{as} \quad |y| \to \infty.$$

Thus, to recover the mass term in the expansion (11.47a) one requires further information about the fields Ω and $\boldsymbol{h}$.

With regards to the second fundamental form, it follows from the transformation rules discussed in Section 11.1.1 that

$$\tilde{K}_{\alpha\beta} = \Omega^{-1} K_{\alpha\beta} = \Omega \psi_{\alpha\beta}.$$

Hence, if the physical field $\tilde{K}_{\alpha\beta}$ satisfies the decay given by condition (11.47b), then

$$\tilde{K}_{\alpha\beta} = O(|x|^0), \qquad \psi_{\alpha\beta} = O(|x|^{-4}), \qquad \text{as } |x| \to 0.$$

Consequently, the decay conditions of Definition 11.1 imply a tensor $\psi_{\alpha\beta}$ which is singular at the points at infinity. To have a regular $\psi_{\alpha\beta}$ one needs the stronger decay condition $\tilde{K}_{\alpha\beta} = O(1/|y|^6)$. This decay excludes the possibility of a non-vanishing *ADM linear momentum* and *ADM angular momentum.*

The regularity at the points at infinity

The regularity requirements on Ω and $\boldsymbol{h}$ of Definition 11.2 are given with respect to some suitable coordinate system. A natural choice is the normal coordinates $\underline{x} = (x^\alpha)$ centred at the point at infinity – intuitively, one expects the regularity with respect to normal coordinates to be optimal. In these coordinates the function $|x|$ is not smooth at i as its second derivative is not well defined there. More generally, even powers of $|x|$ will be smooth, while odd ones will be only C^k, for some k.

Initial data sets for static vacuum spacetimes admit a conformal metric which is, in fact, analytic at the point at infinity; see Beig and Simon (1980b) and Beig and Schmidt (2000). Remarkably, this is not the case for stationary solutions which can be seen to be only C^2 at the point at infinity; see Dain (2001b). More precisely, *any asymptotically Euclidean data set for a stationary spacetime (and in particular for the Kerr solution!) has a conformal metric of the form* $\boldsymbol{h}$ *which, in a suitable neighbourhood* $\mathcal{B}_a(i_k)$ *of infinity, takes the form*

$$\boldsymbol{h} = \boldsymbol{h}' + |x|^3 \boldsymbol{h}'',$$

with $\boldsymbol{h}'$ *and* $\boldsymbol{h}''$ *analytic tensors with respect to normal coordinates.*

11.6.3 Fundamental solutions and punctures

Consider now, for simplicity, an asymptotically Euclidean and regular manifold $(\mathcal{S}, \boldsymbol{h})$ with a single point at infinity i. Suppose, for ease of the presentation, that the conformal factor $\Omega = \vartheta^{-2}$ satisfies the ***Yamabe equation***

$$\Delta_{\boldsymbol{h}}\vartheta - \frac{1}{8}r[\boldsymbol{h}]\vartheta = 0 \qquad \text{on } \mathcal{S}\setminus\{i\}. \tag{11.56}$$

Condition (i) of Definition 11.2 implies a singular behaviour for the conformal factor ϑ. Indeed, from Equation (11.55) it follows that

$$\vartheta|x| \to 1 \qquad \text{as } |x| \to 0. \tag{11.57}$$

In order to develop a better understanding of the singular behaviour at i consider the integral

$$I_\varepsilon \equiv \int_{\mathcal{B}_\varepsilon(i)} \left(\Delta_{\boldsymbol{h}}\vartheta - \frac{1}{8}r[\boldsymbol{h}]\vartheta \right) \mathrm{d}\mu_{\boldsymbol{h}}$$

over an open ball $\mathcal{B}_\varepsilon(i)$ of a suitably small radius $\varepsilon > 0$ centred at i. To simplify the evaluation of the integral it is assumed that the metric $\boldsymbol{h}$ has been chosen such that $r[\boldsymbol{h}] = 0$ on $\mathcal{B}_\varepsilon(i)$; as seen in Section 11.6.2 this is is always possible locally. Using the *divergence theorem* (see the Appendix to this chapter), one has that

$$I_\varepsilon = \int_{\mathcal{B}_\varepsilon(i)} \Delta_{\boldsymbol{h}}\vartheta \mathrm{d}\mu_{\boldsymbol{h}} = -\int_{\partial\mathcal{B}_\varepsilon(i)} \langle \mathbf{d}\vartheta, \boldsymbol{n}\rangle \mathrm{d}S_{\boldsymbol{h}},$$

where $\boldsymbol{n}$ is the outward-pointing unit normal to $\partial\mathcal{B}_\varepsilon(i)$ and $\mathrm{d}\sigma_{\boldsymbol{h}}$ is the surface element of $\partial\mathcal{B}_\varepsilon(i)$ implied by $\boldsymbol{h}$. From the expansion (11.53) it follows for sufficiently small ε that $\mathrm{d}S_{\boldsymbol{h}} = \varepsilon^2\mathrm{d}\sigma + o(\varepsilon^2)$ with $\mathrm{d}\sigma$ the surface element of $\mathbb{S}^2$. Moreover, as a consequence of (11.57) one has

$$\langle \mathbf{d}\vartheta, \boldsymbol{n}\rangle = -\frac{1}{\varepsilon^2} + o(\varepsilon^{-2}).$$

Putting everything together one concludes that

$$I_\varepsilon = -\int_{\partial\mathcal{B}_\varepsilon(i)} \mathrm{d}\sigma + o(\varepsilon) \longrightarrow 4\pi \qquad \text{as } \varepsilon \to 0,$$

so that

$$\int_{\mathcal{S}} \Delta_{\boldsymbol{h}}\vartheta \mathrm{d}\mu_{\boldsymbol{h}} = 4\pi.$$

The latter implies that one can write

$$\Delta_{\boldsymbol{h}}\vartheta = 4\pi\delta(i),$$

where $\delta(i)$ denotes the ***Dirac's delta distribution*** with support at the point i; see the Appendix to this chapter for more details and references. To obtain the expression for a generic metric with a non-vanishing Ricci scalar in a neighbourhood of i one makes use of the transformation law for the Yamabe equation, Equation (11.23), to obtain

$$\left(\Delta_{\boldsymbol{h}'} - \frac{1}{8}r[\boldsymbol{h}']\right)(\phi^{-1}\vartheta) = 4\pi\phi^{-5}\delta(i) \qquad \text{with } \boldsymbol{h}' = \phi^4\boldsymbol{h}.$$

As $\delta(i)$ has support only on i and $\phi(i) = 1$ one finally concludes that

$$\left(\Delta_{\boldsymbol{h}'} - \frac{1}{8}r[\boldsymbol{h}']\right)\vartheta' = 4\pi\delta(i) \qquad \text{with } \vartheta' = \phi^{-1}\vartheta.$$

This expression provides an alternative description of the singular behaviour of solutions to the Yamabe equation which satisfy the boundary condition (i) of Definition 11.2. The previous discussion can be generalised to manifolds $(\mathcal{S}, \boldsymbol{h})$ with several points at infinity. For example, if $\mathcal{S} = \mathbb{S}^3$ and $\boldsymbol{h} = -\boldsymbol{\hbar}$ the standard metric of $\mathbb{S}^3$, then the Yamabe equation

$$\left(\Delta_{-\boldsymbol{\hbar}} - \frac{1}{8}r[-\boldsymbol{\hbar}]\right)\vartheta = 4\pi\Big(\delta(i_N) + \delta(i_S)\Big),$$

where $\delta(i_N)$ and $\delta(i_S)$ are supported, respectively, at the north and south poles of $\mathbb{S}^3$, describes the conformal factor ϑ for time-symmetric data for the Schwarzschild spacetime. Letting $\phi \equiv 1 + m/2r$, it follows from combining the first equation in (11.44) with the conformal factor ω compactifying $\mathbb{R}^3$ into $\mathbb{S}^3$ given in (6.5) that

$$\boldsymbol{h} = -\Omega^2\boldsymbol{\hbar}, \qquad \Omega = \omega\phi^{-2}.$$

Setting $\alpha = 1$ in Equation (6.5), one has that

$$\Omega = \frac{2\sin^2\dfrac{\psi}{2}}{\left(1 + \dfrac{m}{2}\tan\dfrac{\psi}{2}\right)^2}.$$

One can verify that Ω and $\mathbf{d}\Omega$ vanish at $\psi = 0, \pi$ (the north and south poles of $\mathbb{S}^3$). Moreover, one has

$$\Omega = \psi^2 + O(\psi^3), \qquad \Omega = (\psi - \pi)^2 + O((\psi - \pi)^3),$$

so that the fundamental solution $\vartheta = \Omega^{-1/2}$ has the expected singular behaviour.

Conformal decompactification of initial data sets on compact manifolds

From a geometric point of view, the purpose of introducing a conformal factor ϑ which is singular at the point at infinity is to produce a ***conformal decompactification*** of the manifold $\mathcal{S}$. As an example, consider a *vacuum maximal initial data set* $(\tilde{\mathcal{S}}, \tilde{\boldsymbol{h}}, \tilde{\boldsymbol{K}})$ with $\tilde{\mathcal{S}}$ compact. Under these assumptions one has that the Einstein constraints (11.14) reduce to

$$r[\tilde{\boldsymbol{h}}] = -\tilde{K}_{ij}\tilde{K}^{ij}, \qquad \tilde{D}^i\tilde{K}_{ij} = 0, \qquad \tilde{K} = \tilde{h}^{ij}\tilde{K}_{ij} = 0.$$

If given a point $i \in \tilde{\mathcal{S}}$ one can find a solution $\bar{\vartheta}$ to the equation

$$\Delta_{\tilde{\boldsymbol{h}}}\bar{\vartheta} - \frac{1}{8}r[\tilde{\boldsymbol{h}}]\bar{\vartheta} - \frac{1}{8}r[\tilde{\boldsymbol{h}}]\bar{\vartheta}^{-7} = 4\pi\delta(i),$$

it follows from a calculation involving the conformal transformation properties of the various fields that

$$\bar{\boldsymbol{h}} \equiv \bar{\vartheta}\tilde{\boldsymbol{h}}, \qquad \bar{\boldsymbol{K}} \equiv \vartheta^{-2}\tilde{\boldsymbol{K}},$$

gives rise to an asymptotically Euclidean and regular solution to the Einstein constraints

$$r[\bar{\boldsymbol{h}}] = -\bar{K}_{ij}\bar{K}^{ij}, \qquad \bar{D}^i\bar{K}_{ij} = 0, \qquad \bar{K} \equiv \bar{h}^{ij}\bar{K}_{ij} = 0.$$

As pointed out in O'Murchadha (1988), this construction can be used to argue that, in a certain sense, there are more asymptotically flat initial data sets than initial data sets on compact surfaces.

The Yamabe invariant

The possibility of conformally decompactifying a compact Riemannian manifold $(\mathcal{S}, \boldsymbol{h})$ to obtain a physical manifold $(\tilde{\mathcal{S}}, \tilde{\boldsymbol{h}})$ which is asymptotically Euclidean and regular depends on being able to solve the equation

$$\left(\Delta_{\boldsymbol{h}} - \frac{1}{8}r[\boldsymbol{h}]\right)\vartheta = 4\pi\delta(i). \tag{11.58}$$

The discussion in Section 11.5 suggests that this may not be possible for all cases. To explore this further, consider a *test function* $\phi \in \mathfrak{X}(\mathcal{S})$. A calculation shows that

$$\begin{aligned}\int_{\mathcal{S}} |D(\vartheta\phi)|^2 \mathrm{d}\mu_{\boldsymbol{h}} &= \int_{\mathcal{S}} (\vartheta^2|D\phi|^2 + \phi^2|D\vartheta|^2)\mathrm{d}\mu_{\boldsymbol{h}} + \int_{\mathcal{S}} \vartheta D_i\vartheta D^i\phi^2 \mathrm{d}\mu_{\boldsymbol{h}} \\ &= \int_{\mathcal{S}} (\vartheta^2|D\phi|^2 + \phi^2|D\vartheta|^2)\mathrm{d}\mu_{\boldsymbol{h}} - \int_{\mathcal{S}} D^i(\vartheta D_i\vartheta)\phi^2 \mathrm{d}\mu_{\boldsymbol{h}} \\ &= \int_{\mathcal{S}} \vartheta^2|D\phi|^2 \mathrm{d}\mu_{\boldsymbol{h}} - \int_{\mathcal{S}} \vartheta\phi^2\Delta_{\boldsymbol{h}}\vartheta \mathrm{d}\mu_{\boldsymbol{h}},\end{aligned}$$

where the second equality follows by integration by parts on a compact manifold of the last integral in the first line. As $|D\phi|^2 = D_i\phi D^i\phi < 0$ and $\vartheta > 0$ it follows that

$$
\begin{aligned}
-\int_{\mathcal{S}} |D(\vartheta\phi)|^2 \mathrm{d}\mu_{\boldsymbol{h}} &> \int_{\mathcal{S}} \vartheta\phi^2 \Delta_{\boldsymbol{h}}\vartheta \mathrm{d}\mu_{\boldsymbol{h}} \\
&> 4\pi \int_{\mathcal{S}} \vartheta\phi^2\delta(i)\mathrm{d}\mu_{\boldsymbol{h}} + \frac{1}{8}\int_{\mathcal{S}} \vartheta^2\phi^2 r[\boldsymbol{h}] d\mu_{\boldsymbol{h}} \\
&> 4\pi\vartheta(i)\phi^2(i) + \frac{1}{8}\int_{\mathcal{S}} r[\boldsymbol{h}]\vartheta^2\phi^2 \mathrm{d}\mu_{\boldsymbol{h}}, \\
&> \frac{1}{8}\int_{\mathcal{S}} r[\boldsymbol{h}]\vartheta^2\phi^2 \mathrm{d}\mu_{\boldsymbol{h}},
\end{aligned}
$$

where the last inequality follows from the fact that ϕ is an arbitrary test function. Thus setting $\zeta \equiv \vartheta\phi$ one concludes that

$$
-\inf_{\zeta\in\mathfrak{X}(\mathcal{S})} \int_{\mathcal{S}} (8|D\zeta|^2 + r[\boldsymbol{h}]\zeta^2)\mathrm{d}\mu_{\boldsymbol{h}} > 0,
$$

where inf denotes the infimum, that is, the biggest lower bound. *The latter is a necessary condition for the existence of a solution to Equation* (11.58). Under some further technical assumptions, it can be shown to be a sufficient condition; see, for example, Friedrich (2011). The above expression can be reformulated in a conformal way by adding a suitable normalisation factor. Accordingly, one defines the ***Yamabe invariant (number)*** of $\boldsymbol{h}$ as

$$
Y[\boldsymbol{h}] \equiv -\inf_{\zeta\in\mathfrak{X}(\mathcal{S})} \frac{\displaystyle\int_{\mathcal{S}} (8h^{ij}D_i\zeta D_j\zeta + r[\boldsymbol{h}]\zeta^2)\mathrm{d}\mu_{\boldsymbol{h}}}{\left(\displaystyle\int_{\mathcal{S}} \zeta^6 \mathrm{d}\mu_{\boldsymbol{h}}\right)^{1/3}}.
$$

The conformal invariance of the above expression follows from the transformation properties of the three-dimensional Ricci scalar and of the volume element. Accordingly, the Yamabe number is, in fact, a property of the conformal class $[\boldsymbol{h}]$. In particular, if $(\mathcal{S}, \boldsymbol{h})$ is such that $Y[\boldsymbol{h}] > 0$, then there exists $\bar{\boldsymbol{h}} \in [\boldsymbol{h}]$ such that $r[\bar{\boldsymbol{h}}] < 0$ on $\mathcal{S}$; see Lee and Parker (1987).

11.6.4 Constructing solutions to the constraint equations using fundamental solutions

As already mentioned, fundamental functions of the Yamabe equation on compact manifolds allows one to obtain solutions to the Hamiltonian and momentum constraints by means of a procedure of *conformal decompactification*. In this section an overview of some of the technical details of this construction is provided.

In the first instance, attention is restricted to the time-symmetric case. Furthermore, it is assumed that there is only one point at infinity. Given a compact Riemannian manifold $(\mathcal{S},\boldsymbol{h})$ and a point at infinity i, the construction of a time-symmetric initial data set $(\tilde{\mathcal{S}},\tilde{\boldsymbol{h}})$ requires a global solution to Equation (11.58). As already observed, the function $\Gamma = |x|$, defined only in a neighbourhood $\mathcal{B}_a(i)$ with $a>0$, satisfies the required boundary conditions for a solution to Equation (11.58). Indeed, it can be shown that the solution ϑ satisfies

$$\vartheta = \Gamma^{-1} + \frac{m}{2} + O(\Gamma), \qquad \text{near } i,$$

where m is a constant; see Lee and Parker (1987). The above expansion is also valid for any other choice of asymptotic distance function – the constant m is independent of the particular choice. As Γ^2 is a smooth function on its domain of definition, it can be extended to a smooth function (to be denoted again by Γ^2) on the whole of the compact manifold $\mathcal{S}$; see the Appendix to this chapter for further discussion. To obtain the global solution to Equation (11.58), one considers the ansatz

$$\vartheta = \Gamma^{-1} + \frac{m}{2} + W, \tag{11.59}$$

with W some smooth function on $\mathcal{S}$. To make effective use of this ansatz it is assumed, without loss of generality, that the conformal metric $\boldsymbol{h}$ satisfies $r_{\alpha\beta\gamma\delta}(i)=0$ so that one can write

$$h_{\alpha\beta} = -\delta_{\alpha\beta} + \bar{h}_{\alpha\beta}, \qquad \bar{h}_{\alpha\beta} = O(|x|^3).$$

Hence, using the identity

$$\Delta_{\boldsymbol{h}}\vartheta = \frac{1}{\sqrt{-\det \boldsymbol{h}}}\partial_\alpha\left(\sqrt{-\det \boldsymbol{h}}\, h^{\alpha\beta}\partial_\beta\vartheta\right),$$

it follows that

$$\begin{aligned}\mathbf{L}_{\boldsymbol{h}} &= \Delta_{\boldsymbol{h}} - \frac{1}{8}r[\boldsymbol{h}]\\ &= \Delta_{-\boldsymbol{\delta}} + \bar{\mathbf{L}} + r[\boldsymbol{h}],\end{aligned}$$

with

$$\bar{\mathbf{L}} \equiv \bar{h}^{\alpha\beta}\partial_\alpha\partial_\beta + b^\alpha\partial_\alpha, \qquad \bar{h}^{\alpha\beta} = O(|x|^3), \qquad b^\alpha = O(|x|^2)$$

and $r[\boldsymbol{h}] = O(|x|)$. Using the above expressions one can compute that

$$\mathbf{L}_{\boldsymbol{h}}\left(\frac{1}{\Gamma}\right) = \Delta_{-\boldsymbol{\delta}}\left(\frac{1}{\Gamma}\right) + \bar{f} \qquad \text{with } \bar{f} = O(|x|^0).$$

Now, a calculation similar to the one discussed in Section 11.6.3 shows that

$$\Delta_{-\boldsymbol{\delta}}\left(\frac{1}{\Gamma}\right) = 4\pi\delta(i),$$

so that substitution of ansatz (11.59) into $\mathbf{L}_{\boldsymbol{h}}\vartheta = 4\pi\delta(i)$ leads to the *regular equation*

$$\Delta_{\boldsymbol{h}}W - \frac{1}{8}r[\boldsymbol{h}]W = f \qquad \text{with } f = O(|x|^0), \tag{11.60}$$

for which a suitable existence theory is readily available. A unique smooth solution to Equation (11.60) exists if the Yamabe number of the metric $\boldsymbol{h}$ satisfies $Y[\boldsymbol{h}] > 0$; see Beig and O'Murchadha (1991) and Friedrich (1998c). A further argument using the *maximum principle* shows that ϑ – as given by Equation (11.59) – with W solving Equation (11.60) is positive on $\mathcal{S}\setminus\{i\}$ and gives the unique global solution to Equation (11.58). It follows that $(\tilde{\mathcal{S}}, \vartheta^4\boldsymbol{h})$ is an asymptotically Euclidean and regular time-symmetric initial data set.

Data with a non-vanishing extrinsic curvature

The procedure to solve the Yamabe equation described in the previous section can be extended to the case of an initial data set with a trace-free extrinsic curvature. One first needs a solution to the momentum constraint. Several procedures to construct solutions to the maximal momentum constraint (and in particular of the elliptic Equation (11.21)) have been considered in the literature; see, for example, Beig and O'Murchadha (1996), Chaljub-Simon (1982) and Dain and Friedrich (2001). In particular, it is well understood how to specify the free datum ψ'_{ij} in Equation (11.21) so as to ensure non-vanishing ADM linear momentum and ADM angular momentum.

In what follows, assume that Equation (11.21) has been solved for a particular choice of the free datum ψ'_{ij}. Substituting the transverse and trace-free tensor ψ_{ij} obtained from the York splitting (11.20) into the Licnerowicz Equation (11.17) yields the equation

$$\Delta_{\boldsymbol{h}}\vartheta - \frac{1}{8}r[\boldsymbol{h}]\vartheta = \frac{1}{8}\psi_{ij}\psi^{ij}\vartheta^{-7}.$$

As in the case of the Yamabe equation, one can incorporate the singular behaviour of the conformal factor required to decompactify the compact manifold $\mathcal{S}$ via a Dirac's delta. This leads to the equation

$$\Delta_{\boldsymbol{h}}\vartheta - \frac{1}{8}r[\boldsymbol{h}]\vartheta = 4\pi\delta(i) + \frac{1}{8}\psi_{ij}\psi^{ij}\vartheta^{-7}. \tag{11.61}$$

To construct a solution to this equation one first considers a solution $\vartheta_\bullet$ to Equation (11.58) – such solution exists if $Y[\boldsymbol{h}] > 0$. One uses $\vartheta_\bullet$ to write the ansatz

$$\vartheta = \vartheta_\bullet + V$$

with V a smooth function to be determined. Equation (11.61) yields

$$\Delta_{\boldsymbol{h}}V - \frac{1}{8}r[\boldsymbol{h}]V = \frac{1}{8}\psi_{ij}\psi^{ij}\vartheta_\bullet^{-7}(1+\vartheta_\bullet^{-1}V)^{-7}. \tag{11.62}$$

Observe that if $\psi_{ij} = O(|x|^{-4})$, then, in principle, $\frac{1}{8}\psi_{ij}\psi^{ij}\vartheta_{\bullet}^{-7} = O(|x|^{-1})$ so that the right-hand side of Equation (11.62) is still singular. This singularity is, nevertheless, mild, and suitable existence results are available; see theorem 12 in Dain and Friedrich (2001) and also the appendix in Beig and O'Murchadha (1994). The solution ϑ is smooth, and, again, it can be verified that it satisfies $\vartheta > 0$ on $\mathcal{S} \setminus \{i\}$.

11.7 Hyperboloidal manifolds

In certain applications of the conformal field equations it is convenient to consider initial data sets prescribed on hypersurfaces similar to the hyperboloids of the Minkowski spacetime discussed in Section 6.2.4. Hyperboloidal 3-manifolds arise in the construction of asymptotically simple spacetimes with vanishing cosmological constant and in the construction of anti-de Sitter like spacetimes.

11.7.1 Hyperboloidal initial data sets

For the sake of the presentation, the discussion in this section is restricted to the vacuum case with vanishing cosmological constant. Based on the intuition gained through the analysis of hyperboloids in the Minkowski spacetime one has the following definition (see Friedrich (1983) and Kánnár (1996a)):

Definition 11.3 (***hyperboloidal initial data sets***) *A triple $(\tilde{\mathcal{S}}, \tilde{\boldsymbol{h}}, \tilde{\boldsymbol{K}})$ satisfying the vacuum Einstein constraint equations is called a* ***hyperboloidal initial data set*** *if:*

(i) There exists a conformal compactification whereby $\tilde{\mathcal{S}}$ is diffeomorphically identified with the interior of a manifold $\mathcal{S}$ with boundary $\partial\mathcal{S}$ such that $\mathcal{S}$ is diffeomorphic to the closed unit ball in $\mathbb{R}^3$ (whence $\partial\mathcal{S}$ is diffeomorphic to $\mathbb{S}^2$).

(ii) There exist functions Ω and Σ on $\mathcal{S}$ such that $\Omega > 0$ on $\tilde{\mathcal{S}}$ and $\Omega = 0$ and $\Sigma > 0$ on $\partial\mathcal{S}$.

(iii) The conformal fields

$$\boldsymbol{h} = \Omega^2 \tilde{\boldsymbol{h}}, \qquad \boldsymbol{K} = \Omega(\tilde{\boldsymbol{K}} + \Sigma\tilde{\boldsymbol{h}}),$$

extend smoothly to $\mathcal{S}$. Moreover, one has that $\boldsymbol{h}^\sharp(\mathbf{d}\Omega, \mathbf{d}\Omega) = \Sigma^2$ on $\partial\mathcal{S}$.

The simplest type of hyperboloidal initial data sets consists of the case where the physical extrinsic curvature is *pure trace*; that is, one has

$$\tilde{\boldsymbol{K}} = \frac{1}{3}\tilde{K}\tilde{\boldsymbol{h}}. \tag{11.63}$$

As a consequence of the momentum constraint and assuming (11.63) it follows that $\tilde{K}$ must be a constant. From the conformal Hamiltonian constraint (11.15a) one concludes that

$$4\Omega D_i D^i \Omega - 6 D_i \Omega D^i \Omega + 2\Omega^2 r = \tilde{K}^2. \tag{11.64}$$

In order to encode the right behaviour of the conformal factor Ω at $\partial\mathcal{S}$ one introduces a so-called ***boundary defining function*** ρ, that is, a real function over $\mathcal{S}$ satisfying

$$\rho|_{\partial\mathcal{S}} = 0, \qquad \mathbf{d}\rho|_{\partial\mathcal{S}} \neq 0.$$

Given a Riemannian manifold $(\mathcal{S}, \boldsymbol{h})$, such a function can always be constructed. Making use of the ansatz $\Omega = \rho\vartheta^{-2}$ with $\vartheta > 0$ on $\mathcal{S}$, it follows from (11.64) that

$$\rho^2 \Delta_{\boldsymbol{h}} \vartheta - \rho D_i \rho D^i \vartheta + \left(\frac{3}{2} D_i \rho D^i \rho - \frac{1}{8} r[\boldsymbol{h}] \vartheta \rho^2 \theta - \frac{1}{2} \rho \vartheta^2 \Delta_{\boldsymbol{h}} \rho \right) \vartheta = -\frac{1}{8} \tilde{K}^2 \vartheta^{-5}. \tag{11.65}$$

The latter is an elliptic equation for ϑ which becomes singular at $\partial\mathcal{S}$ as its principal part vanishes at this set.

The properties of solutions to Equation (11.65) have been analysed in Andersson et al. (1992). One has the following:

Theorem 11.2 (***existence of hyperboloidal initial data sets***) *Let $(\mathcal{S}, \boldsymbol{h})$ be a smooth Riemannian manifold with boundary $\partial\mathcal{S}$. Then, there exists a unique positive solution ϑ to Equation (11.65). Moreover, the following are equivalent:*

(i) The function ϑ and the tensors

$$L_{ij} \equiv -\frac{1}{\Omega} D_{\{i} D_{j\}} \Omega + \frac{1}{12} \left(r + \frac{2}{3} K^2 \right) h_{ij}, \tag{11.66a}$$

$$d_{ij} \equiv \frac{1}{\Omega^2} D_{\{i} D_{j\}} \Omega + \frac{1}{\Omega} r_{\{ij\}}, \tag{11.66b}$$

determined on $\tilde{\mathcal{S}}$ by $\boldsymbol{h}$ and $\Omega = \rho\vartheta^{-2}$ extend smoothly to all of $\mathcal{S}$.

(ii) The Weyl tensor $C^a{}_{bcd}$ computed from the data on $\mathcal{S}$ vanishes on $\partial\mathcal{S}$.

(iii) The conformal class $[\boldsymbol{h}]$ is such that the extrinsic curvature of $\partial\mathcal{S}$ with respect to its embedding in $(\mathcal{S}, \boldsymbol{h})$ is pure trace.

The expressions for the fields L_{ij} and d_{ij} correspond to the spatial part of the Schouten tensor and the electric part of the rescaled Weyl tensor as determined by the conformal constraint equations of Section 11.4.3. Observe that the expressions for the fields are formally singular at $\Omega = 0$, so that the conclusion of the theorem is non-trivial and ensures the existence of regular hyperboloidal data for the conformal field equations. Extensions of Theorem 11.2 to more general forms of the extrinsic curvature have been analysed in Andersson and Chruściel (1993, 1994).

Initial data for anti-de Sitter-like spacetimes

By making the identification $\tilde{K}^2 \mapsto \lambda$ with $\lambda > 0$ in Equation (11.64), hyperboloidal initial data sets can be interpreted as initial data sets for anti-de Sitter-like spacetimes. Thus, all available knowledge about the existence of hyperboloidal initial data sets can be transferred to this setting. This idea has been investigated for a larger class of data than the one considered in this section in Kánnár (1996a).

11.8 Other methods for solving the constraint equations

The analysis of the Einstein constraint equations carried out in the previous sections relies on a systematic use of the conformal method of Licnerowicz, Choquet-Bruhat and York. There are, however, other alternative procedures, each providing a different insight into the properties of the solutions to the constraint equations; see, for example, Bartnik and Isenberg (2004). In this section, methods of particular relevance for the analysis of the conformal field equations are briefly discussed: the first one based on the so-called *extended constraint equations*, and the second one being the so-called *exterior gluing procedure*.

11.8.1 The extended constraint equations

Given a solution to the conformal constraint equations, Lemma 11.1 shows how to construct initial data for the conformal Einstein field equations. It is, nevertheless, of interest to directly obtain a solution to the conformal constraint equations without having to solve the Einstein constraint equations. A construction of this type is of importance as the expressions for the rescaled Weyl tensor and the Schouten tensor in terms of the conformal factor and the intrinsic 3-geometry of the hypersurface are formally singular at the points where $\Omega = 0$; see, for example, Equations (11.66a) and (11.66b) in Theorem 11.2. Currently available results in this direction are restricted to the case where the $\Omega = 1$; see Butscher (2002, 2007). Despite this limitation, they provide insight into the properties and structure of the conformal constraint equations and lead to a procedure for the construction of initial data sets by perturbative methods.

Assuming that the matter fields vanish, and setting $\Omega = 1$, $\Sigma = 0$ in the conformal constraint equations (11.35a)–(11.35j) one finds that the essential equations of the system can be rewritten in tensorial form as

$$D_j K_{ki} - D_k K_{ji} = \epsilon^l{}_{jk} d^*_{il}, \tag{11.67a}$$

$$D^k d_{ki} = K^{jk} \epsilon^l{}_{ki} d^*_{jl}, \tag{11.67b}$$

$$D^k d^*_{ki} = -\epsilon_i{}^{jl} K_j{}^k r_{kl}, \tag{11.67c}$$

$$r_{ij} = d_{ij} + K K_{ij} - K_i{}^k K_{kj}. \tag{11.67d}$$

These equations are known as the ***extended Einstein constraints*** since a solution thereof implies a solution to the Einstein vacuum constraints; see Lemma 11.1 in this chapter and lemma 1 in Butscher (2007) for a more detailed discussion. The first three equations constitute an underdetermined elliptic system for the fields K_{ij}, d_{ij} and d^*_{ij}.

A direct computation shows that the formal adjoint of the operator in the principal part of Equation (11.67a) is the divergence with respect to the index $_j$. Applying this divergence to the equation and writing

$$K_{ij} = \psi_{ij} - \frac{1}{3}Kh_{ij} \qquad \text{with } \psi_{ij}h^{ij} = 0$$

one obtains the equation

$$D^j D_j \psi_{ki} - D^j D_k \psi_{ji} = \epsilon^l{}_{jk} D^j d^*_{il} + \frac{1}{3}\left(h_{ki} D^j D_j K - D_i D_k K\right).$$

If the fields d^*_{ij} and K are known, this equation can be verified to be an elliptic equation for the trace-free part ψ_{ij}.

Equations (11.67b) and (11.67c) can be transformed into fully elliptic equations by means of a York splitting of the fields d_{ij} and d^*_{ij}; see Section 11.3.3. Hence, writing

$$d_{ij} = D_i v_j + D_j v_i - \frac{2}{3} D_k v^k h_{ij} + d'_{ij},$$
$$d^*_{ij} = D_i u_j + D_j u_i - \frac{2}{3} D_k u^k h_{ij} + d^{*\prime}_{ij},$$

where d'_{ij} and $d^{*\prime}_{ij}$ are freely specifiable symmetric trace-free tensors, one obtains elliptic equations for the fields v_i and u_i whose principal part is identical to that of Equation (11.21). Finally, Equation (11.67d) can be transformed into an elliptic equation for the components of the 3-metric $\boldsymbol{h}$ by introducing ***harmonic coordinates*** $\underline{x} = (x^\alpha)$, $\Delta_{\boldsymbol{h}} x^\alpha = 0$; compare the analogous use of *wave coordinates* in the case of a Lorentzian metric to obtain the reduced Einstein field equation discussed in the Appendix to Chapter 13.

The system of elliptic equations for the fields K_{ij}, v_i, u_i, h_{ij} discussed in the previous paragraphs is called the ***auxiliary system***. Solutions to the auxiliary system could be obtained, in principle, by means of perturbative methods relying on the use of the *implicit function theorem* – see, for example, Ambrosetti and Prodi (1995) – if some background solution is known. The solutions thus obtained are not a priori solutions to the original Equations (11.67a)–(11.67d). Hence, in a second step, one needs to investigate the conditions under which a solution to the auxiliary system implies a solution to the extended Einstein constraints and, consequently, a solution to the vacuum Einstein constraints. This strategy has been investigated in Butscher (2002, 2007) to obtain asymptotically Euclidean solutions to the extended constraints which are close to data for the Minkowski spacetime. The particular details require the use of *weighted Sobolev spaces* to control the decay of the various fields. These methods can be adapted, in principle, to obtain data on $\mathbb{S}^3$ corresponding to perturbations of de Sitter initial data.

11.8.2 Exterior asymptotic gluing

The ***exterior asymptotic gluing*** is a method to construct solutions to the Einstein constraint equations by *gluing the interior region of an asymptotically Euclidean solution to the Einstein vacuum constraints to an asymptotic end of initial data for the Kerr spacetime or, in fact, of any stationary solution*; see Corvino (2000), Chruściel and Delay (2003), Corvino and Schoen (2006) and Corvino (2007). More precisely, given a smooth asymptotically Euclidean initial data set for the vacuum Einstein field equations $(\tilde{\mathcal{S}}, \tilde{\boldsymbol{h}}, \tilde{\boldsymbol{K}})$ and a given compact subset $\mathcal{U} \subset \tilde{\mathcal{S}}$ such that $\tilde{\mathcal{S}} \setminus \mathcal{U}$ is an asymptotic end, it is possible to show that there exists another smooth asymptotically Euclidean solution to the vacuum Einstein constraints $(\tilde{\mathcal{S}}, \bar{\boldsymbol{h}}, \bar{\boldsymbol{K}})$ which is identical to the original solution on $\mathcal{U}$ and coincides with initial data for the Kerr spacetime on $\tilde{\mathcal{S}} \setminus \bar{\mathcal{U}}$ for some $\bar{\mathcal{U}} \subset \tilde{\mathcal{S}}$. In addition, the initial data set $(\tilde{\mathcal{S}}, \bar{\boldsymbol{h}}, \bar{\boldsymbol{K}})$ contains an annular transition region in which the initial data can be controlled. In the case of time-symmetric initial data sets this method glues any interior region to an exterior region of a slice of the Schwarzschild spacetime.

The underlying idea in the asymptotic exterior gluing method is to exploit the underdetermined character of the Einstein constraints as a system of partial differential equations for the fields $(\tilde{\boldsymbol{h}}, \tilde{\boldsymbol{K}})$. Prior to the development of the asymptotic exterior gluing methods Cutler and Wald (1989) have shown that it is possible to make use of the standard conformal method to construct solutions to the time symmetric constraints containing a Minkowskian interior region and a Schwarzschildean exterior region joined together by an annular region containing a *purely magnetic solution to the Einstein-Maxwell constraints.*

As will be discussed in Chapter 20, initial data sets obtained by means of asymptotic exterior gluing play a key role in the construction of Minkowski-like asymptotically simple spacetimes. For simplicity, in the remainder of this section *attention is restricted to the time-symmetric case* for which the Einstein vacuum constraints reduce to $r[\tilde{\boldsymbol{h}}] = 0$. In the present context, one regards the Ricci scalar as a map between the space of Riemannian metrics over $\tilde{\mathcal{S}}$ and $\mathfrak{X}(\tilde{\mathcal{S}})$. Under certain circumstances this mapping is an isomorphism; that is, given a metric $\boldsymbol{h}$ and $f \in \mathfrak{X}(\tilde{\mathcal{S}})$ such that $r[\boldsymbol{h}] = f$ and given a further $g \in \mathfrak{X}(\tilde{\mathcal{S}})$ close enough to f, then there exists another metric $\bar{\boldsymbol{h}}$ close to $\boldsymbol{h}$ such that $r[\bar{\boldsymbol{h}}] = g$. This property of the *Ricci scalar operator* is the essential ingredient in the gluing procedure. As part of the gluing construction, one connects the inner region $(\mathcal{U}, \tilde{\boldsymbol{h}})$ and an asymptotic region $(\mathcal{E}, \tilde{\boldsymbol{h}}_{\mathscr{S}})$ with $\tilde{\boldsymbol{h}}_{\mathscr{S}}$ as given in Equation (11.45) for some choice (so far undetermined) of the constants m and (x_0^α) through an annular region. A positive definite symmetric tensor $\check{\boldsymbol{h}}$ is defined on $\tilde{\mathcal{S}}$ by requiring it to be identical to $\tilde{\boldsymbol{h}}$ on $\mathcal{U}$ and to $\tilde{\boldsymbol{h}}_{\mathscr{S}}$ on $\mathcal{E}$, while on the asymptotic region it is chosen so that it interpolates smoothly between $\tilde{\boldsymbol{h}}$ and $\tilde{\boldsymbol{h}}_{S}$. By construction $r[\breve{\boldsymbol{h}}] = 0$ in both $\mathcal{U}$ and $\mathcal{E}$, while $r[\breve{\boldsymbol{h}}] \neq 0$ in the transitional annular region. Nevertheless, by moving $\mathcal{U}$ suitably into the asymptotic region, one can make $r[\breve{\boldsymbol{h}}]$ small enough so that the isomorphism properties of the Ricci scalar operator can be used to ensure the existence of a tensor $\boldsymbol{k}$ with support on an annular region such that $\bar{\boldsymbol{h}} \equiv \breve{\boldsymbol{h}} + \boldsymbol{k}$ is a Riemannian metric with $r[\bar{\boldsymbol{h}}] = 0$ on $\tilde{\mathcal{S}}$.

The asymptotic exterior gluing construction requires a careful analysis of the properties of the ***linearised Ricci operator***

$$\mathscr{R}_{\boldsymbol{h}}[\bar{\boldsymbol{h}}] \equiv -\Delta_{\boldsymbol{h}}(\mathbf{tr}_{\boldsymbol{h}}(\bar{\boldsymbol{h}})) + \mathbf{div}_{\boldsymbol{h}}(\mathbf{div}_{\boldsymbol{h}}(\bar{\boldsymbol{h}})) - \boldsymbol{h}(\bar{\boldsymbol{h}}, \mathbf{Ric}[\bar{\boldsymbol{h}}]).$$

For a fixed metric $\boldsymbol{h}$, the latter is an underdetermined elliptic operator. It can be transformed into an elliptic system by composition with its formal adjoint

$$\mathscr{R}^*_{\boldsymbol{h}}(f) \equiv -(\Delta_{\boldsymbol{h}} f)\boldsymbol{h} + \mathbf{Hess}(f) - f\mathbf{Ric}[\boldsymbol{h}].$$

The composite elliptic operator $\mathscr{R}^*_{\boldsymbol{h}} \circ \mathscr{R}_{\boldsymbol{h}}$ is a fourth-order partial differential operator. Once the linearised problem is controlled, the non-linear problem is then solved by means of an iteration. To conclude, one needs to show that the metric $\bar{\boldsymbol{h}}$ is indeed a solution to $r[\bar{\boldsymbol{h}}] = 0$. It is in this part of the construction that the value of the constants m and (x_0^α) are fixed. A refined version of the original construction in Corvino (2000) has been given in Corvino (2007), from which the following result has been adapted:

Theorem 11.3 (***exterior asymptotic gluing construction***) *Let* $(\tilde{\mathcal{S}}, \tilde{\boldsymbol{h}})$ *denote an asymptotically Euclidean initial data set for the Einstein vacuum equations. Let* $\mathcal{E} \subset \tilde{\mathcal{S}}$ *be any asymptotically flat end of* $\tilde{\mathcal{S}}$. *Given* $r_0 > 0$ *let* $\mathcal{E}_{r_0} \subset \mathcal{E}$ *be an exterior region in* $\mathcal{E}$ *expressed in asymptotically Cartesian coordinates by* $\mathcal{E}_{r_0} = \{(x^\alpha) \in \mathbb{R}^3 \,|\, |x| > r > r_0\}$. *Suppose, furthermore, that in these coordinates the metric* $\tilde{\boldsymbol{h}}$ *has the form*

$$\tilde{h}_{\alpha\beta} = -\left(1 + \frac{2m}{|x|}\right)\delta_{\alpha\beta} + O_3(|x|^{-2}).$$

Let k *be a non-negative integer. Then for any* $\varepsilon > 0$ *there exists* $r_* > 0$ *and a smooth metric* $\bar{\boldsymbol{h}}$ *satisfying* $r[\bar{\boldsymbol{h}}] = 0$ *and* $||h_{\alpha\beta} - \bar{h}_{\alpha\beta}||_{C^k(\mathcal{E})} < \varepsilon$ *so that* $\bar{\boldsymbol{h}}$ *is equal to* $\tilde{\boldsymbol{h}}$ *on* $\mathcal{U} = \tilde{\mathcal{S}} \setminus \mathcal{E}_{r_*}$ *and identical to an asymptotically flat end of a standard Schwarzschild slice on* $\mathcal{E}_{2r_*}$.

The precise definition of the ***supremum norm*** $|| \;\; ||_{C^k(\mathcal{E})}$ is discussed in the Appendix to this chapter. A schematic depiction of the construction of Theorem 11.3 is given in Figure 11.2. In the applications of this result to the existence of asymptotically simple spacetimes, it is important to control the location of the exterior region $\mathcal{E}_{r_*}$ and to ensure that $r_* \not\to \infty$ as one moves along a family of initial data sets tending, say, to data for the Minkowski spacetime. This possible degeneracy has been dealt with by imposing some reflexion symmetry properties

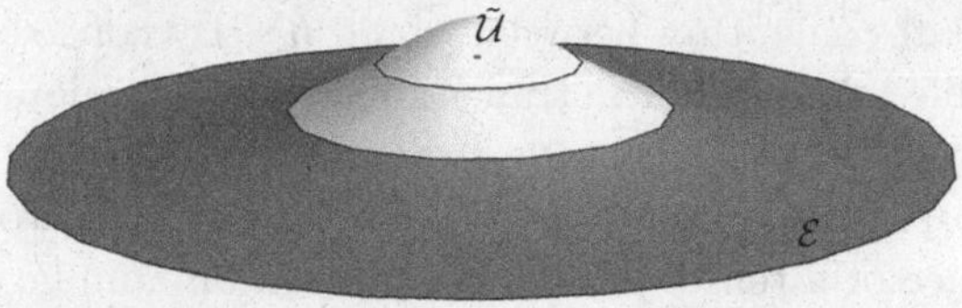

Figure 11.2 Schematic depiction of the exterior gluing construction given by Theorem 11.3. It contains an inner region $\tilde{\mathcal{U}}$ where the 3-metric has a fixed arbitrary value $\tilde{\boldsymbol{h}}$, an annular transition region between $\mathcal{E}_{r_*}$ and $\mathcal{E}_{2r_*}$ and an exterior region $\mathcal{E}$ where it is equal to data for a member of the Schwarzschild family of solutions.

on the metric $\tilde{\boldsymbol{h}}$; see Chruściel and Delay (2003). An alternative solution has been provided in Corvino (2007). This result makes use of symmetric $(0,2)$-tensors $\boldsymbol{k}$ satisfying the condition $\mathscr{R}_{-\boldsymbol{\delta}}(\boldsymbol{k}) = 0$. Making use of a York splitting the tensor $\boldsymbol{k}$ can be decomposed in a unique way into a traceless term with vanishing divergence, a trace part and a part which is the conformal Killing operator of a covector; see Chaljub-Simon (1982). The tensor $\boldsymbol{k}$ is said to be ***non-degenerate*** if its transverse-traceless part is non-zero. Using this terminology one has the following *stability result* (see Corvino (2007) for further details and its proof):

Theorem 11.4 (***stability of the exterior gluing construction***) *Let $\boldsymbol{k}$ be any smooth, compactly supported symmetric $(0,2)$-tensor on $\mathbb{R}^3$ with $\mathscr{R}_{-\boldsymbol{\delta}}(\boldsymbol{k}) = 0$. Moreover, for sufficiently small $\varepsilon > 0$ let*

$$\tilde{\boldsymbol{h}} = -\vartheta^4(\boldsymbol{\delta} + \varepsilon \boldsymbol{k})$$

be asymptotically flat and satisfy $r[\tilde{\boldsymbol{h}}] = 0$. If $\boldsymbol{k}$ is non-degenerate, there exists $r_* > 0$ so that for all ε small enough there is a metric $\bar{\boldsymbol{h}}$ with $r[\bar{\boldsymbol{h}}] = 0$ which agrees with $\tilde{\boldsymbol{h}}$ in the closed ball $\overline{\mathcal{B}}_{r_*}(0)$ and is exactly Schwarzschild on $\mathcal{E}_{2r_*}$. Consequently, the Riemannian manifold $(\mathbb{R}^3, \bar{\boldsymbol{h}})$ admits a smooth conformal point compactification in the sense of Definition 11.2.

This theorem guarantees the existence of time-symmetric solutions to the vacuum Einstein constraint equations which are both close to data for the Minkowski spacetime and exactly Schwarzschildean in a non-trivial exterior region; see Section 20.5.

Versions of the asymptotic exterior gluing construction for initial data sets with non-vanishing extrinsic curvature can be found in Chruściel and Delay (2003) and Corvino and Schoen (2006). There are adaptations of the exterior gluing method to the case of hyperboloidal initial data sets with constant scalar curvature; see Chruściel and Delay (2009).

11.9 Further reading

The best point of entry to the extensive literature on the Einstein constraint equations is through reviews such as those of Bartnik and Isenberg (2004) or Isenberg (2013). An older, classical review on the topic is given in Choquet-Bruhat and York (1980). An alternative review aimed at applications in numerical relativity is Cook (2000). A detailed account of the conformal method to solve the constraint equations, as seen by one of the main contributors of the topic, can be found in Choquet-Bruhat (2008) – this reference contains, in addition, a discussion of the basic aspects of weighted Sobolev spaces. Closely related to the latter is the reference Choquet-Bruhat et al. (2000). A discussion of basic aspects of the theory of elliptic differential equations and its application to the analysis of the Einstein constraints can be found in Rendall (2008). An alternative account of the basic aspects of the analysis of elliptic equations with a number of worked-out examples is Dain (2006). Finally, a detailed account of the conformal equations under the assumption of spherical symmetry is given in Guven and O'Murchadha (1995).

By contrast, the accounts on the conformal Einstein constraints are much more restricted in number. The original references are Friedrich (1983, 1984, 1986a, 1995, 2004); see also the discussion in Frauendiener (2004). A systematic analysis of hyperboloidal initial data sets can be found in Andersson et al. (1992) and Andersson and Chruściel (1993, 1994).

The notion of asymptotically Euclidean and regular manifolds can be traced back to the discussion in Geroch (1972b). These ideas have been further elaborated in Friedrich (1988, 1998c). Accounts of the use of *Dirac's deltas* to represent the points at infinity can be found in Beig and O'Murchadha (1991, 1994). A neat application of this approach to the construction of initial data sets with a conformal toroidal symmetry is given in Beig and Husa (1994). Applications of the method to the construction of initial data for the collision of Kerr-like black holes can be found in Dain (2001a,c). Finally, a detailed construction of initial data sets admitting expansions in powers of the geodesic distance is given in Dain and Friedrich (2001).

Appendix: some results of analysis

As in the main text of this chapter, let $(\mathcal{S}, \boldsymbol{h})$ denote a Riemannian manifold. Moreover, let $p \in \mathcal{S}$ denote a point and consider normal coordinates $\underline{x} = (x^\alpha)$ centred at p; that is, $x^\alpha(0) = 0$.

Order symbols. The behaviour of functions $f : \mathcal{S} \to \mathbb{R}$ near p can be conveniently described by means of the ***big*** O and ***small*** o notations. More precisely, given $f, g : \mathcal{S} \to \mathbb{R}$, if for some $\underline{x} = (x^\alpha)$ sufficiently close to 0 there exists a positive constant M such that

$$|f(\underline{x})| \leq M|g(\underline{x})|,$$

one writes $f(\underline{x}) = O(g(\underline{x}))$, and one says that f *is at most of the order of* g. If, in addition, one has that

$$\partial_\alpha f(\underline{x}) = O(\partial_\alpha g(\underline{x})), \qquad \cdots \qquad \partial_{\alpha_1} \ldots \partial_{\alpha_k} f(\underline{x}) = O(\partial_{\alpha_1} \cdots \partial_{\alpha_k} g(\underline{x})),$$

for some integer k one writes $f(\underline{x}) = O_k(g(\underline{x}))$.

If given f, g one has $f(\underline{x})/g(\underline{x}) \to 0$ as $x^\alpha \to 0$, then one writes

$$f(\underline{x}) = o(g(\underline{x})),$$

and one says that *the order of* f *is bigger than that of* g. Again, if

$$\partial_\alpha f(\underline{x}) = o(\partial_\alpha g(\underline{x})), \qquad \ldots \qquad \partial_{\alpha_1} \cdots \partial_{\alpha_k} f(\underline{x}) = o(\partial_{\alpha_1} \cdots \partial_{\alpha_k} g(\underline{x})),$$

one writes $f(\underline{x}) = o_k(g(\underline{x}))$. For further discussion, see, for example, Courant and John (1989).

Taylor expansions. If a function $f : \mathbb{R}^n \to \mathbb{R}$ is of class C^k on the open ball $\mathcal{B}_a(0) \subset \mathbb{R}^n$ one has that

$$\begin{aligned} f(\underline{x}) = f(0) + \partial_\alpha f(0) x^\alpha + \frac{1}{2!} \partial_{\alpha_1} \partial_{\alpha_2} f(0) x^{\alpha_1} x^{\alpha_2} \\ + \cdots + \frac{1}{(k-1)!} \partial_{\alpha_1} \cdots \partial_{\alpha_{k-1}} f(0) x^{\alpha_1} \cdots x^{\alpha_{k-1}} + O(|x|^k). \end{aligned}$$

For further discussion, see, for example, Courant and John (1989).

Supremum norm. Given $\mathcal{U} \subset \mathbb{R}^n$ and $f \in C^k(\mathcal{U})$, one defines the supremum norm as

$$||f||_{C^k(\mathcal{U})} = \sum_{0 \leq l \leq k} \sup\{|\partial_{\alpha_1} \cdots \partial_{\alpha_l} f(\underline{x})|\,,\, \underline{x} \in \overline{\mathcal{U}}\}$$

where $\overline{\mathcal{U}}$ denotes the closure of $\mathcal{U}$. For further discussion on this and other related norms, see, for example, Ambrosetti and Prodi (1995).

Extension of smooth functions. Let $\mathcal{U} \subset \mathcal{S}$ denote a closed subset and $f : \mathcal{U} \to \mathbb{R}^k$ a smooth function. There exists a smooth function $\tilde{f} : \mathcal{S} \to \mathbb{R}^k$ such that $\tilde{f}|_{\mathcal{U}} = f$ and whose support is contained in $\mathcal{S} \setminus \mathcal{U}$; in other words, $\tilde{f}$ is non-vanishing in $\mathcal{S} \setminus \mathcal{U}$. In a slight abuse of notation $\tilde{f}$ will be denoted, again, by f. For more details on this result, see Lee (2002).

Dirac's delta. Let now $\mathcal{S}$ denote a compact manifold and $p \in \mathcal{S}$ a fixed point within. The Dirac's delta $\delta(p)$ with support on p is the *distribution* (i.e. a linear functional $C^0(\mathcal{S}) \to \mathbb{R}$) satisfying

$$\int_{\mathcal{S}} f(\underline{x}) \delta(p) \mathrm{d}\mu_{\boldsymbol{h}} = f(p), \qquad \text{for all } f \in C^0(\mathcal{S}).$$

In particular, one has that

$$\int_{\mathcal{S}} \delta(p)\, \mathrm{d}\mu_{\boldsymbol{h}} = 1.$$

If $f(p) = 0$, one has the *distributional equality*

$$f(\underline{x})\delta(p) = 0.$$

For further details, the reader is referred to Appel (2007).

Divergence theorem. Given $(\mathcal{M}, \boldsymbol{g})$ a manifold with metric (Riemannian or Lorentzian) and, within, $\mathcal{U} \subset \mathcal{M}$ a compact subset and a smooth covector $\boldsymbol{\omega}$, one has

$$\int_{\mathcal{U}} \mathbf{div}\boldsymbol{\omega}\, \mathrm{d}\mu_{\boldsymbol{h}} = \int_{\partial\mathcal{U}} \langle \boldsymbol{\omega}, \boldsymbol{\nu} \rangle\, \mathrm{d}S_{\boldsymbol{h}},$$

with $\boldsymbol{\nu}$ the outward pointing unit normal to $\partial\mathcal{U}$; see, for example, Frankel (2003) for further details.

12

Methods of the theory of hyperbolic differential equations

This chapter discusses the notions of the theory of hyperbolic differential equations and the existence theorems employed to construct solutions to the conformal Einstein field equations. Conformal methods allow, under suitable circumstances, the use of very general theorems of the theory of partial differential equations (PDEs) to obtain conclusions of a global nature about solutions to the Einstein field equations. The results presented in this chapter have been tailored to fit the general discussion of this book.

The basic result of the theory of hyperbolic PDEs that will be used in this book is Kato's existence, uniqueness and stability result for symmetric hyperbolic systems; see Theorem 12.4. In view of applications to the construction of anti-de Sitter-like spacetimes a basic existence and uniqueness result of the initial boundary value problem of symmetric hyperbolic equations is also discussed; see Theorem 12.6. The chapter concludes with an overview of the basic theory behind characteristic initial value problems; see Theorem 12.7.

12.1 Basic notions

As will be seen in Chapter 13, the conformal Einstein equations give rise to ***quasilinear evolution equations*** which, in local coordinates $x \equiv (x^\mu)$ on an open set $\mathcal{U} \subset \mathcal{M}$ of the spacetime manifold, take the form

$$\mathbf{A}^\mu(x, \mathbf{u})\partial_\mu \mathbf{u} = \mathbf{B}(x, \mathbf{u}) \tag{12.1}$$

where $\mathbf{u}$ is a $\mathbb{C}^N$-valued unknown for some positive integer N and $\mathbf{A}^\mu$, $\mu = 0, \ldots 3$, are $(N \times N)$ matrix-valued functions of the coordinates and of the vector-valued unknown $\mathbf{u}$; thus, one has as many equations as components in the vector $\mathbf{u}$. Finally, $\mathbf{B}(x, \mathbf{u})$ is a vector-valued function of x and $\mathbf{u}$. In what follows, *it will be assumed that the components of* $\mathbf{u}$ *are scalars.* The functions $\mathbf{A}^\mu(x, \mathbf{u})$ and $\mathbf{B}(x, \mathbf{u})$ are, in principle, non-linear functions of the entries of $\mathbf{u}$. If the matrices

$\mathbf{A}^\mu$ do not depend on $\mathbf{u}$, one has a ***semilinear*** system. Without loss of generality, $\mathcal{U}$ can be regarded as some suitable subset of $\mathbb{R}^4$.

Following the terminology of Section 11.2 the term

$$\mathbf{A}^\mu(x, \mathbf{u})\partial_\mu \mathbf{u}$$

is known as the ***principal part*** of Equation (12.1). The ***symbol*** with respect to the unknown $\mathbf{u}$ at the point $p \in \mathcal{U}$ with coordinates $x = x(p)$ for a covector $\boldsymbol{\xi} \in T^*|_p(\mathcal{U})$ is given by the matrix

$$\boldsymbol{\sigma}(x, \mathbf{u}, \boldsymbol{\xi}) \equiv \mathbf{A}^\mu(x, \mathbf{u})\xi_\mu.$$

Under a coordinate transformation $x' = x'(x)$, it follows from Equation (12.1) that

$$\mathbf{A}^{\mu\prime}(x', \mathbf{u})\partial_{\mu'}\mathbf{u} = \mathbf{B}(x', \mathbf{u}),$$

with

$$\mathbf{A}'^\mu(x', \mathbf{u}) = \frac{\partial x'^\mu}{\partial x^\nu}\mathbf{A}^\nu(x(x'), \mathbf{u}). \tag{12.2}$$

It then follows from the transformation law of covectors under coordinate transformations and Equation (12.2) that the symbol of the differential Equation (12.1) is an invariant.

12.1.1 Symmetric hyperbolic systems

The basic properties of the PDE (12.1) and of its solutions depend on the structure of its principal part. Given a matrix $\mathbf{A}$, the operation of taking the transpose of its complex conjugate will be denoted by $\mathbf{A}^*$. One has the following definition:

Definition 12.1 (***symmetric hyperbolic systems***) *Given a solution* $\mathbf{u}(x)$, *the system* (12.1) *is said to be* ***symmetric hyperbolic*** *at* $(x, \mathbf{u})$ *if:*

(i) the matrices $\mathbf{A}^\mu(x, \mathbf{u})$ *are Hermitian; that is* $(\mathbf{A}^\mu)^* = \mathbf{A}^\mu$
(ii) there exists a covector $\boldsymbol{\xi}$ *such that* $\boldsymbol{\sigma}(x, \mathbf{u}, \boldsymbol{\xi}) = \mathbf{A}^\mu(x, \mathbf{u})\xi_\mu$ *is a positive-definite matrix.*

Given two vectors $\mathbf{u}, \mathbf{v} \in \mathbb{C}^N$, their ***inner product*** is defined by

$$\langle \mathbf{u}, \mathbf{v} \rangle \equiv \mathbf{u}^*\mathbf{v}.$$

It follows then that $\langle \mathbf{u}, \mathbf{v} \rangle = \overline{\langle \mathbf{v}, \mathbf{u} \rangle}$ with the overbar denoting the usual complex conjugation of scalars. Moreover, if $\mathbf{A}$ is a Hermitian $N \times N$ matrix, then

$$\langle \mathbf{u}, \mathbf{A}\mathbf{v} \rangle = \langle \mathbf{A}\mathbf{u}, \mathbf{v} \rangle, \qquad \langle \mathbf{u}, \mathbf{A}\mathbf{u} \rangle = \overline{\langle \mathbf{u}, \mathbf{A}\mathbf{u} \rangle}.$$

Spacelike and timelike hypersurfaces with respect to a symmetric hyperbolic system

In what follows, let $\mathcal{S}$ denote a hypersurface on $\mathcal{U} \subset \mathcal{M}$ defined in terms of a smooth scalar $\phi \in \mathcal{X}(\mathcal{M})$ as

$$\mathcal{S} \equiv \{p \in \mathcal{U} \,|\, \phi(p) = 0\}, \tag{12.3}$$

where it is assumed that $\mathbf{d}\phi \neq 0$ so that $\mathcal{S}$ has everywhere a well-defined normal. The positivity condition (i) in Definition 12.1 allows one to define the ***causal nature*** of the hypersurface $\mathcal{S}$ with respect to solutions of Equation (12.1). More precisely, the hypersurface $\mathcal{S}$ is said to be ***spacelike with respect to a solution*** $\mathbf{u}$ ***to the symmetric hyperbolic system*** (12.1) if $\boldsymbol{\sigma}(x, \mathbf{u}, \mathbf{d}\phi)$ is positive definite for $p \in \mathcal{S}$. If $\boldsymbol{\sigma}(x, \mathbf{u}, \mathbf{d}\phi)$ has a non-vanishing determinant and is not positive definite, one says that $\mathcal{S}$ is ***timelike*** for the solution $\mathbf{u}$. Finally, if $\boldsymbol{\sigma}(x, \mathbf{u}, \mathbf{d}\phi)$ has a vanishing determinant, one says that $\mathcal{S}$ is ***null*** – this case is tied to the notion of *characteristics* to be discussed in the next section. These causal definitions are, in principle, independent of the homonymous notion defined in terms of a metric $\boldsymbol{g}$ on $\mathcal{M}$. However, as discussed in Chapter 14, for evolution equations arising from the Einstein field equations, the geometric and PDE notions agree; see Theorem 14.1.

12.1.2 Initial value problems and characteristics

Of particular relevance for a symmetric hyperbolic system of the form (12.1) is the so-called ***initial value problem*** whereby some initial data on a hypersurface $\mathcal{S}$ is prescribed and one purports to obtain the solution to the equation away from the initial hypersurface.

An ***initial data set*** for Equation (12.1) on a hypersurface $\mathcal{S}$ which is spacelike with respect to Equation (12.1) consists of a $\mathbb{C}^N$-valued function $\mathbf{u}_\star$ on $\mathcal{S}$ which is interpreted as the value of the solution $\mathbf{u}$ to Equation (12.1) on $\mathcal{S}$. A question which arises naturally in this context is whether all the components of the vector $\mathbf{u}_\star$ can be specified freely on $\mathcal{S}$.

It is convenient to introduce on $\mathcal{U}$ coordinates $x = (x^0, \underline{x}) = (x^0, x^1, x^2, x^3)$ adapted to $\mathcal{S}$ so that the hypersurface is represented by the condition $x^0 = 0$. Using these adapted coordinates and the initial data $\mathbf{u}_\star$ one can compute the *spatial derivatives* $\partial_\alpha \mathbf{u}_\star$ of $\mathbf{u}$ on $\mathcal{S}$. In order to determine the *time derivatives* $\partial_0 \mathbf{u}$ on $\mathcal{S}$ one substitutes the above into Equation (12.1) to obtain

$$\mathbf{A}^0(0, \underline{x}; \mathbf{u}_\star)(\partial_0 \mathbf{u})_\star + \mathbf{A}^\alpha(0, \underline{x}; \mathbf{u}_\star)\partial_\alpha \mathbf{u}_\star = \mathbf{B}(0, \underline{x}; \mathbf{u}_\star), \tag{12.4}$$

where it is observed that $(\partial_\alpha \mathbf{u})_\star = \partial_\alpha \mathbf{u}_\star$. This equation can be read as a linear algebraic system for $(\partial_0 \mathbf{u})_\star \equiv \partial_0 \mathbf{u}|_\mathcal{S}$ which can be solved if $\mathbf{A}^0(0, \underline{x}, \mathbf{u}_\star)$ can be inverted, that is, if

$$\det\left(\mathbf{A}^0(0, \underline{x}; \mathbf{u}_\star)\right) \neq 0.$$

If $\det \mathbf{A}^0(0,\underline{x};\mathbf{u}_\star) = 0$, then $M \equiv \operatorname{rank} \mathbf{A}^0(0,\underline{x},\mathbf{u}_\star) < N$, and one can make linear combinations of the equations in (12.4) to obtain a new system on $\mathcal{S}$ of the form

$$\bar{\mathbf{A}}^0(0,\underline{x};\mathbf{u}_\star)(\partial_0\mathbf{u})_\star + \bar{\mathbf{A}}^\alpha(0,\underline{x};\mathbf{u}_\star)\partial_\alpha\mathbf{u}_\star = \mathbf{B}(0,\underline{x};\mathbf{u}_\star),$$

where

$$\bar{\mathbf{A}}^0(0,\underline{x},\mathbf{u}_\star) = \begin{pmatrix} a^0_{11}(0,\underline{x};\mathbf{u}_\star) & \cdots & a^0_{1N}(0,\underline{x};\mathbf{u}_\star) \\ \vdots & \ddots & \vdots \\ a^0_{M1}(0,\underline{x};\mathbf{u}_\star) & \cdots & a^0_{MN}(0,\underline{x};\mathbf{u}_\star) \\ 0 & \cdots & 0 \\ \vdots & \ddots & \vdots \\ 0 & \cdots & 0 \end{pmatrix}$$

has $N-M$ rows consisting of zeros. Hence, not all the derivatives $(\partial_0\mathbf{u})_\star$ are determined by the initial data, and one has $N-M$ ***constraint equations*** which have to be satisfied by the initial data $\mathbf{u}_\star$.

The discussion of the previous paragraphs leads to the following general definition which also applies to evolution systems of the form (12.1) which are *not necessarily* symmetric hyperbolic:

Definition 12.2 (***characteristic surfaces of a first order PDE***) *A hypersurface $\mathcal{S}$ defined by a condition of the form* (12.3) *is said to be a* ***characteristic of a solution u of Equation (12.1)*** *if*

$$\det\big(\boldsymbol{\sigma}(x,\mathbf{u},\mathbf{d}\phi)\big) = 0 \quad \textit{for } p \in \mathcal{S}. \tag{12.5}$$

If

$$\det\big(\boldsymbol{\sigma}(x,\mathbf{u},\mathbf{d}\phi)\big) \neq 0 \quad \textit{for } p \in \mathcal{S},$$

then $\mathcal{S}$ is said to be ***nowhere characteristic*** *for the solution* $\mathbf{u}$ *of Equation (12.1).*

On a characteristic, the system (12.1) implies M ***transversal equations*** and $N-M$ ***interior equations*** on $\mathcal{S}$. If $M = 0$, so that the full system (12.1) reduces to interior equations, one says that $\mathcal{S}$ is a ***total characteristic*** of the system. More generally, given a point $p \in \mathcal{U}$, one defines its ***characteristic set*** (or ***Monge cone***) with respect to a solution $\mathbf{u}$ of Equation (12.1) as the subset $C^*_p \subset T^*|_p(\mathcal{U})$ defined by

$$C^*_p \equiv \{\boldsymbol{\xi} \in T^*|_p(\mathcal{U}) \mid \det(\boldsymbol{\sigma}(x,\mathbf{u},\boldsymbol{\xi})) = 0,\ \boldsymbol{\xi} \neq 0\}.$$

That is, the elements of C^*_p are in the kernel of the symbol. The covectors $\boldsymbol{\xi}$ are sometimes called the ***null directions*** at p. The quantity $\det(\boldsymbol{\sigma}(x,\mathbf{u},\boldsymbol{\xi}))$ can be read as a polynomial for the components of the covector $\boldsymbol{\xi}$ – the so-called ***characteristic polynomial***.

Well-posedness

An initial value problem for a system of the form (12.1) (not necessarily symmetric hyperbolic) with data prescribed on a hypersurface $\mathcal{S}$ which is *nowhere characteristic and timelike* with respect to the evolution system at the prescribed data $\mathbf{u}_\star$ will be called a ***Cauchy initial value problem***. If the initial data is prescribed on a hypersurface $\mathcal{N}$ which is characteristic, one speaks of a ***characteristic initial value problem***.

The definitions given in the previous paragraph are motivated by the notion of ***well-posedness***. In broad terms, an initial value problem is ***well posed*** if:

(i) There exist solutions to all initial data.
(ii) The solutions are uniquely determined by the initial data.
(iii) The solutions depend continuously on the initial data.

The first step in the analysis of the well-posedness of an initial value problem for a given class of PDEs is the formulation of the above requirements in a precise manner; see, for example, Rendall (2008) for further discussion on this. Initial value problems which are not well posed are said to be ***ill-posed***.

The Cauchy problem for a symmetric hyperbolic system of the form (12.1) is well-posed. By contrast, an initial value problem with data prescribed on a timelike hypersurface is ill-posed. A further example of an ill-posed problem is the Cauchy problem for elliptic equations. In the case of characteristic initial value problems the well-posedness of the problem depends on the causal relation between the region where one wants to obtain the solution and the initial characteristic surfaces; see Section 12.5.1. Although well-posed initial value problems are of natural importance in general relativity, ill-posed problems also arise in applications such as the uniqueness of stationary black holes; see, for example, Ionescu and Klainerman (2009a,b).

12.1.3 Some examples

The discussion of the previous paragraphs is best illuminated by a couple of explicit examples. Many of the features of these examples are generic and arise in the analysis of the evolution equations implied by the (conformal) Einstein field equations.

In what follows, let $(\mathcal{M}, \boldsymbol{g})$ denote a spacetime. On $\mathcal{U} \subset \mathcal{M}$ consider some local coordinates $x = (x^\mu)$ and a null frame $\{\boldsymbol{e}_{\boldsymbol{AA}'}\}$ with associated cobasis $\{\boldsymbol{\omega}^{\boldsymbol{AA}'}\}$. In terms of the local coordinates one writes $\boldsymbol{e}_{\boldsymbol{AA}'} = e_{\boldsymbol{AA}'}{}^{\mu}\boldsymbol{\partial}_\mu$ and $\boldsymbol{\omega}^{\boldsymbol{AA}'} = \omega^{\boldsymbol{AA}'}{}_{\mu}\mathbf{d}x^\mu$. Moreover, let $\{\epsilon_{\boldsymbol{A}}{}^{A}\}$ be a spinorial frame giving rise to the vector frame $\{\boldsymbol{e}_{\boldsymbol{AA}'}\}$; see the discussion in Section 3.1.9.

A spinorial curl equation

As a first example consider on $\mathcal{U} \subset \mathcal{M}$ a spinorial equation of the form

$$\nabla^{\boldsymbol{Q}}{}_{\boldsymbol{A}'}\varphi_{\boldsymbol{QA}\cdots\boldsymbol{D}} = F_{\boldsymbol{A}'\boldsymbol{A}\cdots\boldsymbol{D}}, \qquad (12.6)$$

for the components $\varphi_{QA\cdots D}$ of a spinor $\varphi_{QA\cdots D}$ with respect to the spin frame $\{\epsilon_{\boldsymbol{A}}{}^{A}\}$. The spinor $\varphi_{QA\cdots D}$ is not assumed to have any particular symmetries and the field $F_{\boldsymbol{A}'\boldsymbol{A}\cdots\boldsymbol{D}}$ may depend on the coordinates or any other field. Notice that the unknowns of Equation (12.6) are scalars.

It is claimed that the combination

$$\nabla^{\boldsymbol{Q}}{}_{\boldsymbol{1}'}\varphi_{\boldsymbol{QA}\cdots\boldsymbol{D}} = F_{\boldsymbol{1}'\boldsymbol{A}\cdots\boldsymbol{D}}, \tag{12.7a}$$

$$-\nabla^{\boldsymbol{Q}}{}_{\boldsymbol{0}'}\varphi_{\boldsymbol{QA}\cdots\boldsymbol{D}} = -F_{\boldsymbol{0}'\boldsymbol{A}\cdots\boldsymbol{D}}, \tag{12.7b}$$

is a symmetric hyperbolic system. In order to see this, observe that

$$\nabla^{\boldsymbol{Q}}{}_{\boldsymbol{A}'}\varphi_{\boldsymbol{QA}\cdots\boldsymbol{D}} = \nabla_{\boldsymbol{1A}'}\varphi_{\boldsymbol{0A}\cdots\boldsymbol{D}} - \nabla_{\boldsymbol{0A}'}\varphi_{\boldsymbol{1A}\cdots\boldsymbol{D}}.$$

Thus, the principal part of the system (12.7a) and (12.7b) can be written in matricial form as

$$\mathbf{A}^{\mu}\partial_{\mu}\boldsymbol{\varphi} \equiv \begin{pmatrix} e_{\boldsymbol{11}'}{}^{\mu} & -e_{\boldsymbol{01}'}{}^{\mu} \\ -e_{\boldsymbol{10}'}{}^{\mu} & e_{\boldsymbol{00}'}{}^{\mu} \end{pmatrix} \partial_{\mu} \begin{pmatrix} \varphi_{\boldsymbol{0A}\cdots\boldsymbol{D}} \\ \varphi_{\boldsymbol{1A}\cdots\boldsymbol{D}} \end{pmatrix}.$$

The matrices $\mathbf{A}^{\mu}$ are Hermitian as $\boldsymbol{e}_{\boldsymbol{00}'}$ and $\boldsymbol{e}_{\boldsymbol{11}'}$ are real vectors and $\boldsymbol{e}_{\boldsymbol{01}'} = \overline{\boldsymbol{e}_{\boldsymbol{10}'}}$. Letting $\xi_{\mu} \equiv \omega^{\boldsymbol{00}'}{}_{\mu} + \omega^{\boldsymbol{11}'}{}_{\mu}$, a calculation shows that

$$\mathbf{A}^{\mu}\xi_{\mu} = \begin{pmatrix} e_{\boldsymbol{11}'}{}^{\mu}\omega^{\boldsymbol{00}'}{}_{\mu} + e_{\boldsymbol{11}'}{}^{\mu}\omega^{\boldsymbol{11}'}{}_{\mu} & -e_{\boldsymbol{01}'}{}^{\mu}\omega^{\boldsymbol{00}'}{}_{\mu} - e_{\boldsymbol{01}'}{}^{\mu}\omega^{\boldsymbol{11}'}{}_{\mu} \\ -e_{\boldsymbol{10}'}{}^{\mu}\omega^{\boldsymbol{00}'}{}_{\mu} - e_{\boldsymbol{10}'}{}^{\mu}\omega^{\boldsymbol{11}'}_{\mu} & e_{\boldsymbol{00}'}{}^{\mu}\omega^{\boldsymbol{00}'}{}_{\mu} + e_{\boldsymbol{00}'}\omega^{\boldsymbol{11}'}{}_{\mu} \end{pmatrix}.$$

Using $e_{\boldsymbol{AA}'}{}^{\mu}\omega^{\boldsymbol{BB}'}{}_{\mu} = \epsilon_{\boldsymbol{A}}{}^{\boldsymbol{B}}\epsilon_{\boldsymbol{A}'}{}^{\boldsymbol{B}'}$ it follows that

$$\mathbf{A}^{\mu}\xi_{\mu} = \begin{pmatrix} 1 & 0 \\ 0 & 1 \end{pmatrix},$$

which is clearly positive definite. Thus, the system (12.7a) and (12.7b) is symmetric hyperbolic as claimed. Given a generic covector $\boldsymbol{\xi}$, the characteristic polynomial is given by

$$\begin{aligned} \det(\mathbf{A}^{\mu}\xi_{\mu}) &= \det\begin{pmatrix} e_{\boldsymbol{11}'}{}^{\mu}\xi_{\mu} & -e_{\boldsymbol{01}'}{}^{\mu}\xi_{\mu} \\ -e_{\boldsymbol{10}'}{}^{\mu}\xi_{\mu} & e_{\boldsymbol{00}'}{}^{\mu}\xi_{\mu} \end{pmatrix} \\ &= (e^{\mu}_{\boldsymbol{11}'}e_{\boldsymbol{00}'}{}^{\nu} - e_{\boldsymbol{01}'}{}^{\mu}e_{\boldsymbol{10}'}{}^{\nu})\xi_{\mu}\xi_{\nu} \\ &= \frac{1}{2}g^{\mu\nu}\xi_{\mu}\xi_{\nu}, \end{aligned}$$

where, in the last equality, Equation (3.30) relating the null frame and the metric has been used. Thus, the characteristics of Equation (12.6) are given by null hypersurfaces with respect to the metric $\boldsymbol{g}$. Furthermore, spacelike hypersurfaces with respect to solutions to the equation coincide with the $\boldsymbol{g}$-spacelike hypersurfaces so that the causal notions given by Equation (12.6) and the background metric $\boldsymbol{g}$ coincide.

The wave equation as a symmetric hyperbolic system

As a second example consider the wave equation

$$\nabla^a \nabla_a \phi = 0 \tag{12.8}$$

on a region $\mathcal{U} \subset \mathcal{M}$. In contrast to the previous example, this equation is second order, and thus, it does not fit into the scheme discussed so far. Nevertheless, the wave equation can be recast as a symmetric hyperbolic system for the scalar field ϕ and some further auxiliary fields.

The spinorial version of Equation (12.8) is given by

$$\nabla^{\boldsymbol{AA'}} \nabla_{\boldsymbol{AA'}} \phi = 0. \tag{12.9}$$

As a first step one introduces the auxiliary variable $\phi_{\boldsymbol{AA'}} \equiv \nabla_{\boldsymbol{AA'}}\phi$. Reading this definition as an equation for the scalar field ϕ and contracting with a spinor $\tau^{\boldsymbol{AA'}}$ representing a timelike vector τ^a, one obtains the evolution equation

$$\mathcal{P}\phi = \varphi, \tag{12.10}$$

where $\varphi \equiv \tau^{\boldsymbol{AA'}}\phi_{\boldsymbol{AA'}}$ and $\mathcal{P} \equiv \tau^{\boldsymbol{AA'}}\nabla_{\boldsymbol{AA'}}$ is the directional derivative along τ^a; see Section 4.3.1. Now, defining $\varphi_{\boldsymbol{AB}} \equiv \tau_{(\boldsymbol{B}}{}^{\boldsymbol{A'}}\phi_{\boldsymbol{A})\boldsymbol{A'}}$ one obtains the decomposition

$$\phi_{\boldsymbol{AA'}} = \frac{1}{2}\varphi\tau_{\boldsymbol{AA'}} - \tau^{\boldsymbol{Q}}{}_{\boldsymbol{A'}}\varphi_{\boldsymbol{AQ}}. \tag{12.11}$$

Having introduced the auxiliary variable $\phi_{\boldsymbol{AA'}}$ one needs to consider a suitable field equation for it. A convenient choice is given by the *no torsion condition*

$$\nabla_{\boldsymbol{AA'}}\nabla_{\boldsymbol{BB'}}\phi - \nabla_{\boldsymbol{BB'}}\nabla_{\boldsymbol{AA'}}\phi = 0,$$

which, in view of the definition of $\phi_{\boldsymbol{AA'}}$, can be rewritten as

$$\nabla_{\boldsymbol{AA'}}\phi_{\boldsymbol{BB'}} - \nabla_{\boldsymbol{BB'}}\phi_{\boldsymbol{AA'}} = 0. \tag{12.12}$$

Contracting the indices $\boldsymbol{A'}$ and $\boldsymbol{B'}$ and using the see-saw rule one obtains

$$\nabla_{(\boldsymbol{A}}{}^{\boldsymbol{Q'}}\phi_{\boldsymbol{B})\boldsymbol{Q'}} = 0,$$

which, as a result of the hermicity of $\phi_{\boldsymbol{AA'}}$ is completely equivalent to Equation (12.12). Finally, using the identity

$$\nabla_{\boldsymbol{A}}{}^{\boldsymbol{Q'}}\phi_{\boldsymbol{BQ'}} = \nabla_{(\boldsymbol{A}}{}^{\boldsymbol{Q'}}\phi_{\boldsymbol{B})\boldsymbol{Q'}} - \frac{1}{2}\epsilon_{\boldsymbol{AB}}\nabla^{\boldsymbol{QQ'}}\phi_{\boldsymbol{QQ'}}$$

and observing that from Equation (12.9) it follows that $\nabla^{\boldsymbol{QQ'}}\phi_{\boldsymbol{QQ'}} = 0$, one concludes that

$$\nabla_{\boldsymbol{A}}{}^{\boldsymbol{Q'}}\phi_{\boldsymbol{BQ'}} = 0.$$

Using Equation (12.11) one can perform a space spinor split of this equation. After some calculations one obtains the pair of equations

$$\mathcal{P}\varphi + 2\mathcal{D}^{\boldsymbol{AB}}\varphi_{\boldsymbol{AB}} = 0, \tag{12.13a}$$

$$\mathcal{P}\varphi_{\boldsymbol{AB}} - \mathcal{D}_{\boldsymbol{AB}}\varphi + 2\mathcal{D}_{(\boldsymbol{A}}{}^{\boldsymbol{Q}}\varphi_{\boldsymbol{B})\boldsymbol{Q}} = 0, \tag{12.13b}$$

where $\mathcal{D}_{\boldsymbol{AB}}$ denotes the directional derivative associated to the Sen connection relative to $\tau^{AA'}$; see Section 4.3.1. Equations (12.10), (12.13a) and (12.13b) are the basic evolution equations. For simplicity of presentation in Equations (12.13a) and (12.13b) the covariant derivatives of $\tau^{AA'}$ have been assumed to vanish. To obtain a system which is symmetric hyperbolic, some normalisation factors have to be added. Some experimentation renders

$$\mathcal{P}\phi = \varphi,$$

$$\mathcal{P}\varphi + 2\mathcal{D}^{\boldsymbol{AB}}\varphi_{\boldsymbol{AB}} = 0,$$

$$\frac{4}{(\boldsymbol{A}+\boldsymbol{B})!(2-\boldsymbol{A}-\boldsymbol{B})!}\left(\mathcal{P}\varphi_{\boldsymbol{AB}} - \mathcal{D}_{\boldsymbol{AB}}\varphi + 2\mathcal{D}_{(\boldsymbol{A}}{}^{\boldsymbol{Q}}\varphi_{\boldsymbol{B})\boldsymbol{Q}}\right) = 0,$$

which is claimed to be symmetric hyperbolic. From these equations, a calculation similar to the one carried out for the Maxwell equations yields the following matricial expression for the principal part:

$$\mathbf{A}^\mu\partial_\mu\boldsymbol{\phi} \equiv \begin{pmatrix} \tau^\mu & 0 & 0 & 0 & 0 \\ 0 & \tau^\mu & 2e_{\boldsymbol{11}}{}^\mu & -4e_{\boldsymbol{01}}{}^\mu & 2e_{\boldsymbol{00}}{}^\mu \\ 0 & -2e_{\boldsymbol{00}}{}^\mu & 2\tau^\mu - 4e_{\boldsymbol{01}}{}^\mu & 4e_{\boldsymbol{00}}{}^\mu & 0 \\ 0 & -4e_{\boldsymbol{01}}{}^\mu & -4e_{\boldsymbol{11}}{}^\mu & 4\tau^\mu & 4e_{\boldsymbol{00}}{}^\mu \\ 0 & -2e_{\boldsymbol{11}}{}^\mu & 0 & -4e_{\boldsymbol{11}}{}^\mu & 2\tau^\mu + 4e_{\boldsymbol{01}}{}^\mu \end{pmatrix} \partial_\mu \begin{pmatrix} \phi \\ \varphi \\ \varphi_0 \\ \varphi_1 \\ \varphi_2 \end{pmatrix},$$

where $\varphi_0 \equiv \varphi_{\boldsymbol{00}}$, $\varphi_1 \equiv \varphi_{\boldsymbol{01}}$ and $\varphi_2 \equiv \varphi_{\boldsymbol{11}}$. Taking into account the reality conditions satisfied by the various frame coefficients one concludes that the matrices are Hermitian. Moreover, a short computation shows that $\mathbf{A}^\mu\tau_\mu$ is positive definite so that, indeed, one has obtained a symmetric hyperbolic system for the wave equation. Finally, a further computation shows that the characteristic polynomial of the system is given by

$$\det(\mathbf{A}^\mu\xi_\mu) = 8(\tau^\mu\xi_\mu)(g^{\nu\lambda}\xi_\nu\xi_\lambda)^2.$$

Accordingly, $\boldsymbol{g}$-null hypersurfaces are characteristics of the system.

12.2 Uniqueness and domains of dependence

An important property of the Cauchy initial value problem for symmetric hyperbolic systems is the ***uniqueness of solutions*** for a given prescription of initial data. The discussion of the uniqueness of solutions is naturally carried out in subsets of $\mathbb{R}^4$ known as lens-shaped domains. A ***lens-shaped domain***

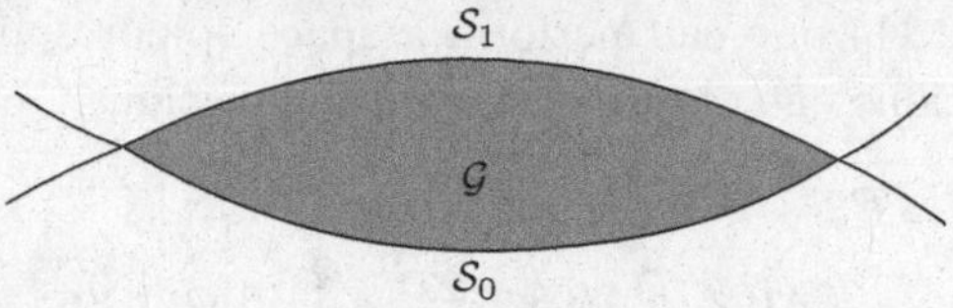

Figure 12.1 Schematic depiction of a lens-shaped domain $\mathcal{G}$. The hypersurfaces $\mathcal{S}_0$ and $\mathcal{S}_1$ are spacelike with respect to a solution $\mathbf{u}$ of a symmetric hyperbolic system of the form (12.1).

with respect to a solution $\mathbf{u}$ to a symmetric hyperbolic system of the form (12.1) is an open subset $\mathcal{G} \subset \mathbb{R}^4$ with compact closure and whose boundary is given by the union of two subsets $\mathcal{S}_0$ and $\mathcal{S}_1$ of hypersurfaces which are spacelike with respect to $\mathbf{u}$; see Figure 12.1. In terms of these domains one has the following result which exploits all the algebraic conditions in Definition 12.1:

Theorem 12.1 (***uniqueness of solutions of symmetric hyperbolic systems***) *Let $\mathcal{G}$ be a lens-shaped domain. If $\mathbf{u}_1$ and $\mathbf{u}_2$ are two solutions to the initial value problem for the symmetric hyperbolic system*

$$\mathbf{A}^\mu(x,\mathbf{u})\partial_\mu \mathbf{u} = \mathbf{B}(x,\mathbf{u}), \qquad \mathbf{u}|_{\mathcal{S}_0} = \mathbf{u}_\star$$

then $\mathbf{u}_1 = \mathbf{u}_2$ on $\mathcal{G}$.

Proof This proof follows closely the discussion in Friedrich and Rendall (2000). Assume one has a symmetric hyperbolic system of the form (12.1) such that the matrices $\mathbf{A}^\mu$ and $\mathbf{B}$ are C^1 functions of their arguments. Moreover, let $\mathbf{u}_1$ and $\mathbf{u}_2$ denote two C^1 solutions. Let $\mathcal{G}$ denote a lens-shaped region with respect to $\mathbf{u}_1$ and $\mathbf{u}_2$ whose boundary is given by the union of two hypersurfaces $\mathcal{S}_0$ and $\mathcal{S}_1$. Using a refined version of the ***mean value theorem*** (see the Appendix to this chapter for further discussion) it follows that there exist continuous functions $\mathbf{M}^\mu$ and $\mathbf{N}$ such that

$$\begin{aligned}
\mathbf{A}^\mu(x,\mathbf{u}_1) - \mathbf{A}^\mu(x,\mathbf{u}_2) &= \mathbf{M}^\mu(x,\mathbf{u}_1,\mathbf{u}_2)(\mathbf{u}_1-\mathbf{u}_2),\\
\mathbf{B}(x,\mathbf{u}_1) - \mathbf{B}(x,\mathbf{u}_2) &= \mathbf{N}(x,\mathbf{u}_1,\mathbf{u}_2)(\mathbf{u}_1-\mathbf{u}_2).
\end{aligned}$$

It follows then from Equation (12.1) that

$$\mathbf{A}^\mu(x,\mathbf{u}_1)\partial_\mu(\mathbf{u}_1-\mathbf{u}_2) + \Big(\mathbf{M}^\mu(x,\mathbf{u}_1,\mathbf{u}_2)\partial_\mu\mathbf{u}_2 + \mathbf{N}(x,\mathbf{u}_1,\mathbf{u}_2)\Big)(\mathbf{u}_1-\mathbf{u}_2) = 0.$$

This equation can be written in a more compact form as

$$\mathbf{A}^\mu(x,\mathbf{u}_1)\partial_\mu(\mathbf{u}_1-\mathbf{u}_2) = \mathbf{Q}(x,\mathbf{u}_1,\mathbf{u}_2,\partial\mathbf{u}_2)(\mathbf{u}_1-\mathbf{u}_2)$$

with $\mathbf{Q}(x,\mathbf{u}_1,\mathbf{u}_2,\partial\mathbf{u}_2)$ a continuous function of its arguments.

Now, choosing coordinates such that $x = (t, \underline{x})$ and using the evolution Equation (12.1) one can verify the identity

$$\partial_\mu\left(e^{-kt}\langle \mathbf{u}_1 - \mathbf{u}_2, \mathbf{A}^\mu(x, \mathbf{u}_1)(\mathbf{u}_1 - \mathbf{u}_2)\rangle\right) = e^{-kt}\langle \mathbf{u}_1 - \mathbf{u}_2, \mathbf{P}(x, \mathbf{u}_1, \mathbf{u}_2, \partial\mathbf{u}_2)(\mathbf{u}_1 - \mathbf{u}_2)\rangle, \tag{12.14}$$

where

$$\mathbf{P}(x, \mathbf{u}_1, \mathbf{u}_2, \partial\mathbf{u}_2) \equiv -k\mathbf{A}^0(x, \mathbf{u}_1) + \partial_\mu\mathbf{A}^\mu(x, \mathbf{u}_1) + \mathbf{Q}(x, \mathbf{u}_1, \mathbf{u}_2, \partial\mathbf{u}_2) + \mathbf{Q}^*(x, \mathbf{u}_1, \mathbf{u}_2, \partial\mathbf{u}_2).$$

Integrating the identity (12.14) over the lens-shaped region $\mathcal{G}$ and using the Gauss theorem one has that

$$\int_{\mathcal{G}} \partial_\mu(e^{-kt}\langle \mathbf{u}_1 - \mathbf{u}_2, \mathbf{A}^\mu(x, \mathbf{u}_1)(\mathbf{u}_1 - \mathbf{u}_2)\rangle)\mathrm{d}^4x = \int_{\mathcal{S}_1} e^{-kt}\langle \mathbf{u}_1 - \mathbf{u}_2, \mathbf{A}^\mu(x, \mathbf{u}_1)(\mathbf{u}_1 - \mathbf{u}_2)\rangle\nu_\mu \mathrm{d}S - \int_{\mathcal{S}_0} e^{-kt}\langle \mathbf{u}_1 - \mathbf{u}_2, \mathbf{A}^\mu(x, \mathbf{u}_1)(\mathbf{u}_1 - \mathbf{u}_2)\rangle\nu_\mu \mathrm{d}S, \tag{12.15}$$

where d^4x is the standard volume element in $\mathbb{R}^4$ and ν_μ denotes the outward pointing unit normal to $\partial\mathcal{G}$.

As $\mathcal{S}_1$ is spatial with respect to the symmetric hyperbolic system, it follows that $\mathbf{A}^\mu(x, \mathbf{u}_1)|_{\mathcal{S}_1}$ is positive definite. Hence, the integral over $\mathcal{S}_1$ in Equation (12.15) is non-negative. By assumption one has that $(\mathbf{u}_1 - \mathbf{u}_2)|_{\mathcal{S}_0} = \mathbf{0}$ so that the integral over $\mathcal{S}_0$ in (12.15) vanishes. Hence, one concludes that

$$\int_{\mathcal{G}} e^{-kt}\langle \mathbf{u}_1 - \mathbf{u}_2, \mathbf{P}(x, \mathbf{u}_1, \mathbf{u}_2, \partial\mathbf{u}_2)(\mathbf{u}_1 - \mathbf{u}_2)\rangle\mathrm{d}^4x \geq 0. \tag{12.16}$$

Finally, as the matrix $\mathbf{A}^0(x, \mathbf{u}_1)$ is positive definite and $\mathcal{G}$ is compact, it follows that the constant $k > 0$ can be chosen so that $\mathbf{P}(x, \mathbf{u}_1, \mathbf{u}_2, \partial\mathbf{u}_2)$ is negative definite uniformly on $\mathcal{G}$. In other words, there exists a positive constant C such that

$$0 > -C\langle \mathbf{u}_1 - \mathbf{u}_2, \mathbf{u}_1 - \mathbf{u}_2\rangle \geq \langle \mathbf{u}_1 - \mathbf{u}_2, \mathbf{P}(x, \mathbf{u}_1, \mathbf{u}_2, \partial\mathbf{u}_2)(\mathbf{u}_1 - \mathbf{u}_2)\rangle.$$

Accordingly, the integral over $\mathcal{G}$ in inequality (12.16) can be made negative by a suitable choice of k. This is a contradiction unless $\mathbf{u}_1 = \mathbf{u}_2$ in $\mathcal{G}$. □

A corollary of the above theorem is the following:

Corollary 12.1 *If* $\mathbf{u}|_{\mathcal{S}_0} = \mathbf{0}$ *and* $\mathbf{B}(x, \mathbf{u})$ *is homogeneous in* $\mathbf{u}$*, then* $\mathbf{u} = \mathbf{0}$ *in* $\mathcal{G}$.

Proof The result follows directly from the previous theorem, observing that $\mathbf{u} = \mathbf{0}$ is a solution. □

The uniqueness Theorem 12.1 shows that, in a neighbourhood of an initial hypersurface $\mathcal{S}$, the solution of a symmetric hyperbolic system is determined by initial data on a compact subset of $\mathcal{S}$ as any point sufficiently close to $\mathcal{S}$ is contained in a lens-shaped region. This consideration leads to the notion of ***domain of dependence.***

Definition 12.3 (***domain of dependence***) *Let $\mathcal{R} \subset \mathcal{S}$. The **domain of dependence** $D(\mathcal{R})$ of $\mathcal{R}$ is the set of all points $p \in \mathcal{U} \subset \mathbb{R}^4$ such that the value of a solution $\mathbf{u}$ to Equation* (12.1) *at p is determined (uniquely) by the restriction of the initial data to $\mathcal{R}$.*

Remark. The term "domain of dependence" is sometimes used in the PDE literature to denote the set of points determining the value of a solution $\mathbf{u}$ at a given point. The notion of domain of dependence used in this book is then called *domain of influence*; see Rendall (2008) for further discussion.

The main property singled out by Definition 12.3 is that the solution of a symmetric hyperbolic system is determined at a given point by data on a proper subset of the initial hypersurface. Thus, the process of solving the Cauchy problem for the symmetric hyperbolic system (12.1) can be *localised in space.* This is a particular property of hyperbolic differential equations which distinguishes them from other types of PDEs. More precisely, if two initial data sets $\mathbf{u}_\star$ and $\bar{\mathbf{u}}_\star$ coincide on an open subset $\mathcal{R} \subset \mathcal{S}$, then the corresponding domains of influence and the solutions $\mathbf{u}$ and $\bar{\mathbf{u}}$ coincide as well. In other words, in the domain of influence $D(\mathcal{R})$ a solution $\mathbf{u}$ is independent of the behaviour of the data $\mathbf{u}_\star$ outside $\mathcal{R}$. In particular, there is no need to impose boundary or fall-off conditions away from $\mathcal{R}$. This observation is usually known as the ***localisability property*** of symmetric hyperbolic systems; that is, the theory does not depend on the global knowledge of the initial data in space. A related observation is that if on $\mathcal{S}$ one has two different intersecting coordinate patches $\mathcal{R}$ and $\mathcal{R}'$ such that on $\mathcal{R} \cap \mathcal{R}'$ one has $x' = x'(x)$, then, as a consequence the transformation rule of Equation (12.1) and the uniqueness of the solution on $D(\mathcal{R} \cap \mathcal{R}')$, one has that $\mathbf{u}'(x') = \mathbf{u}(x(x'))$.

Finite speed of propagation of solutions

A consequence of the existence of a domain of dependence for symmetric hyperbolic systems is the so-called ***finite speed of propagation of their solutions.*** A rough estimate of this phenomenon can be constructed using an argument given in Rendall (2008).

As in previous sections, let $\mathbf{u}$ denote a solution to a symmetric hyperbolic system of the form (12.1) with initial data $\mathbf{u}_\star$ prescribed on the hypersurface

$$\mathcal{S}_0 \equiv \{p \in \mathcal{U} \,|\, t(p) = 0\}.$$

In what follows, assume that the support of $\mathbf{u}_\star$ is contained on a ball of radius $r_\star$ around the origin.

Now, given a point $p \in \mathcal{U}$ with coordinates $(t_\bullet, \underline{x}_\bullet) \equiv (t, x^\alpha_\bullet)$ and a constant $\beta > 0$ consider the paraboloidal hypersurface

$$\mathcal{S}_{\beta;(t_\bullet,\underline{x}_\bullet)} \equiv \{p \in \mathcal{U} \mid t(p) = t_\bullet - \beta\, \delta_{\alpha\beta}(x^\alpha(p) - x^\alpha_\bullet)(x^\beta(p) - x^\beta_\bullet)\}.$$

The normal to these hypersurfaces is given by

$$\boldsymbol{\nu} = \mathbf{d}t - 2\beta\delta_{\alpha\beta}x^\alpha \mathbf{d}x^\beta.$$

Hence, assuming that $\mathbf{A}^0(x, \mathbf{u})$ is positive definite on $\mathcal{S}_0$ it follows that

$$\mathbf{A}^\mu(x, \mathbf{u})\nu_\mu = \mathbf{A}^0(x, \mathbf{u}) + 2\beta\delta_{\alpha\beta}x^\alpha \mathbf{A}^\beta(x, \mathbf{u})$$

can be made positive definite by choosing β sufficiently small, say, $\beta < \beta_0$, so that $\mathcal{S}_{\beta;(t_\bullet,\underline{x}_\bullet)}$ is spacelike with respect to Equation (12.1). For this choice of β the region $\mathcal{G}$ bounded by $\mathcal{S}_0$ and $\mathcal{S}_{\beta;(t_\bullet,\underline{x}_\bullet)}$ is a lens-shaped domain. Now, it can be verified that the intersection of $\mathcal{S}_{\beta;(t_\bullet,\underline{x}_\bullet)}$ with $\mathcal{S}_0$ lies outside a ball of radius

$$r \equiv |x_\bullet| - \sqrt{\frac{t_\bullet}{\beta}}, \qquad |x_\bullet|^2 \equiv \delta_{\alpha\beta}x^\alpha_\bullet x^\beta_\bullet.$$

Thus, if

$$|x_\bullet| - \sqrt{\frac{t_\bullet}{\beta}} > r_\star,$$

then the solution satisfies $\mathbf{u}(t_\bullet, \underline{x}_\bullet) = 0$ as $(t_\bullet, \underline{x}_\bullet)$ lies on the boundary of a lens-shaped region with trivial data. Accordingly, the support of $\mathbf{u}$ on the hypersurface

$$\mathcal{S}_{t_\bullet} = \{p \in \mathcal{U} \mid t(p) = t_\bullet\}$$

must lie within a ball of radius $r_\star + \sqrt{t_\bullet/\beta}$; see Figure 12.2 for further details. Thus, the support of the solution gradually spreads in space at finite speed.

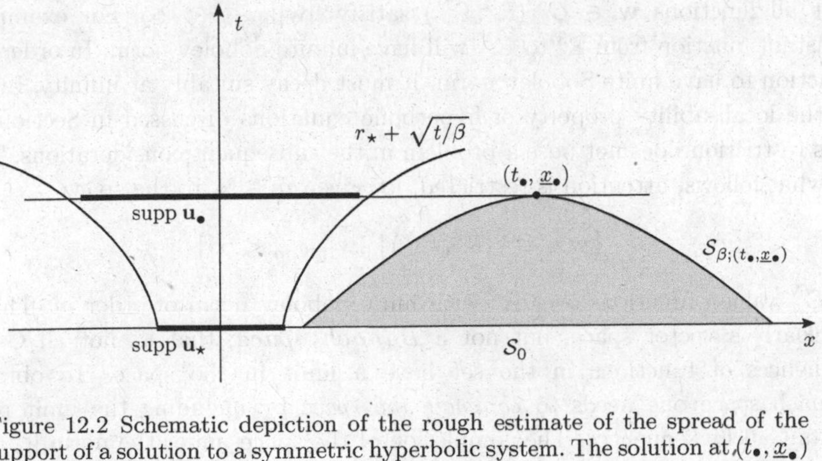

Figure 12.2 Schematic depiction of the rough estimate of the spread of the support of a solution to a symmetric hyperbolic system. The solution at $(t_\bullet, \underline{x}_\bullet)$ is determined by trivial data at the initial hypersurface $\mathcal{S}_0$; see the main text for further details.

12.3 Local existence results for symmetric hyperbolic systems

The purpose of this section is to analyse the basic existence and stability results for symmetric hyperbolic systems of the form (12.1). The precise formulation of existence results is more technical than the one for uniqueness and requires a certain number of notions from the theory of functional analysis. These are discussed in the following subsection.

12.3.1 Sobolev spaces

The precise discussion of existence results for symmetric hyperbolic systems is carried out in terms of Sobolev spaces. The purpose of this section is to introduce some of the basic ideas concerning these function spaces. In a first step, the discussion will consider Sobolev spaces of functions over $\mathbb{R}^3$. These notions can be suitably extended to three-dimensional manifolds with a different topology.

In what follows, let $\underline{x} \equiv (x^\alpha)$ denote some particular choice of Cartesian coordinates and let d^3x be the standard volume element of $\mathbb{R}^3$. The discussion of solutions of symmetric hyperbolic systems of the form (12.1) leads to consider ***$\mathbb{C}^N$-valued functions on*** $\mathbb{R}^3$; that is, $\mathbf{w} : \mathbb{R}^3 \to \mathbb{C}^N$. The space of smooth functions of this type will be denoted by $C^\infty(\mathbb{R}^3, \mathbb{C}^N)$. On $C^\infty(\mathbb{R}^3, \mathbb{C}^N)$ one can introduce, for $m \in \mathbb{N}$, a ***Sobolev norm*** via

$$||\mathbf{w}||_{\mathbb{R}^3,m} \equiv \left(\sum_{k=0}^{m} \sum_{\alpha_1,\ldots\alpha_k=1}^{3} \int_{\mathbb{R}^3} |\partial_{\alpha_k} \cdots \partial_{\alpha_1} \mathbf{w}|^2 \mathrm{d}^3x \right)^{1/2}, \tag{12.17}$$

for $\mathbf{w} \equiv (w_1, \ldots, w_N) \in C^\infty(\mathbb{R}^3, \mathbb{C}^N)$ where $|\mathbf{w}|^2 = \langle \mathbf{w}, \mathbf{w} \rangle$ is the standard norm in $\mathbb{C}^N$. For example, if $\mathbf{u} = (u)$ is a $\mathbb{C}$-valued function, one has that

$$||\mathbf{u}||^2_{\mathbb{R}^3,1} = \int_{\mathbb{R}^3} (u\,\overline{u} + \partial_1 u\, \overline{\partial_1 u} + \partial_2 u\, \overline{\partial_2 u} + \partial_3 u\, \overline{\partial_3 u}) \mathrm{d}^3x.$$

Not all functions $\mathbf{w} \in C^\infty(\mathbb{R}^3, \mathbb{C}^N)$ satisfy $||\mathbf{w}||_{\mathbb{R}^3,m} < \infty$. For example, a constant function from $\mathbb{R}^3$ to $\mathbb{C}^N$ will have infinite Sobolev norm. In order for a function to have finite Sobolev norm, it must decay suitably at infinity. In view of the localisability property of hyperbolic equations discussed in Section 12.2 this restriction does not pose a problem in the subsequent considerations. Thus, in what follows, attention is restricted, for given $m \in \mathbb{N}$, to the space

$$\{\mathbf{w} \in C^\infty(\mathbb{R}^3, \mathbb{C}^N) \,|\, ||\mathbf{w}||_{\mathbb{R}^3,m} < \infty\}$$

of $\mathbb{C}^N$-valued functions over $\mathbb{R}^3$ with finite Sobolev norm of order m. This set is clearly a vector space, but not a ***Banach space***; that is, not all Cauchy sequences of functions in the set have a limit in the space. To obtain a Banach space one needs to *complete the space* by including the limit points of its Cauchy sequences. The completion of the space under the norm $||\ \ ||_{\mathbb{R}^3,m}$ defined by Equation (12.17) is called the ***Sobolev space*** $H^m(\mathbb{R}^3, \mathbb{C}^N)$. Given

$\mathbf{w}_\bullet \in H^m(\mathbb{R}^3, \mathbb{C}^N)$, the (open) ball of radius ε centred at $\mathbf{w}_\bullet$ with respect to the norm $\|\ \|_{\mathbb{R}^3,m}$ is defined as the set

$$B_\varepsilon(\mathbf{w}_\bullet) \equiv \{\mathbf{w} \in H^m(\mathbb{R}^3, \mathbb{C}^N) \,|\, \|\mathbf{w} - \mathbf{w}_\bullet\|_{\mathbb{R}^3,m} < \varepsilon\}.$$

When discussing symmetric hyperbolic systems of the form (12.1), it is convenient to consider their solutions $\mathbf{u}$ as $H^m(\mathbb{R}^3, \mathbb{C}^N)$-valued functions of the time coordinate t. This point of view is expressed by writing

$$\mathbf{u}(t, \cdot) : [0, T] \longrightarrow H^m(\mathbb{R}^3, \mathbb{C}^N).$$

If a $\mathbb{C}^N$-valued function $\mathbf{u}$ is such that for every $t \in [0, T]$, $\mathbf{u}(t, \cdot) \in H^m(\mathbb{R}^3, \mathbb{C}^N)$ with C^k-dependence on t, one writes

$$\mathbf{w} \in C^k([0, T]; H^m(\mathbb{R}^3, \mathbb{C}^N)).$$

For further details on Sobolev spaces, the reader is referred to Evans (1998).

Embedding theorems

Functions in the Sobolev space $H^m(\mathbb{R}^3, \mathbb{C}^N)$ are not necessarily smooth. The reason for this is that by completing the space one has included functions with lower regularity. There is, nevertheless, a relation between functions in H^m and C^k spaces. This relation is expressed in terms of so-called ***embedding theorems***. For the particular case under consideration one has the following:

Proposition 12.1 (***Sobolev embedding theorem***) *If $m \geq 2 + k$, then $H^m(\mathbb{R}^3, \mathbb{C}^N) \subset C^k(\mathbb{R}^3, \mathbb{C}^N)$.*

In other words, if a function belongs to the H^m space, then it has at least $m - 2$ continuous derivatives. A proof of this result can be found in Taylor (1996a), chapter 4, section 1. It follows from Proposition 12.1 that a function over $\mathbb{R}^3$ is smooth (i.e. C^∞) if it belongs to H^m for every m.

Extensions of functions

To exploit the localisability property of hyperbolic equations it is often convenient to extend functions which are defined only on bounded subsets $\mathcal{R} \subset \mathbb{R}^3$ to functions with domain on the whole of $\mathbb{R}^3$. Defining in a natural way the norm $\|\ \|_{\mathcal{R},m}$ and the Sobolev space $H^m(\mathcal{R}, \mathbb{C}^N)$ one has the following result:

Proposition 12.2 (***extension of functions on a compact domain***) *Assume that $\mathcal{R} \subset \mathbb{R}^3$ is bounded with smooth boundary $\partial\mathcal{R}$. Then there exists a linear operator*

$$\mathscr{E} : H^m(\mathcal{R}, \mathbb{C}^N) \longrightarrow H^m(\mathbb{R}^3, \mathbb{C}^N)$$

such that for each $\mathbf{u} \in H^m(\mathcal{R}, \mathbb{C}^N)$:

(i) $\mathscr{E}\mathbf{u} = \mathbf{u}$ *almost everywhere in* $\mathcal{R}$.
(ii) $\mathscr{E}\mathbf{u}$ *has support in an open bounded set* $\mathcal{R}' \supset \mathcal{R}$.
(iii) *There exists a constant* C *depending only on* $\mathcal{U}$ *and* $\mathcal{R}$ *such that*

$$||\mathscr{E}\mathbf{u}||_{\mathbb{R}^3,m} \leq C||\mathbf{u}||_{\mathcal{R},m}.$$

The $\mathbb{C}^N$*-valued function* $\mathscr{E}\mathbf{u}$ *is called an* ***extension of*** $\mathbf{u}$ ***to*** $\mathbb{R}^3$.

A discussion on how to prove this result can be found in Evans (1998).

12.3.2 Kato's existence and stability theorems

Using the terminology introduced in the previous subsections, it is now possible to discuss the basic existence and stability result for quasilinear symmetric hyperbolic systems of the form

$$\mathbf{A}^0(t,\underline{x},\mathbf{u})\partial_t\mathbf{u} + \mathbf{A}^\alpha(t,\underline{x},\mathbf{u})\partial_\alpha\mathbf{u} = \mathbf{B}(t,\underline{x},\mathbf{u}). \tag{12.18}$$

In what follows, it will always be assumed that the matrices $\mathbf{A}^\mu$ are smooth functions of their arguments.

The basic local existence theorem

As it can be seen from the proof of the *Uniqueness Theorem* 12.1, the positive-definiteness of the matrix $\mathbf{A}^0(t,\underline{x},\mathbf{u})$ plays a key role in determining the properties of solutions to the equation. On an initial hypersurface $\mathcal{S}$, this positivity can be set by fiat by choosing suitable initial data. However, in view of the quasilinearity of the equation, the positive-definiteness could be violated at some time as the solution evolves. Intuitively, one would expect this to lead to some sort of problems in the solution. For fixed $(t,\underline{x})$, and given a $\mathbb{C}^N$-valued function $\mathbf{w}$, one says that $\mathbf{A}^0(t,\underline{x},\mathbf{w})$ is ***positive definite and bounded away from zero by*** $\delta > 0$ if

$$\langle \mathbf{z}, \mathbf{A}^0(t,\underline{x},\mathbf{w})\mathbf{z}\rangle > \delta\langle \mathbf{z},\mathbf{z}\rangle$$

for all $\mathbf{z} \in \mathbb{C}^N$.

The basic local existence result for the Cauchy problem of symmetric hyperbolic systems to be considered in this book is the following:

Theorem 12.2 (***local existence of solutions to symmetric hyperbolic systems***) *Consider the Cauchy problem*

$$\begin{aligned}
&\mathbf{A}^0(t,\underline{x},\mathbf{u})\partial_t\mathbf{u} + \mathbf{A}^\alpha(t,\underline{x},\mathbf{u})\partial_\alpha\mathbf{u} = \mathbf{B}(t,\underline{x},\mathbf{u}),\\
&\mathbf{u}(0,\underline{x}) = \mathbf{u}_\star(\underline{x}) \in H^m(\mathbb{R}^3,\mathbb{C}^N) \qquad ,m \geq 4,
\end{aligned}$$

for a quasilinear symmetric hyperbolic system. If $\delta > 0$ *can be found such that* $\mathbf{A}^0(0,\underline{x},\mathbf{u}_\star)$ *is positive definite with lower bound* δ *for all* $p \in \mathbb{R}^3$, *then there* *exists $T > 0$ and a unique solution $\mathbf{u}$ to the Cauchy problem defined on $[0,T] \times \mathbb{R}^3$ such that*

$$\mathbf{u} \in C^{m-2}\big([0,T] \times \mathbb{R}^3, \mathbb{C}^N\big).$$

Moreover, $\mathbf{A}^0(t, \underline{x}, \mathbf{u})$ is positive definite with lower bound δ for $(t, \underline{x}) \in [0,T] \times \mathbb{R}^3$.

This theorem is an adaptation of similar theorems given in Kato (1975a) and Friedrich (1986b). A proof of this result falls beyond the scope of this book. The interested reader is referred to references given above.

Remarks

(a) For convenience, the regularity of the solution has been stated in terms of C^k spaces. However, the conclusions of the theorem can be expressed in a more detailed manner. In particular, one has that the solution satisfies

$$\mathbf{u} \in C^1\big([0,T], H^{m-1}(\mathbb{R}^3, \mathbb{C}^N)\big).$$

The latter can be shown to imply $\mathbf{u} \in H^m([0,T] \times \mathbb{R}^3, \mathbb{C}^N)$ which, in turn, using a Sobolev embedding theorem in four-dimensions gives the regularity stated in the theorem.

(b) In most of the applications given in this book, the initial data $\mathbf{u}_\star$ will be assumed to be smooth, so that $\mathbf{u}_\star \in H^m(\mathbb{R}^3, \mathbb{C}^N)$ for all m. However, as $\mathbb{R}^3$ is an unbounded set, one cannot simply assume that $\mathbf{u}_\star \in C^\infty(\mathbb{R}^3, \mathbb{C}^N)$; compare the remark after Equation (12.17).

(c) As $\mathbf{A}^0$ is a smooth function of its arguments, it follows from the regularity of the solution $\mathbf{u}$ that $\langle \mathbf{z}, \mathbf{A}^0(t, \underline{x}, \mathbf{u})\mathbf{z} \rangle$ for $\mathbf{z} \in \mathbb{C}^N$ depends continuously on $(t, \underline{x})$.

(d) As $\mathbf{A}^0(t, \underline{x}, \mathbf{u})$ is positive definite for $(t, \underline{x}) \in [0,T] \times \mathbb{R}^3$, it follows that the hypersurfaces of constant t are spacelike with respect to the symmetric hyperbolic system (and the solution).

(e) The value of the lower bound δ can often be determined by inspection.

The basic stability result

Of great relevance is the notion of ***Cauchy stability*** – namely, the idea that, given a symmetric hyperbolic system, initial data which are close to each other should lead to solutions which are close in some sense and have a common existence time interval. In view of the inherent error in the physical process of measurement, Cauchy stability is fundamental for the applicability of differential equations to describe physical phenomena. In mathematical terms, the precise formulation of the *closeness* of initial data and solutions is expressed in terms of Sobolev norms.

In the remainder of this section let D be a bounded open subset of $H^m(\mathbb{R}^3, \mathbb{C}^N)$ such that for $\mathbf{w} \in D$ the matrix $\mathbf{A}^0(0, \underline{x}, \mathbf{w})$ is positive definite bounded away from zero by $\delta > 0$ for all $p \in \mathbb{R}^4$. The basic result describing the Cauchy stability of the symmetric hyperbolic system (12.18) is the following theorem, adapted from Kato (1975a):

Theorem 12.3 (***basic Cauchy stability for symmetric hyperbolic systems***) *Let $m \geq 4$. If $\mathbf{u}_\star \in D$ is given as an initial condition for the system (12.18), then:*

(i) There exists $\varepsilon > 0$ such that a common existence time T can be chosen for all initial conditions in the open ball $B_\varepsilon(\mathbf{u}_\star) \subset D$.

(ii) If the solution $\mathbf{u}$ with initial data $\mathbf{u}_\star$ exists on $[0, T]$ for some $T > 0$, then the solutions to all initial conditions in $B_\varepsilon(\mathbf{u}_\star)$ exist on $[0, T]$ if $\varepsilon > 0$ is sufficiently small.

(iii) If ε and T are chosen as in (i) and one has a sequence $\mathbf{u}^n_\star \in B_\varepsilon(\mathbf{u}_\star)$ such that

$$||\mathbf{u}^n_\star - \mathbf{u}_\star||_{\mathbb{R}^3,m} \to 0 \quad \text{as } n \to \infty,$$

then for the solutions $\mathbf{u}^n(t,\cdot)$ with $\mathbf{u}^n(0,\cdot) = \mathbf{u}^n_\star$ it holds that

$$||\mathbf{u}^n(t,\cdot) - \mathbf{u}(t,\cdot)||_{\mathbb{R}^3,m} \to 0, \qquad \text{as } n \to \infty,$$

uniformly for $t \in [0, T]$.

Remarks

(a) Point (i) in the previous theorem essentially states that, given a sufficiently small ball in the space of data on which the *Existence Theorem* 12.2 can be applied, then a common existence time for the solutions arising from this data can be found. Observe, however, that one has no control over the size of the common existence time; one only knows there is one.

(b) If the existence of a particular solution is known, then point (ii) states that, by shrinking the ball on the space of data, one can choose the known existence time as the common existence time.

(c) Point (iii) states that data close to certain reference data give rise to developments which are also close to the reference solution; this is the statement of *Cauchy stability*.

(d) The convergence stated in (iii) is uniform on $[0, T] \times \mathbb{R}^3$.

12.3.3 Localising solutions

The localisability property of hyperbolic equations allows one to apply the existence and stability results discussed in the previous sections to the case of an initial data problem where data are prescribed only on a compact region $\mathcal{R}$. Given smooth initial data $\mathbf{u}_\star$ for a symmetric hyperbolic equation of the form (12.1) on a region $\mathcal{R} \subset \mathbb{R}^3$, one can make use of Proposition 12.2 to extend the

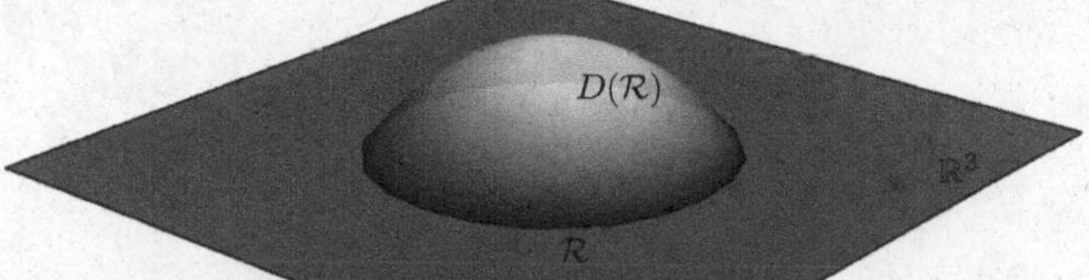

Figure 12.3 Localised solution arising from data prescribed on an open set $\mathcal{R} \subset \mathbb{R}^3$. The associated domain of dependence is denoted by $D(\mathcal{R})$.

initial data $\mathbf{u}_\star$ to the whole of $\mathbb{R}^3$ in a controlled manner. Denoting this extension by $\mathscr{E}\mathbf{u}_\star$, one has that by point (iii) of Proposition 12.2, $\mathscr{E}\mathbf{u}_\star \in H^m(\mathbb{R}^3, \mathbb{C}^N)$.

In order to make use of Theorems 12.2 and 12.3 it is necessary to assume that $\mathbf{A}^0(0, \underline{x}, \mathscr{E}\mathbf{u}_\star)$ is positive definite with some non-zero lower bound uniform on $\mathbb{R}^3$. Thus, one obtains a solution to Equation (12.1) with initial data on $\mathbb{R}^3$ given by $\mathbf{u}(0, \underline{x}) = \mathscr{E}\mathbf{u}_\star(\underline{x})$. As a consequence of the uniqueness of solutions on the domain of dependence, the solution $\mathbf{u}$ on $D(\mathcal{R})$ is independent of the particular extension of the initial data $\mathbf{u}_\star$ on $\mathcal{R}$ to $\mathbb{R}^3$; see Figure 12.3.

12.3.4 Existence and stability result on manifolds with compact spatial sections

The existence and stability Theorems 12.2 and 12.3 can be modified so as to apply to Cauchy problems where data is prescribed on compact, orientable three-dimensional manifolds. In what follows, the main ideas behind this construction are discussed.

Patching together solutions

In the remainder of this section let $\mathcal{S}$ denote an orientable, compact three-dimensional manifold – in most of the applications to be considered in this book one has $\mathcal{S} \approx \mathbb{S}^3$; however, any other compact, orientable topology will work as well. As a result of compactness, there exists a *finite cover* consisting of open sets $\mathcal{R}_1, \ldots, \mathcal{R}_M \subset \mathcal{S}$; that is, one has $\cup_{i=1}^M \mathcal{R}_i = \mathcal{S}$. On each of the open sets $\mathcal{R}_i$, $i = 1, \ldots, M$, one can introduce local coordinates $\underline{x}_i \equiv (x_i^\alpha)$ which allow one to identify $\mathcal{R}_i$ with open subsets $\mathcal{B}_i \subset \mathbb{R}^3$. As $\mathcal{S}$ is assumed to be a smooth manifold, the *coordinate patches* can be chosen so that the change of coordinates on intersecting sets is smooth.

Now, assume that a smooth function $\mathbf{u}_\star : \mathcal{S} \to \mathbb{C}^N$ has been prescribed on $\mathcal{S}$. In what follows, the restriction of $\mathbf{u}_\star$ to a particular open set $\mathcal{R}_i$ will be denoted by $\mathbf{u}_{i\star}$. Using the local coordinates x_i, the function $\mathbf{u}_{i\star}$ can be regarded as a function $\mathbf{u}_{i\star} : \mathcal{B}_i \to \mathbb{C}^N$.

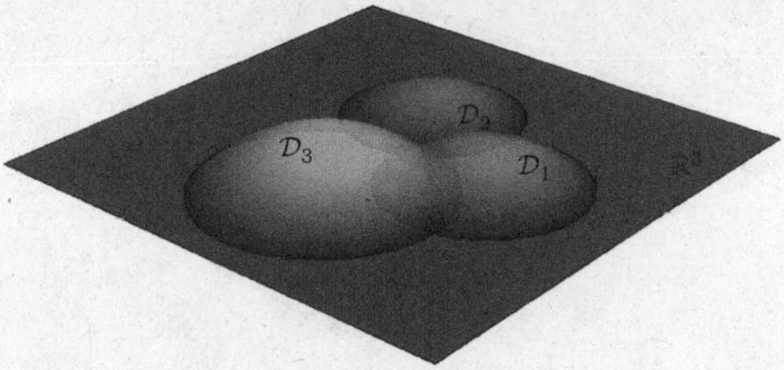

Figure 12.4 Construction of a solution by patching localised solutions to data prescribed on open sets $\mathcal{D}_1$, $\mathcal{D}_2$, $\mathcal{D}_3 \subset \mathbb{R}^3$.

The strategy is now to use the same procedure as described in Section 12.3.3 to ensure the existence of a solution on the domain of dependence of $\mathcal{B}_i$. Accordingly, one makes use of Proposition 12.2 to extend $\mathbf{u}_{i\star}$ to a function $\mathscr{E}\mathbf{u}_{i\star}$ defined on the whole of $\mathbb{R}^3$. Using the extended functions $\mathscr{E}\mathbf{u}_{i\star}$ one defines the norm

$$||\mathbf{u}_\star||_{\mathcal{S},m} \equiv \sum_{i=1}^{M} ||\mathbf{u}_{i\star}||_{\mathbb{R}^3,m}. \tag{12.19}$$

Assuming, as in Section 12.3.3, that $\mathbf{A}^0(0, \underline{x}, \mathscr{E}\mathbf{u}_{i\star})$ is positive definite with lower bound $\delta_i > 0$, one obtains a unique solution $\mathbf{u}_i$ of Equation (12.1) with initial data $\mathbf{u}(0, \underline{x}) = \mathscr{E}\mathbf{u}_{i\star}(\underline{x})$ with existence interval $[0, T_i]$. The solution on $D(\mathcal{B}_i)$ is independent of the particular extension $\mathscr{E}\mathbf{u}_{i\star}$ being used, so that one can speak of a solution $\mathbf{u}_i$ on a domain $\mathcal{D}_i \subset [0, T_i] \times \mathcal{R}_i$; see Figure 12.4.

Now, given two solutions $\mathbf{u}_i$ and $\mathbf{u}_j$ defined, respectively, on intersecting domains $\mathcal{D}_i$ and $\mathcal{D}_j$ one has – following the discussion on the change of coordinates given in Section 12.1 and as a consequence of uniqueness – that $\mathbf{u}_i$ and $\mathbf{u}_j$ must coincide on $\mathcal{D}_i \cap \mathcal{D}_j$. Proceeding in the same manner over the whole finite cover of $\mathcal{S}$, one obtains a unique solution $\mathbf{u}$ on $[0, T] \times \mathcal{S}$ with $T \equiv \min_{i=1,\dots M}\{T_i\}$ which is constructed by *patching together the localised solutions* $\mathbf{u}_1, \dots, \mathbf{u}_M$ defined, respectively on the domains $\mathcal{D}_i, \dots, \mathcal{D}_M$. Observe that the compactness of $\mathcal{S}$ ensures the existence of a minimum non-zero existence time for the whole of the domains $\mathcal{D}_i$.

A general existence and stability result

Using the ideas of the localisation of solutions discussed in the previous subsection, one can formulate a quite general existence and stability result for symmetric hyperbolic systems on manifolds whose spatial sections are given by orientable, compact three-dimensional manifolds. The hypotheses of this theorem are very similar to the ones in Theorems 12.2 and 12.3.

Theorem 12.4 (***existence and stability result for symmetric hyperbolic systems on compact spatial sections***) *Given an orientable, compact, three-dimensional manifold $\mathcal{S}$, consider the Cauchy problem*

$$\mathbf{A}^0(t,\underline{x},\mathbf{u})\partial_t\mathbf{u} + \mathbf{A}^\alpha(t,\underline{x},\mathbf{u})\partial_\alpha\mathbf{u} = \mathbf{B}(t,\underline{x},\mathbf{u}),$$
$$\mathbf{u}(0,\underline{x}) = \mathbf{u}_\star(\underline{x}) \in H^m(\mathcal{S},\mathbb{C}^N) \qquad \textit{for } m \geq 4,$$

for a quasilinear symmetric hyperbolic system. If $\delta > 0$ can be found such that $\mathbf{A}^0(0,\underline{x},\mathbf{u}_\star)$ is positive definite with lower bound δ for all $x \in \mathcal{S}$, then:

(i) There exists $T > 0$ and a unique solution $\mathbf{u}$ to the Cauchy problem defined on $[0,T] \times \mathcal{S}$ such that

$$\mathbf{u} \in C^{m-2}([0,T] \times \mathcal{S},\mathbb{C}^N).$$

Moreover, $\mathbf{A}^0(t,\underline{x},\mathbf{u})$ is positive definite with lower bound δ for $(t,\underline{x}) \in [0,T] \times \mathcal{S}$.

(ii) There exists $\varepsilon > 0$ such that one common existence time T can be chosen for all initial conditions in the open ball $B_\varepsilon(\mathbf{u}_\star)$ and such that $B_\varepsilon(\mathbf{u}_\star) \subset D$.

(iii) If the solution $\mathbf{u}$ with initial data $\mathbf{u}_\star$ exists on $[0,T]$ for some $T > 0$, then the solutions to all initial conditions in $B_\varepsilon(\mathbf{u}_\star)$ exist on $[0,T]$ if $\varepsilon > 0$ is sufficiently small.

(iv) If ε and T are chosen as in (ii) and one has a sequence $\mathbf{u}^n_\star \in B_\varepsilon(\mathbf{u}_\star)$ such that

$$||\mathbf{u}^n_\star - \mathbf{u}_\star||_{\mathcal{S},m} \to 0, \qquad \textit{as } n \to \infty,$$

then for the solutions $\mathbf{u}^n(t,\cdot)$ with $\mathbf{u}^n(0,\cdot) \equiv \mathbf{u}^n_\star$ it holds that

$$||\mathbf{u}^n(t,\cdot) - \mathbf{u}(t,\cdot)||_{\mathcal{S},m} \to 0, \qquad \textit{as } n \to \infty$$

uniformly in $t \in [0,T]$.

Remarks similar to the ones after Theorems 12.2 and 12.3 apply to this result. Further discussion and details can be found in Friedrich (1991).

12.4 Local existence for boundary value problems

As will be seen in Chapter 17, the construction of anti-de Sitter-like spacetimes leads one to consider initial boundary value problems for symmetric hyperbolic systems of the form (12.1). In this type of problem one prescribes initial data on a spacelike hypersurface $\mathcal{S}$ and boundary data on a timelike hypersurface $\mathcal{T}$. These two hypersurfaces intersect on a two-dimensional hypersurface $\mathcal{E} \equiv \mathcal{S} \cap \mathcal{T}$, the ***edge***, on which the initial and the boundary conditions need to satisfy some compatibility conditions; see Figure 12.5. In view of the localisation property of

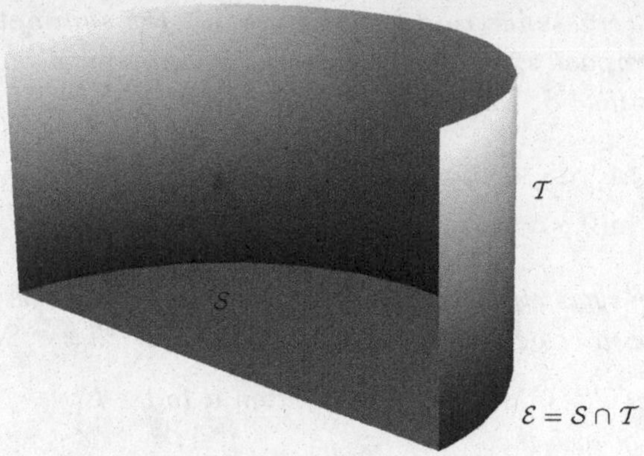

Figure 12.5 Geometric setting of the initial boundary value problem for symmetric hyperbolic systems. The initial data are prescribed on the three-dimensional spacelike hypersurface $\mathcal{S}$; boundary data are prescribed on the three-dimensional timelike hypersurface $\mathcal{T}$. The initial and boundary data must satisfy certain compatibility conditions (corner conditions) on the edge $\mathcal{E} = \mathcal{S} \cap \mathcal{T}$.

symmetric hyperbolic systems, it is sufficient to analyse the problem close to the edge. The solution away from the boundary is obtained by patching domains of dependence.

12.4.1 Basic setting

In a neighbourhood of a point $p \in \mathcal{E}$, one can introduce coordinates $x = (x^\mu)$ such that the domain $\mathcal{U}$ in which the solution to the boundary value problem takes the form

$$\mathcal{U} = \{x \in \mathbb{R}^4 \,|\, x^0 \geq 0,\ x^3 \geq 0\},$$

while the initial hypersurface and the boundary are given, respectively, by

$$\mathcal{S} \equiv \{x \in \mathcal{U} \,|\, x^0 = 0\},$$
$$\mathcal{T} \equiv \{x \in \mathcal{U} \,|\, x^3 = 0\}.$$

The ***normal matrix*** $\mathbf{A}^3(x, \mathbf{u})$ in a symmetric hyperbolic system of the form (12.1) plays a crucial role in the specification of admissible boundary conditions leading to a well-posed initial boundary value problem. Due to the use of coordinates adapted to the boundary, the properties of the matrix $\mathbf{A}^3$ determine the relation between the timelike boundary $\mathcal{T}$ and the characteristics of the hyperbolic evolution equation.

In what follows let $\mathbf{T}(x)$ denote a smooth map from $\mathcal{T}$ to the vector subspaces of $\mathbb{C}^N$ and require as boundary condition that

$$\mathbf{u}(x) \in \mathbf{T}(x), \qquad x \in \mathcal{T}.$$

The map $\mathbf{T}$ is restricted by the requirements:

(i) The set $\mathcal{T}$ is a characteristic of (12.1) of constant multiplicity; that is,

$$\dim \mathrm{Ker}(\mathbf{A}^3) = \text{constant} > 0, \qquad x \in \mathcal{T}.$$

(ii) The map $\mathbf{T}$ satisfies the ***non-positivity condition***

$$\langle \mathbf{u}, \mathbf{A}^3(x,\mathbf{u})\mathbf{u}\rangle \leq 0, \qquad \mathbf{u} \in \mathbf{T}(x),\ x \in \mathcal{T}.$$

(iii) The dimension of the subspace $\mathbf{T}(x)$, $x \in \mathcal{T}$, is equal to the number of non-positive eigenvalues of $\mathbf{A}^3(x,\mathbf{u})$ counting multiplicities.

An important property of Hermitian matrices is that they can be diagonalised by unitary matrices and that all their eigenvalues are real. Accordingly, after a redefinition of the dependent variables one can assume that, at a given point $x \in \mathcal{T}$, the normal matrix $\mathbf{A}^3(x,\mathbf{u})$ has the form

$$\mathbf{A}^3(x,\mathbf{u}) = \kappa \begin{pmatrix} -\mathbf{I}_{j\times j} & \mathbf{0} & \mathbf{0} \\ \mathbf{0} & \mathbf{0}_{k\times k} & \mathbf{0} \\ \mathbf{0} & \mathbf{0} & \mathbf{I}_{l\times l} \end{pmatrix}, \qquad \kappa > 0,$$

where $\mathbf{I}_{j\times j}$ and $\mathbf{I}_{l\times l}$ are, respectively, $j \times j$ and $l \times l$ unit matrices and $\mathbf{0}_{k\times k}$ is the $k \times k$ zero matrix. Moreover, one has that $j + k + l = N$. Writing

$$\mathbf{u}(x) = \begin{pmatrix} \mathbf{a}(x) \\ \mathbf{b}(x) \\ \mathbf{c}(x) \end{pmatrix} \in \mathbb{C}^j \times \mathbb{C}^k \times \mathbb{C}^l,$$

one finds that the linear subspaces admitted by condition (ii) are of the form

$$\mathbf{c} - \mathbf{H}\mathbf{a} = \mathbf{0}$$

with $\mathbf{H} = \mathbf{H}(x)$ an $l \times j$ matrix satisfying

$$-\langle \mathbf{a}, \mathbf{a}\rangle + \langle \mathbf{H}\mathbf{a}, \mathbf{H}\mathbf{a}\rangle \leq 0, \qquad \mathbf{a} \in \mathbb{C}^j.$$

This condition can be reexpressed, alternatively, as $\mathbf{H}^*\mathbf{H} \leq \mathbf{I}_{j\times j}$. The key observation is that the above procedure gives no freedom to prescribe data for the component $\mathbf{b}$ of $\mathbf{u}$ associated with the kernel of the normal matrix $\mathbf{A}^3(x,\mathbf{u})$. In particular, if $\mathbf{A}^3(x,\mathbf{u}) = 0$, one has that the boundary is a total characteristic (see Section 12.1.2) and no boundary conditions can be specified on $\mathcal{T}$ – the solution $\mathbf{u}$ on $\mathcal{T}$ is directly determined by the initial conditions on the edge $\mathcal{E}$. More generally, by a further redefinition of the dependent variables one obtains the ***inhomogeneous maximally dissipative boundary conditions***

$$\mathbf{q}(x) = \mathbf{c}(x) - \mathbf{H}(x)\mathbf{a}(x), \qquad x \in \mathcal{T},$$

with $\mathbf{q}(x)$ a $\mathbb{C}^l$-valued function representing the free boundary data on $\mathcal{T}$.

Corner conditions

To obtain a smooth solution to an initial boundary value problem for a symmetric hyperbolic system of the form (12.18), the initial data prescribed on $\mathcal{S}$ and the boundary data at $\mathcal{T}$ must satisfy certain compatibility conditions at the edge $\mathcal{E} = \partial\mathcal{S} = \mathcal{S} \cap \mathcal{T}$ – the so-called ***corner conditions***. More precisely, if one has initial data of the form

$$\mathbf{u}(0, \underline{x}) = \mathbf{u}_\star(\underline{x}) \quad \text{on } \mathcal{S},$$

with $\mathbf{u}_\star$ smooth and maximally dissipative boundary conditions of the form

$$\mathbf{T}(t, \underline{x})\mathbf{u}(t, \underline{x}) = \mathbf{q}(t, \underline{x}) \quad \text{on } \mathcal{T}, \tag{12.20}$$

then one requires that

$$\mathbf{T}(0, \underline{x})\mathbf{u}_\star|_\mathcal{E} = \mathbf{q}(0, \underline{x}).$$

Higher order corner conditions can be obtained by considering the system (12.18). Evaluating at $\mathcal{E}$ one obtains

$$\mathbf{A}^0(\mathbf{u}_\star)|_\mathcal{E}(\partial_t\mathbf{u})|_\mathcal{E} + \mathbf{A}^\alpha(\mathbf{u}_\star)|_\mathcal{E}(\partial_\alpha\mathbf{u}_\star)|_\mathcal{E} = \mathbf{B}(\mathbf{u}_\star)|_\mathcal{E}.$$

As $\mathbf{A}^0(\mathbf{u}_\star)|_\mathcal{E}$ is positive definite, the above equation can be used to solve for $(\partial_t\mathbf{u})|_\mathcal{E}$. The result should be consistent, upon substitution, with what is obtained from differentiating the boundary condition (12.20). Namely,

$$(\partial_t\mathbf{T})|_\mathcal{E}\mathbf{u}|_\mathcal{E} + \mathbf{T}|_\mathcal{E}(\partial_t\mathbf{u})|_\mathcal{E} = (\partial_t\mathbf{q})|_\mathcal{E}.$$

Further higher order boundary conditions are obtained in an analogous manner by differentiating (12.18) successively with respect to t.

12.4.2 Uniqueness of the solutions to the boundary value problem

Insight into the role of the maximally dissipative boundary conditions can be obtained from the analysis of the uniqueness of solutions to the boundary value problem. The argument follows a strategy similar to the one employed in Theorem 12.1 with a domain $\mathcal{G}$ whose boundary consists of portions of the initial hypersurface $\mathcal{S}_0$, the boundary $\mathcal{T}$ and a hypersurface $\mathcal{S}_1$ which is spacelike with respect to the symmetric hyperbolic system; see Figure 12.6. Set $\mathcal{M} = [0, \infty) \times \mathcal{S}$ such that $\mathcal{S}$ and $\mathcal{T}$ can be identified, in a natural way, as the boundary of $\mathcal{M}$. Define the coordinate $x^0 \equiv t$ in such a way that $\mathcal{S}_0 = \{p \in \mathcal{M} \,|\, t = 0\}$.

Theorem 12.5 (***uniqueness of solutions of the initial boundary problem with maximally dissipative boundary conditions***) *Let $\mathcal{G}$ be a domain as given above. If $\mathbf{u}_1$ and $\mathbf{u}_2$ are two solutions to the initial value problem for the symmetric hyperbolic system*

$$\mathbf{A}^\mu(x, \mathbf{u})\partial_\mu\mathbf{u} = \mathbf{B}(x, \mathbf{u}), \qquad \mathbf{u}|_{\mathcal{S}_0} = \mathbf{u}_\star,$$

with the same maximally dissipative boundary conditions, then $\mathbf{u}_1 = \mathbf{u}_2$ on $\mathcal{G}$.

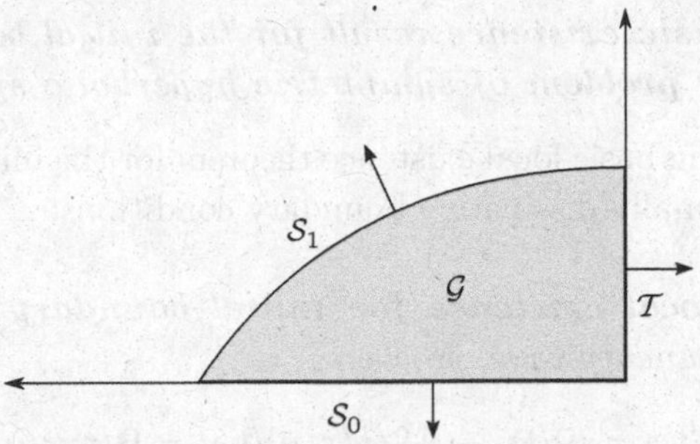

Figure 12.6 Integration domain for the uniqueness argument for initial boundary value problems. The boundary $\partial\mathcal{G}$ consists of portions of the initial hypersurface $\mathcal{S}_0$, the timelike boundary $\mathcal{T}$ and a spacelike hypersurface $\mathcal{S}_1$.

Proof Starting from the identity (12.14) one integrates over a domain $\mathcal{G}$ as depicted in Figure 12.6, where $\mathcal{S}_1$ is spacelike with respect to the symmetric hyperbolic system. Applying the Gauss identity one obtains

$$\begin{aligned}\int_{\mathcal{S}_1} & e^{-kt}\langle \mathbf{u}_1-\mathbf{u}_2, \mathbf{A}^\mu(x,\mathbf{u}_1)(\mathbf{u}_1-\mathbf{u}_2)\rangle \nu_\mu \mathrm{d}S \\ & -\int_{\mathcal{S}_0} e^{-kt}\langle \mathbf{u}_1-\mathbf{u}_2, \mathbf{A}^\mu(x,\mathbf{u}_1)(\mathbf{u}_1-\mathbf{u}_2)\rangle \nu_\mu \mathrm{d}S \\ & -\int_{\mathcal{T}} e^{-kt}\langle \mathbf{u}_1-\mathbf{u}_2, \mathbf{A}^\mu(x,\mathbf{u}_1)(\mathbf{u}_1-\mathbf{u}_2)\rangle \nu_\mu \mathrm{d}S \\ & = \int_{\mathcal{G}} e^{-kt}\langle \mathbf{u}_1-\mathbf{u}_2, \mathbf{P}(x,\mathbf{u}_1,\mathbf{u}_2,\partial\mathbf{u}_2)(\mathbf{u}_1-\mathbf{u}_2)\rangle \mathrm{d}^4x,\end{aligned}$$

with

$$\begin{aligned}\mathbf{P}(x,\mathbf{u}_1,\mathbf{u}_2) \equiv & -k\mathbf{A}^0(x,\mathbf{u}_1) + \partial_\mu \mathbf{A}^\mu(x,\mathbf{u}_1) \\ & + \mathbf{Q}(x,\mathbf{u}_1,\mathbf{u}_2,\partial\mathbf{u}_2) + \mathbf{Q}^*(x,\mathbf{u}_1,\mathbf{u}_2,\partial\mathbf{u}_2)\end{aligned}$$

and $\mathbf{Q}$ obtained as in Theorem 12.1 using the mean value theorem. Exploiting the positive definiteness of $\mathbf{A}^0(x,\mathbf{u}_1)$, one can make the volume integral over $\mathcal{G}$ negative. Moreover, as $\mathbf{u}_1$ and $\mathbf{u}_2$ coincide on $\mathcal{S}_0$ one obtains

$$\begin{aligned}\int_{\mathcal{S}_1} & e^{-kt}\langle \mathbf{u}_1-\mathbf{u}_2, \mathbf{A}^\mu(x,\mathbf{u}_1)(\mathbf{u}_1-\mathbf{u}_2)\rangle \nu_\mu \mathrm{d}S \\ & \leq \int_{\mathcal{T}} e^{-kt}\langle \mathbf{u}_1-\mathbf{u}_2, \mathbf{A}^\mu(x,\mathbf{u}_1)(\mathbf{u}_1-\mathbf{u}_2)\rangle \nu_\mu \mathrm{d}S \leq 0,\end{aligned}$$

where the last inequality follows from the negative definiteness of the maximally dissipative boundary conditions. Thus, one obtains a contradiction with the fact that the surface integral over the spacelike hypersurface $\mathcal{S}_1$ is positive unless $\mathbf{u}_1 = \mathbf{u}_2$. □

12.4.3 The basic existence result for the initial boundary value problem of symmetric hyperbolic systems

One has the following basic local existence theorem for the initial boundary value problem with maximally dissipative boundary conditions:

Theorem 12.6 (***local existence for initial boundary value problems***) *Given the initial boundary value problem*

$$\mathbf{A}^0(t,\underline{x},\mathbf{u})\partial_t\mathbf{u} + \mathbf{A}^\alpha(t,\underline{x},\mathbf{u})\partial_\alpha\mathbf{u} = \mathbf{B}(t,\underline{x},\mathbf{u}), \tag{12.21a}$$

$$\mathbf{T}(t,\underline{x})\mathbf{u} = \mathbf{q}(t,\underline{x}) \quad \text{on } \mathcal{T}, \tag{12.21b}$$

$$\mathbf{u}(0,\underline{x}) = \mathbf{u}_\star(\underline{x}), \quad \text{on } \mathcal{S}, \tag{12.21c}$$

with (12.21a) *symmetric hyperbolic,* $\mathbf{A}^0(0,\underline{x},\mathbf{u}_\star)$ *positive definite and* $\mathbf{q}$, $\mathbf{u}_\star$ *smooth, assume that the boundary condition* (12.21b) *is maximally dissipative with respect to the normal matrix* $\mathbf{A}^3(t,\underline{x},\mathbf{u})$ *and that the boundary data satisfy corner conditions at* $\mathcal{E} = \mathcal{S} \cap \mathcal{T}$ *to all orders. Then, the initial boundary value problem has a unique smooth solution* $\mathbf{u}(t,\underline{x})$ *defined on*

$$\mathcal{M}_T = \{p \in [0,\infty) \times \mathcal{S} \,|\, 0 \leq t(p) < T\},$$

for some $T > 0$.

The reader is refereed to Guès (1990), Friedrich (1995) and Friedrich and Nagy (1999) for details and remarks concerning the proof. As a consequence of the localisability property of hyperbolic equations, the problem can be split into two parts: an interior one away from the boundary in which the standard

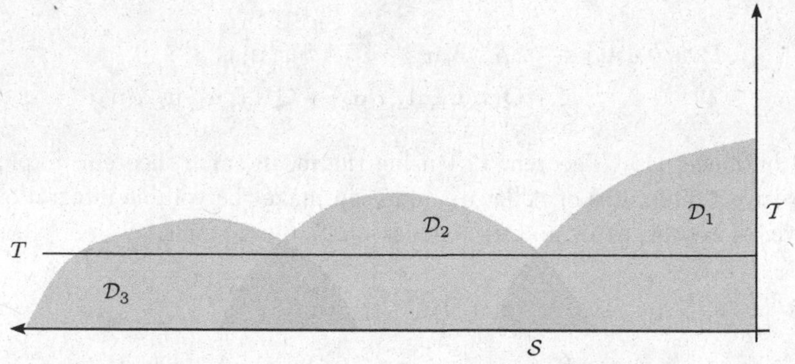

Figure 12.7 Construction of a solution to an initial boundary value problem which is global in space by patching domains. The solution patch $\mathcal{D}_1$ near the boundary $\mathcal{T}$ is obtained using Theorem 12.6. The existence on the domains $\mathcal{D}_2$ and $\mathcal{D}_3$ away from the boundary are obtained by means of Theorem 12.3. The uniqueness of solutions ensures that the solution on the intersections "match together". Due to the compactness of the initial hypersurface it is possible to obtain an existence time T common to all domains.

local existence for the Cauchy problem (as described in Section 12.3) is used, and a boundary part in which the boundary and edge conditions play a role; see Figure 12.7. The local solutions are then patched together to obtain the solution on the whole of $\mathcal{M}_T$.

Remark. The question of the stability of solutions to the initial boundary value problem will not be analysed here. Stability questions for initial boundary value problems are much more complicated than for the Cauchy case. At the time of writing, there are no applications of stability results for boundary value problems involving the conformal field equations.

12.5 Local existence for characteristic initial value problems

Characteristic initial value problems arise naturally in applications to general relativity; see Chapter 18. The purpose of this section is to discuss a method to analyse the local existence of solutions to the characteristic initial value problem for symmetric hyperbolic equations due to Rendall (1990). The idea behind this method is to reduce the characteristic problem to a standard Cauchy problem where the *standard* theory of Section 12.3 can be applied.

12.5.1 General remarks on the characteristic problem

In what follows, consider a quasilinear symmetric hyperbolic system of the form given by Equation (12.1) on $\mathbb{R}^4$. In contrast to the analysis of the Cauchy problem where it is convenient to single out one of the coordinates as a time coordinate, in the characteristic problem it is convenient to make use of coordinates adapted to the characteristic hypersurfaces.

As discussed in Section 12.1.2, for quasilinear equations like (12.1), the notion of characteristic hypersurfaces depends on the solution $\mathbf{u}$. Thus, it is, in principle, unclear on which hypersurfaces one can prescribe the characteristic initial data. There are two approaches to get around this difficulty:

(i) ***Fix the data first, then look for the hypersurface.*** Choose a smooth function $\mathbf{v}$ on $\mathcal{U} \subset \mathbb{R}^4$ such that the matrices $\mathbf{A}^\mu(x, \mathbf{v})$ are defined at each point of $\mathbb{R}^4$, and choose a smooth function $\phi \in \mathcal{X}(\mathcal{U})$, $\mathbf{d}\phi \neq 0$ in $\mathcal{U}$, such that the hypersurface

$$\mathcal{N} \equiv \{x \in \mathcal{U} \mid \phi(x) = 0\} \tag{12.22}$$

is characteristic with respect to $\mathbf{A}^\mu(x, \mathbf{v})$, that is, such that

$$\det(\mathbf{A}^\mu(x, \mathbf{v})\partial_\mu \phi) = 0.$$

The *characteristic data* on $\mathcal{N}$ is then given as the restriction of $\mathbf{v}$ to $\mathcal{N}$; that is, $\mathbf{u}_\star = \mathbf{v}|_\mathcal{N}$.

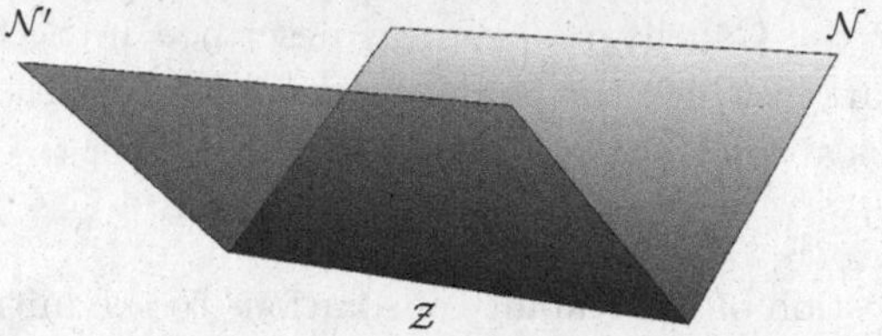

Figure 12.8 Initial hypersurfaces $\mathcal{N}$ and $\mathcal{N}'$ on a characteristic initial value problem. The set $\mathcal{Z} \neq \emptyset$ is the intersection of $\mathcal{N}$ and $\mathcal{N}'$.

(ii) ***Choose the hypersurface first, then look for suitable data.*** Alternatively, one can choose some hypersurface $\mathcal{N}$ in $\mathcal{U} \subset \mathbb{R}^4$ defined as in (12.22), and then consider only those smooth functions $\mathbf{u}_\star$ such that

$$\det(\mathbf{A}^\mu(x, \mathbf{u}_\star)\partial_\mu \phi) = 0.$$

Approach (ii) is more natural in applications where geometric information of the initial hypersurface is available. *This point of view will be adopted in the rest of this section.*

A peculiarity of characteristic initial value problems for the system (12.1) is that data need to be prescribed on *two intersecting characteristic hypersurfaces* $\mathcal{N}$ and $\mathcal{N}'$; see Figure 12.8. Intuitively, this is a consequence of the existence of a subsystem of equations in (12.1) which is intrinsic to the hypersurface $\mathcal{N}$, so that one does not have enough evolution equations transverse to the hypersurface for all the components of $\mathbf{u}$. Alternatively, one can formulate characteristic initial value problems by prescribing initial data on a *cone*. This is a more technically involved problem and will not be discussed here. The interested reader is referred to Cagnac (1981) and Dossa (1997) for further details.

Well- and ill-posed characteristic problems

In what follows, let $\mathcal{N}$ and $\mathcal{N}'$ denote two hypersurfaces on $\mathcal{U} \subset \mathbb{R}^4$ with non-empty intersection $\mathcal{Z} \equiv \mathcal{N} \cap \mathcal{N}'$. One can introduce coordinates u and v such that, at least in a neighbourhood of $\mathcal{N} \cap \mathcal{N}'$, one can write

$$\mathcal{N} \equiv \{p \in \mathcal{U} \,|\, u(p) = 0\}, \qquad \mathcal{N}' \equiv \{p \in \mathcal{U} \,|\, v(p) = 0\}. \tag{12.23}$$

Given suitable initial data on $\mathcal{N} \cup \mathcal{N}'$ one would like to make some statement about the existence and the uniqueness of solutions to Equation (12.1) on some open set

$$\mathcal{V} \subset \{p \in \mathcal{U} \,|\, u(p) \geq 0,\ v(p) \geq 0\}.$$

By symmetry, one could also look for a solution in the region

$$\{p \in \mathcal{U} \,|\, u(p) \leq 0,\ v(p) \leq 0\};$$

see Figure 12.9 (a).

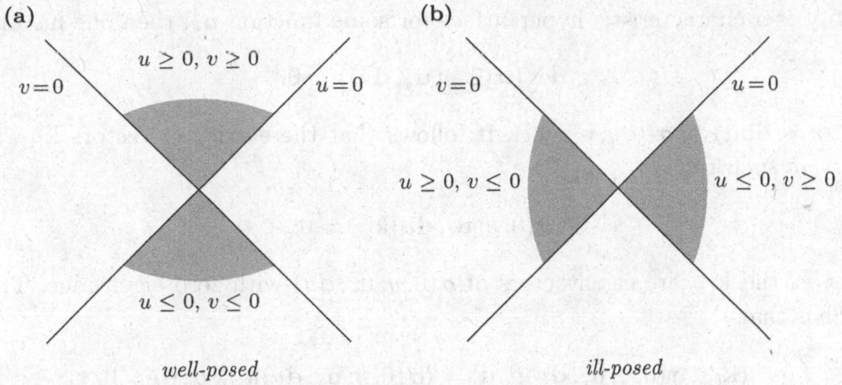

Figure 12.9 Schematic representation of well-posed (a) and ill-posed (b) characteristic initial value problems.

The problem of looking for solutions in domains of the form

$$\bar{\mathcal{V}} \subset \{p \in \mathcal{U} \,|\, u(p) \leq 0,\ v(p) \geq 0\}$$

or

$$\bar{\bar{\mathcal{V}}} \subset \{p \in \mathcal{U} \,|\, u(p) \geq 0,\ v(p) \leq 0\}$$

is *ill-posed* – the reason will become clear once the Rendall's reduction procedure to a Cauchy problem is discussed in Section 12.5.3. Under suitable circumstances, it may be possible to establish uniqueness of a solution – but not existence – for this ill-posed problem. These ideas have been used by Ionescu and Klainerman (2009a,b) to obtain a new strategy to prove the uniqueness of stationary black holes.

12.5.2 Interior equations on the characteristic hypersurfaces

As seen in Section 12.1.2, on a characteristic surface, a system of the form (12.1) implies a subsystem of interior equations on the hypersurface. Assuming that the freely specifiable part of $\mathbf{u}$ is smooth on the characteristic hypersurface, these interior equations can be used to compute the remaining components of $\mathbf{u}$ and their derivatives to any arbitrary order. For conciseness, the subsequent analysis is restricted to the characteristic $\mathcal{N}$ as given by (12.23). The situation on $\mathcal{N}'$ is completely analogous. Letting $x \equiv (u, \underline{y})$ with $\underline{y} \equiv (y^\alpha) = (v, x^2, x^3)$ and using the chain rule, Equation (12.1) can be rewritten as

$$\boldsymbol{\sigma}(u, \underline{y}; \mathbf{u}, \mathbf{d}u)\partial_u \mathbf{u} + \mathbf{A}^\alpha(u, \underline{y}; \mathbf{u})\partial_\alpha \mathbf{u} = \mathbf{B}(u, \underline{y}; \mathbf{u}) \qquad (12.24)$$

with

$$\boldsymbol{\sigma}(u, \underline{y}; \mathbf{u}, \mathbf{d}u) \equiv \mathbf{A}^\mu(u, \underline{y}; \mathbf{u})\frac{\partial u}{\partial x^\mu}.$$

If $\mathcal{N}$ is a characteristic hypersurface for some function $\mathbf{u}_\star$, then one has that

$$\det\big(\boldsymbol{\sigma}(0,\underline{y};\mathbf{u}_\star,\mathbf{d}u)\big)=0.$$

Let $m=\dim\operatorname{Ker}\boldsymbol{\sigma}(0,y;\mathbf{u}_\star,\mathbf{d}u)$. It follows that there exist m vectors $\mathbf{k}_{(i)}$, $i=1,\ldots,m$ such that

$$\boldsymbol{\sigma}(0,\underline{y};\mathbf{u}_\star,\mathbf{d}u)\mathbf{k}_{(i)}=0.$$

That is, the $\mathbf{k}_{(i)}$ are eigenvectors of $\boldsymbol{\sigma}(0,\underline{y};\mathbf{u}_\star,\mathbf{d}u)$ with zero eigenvalue. Thus, one has that

$$\langle\mathbf{k}_{(i)},\boldsymbol{\sigma}(0,\underline{y};\mathbf{u}_\star,\mathbf{d}u)\partial_u\mathbf{u}\rangle=\langle\boldsymbol{\sigma}(0,\underline{y};\mathbf{u}_\star,\mathbf{d}u)\mathbf{k}_{(i)},\partial_u\mathbf{u}\rangle=0,$$

since $\boldsymbol{\sigma}(u,\underline{y};\mathbf{u},\mathbf{d}u)$ is Hermitian as a consequence of the symmetric hyperbolicity of (12.1). Thus, from Equation (12.24) one obtains

$$\langle\mathbf{k}_{(i)},\mathbf{A}^\alpha(0,\underline{y};\mathbf{u}_\star)\partial_\alpha\mathbf{u}_\star\rangle=\langle\mathbf{k}_{(i)},\mathbf{B}(0,\underline{y};\mathbf{u}_\star)\rangle,\qquad i=1,\ldots m,\tag{12.25}$$

a system of m (scalar) interior equations for the components of $\mathbf{u}$. In what follows, it will be assumed that the components of $\mathbf{u}$ have been chosen such that the free data on $\mathcal{N}$ consist of $N-m$ variables $\breve{\mathbf{u}}_\star$ – the so-called ***u*-data**. The remaining m variables, $\bar{\mathbf{u}}_\star$, constrained by the equations in (12.25), are called the ***u*-variables**. Thus, one obtains the split

$$\mathbf{u}_\star=(\breve{\mathbf{u}}_\star,\bar{\mathbf{u}}_\star)\qquad\text{on }\mathcal{N}.\tag{12.26}$$

In terms of this split, the scalar intrinsic equations (12.25) can be rewritten in matricial form as

$$\bar{\mathbf{A}}^\alpha(\underline{y},\breve{\mathbf{u}}_\star,\bar{\mathbf{u}}_\star)\partial_\alpha\bar{\mathbf{u}}=\bar{\mathbf{B}}(\underline{y},\breve{\mathbf{u}}_\star,\bar{\mathbf{u}}_\star),\tag{12.27}$$

for some $(m\times m)$-matrix valued smooth functions $\bar{\mathbf{A}}^\alpha$ and an m-vector valued function $\bar{\mathbf{B}}$. For simplicity, *it will assumed that the system* (12.27) *is a symmetric hyperbolic system on* $\mathcal{N}$ which can be solved, at least locally, in a neighbourhood $\mathcal{W}\subset\mathcal{N}$ of the two-dimensional surface $\mathcal{Z}$ where initial data for the u-variables $\bar{\mathbf{u}}_\star$ is prescribed. In this way, one obtains the value of the whole components of $\mathbf{u}_\star$ on $\mathcal{W}$. Assuming that $\breve{\mathbf{u}}_\star$ is smooth on $\mathcal{N}$, higher intrinsic derivatives can be obtained in a similar manner by formally differentiating Equation (12.27) with respect to ∂_α an arbitrary number of times, say, M. In this manner, one obtains a system of the form

$$\bar{\mathbf{A}}^\alpha(\underline{y},\partial_{\boldsymbol{\alpha}}\breve{\mathbf{u}}_\star,\partial_{\boldsymbol{\alpha}}\bar{\mathbf{u}}_\star)\partial_\alpha\bar{\mathbf{u}}_{\boldsymbol{\alpha}}=\bar{\mathbf{B}}(\underline{y},\partial_{\boldsymbol{\alpha}}\breve{\mathbf{u}}_\star,\partial_{\boldsymbol{\alpha}}\bar{\mathbf{u}}_\star,\bar{\mathbf{u}}_{\boldsymbol{\alpha}}),\tag{12.28}$$

where multi-index notation has been used so that

$$\partial_{\boldsymbol{\alpha}}\breve{\mathbf{u}}_\star\equiv\big(\breve{\mathbf{u}}_\star,\partial_\alpha\breve{\mathbf{u}}_\star,\partial_{\alpha_1}\partial_{\alpha_2}\breve{\mathbf{u}}_\star,\ldots,\partial_{\alpha_1}\cdots\partial_{\alpha_{M-1}}\breve{\mathbf{u}}_\star\big),$$
$$\partial_{\boldsymbol{\alpha}}\bar{\mathbf{u}}_\star\equiv\big(\bar{\mathbf{u}}_\star,\partial_\alpha\bar{\mathbf{u}}_\star,\partial_{\alpha_1}\partial_{\alpha_2}\bar{\mathbf{u}}_\star,\ldots,\partial_{\alpha_1}\cdots\partial_{\alpha_{M-1}}\bar{\mathbf{u}}_\star\big)$$

and

$$\bar{\mathbf{u}}_{\boldsymbol{\alpha}} \equiv \partial_{\alpha_1} \cdots \partial_{\alpha_M} \bar{\mathbf{u}}_\star.$$

By assumption, Equation (12.28) is a symmetric hyperbolic system on $\mathcal{N}$ so that by prescribing initial data for $\bar{\mathbf{u}}_{\boldsymbol{\alpha}}$ on $\mathcal{Z}$ and assuming that the lower order intrinsic derivatives $\partial_{\boldsymbol{\alpha}}\bar{\mathbf{u}}$ have been solved for, one obtains a solution in a neighbourhood of $\mathcal{Z}$ on $\mathcal{N}$. Thus, one can obtain, recursively, the interior partial derivatives

$$\mathbf{u}_\star, \partial_\alpha \mathbf{u}_\star, \partial_{\alpha_1}\partial_{\alpha_2}\mathbf{u}_\star, \ldots, \partial_{\alpha_1} \cdots \partial_{\alpha_M}\mathbf{u}_\star \qquad \text{on } \mathcal{W} \subset \mathcal{N},$$

with $\mathcal{W} \supset \mathcal{Z}$.

Now, not only the interior derivatives on $\mathcal{N}$ can be computed. Using the split (12.26), the subset of $N-m$ equations in (12.1) which are transversal to $\mathcal{N}$ can be written as

$$\mathbf{C}^u(0, \underline{y}, \check{\mathbf{u}}_\star, \bar{\mathbf{u}}_\star)\partial_u \check{\mathbf{u}}_\star + \mathbf{C}^\alpha(0, \underline{y}, \check{\mathbf{u}}_\star, \bar{\mathbf{u}}_\star)\partial_\alpha \check{\mathbf{u}}_\star = \mathbf{D}(0, \underline{y}, \bar{\mathbf{u}}_\star, \bar{\mathbf{u}}), \qquad (12.29)$$

with $\mathbf{C}^\mu$ smooth $(N-m)\times(N-m)$-matrix valued functions and $\mathbf{D}$ an $(N-m)$-vector valued function of their arguments. For clarity of the presentation it is convenient to write $\partial_u \check{\mathbf{u}}_\star \equiv (\partial_u \check{\mathbf{u}})_\star$. By construction, the matrix $\mathbf{C}^\mu$ is invertible, so that Equation (12.29) can be regarded as an algebraic linear system of equations determining the transversal derivatives $\partial_u \check{\mathbf{u}}$ on $\mathcal{N}$ in terms of $\mathbf{u}_\star$ and $\partial_\alpha \mathbf{u}_\star$. To compute the transversal derivatives of the u-variables $\bar{u}_\star$, one differentiates the interior system (12.27) to obtain a system of the form

$$\bar{\mathbf{A}}^\alpha(\underline{y}, \check{\mathbf{u}}_\star, \bar{\mathbf{u}}_\star)\partial_\alpha(\partial_u \bar{\mathbf{u}}_\star) = \bar{\mathbf{B}}(\underline{y}, \check{\mathbf{u}}_\star, \partial_u \check{\mathbf{u}}_\star, \bar{\mathbf{u}}_\star, \partial_u \bar{\mathbf{u}}_\star).$$

As in the case of the system (12.27), the above system can be solved in some neighbourhood of $\mathcal{Z}$ on $\mathcal{N}$ if initial data for $\partial_u \bar{\mathbf{u}}_\star$ are given on $\mathcal{Z}$. This procedure can be repeated to obtain higher order transversal derivatives.

The procedure described in the previous paragraphs can also be implemented on the characteristic hypersurface $\mathcal{N}'$. By analogy to the case of $\mathcal{N}$, one can split the unknown $\mathbf{u}$ as

$$\mathbf{u} = (\check{\mathbf{u}}'_\star, \bar{\mathbf{u}}'_\star), \qquad \text{on } \mathcal{N}',$$

where $\check{\mathbf{u}}'_\star$ are v-***data*** which can be specified freely on $\mathcal{N}'$ and $\bar{\mathbf{u}}'_\star$ are v-***variables*** constrained by interior equations analogous to (12.28). In what follows, these interior equations are assumed to be symmetric hyperbolic on $\mathcal{N}'$. Applying a procedure similar to that used on $\mathcal{N}$, all the derivatives of $\mathbf{u}$ on $\mathcal{N}'$ to any desired order can be computed if $\check{\mathbf{u}}_\star$ is suitably smooth, and the required initial data are supplied on $\mathcal{Z}$.

The discussion described in the previous paragraphs is summarised in the following proposition:

Proposition 12.3 (***evaluation of derivatives on the initial characteristic surface***) *Let $\mathcal{N}$ and $\mathcal{N}'$ denote two characteristic hypersurfaces for the symmetric hyperbolic system* (12.1) *having a non-empty two-dimensional intersection $\mathcal{Z} = \mathcal{N} \cap \mathcal{N}'$. If smooth u-data and v-data are prescribed, respectively, on $\mathcal{N}$ and $\mathcal{N}'$ and the values of the u-variables and v-variables are prescribed on $\mathcal{Z}$ in such a way that the freely specifiable data are smooth on $\mathcal{N} \cup \mathcal{N}'$, then all derivatives of $\mathbf{u}$ on $\mathcal{N} \cup \mathcal{N}'$ to any desired order can be computed in a neighbourhood $\mathcal{W} \subset \mathcal{N}$ of $\mathcal{Z}$.*

12.5.3 Reduction to a standard Cauchy problem

The observations summarised in Proposition 12.3 are the cornerstone of a reduction procedure of the characteristic problem on $\mathcal{N} \cup \mathcal{N}'$ to a standard Cauchy problem for which the theory discussed in Section 12.3 is applicable. This approach to analysing the characteristic initial value problem for hyperbolic equations was originally introduced by Rendall (1990).

In what follows, suppose that characteristic initial data have been prescribed on $\mathcal{N} \cup \mathcal{N}'$ in a manner consistent with Proposition 12.3 so that the values of $\mathbf{u}$ and its derivatives to any order are known in a neighbourhood $\mathcal{W}$ of $\mathcal{Z}$ on $\mathcal{N} \cup \mathcal{N}'$. Rendall's reduction proceeds first by constructing an extension of $\mathbf{u}$ to a neighbourhood $\mathcal{U}$ of $\mathcal{Z}$ in $\mathbb{R}^4$. This type of extension of functions is different from the one discussed in Section 12.3.1 where functions defined on open subsets of a certain space are extended to functions on the whole space. In the present case one needs to extend a function defined on a closed set of $\mathbb{R}^4$. There exists a general result, ***Whitney's extension theorem***, which allows one to obtain the required extension; see the Appendix to this chapter for more details.

To apply Whitney's extension theorem to the collection of fields

$$\{\mathbf{u}_\star, (\partial_\mu \mathbf{u})_\star, (\partial_{\mu_1}\partial_{\mu_2}\mathbf{u})_\star, \ldots, (\partial_{\mu_1} \cdots \partial_{\mu_M}\mathbf{u})_\star\} \tag{12.30}$$

on $\mathcal{W} \subset \mathcal{N} \cup \mathcal{N}'$ for some non-negative integer M, one has to verify that the various fields in this collection are related to each other in the same way as the derivatives of a function are related to each other in a Taylor expansion. The key condition on these *Taylor-like expansions* ensuring the existence of an extension is a requirement on the vanishing rate of the remainder of the expansions. Given two points on $\mathcal{N}$ away from $\mathcal{Z}$ this vanishing of the remainder follows from smoothness of the free data on the characteristic hypersurface, and the fact that the ***derivative candidates*** in (12.30) have been obtained solving hyperbolic differential equations on $\mathcal{N}$ and algebraic equations. The functions thus obtained are smooth on $\mathcal{N}$ and admit a standard Taylor expansion on the characteristic hypersurface. This also holds for two points on $\mathcal{N}'$ away from $\mathcal{Z}$. Thus, the difficulty is to verify Whitney's condition for two points, respectively, on $\mathcal{N}$ and $\mathcal{N}'$, so that one writes

$$x = (0, v, x^2, x^3), \qquad x' = (u, 0, x'^2, x'^3).$$

The complication arises from the fact that the characteristic initial hypersurface $\mathcal{N} \cup \mathcal{N}'$ is continuous only at $\mathcal{Z}$. It is convenient to define a point $x_* \in \mathcal{Z}$ as

$$x_* \equiv \left(0, 0, \tfrac{1}{2}(x^2 + x'^2), \tfrac{1}{2}(x^3 + x'^3)\right).$$

Using the cosines law, it follows that there exists a constant $C > 0$ such that

$$|x - x_*|^2 + |x' - x_*|^2 \leq C|x - x'|^2.$$

To apply Whitney's extension theorem it is necessary to establish that the remainder of the Taylor-like expansion about x vanishes at x' as fast as it would do if the function had an extension as x and x' tend to a common point, say, x_*. The inequality above shows that the points x and x' cannot get closer to each other without getting close to a point on $\mathcal{Z}$. This idea, together with the Cauchy stability of solutions to the interior equations which determine the constrained components of the data on $\mathcal{N} \cup \mathcal{N}'$ yields the required vanishing rate.

Applying Whitney's extension theorem to the collection of derivative candidates (12.30) one obtains a smooth function $\hat{\mathbf{u}}$ in a neighbourhood $\mathcal{U}$ of $\mathcal{Z}$ on $\mathbb{R}^4$. The function $\hat{\mathbf{u}}$ satisfies

$$\hat{\mathbf{u}} = \mathbf{u}_\star, \quad \partial_\mu \hat{\mathbf{u}} = (\partial_\mu \mathbf{u})_\star, \quad \partial_{\mu_2}\partial_{\mu_1}\hat{\mathbf{u}} = (\partial_{\mu_2}\partial_{\mu_1}\mathbf{u})_\star, \ldots,$$

on $\mathcal{W} \subset \mathcal{N} \cup \mathcal{N}'$. In general, $\hat{\mathbf{u}}$ is not a solution to Equation (12.1) away from $\mathcal{N} \cup \mathcal{N}'$. Nevertheless,

$$\Delta \equiv \mathbf{A}^\mu(x, \hat{\mathbf{u}})\partial_\mu \hat{\mathbf{u}} - \mathbf{B}(x, \hat{\mathbf{u}})$$

vanishes to all orders on $\mathcal{W} \subset \mathcal{N} \cup \mathcal{N}'$ and

$$\delta \equiv \begin{cases} 0 & u > 0, \quad v > 0, \\ \Delta & \text{elsewhere,} \end{cases}$$

is smooth in a neighbourhood of $\mathcal{N} \cap \mathcal{N}'$ where $\hat{\mathbf{u}}$ exists.

The desired reduction to a Cauchy problem is now obtained by considering the equation

$$\mathbf{A}^\mu(x, \hat{\mathbf{u}} + \mathbf{v})\partial_\mu(\hat{\mathbf{u}} + \mathbf{v}) - \mathbf{B}(x, \hat{\mathbf{u}} + \mathbf{v}) = \delta, \tag{12.31}$$

for the unknown $\mathbf{v}$ together with the initial data

$$\mathbf{v}_\star = \mathbf{0}, \quad \text{on} \quad \mathcal{S} \equiv \{p \in \mathbb{R}^4 \,|\, u(p) + v(p) = 0\}. \tag{12.32}$$

By assumption, the hypersurface $\mathcal{S}$ has a neighbourhood around $\mathcal{N} \cap \mathcal{N}'$ which is spacelike with respect to Equation (12.31) so that the Cauchy problem given by (12.31) together with the initial data (12.32) is well posed and the theory of Section 12.3 is readily applicable. In particular, one obtains a unique solution $\mathbf{v}$ in a neighbourhood $\mathcal{V}$ of $\mathcal{Z}$ on $\mathbb{R}^4$; see Figure 12.10.

Outside the intersection of $\mathcal{V}$ with the quadrant

$$\{p \in \mathbb{R}^4 \,|\, u(p) \geq 0, \, v(p) \geq 0\},$$

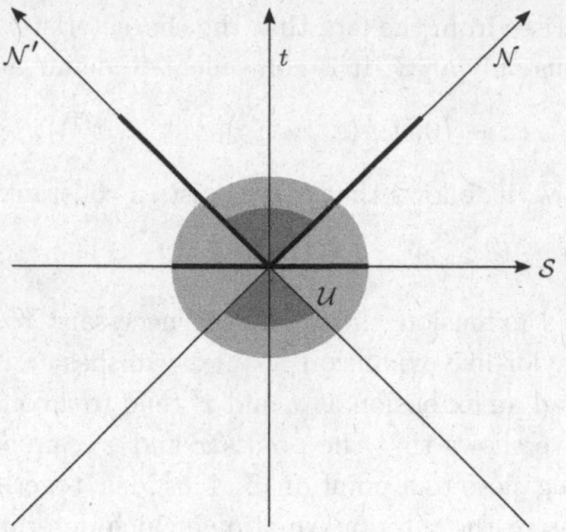

Figure 12.10 Schematic reduction of a characteristic initial value problem to a Cauchy problem. The data on $\mathcal{N} \cup \mathcal{N}'$ (thick line) are extended by means of Whitney's extension theorem to a neighbourhood $\mathcal{U}$ (light gray region) of $\mathcal{Z} \equiv \mathcal{N} \cap \mathcal{N}'$. This extension implies data for an auxiliary initial value problem on the spacelike hypersurface $\mathcal{S}$. The solution to the characteristic problem is found in the upper and lower quadrants (dark gray areas).

Equation (12.31) takes the form

$$\mathbf{A}^\mu(x, \hat{\mathbf{u}} + \mathbf{v})\partial_\mu(\hat{\mathbf{u}} + \mathbf{v}) - \mathbf{B}(x, \hat{\mathbf{u}} + \mathbf{v}) = \mathbf{A}^\mu(x, \hat{\mathbf{u}})\partial_\mu \hat{\mathbf{u}} - \mathbf{B}(x, \hat{\mathbf{u}})$$

so that $\mathbf{v} = 0$ is clearly a solution – by uniqueness, it is the only solution. In contrast, on $\mathcal{V} \cap \{p \in \mathbb{R}^4 \mid u(p) \geq 0,\ v(p) \geq 0\}$ one has the equation

$$\mathbf{A}^\mu(x, \hat{\mathbf{u}} + \mathbf{v})\partial_\mu(\hat{\mathbf{u}} + \mathbf{v}) = \mathbf{B}(x, \hat{\mathbf{u}} + \mathbf{v}).$$

As $\hat{\mathbf{u}} + \mathbf{v}$ coincides with $\mathbf{u}_\star$ on $\mathcal{N} \cup \mathcal{N}'$ one concludes, again by uniqueness of the solution of the reduced Cauchy problem, that

$$\mathbf{u} \equiv \hat{\mathbf{u}} + \mathbf{v}$$

is the required solution to the posed characteristic initial value problem.

The discussion in this section is summarised in the following theorem:

Theorem 12.7 (***local existence for the standard characteristic problem***) *Let $\mathcal{N}$ and $\mathcal{N}'$ denote two characteristic hypersurfaces for the symmetric hyperbolic system* (12.1) *with smooth, freely specifiable data on $\mathcal{N}$ and $\mathcal{N}'$ as given in Proposition 12.3. Then there exists a unique solution $\mathbf{u}$ to the characteristic initial value problem in a neighbourhood $\mathcal{V}$ of $\mathcal{Z}$ with $u \geq 0$, $v \geq 0$.*

Remark. If one were to attempt a similar reduction procedure to construct a solution on the regions for which either $u \geq 0$, $v \leq 0$ or $u \leq 0$, $v \geq 0$,

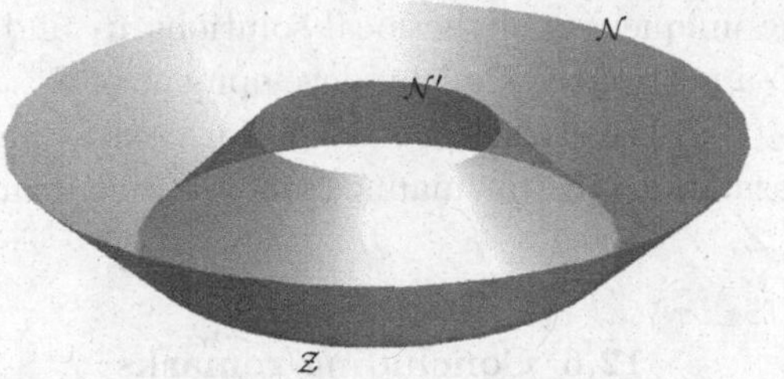

Figure 12.11 Characteristic cones $\mathcal{N}$ and $\mathcal{N}'$ intersecting on a two-dimensional hypersurface $\mathcal{Z}$ which is diffeomorphic to $\mathbb{S}^2$.

one would end up with an initial value problem with data prescribed on a timelike hypersurface. This is an ill-posed problem. Accordingly, the original characteristic problems are, themselves, also ill-posed.

The case $\mathcal{Z} \approx \mathbb{S}^2$

A case that occurs naturally in applications of conformal methods in general relativity is an initial characteristic problem where the intersection $\mathcal{Z} = \mathcal{N} \cap \mathcal{N}'$ is diffeomorphic to the 2-sphere $\mathbb{S}^2$. This is the case, for example, of the intersection of two light cones; see Figure 12.11. The method discussed in the previous section can be adapted to this case; see Kánnár (1996b).

Assuming in what follows that $\mathcal{Z} \approx \mathbb{S}^2$, consider an atlas $\{(U_1, \phi_1), (U_2, \phi_2)\}$ of $\mathcal{Z}$ and closed sets $V_1 \subset U_1$ and $V_2 \subset U_2$ which also cover $\mathcal{Z}$; that is, $\mathcal{Z} = V_1 \cup V_2$. Furthermore, define two smooth functions η_1 and η_2 with compact support on $\mathbb{R}^2$ by

$$\eta_1(x) \equiv \begin{cases} 1 & x \in \phi_1(V_1) \\ 0 & x \in \mathbb{R}^2 \setminus \phi_1(U_1), \end{cases} \qquad \eta_2(x) \equiv \begin{cases} 1 & x \in \phi_2(V_2) \\ 0 & x \in \mathbb{R}^2 \setminus \phi_2(U_2). \end{cases}$$

In what follows, denote by $\breve{\mathbf{u}}_{1\star}$ and $\breve{\mathbf{u}}_{2\star}$, respectively, the restriction to U_1 and U_2 of the freely specifiable data on $\mathcal{N} \cup \mathcal{N}'$. It follows that the functions $\eta_1 \breve{\mathbf{u}}_{1\star}$ and $\eta_2 \breve{\mathbf{u}}_{2\star}$ define a smooth initial value data set on the initial hypersurfaces $\mathcal{N}_1 \equiv \mathbb{R}^+ \times \{0\} \times \mathbb{R}^2$ and $\mathcal{N}_2 \equiv \{0\} \times \mathbb{R}^+ \times \mathbb{R}^2$ which coincides with $\mathbf{u}_{1\star}$ and $\mathbf{u}_{2\star}$ on $\mathbb{R}^+ \times \{0\} \times \phi_1(V_1)$ and $\{0\} \times \mathbb{R}^+ \times \phi_2(V_2)$.

The intrinsic equations on $\mathcal{N}_1$ and $\mathcal{N}_2$ can now be solved in a manner similar to what was done in Section 12.5.2. In this way, one obtains two complete characteristic initial data sets $\mathbf{u}_{1\star}$ and $\mathbf{u}_{2\star}$ on $\mathcal{N}_1 \cup \mathcal{N}_2$. The (interior and transversal) derivatives of these data can be computed to any desired order. Using Theorem 12.7 one obtains two solutions $\mathbf{u}_1$ and $\mathbf{u}_2$ in a neighbourhood $\mathcal{V}$ of $\mathcal{N}_1 \cap \mathcal{N}_2$. Their restrictions to the Cauchy development of $\mathbb{R}^+ \times \mathbb{R}^+ \times \phi_1(V_1)$ and $\mathbb{R}^+ \times \mathbb{R}^+ \times \phi_2(V_2)$ are local solutions to the original problem. The solutions $\mathbf{u}_1$ and $\mathbf{u}_2$ can be glued together to obtain a global solution on $\mathcal{Z} \approx \mathbb{R}^2$. As a consequence of the uniqueness of the local solutions $\mathbf{u}_1$ and $\mathbf{u}_2$, it follows that their restriction to a part of the Cauchy development of $\mathbb{R}^+ \times \mathbb{R}^+ \times \phi_1(V_1 \cap V_2)$ and $\mathbb{R}^+ \times \mathbb{R}^+ \times \phi_2(V_1 \cap V_2)$ – where both solutions exist – must be related by a coordinate transformation. In this manner, one obtains a smooth function in a neighbourhood of $\mathcal{Z}$.

12.6 Concluding remarks

This chapter has provided a succinct discussion of the theory of the local existence and uniqueness of quasilinear symmetric hyperbolic evolution equations. Of course, this is not the only way the subject can be approached. Nor are the issues raised the only relevant ones in the analysis of the evolution problem in general relativity. Thus, it is important to make some remarks concerning some ideas and approaches which have been omitted.

12.6.1 Wave equations

The analysis of Section 12.1.3 gives a hint on how the theory of second-order hyperbolic equations (wave equations) can be reduced to the analysis of symmetric hyperbolic equations. There is, however, a well-developed theory for the local existence and stability of systems of quasilinear equations of the form

$$g^{\mu\nu}(x, \mathbf{u})\partial_\mu\partial_\nu\mathbf{u} = \mathbf{B}(x, \mathbf{u}, \partial\mathbf{u}), \tag{12.33}$$

which does not rely on the reduction to a first order system; see Hughes et al. (1977). Equation (12.33) is quasilinear in the sense that $g^{\mu\nu}$ is the contravariant version of a Lorentzian metric which is allowed to depend not only on the coordinates but also on the unknowns. This "stand alone" theory relaxes the differentiability assumptions made on the equation and data; see, for example, Rendall (2008) for more details.

The results of Hughes et al. (1977) are similar, in spirit, to the results given in Theorems 12.2 and 12.3: given suitably smooth initial data for Equation (12.33) one obtains a unique solution for some existence time T; moreover one also has a Cauchy stability result. It should be pointed out that this theory applies, in fact, to a class of second-order equations more general than that given in Equation (12.33).

Systems of quasilinear wave equations of the form (12.33) arise naturally in the reduction procedure for the Einstein field equations based on the use of wave coordinates; see the Appendix to Chapter 14. Historically, this was the first approach to the Cauchy problem in general relativity; see Fourès-Bruhat (1952).

12.6.2 Global existence of solutions

Conformal methods allow the reformulation of several questions on the global existence of solutions to the Einstein field equations as a local existence question for symmetric hyperbolic systems. Accordingly, the issue of global existence of solutions to symmetric hyperbolic equations has not been addressed in this chapter. Nevertheless, this question is at the heart of current research work in the area; see, for example, Klainerman (2008) for a discussion.

As already pointed out, the local theory of solutions to hyperbolic equations depends solely on the properties of the principal part of the equations. To construct a theory of global existence one has to include the lower order terms of the equations into the analysis. Under certain circumstances, the analysis of the eigenvalues of the matrix arising from the linearisation of the lower order terms of a quasilinear system gives a strong indication of whether one can expect global existence and stability of solutions; see, for example, Kreiss and Lorenz (1998). More generally, one can identify structures in the evolution equations which allow one to prove global existence. One of these structures is the so-called ***null condition***; see, for example, Klainerman (1984).

12.7 Further reading

The theory of hyperbolic differential equations, in general, and their application to the analysis of solutions of the Einstein field equations, in particular, is an extensive area of research so that any list of references can provide only a partial impression of the field. For an overview of the whole field of the theory of PDEs and the interconnection between the various types of equations the reader is referred to Klainerman (2008).

Readers interested in further details of the basic aspects of the theory of PDEs are referred to the classical references by Garabedian (1986) and Courant and Hilbert (1962). A modern introduction to the subject is given in Evans (1998). A comprehensive exposition of the subject is given in the three-volume treatise of Taylor (1996a,b,c). Detailed accounts of the theory of the Cauchy problem for symmetric hyperbolic systems are discussed in the original references by Kato (1975a); Fischer and Marsden (1972) for first-order symmetric hyperbolic systems and Hughes et al. (1977) for second-order equations. A review of the ideas contained in these works can be found in Kato (1975b).

A comprehensive discussion of the role of PDEs in general relativity is given in Rendall (2008). A more compact review is Friedrich and Rendall (2000). Complementary discussions on the topics covered in these references can be found in Rendall (2006) and Reula (1998).

Appendix

A generalised mean value theorem

In the proofs of the uniqueness of solutions for symmetric hyperbolic systems, Theorems 12.2 and 12.5, the following generalisation of the mean value theorem has been used; see Hamilton (1982). In the following result, $M(N\times N,\mathbb{C})$ denotes the set of $N\times N$ matrices with complex entries.

Lemma 12.1 *Let $\mathcal{U}\subset\mathbb{R}^N$, and let $\mathbf{F}:\mathcal{U}\to\mathbb{C}^N$ be a C^1 map. Then there exists a continuous map $\mathbf{M}:\mathcal{U}\times\mathcal{U}\to M(N\times N,\mathbb{C})$ such that*

$$\mathbf{F}(\mathbf{u})-\mathbf{F}(\mathbf{v})=\mathbf{M}(\mathbf{u},\mathbf{v})(\mathbf{u}-\mathbf{v}).$$

The proof is an application of the *fundamental theorem of calculus.*

Whitney's extension theorem

In what follows, let $\boldsymbol{\alpha}=(\alpha_1,\ldots,\alpha_n)$, $\boldsymbol{\beta}=(\beta_1,\ldots,\beta_n)$ denote multi-indices. The factorial $\boldsymbol{\alpha}!$ is defined as $\boldsymbol{\alpha}!\equiv\alpha_1!\cdots\alpha_n!$. Moreover, let

$$|\boldsymbol{\beta}|\equiv\beta_1+\cdots+\beta_n.$$

In terms of this notation one has the following:

Theorem 12.8 (***Whitney's extension theorem***) *Given a non-negative integer k, suppose $\{f_{\boldsymbol{\alpha}}\}$, $|\boldsymbol{\alpha}|<k$ is a collection of real valued functions defined on a closed set $\mathcal{A}\subset\mathbb{R}^n$ satisfying*

$$f_{\boldsymbol{\alpha}}(x)=\sum_{|\boldsymbol{\beta}|\leq k-|\boldsymbol{\alpha}|}\frac{1}{\boldsymbol{\beta}!}f_{\boldsymbol{\alpha}+\boldsymbol{\beta}}(x')(x-x')^{\boldsymbol{\beta}}+R_{\boldsymbol{\alpha}}(x,x'),$$

for every $x,\,x'\in\mathcal{A}$ and each multi-index $\boldsymbol{\alpha}$ with $|\boldsymbol{\alpha}|\leq k$ such that for every $x_0\in\mathcal{A}$

$$R_{\boldsymbol{\alpha}}=o\big(|x-x'|^{k-|\boldsymbol{\alpha}|}\big),\qquad as\qquad x,\,x'\to x_0.$$

Then there exists a C^k function $g:\mathbb{R}^n\to\mathbb{R}$ such that

$$g=f_0,\qquad \partial_{\boldsymbol{\alpha}}g=f_{\boldsymbol{\alpha}}\qquad on\ \mathcal{A}.$$

In other words, for a closed set $\mathcal{A}$, if one is given a function f and candidates $f_{\boldsymbol{\alpha}}$ for its partial derivaties on $\mathcal{A}$, then f can be extended to all of $\mathbb{R}^n$ in such a way that the candidates are indeed the derivatives of f as long as the remainder has a suitable behaviour. A priori, it is not possible to identify the functions $f_{\boldsymbol{\alpha}}$ with the derivatives of f as $\mathcal{A}$ is a closed set and transversal derivatives f to $\partial\mathcal{A}$ may not be defined. For further details on Theorem 12.8 and its proof, see, for example, Abraham and Robbin (1967).

13
Hyperbolic reductions

This chapter discusses several methods for the construction of *symmetric hyperbolic evolution systems* out of the conformal Einstein field equations. Once suitable evolution systems have been obtained, the methods of Chapter 12 allow, in turn, one to make statements about the existence of solutions to the equations. Direct inspection of the conformal field equations reveals that these are overdetermined – there are more equations than unknowns, even if the symmetries of the various tensorial and spinorial fields are taken into account. Thus, the process of hyperbolic reduction for the conformal field equations necessarily requires discarding some of the equations. The discarded equations are then treated as constraints. It is a remarkable structural property of the conformal field equations that these constraints satisfy a system of evolution equations – a so-called ***subsidiary evolution system*** – from where it can be concluded that the constraint equations will be satisfied if they hold at some initial hypersurface and the evolution equations are imposed. This construction is called the ***propagation of the constraints***. The solution of the evolution system together with the propagation of the constraints yields the required solution of the conformal Einstein field equations.

In this chapter, two different procedures for the hyperbolic reduction of the conformal Einstein field equations are considered. The first method, based on the notion of ***gauge source functions***, exploits the fact that certain derivatives of the conformal fields are not directly determined by the equations and, thus, can be freely specified. In the spinorial formulation of the equations, once the required gauge source functions have been specified, the irreducible decomposition of the various zero quantities leads to the required evolution equations. The equations obtained by this procedure include the conformal factor as an unknown.

The second hyperbolic reduction procedure presented in this chapter exploits the properties of congruences of conformal geodesics to construct ***conformal Gaussian gauge systems***. As discussed in Chapter 5, the connection coefficients and components of the Schouten tensor with respect to a frame which is

Weyl propagated along the congruence satisfy certain relations which lead to a particularly simple system of equations in which the evolution of all the geometric unknowns, save for the components of the rescaled Weyl spinor, are either fixed by the gauge or given by transport equations along the congruence. Moreover, as a consequence of the properties of the conformal geodesics one gains an a priori knowledge of the location of the conformal boundary; see Proposition 5.1. Despite these attractive features, this method is less flexible than the one based on the use of gauge functions and may not be readily extended to non-vacuum situations.

13.1 A model problem: the Maxwell equations on a fixed background

To illustrate the various aspects of the construction of evolution equations for the conformal Einstein field equations, it is convenient to analyse the analogous problem for the Maxwell equations on a fixed background.

In the remainder of this section, let $\mathcal{U}$ denote an open region of a spacetime $(\mathcal{M}, \boldsymbol{g})$. It will be assumed that $\mathcal{U}$ is covered by a non-singular congruence of curves with tangent vector $\boldsymbol{\tau}$ satisfying the normalisation condition $\boldsymbol{g}(\boldsymbol{\tau}, \boldsymbol{\tau}) = 2$. The vector $\boldsymbol{\tau}$ does not need to be hypersurface orthogonal. Let $\tau^{AA'}$ denote the spinorial counterpart of τ^a. As discussed in Section 4.2.5, the spinor $\tau^{AA'}$ gives rise to a Hermitian structure, and, accordingly, one can introduce a space spinor formalism. Let $\{\epsilon_{\boldsymbol{A}}{}^A\}$ denote a spin basis such that

$$\tau^{\boldsymbol{AA}'} = \epsilon_{\boldsymbol{0}}{}^{\boldsymbol{A}} \epsilon_{\boldsymbol{0}'}{}^{\boldsymbol{A}'} + \epsilon_{\boldsymbol{1}}{}^{\boldsymbol{A}} \epsilon_{\boldsymbol{1}'}{}^{\boldsymbol{A}'}, \tag{13.1}$$

and with $\{\boldsymbol{e}_{\boldsymbol{AA}'}\}$ its associated null frame. At every point $p \in \mathcal{U}$ a basis of the subspace $\langle \boldsymbol{\tau} \rangle^{\perp}|_p \subset T|_p(\mathcal{U})$ orthogonal to $\boldsymbol{\tau}$ is given by $\boldsymbol{e}_{\boldsymbol{AB}} = \tau_{(\boldsymbol{B}}{}^{\boldsymbol{A}'} \boldsymbol{e}_{\boldsymbol{A})\boldsymbol{A}'}$. In terms of local coordinates $x = (x^\mu)$ in $\mathcal{U}$ one writes

$$\boldsymbol{e}_{\boldsymbol{AB}} = e_{\boldsymbol{AB}}{}^{\mu} \boldsymbol{\partial}_\mu. \tag{13.2}$$

In principle, it is possible for the frame vectors $\boldsymbol{e}_{\boldsymbol{AB}}$ to have components with respect to the time coordinate. The frame components $e_{\boldsymbol{AB}}{}^{\mu}$ satisfy the reality conditions

$$e_{\boldsymbol{01}}{}^{\mu} = \overline{e_{\boldsymbol{01}}{}^{\mu}}, \qquad e_{\boldsymbol{00}}{}^{\mu} = -\overline{e_{\boldsymbol{11}}{}^{\mu}}. \tag{13.3}$$

All spinorial objects will be expressed with respect to the spin basis $\{\epsilon_{\boldsymbol{A}}{}^A\}$. In particular, the spinorial Maxwell Equation (9.15) is written as

$$\nabla^{\boldsymbol{Q}}{}_{\boldsymbol{A}'} \varphi_{\boldsymbol{BQ}} = 0. \tag{13.4}$$

In what follows, it will be convenient to introduce the ***zero quantity***

$$\omega_{\boldsymbol{A}'\boldsymbol{B}} \equiv \nabla^{\boldsymbol{Q}}{}_{\boldsymbol{A}'} \varphi_{\boldsymbol{BQ}},$$

so that (13.4) can be expressed as $\omega_{\boldsymbol{A}'\boldsymbol{B}} = 0$. Here and in the remainder of this chapter, zero quantities such as $\omega_{\boldsymbol{A}'\boldsymbol{B}}$ serve as convenient bookkeeping devices to denote the various field equations.

13.1.1 Space spinor description of the Maxwell equations and hyperbolic reductions

The space spinor version of Equation (13.4) leads to a decomposition into evolution and constraint equations. Following the discussion of Chapter 4 one considers the *unprimed zero quantity* $\omega_{BA} \equiv \tau_B{}^{A'}\omega_{A'A}$. One then has that

$$\begin{aligned}\omega_{BA} &= \nabla^Q{}_B\varphi_{AQ} = \frac{1}{2}\epsilon^Q{}_A\mathcal{P}\varphi_{BQ} + \mathcal{D}^Q{}_A\varphi_{BQ}\\ &= -\frac{1}{2}\mathcal{P}\varphi_{AB} + \mathcal{D}^Q{}_A\varphi_{BQ},\end{aligned}$$

where $\mathcal{P}$ is the covariant directional derivative along τ, $\mathcal{D}_{AB}$ is the Sen covariant derivative implied by $\nabla_{AA'}$ and $\nabla_{AB} \equiv \tau_B{}^{A'}\nabla_{AA'}$. In the above expressions, the decomposition

$$\nabla_{AB} = \frac{1}{2}\epsilon_{AB}\mathcal{P} + \mathcal{D}_{AB} \tag{13.5}$$

has been used; see Section 4.3.1. The spinor ω_{BA} can, in turn, be decomposed in irreducible parts as

$$\omega_{BA} = \frac{1}{2}\epsilon_{BA}\omega + \omega_{(AB)},$$

with

$$\omega \equiv \omega_Q{}^Q = \mathcal{D}^{PQ}\varphi_{PQ}, \qquad \omega_{(AB)} = -\frac{1}{2}\mathcal{P}\varphi_{AB} + \mathcal{D}^Q{}_{(A}\varphi_{B)Q}.$$

Thus, the Maxwell Equations (13.4) imply the equations

$$\omega = \mathcal{D}^{PQ}\varphi_{PQ} = 0, \tag{13.6a}$$

$$-2\omega_{(AB)} = \mathcal{P}\varphi_{AB} - \mathcal{D}^Q{}_{(A}\varphi_{B)Q} = 0. \tag{13.6b}$$

The decomposition of the spinorial Maxwell equation given by (13.6a) and (13.6b) shows that Equation (13.4) is overdetermined. Equation (13.6a) will be interpreted as a constraint equation on the orthogonal subspaces of the distribution generated by the vector field τ, while (13.6b) will be regarded as suitable evolution equations for the symmetric spinorial field φ_{AB}.

13.1.2 The symmetric hyperbolicity of the Maxwell evolution equations

To apply the theory of Chapter 12 one needs to verify that the evolution Equations (13.6b) give rise to a symmetric hyperbolic system for the independent components of φ_{AB}. One considers the slightly modified version

$$\binom{2}{A+B}\left(\mathcal{P}\varphi_{AB} - \mathcal{D}^Q{}_{(A}\varphi_{B)Q}\right) = 0, \tag{13.7}$$

where the binomial coefficient in front of the equation has been included to make the expression manifestly symmetric hyperbolic. The principal part of Equation (13.7) can be written as

$$\binom{2}{\boldsymbol{A}+\boldsymbol{B}}\left(\tau^\mu\partial_\mu\varphi_{\boldsymbol{AB}} - e^{\boldsymbol{Q}}{}_{(\boldsymbol{A}}{}^\mu\partial_{|\mu|}\varphi_{\boldsymbol{B})\boldsymbol{Q}}\right).$$

As a result of the symmetry of $\varphi_{\boldsymbol{AB}}$, the above principal part contains three independent expressions. These can be arranged in the matricial expression

$$\mathbf{A}^\mu\partial_\mu\varphi \equiv \begin{pmatrix} \tau^\mu + e_{\boldsymbol{10}}{}^\mu & -e_{\boldsymbol{00}}{}^\mu & 0 \\ e_{\boldsymbol{11}}{}^\mu & 2\tau^\mu & e_{\boldsymbol{00}}{}^\mu \\ 0 & e_{\boldsymbol{11}}{}^\mu & \tau^\mu - e_{\boldsymbol{01}}{}^\mu \end{pmatrix} \partial_\mu \begin{pmatrix} \varphi_0 \\ \varphi_1 \\ \varphi_2 \end{pmatrix},$$

with

$$\varphi_0 \equiv \varphi_{\boldsymbol{00}}, \qquad \varphi_1 \equiv \varphi_{\boldsymbol{01}}, \qquad \varphi_2 \equiv \varphi_{\boldsymbol{11}}.$$

Thus, making use of the reality conditions (13.3), it follows that the matrices $\mathbf{A}^\mu$ are Hermitian. Moreover, the matrix

$$\mathbf{A}^\mu\tau_\mu = \begin{pmatrix} 2 & 0 & 0 \\ 0 & 4 & 0 \\ 0 & 0 & 2 \end{pmatrix}$$

clearly is positive definite. Thus, Equation (13.7) implies a symmetric hyperbolic system for the independent components of $\varphi_{\boldsymbol{AB}}$. Finally, a direct computation shows that given an arbitrary covector ξ_μ,

$$\begin{aligned} \det(\mathbf{A}^\mu\xi_\mu) &= 2(\tau^\mu\xi_\mu)(\tau^\nu\tau^\lambda + e_{\boldsymbol{00}}{}^\nu e_{\boldsymbol{11}}{}^\lambda - e_{\boldsymbol{01}}{}^\nu e_{\boldsymbol{10}}{}^\lambda)\xi_\nu\xi_\lambda \\ &= 4(\tau^\mu\xi_\mu)(g^{\nu\lambda}\xi_\nu\xi_\lambda), \end{aligned}$$

where in the last line Equation (4.14) for the 1+3 decomposition of the spacetime metric has been used. Thus, $\boldsymbol{g}$-null hypersurfaces are characteristics of Equation (13.7) – these types of characteristics are often called ***physical characteristics***. By contrast, the factor $(\tau^\mu\xi_\mu)$ is associated to ***gauge characteristics***.

For completeness, it is observed that the principal part of the constraint equation is given, explicitly, by

$$e_{\boldsymbol{00}}{}^\mu\partial_\mu\varphi_0 + e_{\boldsymbol{01}}{}^\mu\partial_\mu\varphi_1 + e_{\boldsymbol{11}}{}^\mu\partial_\mu\varphi_2,$$

so that, in general, it will contain derivatives in the time direction. More generally, if the vector $\boldsymbol{\tau}$ is not hypersurface orthogonal, then the constraint equation $\omega = 0$ will not be intrinsic to the leaves of a foliation.

13.1.3 The subsidiary system for the spinorial Maxwell equations

The hyperbolic reduction for the Maxwell equations discussed in Section 13.1.1 splits Equation (13.4) into three evolution equations and one constraint equation. Thus, if one wants to obtain a solution to Equation (13.4) through a Cauchy initial value problem, one uses, in first instance, the theory of Chapter 12 to show the existence of a unique solution to the evolution equations. In a second stage, one has to show that if the constraint equation is satisfied initially, then, by virtue of the evolution equations, it must be satisfied also at later times. This last argument requires the construction of a suitable hyperbolic evolution equation for ω.

To obtain an equation for the zero quantity ω one considers the expression $\nabla^{AA'}\omega_{A'A}$. Using that $\omega_{A'A} = -\tau^{Q}{}_{A'}\omega_{QA}$ one has that

$$\begin{aligned}\nabla^{AA'}\omega_{A'A} &= -\nabla^{AA'}\left(\tau^{Q}{}_{A'}\omega_{QA}\right)\\ &= \nabla^{AQ}\omega_{QA} - (\nabla^{AA'}\tau^{Q}{}_{A'})\omega_{QA}.\end{aligned}$$

Now, using Equation (4.17), a calculation yields

$$\nabla^{AA'}\tau^{Q}{}_{A'} = -\sqrt{2}\chi^{A}{}_{P}{}^{PQ}, \tag{13.8}$$

so that

$$\nabla^{AA'}\omega_{A'A} = \nabla^{AQ}\omega_{QA} + \sqrt{2}\chi^{A}{}_{P}{}^{PQ}\omega_{QA}.$$

Thus, the split (13.5) leads to the expression

$$\mathcal{P}\omega + 2\mathcal{D}^{AB}\omega_{(AB)} + 2\sqrt{2}\chi^{A}{}_{P}{}^{PQ}\omega_{QA} = 2\nabla^{AA'}\omega_{A'A}.$$

If the evolution equations hold – that is, $\omega_{(AB)} = 0$ – then $\omega_{AB} = \frac{1}{2}\epsilon_{AB}\omega$ and one obtains

$$\mathcal{P}\omega + \sqrt{2}\chi_{AB}{}^{AB}\omega = 2\nabla^{AA'}\omega_{A'A}.$$

The next step is to evaluate $\nabla^{AA'}\omega_{A'A}$ in an alternative manner. Using the definition of the zero quantity one has that

$$\nabla^{AA'}\omega_{A'A} = \nabla^{AA'}\nabla^{Q}{}_{A'}\varphi_{AQ}.$$

From the commutator

$$\nabla_{AA'}\nabla_{BB'}\varphi_{CD} - \nabla_{BB'}\nabla_{AA'}\varphi_{CD} = -R^{P}{}_{CAA'BB'}\varphi_{PD} - R^{P}{}_{DAA'BB'}\varphi_{CP},$$

suitably contracting indices one obtains

$$\nabla^{AA'}\nabla^{Q}{}_{A'}\varphi_{AQ} = -2R^{P}{}_{A}{}^{AA'Q}{}_{A'}\varphi_{PQ} - 2R^{P}{}_{Q}{}^{AA'Q}{}_{A'}\varphi_{AP}.$$

Thus, combining the above equation with the decomposition

$$R_{ABCC'DD'} = \Psi_{ABCD}\epsilon_{C'D'} + L_{BC'DD'}\epsilon_{CA} - L_{BD'CC'}\epsilon_{DA}, \tag{13.9}$$

where Ψ_{ABCD} and $L_{BC'DD'}$ denote, respectively, the spinorial counterparts of the Weyl and Schouten tensors, one concludes that $\nabla^{AA'}\nabla^{Q}{}_{A'}\varphi_{AQ} = 0$. Hence, the evolution equation for ω takes the form

$$\mathcal{P}\omega + \sqrt{2}\chi_{AB}{}^{AB}\omega = 0 \qquad \text{if } \omega_{(AB)} = 0.$$

The form of this equation implies, in together with Corollary 12.1, that if $\omega = 0$ on some spacelike hypersurface $\mathcal{S}_\star$ in $\mathcal{U}$, then $\omega = 0$ on lens-shaped domains having $\mathcal{S}_\star$ as base.

13.2 Hyperbolic reductions using gauge source functions

In this section hyperbolic reduction procedures for the conformal Einstein field equations based on the notion of ***gauge source functions*** are considered. Gauge source functions naturally arise in the analysis of frame formulations of the conformal Einstein field equations written in terms of the Levi-Civita connection $\boldsymbol{\nabla}$ of an unphysical metric $\boldsymbol{g}$. The present analysis will be restricted to the spinorial version of the conformal field equations: Equations (8.36a) and (8.36b) or, alternatively, Equations (8.38a) and (8.38b).

Basic set up and assumptions

As in the analysis of the Maxwell equations in Section 13.1, all the calculations will be performed in an open subset $\mathcal{U} \subset \mathcal{M}$ of an unphysical spacetime $(\mathcal{M}, \boldsymbol{g})$ which is conformally related to a spacetime $(\tilde{\mathcal{M}}, \tilde{\boldsymbol{g}})$ satisfying the Einstein field equations. On $\mathcal{U}$ one considers some local coordinates $x = (x^\mu)$ and an arbitrary frame $\{\boldsymbol{c}_{\boldsymbol{a}}\}$ which may or may not be a coordinate frame. Let $\{\boldsymbol{\alpha}^{\boldsymbol{a}}\}$ denote the dual coframe so that $\langle \boldsymbol{\alpha}^{\boldsymbol{a}}, \boldsymbol{c}_{\boldsymbol{b}} \rangle = \delta_{\boldsymbol{b}}{}^{\boldsymbol{a}}$. In what follows, let $\boldsymbol{\nabla}$ denote the Levi-Civita covariant derivative of the metric $\boldsymbol{g}$.

It will be assumed that $\mathcal{U}$ is covered by a non-singular congruence of curves with tangent vector $\boldsymbol{\tau}$ satisfying the normalisation condition $\boldsymbol{g}(\boldsymbol{\tau}, \boldsymbol{\tau}) = 2$. The vector $\boldsymbol{\tau}$ does not need to be hypersurface orthogonal. Let $\tau^{AA'}$ denote the spinorial counterpart of τ^a. In what follows, only spin bases $\{\epsilon_{\boldsymbol{A}}{}^A\}$ satisfying condition (13.1) will be considered. All spinors will be expressed in components with respect to this spin basis.

Let $\{\boldsymbol{e}_{\boldsymbol{AA'}}\}$ and $\{\boldsymbol{\omega}^{\boldsymbol{AA'}}\}$ denote, respectively, the null frame and coframe associated to the spin basis $\{\epsilon_{\boldsymbol{A}}{}^A\}$. By definition, one has that $\langle \boldsymbol{\omega}^{\boldsymbol{AA'}}, \boldsymbol{e}_{\boldsymbol{BB'}} \rangle = \epsilon_{\boldsymbol{B}}{}^{\boldsymbol{A}}\epsilon_{\boldsymbol{B'}}{}^{\boldsymbol{A'}}$. At every point $p \in \mathcal{U}$ a basis of $\langle \boldsymbol{\tau} \rangle^\perp|_p$, the subspace of $T|_p(\mathcal{U})$ orthogonal to $\boldsymbol{\tau}$ is given by $\boldsymbol{e}_{\boldsymbol{AB}} = \tau_{(\boldsymbol{B}}{}^{\boldsymbol{A'}}\boldsymbol{e}_{\boldsymbol{A})\boldsymbol{A'}}$. The *spatial* frame can be expanded in terms of the vectors $\boldsymbol{c}_{\boldsymbol{a}}$ as $\boldsymbol{e}_{\boldsymbol{AB}} = e_{\boldsymbol{AB}}{}^{\boldsymbol{a}}\boldsymbol{c}_{\boldsymbol{a}}$. If the basis $\{\boldsymbol{c}_{\boldsymbol{a}}\}$ is a coordinate basis, the last expression reduces to the one given in Equation (13.2).

A model equation

The general strategy behind the procedure of hyperbolic reduction using gauge functions is best understood through a *model equation.*

In Section 12.1.3 it has been shown that spinorial equations of the form

$$\nabla^{Q}{}_{A'}\varphi_{QB'C\cdots D} = F_{A'B'C\cdots D} \tag{13.10}$$

imply a symmetric hyperbolic system for the components of the field $\varphi_{QB'C\cdots D}$ which is not assumed to have any special symmetries. This equation is now contrasted with the equation

$$\nabla_{AA'}\varphi_{BB'C\cdots D} - \nabla_{BB'}\varphi_{AA'C\cdots D} = F_{AA'BB'C\cdots D}. \tag{13.11}$$

Exploiting the antisymmetry in the pairs AA' and BB' it follows that

$$\nabla^{Q}{}_{(A'}\varphi_{|Q|B')C\cdots D} = \frac{1}{2}F^{Q}{}_{A'QB'C\cdots D}. \tag{13.12}$$

Thus, while Equation (13.10) determines the full derivative $\nabla^{Q}{}_{A'}\varphi_{QB'C\cdots D}$, Equation (13.12) determines only its symmetric part. More precisely, writing

$$\nabla^{Q}{}_{A'}\varphi_{QB'C\cdots D} = \nabla^{Q}{}_{(A'}\varphi_{|Q|B')C\cdots D} - \frac{1}{2}\epsilon_{A'B'}\nabla^{QQ'}\varphi_{QQ'C\cdots D}, \tag{13.13}$$

one has that the first term in the right-hand side is determined by Equation (13.12), while the *divergence* $\nabla^{QQ'}\varphi_{QQ'C\cdots D}$ remains unspecified. Thus, in the absence of other equations providing information about this term, the latter observation suggests completing Equation (13.13) by setting

$$\nabla^{QQ'}\varphi_{QQ'C\cdots D} = f_{C\cdots D}(x),$$

where $f_{C\cdots D} \in \mathcal{X}(\mathcal{M})$ are *smooth freely specifiable functions of the coordinates.* In what follows, functions of this type will be known as ***gauge source functions***. Thus, from (13.13) one obtains the equation

$$\nabla^{Q}{}_{A'}\varphi_{QAC\cdots D} = \frac{1}{2}F^{Q}{}_{A'QB'AC\cdots D} - \frac{1}{2}\epsilon_{A'B'}f_{AC\cdots D}(x),$$

for which one can extract a symmetric hyperbolic evolution system for the components of $\varphi_{AA'C\cdots D}$; see the discussion of Section 12.1.3. In particular, the characteristics of this evolution system are null hypersurfaces of the spacetime metric $\boldsymbol{g}$.

As will be seen in the following subsections, several of the conformal Einstein field equations admit an analysis similar to that of Equation (13.11). A detailed discussion of the resulting evolution equations exploits the particular symmetries of the field appearing in the principal part.

13.2.1 Coordinate gauge source functions

The purpose of this subsection is to analyse the evolution equations arising from the *no-torsion condition* in the frame and spinor formulations of the conformal field equations; see Equations (8.31a), (8.35a), (8.44a) and (8.53a). This leads to the first class of gauge source functions that will be considered in this chapter: the ***coordinate gauge source functions***. Following the general discussion of Chapter 8, the no-torsion condition will be regarded as a differential condition on the coefficients of the frame $\{\boldsymbol{e}_{\boldsymbol{AA'}}\}$. Thus, the ultimate purpose of this section is to derive a symmetric hyperbolic subsystem for these quantities.

In Section 8.3.2 an expression for the spinorial counterpart of the torsion tensor $\Sigma_{\boldsymbol{AA'}}{}^{\boldsymbol{CC'}}{}_{\boldsymbol{BB'}}$ in terms of the spinorial connection coefficients $\Gamma_{\boldsymbol{AA'}}{}^{\boldsymbol{CC'}}{}_{\boldsymbol{BB'}}$ has been given; see Equation (8.35a). In what follows, it is more convenient to make use of an expression involving the reduced spin connection coefficients. Using the relation

$$\Gamma_{\boldsymbol{AA'}}{}^{\boldsymbol{CC'}}{}_{\boldsymbol{BB'}} = \Gamma_{\boldsymbol{AA'}}{}^{\boldsymbol{C}}{}_{\boldsymbol{B}}\epsilon_{\boldsymbol{B'}}{}^{\boldsymbol{C'}} + \bar{\Gamma}_{\boldsymbol{A'A}}{}^{\boldsymbol{C'}}{}_{\boldsymbol{B'}}\epsilon_{\boldsymbol{B}}{}^{\boldsymbol{C}},$$

– compare Equation (3.33) – it can be seen that

$$\begin{aligned}\Sigma_{\boldsymbol{AA'}}{}^{\boldsymbol{QQ'}}{}_{\boldsymbol{BB'}}\boldsymbol{e}_{\boldsymbol{QQ'}} = [\boldsymbol{e}_{\boldsymbol{BB'}}, \boldsymbol{e}_{\boldsymbol{AA'}}] - \Gamma_{\boldsymbol{BB'}}{}^{\boldsymbol{Q}}{}_{\boldsymbol{A}}\boldsymbol{e}_{\boldsymbol{QA'}} - \bar{\Gamma}_{\boldsymbol{BB'}}{}^{\boldsymbol{Q'}}{}_{\boldsymbol{A'}}\boldsymbol{e}_{\boldsymbol{AQ'}}\\ + \Gamma_{\boldsymbol{AA'}}{}^{\boldsymbol{Q}}{}_{\boldsymbol{B}}\boldsymbol{e}_{\boldsymbol{QB'}} + \bar{\Gamma}_{\boldsymbol{AA'}}{}^{\boldsymbol{Q'}}{}_{\boldsymbol{B'}}\boldsymbol{e}_{\boldsymbol{BQ'}}.\end{aligned} \tag{13.14}$$

Using the frame $\{\boldsymbol{c}_{\boldsymbol{a}}\}$ one can write

$$\boldsymbol{e}_{\boldsymbol{AA'}} = e_{\boldsymbol{AA'}}{}^{\boldsymbol{a}}\boldsymbol{c}_{\boldsymbol{a}},$$

so that for fixed frame spinorial indices ${}_{\boldsymbol{AA'}}$, the coefficients $e_{\boldsymbol{AA'}}{}^{\boldsymbol{a}}$ have the natural interpretation of the components of $\boldsymbol{e}_{\boldsymbol{AA'}}$ with respect to $\boldsymbol{c}_{\boldsymbol{a}}$. However, there is an alternative interpretation: for fixed frame index ${}_{\boldsymbol{a}}$, the coefficients $e_{\boldsymbol{AA'}}{}^{\boldsymbol{a}}$ correspond to the components of the covectors $\boldsymbol{\alpha}^{\boldsymbol{a}}$ with respect to the coframe $\boldsymbol{\omega}^{\boldsymbol{AA'}}$. That is, one has

$$\boldsymbol{\alpha}^{\boldsymbol{a}} = e_{\boldsymbol{AA'}}{}^{\boldsymbol{a}}\boldsymbol{\omega}^{\boldsymbol{AA'}},$$

from where it follows that $e_{\boldsymbol{AA'}}{}^{\boldsymbol{a}}\omega^{\boldsymbol{AA'}}{}_{\boldsymbol{b}} = \delta_{\boldsymbol{b}}{}^{\boldsymbol{a}}$. In view of this interpretation, it is convenient to define

$$\nabla_{\boldsymbol{CC'}}e_{\boldsymbol{BB'}}{}^{\boldsymbol{a}} \equiv e_{\boldsymbol{CC'}}{}^{\boldsymbol{b}}\boldsymbol{c}_{\boldsymbol{b}}(e_{\boldsymbol{BB'}}{}^{\boldsymbol{a}}) - \Gamma_{\boldsymbol{CC'}}{}^{\boldsymbol{Q}}{}_{\boldsymbol{B}}e_{\boldsymbol{QB'}}{}^{\boldsymbol{a}} - \bar{\Gamma}_{\boldsymbol{CC'}}{}^{\boldsymbol{Q'}}{}_{\boldsymbol{B'}}e_{\boldsymbol{BQ'}}{}^{\boldsymbol{a}}, \tag{13.15}$$

so that $\nabla_{\boldsymbol{CC'}}\boldsymbol{\alpha}^{\boldsymbol{a}} = (\nabla_{\boldsymbol{CC'}}e_{\boldsymbol{BB'}}{}^{\boldsymbol{a}})\boldsymbol{\omega}^{\boldsymbol{BB'}}$. Expression (13.15) corresponds to the formula one would use to compute the covariant derivative of $e_{\boldsymbol{BB'}}{}^{\boldsymbol{a}}$ if it were the components of a tensor – which, of course, it is not.

Intuition into this general discussion is gained by considering the particular case of a coordinate frame for which $\boldsymbol{e}_{\boldsymbol{AA'}} = e_{\boldsymbol{AA'}}{}^{\mu}\boldsymbol{\partial}_{\mu}$ so that

$$\boldsymbol{e}_{\boldsymbol{AA'}}(x^{\nu}) = e_{\boldsymbol{AA'}}{}^{\mu}\partial_{\mu}(x^{\nu}) = e_{\boldsymbol{AA'}}{}^{\mu}\delta_{\mu}{}^{\nu} = e_{\boldsymbol{AA'}}{}^{\nu}.$$

Moreover, writing $\boldsymbol{\omega}^{\boldsymbol{AA}'} = \omega^{\boldsymbol{AA}'}{}_\mu \mathbf{d}x^\mu$ one has that

$$\mathbf{d}x^\mu = e_{\boldsymbol{AA}'}{}^\mu \boldsymbol{\omega}^{\boldsymbol{AA}'}.$$

That is, for fixed coordinate index $^\mu$, the coefficients $e_{\boldsymbol{AA}'}{}^\mu$ are the components of the coordinate differential $\mathbf{d}x^\mu$ with respect to the coframe $\boldsymbol{\omega}^{\boldsymbol{AA}'}$.

Returning to the general discussion, using the identity

$$[f\boldsymbol{v}, \boldsymbol{u}] = f[\boldsymbol{v}, \boldsymbol{u}] - \boldsymbol{u}(f)\boldsymbol{v}$$

for $\boldsymbol{v}, \boldsymbol{u} \in T(\mathcal{M})$ and $f \in \mathcal{X}(\mathcal{M})$, together with expression (13.15) one can rewrite Equation (13.14) as

$$\Sigma_{\boldsymbol{AA}'}{}^{\boldsymbol{QQ}'}{}_{\boldsymbol{BB}'} e_{\boldsymbol{QQ}'}{}^{\boldsymbol{c}} = \nabla_{\boldsymbol{BB}'} e_{\boldsymbol{AA}'}{}^{\boldsymbol{c}} - \nabla_{\boldsymbol{AA}'} e_{\boldsymbol{BB}'}{}^{\boldsymbol{c}} - e_{\boldsymbol{AA}'}{}^{\boldsymbol{a}} e_{\boldsymbol{BB}'}{}^{\boldsymbol{b}} C_{\boldsymbol{a}}{}^{\boldsymbol{c}}{}_{\boldsymbol{b}}, \tag{13.16}$$

where $C_{\boldsymbol{a}}{}^{\boldsymbol{c}}{}_{\boldsymbol{b}}$ are the ***commutation coefficients*** defined by

$$[\boldsymbol{c_a}, \boldsymbol{c_b}] = C_{\boldsymbol{a}}{}^{\boldsymbol{c}}{}_{\boldsymbol{b}} \boldsymbol{c_c}.$$

In the case of a coordinate frame one obtains the simpler expression

$$\Sigma_{\boldsymbol{AA}'}{}^{\boldsymbol{QQ}'}{}_{\boldsymbol{BB}'} e_{\boldsymbol{QQ}'}{}^\mu = \nabla_{\boldsymbol{BB}'} e_{\boldsymbol{AA}'}{}^\mu - \nabla_{\boldsymbol{AA}'} e_{\boldsymbol{BB}'}{}^\mu,$$

as $[\boldsymbol{\partial}_\mu, \boldsymbol{\partial}_\nu] = 0$.

A final simplification is obtained by exploiting the antisymmetry of Equation (13.16). Contracting the indices $_{\boldsymbol{A}'}$ and $_{\boldsymbol{B}'}$ and symmetrising in $_{\boldsymbol{AB}}$ one concludes that

$$\nabla_{(\boldsymbol{A}}{}^{\boldsymbol{Q}'} e_{\boldsymbol{B})\boldsymbol{Q}'}{}^{\boldsymbol{a}} + \frac{1}{2} e_{\boldsymbol{A}}{}^{\boldsymbol{Q}'\boldsymbol{b}} e_{\boldsymbol{BQ}'}{}^{\boldsymbol{c}} C_{\boldsymbol{b}}{}^{\boldsymbol{a}}{}_{\boldsymbol{c}} = \Sigma_{\boldsymbol{AB}}{}^{\boldsymbol{a}}, \tag{13.17}$$

with

$$\Sigma_{\boldsymbol{AB}}{}^{\boldsymbol{a}} \equiv \frac{1}{2} \Sigma_{\boldsymbol{A}}{}^{\boldsymbol{Q}'\boldsymbol{CC}'}{}_{\boldsymbol{BQ}'} e_{\boldsymbol{CC}'}{}^{\boldsymbol{a}}.$$

As the frame $\boldsymbol{e_{AA'}}$ is Hermitian, that is, $\overline{\boldsymbol{e_{AA'}}} = \boldsymbol{e_{AA'}}$, one has that (13.17) is completely equivalent to Equation (13.16). Moreover, if $\Sigma_{\boldsymbol{AB}}{}^{\boldsymbol{a}} = 0$, then $\Sigma_{\boldsymbol{AA}'}{}^{\boldsymbol{CC}'}{}_{\boldsymbol{BB}'} = 0$ and the connection is torsion free.

The structure of Equation (13.17) is similar to that of the model Equation (13.12), suggesting that by introducing a gauge source function one will obtain a symmetric hyperbolic system for the frame coefficients $e_{\boldsymbol{AA}'}{}^{\boldsymbol{a}}$. Now, Equation (13.17) does not impose restrictions on the divergences $\nabla^{\boldsymbol{QQ}'} e_{\boldsymbol{QQ}'}{}^{\boldsymbol{a}}$ so that one can set

$$\nabla^{\boldsymbol{QQ}'} e_{\boldsymbol{QQ}'}{}^{\boldsymbol{a}} = F^{\boldsymbol{a}}(x), \tag{13.18}$$

where the ***coordinate gauge source functions*** $F^{\boldsymbol{a}}(x)$ are smooth functions of the coordinates $x = (x^\mu)$. In the case of a coordinate frame the above expression reduces to

$$\nabla^{\boldsymbol{QQ}'} \nabla_{\boldsymbol{QQ}'} x^\mu = F^\mu(x), \tag{13.19}$$

the so-called ***generalised wave coordinates condition***.

Combining the identity

$$\nabla_{(\boldsymbol{A}}{}^{\boldsymbol{Q}'} e_{\boldsymbol{B})\boldsymbol{Q}'}{}^{\boldsymbol{a}} = \nabla_{\boldsymbol{A}}{}^{\boldsymbol{Q}'} e_{\boldsymbol{B}\boldsymbol{Q}'}{}^{\boldsymbol{a}} + \frac{1}{2}\epsilon_{\boldsymbol{A}\boldsymbol{B}}\nabla^{\boldsymbol{P}\boldsymbol{P}'} e_{\boldsymbol{P}\boldsymbol{P}'}{}^{\boldsymbol{a}}$$

with Equations (13.17) and (13.18) one finally obtains, for $\Sigma_{\boldsymbol{A}\boldsymbol{B}}{}^{\boldsymbol{a}} = 0$, the equation

$$\nabla_{\boldsymbol{A}}{}^{\boldsymbol{Q}'} e_{\boldsymbol{B}\boldsymbol{Q}'}{}^{\boldsymbol{a}} + \frac{1}{2}\epsilon_{\boldsymbol{A}\boldsymbol{B}} F^{\boldsymbol{a}}(x) + \frac{1}{2} e_{\boldsymbol{A}}{}^{\boldsymbol{Q}'\boldsymbol{b}} e_{\boldsymbol{B}\boldsymbol{Q}'}{}^{\boldsymbol{c}} C_{\boldsymbol{b}}{}^{\boldsymbol{a}}{}_{\boldsymbol{c}} = 0,$$

from which a symmetric hyperbolic system for the frame components of $e_{\boldsymbol{B}\boldsymbol{Q}'}{}^{\boldsymbol{a}}$ can be deduced.

Geometric interpretation

The generalised wave coordinate condition (13.19) shows that a particular choice of coordinate gauge is, implicitly, a choice of coordinates. Equation (13.19) can always be solved locally by choosing some coordinates $x = (x^0, x^\alpha)$ on some fiduciary surface $\mathcal{S}_\star$. If this surface is described by the condition $x^0 = 0$, then it is also natural to require that

$$\frac{\partial x^\alpha}{\partial x^0} = 0, \qquad \text{on } \mathcal{S}_\star.$$

Moreover, one needs the coordinate differentials $\mathbf{d}x^\mu$ to be linearly independent on $\mathcal{S}_\star$. These conditions ensure the existence of a solution to Equation (13.19) close to $\mathcal{S}_\star$.

Conversely, given a particular coordinate choice on a spacetime $(\mathcal{M}, \boldsymbol{g})$, one can use Equation (13.19) to compute the coordinate gauge source function $F^\mu(x)$ associated with the coordinates. Thus, local coordinates and coordinate gauge source functions are in a one-to-one correspondence.

Construction of coordinates in perturbations of spacetimes

The discussion of the previous subsection can be applied to the construction of coordinates in spacetimes $(\mathcal{M}, \boldsymbol{g})$ which are *perturbations* of a certain *exact background spacetime* $(\mathring{\mathcal{M}}, \mathring{\boldsymbol{g}})$. In this situation, one would expect the spacetime manifolds $\mathcal{M}$ and $\mathring{\mathcal{M}}$ to be diffeomorphic to each other so that coordinates in the background spacetime could be used as coordinates in the perturbed spacetime. This does not mean that the spacetimes $(\mathcal{M}, \boldsymbol{g})$ and $(\mathring{\mathcal{M}}, \mathring{\boldsymbol{g}})$ are isometric! The intuition expressed in this paragraph will now be formalised.

In what follows, assume that one has two spacetimes $(\mathcal{M}, \boldsymbol{g})$ and $(\mathring{\mathcal{M}}, \mathring{\boldsymbol{g}})$ such that the manifolds $\mathcal{M}$ and $\mathring{\mathcal{M}}$ are diffeomorphic. Let $\varphi : \mathcal{M} \to \mathring{\mathcal{M}}$ denote a diffeomorphism between them. This choice is clearly not unique. The subsequent discussion will single out a particular type of diffeomorphism between $\mathcal{M}$ and $\mathring{\mathcal{M}}$.

Let $x = (x^\mu)$ and $\mathring{x} = (\mathring{x}^\mu)$ denote, respectively, local coordinates on $\mathcal{M}$ and $\mathring{\mathcal{M}}$. In terms of these local coordinates the diffeomorphism φ is given by $\mathring{x}^\mu = \mathring{x}^\mu(x)$ and its inverse by $x^\mu = x^\mu(\mathring{x})$. On $\mathring{\mathcal{M}}$ consider a frame $\{\mathring{\boldsymbol{c}}_{\boldsymbol{a}}\}$ and its dual coframe $\{\mathring{\boldsymbol{\alpha}}^{\boldsymbol{a}}\}$. The frame is not necessarily assumed to be $\mathring{\boldsymbol{g}}$-orthonormal. From this frame and coframe one can introduce a frame $\{\boldsymbol{c}_{a}\}$ and a coframe $\{\boldsymbol{\alpha}^{\boldsymbol{a}}\}$ on $\mathcal{M}$ using, respectively, the push-forward and the pull-back implied by $\varphi : \mathcal{M} \rightarrow \mathring{\mathcal{M}}$. More precisely,

$$\mathring{\boldsymbol{c}}_{\boldsymbol{a}} = (\varphi)_* \boldsymbol{c}_{\boldsymbol{a}}, \qquad \mathring{\boldsymbol{\alpha}}^{\boldsymbol{a}} = (\varphi^{-1})^* \boldsymbol{\alpha}^{\boldsymbol{a}}.$$

Thus, writing

$$\boldsymbol{\alpha}^{\boldsymbol{a}} = \alpha^{\boldsymbol{a}}{}_{\mu} \mathbf{d}x^{\mu}, \qquad \mathring{\boldsymbol{\alpha}}^{\boldsymbol{a}} = \mathring{\alpha}^{\boldsymbol{a}}{}_{\mu} \mathbf{d}\mathring{x}^{\mu},$$

one concludes that

$$\alpha^{\boldsymbol{a}}{}_{\mu} = \mathring{\alpha}^{\boldsymbol{a}}{}_{\nu} \frac{\partial \mathring{x}^{\nu}}{\partial x^{\mu}}.$$

Now, observing that $\langle \boldsymbol{\alpha}^{\boldsymbol{a}}, \boldsymbol{e}_{\boldsymbol{b}} \rangle = \delta_{\boldsymbol{b}}{}^{\boldsymbol{a}}$, it follows that $\nabla_{\boldsymbol{c}} \boldsymbol{\alpha}^{\boldsymbol{a}} = (\nabla_{\boldsymbol{c}} e_{\boldsymbol{b}}{}^{\boldsymbol{a}}) \boldsymbol{\alpha}^{\boldsymbol{b}}$ and, consequently,

$$\nabla^{\boldsymbol{b}} e_{\boldsymbol{b}}{}^{\boldsymbol{a}} = \eta^{\boldsymbol{cd}} \langle \nabla_{\boldsymbol{c}} \boldsymbol{\alpha}^{\boldsymbol{a}}, \boldsymbol{e}_{\boldsymbol{d}} \rangle = e_{\boldsymbol{b}}{}^{\mu} \nabla^{\boldsymbol{b}} \alpha^{\boldsymbol{a}}{}_{\mu} = \nabla^{\mu} \alpha^{\boldsymbol{a}}{}_{\mu}.$$

The above expression can be used to write the divergence $\nabla^{\boldsymbol{QQ'}} e_{\boldsymbol{QQ'}}{}^{\boldsymbol{a}}$ appearing in Equation (13.18) in terms of quantities associated to the diffeomorphism $\varphi : \mathcal{M} \rightarrow \mathring{\mathcal{M}}$.

Treating the coordinates $\mathring{x} = (\mathring{x}^{\mu})$ as scalars and recalling that $\mathring{\alpha}^{\boldsymbol{a}}{}_{\nu} = \langle \mathring{\boldsymbol{\alpha}}^{\boldsymbol{a}}, \boldsymbol{\partial}/\boldsymbol{\partial}\mathring{x}^{\nu} \rangle$ so that the coefficients $\mathring{\alpha}^{\boldsymbol{a}}{}_{\nu}$ are also scalars, one finds that

$$\begin{aligned}\nabla_{\nu} \alpha^{\boldsymbol{a}}{}_{\mu} &= \mathring{\alpha}^{\boldsymbol{a}}{}_{\lambda} \nabla_{\nu} \left(\frac{\partial \mathring{x}^{\lambda}}{\partial x^{\mu}} \right) + \frac{\partial \mathring{\alpha}^{\boldsymbol{a}}{}_{\lambda}}{\partial x^{\nu}} \frac{\partial \mathring{x}^{\lambda}}{\partial x^{\mu}} \\ &= \mathring{\alpha}^{\boldsymbol{a}}{}_{\lambda} \nabla_{\nu} \nabla_{\mu} \mathring{x}^{\lambda} + \mathring{\nabla}_{\rho} \mathring{\alpha}^{\boldsymbol{a}}{}_{\lambda} \frac{\partial \mathring{x}^{\rho}}{\partial x^{\nu}} \frac{\partial \mathring{x}^{\lambda}}{\partial x^{\mu}},\end{aligned}$$

where in the last equality the chain rule has been used. Consequently, one has

$$\nabla^{\mu} \alpha^{\boldsymbol{a}}{}_{\mu} = \mathring{\alpha}^{\boldsymbol{a}}{}_{\lambda} \nabla^{\mu} \nabla_{\mu} \mathring{x}^{\lambda} + g^{\mu\nu} \mathring{\nabla}_{\rho} \mathring{\alpha}^{\boldsymbol{a}}{}_{\lambda} \frac{\partial \mathring{x}^{\rho}}{\partial x^{\nu}} \frac{\partial \mathring{x}^{\lambda}}{\partial x^{\mu}} = F^{\boldsymbol{a}}(x),$$

or, more suggestively,

$$\nabla^{\mu} \nabla_{\mu} \mathring{x}^{\sigma} + \mathring{c}_{\boldsymbol{a}}{}^{\sigma} \left(g^{\mu\nu} \mathring{\nabla}_{\rho} \mathring{\alpha}^{\boldsymbol{a}}{}_{\lambda} \frac{\partial \mathring{x}^{\rho}}{\partial x^{\nu}} \frac{\partial \mathring{x}^{\lambda}}{\partial x^{\mu}} - F^{\boldsymbol{a}}(x) \right) = 0.$$

So far, the diffeomorphism $\varphi : \mathcal{M} \rightarrow \mathring{\mathcal{M}}$ has been kept completely general. However, if one sets

$$g^{\mu\nu} \mathring{\nabla}_{\rho} \mathring{\alpha}^{\boldsymbol{a}}{}_{\lambda} \frac{\partial \mathring{x}^{\rho}}{\partial x^{\nu}} \frac{\partial \mathring{x}^{\lambda}}{\partial x^{\mu}} = F^{\boldsymbol{a}}(x), \tag{13.20}$$

one finds that

$$\nabla^{\mu} \nabla_{\mu} \mathring{x}^{\sigma} = 0.$$

That is, under condition (13.20), the diffeomorphism $\varphi : \mathcal{M} \to \mathring{\mathcal{M}}$ given by $\mathring{x}^\mu = \mathring{x}^\mu(x)$ is a ***wave map***. Wave maps can be regarded as a generalisation of the geodesic equation. Further discussion on this notion, which plays an important role in modern research in PDE theory and geometric analysis, can be found in the review by Tataru (2004).

Now, it is convenient to regard the manifolds $\mathcal{M}$ and $\mathring{\mathcal{M}}$ as being the same and let $\mathring{x}^\mu = \mathring{x}^\mu(x)$ be the ***identity map*** so that $\partial\mathring{x}^\rho/\partial x^\nu = \delta_\nu{}^\rho$. *This amounts to saying that the coordinates* $\mathring{x} = (\mathring{x}^\mu)$ *are used as coordinates of the perturbed spacetime* $(\mathcal{M}, \boldsymbol{g})$. In this case condition (13.20) reduces to

$$\mathring{\nabla}^{\boldsymbol{b}}\mathring{\alpha}^{\boldsymbol{a}}{}_{\boldsymbol{b}} = F^{\boldsymbol{a}}(x).$$

If in the reference spacetime one has $\mathring{\boldsymbol{\omega}}^{\boldsymbol{a}} = \mathring{\boldsymbol{\alpha}}^{\boldsymbol{a}}$ so that $\mathring{\alpha}^{\boldsymbol{a}}{}_{\boldsymbol{b}} \equiv \langle \mathring{\boldsymbol{\alpha}}^{\boldsymbol{a}}, \mathring{\boldsymbol{c}}_{\boldsymbol{b}} \rangle = \delta_{\boldsymbol{b}}{}^{\boldsymbol{a}}$, then

$$\mathring{\nabla}^{\boldsymbol{b}}\mathring{\alpha}^{\boldsymbol{a}}{}_{\boldsymbol{b}} = -\eta^{\boldsymbol{bc}}\mathring{\Gamma}_{\boldsymbol{b}}{}^{\boldsymbol{a}}{}_{\boldsymbol{c}}.$$

Accordingly, the coordinate gauge source function $F^{\boldsymbol{a}}(x)$ can be expressed in terms of the connection of the background spacetime via

$$F^{\boldsymbol{a}}(x) = -\eta^{\boldsymbol{bc}}\mathring{\Gamma}_{\boldsymbol{b}}{}^{\boldsymbol{a}}{}_{\boldsymbol{c}},$$

or, in spinorial terms

$$F^{\boldsymbol{a}}(x) = -\epsilon^{\boldsymbol{AB}}\epsilon^{\boldsymbol{A'B'}}e_{\boldsymbol{AA'}}{}^{\boldsymbol{b}}e_{\boldsymbol{BB'}}{}^{\boldsymbol{c}}\mathring{\Gamma}_{\boldsymbol{b}}{}^{\boldsymbol{a}}{}_{\boldsymbol{c}}.$$

Space spinor decomposition of the equation for the frame coefficients

The space spinor decomposition of Equation (13.17) provides a systematic approach to the extraction of the required symmetric hyperbolic system. Accordingly, one considers the space spinor split of the frame fields given by

$$e_{\boldsymbol{AA'}}{}^{\boldsymbol{a}} = \frac{1}{2}\tau_{\boldsymbol{AA'}}e^{\boldsymbol{a}} - \tau^{\boldsymbol{Q}}{}_{\boldsymbol{A'}}e_{\boldsymbol{AQ}}{}^{\boldsymbol{a}}$$

with

$$e^{\boldsymbol{a}} \equiv \tau^{\boldsymbol{AA'}}e_{\boldsymbol{AA'}}{}^{\boldsymbol{a}}, \qquad e_{\boldsymbol{AB}}{}^{\boldsymbol{a}} \equiv \tau_{(\boldsymbol{A}}{}^{\boldsymbol{A'}}e_{\boldsymbol{B})\boldsymbol{A'}}{}^{\boldsymbol{a}}.$$

Alternatively, one can write

$$\tau_{\boldsymbol{B}}{}^{\boldsymbol{Q'}}e_{\boldsymbol{AQ'}}{}^{\boldsymbol{a}} = \frac{1}{2}\epsilon_{\boldsymbol{AB}}e^{\boldsymbol{a}} + e_{\boldsymbol{AB}}{}^{\boldsymbol{a}}.$$

Using

$$\nabla_{\boldsymbol{AB}}\tau_{\boldsymbol{CD'}} = -\sqrt{2}\tau^{\boldsymbol{D}}{}_{\boldsymbol{D'}}\chi_{\boldsymbol{ABCD}},$$

– compare Equation (4.17) – together with the decomposition of $\nabla_{\boldsymbol{AB}}$ given in Equation (13.5), it follows from Equation (13.18) that

$$\mathcal{P}e^{\boldsymbol{a}} + 2\mathcal{D}^{\boldsymbol{PQ}}e_{\boldsymbol{PQ}}{}^{\boldsymbol{a}} + \sqrt{2}e^{\boldsymbol{a}}\chi_{\boldsymbol{PQ}}{}^{\boldsymbol{PQ}} + 2\sqrt{2}e_{\boldsymbol{PQ}}{}^{\boldsymbol{a}}\chi^{\boldsymbol{P}}{}_{\boldsymbol{C}}{}^{\boldsymbol{CQ}} - 2F^{\boldsymbol{a}}(x) = 0. \tag{13.21}$$

A similar computation for Equation (13.17) yields

$$\Sigma_{AB}{}^{a} = \frac{1}{2}\mathcal{P}e_{AB}{}^{a} - \frac{1}{2}\mathcal{D}_{AB}e^{a} + \mathcal{D}_{(A}{}^{Q}e_{B)Q}{}^{a} - \frac{1}{\sqrt{2}}e^{a}\chi_{(A|Q|}{}^{Q}{}_{B)}$$
$$+ \sqrt{2}e_{P(A}{}^{a}\chi_{B)Q}{}^{QP} - \frac{1}{2}(e^{b}e_{AB}{}^{c} + e_{AQ}{}^{b}e_{B}{}^{Qc})C_{b}{}^{a}{}_{c}.$$

A further independent equation can be obtained from the Hermitian conjugate

$$\Sigma^{+}_{AB}{}^{a} \equiv \tau_{A}{}^{A'}\tau_{B}{}^{B'}\bar{\Sigma}_{A'B'}{}^{a}.$$

Exploiting the identity

$$\tau_{A}{}^{A'}\tau_{B}{}^{B'}\nabla^{Q}{}_{A'}e_{QB'}{}^{a} = \nabla^{Q}{}_{A}\big(\tau_{B}{}^{B'}e_{QB'}{}^{a}\big) - e_{QB'}{}^{a}\nabla^{Q}{}_{A}\tau_{B}{}^{B'},$$

one arrives at

$$\Sigma^{+}_{AB}{}^{a} = -\frac{1}{2}\mathcal{P}e_{AB}{}^{a} + \frac{1}{2}\mathcal{D}_{AB}e^{a} + \mathcal{D}_{(A}{}^{Q}e_{B)Q}{}^{a} + \frac{1}{\sqrt{2}}e^{a}\chi^{Q}{}_{(AB)Q}$$
$$- \sqrt{2}e_{PQ}{}^{a}\chi^{P}{}_{(AB)}{}^{Q} - \frac{1}{2}(e^{b}e_{AB}{}^{c} + e_{AQ}{}^{b}e_{B}{}^{Qc})C_{b}{}^{a}{}_{c}.$$

The required evolution equation complementing (13.21) is then obtained from

$$\Sigma_{AB}{}^{a} - \Sigma^{+}_{AB}{}^{a} = 0,$$

where

$$\Sigma_{AB}{}^{a} - \Sigma^{+}_{AB}{}^{a} = \mathcal{P}e_{AB}{}^{a} - \mathcal{D}_{AB}e^{a} - \frac{1}{\sqrt{2}}e^{a}\big(\chi_{(A|Q|}{}^{Q}{}_{B)} + \chi_{Q(AB)}{}^{Q}\big)$$
$$+ \sqrt{2}e_{P(A}{}^{a}\chi_{B)Q}{}^{QP} + \sqrt{2}e_{PQ}{}^{a}\chi^{P}{}_{(AB)}{}^{Q}$$
$$- e^{c}e_{AB}{}^{b}C_{b}{}^{a}{}_{c}. \qquad (13.22)$$

A direct inspection shows that Equations (13.21) and (13.22) imply, for fixed frame index ^{a}, a symmetric hyperbolic system of four equations for e^{a} and the independent components of $e_{AB}{}^{a}$. A further computation shows that the characteristic polynomial of the system is given by

$$-4(\tau^{\mu}\xi_{\mu})^{2}(g^{\lambda\rho}\xi_{\lambda}\xi_{\rho}).$$

As a by-product of the analysis one obtains the constraint equations implied by (13.17) from

$$\Sigma_{AB}{}^{a} + \Sigma^{+}_{AB}{}^{a} = 0,$$

where

$$\Sigma_{AB}{}^{a} + \Sigma^{+}_{AB}{}^{a} = 2\mathcal{D}^{Q}{}_{(A}e_{B)Q}{}^{a} + \frac{1}{\sqrt{2}}e^{a}\big(\chi_{(A|Q|}{}^{Q}{}_{B)} + \chi^{Q}{}_{(AB)Q}\big)$$
$$+ \sqrt{2}e_{P(A}{}^{a}\chi_{B)Q}{}^{QP} - \sqrt{2}e_{PQ}{}^{a}\chi^{P}{}_{(AB)}{}^{Q}$$
$$- (e^{b}e_{AB}{}^{c} + e_{AQ}{}^{b}e_{B}{}^{Qc})C_{b}{}^{a}{}_{c}.$$

Expanding the principal part of this constraint equation, one finds it contains derivatives in the time direction.

13.2.2 Frame gauge source functions

After having analysed the gauge source conditions arising from the no-torsion condition, one can now consider the gauge source functions associated to the Ricci identity – that is, the condition requiring that the geometric and the algebraic curvatures coincide. As with the no-torsion condition, the equality between the two expressions for the curvature is part of the frame and spinorial formulations of the conformal field equations; compare Equations (8.31b), (8.35b), (8.44b) and (8.53b).

Rather than working with the full expressions for the curvature spinors, in the subsequent discussion it will be convenient to make use of the reduced spinorial counterpart of the Riemann tensor in terms of the reduced connection coefficients:

$$\begin{aligned} R_{ABCC'DD'} + \Sigma_{CC'}{}^{QQ'}{}_{DD'}\Gamma_{QQ'AB} &= \nabla_{DD'}\Gamma_{CC'AB} - \nabla_{CC'}\Gamma_{DD'AB} \\ &\quad - \Gamma_{DD'}{}^{Q}{}_{A}\Gamma_{CC'QB} - \Gamma_{CC'}{}^{Q}{}_{A}\Gamma_{DD'QB}, \end{aligned} \tag{13.23}$$

where the definition

$$\begin{aligned} \nabla_{DD'}(\Gamma_{CC'AB}) \equiv e_{DD'}(\Gamma_{CC'AB}) &- \Gamma_{DD'}{}^{Q}{}_{C}\Gamma_{QC'AB} \\ &- \bar{\Gamma}_{DD'}{}^{Q}{}_{C}\Gamma_{QC'AB} - \Gamma_{DD'}{}^{Q}{}_{B}\Gamma_{CC'AQ} \end{aligned}$$

has been used in order to obtain a more concise expression; see Section 8.3.2 for further details. This last expression is formally the same as the one for the covariant derivative of a spinor field with the same index structure as $\Gamma_{CC'AB}$. Equation (13.23) is encoded in the zero quantity

$$\Xi_{ABCC'DD'} \equiv R_{ABCC'DD'} - \rho_{ABCC'DD'},$$

where $R_{ABCC'DD'}$ and $\rho_{ABCC'DD'}$ denote, respectively, the geometric and algebraic curvatures. One has the symmetries

$$\Xi_{ABCC'DD'} = \Xi_{(AB)CC'DD'} = -\Xi_{ABDD'CC'}.$$

Exploiting the antisymmetry of Equation (13.23) one obtains the pair of equations

$$\nabla_{(C}{}^{Q'}\Gamma_{D)Q'AB} + \Gamma_{(C}{}^{Q'Q}{}_{|A|}\Gamma_{D)Q'QB} = R_{ABCD} + \Sigma_{C}{}^{QQ'}{}_{D}\Gamma_{QQ'AB}, \tag{13.24a}$$

$$\nabla^{P}{}_{(C'}\Gamma_{|P|D')AB} + \Gamma^{P}{}_{(C'}{}^{Q}{}_{|A}\Gamma_{P|D')QB} = R_{ABC'D'} + \Sigma_{C'}{}^{QQ'}{}_{D'}\Gamma_{QQ'AB}, \tag{13.24b}$$

where

$$R_{ABCD} \equiv \frac{1}{2}R_{ABCQ'D}{}^{Q'}, \qquad R_{ABC'D'} \equiv \frac{1}{2}R_{ABQC'}{}^{Q}{}_{D'},$$

$$\Sigma_{C}{}^{QQ'}{}_{D} \equiv \frac{1}{2}\Sigma_{CP'}{}^{QQ'}{}_{D}{}^{P'}, \qquad \Sigma_{C'}{}^{QQ'}{}_{D'} \equiv \frac{1}{2}\Sigma_{PC'}{}^{QQ'P}{}_{D'}.$$

As the field $\Gamma_{AA'BC}$ is not Hermitian, two reduced equations are necessary to encode the content of (13.23) – contrast this with the analysis of the no-torsion Equation (13.14).

From the structure of Equations (13.24a) and (13.24b) one concludes that the derivative $\nabla^{QQ'}\Gamma_{QQ'AB}$ is not determined by the equations. Accordingly, one can set

$$\nabla^{QQ'}\Gamma_{QQ'AB} = F_{AB}(x), \tag{13.25}$$

where $F_{AB} = F_{(AB)}$ are smooth arbitrary functions of the coordinates – the ***frame gauge source functions.***

Geometric interpretation

To gain intuition on the role played by the frame gauge source functions recall that $\Gamma_{AA'}{}^{B}{}_{C} = \epsilon^{B}{}_{B}\nabla_{AA'}\epsilon_{C}{}^{B}$; see Equation (3.32). Equation (13.25) can be rewritten as

$$\epsilon^{A}{}_{B}\nabla^{PP'}\nabla_{PP'}\epsilon_{B}{}^{B} + \nabla^{PP'}\epsilon^{A}{}_{B}\nabla_{PP'}\epsilon_{B}{}^{B} = F^{A}{}_{B}(x). \tag{13.26}$$

This is to be read as a quasilinear wave equation for the spin frame $\{\epsilon_{B}{}^{B}\}$. Using the symmetry of F_{AB} and the wave Equation (13.26) one obtains

$$\nabla^{PP'}\nabla_{PP'}\left(\epsilon_{B}{}^{B}\epsilon^{A}{}_{B}\right) = 0,$$

so that by choosing

$$\epsilon_{B}{}^{B}\epsilon^{A}{}_{B} = \delta_{B}{}^{A}, \qquad \nabla_{PP'}\left(\epsilon_{B}{}^{B}\epsilon^{A}{}_{B}\right) = 0,$$

on some fiduciary hypersurface $\mathcal{S}_{\star}$ one obtains a spin frame which is normalised at later times.

Space spinor decomposition of the equation for the spin connection coefficients

To obtain a suitable space spinor decomposition of Equations (13.24a), (13.24b) and (13.25), one defines

$$\Gamma_{ABCD} \equiv \tau_{B}{}^{A'}\Gamma_{AA'CD}$$

and considers the split

$$\Gamma_{ABCD} = \frac{1}{2}\epsilon_{AB}\Gamma_{CD} + \Gamma_{(AB)CD}, \qquad \Gamma_{CD} \equiv \Gamma_{Q}{}^{Q}{}_{CD}.$$

Now, from

$$\begin{aligned}
\nabla^{QQ'}\Gamma_{QQ'AB} &= -\nabla^{QQ'}\left(\tau^{P}{}_{Q'}\Gamma_{QPAB}\right)\\
&= \tau^{SQ'}\nabla^{Q}{}_{Q'}\Gamma_{QSAB} - (\nabla^{QQ'}\tau^{S}{}_{Q'})\Gamma_{QSAB}\\
&= \nabla^{PQ}\Gamma_{PQAB} + \sqrt{2}\chi^{P}{}_{R}{}^{QR}\Gamma_{PQAB},
\end{aligned}$$

it follows, using the split of ∇_{AB}, that

$$\mathcal{P}\Gamma_{AB} + 2\mathcal{D}^{PQ}\Gamma_{(PQ)AB} + 2\sqrt{2}\chi^{P}{}_{R}{}^{QR}\Gamma_{PQAB} = 2F_{AB}(x). \qquad (13.27)$$

In view of its symmetries, the zero quantity $\Xi_{ABCC'EE'}$ is decomposed as

$$\Xi_{ABCC'EE'} = \Xi_{ABCE}\epsilon_{C'E'} + \Xi_{ABC'E'}\epsilon_{CE},$$

with

$$\Xi_{ABCE} \equiv \frac{1}{2}\Xi_{ABCQ'E}{}^{Q'}, \qquad \Xi_{ABC'E'} \equiv \frac{1}{2}\Xi_{ABQC'}{}^{Q}{}_{E'}.$$

In terms of space spinors the latter decomposition can be rewritten as

$$\Xi_{ABCDEF} = \Xi_{ABCE}\epsilon_{DF} + \Xi^{*}_{ABDF}\epsilon_{CE},$$

where

$$\Xi_{ABCDEF} \equiv \tau_{D}{}^{C'}\tau_{F}{}^{E'}\Xi_{ABCC'EE'},$$
$$\Xi_{ABDF} \equiv \tau_{D}{}^{C'}\tau_{F}{}^{E'}\Xi_{ABC'E'}, \qquad \Xi^{*}_{ABDF} \equiv \tau_{D}{}^{C'}\tau_{F}{}^{E'}\Xi_{ABC'E'}.$$

To expand Ξ_{ABCE} and Ξ^{*}_{ABDF} it is observed that

$$\begin{aligned}
\nabla_{(C}{}^{Q'}\Gamma_{D)Q'AB} &= -\nabla_{(C}{}^{Q'}\left(\Gamma_{D)SAB}\tau^{S}{}_{Q'}\right)\\
&= -\tau^{S}{}_{Q'}\nabla_{(C}{}^{Q'}\Gamma_{D)SAB} - \nabla_{(C}{}^{Q'}\tau^{S}{}_{|Q'|}\Gamma_{D)SAB}\\
&= \nabla_{(C}{}^{S}\Gamma_{D)SAB} + \sqrt{2}\chi_{(C|Q|}{}^{SQ}\Gamma_{D)SAB}\\
&= \frac{1}{2}\mathcal{P}\Gamma_{(CD)AB} + \mathcal{D}_{(C}{}^{S}\Gamma_{D)SAB} + \sqrt{2}\chi_{(C|Q|}{}^{SQ}\Gamma_{D)SAB}
\end{aligned}$$

and that

$$\begin{aligned}
\tau_{C}{}^{C'}\tau_{D}{}^{D'}\nabla^{P}{}_{(C'}\Gamma_{|P|D')AB} &= \nabla^{P}{}_{(C}\Gamma_{|P|D)AB}\\
&= -\frac{1}{2}\mathcal{P}\Gamma_{(CD)AB} + \mathcal{D}^{P}{}_{(C}\Gamma_{|P|D)AB}.
\end{aligned}$$

From the above expressions it follows that

$$\begin{aligned}
\Xi_{ABCD} = {} & \frac{1}{2}\mathcal{P}\Gamma_{(CD)AB} - \frac{1}{2}\mathcal{D}_{CD}\Gamma_{AB} + \frac{1}{2}\left(\mathcal{D}_{C}{}^{S}\Gamma_{(DS)AB} + \mathcal{D}_{D}{}^{S}\Gamma_{(CS)AB}\right)\\
& + \Gamma_{(C}{}^{PQ}{}_{|A|}\Gamma_{D)PQB} - \Sigma_{C}{}^{PQ}{}_{D}\Gamma_{PQAB} - R_{ABCD},\\
\Xi^{*}_{ABCD} = {} & -\frac{1}{2}\mathcal{P}\Gamma_{(CD)AB} + \frac{1}{2}\mathcal{D}_{CD}\Gamma_{AB} + \frac{1}{2}\left(\mathcal{D}^{P}{}_{C}\Gamma_{(PD)AB} + \mathcal{D}^{P}{}_{D}\Gamma_{(PC)AB}\right)\\
& + \Gamma^{P}{}_{(C}{}^{Q}{}_{|A}\Gamma_{P|D)QB} + \Sigma^{+}_{C}{}^{PQ}{}_{D}\Gamma_{PQAB} - R^{*}_{ABCD}.
\end{aligned}$$

Constraint equations are obtained from the combination

$$\Xi_{ABCD} + \Xi^{*}_{ABCD} = 0,$$

where

$$\begin{aligned}\Xi_{ABCD} + \Xi^{*}_{ABCD} = {} & \mathcal{D}^{P}{}_{C}\Gamma_{(PD)AB} + \mathcal{D}^{P}{}_{D}\Gamma_{(PC)AB} \\ & + \Gamma_{(C}{}^{PQ}{}_{|A|}\Gamma_{D)PQB} + \Gamma^{P}{}_{(C}{}^{Q}{}_{|A}\Gamma_{P|D)QB} \\ & + \Sigma^{+}_{C}{}^{PQ}{}_{D}\Gamma_{PQAB} - \Sigma_{C}{}^{PQ}{}_{D}\Gamma_{PQAB} \\ & - R_{ABCD} - R^{*}_{ABCD},\end{aligned}$$

while the required evolution equations arise from

$$\Xi_{ABCD} - \Xi^{*}_{ABCD} = 0,$$

with

$$\begin{aligned}\Xi_{ABCD} - \Xi^{*}_{ABCD} = {} & \mathcal{P}\Gamma_{(CD)AB} - \mathcal{D}_{CD}\Gamma_{AB} \\ & + \Gamma_{(C}{}^{PQ}{}_{|A|}\Gamma_{D)PQB} - \Gamma^{P}{}_{(C}{}^{Q}{}_{|A}\Gamma_{P|D)QB} \\ & - \Sigma^{+}_{C}{}^{PQ}{}_{D}\Gamma_{PQAB} - \Sigma_{C}{}^{PQ}{}_{D}\Gamma_{PQAB} \\ & - R_{ABCD} + R^{*}_{ABCD}.\end{aligned} \tag{13.28}$$

It can be verified that the system composed by (13.27) and (13.28) leads to a symmetric hyperbolic system for the independent components of Γ_{AB} and $\Gamma_{(CD)AB}$ – up to a suitable normalisation factor. A simple counting argument shows that the system consists of 12 equations, three coming from Equation (13.27) and nine from Equation (13.28). The characteristic polynomial of the system is given by

$$-64(\tau^{\mu}\xi_{\mu})^{6}(g^{\nu\lambda}\xi_{\nu}\xi_{\lambda})^{3}.$$

13.2.3 The conformal gauge source function

The third type of gauge source function to be considered arises from the analysis of the Cotton equation; see Equations (8.31e) and (8.35f). The starting point of the analysis is the spinorial counterpart, Equation (8.37a), associated with the zero quantity

$$\Delta_{CDBB'} \equiv \nabla_{(C}{}^{Q'}L_{D)Q'BB'} + \nabla^{Q}{}_{B'}\Xi\phi_{CDBQ} + \Xi T_{CDBB'}.$$

To deduce a symmetric hyperbolic system from this equation one needs to complete the symmetrised derivative $\nabla_{(C}{}^{Q'}L_{D)Q'BB'}$ with the divergence $\nabla^{QQ'}L_{QQ'BB'}$. Information about this derivative is provided by the contracted Bianchi identity for the Schouten tensor; compare Equation (8.17). In spinorial notation one has

$$\nabla^{QQ'}L_{QQ'BB'} = \frac{1}{6}\nabla_{BB'}R. \tag{13.29}$$

Thus, using

$$\nabla_{(C}{}^{Q'}L_{D)Q'BB'} = \nabla_{C}{}^{Q'}L_{DQ'BB'} + \frac{1}{2}\epsilon_{CD}\nabla^{QQ'}L_{QQ'BB'},$$

one can rewrite the zero quantity $\Delta_{CDBB'}$ as

$$\begin{aligned}\Delta_{CDBB'} = \nabla_C{}^{Q'} L_{DQ'BB'} + \frac{1}{12}\epsilon_{CD}\nabla_{BB'}R \\ + \Sigma^{Q}{}_{B'}\phi_{CDBQ} + \Xi T_{CDBB'},\end{aligned} \tag{13.30}$$

where $\Sigma_{AA'} \equiv \nabla_{AA'}\Xi$.

As discussed in Chapter 8, the conformal field equations impose no differential condition on the unphysical Ricci scalar R. Accordingly, R can be specified freely as a function of the coordinates. Thus, if the reduced rescaled Cotton spinor $T_{CDBB'}$ can be rewritten so that it does not explicitly contain derivatives of the matter fields, one can deduce a symmetric hyperbolic system for the components of $L_{AA'BB'}$ from Equation (13.30).

Geometric interpretation

The particular choice of the Ricci scalar fixes the conformal gauge freedom. Thus, it is natural to call $R(x)$ the ***conformal gauge source function***. Given a particular choice of $R(x)$, the transformation law for the Ricci scalar implies a wave equation for the conformal factor realising the prescribed Ricci scalar; see Equation (8.30). This equation can always be solved locally if initial data on a fiduciary hypersurface $\mathcal{S}_\star$ is provided – namely, the values of the conformal factor and its normal derivative on the hypersurface. Conversely, given an unphysical spacetime $(\mathcal{M}, \boldsymbol{g})$ and a conformal factor Ξ linking it to a physical spacetime $(\tilde{\mathcal{M}}, \tilde{\boldsymbol{g}})$ via the standard relation $\boldsymbol{g} = \Xi^2\tilde{\boldsymbol{g}}$, one can compute the corresponding conformal gauge source function $R(x)$.

Space spinor decomposition of the equation for the components of the Schouten tensor

The space spinor decomposition of the equations for the Schouten tensor is based on the expression

$$L_{AA'CC'} = \Phi_{AA'CC'} + \frac{1}{24}\epsilon_{AC}\epsilon_{A'C'}R(x), \tag{13.31}$$

where $\Phi_{AA'CC'}$ denotes the spinorial counterpart of the trace-free part of the Ricci tensor; see Section 3.2.4. The space spinor counterpart of $L_{AA'CC'}$ is defined as

$$\begin{aligned}L_{ABCD} &\equiv \tau_B{}^{A'}\tau_D{}^{C'}L_{AA'CC'}, \\ &= \Phi_{ABCD} + \frac{1}{24}\epsilon_{AC}\epsilon_{BD}R(x),\end{aligned}$$

where $\Phi_{ABCD} \equiv \tau_B{}^{A'}\tau_D{}^{C'}\Phi_{AA'CC'}$ so that

$$\Phi_{ABCD} = \Phi_{CBAD} = \Phi_{ADCB},$$

as a consequence of the symmetries of $\Phi_{AA'CC'}$; see Equation (3.44). A spinor with these symmetries can be decomposed as

$$\Phi_{ABCD} = \Phi_{(ABCD)} + \frac{1}{2}\left(\epsilon_{A(B}\Phi_{D)C} + \epsilon_{C(B}\Phi_{D)A}\right) + \frac{1}{3}\Phi h_{ACBD}, \quad (13.32)$$

where

$$\Phi_{AB} \equiv \Phi_{(AB)Q}{}^{Q}, \qquad \Phi \equiv \Phi_{ABCD}h^{ACBD}.$$

Now, using that

$$\nabla_{A}{}^{Q'}L_{BQ'CC'} = \nabla_{(A}{}^{Q'}L_{B)Q'CC'} - \frac{1}{2}\epsilon_{AB}\nabla^{QQ'}L_{QQ'CC'},$$

together with the contracted Bianchi identity (13.29) one can rewrite the zero quantity $\Delta_{ABCC'}$ as

$$\Delta_{ABCC'} = \nabla_{A}{}^{Q'}L_{BQ'CC'} + \frac{1}{12}\epsilon_{AB}\nabla_{CC'}R(x) + \Sigma^{Q}{}_{C'}\phi_{ABCQ} + \Xi T_{ABCC'}. \quad (13.33)$$

Defining

$$\Delta_{ABCD} \equiv \tau_{D}{}^{C'}\Delta_{ABCC'},$$

a calculation using (13.33) together with the definitions of the spinors L_{ABCD} and χ_{ABCD}, yields

$$\begin{aligned}\Delta_{ABCD} = &\nabla_{A}{}^{Q}L_{BQCD} + \sqrt{2}\chi_{AP}{}^{QP}L_{BQCD} - \sqrt{2}\chi_{A}{}^{QP}{}_{D}L_{BQCP}\\ &+ \frac{1}{2}\epsilon_{AB}\nabla_{CD}R(x) + \Sigma^{Q}{}_{D}\phi_{ABCQ} + \Xi T_{ABCD},\end{aligned}$$

where $\Sigma_{AB} \equiv \tau_{B}{}^{Q'}\Sigma_{AQ'}$. Thus, using the decomposition of the operator ∇_{AB} one obtains

$$\begin{aligned}\Delta_{ABCD} = &\frac{1}{2}\mathcal{P}L_{BACD} + \mathcal{D}_{A}{}^{Q}L_{BQCD} + \sqrt{2}\chi_{AP}{}^{QP}L_{BQCD}\\ &- \sqrt{2}\chi_{A}{}^{PQ}{}_{D}L_{BPCQ} + \frac{1}{2}\epsilon_{AB}\nabla_{CD}R(x) + \Sigma^{Q}{}_{D}\phi_{ABCQ} + \Xi T_{ABCD}.\end{aligned}$$

To extract the full information of $\Delta_{ABCC'}$ one also needs to consider

$$\Delta^{+}_{ABCD} \equiv \tau_{A}{}^{P'}\tau_{B}{}^{Q'}\tau_{C}{}^{R'}\tau_{D}{}^{S'}\bar{\Delta}_{P'Q'R'S'}.$$

Proceeding as with Δ_{ABCD} one finds that

$$\begin{aligned}\Delta^{+}_{ABCD} = &\frac{1}{2}\mathcal{P}L_{ABCD} - \mathcal{D}^{Q}{}_{A}L_{QBDC} + \sqrt{2}\chi^{Q}{}_{A}{}^{R}{}_{B}L_{QRDC}\\ &+ \sqrt{2}\chi^{Q}{}_{A}{}^{P}{}_{C}L_{QBDP} - \frac{1}{2}\epsilon_{AB}\nabla_{CD}R(x) + \Sigma^{+R}{}_{D}\phi^{+}_{ABCR} + \Xi T^{+}_{ABCD}.\end{aligned}$$

Given the above expressions for Δ_{ABCD} and Δ^+_{ABCD}, suitable symmetric hyperbolic evolution equations for the independent components of the fields $\Phi_{(ABCD)}$, Φ_{AB} and Φ can be found from the combinations

$$\Delta_{(ABCD)} + \Delta^+_{(ABCD)} = 0, \tag{13.34a}$$
$$\Delta^{+}_{Q}{}^{Q}{}_{(CD)} - \Delta_{Q}{}^{Q}{}_{(CD)} = 0, \tag{13.34b}$$
$$\Delta_{Q}{}^{Q}{}_{P}{}^{P} + \Delta^{+}_{Q}{}^{Q}{}_{P}{}^{P} = 0, \tag{13.34c}$$

while constraint equations arise from

$$\Delta_{ABCD} - \Delta^+_{ABCD} = 0,$$
$$\Delta^{+}_{Q}{}^{Q}{}_{(CD)} + \Delta_{Q}{}^{Q}{}_{(CD)} = 0,$$
$$\Delta_{Q}{}^{Q}{}_{P}{}^{P} - \Delta^{+}_{Q}{}^{Q}{}_{P}{}^{P} = 0.$$

The principal parts of Equations (13.34a)–(13.34c) are given, respectively, by

$$\mathcal{P}\Phi_{(ABCD)} - \mathcal{D}_{(AB}\Phi_{CD)},$$
$$\mathcal{P}\Phi_{AB} + 2\mathcal{D}^{PQ}\Phi_{PQAB} - \frac{1}{3}\mathcal{D}_{AB}\Phi,$$
$$\mathcal{P}\Phi + \mathcal{D}^{PQ}\Phi_{PQ}.$$

The above expressions imply a symmetric hyperbolic system for the independent components of the fields $\Phi_{(ABCD)}$, Φ_{AB} and Φ. The explicit form of this system will not be required in the subsequent discussion but can be readily computed.

13.2.4 The hyperbolic reduction of the Bianchi equation

This section discusses the hyperbolic reduction of the spinorial Bianchi identity. This procedure leads to evolution equations for the components of the rescaled Weyl spinor and is completely analogous to that for the Maxwell equations; see Section 13.1.1. In particular, no gauge source functions are required for this subsystem.

The spinorial Bianchi equation is encoded in the zero quantity

$$\Lambda_{A'BCD} \equiv \nabla^{Q}{}_{A'}\phi_{BCDQ} + T_{CDBA'}.$$

In the following it will be convenient to work with a space spinor version of this zero quantity, namely,

$$\Lambda_{ABCD} \equiv \nabla^{Q}{}_{A}\phi_{BCDQ} + T_{CDBA}, \qquad T_{CDBA} \equiv \tau_{A}{}^{A'}T_{CDBA'}.$$

Using the decomposition (13.5) one can compute that

$$\Lambda_{ABCD} = -\frac{1}{2}\mathcal{P}\phi_{ABCD} + \mathcal{D}^{Q}{}_{A}\phi_{BCDQ} + T_{CDBA}. \tag{13.35}$$

Suitable evolution equations are obtained from the above expression by considering

$$-2\Lambda_{(\boldsymbol{ABCD})} = \mathcal{P}\phi_{\boldsymbol{ABCD}} - 2\mathcal{D}^{\boldsymbol{Q}}{}_{(\boldsymbol{A}}\phi_{\boldsymbol{BCD})\boldsymbol{Q}} + T_{(\boldsymbol{ABCD})} = 0. \qquad (13.36)$$

In what follows, this system of evolution equations will be known as the ***standard system***. It gives rise to five independent equations for the five independent components of $\phi_{\boldsymbol{ABCD}}$. Contracting the indices $\boldsymbol{A}$ and $\boldsymbol{B}$ in Equation (13.35) one obtains

$$\Lambda_{\boldsymbol{CD}} \equiv \Lambda^{\boldsymbol{Q}}{}_{\boldsymbol{QCD}} = \mathcal{D}^{\boldsymbol{PQ}}\phi_{\boldsymbol{PQCD}} + T_{\boldsymbol{CDQ}}{}^{\boldsymbol{Q}} = 0,$$

the so-called ***Bianchi constraints***. As in the case of the other constraint equations discussed in the previous sections, the Bianchi constraints may contain derivatives in the time direction.

The hyperbolicity of the standard system

The overall structure of Equation (13.36) suggests that it should imply a symmetric hyperbolic system. In analogy to the Maxwell equations, one considers a slightly modified version of Equation (13.36) given by

$$-2\left(\frac{4}{\boldsymbol{A}+\boldsymbol{B}+\boldsymbol{C}+\boldsymbol{D}}\right)\Lambda_{(\boldsymbol{ABCD})} = 0.$$

The principal part of this equation can be written in matricial form as

$$\mathbf{A}^\mu\partial_\mu\phi \equiv \begin{pmatrix} \tau^\mu + 2e_{\boldsymbol{01}}{}^\mu & -2e_{\boldsymbol{00}}{}^\mu & 0 & 0 & 0 \\ 2e_{\boldsymbol{11}}{}^\mu & 4\tau^\mu + 4e_{\boldsymbol{01}}{}^\mu & -6e_{\boldsymbol{00}}{}^\mu & 0 & 0 \\ 0 & 6e_{\boldsymbol{11}}{}^\mu & 6\tau^\mu & -6e_{\boldsymbol{00}}{}^\mu & 0 \\ 0 & 0 & 6e_{\boldsymbol{11}}{}^\mu & 4\tau^\mu - 4e_{\boldsymbol{01}}{}^\mu & -2e_{\boldsymbol{00}}{}^\mu \\ 0 & 0 & 0 & 2e_{\boldsymbol{11}}{}^\mu & \tau^\mu - 2e_{\boldsymbol{01}}{}^\mu \end{pmatrix}$$

$$\times\,\partial_\mu \begin{pmatrix} \phi_0 \\ \phi_1 \\ \phi_2 \\ \phi_3 \\ \phi_4 \end{pmatrix},$$

with

$$\phi_0 \equiv \phi_{\boldsymbol{0000}}, \quad \phi_1 \equiv \phi_{\boldsymbol{0001}}, \quad \phi_2 \equiv \phi_{\boldsymbol{0011}}, \quad \phi_3 \equiv \phi_{\boldsymbol{0111}}, \quad \phi_4 \equiv \phi_{\boldsymbol{1111}}.$$

Using the reality conditions satisfied by the vectors $\boldsymbol{e_{AB}}$, it follows that the matrices of the system are Hermitian. Moreover, one has that $\mathbf{A}^\mu\tau_\mu$ is positive definite. Thus, the standard evolution system implies a symmetric hyperbolic system for the independent components of $\phi_{\boldsymbol{ABCD}}$. The characteristic matrix of the system is given by

$$\det(\mathbf{A}^\mu \xi_\mu) = 36(\tau^\mu \xi_\mu)(g^{\nu\lambda}\xi_\nu\xi_\lambda)\left(\tau^\rho\tau^\sigma + \frac{2}{3}g^{\rho\sigma}\right)\xi_\rho\xi_\sigma.$$

Thus, $\boldsymbol{g}$-null hypersurfaces are characteristics of the standard system.

13.2.5 The hyperbolic reduction of the equations for the conformal factor and its concomitants

Finally, one requires evolution equations for the conformal factor Ξ and its concomitants $\Sigma_{AA'}$ and s. The relevant zero quantities are given by

$$Q_{AA'} \equiv \Sigma_{AA'} - \nabla_{AA'}\Xi, \tag{13.37a}$$

$$Z_{AA'BB'} \equiv \nabla_{AA'}\Sigma_{BB'} + \Xi L_{AA'BB'} - s\epsilon_{AB}\epsilon_{A'B'} - \frac{1}{2}\Xi^3 T_{AA'BB'}, \tag{13.37b}$$

$$Z_{AA'} \equiv \nabla_{AA'}s + L_{AA'CC'}\nabla^{CC'}\Xi - \frac{1}{2}\Xi^2\nabla^{CC'}\Xi T_{AA'CC'}. \tag{13.37c}$$

Their space spinor counterparts are defined by

$$Q_{AB} \equiv \tau_B{}^{A'}Q_{AA'}, \qquad Z_{ABCD} \equiv \tau_B{}^{A'}\tau_D{}^{C'}Z_{AA'CC'}, \qquad Z_{AB} \equiv \tau_B{}^{A'}Z_{AA'}.$$

It is also convenient to make use of the split

$$\Sigma_{AB} \equiv \tau_B{}^{A'}\Sigma_{AA'} = \frac{1}{2}\epsilon_{AB}\Sigma + \Sigma_{(AB)}, \qquad \Sigma \equiv \Sigma_Q{}^Q.$$

From the condition $Q_{AB} = 0$ one obtains the equations

$$\mathcal{P}\Xi = \Sigma, \qquad \mathcal{D}_{AB}\Sigma = \Sigma_{(AB)},$$

which are, respectively, an evolution equation for Ξ and a constraint equation. Next, using the identity

$$\tau_B{}^{A'}\tau_D{}^{C'}\nabla_{AA'}\Sigma_{CC'} = \nabla_{AB}(\tau_D{}^{C'}\Sigma_{CC'}) - \sqrt{2}\Sigma_{CP}\chi_{AB}{}^P{}_D$$

and the split of ∇_{AB} it follows that

$$\begin{aligned} Z_{ABCD} = {} & \frac{1}{4}\epsilon_{AB}\epsilon_{CD}\mathcal{P}\Sigma + \frac{1}{2}\epsilon_{AB}\mathcal{P}\Sigma_{(CD)} + \frac{1}{2}\epsilon_{CD}\mathcal{D}_{AB}\Sigma + \mathcal{D}_{AB}\Sigma_{(CD)} \\ & + \frac{1}{\sqrt{2}}\chi_{ABCD} - \sqrt{2}\Sigma_{CP}\chi_{AB}{}^P{}_D \\ & + \Xi L_{ABCD} - s\epsilon_{AB}\epsilon_{CD} - \frac{1}{2}T_{ABCD}. \end{aligned}$$

Evolution equations for Σ and $\Sigma_{(AB)}$ are obtained from

$$2Z_{AB}{}^{AB} = 0, \qquad Z_A{}^A{}_{(CD)} = 0,$$

where

$$2Z_{AB}{}^{AB} = \mathcal{P}\Sigma + \sqrt{2}\chi_{AB}{}^{AB}\Sigma - 2\sqrt{2}\chi^{ABP}{}_{B}\Sigma_{(AP)} + \Xi L_{AB}{}^{AB} - 4s,$$

$$Z_{A}{}^{A}{}_{(CD)} = \mathcal{P}\Sigma_{(CD)} + \frac{1}{\sqrt{2}}\chi_{A}{}^{A}{}_{(CD)} - \sqrt{2}\Sigma_{(C|P}\chi_{A}{}^{AP}{}_{|D)}$$
$$+ \Xi L_{A}{}^{A}{}_{(CD)} - \frac{1}{2}\Xi^{3}T_{A}{}^{A}{}_{(CD)}.$$

The corresponding constraints arise from

$$Z_{(ABCD)} = 0, \qquad Z_{(AB)C}{}^{C} = 0,$$

with

$$Z_{(ABCD)} = \mathcal{D}_{(AB}\Sigma_{CD)} + \frac{1}{\sqrt{2}}\chi_{(ABCD)}\Sigma - \sqrt{2}\Sigma_{(C|P|}\chi_{|AB}{}^{P}{}_{D)}$$
$$+ \Xi L_{(ABCD)} - \frac{1}{2}\Xi^{3}T_{(ABCD)},$$

$$Z_{(AB)C}{}^{C} = \mathcal{D}_{AB}\Sigma + \frac{1}{\sqrt{2}}\chi_{(AB)C}{}^{C}\Sigma - \sqrt{2}\Sigma_{PQ}\chi_{(AB)}{}^{PQ}$$
$$+ \Xi L_{(AB)C}{}^{C} - \frac{1}{2}\Xi^{3}T_{(AB)C}{}^{C}.$$

Finally, similar calculations lead to the expression

$$Z_{AB} = \frac{1}{2}\epsilon_{AB}\mathcal{P}s + \mathcal{D}_{AB}s - \frac{1}{2}L_{ABC}{}^{C}\Sigma + L_{ABCD}\Sigma^{CD}$$
$$+ \frac{1}{4}\Xi^{2}T_{ABC}{}^{C} - \frac{1}{2}\Xi^{2}\Sigma^{CD}T_{ABCD}.$$

The evolution and constraint equations for s are then given, respectively, by

$$Z_{A}{}^{A} = 0, \qquad Z_{(AB)} = 0,$$

with

$$Z_{A}{}^{A} = \mathcal{P}s - \frac{1}{2}L_{A}{}^{A}{}_{C}{}^{C} + L_{A}{}^{A}{}_{CD}\Sigma^{CD} + \frac{1}{4}\Xi^{2}\Sigma T_{A}{}^{A}{}_{C}{}^{C}$$
$$- \frac{1}{2}\Xi^{2}\Sigma^{CD}T_{A}{}^{A}{}_{CD},$$

$$Z_{(AB)} = \mathcal{D}_{AB}s - \frac{1}{2}L_{(AB)C}{}^{C}\Sigma + L_{(AB)CD}\Sigma^{CD} + \frac{1}{4}\Xi^{2}\Sigma T_{(AB)C}{}^{C}$$
$$- \frac{1}{2}\Xi^{2}\Sigma^{CD}T_{(AB)CD}.$$

Remark. It should be observed that all the evolution equations obtained in this section are transport equations – that is, they involve only the directional derivative $\mathcal{P}$. Accordingly the characteristic polynomial of each of them is just $\tau^{\mu}\xi_{\mu}$.

13.3 The subsidiary equations for the standard conformal field equations

After having discussed a set of evolution equations implied by the conformal field Equations (8.38a) and (8.38b), one is now in the position of analysing the construction of the associated subsidiary system. The subsidiary equations constitute a system of evolution equations for the zero quantities encoding the conformal field equations. To prove the ***propagation of the constraints*** it is necessary that these subsidiary evolution equations are homogeneous in the various zero quantities. If this is the case, then Corollary 12.1 implies a unique vanishing solution to the subsidiary equations if the zero quantities are zero initially. The construction of the subsidiary system involves lengthy computations, parts of which are best carried out with spinorial expressions, while others are more conveniently described in tensorial terms. The basic strategy behind the analysis can be understood by first discussing some model equations.

General setup

The general setup for the construction of the subsidiary equations for the conformal field equations is similar to the one for the construction of the evolution equations: one works in an open subset $\mathcal{U} \subset \mathcal{M}$ of the unphysical spacetime manifold; vector and spinor bases are introduced in a similar manner. The key difference lies in the fact that the covariant derivative $\boldsymbol{\nabla}$ is, a priori, not assumed to be the Levi-Civita connection of the metric $\boldsymbol{g}$. Thus, when considering the commutator of covariant derivatives, one has to make use of the general expression involving a non-vanishing torsion tensor. This is because the torsion tensor is, in itself, a zero quantity of the conformal field equations. On similar grounds, one cannot regard the algebraic and geometric curvatures as being equal to each other.

13.3.1 Hyperbolic reduction of model equations

The construction of a system of subsidiary equations for the conformal Einstein equations leads to spinorial equations whose tensorial counterparts are of one of the following forms

$$\nabla_{[\boldsymbol{a}} M_{\boldsymbol{b}]\mathcal{K}} = N_{\boldsymbol{a}\boldsymbol{b}\mathcal{K}}, \tag{13.38a}$$

$$\nabla_{[\boldsymbol{a}} P_{\boldsymbol{b}\boldsymbol{c}]\mathcal{L}} = Q_{\boldsymbol{a}\boldsymbol{b}\boldsymbol{c}\mathcal{L}}, \tag{13.38b}$$

where $M_{\boldsymbol{a}\mathcal{K}}$ and $P_{\boldsymbol{a}\boldsymbol{b}\mathcal{L}}$ are some zero quantities with

$$N_{\boldsymbol{a}\boldsymbol{b}\mathcal{K}} = N_{[\boldsymbol{a}\boldsymbol{b}]\mathcal{K}}, \qquad P_{\boldsymbol{a}\boldsymbol{b}\mathcal{L}} = P_{[\boldsymbol{a}\boldsymbol{b}]\mathcal{L}}, \qquad Q_{\boldsymbol{a}\boldsymbol{b}\boldsymbol{c}\mathcal{L}} = Q_{[\boldsymbol{a}\boldsymbol{b}\boldsymbol{c}]\mathcal{L}},$$

and $_{\mathcal{K}}$ and $_{\mathcal{L}}$ denote an arbitrary string of indices.

Equations (13.38a) and (13.38b) arise from the following observations concerning differential forms; see the Appendix to this chapter for a brief discussion on this and related notions. The fields $M_{a\mathcal{K}}$ and $P_{ab\mathcal{K}}$ can be regarded as the components, respectively, of the *1-form* and *2-form*

$$\boldsymbol{M}_{\mathcal{K}} \equiv M_{a\mathcal{K}}\,\boldsymbol{\omega}^a, \qquad \boldsymbol{P}_{\mathcal{L}} \equiv P_{ab\mathcal{L}}\,\boldsymbol{\omega}^a \wedge \boldsymbol{\omega}^b.$$

Accordingly, Equations (13.38a) and (13.38b) can be written as

$$\mathbf{d}\boldsymbol{M}_{\mathcal{K}} = N_{ab\mathcal{K}}\,\boldsymbol{\omega}^a \wedge \boldsymbol{\omega}^b, \qquad \mathbf{d}\boldsymbol{P}_{\mathcal{L}} = Q_{abc\mathcal{L}}\,\boldsymbol{\omega}^a \wedge \boldsymbol{\omega}^b \wedge \boldsymbol{\omega}^c.$$

If $\boldsymbol{\tau}$ denotes a timelike vector field, then the Lie derivatives of $\boldsymbol{M}_{\mathcal{K}}$ and $\boldsymbol{P}_{\mathcal{L}}$ along the direction of $\boldsymbol{\tau}$ are given by the so-called ***Cartan's formula***

$$\pounds_{\boldsymbol{\tau}}\boldsymbol{M}_{\mathcal{K}} = i_{\boldsymbol{\tau}}\mathbf{d}\boldsymbol{M}_{\mathcal{K}} + \mathbf{d}(i_{\boldsymbol{\tau}}\boldsymbol{M}_{\mathcal{K}}), \qquad \pounds_{\boldsymbol{\tau}}\boldsymbol{P}_{\mathcal{L}} = i_{\boldsymbol{\tau}}\mathbf{d}\boldsymbol{P}_{\mathcal{L}} + \mathbf{d}(i_{\boldsymbol{\tau}}\boldsymbol{P}_{\mathcal{L}}),$$

where $i_{\boldsymbol{\tau}}$ denotes the operation of contraction between the vector $\boldsymbol{\tau}$ and a differential form; see Frankel (2003). In terms of this notation the evolution equations are given, respectively, by

$$i_{\boldsymbol{\tau}}\boldsymbol{M}_{\mathcal{K}} = 0, \qquad i_{\boldsymbol{\tau}}\boldsymbol{P}_{\mathcal{L}} = 0,$$

so that

$$\pounds_{\boldsymbol{\tau}}\boldsymbol{M}_{\mathcal{K}} = i_{\boldsymbol{\tau}}\mathbf{d}\boldsymbol{M}_{\mathcal{K}}, \qquad \pounds_{\boldsymbol{\tau}}\boldsymbol{P}_{\mathcal{L}} = i_{\boldsymbol{\tau}}\mathbf{d}\boldsymbol{P}_{\mathcal{L}}.$$

The latter can be read as suitable evolution equations for the zero quantities $M_{a\mathcal{K}}$ and $P_{ab\mathcal{L}}$. Their frame component version is given by

$$\nabla_{[\mathbf{0}}M_{\mathbf{b}]\mathcal{K}} = N_{\mathbf{0b}\mathcal{K}}, \qquad \nabla_{[\mathbf{0}}P_{\mathbf{bc}]\mathcal{L}} = Q_{\mathbf{0bc}\mathcal{L}}.$$

Detailed analysis of the first model equation

The spinorial analogue of Equation (13.38a) is given by

$$\nabla_{\boldsymbol{AA'}}M_{\boldsymbol{BB'}\mathcal{K}} - \nabla_{\boldsymbol{BB'}}M_{\boldsymbol{AA'}\mathcal{K}} = 2N_{\boldsymbol{AA'BB'}\mathcal{K}}.$$

Exploiting the antisymmetry one obtains the equivalent expression

$$\nabla_{(\boldsymbol{A}}{}^{\boldsymbol{Q'}}M_{\boldsymbol{B})\boldsymbol{Q'}\mathcal{K}} = N_{\boldsymbol{A}}{}^{\boldsymbol{Q'}}{}_{\boldsymbol{BQ'}\mathcal{K}}, \qquad N_{\boldsymbol{A}}{}^{\boldsymbol{Q'}}{}_{\boldsymbol{BQ'}\mathcal{K}} = N_{\boldsymbol{B}}{}^{\boldsymbol{Q'}}{}_{\boldsymbol{AQ'}\mathcal{K}}.$$

Defining the space spinor counterpart $M_{\boldsymbol{AB}\mathcal{K}} \equiv \tau_{\boldsymbol{B}}{}^{\boldsymbol{A'}}M_{\boldsymbol{AA'}\mathcal{K}}$ and using the definition of the spinor $\chi_{\boldsymbol{ABCD}}$ together with the decomposition (13.5) of $\nabla_{\boldsymbol{AB}}$ one obtains the expression

$$\mathcal{P}M_{(\boldsymbol{AB})\mathcal{K}} + 2\mathcal{D}_{(\boldsymbol{A}}{}^{\boldsymbol{P}}M_{\boldsymbol{B})\boldsymbol{P}\mathcal{K}} + 2\sqrt{2}\chi_{(\boldsymbol{A}|\boldsymbol{Q}|}{}^{\boldsymbol{PQ}}M_{\boldsymbol{B})\boldsymbol{P}\mathcal{K}} = N_{\boldsymbol{A}}{}^{\boldsymbol{Q'}}{}_{\boldsymbol{BQ'}\mathcal{K}}.$$

Finally, assuming that the evolution equations implied by the zero quantity $M_{\boldsymbol{AA'}\mathcal{K}}$ are given by $M_{\boldsymbol{Q}}{}^{\boldsymbol{Q}}{}_{\mathcal{K}} = 0$, it follows that $M_{\boldsymbol{BP}\mathcal{K}} = M_{(\boldsymbol{BP})\mathcal{K}}$ and, moreover, that

$$\mathcal{P}M_{(\boldsymbol{AB})\mathcal{K}} + \mathcal{D}_{\boldsymbol{A}}{}^{\boldsymbol{P}}M_{(\boldsymbol{BP})\mathcal{K}} + \mathcal{D}_{\boldsymbol{B}}{}^{\boldsymbol{P}}M_{(\boldsymbol{AP})\mathcal{K}} + 2\sqrt{2}\chi_{(\boldsymbol{A}|\boldsymbol{Q}|}{}^{\boldsymbol{PQ}}M_{\boldsymbol{B})\boldsymbol{P}\mathcal{K}} = N_{\boldsymbol{A}}{}^{\boldsymbol{Q'}}{}_{\boldsymbol{BQ'}\mathcal{K}}.$$

This last expression is a suitable evolution equation for $M_{(AB)\mathcal{K}}$ if $N_A{}^{Q'}{}_{BQ'\mathcal{K}}$ can be expressed as a linear combination of zero quantities. This computation depends on the particular structure of the conformal equation under consideration.

Detailed analysis of the second model equation

In what follows, let $P_{AA'BB'\mathcal{L}}$ and $Q_{AA'BB'CC'\mathcal{L}}$ denote, respectively, the spinorial counterparts of the fields $P_{ab\mathcal{L}}$ and $Q_{abc\mathcal{L}}$. The spinorial counterpart of Equation (13.38b) can be conveniently written using the spinorial counterpart of the volume form as

$$\epsilon^{AA'BB'CC'}{}_{DD'}\nabla_{AA'}P_{BB'CC'\mathcal{L}} = \epsilon^{AA'BB'CC'}{}_{DD'}Q_{AA'BB'CC'\mathcal{L}}. \quad (13.39)$$

A convenient way of obtaining the space spinor version of this last equation is to consider, alternatively, the expression

$$\epsilon^{EFCDGH}{}_{AB}\nabla_{EF}P_{CDGH\mathcal{L}},$$

where, following standard conventions, one defines

$$P_{CDGH\mathcal{L}} \equiv \tau_D{}^{C'}\tau_H{}^{G'}P_{CC'GG'\mathcal{L}},$$
$$\epsilon_{EFCDGHAB} \equiv \tau_F{}^{F'}\tau_D{}^{D'}\tau_H{}^{H'}\tau_B{}^{B'}\epsilon_{EF'CD'GH'AB'}.$$

A short computation using the expression of the volume form in terms of ϵ-spinors yields

$$\epsilon_{EFCDGHAB} = \mathrm{i}(\epsilon_{EG}\epsilon_{CA}\epsilon_{FB}\epsilon_{DH} - \epsilon_{EA}\epsilon_{CG}\epsilon_{FH}\epsilon_{DB}).$$

Now, exploiting the symmetries of $P_{CDGH\mathcal{L}}$ one can write

$$P_{CDGH\mathcal{L}} = P_{CG\mathcal{L}}\epsilon_{DH} + P^*_{DH\mathcal{L}}\epsilon_{CG},$$

where

$$P_{CG\mathcal{L}} \equiv \frac{1}{2}P_{CQG}{}^{Q}{}_{\mathcal{L}}, \qquad P^*_{DH\mathcal{L}} \equiv \frac{1}{2}P_{QD}{}^{Q}{}_{H\mathcal{L}}.$$

A calculation shows that the above expressions lead to

$$\begin{aligned}\epsilon^{EFCDGH}{}_{AB}\nabla_{EF}P_{CDGH\mathcal{L}} &\\ &= 2\mathrm{i}\left(\nabla_A{}^Q P^*_{BQ\mathcal{L}} - \nabla^Q{}_B P_{AQ\mathcal{L}}\right)\\ &= \mathrm{i}\mathcal{P}\left(P^*_{AB\mathcal{L}} + P_{AB\mathcal{L}}\right) + 2\mathrm{i}\mathcal{D}^Q{}_A P^*_{BQ\mathcal{L}} - 2\mathrm{i}\mathcal{D}^Q{}_B P_{AQ\mathcal{L}}.\end{aligned}$$

If the evolution equations associated with the zero quantity $P_{ABCD\mathcal{L}}$ are given by the condition

$$P_{AB\mathcal{L}} - P^*_{AB\mathcal{L}} = 0,$$

it follows that

$$\mathcal{P}P_{AB\mathcal{L}} = -\frac{\mathrm{i}}{2}\epsilon^{EFCDGH}{}_{(AB)}\nabla_{EF}P_{CDGH\mathcal{L}}.$$

It can be verified that the expression one obtains by working directly with the left-hand side of Equation (13.39) differs from the above expression by homogeneous terms involving $P_{CDGH\mathcal{L}}$ and χ_{ABCD}. To complete the construction of a suitable subsidiary equation for $P_{ABCD\mathcal{L}}$ it is necessary to show that the right-hand side of Equation (13.39) can be expressed as a linear combination of zero quantities – this computation is specific to each zero quantity.

13.3.2 The subsidiary equations for the equations governing the conformal factor and its concomitants

The zero quantities $Q_{AA'}$, $Z_{AA'BB'}$ and $Z_{AA'}$ – see Equations (13.37a)–(13.37c) – lead to subsidiary equations which fall into the class described by the model Equation (13.38a). Accordingly, one will have suitable subsidiary evolution equations for the zero quantities $Q_{AA'}$, $Z_{AA'BB'}$ and $Z_{AA'}$ if the derivatives

$$\nabla_{(A}{}^{Q'}Q_{B)Q'}, \qquad \nabla_{(A}{}^{Q'}Z_{B)Q'CC'}, \qquad \nabla_{(A}{}^{Q'}Z_{B)Q'},$$

can be expressed as linear combinations of other zero quantities.

The subsidiary equation for $Q_{AA'}$

A direct computation using the definition of $Q_{AA'}$ shows that

$$\nabla_{(A}{}^{Q'}Q_{B)Q'} = \nabla_{(A}{}^{Q'}\Sigma_{B)Q'} - \Sigma_{A}{}^{QQ'}{}_{B}\Sigma_{QQ'},$$

where the definition of the torsion spinor – see Equation (8.35a) – has been used to write

$$\nabla_{(A}{}^{Q'}\nabla_{B)Q'}\Xi = \Sigma_{A}{}^{QQ'}{}_{B}\Sigma_{QQ'}, \qquad \Sigma_{A}{}^{QQ'}{}_{B} \equiv \frac{1}{2}\Sigma_{A}{}^{P'QQ'}{}_{BP'}.$$

Finally, using the definition of the zero quantity $Z_{AA'BB'}$ one can eliminate the term $\nabla_{(A}{}^{Q'}\Sigma_{B)Q'}$. Observing that $L_{(A}{}^{Q'}{}_{B)Q'} = T_{(A}{}^{Q'}{}_{B)Q'} = 0$ – as these are the spinorial counterparts of symmetric rank-2 tensors – one finds

$$\nabla_{(A}{}^{Q'}\Sigma_{B)Q'} = Z_{(A}{}^{Q'}{}_{B)Q'},$$

so that one concludes that

$$\nabla_{(A}{}^{Q'}Q_{B)Q'} = Z_{(A}{}^{Q'}{}_{B)Q'} - \Sigma_{A}{}^{QQ'}{}_{B}\Sigma_{QQ'},$$

which is a linear combination of zero quantities as required.

The subsidiary equation for $Z_{AA'BB'}$

A direct computation starting from the definition of $Z_{AA'BB'}$ yields the expression

$$\nabla_{(A}{}^{Q'}Z_{B)Q'CC'} = \nabla_{(A}{}^{Q'}\nabla_{B)Q'}\Sigma_{CC'} + \Sigma_{(A}{}^{Q'}L_{B)Q'CC'} + \Xi\nabla_{(A}{}^{Q'}L_{B)Q'CC'} + \epsilon_{C(A}\nabla_{B)C'}s - \frac{3}{2}\Xi^2\Sigma_{(A}{}^{Q'}T_{B)Q'CC'} - \Xi^3\nabla_{(A}{}^{Q'}T_{B)Q'CC'}.$$

Using the commutator

$$\nabla_{AA'}\nabla_{BB'}\Sigma_{CC'} - \nabla_{BB'}\nabla_{AA'}\Sigma_{CC'} = -R^{P}{}_{CAA'BB'}\Sigma_{PC'} - \bar{R}^{P'}{}_{C'A'ABB'}\Sigma_{CP'} - \Sigma_{AA'}{}^{QQ'}{}_{BB'}\nabla_{QQ'}\Sigma_{CC'},$$

one finds that

$$\nabla_{(A}{}^{Q'}\nabla_{B)Q'}\Sigma_{CC'} = -R^{P}{}_{C(AB)}\Sigma_{PC'} - \bar{R}^{P'}{}_{C'(AB)}\Sigma_{CP'} - \Sigma_{A}{}^{QQ'}{}_{B}\nabla_{QQ'}\Sigma_{CC'},$$

where

$$R_{ABCD} \equiv \frac{1}{2}R_{ABCQ'D}{}^{Q'}, \qquad \bar{R}_{A'B'CD} \equiv \frac{1}{2}\bar{R}_{A'B'Q'CD}{}^{Q'}.$$

Using the definitions of the zero quantities $\Delta_{CDBB'}$ and $Z_{AA'}$ to eliminate, respectively, $\nabla_{(A}{}^{Q'}L_{B)Q'CC'}$ and $\nabla_{BC'}s$, one obtains

$$\begin{aligned}\nabla_{(A}{}^{Q'}Z_{B)Q'CC'} = {} & -R^{P}{}_{C(AB)}\Sigma_{PC'} - \bar{R}^{P'}{}_{C'(AB)}\Sigma_{CP'} - \Sigma_{A}{}^{QQ'}{}_{B}\nabla_{QQ'}\Sigma_{CC'} \\ & + \Sigma_{(A}{}^{Q'}L_{B)Q'CC'} + \Xi\Delta_{ABCC'} - \Xi\Sigma^{Q}{}_{C'}\phi_{ABCQ} \\ & - \Xi^2 T_{ABCC'} + \epsilon_{C(A}Z_{B)C'} - \epsilon_{C(A}L_{B)C'QQ'}\Sigma^{QQ'} \\ & - \frac{1}{2}\Xi^2\Sigma^{QQ'}\epsilon_{C(A}T_{B)C'QQ'} - \frac{3}{2}\Xi^2\Sigma_{(A}{}^{Q'}T_{B)Q'CC'} \\ & - \frac{1}{2}\Xi^3\nabla_{(A}{}^{Q'}T_{B)Q'CC'}.\end{aligned}$$

Next, one uses the zero quantity $\Xi_{ABCC'DD'}$ to eliminate the geometric curvature terms $R^{P}{}_{C(AB)}$ and $\bar{R}^{P'}{}_{C'(AB)}$. Taking into account the expression of the algebraic curvature in terms of the Schouten tensor and the rescaled Weyl tensor one obtains

$$\begin{aligned}\nabla_{(A}{}^{Q'}Z_{B)Q'CC'} = {} & -\Xi^{P}{}_{C(AB)}\Sigma_{PC'} - \bar{\Xi}^{P'}{}_{C'(AB)}\Sigma_{CP'} - \Sigma_{A}{}^{QQ'}{}_{B}\nabla_{QQ'}\Sigma_{CC'} \\ & + \Xi\Delta_{ABCC'} - \Xi^2 T_{ABCC'} + \epsilon_{C(A}Z_{B)C'} \\ & - \frac{1}{2}\Xi^2\Sigma^{QQ'}\epsilon_{C(A}T_{B)C'QQ'} - \frac{3}{2}\Xi^2\Sigma_{(A}{}^{Q'}T_{B)Q'CC'} \\ & - \frac{1}{2}\Xi^3\nabla_{(A}{}^{Q'}T_{B)Q'CC'},\end{aligned}$$

where

$$\Xi_{ABCD} \equiv \frac{1}{2}\Xi_{ABCQ'D}{}^{Q'}, \qquad \Xi_{A'B'CD} \equiv \frac{1}{2}\Xi_{A'B'Q'CD}{}^{Q'}.$$

Finally, observing that the definition of the rescaled Cotton tensor implies that $T_{ABCC'} = -\frac{1}{2}\Xi\nabla_{(A}{}^{Q'}T_{B)Q'CC'}$ and exploiting the trace-freeness of the energy-momentum tensor one ends up with the expression

$$\begin{aligned}\nabla_{(A}{}^{Q'}Z_{B)Q'CC'} = &-\Xi^{P}{}_{C(AB)}\Sigma_{PC'} - \Xi^{P'}{}_{C'(AB)}\Sigma_{CP'} - \Sigma_{A}{}^{QQ'}{}_{B}\nabla_{QQ'}\Sigma_{CC'}\\ &+ \Xi\Delta_{ABCC'} + \epsilon_{C(A}Z_{B)C'},\end{aligned}$$

which, as required, is a linear combination of zero quantities.

The subsidiary equation for $Z_{AA'}$

In this case one needs to evaluate $\nabla_{(A}{}^{Q'}Z_{B)Q'}$. Making use of the definition of $Z_{AA'}$ one finds that

$$\begin{aligned}\nabla_{(A}{}^{Q'}Z_{B)Q'} = &\nabla_{(A}{}^{Q'}\nabla_{B)Q'}s + \nabla_{(A}{}^{Q'}L_{B)Q'PP'}\Sigma^{PP'} + \nabla_{(A}{}^{Q'}\Sigma^{PP'}L_{B)Q'PP'}\\ &- \Xi\Sigma_{(A}{}^{Q'}\Sigma^{PP'}T_{B)Q'PP'} - \frac{1}{2}\Xi^2\nabla_{(A}{}^{Q'}\Sigma^{PP'}T_{B)Q'PP'}\\ &- \frac{1}{2}\Xi^2\Sigma^{PP'}\nabla_{(A}{}^{Q'}T_{B)Q'PP'}.\end{aligned}$$

Using the definition of the torsion tensor in the form

$$\nabla_{(A}{}^{Q'}\nabla_{B)Q'}s = \Sigma_{A}{}^{QQ'}{}_{B}\nabla_{QQ'}s,$$

and the definitions of $\Delta_{ABCC'}$ and $Z_{AA'BB'}$ to eliminate, respectively, $\nabla_{(A}{}^{Q'}L_{B)Q'PP'}$ and $\nabla_{A}{}^{Q'}\Sigma^{PP'}$ one obtains – after some simplifications involving the symmetries of $L_{AA'BB'}$ and $T_{AA'BB'}$ –

$$\begin{aligned}\nabla_{(A}{}^{Q'}Z_{B)Q'} = &\Sigma_{(A}{}^{QQ'}{}_{B)}\nabla_{QQ'}s + \Delta_{ABPP'}\Sigma^{PP'} + Z_{(A}{}^{Q'PP'}L_{B)Q'PP'}\\ &- \frac{1}{2}\Xi^2 Z_{(A}{}^{Q'PP'}T_{B)Q'PP'}.\end{aligned}$$

This expresion is a linear combination of zero quantities.

13.3.3 Subsidiary equation for the no-torsion condition

Following the general discussion of Section 13.3.1, one defines

$$\Sigma_{ABCD}{}^{a} \equiv \tau_{B}{}^{A'}\tau_{D}{}^{C'}\Sigma_{AA'}{}^{QQ'}{}_{CC'}e_{QQ'}{}^{a}.$$

One can write

$$\Sigma_{ABCD}{}^{a} = -\Sigma_{AC}{}^{a}\epsilon_{BD} - \Sigma^{+}{}_{BD}{}^{a}\epsilon_{AC},$$

so that, if the evolution equation $\Sigma_{AB}{}^{a} - \Sigma^{+}{}_{AB}{}^{a} = 0$ holds, then

$$\mathcal{P}\Sigma_{AB}{}^{a} = -\frac{\mathrm{i}}{2}\nabla_{EF}\Sigma_{CDGH}{}^{a}\epsilon^{EFCDGH}{}_{(AB)}. \qquad (13.40)$$

To conclude the argument one needs to express the right-hand side of the above equation as a linear combination of zero quantities. To this end, one makes use of the first Bianchi identity (2.10) to write

$$\nabla_{[a}\Sigma_{b}{}^{d}{}_{c]} = -\Xi^{d}{}_{[cab]} - \rho^{d}{}_{[cab]} - \Sigma_{[a}{}^{e}{}_{b}\Sigma_{c]}{}^{d}{}_{e},$$

where the zero quantity $\Xi^{d}{}_{cab} \equiv R^{d}{}_{cab} - \rho^{d}{}_{cab}$. By construction, the algebraic curvature has the same algebraic symmetries as the Riemann tensor of a Levi-Civita connection so that, in particular, $\rho^{d}{}_{[cab]} = 0$ and one has that

$$\nabla_{[a}\Sigma_{b}{}^{d}{}_{c]} = -\Xi^{d}{}_{[cab]} - \Sigma_{[a}{}^{e}{}_{b}\Sigma_{c]}{}^{d}{}_{e}.$$

This expression shows that the right-hand side of Equation (13.40) can be written as a linear combination of zero quantities.

13.3.4 Subsidiary equation for the Ricci identity

It follows from the general discussion of Section 13.3.1 that, if the evolution equations $\Xi_{ABCD} - \Xi^{*}_{ABCD} = 0$ are satisfied, then

$$\mathcal{P}\Xi_{ABCD} = -\frac{\mathrm{i}}{2}\nabla_{EF}\Xi_{CDGH}\epsilon^{EFCDGH}{}_{(AB)}. \tag{13.41}$$

To express the right-hand side of this last equation as a linear combination of zero quantities one makes use of the second Bianchi identity (2.11) to obtain

$$\begin{aligned}\epsilon_{f}{}^{abc}\nabla_{[a}\Xi^{d}{}_{|e|bc]} &= \epsilon_{f}{}^{abc}\nabla_{[a}R^{d}{}_{|e|bc]} - \epsilon_{f}{}^{abc}\nabla_{[a}\rho^{d}{}_{|e|bc]} \\ &= -\epsilon_{f}{}^{abc}\Sigma_{[a}{}^{g}{}_{b}R^{d}{}_{|e|c]g} - \epsilon_{f}{}^{abc}\nabla_{[a}\rho^{d}{}_{|e|bc]}.\end{aligned} \tag{13.42}$$

The first term in the right-hand side of the last equation already has the desired form. The second term needs to be examined in more detail. One considers

$$\begin{aligned}\epsilon_{f}{}^{abc}\nabla_{[a}\rho^{d}{}_{|e|bc]} &= \epsilon_{f}{}^{abc}\nabla_{a}\rho^{d}{}_{ebc} \\ &= \Xi\epsilon_{f}{}^{abc}\nabla_{a}d^{d}{}_{ebc} + \epsilon_{f}{}^{abc}\nabla_{a}\Xi d^{d}{}_{ebc} + 2\epsilon_{f}{}^{abc}S_{eb}{}^{dh}\nabla_{a}L_{ch},\end{aligned}$$

where in the last line the expression of the algebraic curvature in terms of the Weyl tensor and the Schouten tensor has been used. Now, a computation using the properties of the Hodge dual and the definition of the zero quantity Λ_{abc} shows that

$$\begin{aligned}\epsilon_{f}{}^{abc}\nabla_{a}d^{d}{}_{ebc} &= -\epsilon_{f}{}^{abc}\nabla_{a}{}^{*}d^{*d}{}_{ebc} = -2\nabla_{a}{}^{*}d^{**d}{}_{ef}{}^{a} \\ &= 2\nabla_{a}{}^{*}d^{d}{}_{ef}{}^{a} = 2\nabla_{a}d^{*}{}_{f}{}^{ad}{}_{e} \\ &= \epsilon_{e}{}^{dgh}\nabla_{a}d^{a}{}_{fgh} = \epsilon_{e}{}^{dgh}(\Lambda_{fgh} + T_{fgh}).\end{aligned} \tag{13.43}$$

Using the above expression together with the definition of the zero quantities Δ_{abc} and Q_{a} to eliminate $\nabla_{[a}L_{c]f}$ and $\nabla_{a}\Xi$, respectively, one finds that

$$\epsilon_{f}{}^{abc}\nabla_{[a}\rho^{d}{}_{|e|bc]} = \epsilon_{f}{}^{abc}Q_{a}d^{d}{}_{ebc} + \Xi\epsilon_{e}{}^{dgh}\Lambda_{fgh} + \epsilon_{f}{}^{abc}S_{eb}{}^{dh}\Delta_{ach}.$$

Substituting this last expression into Equation (13.42) one obtains the required expression for $\nabla_{[a}\Xi^{d}{}_{|e|bc]}$ as a linear combination of zero quantities. The spinorial counterpart of this expression differs from the right-hand side of Equation (13.41) by terms homogeneous in zero quantities involving the spinor χ_{ABCD}.

13.3.5 The subsidiary equations for the Cotton equation

Applying the general discussion of Section 13.3.1 to the zero quantity $\Delta_{AA'BB'CC'}$ associated with the unphysical Bianchi identity leads to the expression

$$\frac{1}{2}\mathcal{P}(\Delta^{+}_{ABKL} + \Delta_{ABKL}) + \mathcal{D}^{Q}{}_{A}\Delta^{+}_{BQKL} - \mathcal{D}^{Q}{}_{B}\Delta_{AQKL} = -\frac{\mathrm{i}}{2}\epsilon^{EFCDGH}{}_{AB}\nabla_{EF}\Delta_{CDGHKL}. \quad (13.44)$$

To make use of the evolution equations in the above expression it is observed that

$$\Delta_{ABCD} = \Delta_{(ABCD)} + \frac{1}{3}h_{ABCD}\Delta_{PQ}{}^{PQ} + \frac{1}{2}\epsilon_{CD}\Delta_{ABQ}{}^{Q}.$$

Thus, using the evolution equations for the various components of the Schouten tensor one obtains

$$\mathcal{P}\Delta_{(ABKL)} = -\frac{\mathrm{i}}{2}\nabla_{EF}\Delta_{CDGH(AB}\epsilon^{EFCDGH}{}_{KL)},$$

$$\mathcal{P}\Delta_{PQ}{}^{PQ} = -\frac{\mathrm{i}}{2}\epsilon^{EFCDGHKL}\nabla_{EF}\Delta_{CDGHKL},$$

$$\mathcal{P}\Delta_{ABQ}{}^{Q} = -\frac{\mathrm{i}}{2}\epsilon^{EFCDGH}{}_{AB}\nabla_{EF}\Delta_{CDGHQ}{}^{Q}.$$

To analyse the right-hand sides of the above equations it is more convenient to analyse $\epsilon_{AA'}{}^{BB'CC'DD'}\nabla_{BB'}\Delta_{CC'DD'EE'}$. This expression differs from the right-hand side of Equation (13.44) by terms involving the spinor χ_{ABCD}. For conciseness, the analysis is carried out using tensorial notation. One has that

$$\epsilon_{f}{}^{ecd}\nabla_{e}\Delta_{cdb} = \epsilon_{f}{}^{ecd}\big(2\nabla_{e}\nabla_{[c}L_{d]b} - \nabla_{e}\Sigma_{a}d^{a}{}_{bcd} - \Sigma_{a}\nabla_{e}d^{a}{}_{bcd} - \nabla_{e}\Xi T_{cdb} - \Xi\nabla_{e}T_{cdb}\big). \quad (13.45)$$

The first term on the right-hand side of the above equation is manipulated using the commutator of covariant derivatives by observing that

$$\begin{aligned}2\epsilon_{f}{}^{ecd}\nabla_{e}\nabla_{[c}L_{d]b} &= 2\epsilon_{f}{}^{ecd}\nabla_{[e}\nabla_{c]}L_{db}\\ &= -2\epsilon_{f}{}^{ecd}\big(2R^{s}{}_{(d|ec|}L_{b)s} - \Sigma_{e}{}^{s}{}_{c}\nabla_{s}L_{db}\big)\\ &= -2\epsilon_{f}{}^{ecd}\big(2\Xi^{s}{}_{(d|ec|}L_{b)s} + \epsilon_{f}{}^{ecd}\rho^{s}{}_{bec}L_{ds} + \Sigma_{e}{}^{s}{}_{c}\nabla_{s}L_{db}\big),\end{aligned}$$

where in the third line the identity $\rho^{a}{}_{[bcd]} = 0$ has been used. The second term in the last equation does not contain zero quantities. The third term in

Equation (13.45) is now cast in a suitable form using the properties of Hodge dual; compare the analogous argument leading to (13.43). One finds that

$$\epsilon_f{}^{ecd}\nabla_e d_{abcd} = -\epsilon_{ab}{}^{gh}(\Lambda_{fgh} + T_{fgh}).$$

Substituting the above two identities into Equation (13.45) and using the zero quantity Z_{ab} to eliminate $\nabla_a\Sigma_b$ one obtains

$$\begin{aligned}
2\epsilon_f{}^{ecd}\nabla_e\nabla_{[c}L_{d]b} \\
&= -4\epsilon_f{}^{ecd}\Xi^s{}_{(d|ec|}L_{b)s} - 2\Sigma_e{}^s{}_c\nabla_s L_{fb} \\
&\quad - \Sigma^a\epsilon_{ab}{}^{gh}\Lambda_{fgh} - \epsilon_f{}^{ecd}Z_{ea}d^a{}_{bcd} \\
&\quad - \left\{\epsilon_f{}^{ecd}\left(\frac{1}{2}\Xi^3 T_{ea}d^a{}_{bcd} + \nabla_e\Xi T_{cdb} - \Xi\nabla_e T_{cdb}\right) - \epsilon_{ab}{}^{gh}\Sigma^a T_{fgh}\right\},
\end{aligned} \tag{13.46}$$

where the explicit expression of the algebraic curvature has been used to show that

$$\epsilon_f{}^{ecd}\left(\Xi d^a{}_{bcd}L_{ea} - 2\rho^s{}_{bec}L_{ds}\right) = 0.$$

Expression (13.46) is, up to the *matter terms* in curly brackets, a linear combination of zero quantities. Whether the terms in curly brackets can be expressed as a linear combination of (matter) zero quantities depends on the particular features of the matter model under consideration.

13.3.6 The subsidiary equations for the Bianchi identity

The construction of the subsidiary equation for the Bianchi identity is similar to that of the subsidiary equation for the Maxwell equations. In this case the relevant zero quantity is given by

$$\Lambda_{A'BCD} \equiv \nabla^Q{}_{A'}\phi_{BCDQ} + T_{CDBA'},$$

for which one computes $\nabla^{BB'}\Lambda_{B'BCD}$ in two different manners.

First, making use of the space spinor zero quantity $\Lambda_{ABCD} \equiv \tau_A{}^{A'}\Lambda_{A'BCD}$ one has that

$$\begin{aligned}
\nabla^{BB'}\Lambda_{B'BCD} &= -\nabla^{BB'}\left(\tau^P{}_{B'}\Lambda_{PBCD}\right) \\
&= \nabla^{AB}\Lambda_{ABCD} - \left(\nabla^{BA'}\tau^A{}_{A'}\right)\Lambda_{ABCD}.
\end{aligned}$$

Using Equation (13.8) for the derivative of the spinor $\tau_{AA'}$ and the split of ∇_{AB} one obtains

$$\mathcal{P}\Lambda_{CD} - 2\mathcal{D}^{AB}\Lambda_{ABCD} - 2\sqrt{2}\chi^B{}_P{}^{PA}\Lambda_{ABCD} = -2\nabla^{BB'}\Lambda_{B'BCD},$$

where it is recalled that $\Lambda_{CD} \equiv \Lambda^Q{}_{QCD}$. Now, a calculation shows that the symmetry $\Lambda_{ABCD} = \Lambda_{A(BCD)}$ implies the decomposition

$$\Lambda_{ABCD} = \Lambda_{(ABCD)} - \frac{3}{4}\epsilon_{A(B}\Lambda_{CD)},$$

so that

$$\begin{aligned}\mathcal{P}\Lambda_{CD} - 2\mathcal{D}^P{}_{(C}\Lambda_{D)P} - 2\mathcal{D}^{AB}\Lambda_{(ABCD)} - 2\sqrt{2}\chi^B{}_P{}^{PA}\Lambda_{(ABCD)}\\ + \frac{3}{\sqrt{2}}\chi^B{}_P{}^{PA}\epsilon_{A(B}\Lambda_{CD)} = -2\nabla^{BB'}\Lambda_{B'BCD}.\end{aligned} \tag{13.47}$$

As a second way of evaluating $\nabla^{BB'}\Lambda_{B'BCD}$ one makes use of the definition of the zero quantity so that

$$\nabla^{BB'}\Lambda_{B'BCD} = \nabla^{BB'}\nabla^Q{}_{B'}\phi_{BCDQ} + \nabla^{BB'}T_{CDBB'}.$$

The first term of the right-hand side is manipulated using the commutator

$$\begin{aligned}&\nabla_{AA'}\nabla_{BB'}\phi_{CDEF} - \nabla_{BB'}\nabla_{AA'}\phi_{CDEF}\\ &\quad = -R^S{}_{CAA'BB'}\phi_{SDEF} - R^S{}_{DAA'BB'}\phi_{SCEF} - R^S{}_{EAA'BB'}\phi_{SCDF}\\ &\qquad - R^S{}_{FAA'BB'}\phi_{SCDE} + \Sigma_{AA'}{}^{PP'}{}_{BB'}\nabla_{PP'}\phi_{CDEF}.\end{aligned}$$

Observe that the torsion $\Sigma_{AA'}{}^{PP'}{}_{BB'}$, being one of the unknowns in the subsidiary system, needs to be included in the commutator. Also, the curvature terms in the above expression are understood to be those of the *geometric curvature*. Contracting the expression of the commutator leads to

$$\begin{aligned}2\nabla^{BB'}\nabla^Q{}_{B'}\phi_{BCDQ} = &-R^S{}_C{}^{BA'Q}{}_{A'}\phi_{SDBQ} - R^S{}_D{}^{BA'Q}{}_{A'}\phi_{SCBQ}\\ &- R^S{}_B{}^{BA'Q}{}_{A'}\phi_{SCDQ} - R^S{}_Q{}^{BA'Q}{}_{A'}\phi_{SCDB}\\ &+ \Sigma^{AA'SS'Q}{}_{A'}\nabla_{SS'}\phi_{CDAQ}.\end{aligned}$$

Using the zero quantity

$$\Xi^C{}_{DAA'BB'} = R^C{}_{DAA'BB'} - \rho^C{}_{DAA'BB'},$$

to eliminate the geometric curvature and the decomposition (13.9) one obtains an expression which is homogenous in zero quantities:

$$\begin{aligned}2\nabla^{BB'}\nabla^Q{}_{B'}\phi_{BCDQ} = &-\Xi^S{}_C{}^{BA'Q}{}_{A'}\phi_{SDBQ} - \Xi^S{}_D{}^{BA'Q}{}_{A'}\phi_{SCBQ}\\ &- \Xi^S{}_B{}^{BA'Q}{}_{A'}\phi_{SCDQ} - \Xi^S{}_Q{}^{BA'Q}{}_{A'}\phi_{SCDB}\\ &+ \Sigma^{AA'SS'Q}{}_{A'}\nabla_{SS'}\phi_{CDAQ}.\end{aligned} \tag{13.48}$$

In particular, all the terms coming from the algebraic curvature cancel out.

Combining Equations (13.47) and (13.48) one obtains the required subsidiary equation. Namely, one has that if $\Lambda_{(ABCD)} = 0$, then

$$\begin{aligned}\mathcal{P}\Lambda^{Q}{}_{QCD} - 2\mathcal{D}^{P}{}_{(C}\Lambda^{Q}{}_{D)PQ} + \frac{3}{\sqrt{2}}\chi^{B}{}_{P}{}^{PA}\epsilon_{A(B}\Lambda^{Q}{}_{CD)Q} \\ = \Xi^{S}{}_{C}{}^{BA'Q}{}_{A'}\phi_{SDBQ} + \Xi^{S}{}_{D}{}^{BA'Q}{}_{A'}\phi_{SCBQ} \\ + \Xi^{S}{}_{B}{}^{BA'Q}{}_{A'}\phi_{SCDQ} + \Xi^{S}{}_{Q}{}^{BA'Q}{}_{A'}\phi_{SCDB} \\ - 2\Sigma^{AA'SS'Q}{}_{A'}\nabla_{SS'}\phi_{CDAQ} - 2\nabla^{BB'}T_{CDBB'},\end{aligned}$$

which is homogeneous in zero quantities if the matter term $\nabla^{BB'}T_{CDBB'}$ can be expressed, in turn, as a homogeneous expression of matter zero quantities.

Alternatively, one can perform the computation with tensorial objects. In this case one looks at

$$\nabla^{b}\Lambda_{bcd} = \nabla^{b}\nabla_{a}d^{a}{}_{bcd} - \nabla^{b}T_{cdb}.$$

Again, using the properties of the Hodge dual one can write

$$\begin{aligned}\nabla^{b}\nabla_{a}d^{a}{}_{bcd} &= -\nabla^{a}\nabla^{b}d_{abcd} \\ &= \nabla^{a}\nabla^{b*}d^{*}_{abcd} = \frac{1}{4}\epsilon^{abef}\epsilon^{gh}{}_{cd}\nabla_{b}\nabla_{a}d_{efgh} \\ &= \frac{1}{4}\epsilon^{abef}\epsilon^{gh}{}_{cd}\big(R^{s}{}_{eab}d_{fsgh} + R^{s}{}_{gab}d_{hsef} - \frac{1}{2}\Sigma_{a}{}^{s}{}_{b}\nabla_{s}d_{efgh}\big) \\ &= \frac{1}{4}\epsilon^{abef}\epsilon^{gh}{}_{cd}\big(\Xi^{s}{}_{eab}d_{fsgh} + \Xi^{s}{}_{gab}d_{hsef} - \frac{1}{2}\Sigma_{a}{}^{s}{}_{b}\nabla_{s}d_{efgh}\big).\end{aligned}$$

Hence, one concludes that $\nabla^{b}\Lambda_{bcd}$, except for the *matter term* $\nabla^{b}T_{cdb}$ can be written as a linear combination of zero quantities.

13.3.7 Summary

In most applications, the detailed form of the evolution and subsidiary equations is not required; general structural properties suffice. These properties are summarised in the following propositions.

It is convenient to group the independent components of the unknowns appearing in the spinorial formulation of the conformal field equations in the following manner:

$\boldsymbol{\sigma}$ independent components of Ξ, $\Sigma_{AA'}$, s;

$\boldsymbol{\upsilon}$ independent components of $e^{\mu}_{AA'}$, $\Gamma_{AA'BC}$, $\Phi_{AA'BB'}$;

$\boldsymbol{\phi}$ independent components of ϕ_{ABCD};

$\boldsymbol{\varphi}$ independent components of matter fields.

Moreover, let $\boldsymbol{e}$ and $\boldsymbol{\Gamma}$ denote, respectively, the independent components of the frame components and the connection coefficients. In terms of these objects one has the following:

Proposition 13.1 (***properties of the conformal evolution equations***) *Given arbitrary smooth gauge source functions*

$$F^{a}(x), \qquad F_{\boldsymbol{AB}}(x), \qquad R(x),$$

such that

$$\nabla^{\boldsymbol{QQ}'} e_{\boldsymbol{QQ}'}{}^{\boldsymbol{a}} = F^{\boldsymbol{a}}(x), \qquad \nabla^{\boldsymbol{QQ}'}\Gamma_{\boldsymbol{QQ}'\boldsymbol{AB}} = F_{\boldsymbol{AB}}(x),$$
$$\nabla^{\boldsymbol{QQ}'} L_{\boldsymbol{QQ}'\boldsymbol{BB}'} = \frac{1}{6}\nabla_{\boldsymbol{BB}'}R(x),$$

and assuming that the components of the matter tensors $T_{\boldsymbol{ab}}$ and $T_{\boldsymbol{abc}}$ can be written in such a way that they do not contain derivatives of the matter fields, then the conformal Einstein field equations

$$Q_{\boldsymbol{a}} = 0, \qquad Z_{\boldsymbol{ab}} = 0, \qquad Z_{\boldsymbol{a}} = 0, \qquad \Sigma_{\boldsymbol{a}}{}^{\boldsymbol{c}}{}_{\boldsymbol{b}} = 0, \qquad \Xi^{\boldsymbol{c}}{}_{\boldsymbol{dab}} = 0,$$
$$\Delta_{\boldsymbol{abc}} = 0, \qquad \Lambda_{\boldsymbol{abc}} = 0,$$

imply a symmetric hyperbolic system of equations for the independent components of the geometric fields $(\boldsymbol{\sigma}, \boldsymbol{\upsilon}, \boldsymbol{\phi})$ of the form

$$(\mathbf{I} + \mathbf{A}^0(\boldsymbol{e}))\partial_\tau \boldsymbol{\phi} + \mathbf{A}^\alpha(\boldsymbol{e})\partial_\alpha \boldsymbol{\phi} = \mathbf{B}(\boldsymbol{\Gamma})\boldsymbol{\phi} + \mathbf{C}(\boldsymbol{\sigma}, \boldsymbol{\upsilon}, \boldsymbol{\varphi}),$$
$$(\mathbf{I} + \mathbf{D}^0(\boldsymbol{e}))\partial_\tau \boldsymbol{\upsilon} + \mathbf{D}^\mu(\boldsymbol{e})\partial_\mu \boldsymbol{\upsilon} = \mathbf{E}(\boldsymbol{\Gamma})\boldsymbol{\upsilon} + \mathbf{F}(\boldsymbol{\sigma}, \boldsymbol{\upsilon}, \boldsymbol{\phi}, \boldsymbol{\varphi}),$$
$$\partial_\tau \boldsymbol{\sigma} = \mathbf{G}(\boldsymbol{\Gamma})\boldsymbol{\sigma} + \mathbf{H}(\boldsymbol{\sigma}, \boldsymbol{\upsilon}, \boldsymbol{\varphi}),$$

where **I** *denotes the identity matrix of the required dimensions,*

$$\mathbf{A}^\mu(\boldsymbol{e}), \qquad \mathbf{D}^\mu(\boldsymbol{e})$$

are smooth matrix-valued functions of the components of the frame components,

$$\mathbf{B}(\boldsymbol{\Gamma}), \qquad \mathbf{E}(\boldsymbol{\Gamma}), \qquad \mathbf{G}(\boldsymbol{\Gamma})$$

are smooth matrix-valued functions of the connection coefficients and

$$\mathbf{C}(\boldsymbol{\sigma}, \boldsymbol{\upsilon}, \boldsymbol{\varphi}), \qquad \mathbf{F}(\boldsymbol{\sigma}, \boldsymbol{\upsilon}, \boldsymbol{\phi}, \boldsymbol{\varphi}), \qquad \mathbf{H}(\boldsymbol{\sigma}, \boldsymbol{\upsilon}, \boldsymbol{\varphi})$$

are smooth vector-valued functions with polynomial dependence on their arguments. The characteristics of this system satisfy a characteristic polynomial involving factors of the form

$$\tau^\mu \xi_\mu, \qquad g^{\nu\lambda}\xi_\nu\xi_\lambda, \qquad \left(\tau^\rho\tau^\sigma + \frac{2}{3}g^{\rho\sigma}\right)\xi_\rho\xi_\sigma.$$

Remarks

(i) In the presence of matter, the symmetric hyperbolic system given in the above proposition needs to the supplemented by a symmetric hyperbolic system for the matter fields. As the rescaled Cotton tensor $T_{\boldsymbol{abc}}$ (and hence also the spinor $T_{\boldsymbol{ABCC}'}$) is made up of derivatives of the energy-momentum tensor, the *matter evolution equations* will need to include equations for the matter field derivatives appearing in the *geometric evolution equations.*

(ii) The choice of the gauge source functions is dictated by the particular analysis under consideration.

With regards to the subsidiary system one has the following:

Proposition 13.2 (***properties of the subsidiary evolution system***) *Assume that the evolution equations implied by the conformal Einstein field equations are satisfied and that the energy-momentum tensor T_{ab} is such that the quantities*

$$M_{cd} \equiv \nabla^b T_{cdb},$$

$$N_{bf} \equiv \epsilon_f{}^{ecd}\left(\frac{1}{2}\Xi^3 T_{ea} d^a{}_{bcd} + \nabla_e \Xi T_{cdb} - \Xi \nabla_e T_{cdb}\right) - \epsilon_{ab}{}^{gh}\Sigma^a T_{fgh},$$

can be written as homogeneous expressions of the geometric and matter zero quantities. Then the zero quantities encoding the constraint equations implied by the conformal Einstein equations under the hyperbolic reduction procedure leading to Proposition 13.1 satisfy a symmetric hyperbolic system which is a homogeneous expression of zero quantities.

13.4 Hyperbolic reductions using conformal Gaussian systems

This section discusses a hyperbolic reduction procedure based on the properties of congruences of conformal geodesics. The approach discussed in this section makes use of the formulation of the conformal field equations in terms of Weyl connections – the so-called extended conformal field equations. As will be seen, this procedure leads to simpler evolution equations than the ones obtained by the reduction procedure discussed in Section 13.2.

For conciseness of the presentation, the discussion in the rest of this section is restricted to the vacuum case.

13.4.1 Basic set up

In what follows, it is assumed one has a region $\mathcal{U}$ of a spacetime $(\tilde{\mathcal{M}}, \tilde{g})$ which is covered by a congruence of conformal geodesics $(\dot{x}(\tau), \beta(\tau))$. For convenience, the vector field tangent to the congruence will be denoted by $\boldsymbol{\tau}$. As discussed in Section 5.5, a canonical representative $\boldsymbol{g}$ of the conformal class $[\tilde{\boldsymbol{g}}]$ is singled out by the requirement

$$\boldsymbol{g}(\boldsymbol{\tau}, \boldsymbol{\tau}) = 1,$$

so that $\boldsymbol{g} = \Theta^2 \tilde{\boldsymbol{g}}$ where the conformal factor Θ satisfies a third-order ordinary differential equation along the congruence of conformal geodesics; see Equation (5.53b). In the case of a vacuum spacetime this equation can be explicitly solved yielding a formula for Θ as a quadratic polynomial in the parameter τ. The conformal factor is completely determined by the three coefficients $\Theta_\star$, $\dot{\Theta}_\star$, $\ddot{\Theta}_\star$ specified, say, on an initial hypersurface $\tilde{\mathcal{S}}_\star$.

In the following, $\{\boldsymbol{e_a}\}$ will denote a $\boldsymbol{g}$-orthonormal frame which is Weyl-propagated along the conformal geodesics and such that $\boldsymbol{e_0} = \boldsymbol{\tau}$. As discussed in Section 5.5, to every congruence of conformal geodesics one can associate a Weyl connection $\hat{\boldsymbol{\nabla}}$. This Weyl connection satisfies the relations

$$\hat{\nabla}_{\boldsymbol{\tau}} \boldsymbol{e_a} = 0, \qquad \hat{\boldsymbol{L}}(\boldsymbol{\tau}, \cdot) = 0,$$

where $\hat{\boldsymbol{L}}$ denotes the Schouten tensor of $\hat{\boldsymbol{\nabla}}$; see Equation (5.41). In terms of frame components, the above conditions can be rewritten as

$$\hat{\Gamma}_{\boldsymbol{0}}{}^{\boldsymbol{a}}{}_{\boldsymbol{b}} = 0, \qquad \hat{L}_{\boldsymbol{0a}} = 0. \tag{13.49}$$

In particular, it follows that the covector $\boldsymbol{f}$ which defines the Weyl connection $\hat{\boldsymbol{\nabla}}$ satisfies

$$f_{\boldsymbol{0}} = 0.$$

The gauge choice can be refined further by choosing the parameter of the conformal geodesics τ as the time coordinate. Thus, one has the additional gauge condition

$$\boldsymbol{e_0} = \boldsymbol{\partial}_\tau, \qquad \text{so that} \qquad \boldsymbol{e_0}^\mu = \delta_0{}^\mu. \tag{13.50}$$

In most applications, initial data for the congruence of conformal geodesics will be prescribed on the initial hypersurface $\mathcal{S}_\star$. On $\mathcal{S}_\star$ choose some local coordinates (x^α). Assuming that each curve of the congruence of conformal geodesics intersects $\mathcal{S}_\star$ only once, one can extend coordinates on $\mathcal{S}_\star$ off the hypersurface by requiring them to be constant along the conformal geodesic which intersects $\mathcal{S}_\star$ at the point with coordinates (x^α); see Figure 13.1. The spacetime coordinates (τ, x^α) one obtains by this procedure are known as ***conformal Gaussian coordinates***. More generally, the collection of the conformal factor Θ, Weyl-propagated frame vectors $\{\boldsymbol{e_a}\}$ and coordinates (τ, x^α) extended off some initial hypersurface $\mathcal{S}_\star$ using a congruence of conformal geodesics will be known as a ***conformal Gaussian gauge system***.

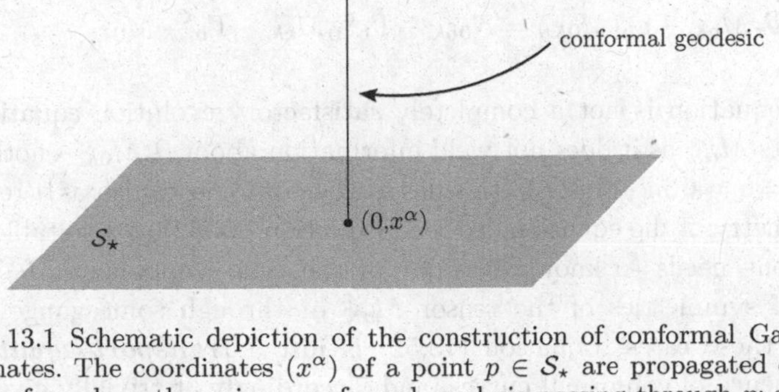

Figure 13.1 Schematic depiction of the construction of conformal Gaussian coordinates. The coordinates (x^α) of a point $p \in \mathcal{S}_\star$ are propagated off the hypersurface along the unique conformal geodesic passing through p; see the main text for further details.

Remarks

(i) The specific choice of the data for the conformal Gaussian gauge system on an initial hypersurface $\mathcal{S}$ is dictated by the particular geometric setting under consideration.

(ii) The discussion in this section can be adapted, with minor changes, to the case of a congruence of so-called *conformal curves* in non-vacuum spacetimes; see, for example, Lübbe and Valiente Kroon (2012) for further details.

13.4.2 Hyperbolic reduction of a model equation

The general ideas behind the procedure of hyperbolic reduction using conformal Gaussian systems are best illustrated with a model equation. All the extended conformal equations, except for the unphysical Bianchi identity, are of the form

$$\hat{\nabla}_{\boldsymbol{a}} M_{\boldsymbol{b}\mathcal{K}} - \hat{\nabla}_{\boldsymbol{b}} M_{\boldsymbol{a}\mathcal{K}} = N_{\boldsymbol{a}\boldsymbol{b}\mathcal{K}}, \tag{13.51}$$

where $M_{\boldsymbol{a}\mathcal{K}}$ and $N_{\boldsymbol{a}\boldsymbol{b}\mathcal{K}} = N_{[\boldsymbol{a}\boldsymbol{b}]\mathcal{K}}$ denote the components of some tensorial quantities with respect to the frame $\{\boldsymbol{e_a}\}$ and $\mathcal{K}$ denotes an arbitrary set of tensor indices. To derive an evolution equation along the direction given by the congruence of curves, one sets $\boldsymbol{a} = \boldsymbol{0}$ so that

$$\hat{\nabla}_{\boldsymbol{0}} M_{\boldsymbol{b}\mathcal{K}} - \hat{\nabla}_{\boldsymbol{b}} M_{\boldsymbol{0}\mathcal{K}} = N_{\boldsymbol{0}\boldsymbol{b}\mathcal{K}},$$

or, more explicitly,

$$\boldsymbol{e_0}(M_{\boldsymbol{b}\mathcal{K}}) - \boldsymbol{e_b}(M_{\boldsymbol{0}\mathcal{K}}) = N_{\boldsymbol{0}\boldsymbol{b}\mathcal{K}} + \hat{\Gamma}_{\boldsymbol{0}}{}^{\boldsymbol{c}}{}_{\boldsymbol{b}} M_{\boldsymbol{c}\mathcal{K}} + \hat{\Gamma}_{\boldsymbol{0}}{}^{\mathcal{L}}{}_{\mathcal{K}} M_{\boldsymbol{b}\mathcal{L}} - \hat{\Gamma}_{\boldsymbol{b}}{}^{\boldsymbol{c}}{}_{\boldsymbol{0}} M_{\boldsymbol{c}\mathcal{K}} - \hat{\Gamma}_{\boldsymbol{b}}{}^{\mathcal{L}}{}_{\mathcal{K}} M_{\boldsymbol{0}\mathcal{L}}.$$

If the gauge conditions (13.49) are taken into account and coordinates are chosen such that $\boldsymbol{e_0} = \boldsymbol{\partial}_\tau$, then the above equation reduces to

$$\partial_\tau M_{\boldsymbol{b}\mathcal{K}} - \boldsymbol{e_b}(M_{\boldsymbol{0}\mathcal{K}}) = N_{\boldsymbol{0}\boldsymbol{b}\mathcal{K}} - \hat{\Gamma}_{\boldsymbol{b}}{}^{\boldsymbol{c}}{}_{\boldsymbol{0}} M_{\boldsymbol{c}\mathcal{K}} - \hat{\Gamma}_{\boldsymbol{b}}{}^{\mathcal{L}}{}_{\mathcal{K}} M_{\boldsymbol{0}\mathcal{L}}. \tag{13.52}$$

This last equation is not a completely satisfactory evolution equation for the components $M_{\boldsymbol{a}\mathcal{K}}$ as it does not yield information about $\partial_\tau M_{\boldsymbol{0}\mathcal{K}}$ – notice that by setting $\boldsymbol{a} = \boldsymbol{b} = \boldsymbol{0}$ in (13.51) both sides of the equation vanish as a result of the skew symmetry of the equation. To read Equation (13.52) as a suitable evolution equation one needs to know the value of the time component $M_{\boldsymbol{0}\mathcal{K}}$ either as a result of symmetries of the tensor $M_{\boldsymbol{a}\mathcal{K}}$ or through some gauge condition. In any of these cases, Equation (13.52) is just a ***transport equation*** along the congruence of conformal curves, and, accordingly, it trivially gives rise to a symmetric hyperbolic subsystem of equations.

Analysis in terms of spinors

In view of subsequent applications, the properties of the spinorial counterpart of Equation (13.51) are now analysed. In this case one has

$$\hat{\nabla}_{AA'}M_{BB'\mathcal{K}} - \hat{\nabla}_{BB'}M_{AA'\mathcal{K}} = N_{AA'BB'\mathcal{K}}, \tag{13.53}$$

where $\mathcal{K}$ denotes an arbitrary string of spinorial indices. In view of its antisymmetry, Equation (13.53) is completely equivalent to the pair of contracted equations

$$\hat{\nabla}_{(A|P'|}M_{B)}{}^{P'}{}_{\mathcal{K}} = \frac{1}{2}N_{(A|P'|B)}{}^{P'}{}_{\mathcal{K}}, \tag{13.54a}$$

$$\hat{\nabla}_{P(A'}M_{B')}{}^{P}{}_{\mathcal{K}} = \frac{1}{2}N_{P(A'}{}^{P}{}_{B')\mathcal{K}}. \tag{13.54b}$$

Thus, not unsurprisingly, one has arrived at a situation similar to the one analysed in Section 13.2. Namely, one has equations containing a symmetrised spinorial curl. A symmetric hyperbolic system can then be obtained if suitable information about the divergence $\hat{\nabla}^{PP'}M_{PP'\mathcal{K}}$ is available.

The next step in the procedure consists of introducing the space spinor version of $M_{AA'\mathcal{K}}$, namely, $M_{BB'\mathcal{K}} = -\tau^{P}{}_{B'}M_{BP\mathcal{K}}$ so that

$$\begin{aligned}
\tau_P{}^{A'}\tau_Q{}^{B'}\hat{\nabla}_{AA'}M_{BB'\mathcal{K}} &= -\tau_P{}^{A'}\tau_Q{}^{B'}\left(\tau^{R}{}_{B'}\hat{\nabla}_{AA'}M_{BR\mathcal{K}} + M_{BR\mathcal{K}}\hat{\nabla}_{AA'}\tau^{R}{}_{B'}\right)\\
&= \hat{\nabla}_{AP}M_{BQ\mathcal{K}} - \sqrt{2}M_{BR\mathcal{K}}\,\hat{\chi}_{AP}{}^{R}{}_{Q},
\end{aligned}$$

where it has been used that $\sqrt{2}\hat{\chi}_{ABCD} \equiv \tau_B{}^{A'}\tau_D{}^{C'}\hat{\nabla}_{AA'}\tau_{CC'}$ consistent with formula (4.17). From the above identity together with Equations (13.54a) and (13.54b) one obtains

$$\hat{\nabla}_{(A|P|}M_{B)}{}^{P}{}_{\mathcal{K}} = \frac{1}{2}N_{(A|P|B)}{}^{P}{}_{\mathcal{K}} - \sqrt{2}M_{(A}{}^{R}{}_{|\mathcal{K}|}\,\hat{\chi}_{B)PR}{}^{P},$$

$$\hat{\nabla}_{A(P}M^{A}{}_{Q)\mathcal{K}} = \frac{1}{2}N_{A(P}{}^{A}{}_{Q)\mathcal{K}} + \sqrt{2}M^{A}{}_{R\mathcal{K}}\,\hat{\chi}_{A(P}{}^{R}{}_{Q)}.$$

Using the decomposition

$$\hat{\nabla}_{AB} = \frac{1}{2}\epsilon_{AB}\hat{\mathcal{P}} + \hat{\mathcal{D}}_{AB},$$

with $\hat{\mathcal{P}} \equiv \tau^{AA'}\hat{\nabla}_{AA'}$ and $\hat{\mathcal{D}}_{AB} \equiv \tau_{(A}{}^{A'}\hat{\nabla}_{B)A'}$ – compare Equation (4.16) – and writing $M_{AB\mathcal{K}}$ as

$$M_{AB\mathcal{K}} = \frac{1}{2}\epsilon_{AB}m_{\mathcal{K}} + m_{AB\mathcal{K}}$$

where $m_{\mathcal{K}} \equiv M_Q{}^{Q}{}_{\mathcal{K}}$ and $m_{AB\mathcal{K}} \equiv M_{(AB)\mathcal{K}}$, one obtains

$$\begin{aligned}
\frac{1}{2}\hat{\mathcal{P}}m_{AB\mathcal{K}} - \frac{1}{2}\hat{\mathcal{D}}_{AB}m_{\mathcal{K}} - \hat{\mathcal{D}}_{P(A}m_{B)}{}^{P}{}_{\mathcal{K}}&\\
= -\frac{1}{2}N_{(A|P|B)}{}^{P}{}_{\mathcal{K}} &+ \sqrt{2}M_{(A}{}^{R}{}_{|\mathcal{K}|}\,\hat{\chi}_{B)PR}{}^{P},
\end{aligned}$$

$$\frac{1}{2}\hat{\mathcal{P}}m_{PQ\mathcal{K}} - \frac{1}{2}\hat{\mathcal{D}}_{PQ}m_{\mathcal{K}} + \hat{\mathcal{D}}_{A(P}m^{A}{}_{Q)\mathcal{K}}$$
$$= \frac{1}{2}N_{A(P}{}^{A}{}_{Q)\mathcal{K}} + \sqrt{2}M^{A}{}_{R\mathcal{K}}\,\hat{\chi}_{A(P}{}^{R}{}_{Q)}.$$

Taking linear combinations of the latter equations one finally arrives at

$$\hat{\mathcal{P}}m_{AB\mathcal{K}} - \hat{\mathcal{D}}_{AB}m_{\mathcal{K}} = E_{AB\mathcal{K}}, \tag{13.55a}$$
$$\hat{\mathcal{D}}_{P(A}m^{P}{}_{B)\mathcal{K}} = C_{AB\mathcal{K}}, \tag{13.55b}$$

where $E_{AB\mathcal{K}}$ and $C_{AB\mathcal{K}}$ are some expressions not involving derivatives of $M_{AB\mathcal{K}}$ whose precise form is not relevant for the subsequent discussion. Equation (13.55a) can be read as an evolution equation for the *spatial components* $m_{AB\mathcal{K}}$ if the *time component* $m_{\mathcal{K}}$ is known. Observe that the reduction procedure does not produce an equivalent equation for $m_{\mathcal{K}}$ consistent with the discussion of Equation (13.51).

13.4.3 The evolution equations in the frame formalism

To obtain some intuition into the structural properties of the evolution equations, it is convenient to look first at the form of the equations in a tensor frame formalism. Accordingly, one considers the vacuum extended conformal field equations as given in Section 8.4.1; see Equations (8.46).

The required evolution equations for the frame components, connection coefficients and components of the Schouten tensor are obtained from the conditions

$$\hat{\Sigma}_{\mathbf{0}\mathbf{b}} = 0, \qquad \hat{\Xi}^{\mathbf{c}}{}_{\mathbf{d}\mathbf{0}\mathbf{b}} = 0, \qquad \hat{\Delta}_{\mathbf{0}\mathbf{b}\mathbf{c}} = 0.$$

In particular, the evolution equation for the covector $\boldsymbol{f}$ defining the Weyl connection is given by

$$\hat{\Xi}^{\mathbf{c}}{}_{\mathbf{c}\mathbf{0}\mathbf{b}} = 0.$$

Using the definitions of the zero quantities given in Equations (8.44a)–(8.44c), recalling that in the vacuum case $T_{\mathbf{c}\mathbf{d}\mathbf{b}} = 0$, and making use of the gauge conditions (13.49) and (13.50), one obtains the evolution equations

$$\partial_\tau e_{\mathbf{b}}{}^{\mu} = -\hat{\Gamma}_{\mathbf{b}}{}^{\mathbf{f}}{}_{\mathbf{0}} e_{\mathbf{f}}{}^{\mu},$$
$$\partial_\tau \hat{\Gamma}_{\mathbf{b}}{}^{\mathbf{c}}{}_{\mathbf{d}} = -\hat{\Gamma}_{\mathbf{f}}{}^{\mathbf{c}}{}_{\mathbf{d}}\hat{\Gamma}_{\mathbf{b}}{}^{\mathbf{f}}{}_{\mathbf{0}} + \delta_{\mathbf{0}}{}^{\mathbf{c}}\hat{L}_{\mathbf{b}\mathbf{d}} + \delta_{\mathbf{d}}{}^{\mathbf{c}}\hat{L}_{\mathbf{b}\mathbf{0}} - \eta_{\mathbf{0}\mathbf{d}}\eta^{\mathbf{f}\mathbf{c}}\hat{L}_{\mathbf{b}\mathbf{f}} + \Theta d^{\mathbf{c}}{}_{\mathbf{d}\mathbf{0}\mathbf{b}},$$
$$\partial_\tau \hat{L}_{\mathbf{b}\mathbf{c}} = -\hat{\Gamma}_{\mathbf{b}}{}^{\mathbf{f}}{}_{\mathbf{0}}\hat{L}_{\mathbf{f}\mathbf{c}} + d_{\mathbf{f}} d^{\mathbf{f}}{}_{\mathbf{c}\mathbf{0}\mathbf{b}}.$$

These equations contain derivatives only in the τ direction – that is, they are transport equations along the conformal geodesics.

The evolution equations for the components of the rescaled Weyl tensor are obtained by resorting to an electric-magnetic decomposition; see Section 11.1.2. Using Equations (11.9) and (11.10) for the decomposition of a Weyl candidate in the equations

$$\nabla^a d_{abcd} = 0, \qquad \nabla^a d^*_{abcd} = 0,$$

one obtains the expressions

$$\begin{aligned}
\Lambda^*_{(b|0|d)} &= e_0(E_{bd}) + D_a B_{c(b}\epsilon_{d)}{}^{ac} + 2a_a\epsilon^{ac}{}_{(b}B_{d)c} - 3\chi_{(b}{}^c E_{d)c} \\
&\quad - \epsilon_b{}^{ac}\epsilon_d{}^{ef} E_{ac}\chi_{cf} + \chi E_{bd} = 0, \\
\Lambda_{(b|0|d)} &= e_0(B_{bd}) - D_a E_{c(b}\epsilon_{d)}{}^{ac} - 2a_a\epsilon^{ac}{}_{(b}E_{d)c} - 3\chi^a{}_{(b}B_{d)a} \\
&\quad - \epsilon_b{}^{ac}\epsilon_d{}^{ef} B_{ac}\chi_{cf} + \chi B_{bd} = 0,
\end{aligned}$$

where

$$h_{ab} \equiv g_{ab} - \tau_a\tau_b, \qquad \chi_{ab} = h_a{}^c\nabla_c\tau_b, \qquad \chi = h^{ab}\chi_{ab}, \qquad a_a \equiv \tau^b\nabla_b\tau_a.$$

The above form of the equations is completely general. In the particular case of a conformal Gaussian gauge system one has $e_0 = \partial_\tau$.

13.4.4 The evolution equations in the spinorial formalism

To discuss the spinorial version of the evolution equations one makes use of the extended conformal field equations

$$\hat{\Sigma}_{AA'BB'} = 0, \qquad \hat{\Xi}^C{}_{DAA'BB'} = 0, \qquad \hat{\Delta}_{CC'DD'BB'} = 0, \qquad \hat{\Lambda}_{BB'CD} = 0,$$

with the zero quantities as given in (8.53a)–(8.53e). These equations are regarded as differential conditions on the fields

$$e_{AA'}{}^\mu, \qquad \hat{\Gamma}_{AA'BC}, \qquad \hat{L}_{AA'BB'}, \qquad \phi_{ABCD}.$$

Moreover, one considers the spinor $\tau^{AA'}$ – the counterpart of the vector $\boldsymbol{\tau}$, with normalisation $\tau_{AA'}\tau^{AA'} = 2$. In terms of a spinor dyad $\{\epsilon_A{}^A\}$ adapted to $\tau^{AA'}$ one has

$$\tau^{AA'} = \epsilon_0{}^A\epsilon_{0'}{}^{A'} + \epsilon_1{}^A\epsilon_{1'}{}^{A'}.$$

In what follows, all spinorial objects will be expressed with respect to this basis. In particular, the components of $\tau^{AA'}$ with respect to $\{\epsilon_A{}^A\}$ will be denoted by $\tau^{\boldsymbol{AA'}}$.

The gauge conditions (13.49) and (13.50) in the spinorial formalism take the form

$$\tau^{AA'}e_{AA'} = \sqrt{2}\partial_\tau, \qquad \tau^{AA'}\hat{\Gamma}_{AA'}{}^B{}_C = 0, \qquad \tau^{AA'}\hat{L}_{AA'BB'} = 0. \quad (13.56)$$

For future use, it is recalled that the reduced spin Weyl connection coefficients $\hat{\Gamma}_{CC'AB}$ can be written in terms of the unphysical Levi-Civita connection coefficients $\Gamma_{CC'AB}$ and the covector $f_{AA'}$ as

$$\hat{\Gamma}_{CC'AB} = \Gamma_{CC'AB} - \epsilon_{AC}f_{BC'}.$$

Combining the above with the gauge conditions one obtains

$$\tau^{CC'}\Gamma_{CC'AB} = -\tau_A{}^{C'} f_{BC'} \tag{13.57}$$

and, furthermore, that

$$\hat{\Gamma}_{CC'AB} = \Gamma_{CC'AB} - \epsilon_{AC}\tau^{QQ'}\Gamma_{QQ'PB}\tau^P{}_{C'}.$$

In the gauge given by conditions (13.56) *the connection coefficients* $\hat{\Gamma}_{CC'AB}$ *can be fully expressed in terms of the coefficients* $\Gamma_{CC'AB}$, *and vice versa.* Comparing Equation (13.57) with the definition in Equation (4.17), one sees that the spinor $f_{AA'}$ encodes the acceleration of the congruence of conformal geodesics. In particular, if $f_{AA'} = 0$, then the congruence consists of standard geodesics and one obtains a Gaussian gauge system.

The reduced symmetric hyperbolic system of evolution equations can be deduced from the following contractions of the conformal field equations

$$\tau^{AA'}\hat{\Sigma}_{AA'}{}^{PP'}{}_{BB'}e_{PP'}{}^{\mu} = 0, \qquad \tau^{CC'}\hat{\Xi}_{ABCC'DD'} = 0,$$
$$\tau^{AA'}\hat{\Delta}_{AA'BB'CC'} = 0, \qquad \tau_{(A}{}^{A'}\hat{\Lambda}_{|A'|BCD)} = 0.$$

Explicitly, for the first three equations one has

$$\begin{aligned}
\sqrt{2}\partial_\tau e_{AA'}{}^{\mu} &= -\left(\hat{\Gamma}_{AA'}{}^{Q}{}_{B}\tau^{BQ'} + \bar{\hat{\Gamma}}_{A'A}{}^{Q'}{}_{B'}\tau^{QB'}\right)e_{QQ'}{}^{\mu},\\
\sqrt{2}\partial_\tau \hat{\Gamma}_{AA'}{}^{B}{}_{C} &= -\left(\hat{\Gamma}_{AA'}{}^{P}{}_{Q}\hat{\Gamma}_{PQ'}{}^{B}{}_{C} + \bar{\hat{\Gamma}}_{AA'}{}^{P'}{}_{Q'}\hat{\Gamma}_{QP'}{}^{B}{}_{C}\right)\tau^{QQ'}\\
&\quad + \hat{L}_{AA'CQ'}\tau^{BQ'} + \Theta\phi^{B}{}_{CQA}\tau^{Q}{}_{A'},\\
\sqrt{2}\partial_\tau \hat{L}_{AA'BB'} &= -\left(\hat{\Gamma}_{AA'}{}^{P}{}_{Q}\hat{L}_{PQ'BB'} + \bar{\hat{\Gamma}}_{A'A}{}^{P'}{}_{Q'}\hat{L}_{QP'BB'}\right)\tau^{QQ'}\\
&\quad - d^{PP'}\left(\phi_{PAQB}\epsilon_{P'B'}\tau^{Q}{}_{A'} + \bar{\phi}_{P'A'Q'B'}\epsilon_{PB}\tau_{A}{}^{Q'}\right).
\end{aligned}$$

Following the same procedure discussed in Section 13.2.4 one finds, for the Bianchi identity, that

$$\mathcal{P}\phi_{ABCD} - 2\mathcal{D}_{(A}{}^{Q}\phi_{BCD)Q} = 0. \tag{13.58}$$

Observe that this last expression is, for convenience, expressed in terms of the Levi-Civita connection $\boldsymbol{\nabla}$.

The space spinor split of the evolution equations

A more detailed version of the evolution equations is obtained by resorting to the space spinor formalism, and, in particular, to the split of the connection coefficients as given in Section 4.3.1.

Following the general strategy behind the space spinor formalism, it is convenient to define

$$\hat{\Gamma}_{ABCD} \equiv \tau_B{}^{A'}\hat{\Gamma}_{AA'CD}, \qquad \Gamma_{ABCD} \equiv \tau_B{}^{A'}\Gamma_{AA'CD}, \qquad f_{AB} \equiv \tau_B{}^{A'} f_{AA'},$$
$$\Theta_{ABCD} \equiv \tau_B{}^{A'}\tau_D{}^{C'}\hat{L}_{AA'CC'}$$

In particular, one has

$$\hat{\Gamma}_{ABCD} = \Gamma_{ABCD} - \epsilon_{CA} f_{DB}.$$

As a consequence of the gauge conditions (13.56) it follows that

$$f_{AB} = f_{(AB)}, \qquad \Gamma_Q{}^Q{}_{AB} = -f_{AB}, \qquad \hat{L}_Q{}^Q{}_{AB} = 0.$$

Defining, as in Section 4.3.1, the spinors χ_{ABCD} and ξ_{ABCD} via

$$\chi_{ABCD} \equiv -\frac{1}{\sqrt{2}}(\Gamma_{ABCD} + \Gamma^{+}_{ABCD}), \qquad \xi_{ABCD} \equiv \frac{1}{\sqrt{2}}(\Gamma_{ABCD} - \Gamma^{+}_{ABCD}),$$

one obtains from the metricity of the connection $\boldsymbol{\nabla}$ that

$$\begin{aligned}\Gamma_{ABCD} &= \frac{1}{\sqrt{2}}(\xi_{ABCD} - \chi_{ABCD}) \\ &= \frac{1}{\sqrt{2}}(\xi_{ABCD} - \chi_{(AB)CD}) - \frac{1}{2}\epsilon_{AB} f_{CD}.\end{aligned}$$

Exploiting the gauge conditions, the spinor Θ_{ABCD} can be decomposed into

$$\Theta_{ABCD} = \Theta_{AB(CD)} + \frac{1}{2}\epsilon_{CD}\Theta_{ABQ}{}^{Q}.$$

In addition, it is convenient to introduce the ***electric and magnetic parts of the rescaled Weyl spinor*** ϕ_{ABCD} via

$$\eta_{ABCD} \equiv \frac{1}{2}(\phi_{ABCD} + \phi^{+}_{ABCD}), \qquad \mu_{ABCD} \equiv -\frac{\mathrm{i}}{2}(\phi_{ABCD} - \phi^{+}_{ABCD}).$$

A calculation using the above definitions yields the detailed system:

$$\partial_\tau e_{AB}{}^0 = -\chi_{(AB)}{}^{PQ} e_{PQ}{}^0 - f_{AB}, \tag{13.59a}$$

$$\partial_\tau e_{AB}{}^\alpha = -\chi_{(AB)}{}^{PQ} e_{PQ}{}^\alpha, \tag{13.59b}$$

$$\begin{aligned}\partial_\tau \xi_{ABCD} = &-\chi_{(AB)}{}^{PQ}\xi_{PQCD} + \frac{1}{\sqrt{2}}(\epsilon_{AC}\chi_{(BD)PQ} + \epsilon_{BD}\chi_{(AC)PQ}) f^{PQ} \\ &- \sqrt{2}\chi_{(AB)(C}{}^{E} f_{D)E} - \frac{1}{2}(\epsilon_{AC}\Theta_{BDQ}{}^{Q} + \epsilon_{BD}\Theta_{ACQ}{}^{Q}) \\ &- \mathrm{i}\Theta\mu_{ABCD},\end{aligned} \tag{13.59c}$$

$$\partial_\tau f_{AB} = -\chi_{(AB)}{}^{PQ} f_{PQ} + \frac{1}{\sqrt{2}}\Theta_{ABQ}{}^{Q}, \tag{13.59d}$$

$$\partial_\tau \chi_{(AB)CD} = -\chi_{(AB)}{}^{PQ}\chi_{PQCD} - \Theta_{AB(CD)} + \Theta\eta_{ABCD}, \tag{13.59e}$$

$$\begin{aligned}\partial_\tau \Theta_{CD(AB)} = &-\chi_{(CD)}{}^{PQ}\Theta_{PQ(AB)} - \partial_\tau\Theta\eta_{ABCD} \\ &+ \mathrm{i}\sqrt{2} d^P{}_{(A}\mu_{B)CDP},\end{aligned} \tag{13.59f}$$

$$\partial_\tau \Theta_{ABQ}{}^{Q} = -\chi_{(AB)}{}^{EF}\Theta_{EFQ}{}^{Q} + \sqrt{2} d^{PQ}\eta_{ABPQ}. \tag{13.59g}$$

Remark. The term $\partial_\tau\Theta$ in the second term of the left-hand side of Equation (13.59f) arises from the fact that, in a conformal Gaussian system, the time component of the covector $\boldsymbol{d}$ is given by $\dot{\Theta}$; see Proposition 5.1.

Setting

$$\phi_0 \equiv \phi_{\mathbf{0000}}, \quad \phi_1 \equiv \phi_{\mathbf{0001}}, \quad \phi_2 \equiv \phi_{\mathbf{0011}}, \quad \phi_3 \equiv \phi_{\mathbf{0111}}, \quad \phi_4 \equiv \phi_{\mathbf{1111}},$$

the standard Bianchi system, Equation (13.58), explicitly reads

$$\begin{aligned}
&(\sqrt{2} + 2e_{\mathbf{01}}{}^0)\partial_\tau\phi_0 - 2e_{\mathbf{11}}{}^0\partial_\tau\phi_1 + 2e_{\mathbf{01}}{}^\alpha\partial_\alpha\phi_0 - 2e_{\mathbf{11}}{}^\alpha\partial_\alpha\phi_1\\
&\qquad = -6\Gamma_{\mathbf{1111}}\phi_2 + (4\Gamma_{\mathbf{1110}} + 8\Gamma_{\mathbf{0111}})\phi_1 + (2\Gamma_{\mathbf{1100}} - 8\Gamma_{\mathbf{0101}})\phi_0,\\
&(\sqrt{2} + 2e_{\mathbf{01}}{}^0)\partial_\tau\phi_1 - 2e_{\mathbf{11}}{}^0\partial_\tau\phi_2 - 2e_{\mathbf{11}}{}^\alpha\partial_\alpha\phi_2 + 2e_{\mathbf{01}}{}^\alpha\partial_\alpha\phi_1\\
&\qquad = -4\Gamma_{\mathbf{1111}}\phi_3 + (6\Gamma_{(\mathbf{01})\mathbf{11}} - 3f_{\mathbf{11}})\phi_2\\
&\qquad\quad + (4\Gamma_{\mathbf{1100}} - 4\Gamma_{(\mathbf{01})\mathbf{01}} + 2f_{\mathbf{01}})\phi_1 - (2\Gamma_{(\mathbf{01})\mathbf{00}} + f_{\mathbf{00}})\phi_0,\\
&\sqrt{2}\partial_\tau\phi_2 - e_{\mathbf{11}}{}^0\partial_\tau\phi_3 + e_{\mathbf{00}}{}^0\partial_\tau\phi_1 - e_{\mathbf{11}}{}^\alpha\partial_\alpha\phi_3 + e_{\mathbf{00}}{}^\alpha\partial_\alpha\phi_1\\
&\qquad = -\Gamma_{\mathbf{1111}}\phi_4 - 2(\Gamma_{\mathbf{1101}} + f_{\mathbf{11}})\phi_3 + 3(\Gamma_{\mathbf{0011}} + \Gamma_{\mathbf{1100}})\phi_2\\
&\qquad\quad - 2(\Gamma_{\mathbf{0001}} - f_{\mathbf{00}})\phi_1 - \Gamma_{\mathbf{0000}}\phi_0,\\
&(\sqrt{2} - 2e_{\mathbf{01}}{}^0)\partial_\tau\phi_3 + 2e_{\mathbf{00}}{}^0\partial_\tau\phi_2 - 2e_{\mathbf{01}}{}^\alpha\partial_\alpha\phi_3 + 2e_{\mathbf{00}}{}^\alpha\partial_\alpha\phi_2\\
&\qquad = -(2\Gamma_{(\mathbf{01})\mathbf{11}} + f_{\mathbf{11}})\phi_4 + (2\Gamma_{\mathbf{0011}} - 4\Gamma_{(\mathbf{01})\mathbf{01}} - 2f_{\mathbf{01}})\phi_3\\
&\qquad\quad + (6\Gamma_{(\mathbf{01})\mathbf{00}} + 3f_{\mathbf{00}})\phi_2 - 4\Gamma_{\mathbf{0000}}\phi_1,\\
&(\sqrt{2} - 2e_{\mathbf{01}}{}^0)\partial_\tau\phi_4 + 2e_{\mathbf{00}}{}^0\partial_\tau\phi_3 - 2e_{\mathbf{01}}{}^\alpha\partial_\alpha\phi_4 + 2e_{\mathbf{00}}{}^\alpha\partial_\alpha\phi_3\\
&\qquad = (2\Gamma_{\mathbf{0011}} - 8\Gamma_{\mathbf{1010}})\phi_4 + (4\Gamma_{\mathbf{0001}} + 8\Gamma_{\mathbf{1000}})\phi_3 - 6\Gamma_{\mathbf{0000}}\phi_2.
\end{aligned}$$

For completeness, the constraints

$$\Lambda_{\mathbf{AB}} \equiv \mathcal{D}^{\mathbf{PQ}}\phi_{\mathbf{PQAB}} = 0$$

are also given in explicit form:

$$\begin{aligned}
&e_{\mathbf{11}}{}^0\partial_\tau\phi_4 - 2e_{\mathbf{01}}{}^0\partial_\tau\phi_3 + e_{\mathbf{00}}{}^0\partial_\tau\phi_2 + e_{\mathbf{11}}{}^\alpha\partial_\alpha\phi_4 - 2e_{\mathbf{01}}{}^\alpha\partial_\alpha\phi_3 + e_{\mathbf{00}}{}^\alpha\partial_\alpha\phi_2\\
&\qquad = -(2\Gamma_{(\mathbf{01})\mathbf{11}} - 4\Gamma_{\mathbf{1110}})\phi_4 + (2\Gamma_{\mathbf{0011}} - 4\Gamma_{(\mathbf{01})\mathbf{01}} - 4\Gamma_{\mathbf{1100}})\phi_3\\
&\qquad\quad + 6\Gamma_{(\mathbf{01})\mathbf{00}}\phi_2 - 2\Gamma_{\mathbf{0000}}\phi_1,\\
&e_{\mathbf{11}}{}^0\partial_\tau\phi_3 - 2e_{\mathbf{01}}{}^0\partial_\tau\phi_2 + e_{\mathbf{00}}{}^0\partial_\tau\phi_1 + e_{\mathbf{11}}{}^\alpha\partial_\alpha\phi_3 - 2e_{\mathbf{01}}{}^\alpha\partial_\alpha\phi_2 + e_{\mathbf{00}}{}^\alpha\partial_\alpha\phi_1\\
&\qquad = \Gamma_{\mathbf{1111}}\phi_4 - (4\Gamma_{(\mathbf{01})\mathbf{11}} - 2\Gamma_{\mathbf{1101}})\phi_3 + 3(\Gamma_{\mathbf{0011}} - \Gamma_{\mathbf{1100}})\phi_2\\
&\qquad\quad - (2\Gamma_{\mathbf{0001}} - 4\Gamma_{(\mathbf{01})\mathbf{00}})\phi_1 - \Gamma_{\mathbf{0000}}\phi_0,\\
&e_{\mathbf{11}}{}^0\partial_\tau\phi_2 - 2e_{\mathbf{01}}{}^0\partial_\tau\phi_1 + e_{\mathbf{00}}{}^0\partial_\tau\phi_0 + e_{\mathbf{11}}{}^\alpha\partial_\alpha\phi_2 - 2e_{\mathbf{01}}{}^\alpha\partial_\alpha\phi_1 + e_{\mathbf{00}}{}^\alpha\partial_\alpha\phi_0\\
&\qquad = 2\Gamma_{\mathbf{1111}}\phi_3 - 6\Gamma_{(\mathbf{01})\mathbf{11}}\phi_2 + (4\Gamma_{\mathbf{0011}} + 4\Gamma_{(\mathbf{01})\mathbf{01}} - 2\Gamma_{\mathbf{1100}})\phi_1\\
&\qquad\quad - (4\Gamma_{\mathbf{0001}} - 2\Gamma_{(\mathbf{01})\mathbf{00}})\phi_0.
\end{aligned}$$

These *constraint equations* contain time derivatives of the components of the Weyl spinor. Furthermore, as the congruence of conformal curves is, in general, not hypersurface orthogonal, the constraint equations are not intrinsic to the leaves of a foliation.

The boundary adapted system

The standard system (13.36) is not the only symmetric hyperbolic evolution system that can be extracted from the Bianchi equation. In certain applications, such as the ones involving evolution domains with a timelike boundary, another form of the evolution equations is more convenient. In what follows, the system extracted from

$$-2\Lambda_{(0000)} = 0, \quad -2\Lambda_{(0001)} - \frac{1}{2}C_{00} = 0, \quad -2\Lambda_{(0011)} = 0, \tag{13.60a}$$

$$-2\Lambda_{(0111)} + \frac{1}{2}C_{11} = 0, \quad -2\Lambda_{1111} = 0, \tag{13.60b}$$

will be known as the ***boundary adapted system***. In the following, it will be shown that it is, indeed, symmetric hyperbolic. The principal part of the boundary adapted system can be written as

$$\mathbf{A}^\mu\partial_\mu\boldsymbol{\phi} = \begin{pmatrix} \tau^\mu + 2e_{01}{}^\mu & -2e_{00}{}^\mu & 0 & 0 & 0 \\ 2e_{11}{}^\mu & 2\tau^\mu & -2e_{00}{}^\mu & 0 & 0 \\ 0 & 2e_{11}{}^\mu & 2\tau^\mu & -2e_{00}{}^\mu & 0 \\ 0 & 0 & 2e_{11}{}^\mu & 2\tau^\mu & -2e_{00}{}^\mu \\ 0 & 0 & 0 & 2e_{11}{}^\mu & \tau^\mu - 2e_{01}{}^\mu \end{pmatrix} \partial_\mu \begin{pmatrix} \phi_0 \\ \phi_1 \\ \phi_2 \\ \phi_3 \\ \phi_4 \end{pmatrix}, \tag{13.61}$$

so that the matrices $\mathbf{A}^\mu$ are Hermitian, and, in particular, $\mathbf{A}^\mu\tau_\mu$ is positive definite. The characteristic polynomial is given by

$$\det(\mathbf{A}^\mu\xi_\mu) = 4(\tau^\mu\xi_\mu)(g^{\nu\lambda}\xi_\nu\xi_\lambda)(l^{\rho\sigma}\xi_\rho\xi_\sigma),$$

where $l^{\rho\sigma} \equiv \tau^\rho\tau^\sigma + e_{00}{}^{(\rho}e_{11}{}^{\sigma)}$. In Chapter 14, it will be seen that when τ^μ is tangent to a timelike hypersurface, then the pull-back of $l_{\mu\nu}$ gives the components of the intrinsic three-dimensional Lorentzian metric implied by $\boldsymbol{g}$ on the hypersurface.

Explicitly, the boundary adapted system takes the form

$$\begin{aligned}
&(\sqrt{2} + 2e_{01}{}^0)\partial_\tau\phi_0 - 2e_{11}{}^0\partial_\tau\phi_1 + 2e_{01}{}^\alpha\partial_\alpha\phi_0 - 2e_{11}{}^\alpha\partial_\alpha\phi_1 \\
&\qquad = -6\Gamma_{1111}\phi_2 + (4\Gamma_{1110} + 8\Gamma_{0111})\phi_1 + (2\Gamma_{1100} - 8\Gamma_{0101})\phi_0, \\
&\sqrt{2}\partial_\tau\phi_1 - e_{11}{}^0\partial_\tau\phi_2 + e_{00}{}^0\partial_\tau\phi_0 - e_{11}{}^\alpha\partial_\alpha\phi_2 + e_{00}{}^\alpha\partial_\alpha\phi_4 \\
&\qquad = -2\Gamma_{1111}\phi_3 - 3f_{11}\phi_2 + (2\Gamma_{1100} + 4\Gamma_{0011} + 2f_{01})\phi_1 - (4\Gamma_{0001} - f_{00})\phi_0, \\
&\sqrt{2}\partial_\tau\phi_2 - e_{11}{}^0\partial_\tau\phi_3 + e_{00}{}^0\partial_\tau\phi_1 - e_{11}{}^\alpha\partial_\alpha\phi_3 + e_{00}{}^\alpha\partial_\alpha\phi_1 \\
&\qquad = -\Gamma_{1111}\phi_4 - 2(\Gamma_{1101} + f_{11})\phi_3 + 3(\Gamma_{0011} + \Gamma_{1100})\phi_2 \\
&\qquad\quad - 2(\Gamma_{0001} - f_{00})\phi_1 - \Gamma_{0000}\phi_0, \\
&\sqrt{2}\partial_\tau\phi_3 - e_{11}{}^0\partial_\tau\phi_4 + e_{00}{}^0\partial_\tau\phi_2 - e_{11}{}^\alpha\partial_\alpha\phi_4 + e_{00}{}^\alpha\partial_\alpha\phi_2 \\
&\qquad = -(4\Gamma_{1110} + f_{11})\phi_2 + (2\Gamma_{0011} + 4\Gamma_{1100} - 2f_{01})\phi_3 + 3f_{00}\phi_2 - 2\Gamma_{0000}\phi_1, \\
&(\sqrt{2} - 2e_{01}{}^0)\partial_\tau\phi_4 + 2e_{00}{}^0\partial_\tau\phi_3 - 2e_{01}{}^\alpha\partial_\alpha\phi_4 + 2e_{00}{}^\alpha\partial_\alpha\phi_3 \\
&\qquad = (2\Gamma_{0011} - 8\Gamma_{1010})\phi_4 + (4\Gamma_{0001} + 8\Gamma_{1000})\phi_3 - 6\Gamma_{0000}\phi_2.
\end{aligned}$$

13.4.5 The construction of a subsidiary system

This section addresses the construction of a system of subsidiary equations for the evolution equations discussed in the previous section. The particular problem at hand consists of constructing evolution equations for the zero quantities

$$\hat{\Sigma}_{\boldsymbol{a}}{}^{\boldsymbol{c}}{}_{\boldsymbol{b}}, \qquad \hat{\Xi}^{\boldsymbol{c}}{}_{\boldsymbol{dab}}, \qquad \hat{\Delta}_{\boldsymbol{abc}}, \qquad \Lambda_{\boldsymbol{abc}},$$

encoding the extended conformal field equations. In addition, in the present hyperbolic reduction procedure, one also needs to construct evolution equations for the additional zero quantities

$$\delta_{\boldsymbol{a}}, \qquad \gamma_{\boldsymbol{ab}}, \qquad \varsigma_{\boldsymbol{ab}},$$

which play the role of constraints of the conformal equations; see Equations (8.47a)–(8.47c) for their definitions. The necessity of these extra zero quantities can be traced back to Proposition 8.3.

As in the case of the analysis of the subsidiary equations for the hyperbolic reduction procedure using gauge source functions, the subsidiary equations need to be homogeneous in zero quantities so that the vanishing of the latter on an initial hypersurface readily implies a unique vanishing solution. The basic assumption in the construction of the subsidiary system is that the evolution equations associated to the extended conformal field equations are satisfied. That is, one assumes that

$$\hat{\Sigma}_{\boldsymbol{0}}{}^{\boldsymbol{c}}{}_{\boldsymbol{b}} = 0, \qquad \hat{\Xi}^{\boldsymbol{c}}{}_{\boldsymbol{d0b}} = 0, \qquad \hat{\Delta}_{\boldsymbol{0bc}} = 0,$$

hold, together with either the standard or the boundary adapted system for the components of the Weyl spinor. The aforementioned evolution equations have been constructed using the gauge conditions

$$f_{\boldsymbol{0}} = 0, \qquad \hat{\Gamma}_{\boldsymbol{0}}{}^{\boldsymbol{b}}{}_{\boldsymbol{c}} = 0, \qquad \hat{L}_{\boldsymbol{0b}} = 0,$$

which, therefore, can also be used in the construction of the subsidiary system. Note also, that in the present gauge $d_{\boldsymbol{0}} = \Theta\beta_{\boldsymbol{0}} = \nabla_{\boldsymbol{0}}\Theta$ so that one has

$$\delta_{\boldsymbol{0}} = 0.$$

Similarly,

$$\gamma_{\boldsymbol{0b}} = \hat{L}_{\boldsymbol{0b}} - \hat{\nabla}_{\boldsymbol{0}}\beta_{\boldsymbol{b}} - \frac{1}{2}S_{\boldsymbol{0b}}{}^{\boldsymbol{ef}}\beta_{\boldsymbol{e}}\beta_{\boldsymbol{f}} + \lambda\Theta^{-2}\eta_{\boldsymbol{0b}} = 0$$

by virtue of the gauge conditions and the evolution equation

$$\tilde{\nabla}_{\boldsymbol{0}}\beta_{\boldsymbol{a}} + \beta_{\boldsymbol{0}}\beta_{\boldsymbol{a}} - \frac{1}{2}\eta_{\boldsymbol{0a}}(\beta_{\boldsymbol{e}}\beta^{\boldsymbol{e}} - 2\lambda\Theta^{-2}) = 0, \tag{13.62}$$

for the covector $\beta_{\boldsymbol{a}}$. Finally, one has

$$\varsigma_{\boldsymbol{0b}} = -\hat{L}_{\boldsymbol{b0}} - \hat{\nabla}_{\boldsymbol{0}}f_{\boldsymbol{b}} + \hat{\Gamma}_{\boldsymbol{b}}{}^{\boldsymbol{e}}{}_{\boldsymbol{0}}f_{\boldsymbol{e}} = 0,$$

as a result of the evolution equation for the covector $\boldsymbol{f}$.

The construction of subsidiary equations is similar in spirit to the one discussed in Section 13.3. There are, however, certain differences. The most conspicuous one is the fact that one is now working with a connection which is non-metric.

The subsidiary equation for the no-torsion condition

To construct the subsidiary equation for the no-torsion condition one considers the totally antisymmetric covariant derivative $\hat{\nabla}_{[a}\hat{\Sigma}_b{}^d{}_{c]}$ and observes that

$$\begin{aligned}3\hat{\nabla}_{[0}\hat{\Sigma}_b{}^d{}_{c]} &= \hat{\nabla}_0\hat{\Sigma}_b{}^d{}_c + \hat{\nabla}_b\hat{\Sigma}_c{}^d{}_0 + \hat{\nabla}_c\hat{\Sigma}_0{}^d{}_b \\ &= \hat{\nabla}_0\hat{\Sigma}_b{}^d{}_c - \hat{\Gamma}_b{}^e{}_0\hat{\Sigma}_c{}^d{}_e - \hat{\Gamma}_c{}^e{}_0\hat{\Sigma}_e{}^d{}_b. \end{aligned} \tag{13.63}$$

On the other hand, from the first Bianchi identity, Equation (2.10), and the definition of $\hat{\Xi}^d{}_{cab}$ one obtains

$$\hat{\nabla}_{[a}\hat{\Sigma}_b{}^d{}_{c]} = -\hat{\Xi}^d{}_{[cab]} - \hat{\Sigma}_{[a}{}^e{}_b\hat{\Sigma}_{c]}{}^d{}_e, \tag{13.64}$$

where it has been used that $\hat{\rho}^d{}_{[cab]} = 0$ by construction. The desired evolution equation is obtained from combining Equations (13.63) and (13.64). More precisely, one has

$$\hat{\nabla}_0\hat{\Sigma}_b{}^d{}_c = -\frac{1}{3}\hat{\Gamma}_c{}^e{}_0\hat{\Sigma}_e{}^d{}_b - \frac{1}{3}\hat{\Gamma}_c{}^e{}_0\hat{\Sigma}_e{}^d{}_b - \hat{\Xi}^d{}_{0bc}.$$

This evolution equation has the required homogeneous form.

The subsidiary equation for the Ricci identity

In this case, one considers the totally symmetrised covariant derivative $\hat{\nabla}_{[a}\hat{\Xi}^d{}_{|e|bc]}$. A direct computation shows that

$$\begin{aligned}3\hat{\nabla}_{[0}\hat{\Xi}^d{}_{|e|bc]} &= \hat{\nabla}_0\hat{\Xi}^d{}_{ebc} + \hat{\nabla}_b\hat{\Xi}^d{}_{ec0} + \hat{\nabla}_c\hat{\Xi}^d{}_{e0b} \\ &= \hat{\nabla}_0\hat{\Xi}^d{}_{ebc} - \hat{\Gamma}_b{}^f{}_0\hat{\Xi}^d{}_{ecf} - \hat{\Gamma}_c{}^f{}_0\hat{\Xi}^d{}_{efb}.\end{aligned}$$

Using the second Bianchi identity, Equation (2.11), and the definition of $\hat{\Xi}^d{}_{ebc}$ one arrives at the expression

$$\hat{\nabla}_{[a}\hat{\Xi}^d{}_{|e|bc]} = -\hat{\Sigma}_{[a}{}^f{}_b\hat{R}^d{}_{|e|c]f} - \hat{\nabla}_{[a}\hat{\rho}^d{}_{|e|bc]}.$$

The first term on the right-hand side is already of the required form. The second one needs to be analysed in more detail. It is recalled that

$$\hat{\rho}^d{}_{ebc} \equiv C^d{}_{ebc} + 2S_{e[b}{}^{df}\hat{L}_{c]f}.$$

Thus,

$$\hat{\nabla}_{[a}\hat{\rho}^d{}_{|e|bc]} = \hat{\nabla}_{[a}C^d{}_{|e|bc]} + 2S_{e[b}{}^{df}\hat{\nabla}_a\hat{L}_{c]f}.$$

In order to further expand this expression one considers $\epsilon_f{}^{abc}\hat{\nabla}_a\hat{\rho}^d{}_{ebc}$. A direct calculation shows that

$$\hat{\nabla}_{[a}C^d{}_{|e|bc]} = \nabla_{[a}C^d{}_{|e|bc]} + \delta_{[a}{}^d f_{|f}C^f{}_{e|bc]} + \eta_{e[a}f^f C^d{}_{|f|bc]}. \qquad (13.65)$$

Moreover, one has

$$\begin{aligned}\epsilon_f{}^{abc}\nabla_a C^d{}_{ebc} &= -\epsilon_f{}^{abc}\nabla_a{}^*C^{*d}{}_{ebc}\\ &= -2\nabla_a{}^*C^d{}_{ef}{}^a = 2\nabla_a C^{*a}{}_f{}^d{}_e\\ &= -\epsilon_e{}^{dgh}\nabla_a C^a{}_{fgh}.\end{aligned}$$

Thus, using that $C^c{}_{dab} = \Theta d^c{}_{dab}$ and the definition of the zero quantity Λ_{abc} one concludes that

$$\epsilon_f{}^{abc}\hat{\nabla}_a C^d{}_{ebc} = \Theta\epsilon_e{}^{dgh}\Lambda_{fgh} + 2\nabla^g\Theta d^{*d}{}_{efg} + 2\Theta f^g d^*_{gef}{}^d + 2\Theta f^g d^{*d}{}_{gfe}.$$

A similar computation using the definition of $\hat{\Delta}_{abc}$ yields

$$2\epsilon_f{}^{abc}S_{eb}{}^{dg}\hat{\Delta}_{acg} = 2\Theta\beta_g d^{*g}{}_{ef}{}^d - 2\Theta\beta_g d^{*gd}{}_{fe}.$$

Thus, using the symmetries of $d^{*c}{}_{dab}$ and the definition of δ_a one concludes that

$$\epsilon_f{}^{abc}\hat{\nabla}_a\hat{\rho}^d{}_{ebc} = \Theta\epsilon_e{}^{dgh}\Lambda_{fgh} - 2\Theta\delta^g d^{*d}{}_{efg} + \epsilon_f{}^{abc}S_{eb}{}^{dg}\hat{\Delta}_{acg}.$$

Alternatively, using the properties of the generalised Hodge duals † and ‡ defined in Equation (2.24), one can write

$$\hat{\nabla}_{[a}\hat{\rho}^d{}_{|e|bc]} = \frac{1}{6}\Theta\epsilon^f{}_{abc}\epsilon_e{}^{dgh}\Lambda_{fgh} - \frac{1}{3}\Theta\epsilon^f{}_{abc}\delta^g d^{*d}{}_{efg} - S_{e[b}{}^{dg}\hat{\Delta}_{ac]g}.$$

Combining the expressions, one obtains the required evolution equation. Namely, one has

$$\begin{aligned}\hat{\nabla}_0\hat{\Xi}^d{}_{ebc} = \hat{\Gamma}_b{}^f{}_0\hat{\Xi}^d{}_{ecf} + \hat{\Gamma}_c{}^f{}_0\hat{\Xi}^d{}_{efb} - \hat{\Sigma}_b{}^f{}_c\hat{R}^d{}_{e0f} - \frac{1}{2}\Theta\epsilon^f{}_{0bc}\epsilon_e{}^{dgh}\Lambda_{fgh}\\ + \epsilon^f{}_{0bc}\delta^g d^{*d}{}_{efg} + 3S_{e0}{}^{dg}\hat{\Delta}_{cbg},\end{aligned}$$

which is homogeneous in the zero quantities.

The subsidiary equation for the Cotton equation

In this case one considers the skew derivative $\hat{\nabla}_{[a}\hat{\Delta}_{bc]d}$. A direct computation yields

$$\begin{aligned}3\hat{\nabla}_{[0}\hat{\Delta}_{bc]d} &= \hat{\nabla}_0\hat{\Delta}_{bcd} + \hat{\nabla}_b\hat{\Delta}_{c0d} + \hat{\nabla}_c\hat{\Delta}_{0bd}\\ &= \hat{\nabla}_0\hat{\Delta}_{bcd} - \hat{\Gamma}_b{}^e{}_0\hat{\Delta}_{ced} - \hat{\Gamma}_c{}^e{}_0\hat{\Delta}_{ebd}.\end{aligned}$$

On the other hand, using the definition of $\hat{\Xi}^e{}_{cab}$ and the symmetries of $\hat{\rho}^e{}_{cab}$ one obtains

$$\begin{aligned}\hat{\nabla}_{[a}\hat{\Delta}_{bc]d} &= 2\hat{\nabla}_{[a}\hat{\nabla}_b\hat{L}_{c]d} - \hat{\nabla}_{[a}d_{|e}d^e{}_{d|bc]} - d_e\hat{\nabla}_{[a}d^e{}_{|d|bc]} \\ &= -\hat{\Xi}^e{}_{[cab]}\hat{L}_{ed} - \hat{\Xi}^e{}_{d[ab}\hat{L}_{c]e} - \hat{\rho}^e{}_{d[ab}\hat{L}_{c]e} + \hat{\Sigma}_{[a}{}^e{}_b\hat{\nabla}_{|e|}\hat{L}_{c]d} \\ &\quad - \hat{\nabla}_{[a}d_{|e}d^e{}_{d|bc]} - d_e\hat{\nabla}_{[a}d^e{}_{|d|bc]}.\end{aligned}$$

Using the definition of δ_a and γ_{ab} one finds that

$$\hat{\nabla}_{[a}d_{|e}d^e{}_{d|bc]} = -\Theta\delta_{[a}\beta_{|e}d^e{}_{d|bc]} - \Theta\gamma_{[a|e}d^e{}_{d|bc]} - \Theta f_{[a}\beta_{|e}d^e{}_{d|bc]} + \Theta\hat{L}_{[a|e}d^e{}_{d|bc]}.$$

Finally, a calculation similar to the one carried out in the previous subsection shows that

$$\epsilon_f{}^{abc}\nabla_a d^e{}_{dbc} = \epsilon_d{}^{egh}\nabla_a d^a{}_{fgh},$$

so that using Equation (13.65) and the properties of the generalised duals $\dagger$ and $\ddagger$ – see Equation (2.24) – one finds that

$$\hat{\nabla}_{[a}d^e{}_{|d|bc]} = \frac{1}{6}\epsilon_{abc}{}^f\epsilon_d{}^{egh}\Lambda_{fgh} + \delta_{[a}{}^e f_{|f}d^f{}_{d|bc]} + \eta_{d[a}f^f d^e{}_{|f|bc]}.$$

Combining the above expressions and using the properties of the decomposition of $\hat{\rho}^d{}_{dab}$ one obtains the expression

$$\begin{aligned}\hat{\nabla}_{[a}\hat{\Delta}_{bc]d} &= -\hat{\Xi}^e{}_{[cab]}\hat{L}_{ed} - \hat{\Xi}^e{}_{d[ab}\hat{L}_{c]e} + \hat{\Sigma}_{[a}{}^e{}_b\hat{\nabla}_{|e|}\hat{L}_{c]d} \\ &\quad + \Theta\delta_{[a}\beta_{|e}d^e{}_{d|bc]} + \Theta\gamma_{[a|e}d^e{}_{d|bc]} - \frac{1}{6}\epsilon_{abc}{}^f\epsilon_d{}^{egh}\Lambda_{fgh}\beta_e,\end{aligned}$$

and, finally, the evolution equation

$$\begin{aligned}\hat{\nabla}_0\hat{\Delta}_{bcd} &= \hat{\Gamma}_b{}^e{}_0\hat{\Delta}_{ced} + \hat{\Gamma}_c{}^e{}_0\hat{\Delta}_{ebd} - \hat{\Xi}^e{}_{0bc}\hat{L}_{ed} + \delta_b d_e d^e{}_{dc0} + \delta_c d_e d^e{}_{d0b} \\ &\quad + \Theta\gamma_{be}d^e{}_{dc0} + \Theta\gamma_{ce}d^e{}_{d0b} - \frac{1}{2}\epsilon_{0bc}{}^f\epsilon_d{}^{egh}\Lambda_{fgh}\beta_e,\end{aligned}$$

which is homogeneous in zero quantities as required.

The subsidiary equations for the physical Bianchi identity

The argument to show the propagation of the Bianchi identity in the present context is similar to the one discussed in Section 13.3.6. In particular, the zero quantity Λ_{ABCD} satisfies Equation (13.47). It remains to compute $\nabla^b\Lambda_{bcd}$ and express it in terms of zero quantities associated with the extended conformal field equations. A calculation using the commutator of the covariant derivative $\boldsymbol{\nabla}$ yields

$$\begin{aligned}2\nabla^b\Lambda_{bcd} &= 2\nabla^b\nabla^a d_{abcd} = 2\nabla^{[b}\nabla^{a]}d_{abcd} \\ &= 2R^e{}_{[c}{}^{ba}d_{d]eab} - 2R^e{}_a{}^{ba}d_{ebcd} + \Sigma_b{}^e{}_a\nabla_e d^{ab}{}_{cd}.\end{aligned}$$

Now, it is recalled that if $\hat{\boldsymbol{\nabla}} - \boldsymbol{\nabla} = \boldsymbol{S}(\boldsymbol{f})$, then $\hat{\Sigma}_a{}^c{}_b = \Sigma_a{}^c{}_b$. Moreover, using the formula relating the curvature tensors of the connections $\hat{\boldsymbol{\nabla}}$ and $\boldsymbol{\nabla}$, Equation (5.25b), the definitions of the zero quantities $\hat{\Xi}^c{}_{dab}$ and ς_{ab} and the symmetries of d_{abcd}, one concludes that

$$\nabla^b \Lambda_{bcd} = \hat{\Xi}^e{}_{[c}{}^{ba} d_{d]eab} - \hat{\Xi}^e{}_a{}^{ba} d_{ebcd} + \frac{1}{2}\hat{\Sigma}_b{}^e{}_a \nabla_e d^{ab}{}_{cd} + \varsigma^{ab} d_{abcd}.$$

This expression is homogeneous in zero quantities and, thus, also its spinorial counterpart $\nabla^{AA'}\Lambda_{A'ACD}$. Consequently, if the *standard evolution equations* hold, it follows from Equation (13.47) and the calculations in the previous paragraph that

$$\mathcal{P}\Lambda_{AB} - \mathcal{D}_{(A}{}^P \Lambda_{B)P} - \frac{3}{\sqrt{2}}\chi^P{}_Q{}^{SQ}\epsilon_{S(A}\Lambda_{BP)} = 2\nabla^{QQ'}\Lambda_{QQ'AB}$$

is homogeneous in zero quantities.

Finally, in the case of the *boundary adapted system*, one obtains a symmetric hyperbolic system of evolution equations of the form

$$\mathcal{P}\Lambda_{00} + \mathcal{D}_{00}\Lambda_{01} = U_{00}, \tag{13.66a}$$
$$\mathcal{P}\Lambda_{01} + \mathcal{D}_{00}\Lambda_{11} - \mathcal{D}_{11}\Lambda_{00} = U_{01}, \tag{13.66b}$$
$$\mathcal{P}\Lambda_{11} - \mathcal{D}_{11}\Lambda_{01} = U_{11}, \tag{13.66c}$$

where U_{00}, U_{01} and U_{11} are expressions homogeneous in zero quantities.

The subsidiary equations for the auxiliary zero quantities

Direct computations show that

$$2\hat{\nabla}_{[0}\delta_{b]} = \hat{\nabla}_0\delta_b + \hat{\Gamma}_b{}^e{}_0\delta_e,$$
$$2\hat{\nabla}_{[0}\gamma_{b]c} = \hat{\nabla}_0\gamma_{bc} + \hat{\Gamma}_b{}^e{}_0\gamma_{ec},$$
$$3\hat{\nabla}_{[0}\varsigma_{bc]} = \hat{\nabla}_0\varsigma_{bc} - \hat{\Gamma}_b{}^e{}_0\varsigma_{ce} - \hat{\Gamma}_c{}^e{}_0\varsigma_{eb}.$$

For δ_a one finds, using the definitions of the various zero quantities, that

$$\begin{aligned}\hat{\nabla}_{[a}\delta_{b]} &= \hat{\nabla}_a\beta_b - \hat{\nabla}_a f_b - \Theta^{-1}\hat{\nabla}_{[a}\hat{\nabla}_{b]}\Theta \\ &= -\gamma_{[ab]} + \varsigma_{ab} - \frac{1}{2}\Theta^{-1}\Sigma_a{}^e{}_b\hat{\nabla}_e\Theta.\end{aligned}$$

A lengthier computation yields

$$\begin{aligned}2\hat{\nabla}_{[a}\gamma_{b]c} &= 2\hat{\nabla}_{[a}\hat{L}_{b]c} - 2\hat{\nabla}_{[a}\hat{\nabla}_{b]}\beta_c + 2S_{c[a}{}^{ef}\beta_{|e|}\hat{\nabla}_{b]}\beta_f \\ &\quad - 2\lambda\Theta^{-3}\hat{\nabla}_{[a}\Theta\eta_{b]c} - 2\lambda\Theta^{-2}f_{[a}\eta_{b]c} \\ &= \hat{\Delta}_{abc} + \beta_e\hat{\Xi}^e{}_{cab} - \hat{\Sigma}_a{}^e{}_b\hat{\nabla}_e\beta_c + 2\beta_c\gamma_{[ab]} - 2\beta_{[a}\gamma_{b]c} + \eta_{c[a}\beta^e\gamma_{b]e} \\ &\quad + 2\lambda\Theta^{-2}\delta_{[a}\eta_{b]c} + \beta_{[a}\eta_{b]c}\beta_e\beta^e - 2\lambda\Theta^{-2}\eta_{c[a}\beta_{b]}.\end{aligned}$$

Similarly, using $d^e{}_{[cab]} = 0$, one obtains

$$\begin{aligned}\hat{\nabla}_{[a}\varsigma_{bc]} &= \hat{\nabla}_{[[a}\hat{L}_{b]c]} - \hat{\nabla}_{[[a}\hat{\nabla}_{b]}f_{c]} \\ &= \frac{1}{2}\hat{\Delta}_{[abc]} - \frac{1}{2}d_e d^e{}_{[cab]} + \frac{1}{2}\hat{R}^e{}_{[cab]}f_e - \frac{1}{2}\hat{\Sigma}_{[a}{}^e{}_b\hat{\nabla}_{|e|}f_{c]} \\ &= \frac{1}{2}\hat{\Delta}_{[abc]} + \frac{1}{2}\hat{\Xi}^e{}_{[cab]}f_e - \frac{1}{2}\hat{\Sigma}_{[a}{}^e{}_b\hat{\nabla}_{|e|}f_{c]}.\end{aligned}$$

Hence, one obtains the evolution equations

$$\begin{aligned}&\hat{\nabla}_0\delta_i = \gamma_{i0} - \hat{\Gamma}_i{}^e{}_0\delta_e, \\ &\hat{\nabla}_0\gamma_{ic} = -\gamma_{jc}\hat{\Gamma}_i{}^j{}_0 - \beta_0\gamma_{ic} - \beta_c\gamma_{i0} + \eta_{0c}(\beta^e\gamma_{ie} - 2\lambda\Theta^{-2}\delta_i), \\ &\hat{\nabla}_0\varsigma_{jk} = \hat{\Gamma}_j{}^e{}_0\varsigma_{ke} + \hat{\Gamma}_k{}^e{}_0\varsigma_{ej} + \frac{1}{2}\hat{\Delta}_{jk0} + \frac{1}{2}\hat{\Xi}^e{}_{0jk}f_e + \frac{1}{2}\hat{\Sigma}_j{}^e{}_k\hat{\Gamma}_e{}^f{}_0 f_f,\end{aligned}$$

where, in the last equation relation, (13.62) has been used to get further cancellation of terms.

13.4.6 Summary of the analysis

It is convenient to group the independent components of the spinorial fields in the extended conformal field equations as:

$$\begin{aligned}&\hat{\upsilon} \text{ independent components of } e_{AA'}{}^\mu,\ \hat{\Gamma}_{AA'BC},\ \hat{L}_{AA'BB'}, \\ &\phi \text{ independent components of } \phi_{ABCD}.\end{aligned}$$

Also, let e and $\hat{\Gamma}$ denote, respectively, the independent components of the frame and connection coefficients. In terms of the above definitions one has:

Proposition 13.3 (***properties of the conformal evolution equations***) *The extended conformal vacuum Einstein field equations*

$$\hat{\Sigma}_a{}^c{}_b = 0, \qquad \hat{\Xi}^c{}_{dab} = 0, \qquad \hat{\Delta}_{abc} = 0, \qquad \Lambda_{abc} = 0,$$

expressed in terms of a conformal Gaussian gauge imply a symmetric hyperbolic system for the components of $(\hat{\upsilon}, \phi)$ *of the form*

$$\begin{aligned}&\partial_\tau\hat{\upsilon} = \mathbf{K}\hat{\upsilon} + \mathbf{Q}(\hat{\Gamma})\hat{\upsilon} + \mathbf{L}(x)\phi, \\ &(\mathbf{I} + \mathbf{A}^0(e))\partial_\tau\phi + \mathbf{A}^\alpha(e)\partial_\alpha\phi = \mathbf{B}(\hat{\Gamma})\phi,\end{aligned}$$

where $\mathbf{I}$ *is the* 5×5 *unit matrix,* $\mathbf{K}$ *is a constant matrix,* $\mathbf{Q}(\hat{\Gamma})$ *is a smooth matrix-valued function,* $\mathbf{L}(x)$ *is a smooth matrix-valued function of the coordinates,* $\mathbf{A}^\mu(e)$ *are Hermitian matrices depending smoothly on the frame coefficients* e *and* $\mathbf{B}(\hat{\Gamma})$ *is a smooth matrix-valued function of the connection coefficients. In the case of the standard Bianchi system, the characteristic polynomial consists of the factors*

$$\tau^\mu\xi_\mu, \qquad g^{\mu\nu}\xi_\mu\xi_\nu, \qquad \left(\tau^\mu\tau^\nu + \frac{2}{3}g^{\mu\nu}\right)\xi_\mu\xi_\nu,$$

while for the boundary-adapted Bianchi system one has the factors

$$\tau^\mu \xi_\mu, \qquad g^{\mu\nu}\xi_\mu\xi_\nu, \qquad (\tau^\mu\tau^\nu + e_{\mathbf{00}}{}^{(\mu}e_{\mathbf{11}}{}^{\nu)})\xi_\mu\xi_\nu.$$

Remark. It is important to emphasise the relative simplicity of the evolution system provided by Proposition 13.3 compared with the one given in Proposition 13.1. This structure reinforces the intuition that the Weyl tensor encodes the degrees of freedom of the gravitational field.

With regards to the subsidiary system one obtains a result analogous to Proposition 13.2:

Proposition 13.4 (***properties of the subsidiary evolution***) *Assume that the conditions*

$$\hat{\Sigma}_{\mathbf{0}}{}^{\mathbf{c}}{}_{\mathbf{b}} = 0, \qquad \hat{\Xi}^{\mathbf{c}}{}_{\mathbf{d0b}} = 0, \qquad \hat{\Delta}_{\mathbf{0bc}} = 0,$$

hold and that the associated evolution equations are expressed in terms of a conformal Gaussian gauge system. Moreover, let the independent components of the rescaled Weyl spinor satisfy either the standard or the boundary-adapted evolution system. Then, the independent components of the zero quantities

$$\hat{\Sigma}_{\mathbf{a}}{}^{\mathbf{c}}{}_{\mathbf{b}}, \qquad \hat{\Xi}^{\mathbf{c}}{}_{\mathbf{dab}}, \qquad \hat{\Delta}_{\mathbf{abc}}, \qquad \Lambda_{\mathbf{abc}}, \qquad \delta_{\mathbf{a}}, \qquad \gamma_{\mathbf{ab}}, \qquad \varsigma_{\mathbf{ab}},$$

which are not determined by either the evolution equations or gauge conditions satisfy a symmetric hyperbolic system which is homogeneous in zero quantities.

Controlling the conformal Gaussian gauge

The *conformal Gaussian* hyperbolic reduction procedure is based on the assumption of the existence of a non-singular (i.e. non-intersecting) congruence of conformal geodesics. While this assumption may be valid close to an initial hypersurface, it may fail at later times. To analyse the potential breakdown of the gauge, one appends to the evolution system given in Proposition 13.4 an evolution equation for the components of the deviation vector of the congruence; see Section 5.5.7.

In what follows, let $\boldsymbol{z}$ denote a ***separation vector*** for the congruence of conformal geodesics. Accordingly, one has

$$[\dot{\boldsymbol{x}}, \boldsymbol{z}] = 0.$$

Thus, writing $\boldsymbol{z} = z^{\mathbf{a}}\boldsymbol{e}_{\mathbf{a}}$ where $\{\boldsymbol{e}_{\mathbf{a}}\}$ is a Weyl propagated frame such that $\boldsymbol{e}_{\mathbf{0}} = \dot{\boldsymbol{x}}$, it follows that

$$\boldsymbol{e}_{\mathbf{0}}(z^{\mathbf{a}})\boldsymbol{e}_{\mathbf{a}} = z^{\mathbf{a}}[\boldsymbol{e}_{\mathbf{a}}, \boldsymbol{e}_{\mathbf{0}}].$$

Using the conformal field equation $\hat{\Sigma}_{\mathbf{a}}{}^{\mathbf{c}}{}_{\mathbf{b}} = 0$ and using that, in the present gauge, $\boldsymbol{e}_{\mathbf{0}} = \boldsymbol{\partial}_\tau$ and $\hat{\Gamma}_{\mathbf{0}}{}^{\mathbf{c}}{}_{\mathbf{b}} = 0$, the above expression can be rewritten as

$$\partial_\tau z^{\mathbf{a}} = \hat{\Gamma}_{\mathbf{b}}{}^{\mathbf{a}}{}_{\mathbf{0}} z^{\mathbf{b}}.$$

Now, let $z_{AA'}$ denote the spinorial counterpart of z^a. Defining the space spinor counterpart $z_{AB} \equiv \tau_B{}^{A'} z_{AA'}$ and using the split

$$z_{AB} = \frac{1}{2} z \epsilon_{AB} + z_{(AB)},$$

a computation similar to the one used to derive the evolution equations yields the following evolution equations for the irreducible components of z_{AB}:

$$\partial_\tau z = f_{AB} z^{(AB)}, \tag{13.67a}$$

$$\partial_\tau z_{(AB)} = \chi_{CD(AB)} z^{(CD)}. \tag{13.67b}$$

The congruence of conformal geodesics will be non-intersecting as long as $z_{(AB)} \neq 0$.

13.5 Other hyperbolic reduction strategies

The hyperbolic reduction procedures discussed in Sections 13.2 and 13.4 do not exhaust the possible strategies to extract evolution equations from the conformal Einstein field equations. Indeed, other approaches have been put forward in the literature.

13.5.1 Hyperbolic reductions for the metric conformal field equations

Numerical evaluations of solutions to the vacuum conformal Einstein field equations have been carried out in Hübner (1999a,b, 2001a) using the metric formulation of the equations; see Equations (8.28a)–(8.28e). As the metric conformal field equations contain no equation which can be read as a differential equation for the components of the unphysical metric $\boldsymbol{g}$, one needs to supplement the equations in some manner. Assuming that suitable evolution equations can be found for the components of the conformal fields Ξ, Σ_a, s, L_{ab} and $d^a{}_{bcd}$ in some local coordinates $x = (x^\mu)$, the conformal metric $\boldsymbol{g}$ can be computed from the components of the Schouten tensor, $L_{\mu\nu}$, using ***generalised wave coordinates***; see the Appendix to this chapter for the vacuum Einstein field equations and the remark at the end of Section 8.2.5. More precisely, the components $g_{\mu\nu}$ of $\boldsymbol{g}$ are given as the solutions to the equations

$$\begin{aligned} \square g_{\mu\nu} - 2\nabla_{(\mu}F_{\nu)} - 2g_{\lambda\rho}g^{\sigma\tau}\Gamma_\sigma{}^\lambda{}_\mu\Gamma_\tau{}^\rho{}_\nu - 4\Gamma_\lambda{}^\sigma{}_\rho g^{\lambda\tau}g_{\sigma(\mu}\Gamma_{\nu)}{}^\rho{}_\tau \\ = -4L_{\mu\nu} - \frac{1}{3}R(x)g_{\mu\nu}, \\ \square x^\mu = -F^\mu(x), \qquad \text{that is,} \qquad \Gamma^\mu = F^\mu(x), \end{aligned}$$

where $F^\mu(x)$ are some suitable ***coordinate gauge source functions*** and it has been used that $R_{ab} = 2L_{ab} + \frac{1}{6}R(x)g_{ab}$. Observe that in the right-hand side of the first of the above equations one has the Ricci scalar R, which, following the discussion from previous sections, is to be treated as a further gauge source function.

From an analytic point of view, the approach described in the previous paragraphs leads to an evolution system with equations of mixed order. This type of system requires a more general notion of hyperbolicity than the one discussed in Chapter 12: the so-called ***Leray hyperbolicity***; see, for example, Choquet-Bruhat (2008) and Rendall (2008).

13.5.2 Wave equations for the conformal fields

One way of avoiding mixed-order evolution systems is to construct wave equations for the components of the conformal fields Ξ, Σ_a, s, L_{ab} and $d^a{}_{bcd}$. This strategy has been pursued in Paetz (2015) for the metric (vacuum) conformal Einstein field equations. More precisely, it has been shown that by introducing suitable gauge source functions, the conformal field equations can be rewritten as a system of quasilinear wave equations for the conformal fields. An alternative reformulation can be obtained using spinors; see Gasperín and Valiente Kroon (2015). This approach is briefly discussed in the remainder of this section.

Wave equations for the concomitants of the conformal factor

Wave equations for the fields Ξ, $\Sigma_{AA'}$ and s can be obtained from the following derivatives of the relevant zero quantities:

$$\nabla^{AA'} Q_{AA'} = 0, \qquad \nabla^{AA'} Z_{AA'BB'} = 0, \qquad \nabla^{AA'} Z_{AA'} = 0.$$

A direct computation renders the equations

$$\begin{aligned}
&\square \Xi - \nabla^{AA'} \Sigma_{AA'} = 0, \\
&\square \Sigma_{BB'} + \Sigma^{AA'} L_{AA'BB'} + \Xi \nabla^{AA'} L_{AA'BB'} - \nabla_{BB'} s = 0, \\
&\square s + \Sigma^{CC'} \nabla^{AA'} L_{AA'CC'} + \nabla^{AA'} \Sigma^{CC'} L_{AA'CC'} = 0.
\end{aligned}$$

The wave equation satisfied by the rescaled Weyl spinor

Recalling the definition of the zero quantity $\Lambda_{B'BCD}$, one has

$$\begin{aligned}
\nabla_A{}^{B'} \Lambda_{B'BCD} &= \nabla_A{}^{B'} \nabla^Q{}_{B'} \phi_{BCDQ} \\
&= -\nabla_{(A}{}^{B'} \nabla_{Q)B'} \phi_{BCD}{}^Q + \frac{1}{2} \epsilon_{AQ} \nabla^{PP'} \nabla_{PP'} \phi_{BCD}{}^Q \\
&= \square_{AQ} \phi_{BCD}{}^Q - \frac{1}{2} \square \phi_{ABCD},
\end{aligned}$$

where $\square_{AB}$ denotes the box operator discussed in Section 3.2.5. A further calculation shows that

$$\square_{AQ} \phi_{BCD}{}^Q = 6 \Xi \phi^{PQ}{}_{(AB} \phi_{CD)PQ} - \frac{1}{4} R(x) \phi_{ABCD}.$$

Thus, the condition $-2\nabla_A{}^{Q'}\Lambda_{BQ'CD} = 0$ implies the wave equation

$$\square\phi_{ABCD} - 12\Xi\phi^{PQ}{}_{(AB}\phi_{CD)PQ} + \frac{1}{2}R(x)\phi_{ABCD} = 0$$

for the components of the rescaled Weyl spinor as long as the conformal gauge source function $R(x)$ is explicitly provided.

The wave equation satisfied by the components of the Schouten spinor

To construct an equation for the Schouten spinor, one considers the expression

$$-2\nabla^C{}_{C'}\Delta_{CDBB'} = 0,$$

as given by Equation (13.30) together with the decomposition (13.31) for the Schouten tensor in terms of the spinor $\Phi_{AA'BB'}$ and the Ricci scalar. Accordingly, one has

$$\begin{aligned}2\nabla^C{}_{C'}\Delta_{CDBB'} &= \nabla^C{}_{C'}\nabla_C{}^{Q'}\Phi_{DQ'BB'} + \frac{1}{2}\epsilon_{DB}\nabla^C{}_{C'}\nabla_{CB'}R(x)\\ &\quad + \nabla^C{}_{C'}\Sigma^Q{}_{B'}\phi_{CDBQ} + \Sigma^Q{}_{B'}\nabla^C{}_{C'}\phi_{CDBQ},\end{aligned}$$

where

$$\begin{aligned}\nabla^C{}_{C'}\nabla_C{}^{Q'}\Phi_{DQ'BB'} &= -\nabla^C{}_{(C'}\nabla_{|C|Q')}\Phi_D{}^{Q'}{}_{BB'}\\ &\quad - \frac{1}{2}\epsilon_{C'Q'}\nabla^C{}_{P'}\nabla_C{}^{P'}\Phi_D{}^{Q'}{}_{BB'}\\ &= -\bar{\square}_{C'Q'}\Phi_D{}^{Q'}{}_{BB'} - \frac{1}{2}\square\Phi_{DC'BB'},\end{aligned}$$

$$\nabla^C{}_{C'}\nabla_{CB'}R(x) = \frac{1}{2}\epsilon_{C'B'}\square R(x).$$

Thus, using that

$$\begin{aligned}\bar{\square}_{C'Q'}\Phi_D{}^{Q'}{}_{BB'} &= \Phi^{PQ'}{}_{BB'}\Phi_{DC'PQ'} + \Phi_D{}^{Q'P}{}_{B'}\Phi_{BC'PQ'}\\ &\quad + \Xi\bar{\phi}_{C'Q'B'S'}\Phi_D{}^{Q'}{}_B{}^{S'}\\ &\quad - \frac{1}{8}R(x)\Phi_{DC'BB'} - \frac{1}{24}R(x)\Phi_{DB'BC'},\end{aligned}$$

one obtains the desired wave equation for the components of $\Phi_{AA'BB'}$. Finally, a suitable subsidiary equation to ensure that the conformal gauge source function $R(x)$ is, indeed, the Ricci scalar of the connection $\boldsymbol{\nabla}$ can be obtained from the contracted Bianchi identity (13.29).

13.6 Further reading

The original references for the hyperbolic reduction procedure based on the use of spinors and gauge source functions are Friedrich (1983, 1991) – in particular, the latter reference contains a discussion of the hyperbolic reduction of the Einstein-Yang-Mills equations. The hyperbolic reduction procedure using a conformal Gaussian system was first discussed in Friedrich (1995, 1998c); extensions of these ideas to the non-vacuum case using conformal curves have been given in Lübbe and Valiente Kroon (2012). An alternative discussion of hyperbolic reductions of the conformal field equations using space spinors can be found in Frauendiener (1998a,b). A gauge source function-based hyperbolic reduction of the conformal Einstein-Euler system for a perfect fluid with a radiation equation of state has been described in Lübbe and Valiente Kroon (2013b). A discussion of the hyperbolic reduction of the conformal Einstein-scalar field system using gauge source functions is given in Hübner (1995).

A general discussion of the procedure of hyperbolic reduction of the *standard* Einstein field equations in the vacuum and matter case can be found in Friedrich and Rendall (2000), where, for example, the case of the Einstein-Dirac system is discussed. A related reference is Reula (1998). More specific discussions of hyperbolic reductions for the vacuum Einstein field equations and their associated subsidiary evolution systems can be found in Friedrich (1996, 2005). A *Lagrangian* hyperbolic reduction for the Einstein-Euler system has been discussed in Friedrich (1998b). Extensions of this Lagrangian approach have been obtained for the equations of relativistic magnetohydrodynamics coupled to gravity – the so-called Einstein-Euler-Maxwell system – in Pugliese and Valiente Kroon (2012) and for the Einstein-charged scalar field system in Pugliese and Valiente Kroon (2013).

Readers interested in the hyperbolic reductions of the Einstein field equations used in numerical relativity are referred to the monographs by Alcubierre (2008) and Baumgarte and Shapiro (2010) as an entry point to the extensive literature.

Appendix A.1: the reduced Einstein field equations

This chapter has been primarily focused on hyperbolic reductions for the conformal Einstein field equations in their spinorial formulation. In order to put the discussion into a more general context, it is useful to briefly consider the hyperbolic reduction procedure of the (standard) Einstein field equations using ***generalised wave coordinates***. This procedure is essentially the one used in the seminal work by Fourès-Bruhat (1952) where the well-posedness of the Cauchy problem in general relativity was first established.

For simplicity, in the following, the discussion is restricted to the vacuum case so that the Einstein field equations are equivalent to

$$\tilde{R}_{ab} = 0. \tag{13.68}$$

Given general coordinates $x = (x^\mu)$, the Ricci tensor can be explicitly written in terms of the components of the metric tensor $\tilde{\boldsymbol{g}}$ and its first and second partial derivatives as

$$\begin{aligned}\tilde{R}_{\mu\nu} = &-\frac{1}{2}\tilde{g}^{\lambda\rho}\partial_\lambda\partial_\rho\tilde{g}_{\mu\nu} + \tilde{\nabla}_{(\mu}\tilde{\Gamma}_{\nu)} \\ &+ \tilde{g}_{\lambda\rho}\tilde{g}^{\sigma\tau}\tilde{\Gamma}_\sigma{}^\lambda{}_\mu\tilde{\Gamma}_\tau{}^\rho{}_\nu + 2\tilde{\Gamma}_\lambda{}^\sigma{}_\rho\tilde{g}^{\lambda\tau}\tilde{g}_{\sigma(\mu}\tilde{\Gamma}_{\nu)}{}^\rho{}_\tau,\end{aligned} \tag{13.69}$$

where it is recalled that the Christoffel symbols $\tilde{\Gamma}_\mu{}^\nu{}_\lambda$ can be written in terms of partial derivatives of the metric tensor as

$$\tilde{\Gamma}_\mu{}^\nu{}_\lambda = \frac{1}{2}\tilde{g}^{\nu\rho}(\partial_\mu \tilde{g}_{\rho\lambda} + \partial_\lambda \tilde{g}_{\mu\rho} - \partial_\rho \tilde{g}_{\mu\lambda}),$$

and one has defined

$$\tilde{\Gamma}^\nu \equiv \tilde{g}^{\mu\lambda}\tilde{\Gamma}_\mu{}^\nu{}_\lambda,$$

the so-called ***contracted Christoffel symbols***. The principal part of the vacuum Einstein field equation (13.68) is given by the terms

$$-\frac{1}{2}\tilde{g}^{\lambda\rho}\partial_\lambda\partial_\rho \tilde{g}_{\mu\nu} + \tilde{\nabla}_{(\mu}\tilde{\Gamma}_{\nu)}.$$

The first term in the above expression is hyperbolic as it coincides with the principal part of the D'Alambertian operator $\tilde{\square} \equiv \tilde{\nabla}^\mu \tilde{\nabla}_\mu$ acting on the components $\tilde{g}_{\mu\nu}$. If the second term in the principal part can be removed one would obtain a system of non-linear wave equations for $\tilde{g}_{\mu\nu}$.

Generalised wave coordinates

A systematic approach to the construction of coordinates $x = (x^\mu)$ is to require the coordinates to satisfy the equation

$$\tilde{\square} x^\mu = -F^\mu(x), \tag{13.70}$$

where the ***coordinate gauge source functions*** $F^\mu(x)$ are arbitrary smooth functions of the coordinates x. In the particular case where $F^\mu(x) = 0$ one talks of ***wave coordinates***, called ***harmonic coordinates*** in older accounts. In order to unravel the consequences of Equation (13.70), one treats the coordinates x^μ as scalar fields over $\tilde{\mathcal{M}}$. Accordingly, a direct computation gives

$$\tilde{\nabla}_\nu x^\mu = \partial_\nu x^\mu = \delta_\nu{}^\mu,$$
$$\tilde{\nabla}_\lambda \tilde{\nabla}_\nu x^\mu = \partial_\lambda \delta_\nu{}^\mu - \tilde{\Gamma}_\lambda{}^\rho{}_\nu \delta_\rho{}^\mu = -\tilde{\Gamma}_\nu{}^\mu{}_\lambda,$$

so that

$$\tilde{\square} x^\mu = \tilde{g}^{\nu\lambda}\tilde{\Gamma}_\nu{}^\mu{}_\lambda = -\tilde{\Gamma}^\mu. \tag{13.71}$$

A natural way of prescribing initial conditions for Equation (13.70) on a hypersurface $\tilde{\mathcal{S}}_\star$ with normal ν^a is to set $x^0 = 0$ with $\nu^\mu \partial_\mu x^0 = 1$ while setting the spatial coordinates (x^α) to be equal to some given coordinates on $\tilde{\mathcal{S}}_\star$ and requiring that $\nu^\mu \partial_\mu x^\alpha = 0$. Given this data, the general theory of hyperbolic differential equations ensures the existence of a solution to Equation (13.70), and as a result of Equation (13.71), one concludes that

$$\tilde{\Gamma}^\mu = F^\mu(x). \tag{13.72}$$

Moreover, if the coordinate differentials $\mathbf{d}x^\mu$ are chosen initially to be pointwise independent on the initial hypersurface $\tilde{\mathcal{S}}_\star$, then they will also remain pointwise independent close to $\tilde{\mathcal{S}}_\star$. Thus, by a suitable choice of coordinates, the contracted Christoffel symbols can be made to agree, locally, with any prescribed set of functions $F^\mu(x)$. These coordinate gauge source functions and the data for Equation (13.71) uniquely determine the coordinates. Conversely, given any metric $\tilde{\boldsymbol{g}}$, any coordinate system is characterised by some suitable gauge source function and initial data. The domain on which the coordinates $x = (x^\mu)$ form a good coordinate system depends on the initial data, the coordinate gauge source functions and the metric $\tilde{\boldsymbol{g}}$ itself. Consequently, there is little that can be said, a priori, about the domain of existence of the coordinates.

The reduced Einstein equation and the subsidiary evolution equation

Substituting Equation (13.72) into the Einstein field equations in the form given by (13.69) one finds that

$$
\begin{aligned}
-\frac{1}{2}\tilde{g}^{\lambda\rho}\partial_\lambda\partial_\rho\tilde{g}_{\mu\nu} + \tilde{\nabla}_{(\mu}F_{\nu)}(x) + \tilde{g}_{\lambda\rho}\tilde{g}^{\sigma\tau}\tilde{\Gamma}_\sigma{}^\lambda{}_\mu\tilde{\Gamma}_\tau{}^\rho{}_\nu& \\
+ 2\tilde{\Gamma}_\lambda{}^\sigma{}_\rho\tilde{g}^{\lambda\tau}\tilde{g}_{\sigma(\mu}\tilde{\Gamma}_{\nu)}{}^\rho{}_\tau = 0,& \qquad (13.73)
\end{aligned}
$$

where $F_\mu(x) \equiv g_{\mu\nu}F^\nu(x)$. This equation is a system of quasilinear wave equations for the components of the metric tensor $\tilde{\boldsymbol{g}}$. For this system, the local Cauchy problem with data on a spacelike hypersurface $\tilde{\mathcal{S}}_\star$ is well posed – one can show the existence and uniqueness of solutions and their continuous dependence on the data; see, for example, Friedrich and Rendall (2000). Equation (13.73) is known as the ***reduced Einstein field equation***.

The introduction of a specific system of coordinates via the gauge source functions $F^\mu(x)$ breaks the tensoriality of the Einstein field equation (13.68). Given a solution to the reduced Einstein field equation (13.73) the latter will imply a solution to the actual Einstein field equations as long as the coordinates $x = (x^\mu)$ satisfy Equation (13.71) for the chosen coordinate source function $F^\mu(x)$ appearing in the reduced equation. To prove that this is the case one needs to construct a suitable *subsidiary evolution equation.*

A suitable subsidiary equation for the hyperbolic reduction procedure under consideration can be obtained by observing that the reduced Einstein field equation, Equation (13.73), can be written as

$$\tilde{R}_{\mu\nu} = \tilde{\nabla}_{(\mu}Q_{\nu)}, \qquad Q_\mu \equiv \tilde{\Gamma}_\mu - F_\mu(x), \qquad (13.74)$$

where $\tilde{\Gamma}_\mu = \tilde{g}_{\mu\nu}\tilde{\Gamma}^\nu$. Now, from the contracted Bianchi identity in the form

$$\tilde{\nabla}^\mu\left(\tilde{R}_{\mu\nu} - \frac{1}{2}\tilde{R}\tilde{g}_{\mu\nu}\right) = 0,$$

it follows, by substituting Equation (13.74), that

$$\tilde{\square}Q_\nu + \tilde{R}^\mu{}_\nu Q_\mu = 0.$$

From the homogeneity on Q_μ of this wave equation, it follows that if $Q_\nu = 0$ and $\tilde{\nabla}_\mu Q_\nu = 0$ on some initial hypersurface and if $\tilde{g}_{\mu\nu}$ satisfies the reduced Einstein field equations, then $\tilde{\Gamma}^\mu = F^\mu(x)$ at later times.

Appendix A.2: differential forms

Let $\mathcal{M}$ be a four-dimensional manifold. A ***p-form*** $\boldsymbol{\alpha}$ on $\mathcal{M}$ is a totally antisymmetric covariant tensor of rank p. Thus, if $\alpha_{i_1\cdots i_p}$ is the abstract index version of $\boldsymbol{\alpha}$, one has that

$$\alpha_{i_1\cdots i_p} = \alpha_{[i_1\cdots i_p]}.$$

Given $q \in \mathcal{M}$, the space of p-forms at q is denoted by $\Lambda^p|_q(\mathcal{M})$, while the bundle of p-forms over $\mathcal{M}$ is denoted by $\Lambda^p(\mathcal{M})$. In particular, 0-forms are scalar fields so that $\Lambda^0(\mathcal{M}) = \mathcal{X}(\mathcal{M})$ and 1-forms are covectors – accordingly, $\Lambda^1(\mathcal{M}) = T^*(\mathcal{M})$. A counting argument readily shows that $\dim \Lambda^p|_q(\mathcal{M}) = 4!/p!(4-p)!$ – thus, in four dimensions any 4-form is proportional to the volume form. Given a p-form $\boldsymbol{\alpha}$ and a q-form $\boldsymbol{\beta}$, their ***wedge product*** $\boldsymbol{\alpha} \wedge \boldsymbol{\beta}$ is defined, using abstract index notation, as

$$(\alpha \wedge \beta)_{a_1\cdots a_p b_1\cdots b_q} \equiv \frac{(p+q)!}{p!q!}\alpha_{[a_1\cdots a_p}\beta_{b_1\cdots b_q]}.$$

Given local coordinates $x = (x^\mu)$ in $\mathcal{M}$, a 1-form $\boldsymbol{\alpha}$ can be written as $\boldsymbol{\alpha} = \alpha_\mu \mathbf{d}x^\mu$. More generally, for a p-form one has the expansion

$$\boldsymbol{\alpha} = \alpha_{\mu_1\cdots\mu_p}\mathbf{d}x^{\mu_1} \wedge \cdots \wedge \mathbf{d}x^{\mu_p}.$$

It can be verified that

$$\mathbf{d}x^\mu \wedge \mathbf{d}x^\nu = \mathbf{d}x^\mu \otimes \mathbf{d}x^\nu - \mathbf{d}x^\nu \otimes \mathbf{d}x^\mu.$$

Given a p-form $\boldsymbol{\alpha}$ and a vector $\boldsymbol{v} = v^\mu \boldsymbol{\partial}_\mu$, one defines the ***contraction*** $i_{\boldsymbol{v}}\boldsymbol{\alpha}$ as the $(p-1)$-form

$$i_{\boldsymbol{v}}\boldsymbol{\alpha} \equiv v^\nu \alpha_{\nu\mu_1\cdots\mu_{p-1}}\mathbf{d}x^{\mu_1} \wedge \cdots \wedge \mathbf{d}x^{\mu_{p-1}}.$$

The ***exterior derivative*** $\mathbf{d}\boldsymbol{\alpha}$ is the $(p+1)$-form defined via the relation

$$\mathbf{d}\boldsymbol{\alpha} \equiv \partial_{[\mu_1}\alpha_{\mu_2\cdots\mu_{p+1}]}\mathbf{d}x^{\mu_1} \wedge \cdots \wedge \mathbf{d}x^{\mu_{p+1}}.$$

It follows from the commutativity of partial derivatives that $\mathbf{d}^2\boldsymbol{\alpha} = 0$.

Finally, it observed that the Lie derivative of a p-form can be computed using ***Cartan's formula***:

$$\pounds_{\boldsymbol{v}}\boldsymbol{\alpha} = i_{\boldsymbol{v}}\mathbf{d}\boldsymbol{\alpha} + \mathbf{d}i_{\boldsymbol{v}}\boldsymbol{\alpha}.$$

Further details on the above expressions can be found in, for example, Frankel (2003).

14

Causality and the Cauchy problem in general relativity

The study of the Cauchy problem in general relativity was initiated by the seminal work by Fourès-Bruhat (1952). The extensions and refinements of this work and, in particular, the analysis of the existence of maximal Cauchy developments by Choquet-Bruhat and Geroch (1969) bring to the forefront the delicate interplay between geometry and the theory of partial differential equations arising in Einstein's theory of general relativity.

This chapter provides a discussion of two aspects of the Cauchy problem in general relativity: (i) the connection between the notions of causality originating from the theory of symmetric hyperbolic equations and those derived from the existence of a Lorentzian metric on the underlying spacetime manifold – the so-called *Lorentzian causality*, and (ii) the existence and uniqueness of a so-called *maximal Cauchy development* of an initial value problem for the Einstein field equations. This chapter sets the context for the discussion in Part IV of this book where asymptotically simple spacetimes are constructed by means of suitably posed initial value problems.

14.1 Basic elements of Lorentzian causality

In Section 2.5 some basic notions of Lorentzian geometry have already been introduced. These ideas are now further elaborated to present the notions of ***Lorentzian causal theory***. The summary presented here is adapted from the discussion in Ringström (2009).

In what follows, the discussion is restricted to four-dimensional Lorentzian manifolds $(\tilde{\mathcal{M}}, \tilde{\boldsymbol{g}})$ which are *orientable and time orientable*. In particular, *time orientability is equivalent to the existence of a smooth timelike vector* $\boldsymbol{t}$; see Section 2.1. The Lorentzian manifold $(\tilde{\mathcal{M}}, \tilde{\boldsymbol{g}})$ is not assumed to satisfy the Einstein field equations.

Chronological future, causal future, and so on

A vector $\boldsymbol{v} \in T(\tilde{\mathcal{M}})$ is said to be ***causal*** if $\boldsymbol{v} \neq 0$ and $\boldsymbol{v}$ is either timelike or null. Consistent with the discussion of Section 2.5, $\boldsymbol{v}$ is said to be ***future pointing*** if $\tilde{\boldsymbol{g}}(\boldsymbol{v}, \boldsymbol{t}) > 0$ and ***past pointing*** if $\tilde{\boldsymbol{g}}(\boldsymbol{v}, \boldsymbol{t}) < 0$. A ***future-pointing causal curve*** on $(\tilde{\mathcal{M}}, \tilde{\boldsymbol{g}})$ is one for which its tangent vector is everywhere future pointing causal. The notion of ***past-pointing causal curve*** is defined in an analogous manner.

Causal curves can be used to define *order relations* between points of the manifold $\tilde{\mathcal{M}}$. Given p, $q \in \tilde{\mathcal{M}}$, one writes $p \prec\!\prec q$ if there is a *future-pointing timelike curve* in $\tilde{\mathcal{M}}$ from p to q; $p \prec q$ if there is a *future causal curve* from p to q; and $p \preceq q$ if either $p = q$ or $p \prec q$. Given a subset $\mathcal{U} \subset \tilde{\mathcal{M}}$ one defines the ***chronological future*** and ***chronological past*** of $\mathcal{U}$, respectively, as

$$I^+(\mathcal{U}) \equiv \{p \in \tilde{\mathcal{M}} \mid q \prec\!\prec p \text{ for some } q \in \mathcal{U}\},$$
$$I^-(\mathcal{U}) \equiv \{p \in \tilde{\mathcal{M}} \mid p \prec\!\prec q \text{ for some } q \in \mathcal{U}\}.$$

Moreover, the ***causal future*** and ***causal past*** of $\mathcal{U}$ are defined, respectively, as

$$J^+(\mathcal{U}) \equiv \{p \in \tilde{\mathcal{M}} \mid q \preceq p \text{ for some } q \in \mathcal{U}\},$$
$$J^-(\mathcal{U}) \equiv \{p \in \tilde{\mathcal{M}} \mid p \preceq q \text{ for some } q \in \mathcal{U}\}.$$

A schematic depiction of the sets $I^\pm(\mathcal{U})$ and $J^\pm(\mathcal{U})$ is given in Figure 14.1. The sets $I^+(\mathcal{U})$ and $I^-(\mathcal{U})$ can be shown to be open. No general statements of this type can be made about $J^+(\mathcal{U})$ and $J^-(\mathcal{U})$. However, one has that $I^+(\mathcal{U}) \subseteq J^+(\mathcal{U})$ and $I^-(\mathcal{U}) \subseteq J^-(\mathcal{U})$.

Global hyperbolicity

The natural class of spacetimes for which an initial value problem can be formulated is that of globally hyperbolic ones.

A spacetime $(\tilde{\mathcal{M}}, \tilde{\boldsymbol{g}})$ without closed timelike curves is called ***causal***. A causal spacetime $(\tilde{\mathcal{M}}, \tilde{\boldsymbol{g}})$ is said to be ***globally hyperbolic*** if for any pair of points p, $q \in \tilde{\mathcal{M}}$ with $p \prec q$ the ***causal diamond*** $J^+(p) \cap J^-(q)$ is compact; see Figure 14.2. The classical definition of global hyperbolicity as given, for example, in Wald

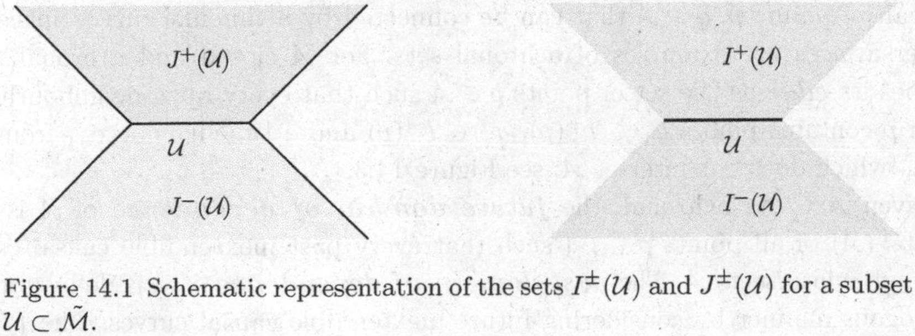

Figure 14.1 Schematic representation of the sets $I^\pm(\mathcal{U})$ and $J^\pm(\mathcal{U})$ for a subset $\mathcal{U} \subset \tilde{\mathcal{M}}$.

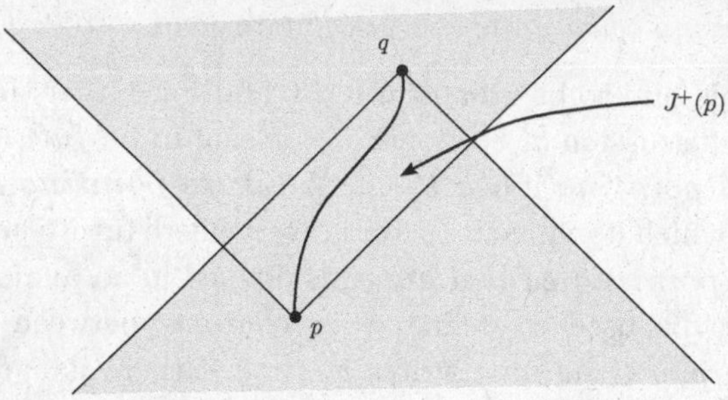

Figure 14.2 The causal diamond $J^+(p) \cap J^-(q)$: the points $p, q \in \tilde{\mathcal{M}}$ satisfy $p \prec q$. In a globally hyperbolic spacetime any such diamond is compact.

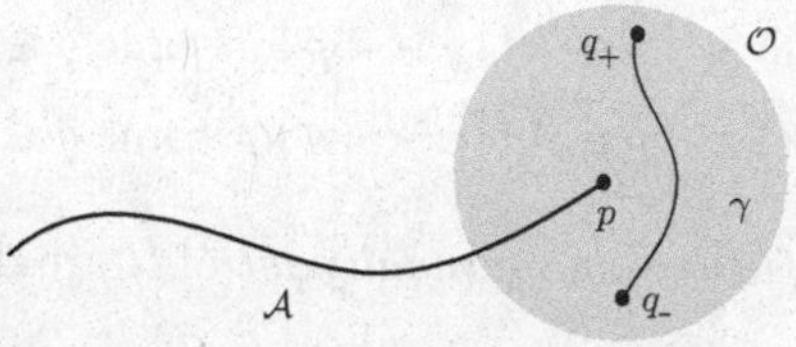

Figure 14.3 The edge of a closed achronal set $\mathcal{A}$: for the point $p \in \mathcal{A}$ there exists an open neighbourhood $\mathcal{O}$ containing p and $q_+ \in I^+(p)$, $q_- \in I^-(p)$ such that q_- and q_+ can be joined by a timelike curve γ not intersecting $\mathcal{A}$.

(1984), makes use of the *stronger* notion of ***strongly causal spacetimes***, that is, the non-existence of "almost closed" causal curves. The classical definition and the one given here have been shown to be equivalent in Bernal and Sánchez (2007).

In physical terms, global hyperbolicity is closely connected to the idea of ***classical determinism***, that is, the prediction or retrodiction of future or past states, respectively, from a set of initial conditions. Pathologies like the existence of closed timelike curves are not present in globally hyperbolic spacetimes.

Cauchy surfaces

A subset $\mathcal{A}$ of a Lorentzian manifold $(\tilde{\mathcal{M}}, \tilde{\boldsymbol{g}})$ is said to be ***achronal*** if there is no pair of points p, $q \in \mathcal{A}$ that can be connected by a timelike curve. Spacelike hypersurfaces are examples of achronal sets. For $\mathcal{A}$ closed and achronal, one defines its ***edge*** as the set of points $p \in \mathcal{A}$ such that every open neighbourhood $\mathcal{O}$ of p contains points $q_+ \in I^+(p)$, $q_- \in I^-(p)$ and a timelike curve γ from q_- to q_+ which does not interset $\mathcal{A}$; see Figure 14.3.

Given $\mathcal{A} \subset \tilde{\mathcal{M}}$ achronal, the ***future domain of dependence*** of $\mathcal{A}$ is the set $D^+(\mathcal{A})$ of all points $p \in \tilde{\mathcal{M}}$ such that every past inextendible causal curve through p intersects $\mathcal{A}$. The ***past domain of dependence*** of $\mathcal{A}$ is defined in an analogous manner by considering future inextendible causal curves. The ***(full)***

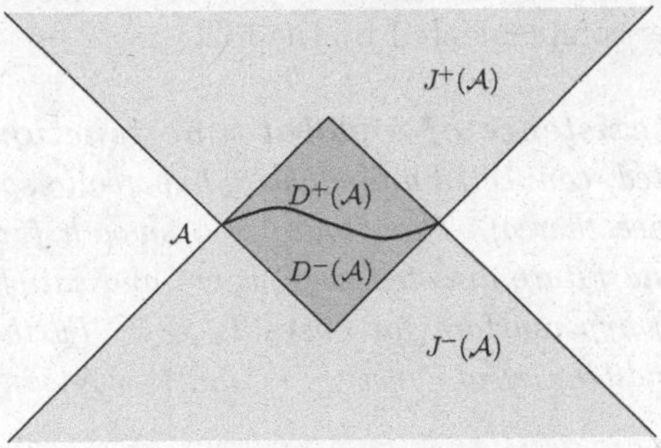

Figure 14.4 Domain of dependence of an achronal set $\mathcal{A}$ and its relation to the causal past and future $J^{\pm}(\mathcal{A})$.

domain of dependence of $\mathcal{A}$ is then defined as

$$D(\mathcal{A}) \equiv D^{+}(\mathcal{A}) \cup D^{-}(\mathcal{A}).$$

In some accounts, the sets $D^{+}(\mathcal{A})$ and $D^{-}(\mathcal{A})$ are called, respectively, the ***future*** and ***past Cauchy development of*** $\mathcal{A}$. The reason for these alternative names is clarified by the discussion in Section 14.2. It can be verified that $\mathcal{A} \subset D^{+}(\mathcal{A}) \subset J^{+}(\mathcal{A})$; see Figure 14.4. From the achronality of $\mathcal{A}$ it follows that $D^{+}(\mathcal{A}) \cap I^{-}(\mathcal{A}) = \emptyset$.

Since information travels along causal curves, a point $p \in D^{+}(\mathcal{S})$ receives information only from $\mathcal{S}$. Accordingly, if *physical laws are causal* – as in the case of general relativity – initial data should determine the physics in $D^{+}(\mathcal{S})$ – and, in fact, in all of $D(\mathcal{S})$.

A ***Cauchy hypersurface*** in $\tilde{\mathcal{M}}$ is a hypersurface $\tilde{\mathcal{S}}$ such that

$$D(\tilde{\mathcal{S}}) = \tilde{\mathcal{M}}.$$

Cauchy hypersurfaces are characterised by the fact that they are intersected exactly once by every inextendible timelike curve in $\tilde{\mathcal{M}}$; see, for example, Ringström (2009). Cauchy hypersurfaces are continuous three-dimensional submanifolds of the spacetime manifold $\tilde{\mathcal{M}}$; see, for example, Wald (1984). Cauchy hypersurfaces provide an alternative description of globally hyperbolic spacetimes: *any globally hyperbolic spacetime possesses a Cauchy hypersurface.* Global hyperbolicity restricts the topology of a spacetime. More precisely, one has that:

Proposition 14.1 (***topology of globally hyperbolic spacetimes***) *Let* $(\tilde{\mathcal{M}}, \tilde{\boldsymbol{g}})$ *denote a connected, time-oriented globally hyperbolic Lorentzian manifold and let* $\tilde{\mathcal{S}}$ *be a Cauchy hypersurface thereof. Then*

$$\tilde{\mathcal{M}} \approx \mathbb{R} \times \tilde{\mathcal{S}}.$$

In other words, $\tilde{\mathcal{M}}$ *can be foliated by Cauchy hypersurfaces. Moreover, if* $\tilde{\mathcal{S}}'$ *is another Cauchy hypersurface, then* $\tilde{\mathcal{S}} \approx \tilde{\mathcal{S}}'$.

The above result is complemented by the following:

Proposition 14.2 (***existence of a global time function***) *Let $(\tilde{\mathcal{M}}, \tilde{g})$ be an oriented, time-oriented, connected and globally hyperbolic spacetime and let $\tilde{\mathcal{S}}$ be a Cauchy hypersurface thereof. Then there is a smooth function t on $\tilde{\mathcal{M}}$ such that $\mathrm{d}t$ is timelike and future directed everywhere and satisfies the property that $t^{-1}(T_\bullet)$ is a Cauchy hypersurface for every $T_\bullet \in \mathbb{R}$. Furthermore, $t^{-1}(0) = \tilde{\mathcal{S}}$ and for every inextendible causal curve $\gamma : (\mathrm{s}_-, \mathrm{s}_+) \to \tilde{\mathcal{M}}$ one has $t(\gamma(\mathrm{s})) \to \pm\infty$ as $\mathrm{s} \to \mathrm{s}_\pm$.*

For a proof of these results, see, for example, Ringström (2009), proposition 11.3 and theorem 11.27. Finally, one has the following:

Proposition 14.3 (***asymptotic simplicity and global hyperbolicity***) *An asymptotically simple and empty spacetime $(\tilde{\mathcal{M}}, \tilde{\boldsymbol{g}})$ is globally hyperbolic.*

The reader interested in a proof is referred to Hawking and Ellis (1973), proposition 6.9.2.

Cauchy horizons

In what follows let $\overline{D^+(\mathcal{A})}$ denote the *closure of the future domain of dependence* of an achronal set $\mathcal{A} \subset \tilde{\mathcal{M}}$. This set is characterised by the fact that for $p \in \overline{D^+(\mathcal{A})}$ every past inextendible timelike curve from p intersects $\mathcal{A}$; see proposition 8.3.2 in Wald (1984). The achronal set $\mathcal{A}$ is not necessarily a Cauchy hypersurface. To characterise how much $\mathcal{A}$ deviates from being a Cauchy hypersurface, it is convenient to introduce the set

$$H^+(\mathcal{A}) \equiv \overline{D^+(\mathcal{A})} \setminus I^-\big(D^+(\mathcal{A})\big),$$

the so-called ***future Cauchy horizon*** of $\mathcal{A}$. The ***past Cauchy horizon*** is defined in an analogous manner as $H^-(\mathcal{A}) \equiv \overline{D^-(\mathcal{A})} \setminus I^+\big(D^-(\mathcal{A})\big)$. It can be shown that $H^+(\mathcal{A})$ is achronal. Moreover, one has that $\mathcal{A} \subset D^+(\mathcal{A})$ and $\partial D^+(\mathcal{A}) = H^+(\mathcal{A}) \cup \mathcal{A}$; see Figure 14.5. Similar properties hold for $D^-(\mathcal{A})$.

The ***(full) Cauchy horizon*** is then defined as $H(\mathcal{A}) \equiv H^+(\mathcal{A}) \cup H^-(\mathcal{A})$. It can be proved that $H(\mathcal{A}) = \partial\big(D(\mathcal{A})\big)$ and that the achronal set $\mathcal{A}$ is a Cauchy surface for $(\tilde{\mathcal{M}}, \boldsymbol{g})$ if and only if $H(\mathcal{A}) = \emptyset$; see proposition 8.3.6 and its corollary in Wald (1984).

The following property of Cauchy horizons will be used at various points in this book (cf. theorem 8.3.5 in Wald (1984)):

Proposition 14.4 (***structure of Cauchy horizons***) *Every point $p \in H^+(\mathcal{A})$ lies on a null geodesic contained entirely in $H^+(\mathcal{A})$ which is either inextendible or has a past endpoint on the edge of $\mathcal{A}$.*

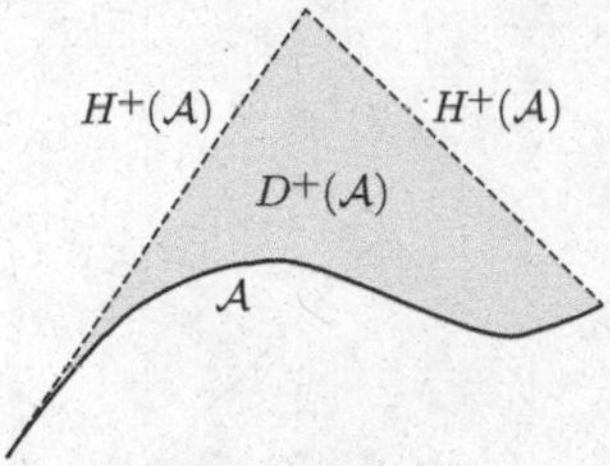

Figure 14.5 The Cauchy horizon of an achronal set $\mathcal{A}$. Observe that $\partial D^+(\mathcal{A}) = H^+(\mathcal{A}) \cup \mathcal{A}$.

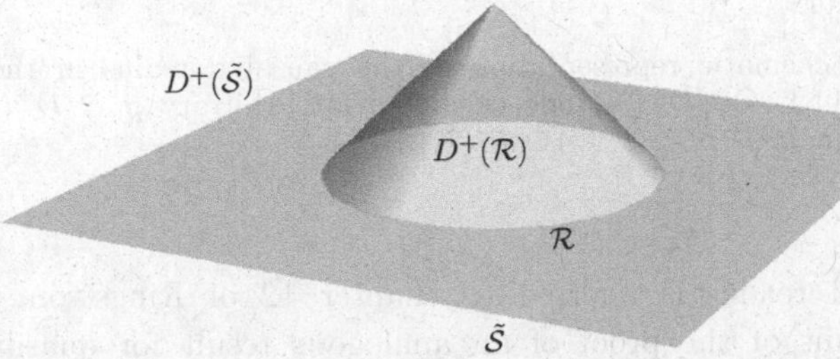

Figure 14.6 Schematic representation of the causal domains of Theorem 14.1. The hypersurface $\tilde{\mathcal{S}}$ is a Cauchy hypersurface and $\mathcal{R} \subseteq \tilde{\mathcal{S}}$ is a region within such that $\mathbf{u} = 0$.

14.2 PDE causality versus Lorentzian causality

Two different notions of causality have been discussed so far in this book: *partial differential equation (PDE) causality* based on the uniqueness of solutions to symmetric hyperbolic systems – Theorem 12.1 – and *Lorentzian causality*, discussed in the first sections of this chapter. These notions of causality are conceptually different from each other. However, they are linked by the following result (see also Figure 14.6):

Theorem 14.1 (***the relation between PDE and Lorentzian causalities***) *Let $(\tilde{\mathcal{M}}, \tilde{\mathbf{g}})$ be a connected, oriented, time-oriented, globally hyperbolic spacetime and let $\tilde{\mathcal{S}}$ be a smooth spacelike Cauchy hypersurface. Let $\mathcal{R} \subseteq \tilde{\mathcal{S}}$ and let $\mathcal{U}$ be an open set containing $D^+(\mathcal{R})$. Assume that $\mathbf{u} : \mathcal{U} \to \mathbb{C}^N$ solves the symmetric hyperbolic system*

$$\mathbf{A}^\mu(x, \mathbf{u})\partial_\mu \mathbf{u} + \mathbf{B}(x, \mathbf{u}) = 0.$$

Moreover, assume that the above equation has a characteristic polynomial which contains the factor $(\tilde{g}^{\mu\nu}\xi_\mu\xi_\nu)$ where $\tilde{g}^{\mu\nu}$ denotes the contravariant components of the metric $\tilde{\mathbf{g}}$. If $\mathbf{u}$ vanishes on $\mathcal{R}$, then $\mathbf{u}$ vanishes on $D^+(\mathcal{R})$. There is a similar statement for $D^-(\mathcal{R})$.

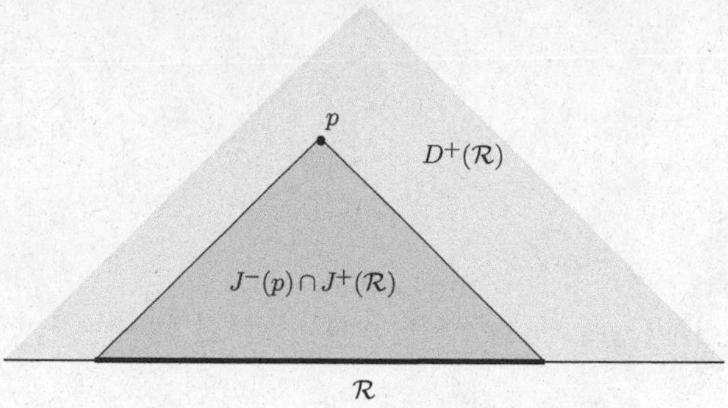

Figure 14.7 Schematic representation of the causal domains in the proof of Theorem 14.1. On $D^+(\mathcal{R})$ one considers for arbitrary $p \in D^+(\mathcal{R})$ the associated domain $\mathcal{D}(p) \equiv J^-(p) \cap J^+(\mathcal{R})$.

The interested reader is referred to chapter 12 of Ringström (2009) for a detailed account of the proof of an analogous result for quasilinear wave equations. It is, nevertheless, useful to discuss some of the ideas behind the proof.

Theorem 12.1 and its Corollary 12.1 – ensuring the uniqueness of solutions to symmetric hyperbolic systems – can be applied only on *lens-shaped domains*. The main idea behind Theorem 14.1 is then to construct a cover of $D^+(\mathcal{R})$ consisting of lens-shaped domains. The metric $\tilde{\boldsymbol{g}}$ provides a natural way of constructing the required cover. Accordingly, the Lorentzian metric allows the introduction of the notions of Lorentzian causality discussed in the first sections of this chapter.

One begins by considering points $p \in D^+(\mathcal{R})$ which are *suitably close* to $\mathcal{R}$ and aims to conclude that $\mathbf{u} = 0$ on $\mathcal{D}(p) \equiv J^-(p) \cap J^+(\mathcal{R})$;—see Figure 14.7. By means of the exponential map $\exp_p : T|_p(\tilde{\mathcal{M}}) \supset \bar{\mathcal{V}} \to \mathcal{V} \subset \tilde{\mathcal{M}}$ – see Section 11.6.2 – the metric $\tilde{\boldsymbol{g}}$ allows the introduction of normal coordinates in some neighbourhood of p – these coordinates can be seen as providing a diffeomorphism between a neighbourhood $\bar{\mathcal{V}}$ of the origin in $T|_p(\tilde{\mathcal{M}})$ to a neighbourhood $\mathcal{V}$ of p. By considering p sufficiently close to $\mathcal{R}$ one can ensure that $\mathcal{D}(p)$ is compact and completely contained in $\mathcal{V}$. On $T|_p(\tilde{\mathcal{M}})$ one can define a function $\bar{f} : T|_p(\tilde{\mathcal{M}}) \to \mathbb{R}$ via $\bar{f}(\boldsymbol{v}) = \tilde{\boldsymbol{g}}(\boldsymbol{v}, \boldsymbol{v})$. Hence, for the present purposes, the neighbourhood $\mathcal{V}$ can be regarded as a subset of the Minkowski spacetime coordinatised by standard Cartesian coordinates. One also defines $f : \mathcal{V} \to \mathbb{R}$ such that $f \equiv \bar{f} \circ \exp_p^{-1}$. Now, given a constant $c > 0$, the condition $\tilde{\boldsymbol{g}}(\boldsymbol{v}, \boldsymbol{v}) = c$ defines (spacelike) hyperboloids on $\mathcal{V}$. More precisely, for given c, the locus of points in $\mathcal{V}$ corresponding to the hyperboloid is given by $f^{-1}(c) \equiv \{q \in \mathcal{V} \,|\, \tilde{\boldsymbol{g}}(\exp_p^{-1}(q), \exp_p^{-1}(q)) = c\}$ – observe that both $\bar{f}$ and f are not injective so that f^{-1} is a set consisting of more than a single point. Now, $f^{-1}(c)$ has two components: one associated to future-directed vectors and the other associated to past-directed vectors. For $c > 0$, let $\mathcal{Q}_c(p)$ denote the component of $f^{-1}(c)$ associated to past-directed vectors and let $\mathcal{Q}_0(p)$ denote the past null cone through p. One can use the hyperboloids

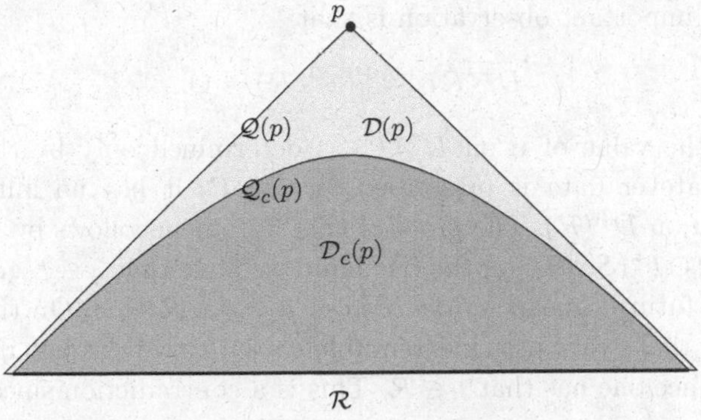

Figure 14.8 Schematic representation of the causal domains in the proof of Theorem 14.1. Given $p \in D^+(\mathcal{R})$, the sets $\mathcal{Q}(p)$ and $\mathcal{D}(p)$ describe, respectively, the past light cone through p and the region between the light cone and $\mathcal{R}$. For $c > 0$, the set $\mathcal{Q}_c(p)$ describes a hyperboloid inside the past light cone of p, while $\mathcal{D}_c(p)$ is the region between the hyperboloid and $\mathcal{R}$. The set $\mathcal{D}_c(p)$ is a lens-shaped domain. See the main text for further details.

$\mathcal{Q}_c(p)$ to foliate the interior of the past light cone passing through p. One defines $\mathcal{D}_c(p) \equiv J^-(\mathcal{Q}_c(p)) \cap J^+(\mathcal{R})$. The domain $\mathcal{D}_c(p)$ describes the region between the hyperboloid $Q_c(p)$ and $\mathcal{R}$, while $\mathcal{D}(p)$ describes the region between the past light cone and $\mathcal{R}$; see Figure 14.8. Now, $\mathcal{D}(p)$ can be shown to be compact. Moreover, it can be seen that $\mathcal{Q}_c(p) \subset I^-(p)$ for $c > 0$ so that $J^-(\mathcal{Q}_c(p)) \subset J^-(p)$ and, in addition, that $\mathcal{D}_c(p) \subset \mathcal{D}(p)$. A further argument allows one to verify that $\mathcal{Q}_c(p)$ for $c > 0$ is a lens-shaped domain on which, modulo some technical details, Corollary 12.1 can be applied. Thus, if $\mathbf{u} = 0$ on $\mathcal{R}$, one concludes that $\mathbf{u} = 0$ on $\mathcal{D}_c(p)$.

To show that $\mathbf{u} = 0$ on $\mathcal{D}(p)$ one now considers a sequence $\{c_l\}$, $l \in \mathbb{N}$, of positive numbers converging to zero. It can then be shown that

$$\operatorname{int} \mathcal{D}(p) \subset \cup_l \mathcal{D}_{c_l}(p) \subset \mathcal{D}(p)$$

– intuitively, by choosing smaller and smaller c_l's one obtains hyperboloids which are, successively, “closer” to the light cone $\mathcal{Q}_p$ thus “filling” $\mathcal{D}(p)$. From this observation and given that $\mathbf{u} = 0$ on each of the $\mathcal{D}_{c_l}(p)$ one can conclude that, indeed, $\mathbf{u} = 0$ on $\mathcal{D}(p)$.

Now, an adaptation of Proposition 14.2 ensures the existence of a time function t on $D^+(\mathcal{R})$. Given $c > 0$, $t^{-1}([0, c])$ denotes a ***slab*** in $D^+(\tilde{\mathcal{S}}) \supset D^+(\mathcal{R})$. Considering points suitably close to $\tilde{\mathcal{S}}$, it is possible to construct a slab $K_\varepsilon \equiv t^{-1}([0, \varepsilon]) \cap D^+(\mathcal{R})$ in $D^+(\mathcal{R})$ for some $\varepsilon > 0$ – this slab can be thought of as the union of domains of the type $\mathcal{D}(p)$ on each of which one already knows that $\mathbf{u} = 0$. The ***top*** of the slab, $t^{-1}(\varepsilon) \cap D^+(\mathcal{R})$ – on which $\mathbf{u} = 0$ – can now be used as a new initial surface from which one constructs a further slab. The rest of the proof consists of showing that $D^+(\mathcal{R})$ can be fully covered by slabs of the type described above so that $\mathbf{u} = 0$ everywhere on $D^+(\mathcal{R})$.

Remark. An important observation is that

$$D^+(\mathcal{R}) \cap I^+(\tilde{\mathcal{S}} \setminus \mathcal{R}) = \emptyset.$$

Accordingly, the value of $\mathbf{u}$ on $D^+(\mathcal{R})$ is determined only by the data on $\mathcal{R}$ – that is, whatever data is prescribed on $\tilde{\mathcal{S}} \setminus \mathcal{R}$, it has no influence on the behaviour of $\mathbf{u}$ on $D^+(\mathcal{R})$. The proof of this statement follows by contradiction: let $q \in D^+(\mathcal{R}) \cap I^+(\tilde{\mathcal{S}} \setminus \mathcal{R})$; on the one hand we have that $q \in I^+(\tilde{\mathcal{S}} \setminus \mathcal{R})$ so that there exists a future timelike curve γ from $p \in \tilde{\mathcal{S}} \setminus \mathcal{R}$ to q. On the other hand $q \in D^+(\mathcal{R})$ so that every past inextendible causal curve through q intersects $\mathcal{R}$. As a consequence one has that $p \in \mathcal{R}$. This is a contradiction since $p \in \tilde{\mathcal{S}} \setminus \mathcal{R}$.

14.3 Cauchy developments and maximal Cauchy developments

For ease of presentation, the subsequent discussion is restricted to the case of *standard Cauchy initial value problems* where initial data is prescribed on a Cauchy hypersurface $\tilde{\mathcal{S}}$. For a detailed account of the Cauchy problem in general relativity the reader is referred to the monograph by Ringström (2009).

As discussed in Section 11.3, the (say, vacuum) Einstein field equations imply on $\tilde{\mathcal{S}}$ a set of constraint equations: the so-called Hamiltonian and momentum constraints for a Riemannian metric $\tilde{\boldsymbol{h}}$ and a symmetric trace-free tensor $\tilde{\boldsymbol{K}}$. Assume one is given a solution $(\tilde{\boldsymbol{h}}, \tilde{\boldsymbol{K}})$ to the Hamiltonian and momentum constraint Equations (11.13a) and (11.13b) on $\tilde{\mathcal{S}}$. To discuss the relation between a solution to the Einstein constraint equations and a solution to the Einstein field equations one needs to introduce the notion of a ***Cauchy development***:

Definition 14.1 (***Cauchy development***) *A* ***Cauchy development*** *of the initial data set $(\tilde{\mathcal{S}}, \tilde{\boldsymbol{h}}, \tilde{\boldsymbol{K}})$ consists of a solution $(\tilde{\mathcal{M}}, \tilde{\boldsymbol{g}})$ of the vacuum Einstein field equations, an embedding φ of $\tilde{\mathcal{S}}$ into $\tilde{\mathcal{M}}$ and a choice of a unit normal vector such that $\varphi(\tilde{\mathcal{S}})$ is a Cauchy hypersurface and the pull-backs by φ of the induced metric and the second fundamental form for the prescribed unit normal coincide with $\tilde{\boldsymbol{h}}$ and $\tilde{\boldsymbol{K}}$.*

The seminal work in Fourès-Bruhat (1952) has shown that, given a solution to the constraint equations on $\tilde{\mathcal{S}}$, it is always possible to obtain a Cauchy development. More precisely, one has the following (see also Figure 14.9):

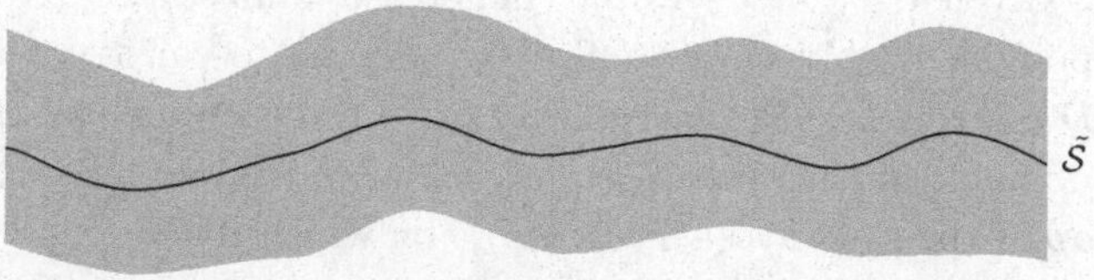

Figure 14.9 Schematic representation of a Cauchy development (in gray) of some initial data set $(\tilde{\mathcal{S}}, \tilde{\boldsymbol{h}}, \tilde{\boldsymbol{K}})$ for the Einstein field equations.

Theorem 14.2 (***existence of a development of an initial data set***) *Given an initial data set $(\tilde{\mathcal{S}}, \tilde{\boldsymbol{h}}, \tilde{\boldsymbol{K}})$ for the Einstein field equations it is always possible to find a corresponding development.*

The above result is a cornerstone of the mathematical study of the Einstein field equations as it shows that it is meaningful to formulate a Cauchy problem for the Einstein field equations. The original proof of the theorem used the hyperbolic reduction of the Einstein field equations based on *wave coordinates*; see the Appendix to Chapter 13. The hyperbolic reductions for the conformal Einstein field equations discussed in Chapter 13 readily lead to an alternative proof which is briefly sketched for completeness.

Proof The proof of the theorem amounts to a local existence result for the Cauchy problem for the Einstein field equations. For convenience, consider the spinorial version of the standard conformal Einstein field equations; see Section 8.3.2. *Setting $\Xi = 1$ and $\Phi_{\boldsymbol{AA'BB'}} = 0$ one obtains a spinorial representation of the vacuum Einstein field equations.* For these equations, the hyperbolic reduction procedure summarised in Proposition 13.1 shows that given a choice of coordinate and frame gauge source functions $F^{\boldsymbol{a}}(x)$ and $F_{\boldsymbol{AB}}(x)$, the Einstein field equations imply a symmetric hyperbolic system for the frame coefficients, connection coefficients and the Weyl spinor. Smooth initial data $\mathbf{u}_\star$ for these evolution equations can be obtained from the pair $(\tilde{\boldsymbol{h}}, \tilde{\boldsymbol{K}})$ using the procedure leading to Lemma 11.1. The basic existence and uniqueness result for symmetric hyperbolic systems given in Theorem 12.2 ensures the existence of a solution $\mathbf{u}$ to the *evolution equations* in a slab of the form $\tilde{\mathcal{M}}_T \equiv (-T, T) \times \tilde{\mathcal{S}}$ for some $T > 0$. In what follows, for conceptual clarity, the Riemannian 3-manifold $\tilde{\mathcal{S}}$ regarded as a submanifold of $\tilde{\mathcal{M}}_T$ will be denoted as $\tilde{\mathcal{S}}_\star$; that is, one has $\tilde{\mathcal{S}}_\star = \varphi(\tilde{\mathcal{S}})$. On $\tilde{\mathcal{S}}_\star$ the solution $\mathbf{u}$ coincides with the initial data $\mathbf{u}_\star$. In view of the homogeneous structure of the subsidiary evolution equations as described in Proposition 13.2, the solution $\mathbf{u}$ implies a solution to the *conformal Einstein field equations* with $\Xi = 1$ and $\Phi_{\boldsymbol{AA'BB'}} = 0$. From the components of $\mathbf{u}$ one can construct a Lorentzian metric $\tilde{\boldsymbol{g}}$ which will be a solution to the Einstein field equations on $\tilde{\mathcal{M}}_T$; compare Proposition 8.1. To conclude, it is observed that the hyperbolic procedure leading to the evolution equations is based on an adapted frame tetrad $\{\boldsymbol{e_a}\}$ such that $\boldsymbol{e_0}$ on $\tilde{\mathcal{S}}_\star$ gives the unit normal of the initial hypersurface; see Section 11.4. From this observation it follows that the pull-back of $\tilde{\boldsymbol{g}}$ to $\tilde{\mathcal{S}}_\star$ coincides with the Riemannian metric $\tilde{\boldsymbol{h}}$. Moreover, by construction, the extrinsic curvature of $\tilde{\mathcal{S}}_\star$ coincides with the tensor $\tilde{\boldsymbol{K}}$. Accordingly $(\tilde{\mathcal{M}}_T, \tilde{\boldsymbol{g}})$ provides the required Cauchy development. □

An important aspect of the notion of a Cauchy development is its non-uniqueness. A different choice of gauge source functions will, in general, lead to *a different Cauchy development for the same initial data.* Observe, however, that as one is constructing solutions to tensorial equations in the regions where two different Cauchy developments $(\tilde{\mathcal{M}}, \tilde{\boldsymbol{g}})$ and $(\tilde{\mathcal{M}}', \tilde{\boldsymbol{g}}')$ overlap $\tilde{\mathcal{M}} \cap \tilde{\mathcal{M}}'$, these must be related to each other by a diffeomorphism, that is, a coordinate transformation. This non-uniqueness of Cauchy developments of a given initial data creates a tension with the notion of ***geometric uniqueness***, that is, the expectation that a given initial data set should give rise to a unique solution to the Einstein field equations. To deal with this issue one introduces the notion of a ***maximal Cauchy development***:

Definition 14.2 (***maximal Cauchy development***) *Let* $(\tilde{\mathcal{S}}, \tilde{\boldsymbol{h}}, \tilde{\boldsymbol{K}})$ *be an initial data set for the vacuum Einstein equations. A Cauchy development* $(\tilde{\mathcal{M}}, \tilde{\boldsymbol{g}})$ *with embedding* $\varphi : \tilde{\mathcal{S}} \to \tilde{\mathcal{M}}$ *of this data is said to be* ***maximal*** *if for any other Cauchy development* $(\tilde{\mathcal{M}}', \tilde{\boldsymbol{g}}')$ *with embedding* $\varphi' : \tilde{\mathcal{S}} \to \tilde{\mathcal{M}}'$, *there is a smooth map* $\psi : \tilde{\mathcal{M}}' \to \tilde{\mathcal{M}}$ *which is a diffeomorphism onto its image such that* $\varphi = \psi \circ \varphi'$ *and* $\psi^* \tilde{\boldsymbol{g}} = \tilde{\boldsymbol{g}}'$.

The maximal Cauchy development describes the biggest spacetime that can be recovered from a given initial data set for the Einstein field equations. Any other Cauchy development must be contained in it. For this notion to be of utility it should satisfy some existence and uniqueness properties. Indeed, one has the following result, first proven in Choquet-Bruhat and Geroch (1969):

Theorem 14.3 (***existence of a maximal Cauchy development***) *Given some initial data* $(\tilde{\mathcal{S}}, \tilde{\boldsymbol{h}}, \tilde{\boldsymbol{K}})$ *for the Einstein field equations, there exists a maximal Cauchy development which is unique up to isometries.*

The original proof of this theorem famously relies on *Zorn's lemma*. Alternative proofs not depending on this axiom of set theory have been given more recently in Sbierski (2013) and Wong (2013).

Remark. The maximal Cauchy development of an initial data set is, in general, different from the so-called ***maximal analytic extension*** of the solution to the initial value problem, that is, the biggest spacetime that can be associated to a given metric allowing for analytic changes of coordinates. As an example, compare the Penrose diagram of the maximal analytical extension of the Reissner-Nordström spacetime given in Figure 6.14 and the Penrose diagram of its maximal Cauchy development in Figure 14.10.

The characterisation and construction of the maximal Cauchy development of an arbitrary initial data set $(\tilde{\mathcal{S}}, \tilde{\boldsymbol{h}}, \tilde{\boldsymbol{K}})$ is a challenging endeavour. It requires controlling the evolution dictated by the Einstein field equations under very general conditions. Generically, one expects the following to be true:

Conjecture 14.1 (***strong cosmic censorship***) *The maximal Cauchy development of generic initial data for the vacuum Einstein field equations is inextendible.*

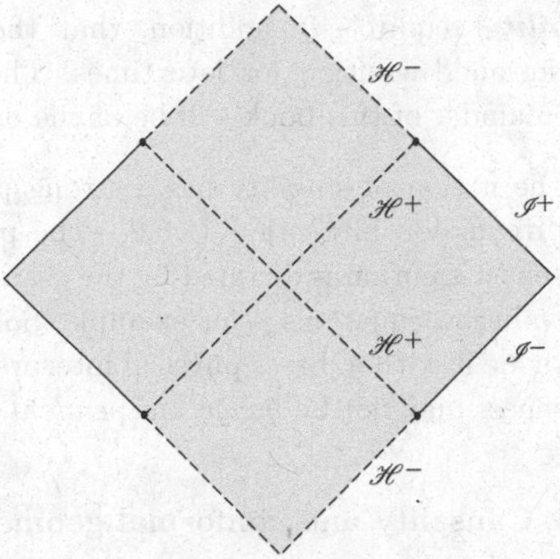

Figure 14.10 Maximal Cauchy development of the Reissner-Nordström spacetime. In this case, the spacetime extends only up to the Cauchy horizon $\mathscr{H}^-$. Notice that the timelike singularities of the spacetime do not appear in the diagram.

A concise discussion of the above conjecture and its various caveats can be found in Chruściel (1991), Rendall (2008) and Ringström (2009).

14.4 Stability of solutions

A problem simpler than cosmic censorship is the construction of the development of initial data sets which are, in some sense, close to initial data for some exact solution (the ***background solution***) whose global structure is well known. Such initial data are called a ***perturbation of the initial data for the exact solution***. Under suitable circumstances one expects the maximal Cauchy development of the perturbed initial data to have a global structure similar to that of the maximal Cauchy development of the exact solution. The resulting spacetime is called a ***perturbation of the exact solution***. This notion of perturbations is a non-linear one: the perturbed solutions are required to satisfy the Einstein field equations without any approximation – as opposed, say, to linearised perturbations where one considers solutions to evolution equations which are linearised with respect to some background exact solution. The underlying strategy behind the analysis of non-linear perturbations is to use the knowledge of the global properties of a solution to the equations of general relativity to infer the existence of other solutions with analogous properties. This point of view leads to the notion of ***stability***.

When discussing the stability of solutions to the Einstein field equations one typically distinguishes between the notions of orbital and asymptotical stability. A solution is said to be ***orbitally stable*** if the global geometry of the perturbed evolution exhibits the same features as the original (background) solution – for example, the existence of a complete null infinity. The stronger notion of ***asymptotic stability*** requires, in addition, that the perturbed solution converges to the background solution for late times. The stability results to be discussed in the remainder of this book will be of the orbital type.

Remark. Although the notion of stability has a strong physical motivation – see, for example, the discussion in Section 12.3.2 – the precise formulation of *closeness* to a certain exact solution is dictated by the details of the PDE theory used to analyse the evolution equations – for example, Sobolev norms – and it may be difficult to provide it with a direct physical interpretation. In particular, statements about closeness may not be *gauge independent.*

14.5 Causality and conformal geometry

Let $(\mathcal{M}, \boldsymbol{g})$ denote a conformal extension of a *physical spacetime* $(\tilde{\mathcal{M}}, \tilde{\boldsymbol{g}})$ with $\boldsymbol{g} = \Xi^2 \tilde{\boldsymbol{g}}$. As a Lorentzian manifold in its own right, the unphysical spacetime $(\mathcal{M}, \boldsymbol{g})$ gives rise to its own causal notions. The causal notions in $(\tilde{\mathcal{M}}, \tilde{\boldsymbol{g}})$ and $(\mathcal{M}, \boldsymbol{g})$ are, however, related to each other – it is not hard to see that *the causal notions introduced in Section 14.1 are conformally invariant.* More precisely, if $p, q \in \tilde{\mathcal{M}}$ are connected to each other via some particular causal relation with respect to the metric $\tilde{\boldsymbol{g}}$ (e.g. $p \prec q$, $p \preceq q$ or $p \prec\!\prec q$), then they are also causally related in the same way with respect to the metric $\boldsymbol{g}$. Special care is needed, however, when discussing points which lie on the conformal boundary of the conformal extension $(\mathcal{M}, \boldsymbol{g})$ as these points do not exist in the physical spacetime manifold $\tilde{\mathcal{M}}$. Moreover, any compact set in the unphysical manifold $(\mathcal{M}, \boldsymbol{g})$ which intersects the conformal boundary will be, from the perspective of the physical manifold $\tilde{\mathcal{M}}$, non-compact. This observation is of importance for the notion of global hyperbolicity as it is formulated in terms of compactness of domains in the physical spacetime $(\tilde{\mathcal{M}}, \tilde{\boldsymbol{g}})$.

A further cautionary note concerns Cauchy horizons in the unphysical spacetime $(\mathcal{M}, \boldsymbol{g})$ which may not correspond to domains in $\tilde{\mathcal{M}}$. The prototypical case of this situation arises in the discussion of Minkowski-like spacetimes. From the point of view of $(\mathcal{M}, \boldsymbol{g})$, the conformal boundary of these spacetimes corresponds to the Cauchy horizon of hyperboloidal hypersurfaces – which from the conformal point of view are compact domains. From the physical perspective of $(\tilde{\mathcal{M}}, \tilde{\boldsymbol{g}})$ the hyperboloids are non-compact and there are no Cauchy horizons. The correspondence between the conformal boundary and Cauchy horizons for Minkowski-like spacetimes is analysed in some detail in Section 16.3.

Penrose diagrams provide a convenient way of visualising the causal properties of spacetimes. For example, an inspection of the Penrose diagram of the anti-de Sitter spacetime, Figure 14.11, readily shows that the spacetime cannot be globally hyperbolic: causal diamonds intersecting the conformal boundary of the conformal representation correspond to non-compact causal diamonds in the physical spacetime. Alternatively, by looking at the Penrose diagram it is easy to draw a timelike curve which does not intersect a putative Cauchy

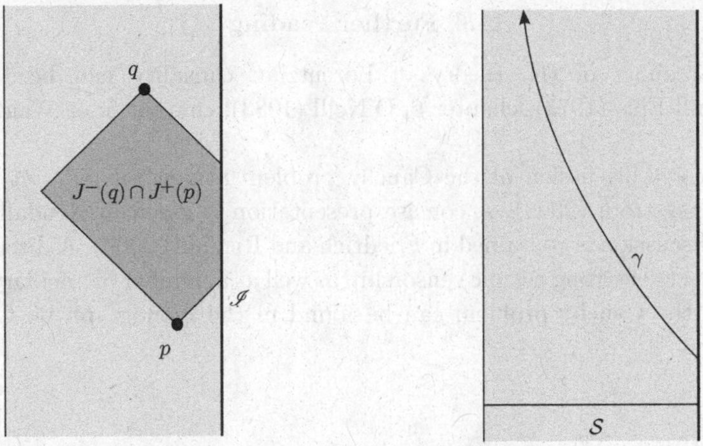

Figure 14.11 Non-global hyperbolicity of the anti-de Sitter spacetime. To the left: causal diamonds intersecting the conformal boundary are non-compact in the physical picture. To the right: given a putative Cauchy hypersurface $\mathcal{S}$, it is always possible to find a timelike curve γ not intersecting $\mathcal{S}$.

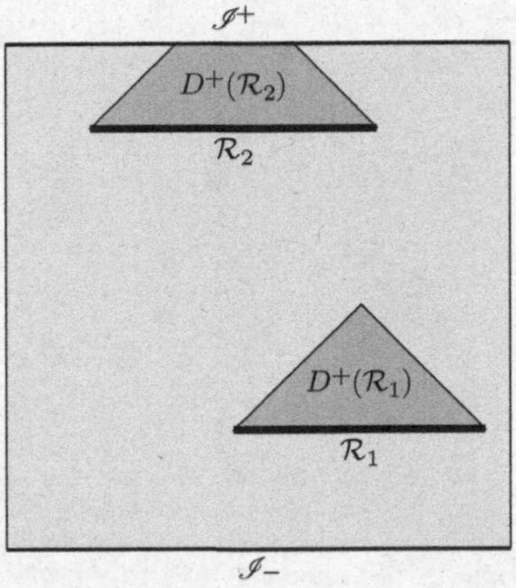

Figure 14.12 Examples of some domains of dependence in the de Sitter spacetime.

hypersurface $\mathcal{S}$ – it is only necessary that in the conformal picture the curve starts at some point of the conformal boundary which lies in the future of $\mathcal{S}$. A second example of the insights provided by the inspection of the Penrose diagrams involves the de Sitter spacetime; see Figure 14.12. A peculiarity of this spacetime is that there exist regions in the spacetime whose domain of dependence is non-compact – to see this, it is only necessary to consider domains which are, from the conformal point of view, sufficiently close to the conformal boundary $\mathscr{I}$.

14.6 Further reading

Detailed accounts of the theory of Lorentzian causality can be found in Hawking and Ellis (1973), chapter 6; O'Neill (1983), chapter 5; or Wald (1984), chapter 8.

An extensive discussion of the Cauchy problem in general relativity can be found in Ringström (2009). A concise presentation is given in Rendall (2008). A related discussion is contained in Friedrich and Rendall (2000). A discussion of various aspects of strong cosmic censorship as well as a number of ancillary results concerning the Cauchy problem can be found in the monograph by Chruściel (1991).

Part IV

Applications

15

De Sitter-like spacetimes

This chapter discusses the global existence and stability of de Sitter-like spacetimes, that is, vacuum spacetimes with a de Sitter-like value of the cosmological constant. This class of spacetimes admits a conformal extension with a spacelike conformal boundary; see Theorem 10.1. The construction of de Sitter-like spacetimes provides, arguably, the simplest application of the conformal field equations to the analysis of global properties of spacetimes. The original discussion of the analysis presented in this chapter was given in Friedrich (1986b). The results of this seminal analysis were subsequently generalised to the case of Einstein equations coupled to the Yang-Mills field in Friedrich (1991). The methods used in the proof of the stability of the de Sitter spacetime can be adapted to analyse the future non-linear stability of Friedman-Robertson-Walker cosmologies with a perfect fluid satisfying the equation of state of radiation; see Lübbe and Valiente Kroon (2013b).

The global existence and stability theorem proven in this chapter can be formulated as follows:

Theorem *(global existence and stability of de Sitter-like spacetimes).* *Small enough perturbations of initial data for the de Sitter spacetime give rise to solutions of the vacuum Einstein field equations which exist globally towards the past and the future. The solutions have the same global structure as the de Sitter spacetime. Thus, perturbations of the de Sitter spacetime are asymptotically simple.*

Intuitively, the last statement in the theorem can be read as saying that the resulting spacetimes have a Penrose diagram similar to the one of the de Sitter spacetime; see Figure 15.1. Accordingly, these spacetimes provide non-trivial (i.e. dynamic) examples of asymptotically simple spacetimes. A detailed formulation of the above result is given in the main text of the chapter; see Theorem 15.1.

To illustrate the comparative advantages of the hyperbolic reduction procedures discussed in Chapter 13, two versions of the proof are provided. The first one makes use of *gauge source functions* and follows the original proofs

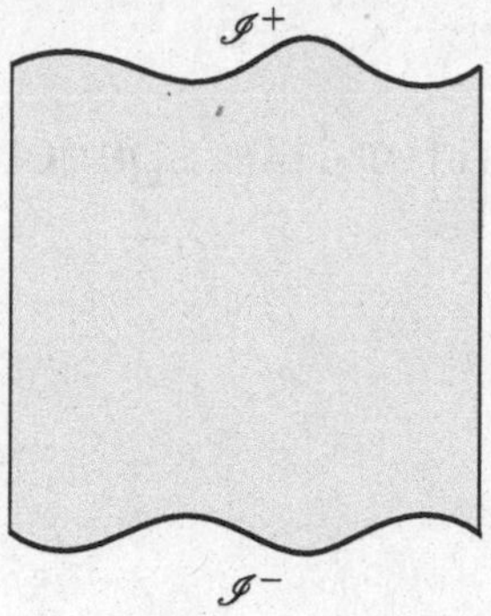

Figure 15.1 Penrose diagram of a de Sitter-like spacetime.

in Friedrich (1986b, 1991). The second proof makes use of *conformal Gaussian systems* and is based on the analysis given in Lübbe and Valiente Kroon (2009). In both approaches, and as a consequence of the use of the conformal field equations, it is possible to formulate initial value problems for the perturbed de Sitter-like spacetime not only on a standard initial hypersurface at a fiduciary finite time, but also on a hypersurface corresponding to the conformal boundary of the spacetime.

The basic strategy used in this chapter to analyse the global existence of solutions to the Einstein field equations had been previously used in Choquet-Bruhat and Christodoulou (1981) to establish the global existence of solutions to the Yang-Mills equations.

15.1 The de Sitter spacetime as a solution to the conformal field equations

The basic conformal properties of the de Sitter spacetime have already been discussed in Section 6.3. In this section the de Sitter spacetime is recast as a solution to the conformal field equations. This is a first step in the construction of an existence and global stability result.

15.1.1 Basic representation in the Einstein cosmos

For simplicity of the exposition, the cosmological constant will be assumed to take the value $\lambda = -3$ so that the conformal factor realising the embedding of the de Sitter spacetime (given in standard coordinates) into the Einstein static universe is given by

$$\mathring{\Xi} \equiv \Xi_{dS} = \cos\tau, \tag{15.1}$$

where τ, the affine parameter of the geodesics introduced in Equation (6.9), is used as a time coordinate. Here, and in the rest of the chapter, the symbol $\mathring{}$ is used to indicate that the associated object is treated as a *background field.*

As a consequence of the conformal embedding, the geometry of the conformal de Sitter spacetime is given by the corresponding expressions for the Einstein static universe as discussed in Section 6.1.3. The various conformal fields on the Einstein cylinder $(\mathbb{R} \times \mathbb{S}^3, \boldsymbol{g}_{\mathscr{E}})$, with

$$\boldsymbol{g}_{\mathscr{E}} = \mathbf{d}\tau \otimes \mathbf{d}\tau - \boldsymbol{\hbar},$$

will be expressed in terms of an orthonormal frame $\{\mathring{\boldsymbol{e}}_{\boldsymbol{a}}\}$ such that

$$\mathring{\boldsymbol{e}}_0 = \boldsymbol{\partial}_\tau, \qquad \mathring{\boldsymbol{e}}_{\boldsymbol{i}} = \boldsymbol{c}_{\boldsymbol{i}}, \tag{15.2}$$

where $\{\boldsymbol{c}_{\boldsymbol{i}}\}$ denotes the globally defined frame on $\mathbb{S}^3$ discussed in Section 6.1.2; see Equations (6.2a)–(6.2c).

In what follows, the manifold $\mathcal{M}_{\mathscr{E}} \equiv \mathbb{R} \times \mathbb{S}^3$ will be described locally in terms of ***Gaussian coordinates*** (τ, x^α) where (x^α) are some local coordinates on $\mathbb{S}^3$ which are extended to coordinates on a subset $\mathcal{U} \subset \mathcal{M}_{\mathscr{E}}$ by requiring them to remain constant along the geodesics parametrised by τ. As a consequence of the $\boldsymbol{g}_{\mathscr{E}}$-orthonormality of the vector fields $\{\boldsymbol{\partial}_\tau, \boldsymbol{c}_{\boldsymbol{i}}\}$, it follows that

$$\mathring{\boldsymbol{e}}_{\boldsymbol{a}} = \delta_{\boldsymbol{a}}{}^{\boldsymbol{b}} \boldsymbol{c}_{\boldsymbol{b}} \equiv \mathring{e}_{\boldsymbol{a}}{}^{\boldsymbol{b}} \boldsymbol{c}_{\boldsymbol{b}}. \tag{15.3}$$

Using the structure equations – see Section 2.7.3 – on $\mathbb{S}^3$, it can be verified that the connection coefficients $\mathring{\gamma}_{\boldsymbol{i}}{}^{\boldsymbol{j}}{}_{\boldsymbol{k}}$ of the Levi-Civita connection $\boldsymbol{D}$ of the standard metric of $\mathbb{S}^3$, $\boldsymbol{\hbar}$, with respect to the *spatial frame* $\{\boldsymbol{c}_{\boldsymbol{i}}\}$ are given by

$$\mathring{\gamma}_{\boldsymbol{i}}{}^{\boldsymbol{j}}{}_{\boldsymbol{k}} = -\epsilon_{\boldsymbol{i}}{}^{\boldsymbol{j}}{}_{\boldsymbol{k}}, \tag{15.4}$$

where $\epsilon_{\boldsymbol{ijk}}$ denotes the components of the volume form on $\mathbb{S}^3$; see Section 6.1.2. Now, observing that $\mathring{\boldsymbol{e}}_0 = \boldsymbol{\partial}_\tau$ is a Killing vector of the Einstein cylinder, it follows that the connection coefficients associated to the frame $\{\mathring{\boldsymbol{e}}_{\boldsymbol{a}}\}$ are given by

$$\mathring{\Gamma}_{\boldsymbol{a}}{}^{\boldsymbol{b}}{}_{\boldsymbol{c}} = \epsilon_{0\boldsymbol{a}}{}^{\boldsymbol{b}}{}_{\boldsymbol{c}}. \tag{15.5}$$

The sign difference between Equations (15.4) and (15.5) arises from the fact that the Riemannian metric implied by $\boldsymbol{g}_{\mathscr{E}}$ on $\mathbb{S}^3$ is negative definite.

Using the expressions (6.8a) and (6.8b) for the Schouten tensor of the Einstein cylinder, it follows that, in terms of the frame $\{\mathring{\boldsymbol{e}}_{\boldsymbol{a}}\}$ described above, one has

$$\mathring{L}_{\boldsymbol{ab}} = \delta_{\boldsymbol{a}}{}^{\boldsymbol{0}} \delta_{\boldsymbol{b}}{}^{\boldsymbol{0}} - \frac{1}{2}\eta_{\boldsymbol{ab}}, \tag{15.6a}$$

$$\mathring{d}^{\boldsymbol{a}}{}_{\boldsymbol{bcd}} = 0. \tag{15.6b}$$

For later use, it is also observed that the components of the trace-free Ricci tensor are given by

$$\mathring{\Phi}_{\boldsymbol{ab}} = \delta_{\boldsymbol{a}}{}^{\boldsymbol{0}} \delta_{\boldsymbol{b}}{}^{\boldsymbol{0}} - \frac{1}{4}\eta_{\boldsymbol{ab}}.$$

Finally, a further computation yields that

$$\mathring{\Sigma} = -\sin\tau, \qquad \mathring{\Sigma}_i = 0, \tag{15.7a}$$

$$\mathring{s} = -\frac{1}{4}\cos\tau. \tag{15.7b}$$

Spinorial expressions

To compute the spinorial counterpart of the fields discussed in the previous section let $\tau^{AA'}$ denote the spinorial counterpart of the vector $\sqrt{2}\boldsymbol{\partial}_\tau$ so that one has the normalisation $\tau_{\boldsymbol{AA}'}\tau^{\boldsymbol{AA}'} = 2$.

The spinorial counterpart of the frame coefficients $\mathring{e}_{\boldsymbol{a}}{}^{\boldsymbol{b}} = \delta_{\boldsymbol{a}}{}^{\boldsymbol{b}}$ – compare Equation (15.3) – is given by

$$\mathring{e}_{\boldsymbol{AA}'}{}^{\boldsymbol{b}} = \sigma_{\boldsymbol{AA}'}{}^{\boldsymbol{b}},$$

where $\sigma_{\boldsymbol{AA}'}{}^{\boldsymbol{b}}$ denotes the Infeld-van der Waerden symbols; see Section 3.1.9. In general, the coefficients $\mathring{e}_{\boldsymbol{AA}'}{}^{\boldsymbol{a}}$ can be decomposed as

$$\mathring{e}_{\boldsymbol{AA}'}{}^{\boldsymbol{a}} = \frac{1}{2}\tau_{\boldsymbol{AA}'}\mathring{e}^{\boldsymbol{a}} - \tau^{\boldsymbol{Q}}{}_{\boldsymbol{A}'}\mathring{e}_{(\boldsymbol{AQ})}{}^{\boldsymbol{a}},$$

with

$$\mathring{e}^{\boldsymbol{a}} \equiv \tau^{\boldsymbol{AA}'}\mathring{e}_{\boldsymbol{AA}'}{}^{\boldsymbol{a}}, \qquad \mathring{e}_{\boldsymbol{AB}}{}^{\boldsymbol{a}} \equiv \tau_{\boldsymbol{B}}{}^{\boldsymbol{A}'}\mathring{e}_{\boldsymbol{AA}'}{}^{\boldsymbol{a}}.$$

By construction it follows that

$$\mathring{e}^{\boldsymbol{0}} = 1, \qquad \mathring{e}_{(\boldsymbol{AB})}{}^{\boldsymbol{0}} = 0,$$
$$\mathring{e}^{\boldsymbol{i}} = 0, \qquad \mathring{e}_{(\boldsymbol{AB})}{}^{\boldsymbol{i}} = \sigma_{\boldsymbol{AB}}{}^{\boldsymbol{i}},$$

with $\sigma_{\boldsymbol{AB}}{}^{\boldsymbol{i}}$ the spatial Infeld-van der Waerden symbols; see Section 4.2.2.

The spinorial counterpart of the trace-free Ricci tensor is given by

$$\mathring{\Phi}_{\boldsymbol{AA}'\boldsymbol{BB}'} = \frac{1}{2}\tau_{\boldsymbol{AA}'}\tau_{\boldsymbol{BB}'} - \frac{1}{4}\epsilon_{\boldsymbol{AB}}\epsilon_{\boldsymbol{A}'\boldsymbol{B}'},$$

so that its space spinor version $\mathring{\Phi}_{\boldsymbol{ABCD}} = \tau_{\boldsymbol{B}}{}^{\boldsymbol{A}'}\tau_{\boldsymbol{D}}{}^{\boldsymbol{C}'}\mathring{\Phi}_{\boldsymbol{AA}'\boldsymbol{CC}'}$ is given by

$$\mathring{\Phi}_{\boldsymbol{ABCD}} = \frac{1}{2}\epsilon_{\boldsymbol{AB}}\epsilon_{\boldsymbol{CD}} - \frac{1}{4}\epsilon_{\boldsymbol{AC}}\epsilon_{\boldsymbol{BD}}.$$

From this last expression is can be verified that the irreducible components of $\mathring{\Phi}_{\boldsymbol{ABCD}}$ are given by

$$\mathring{\Phi}_{(\boldsymbol{ABCD})} = 0, \qquad \mathring{\Phi}_{\boldsymbol{AB}} = 0, \qquad \mathring{\Phi} = -\frac{3}{4};$$

compare the expressions in Equation (13.32). The rescaled Weyl spinor is trivially given by

$$\mathring{\phi}_{\boldsymbol{ABCD}} = 0.$$

Let $\mathring{\Gamma}_{AA'}{}^{BB'}{}_{CC'}$ denote the spinorial counterpart of the connection coefficients $\mathring{\Gamma}_a{}^b{}_c$. Using the spinorial expression of the spacetime volume form, Equation (3.25), one finds that

$$\begin{aligned}\mathring{\Gamma}_{AA'}{}^{BB'}{}_{CC'} &= \frac{1}{\sqrt{2}}\tau^{DD'}\epsilon_{DD'AA'}{}^{BB'}{}_{CC'},\\ &= \frac{\mathrm{i}}{\sqrt{2}}\big(\tau_C{}^{B'}\epsilon_{C'A'}\delta_A{}^B - \tau^B{}_{C'}\epsilon_{CA}\delta_{A'}{}^{B'}\big).\end{aligned}$$

The reduced spin connection coefficients can be obtained from the expression $\mathring{\Gamma}_{AA'}{}^B{}_C = \frac{1}{2}\mathring{\Gamma}_{AA'}{}^{BQ'}{}_{CQ'}$. One obtains

$$\mathring{\Gamma}_{AA'BC} = \frac{\mathrm{i}}{\sqrt{2}}\epsilon_{A(B}\tau_{C)A'}.$$

The space spinor version of the above expression is given by

$$\mathring{\Gamma}_{ABCD} = \tau_B{}^{A'}\mathring{\Gamma}_{AA'CD} = -\frac{\mathrm{i}}{\sqrt{2}}h_{ABCD},$$

where it is recalled that $h_{ABCD} \equiv -\epsilon_{A(C}\epsilon_{D)B}$. It can be verified that $\mathring{\Gamma}^\dagger_{ABCD} = -\mathring{\Gamma}_{ABCD}$; that is, the spin connection coefficients are the components of an *imaginary spinor.* From here it follows that

$$\mathring{\xi}_{ABCD} = -\mathrm{i}h_{ABCD}, \qquad \mathring{\chi}_{ABCD} = 0.$$

Gauge source functions

The gauge source functions associated to the considered conformal representation of the de Sitter spacetime can be computed from the expressions given in the previous section.

Treating the frame component $\mathring{e}_a{}^b$ as the component of a covariant tensor one finds that

$$\begin{aligned}\mathring{\nabla}^b\mathring{e}_b{}^a &= \eta^{cb}\mathring{e}_c(\delta_b{}^a) - \eta^{cb}\mathring{\Gamma}_c{}^e{}_b\delta_e{}^a\\ &= -\eta^{cb}\mathring{\Gamma}_c{}^a{}_b = -\eta^{cb}\epsilon_{0c}{}^a{}_b = 0.\end{aligned}$$

It follows that the *coordinate gauge source function* is given by

$$\mathring{F}^a(x) = \mathring{\nabla}^{AA'}\mathring{e}_{AA'}{}^a = 0.$$

Similarly, treating the connection coefficients as the components of a $(1,2)$-tensor one has

$$\begin{aligned}\eta^{da}\mathring{\nabla}_d\mathring{\Gamma}_a{}^b{}_c &= \eta^{da}e_d(\mathring{\Gamma}_a{}^b{}_c) + \eta^{da}\mathring{\Gamma}_d{}^b{}_e\mathring{\Gamma}_a{}^e{}_c - \eta^{da}\mathring{\Gamma}_d{}^e{}_a\mathring{\Gamma}_e{}^b{}_c - \eta^{da}\mathring{\Gamma}_d{}^e{}_c\mathring{\Gamma}_a{}^b{}_e\\ &= \eta^{da}\epsilon_{0d}{}^b{}_e\epsilon_{0a}{}^e{}_c - \eta^{da}\epsilon_{0d}{}^e{}_c\epsilon_{0a}{}^b{}_e - \eta^{da}\epsilon_{0d}{}^e{}_a\epsilon_{0e}{}^b{}_c = 0.\end{aligned}$$

It follows that the *frame gauge source functions* for the present representation of the de Sitter spacetime are given by

$$\mathring{F}_{BC}(x) = \mathring{\nabla}^{AA'}\mathring{\Gamma}_{AA'BC} = 0.$$

Finally, the *conformal gauge source function* is given by the value of the Ricci scalar. That is, one has

$$\mathring{R}(x) = -6.$$

Summary

The results from the previous analysis are summarised in the following:

Lemma 15.1 (***de Sitter spacetime as a solution to the conformal Einstein field equations***) *The fields*

$$(\mathring{\Xi},\, \mathring{\Sigma},\, \mathring{\Sigma}_{\boldsymbol{i}},\, \mathring{e}_{\boldsymbol{a}}{}^{\boldsymbol{b}},\, \mathring{\Gamma}_{\boldsymbol{a}}{}^{\boldsymbol{b}}{}_{\boldsymbol{c}},\, \mathring{L}_{\boldsymbol{ab}},\, \mathring{d}^{\boldsymbol{a}}{}_{\boldsymbol{bcd}})$$

as given by Equations (15.1)–(15.5), (15.6a), (15.6b), (15.7a) *and* (15.7b) *or, respectively, their spinorial counterparts*

$$(\mathring{\Xi},\, \mathring{\Sigma},\, \mathring{\Sigma}_{\boldsymbol{AA'}},\, \mathring{e}_{\boldsymbol{AA'}}{}^{\boldsymbol{b}},\, \mathring{\Gamma}_{\boldsymbol{AA'}}{}^{\boldsymbol{B}}{}_{\boldsymbol{C}},\, \mathring{\Phi}_{\boldsymbol{AA'BB'}},\, \mathring{\phi}_{\boldsymbol{ABCD}})$$

defined over the Einstein cylinder $\mathbb{R}\times\mathbb{S}^3$ *constitute a solution to the standard frame vacuum conformal Einstein field Equations* (8.32a) *and* (8.32b) *and, respectively, the spinorial vacuum conformal Einstein field Equations* (8.38a) *and* (8.38b). *The gauge source functions associated to this solution are given by*

$$\mathring{F}^{\boldsymbol{a}}(x) = 0, \qquad \mathring{F}_{\boldsymbol{AB}}(x) = 0, \qquad \mathring{R}(x) = -6.$$

15.1.2 Representation using conformal Gaussian systems

In Section 6.1.3 it has been shown that an alternative conformal representation of the de Sitter spacetime is given by the conformal metric

$$\bar{\boldsymbol{g}}_{\mathscr{E}} = \mathbf{d}\bar{\tau}\otimes\mathbf{d}\bar{\tau} - \left(1+\frac{\bar{\tau}^2}{4}\right)^2 \boldsymbol{\hbar},$$

where $\bar{\tau}$ is an affine parameter of the $\tilde{\boldsymbol{g}}_{dS}$-conformal geodesics as in Equation (6.32); that is, $\dot{\boldsymbol{x}} = \boldsymbol{\partial}_{\bar{\tau}}$. The associated covector is given by

$$\boldsymbol{\beta}_{dS}(\bar{\tau}) = -\frac{2\bar{\tau}}{4-\bar{\tau}^2}\mathbf{d}\bar{\tau}.$$

This covector is exact, thus indicating that the Weyl connection $\hat{\boldsymbol{\nabla}} = \tilde{\boldsymbol{\nabla}} + \boldsymbol{S}(\boldsymbol{\beta}_{dS})$ is, in fact, a Levi-Civita connection. Now, the metric $\bar{\boldsymbol{g}}_{\mathscr{E}}$ is related to the physical de Sitter metric $\tilde{\boldsymbol{g}}_{dS}$ via

$$\bar{\boldsymbol{g}}_{\mathscr{E}} = \Theta_{dS}^2\tilde{\boldsymbol{g}}_{dS}, \qquad \Theta_{dS} = 1 - \frac{\bar{\tau}^2}{4},$$

so that a calculation shows that

$$\boldsymbol{\beta}_{dS} = \Theta_{dS}^{-1}\mathbf{d}\Theta_{dS}.$$

That is, *the Weyl connection associated to the congruence of conformal geodesics* (6.32) *coincides with the Levi-Civita connection* $\bar{\boldsymbol{\nabla}}$ *of the metric* $\bar{\boldsymbol{g}}_{\mathscr{E}}$. Recalling that $\boldsymbol{g}_{\mathscr{E}}$ and $\bar{\boldsymbol{g}}_{\mathscr{E}}$ are related to each other by $\bar{\boldsymbol{g}}_{\mathscr{E}} = \bar{\Theta}^2\boldsymbol{g}_{\mathscr{E}}$ with

$$\bar{\Theta} \equiv 1 + \frac{\bar{\tau}^2}{4},$$

one finds that an adapted $\bar{\boldsymbol{g}}_{\mathscr{E}}$-orthonormal frame $\{\bar{\boldsymbol{e}}_{\boldsymbol{a}}\}$ is given by

$$\bar{\boldsymbol{e}}_{\boldsymbol{0}} = \boldsymbol{\partial}_{\bar{\tau}} = \bar{\Theta}^{-1}\boldsymbol{\partial}_{\tau}, \qquad \bar{\boldsymbol{e}}_{\boldsymbol{i}} = \bar{\Theta}^{-1}\boldsymbol{c}_i.$$

This frame can be verified to be *Weyl propagated.* It follows that the frame coefficients $\bar{e}_{\boldsymbol{i}}{}^{\boldsymbol{b}}$, with $\bar{\boldsymbol{e}}_{\boldsymbol{i}} = \bar{e}_{\boldsymbol{i}}{}^{\boldsymbol{b}}\boldsymbol{c}_{\boldsymbol{b}}$, are given by

$$\bar{e}_{\boldsymbol{i}}{}^{\boldsymbol{b}} = \frac{4}{4+\bar{\tau}^2}\delta_{\boldsymbol{i}}{}^{\boldsymbol{b}}. \tag{15.8}$$

In terms of the above, the components of the covector $\bar{\boldsymbol{d}} = \Theta_{dS}\boldsymbol{\beta}_{dS}$ with respect to the frame $\{\bar{\boldsymbol{e}}_{\boldsymbol{a}}\}$ are given by

$$\bar{d}_{\boldsymbol{0}} = \dot{\Theta}_{dS} = -\frac{\bar{\tau}}{2}, \qquad \bar{d}_{\boldsymbol{i}} = 0.$$

The computation of the connection coefficients $\bar{\Gamma}_{\boldsymbol{a}}{}^{\boldsymbol{b}}{}_{\boldsymbol{c}}$ requires a certain amount of care. Recalling that the connections $\hat{\boldsymbol{\nabla}} = \bar{\boldsymbol{\nabla}}$ and $\boldsymbol{\nabla}$ are related to each other via $\hat{\boldsymbol{\nabla}} - \boldsymbol{\nabla} = \boldsymbol{S}(\boldsymbol{\Upsilon})$ with $\boldsymbol{\Upsilon} \equiv \bar{\Theta}^{-1}\mathbf{d}\bar{\Theta}$, it follows by definition that

$$\begin{aligned}
\bar{\Gamma}_{\boldsymbol{a}}{}^{\boldsymbol{b}}{}_{\boldsymbol{c}} &= \bar{\omega}^{\boldsymbol{b}}{}_c\bar{e}_{\boldsymbol{a}}{}^a\bar{\nabla}_a\bar{e}_{\boldsymbol{c}}{}^c \\
&= \bar{\omega}^{\boldsymbol{b}}{}_c\bar{e}_{\boldsymbol{a}}{}^a\nabla_a\bar{e}_{\boldsymbol{c}}{}^c + \bar{\omega}^{\boldsymbol{b}}{}_c\bar{e}_{\boldsymbol{a}}{}^a\bar{e}_{\boldsymbol{c}}{}^d S_{ad}{}^{ec}\Upsilon_e,
\end{aligned}$$

where $\bar{\Upsilon}_{\boldsymbol{a}} \equiv \langle \bar{\boldsymbol{\Upsilon}}, \bar{\boldsymbol{e}}_{\boldsymbol{a}}\rangle$. Using that $\bar{e}_{\boldsymbol{a}}{}^a = \bar{\Theta}^{-1}e_{\boldsymbol{a}}{}^a$ one computes

$$\begin{aligned}
\bar{\omega}^{\boldsymbol{b}}{}_c\bar{e}_{\boldsymbol{a}}{}^a\nabla_a\bar{e}_{\boldsymbol{c}}{}^c &= -\bar{\Theta}^{-1}\omega^{\boldsymbol{b}}{}_c e_{\boldsymbol{a}}{}^a e_{\boldsymbol{c}}{}^c\nabla_a\bar{\Theta} + \bar{\Theta}^{-1}\omega^{\boldsymbol{b}}{}_c e_{\boldsymbol{a}}{}^a\nabla_a e_{\boldsymbol{c}}{}^c \\
&= -\bar{\Upsilon}_{\boldsymbol{a}}\delta_{\boldsymbol{c}}{}^{\boldsymbol{b}} + \bar{\Theta}^{-1}\Gamma_{\boldsymbol{a}}{}^{\boldsymbol{b}}{}_{\boldsymbol{c}}
\end{aligned}$$

and

$$\bar{\Theta}^{-1}\bar{\omega}^{\boldsymbol{b}}{}_c\bar{e}_{\boldsymbol{a}}{}^a\bar{e}_{\boldsymbol{c}}{}^d S_{ad}{}^{ec}\nabla_e\bar{\Theta} = \bar{\Upsilon}_{\boldsymbol{a}}\delta_{\boldsymbol{c}}{}^{\boldsymbol{b}} + \bar{\Upsilon}_{\boldsymbol{c}}\delta_{\boldsymbol{a}}{}^{\boldsymbol{b}} - \eta_{\boldsymbol{ac}}\bar{\Upsilon}^{\boldsymbol{b}}.$$

Accordingly, one concludes that

$$\bar{\Gamma}_{\boldsymbol{a}}{}^{\boldsymbol{b}}{}_{\boldsymbol{c}} = \bar{\Theta}^{-1}\Gamma_{\boldsymbol{a}}{}^{\boldsymbol{b}}{}_{\boldsymbol{c}} + \left(\Upsilon_{\boldsymbol{c}}\delta_{\boldsymbol{a}}{}^{\boldsymbol{b}} - \eta_{\boldsymbol{ac}}\Upsilon^{\boldsymbol{b}}\right).$$

Using

$$\Upsilon_{\boldsymbol{a}} = \frac{2\bar{\tau}}{4+\bar{\tau}^2}\delta_{\boldsymbol{a}}{}^{\boldsymbol{0}},$$

it can be verified that $\bar{\Gamma}_{\boldsymbol{0}}{}^{\boldsymbol{b}}{}_{\boldsymbol{c}} = 0$, as one would expect from the Weyl connection associated to a congruence of conformal geodesics. Moreover,

$$\bar{f}_{\boldsymbol{a}} = \frac{1}{4}\hat{\Gamma}_{\boldsymbol{a}}{}^{\boldsymbol{b}}{}_{\boldsymbol{b}} = 0. \tag{15.9}$$

A direct computation shows that

$$\begin{aligned} R[\bar{\boldsymbol{g}}_{\mathscr{E}}] &= -\frac{36}{4+\bar{\tau}^2}, \\ \boldsymbol{Schouten}[\bar{\boldsymbol{g}}_{\mathscr{E}}] &= \frac{1}{2}\left(1+\frac{1}{4}\bar{\tau}^2\right)\boldsymbol{\hbar}, \\ \boldsymbol{Weyl}[\bar{\boldsymbol{g}}_{\mathscr{E}}] &= 0, \end{aligned}$$

where the last expression follows simply by the conformal invariance of the Weyl tensor. The components of the Schouten tensor with respect to the frame $\{\bar{\boldsymbol{e}}_{\boldsymbol{a}}\}$ are given by

$$\bar{L}_{\boldsymbol{0a}} = 0, \qquad \bar{L}_{\boldsymbol{ij}} = \frac{2}{4+\bar{\tau}^2}\delta_{\boldsymbol{ij}}. \tag{15.10}$$

Furthermore, one has that

$$\bar{d}^{\boldsymbol{a}}{}_{\boldsymbol{bcd}} = 0. \tag{15.11}$$

Spinorial expressions

In what follows, let $\bar{\tau}^{AA'}$ denote the spinorial counterpart of the vector $\sqrt{2}\boldsymbol{\partial}_{\bar{\tau}}$. One has the normalisation $\bar{\tau}_{AA'}\bar{\tau}^{AA'} = 2$. Denoting the spinorial counterpart of the frame coefficients by $\bar{e}_{\boldsymbol{AA'}}{}^{\boldsymbol{a}}$ and making use of the standard space spinor decomposition

$$\bar{e}_{\boldsymbol{AA'}}{}^{\boldsymbol{a}} = \frac{1}{2}\bar{\tau}_{\boldsymbol{AA'}}\bar{e}^{\boldsymbol{a}} - \bar{\tau}^{\boldsymbol{Q}}{}_{\boldsymbol{A'}}\bar{e}_{(\boldsymbol{AQ})}{}^{\boldsymbol{a}},$$

one obtains

$$\bar{e}^{\boldsymbol{0}} = 1, \qquad \bar{e}_{(\boldsymbol{AB})}{}^{\boldsymbol{0}} = 0, \tag{15.12a}$$

$$\bar{e}^{\boldsymbol{i}} = 0, \qquad \bar{e}_{(\boldsymbol{AB})}{}^{\boldsymbol{i}} = \frac{4}{4+\bar{\tau}^2}\sigma_{\boldsymbol{AB}}{}^{\boldsymbol{i}}. \tag{15.12b}$$

Now, let $\bar{L}_{\boldsymbol{AA'BB'}}$ denote the spinorial counterpart of the components of the Schouten tensor $\bar{L}_{\boldsymbol{ab}}$. Setting $\bar{L}_{\boldsymbol{ABCD}} \equiv \bar{\tau}_{\boldsymbol{B}}{}^{\boldsymbol{A'}}\bar{\tau}_{\boldsymbol{D}}{}^{\boldsymbol{C'}}\bar{L}_{\boldsymbol{AA'CC'}}$ one finds

$$\bar{\Theta}_{\boldsymbol{CD}} = 0, \qquad \bar{\Theta}_{\boldsymbol{ABCD}} = -\frac{2}{4+\bar{\tau}^2}h_{\boldsymbol{ABCD}},$$

with $\bar{\Theta}_{\boldsymbol{CD}} \equiv \bar{L}_{\boldsymbol{Q}}{}^{\boldsymbol{Q}}{}_{\boldsymbol{CD}}$ and $\bar{\Theta}_{\boldsymbol{ABCD}} \equiv \bar{L}_{(\boldsymbol{AB})(\boldsymbol{CD})}$. For the rescaled Weyl spinor one has that

$$\bar{\phi}_{\boldsymbol{ABCD}} = 0.$$

To obtain a spinorial expression for the connection coefficients one observes that the spinorial counterpart of $\zeta_a{}^b{}_c = \delta_c{}^0\delta_a{}^b - \eta_{ac}\delta_0{}^b$ is given by

$$\zeta_{AA'}{}^{BB'}{}_{CC'} = \frac{1}{\sqrt{2}}\big(\delta_A{}^B\delta_{A'}{}^{B'}\tau_{CC'} - \epsilon_{AC}\epsilon_{A'C'}\tau^{BB'}\big),$$

so that the associated reduced coefficients are

$$\begin{aligned}\zeta_{AA'}{}^B{}_C &\equiv \frac{1}{2}\zeta_{AA'}{}^{BQ'}{}_{CQ'}\\ &= \frac{1}{2\sqrt{2}}\big(\delta_A{}^B\tau_{CA'} + \epsilon_{AC}\tau^B{}_{A'}\big).\end{aligned}$$

The space spinor version $\zeta_{ABCD} \equiv \tau_B{}^{A'}\zeta_{AA'CD}$ takes the form

$$\zeta_{ABCD} = -\frac{1}{\sqrt{2}}h_{ABCD}.$$

From the expressions computed in the previous paragraph it follows that

$$\bar{\Gamma}_{ABCD} = -\frac{2(\bar{\tau}+2\mathrm{i})}{\sqrt{2}(1+4\bar{\tau}^2)}h_{ABCD}$$

and, consequently,

$$\bar{\xi}_{ABCD} = -\frac{4\mathrm{i}}{4+\bar{\tau}^2}h_{ABCD}, \tag{15.13a}$$

$$\bar{\chi}_{ABCD} = \frac{2\bar{\tau}}{4+\bar{\tau}^2}h_{ABCD}, \tag{15.13b}$$

$$f_{AB} = 0. \tag{15.13c}$$

To keep track of the behaviour of the conformal Gaussian gauge system, one considers separation fields measuring the deviation of the congruence of conformal geodesics. The separation fields are governed by Equations (13.67a) and (13.67b). Assume, without loss of generality, a separation vector field $\boldsymbol{z}$ that is spatial on the fiduciary hypersurface $\mathcal{S}_\star$ described by the condition $\bar{\tau} = 0$, so that

$$z_{AA'\star} = -\tau^Q{}_{A'}z_{(AQ)\star}.$$

Using Equations (15.13b) and (15.13c) one can integrate Equations (13.67a) and (13.67b) to find

$$z = 0, \qquad z_{(AB)} = \left(1 + \frac{1}{4}\bar{\tau}^2\right)z_{(AB)\star}. \tag{15.14}$$

Observe that $z_{(AB)} \neq 0$ for all $\bar{\tau}$. Thus, the congruence of conformal geodesics remains non-singular. This observation is key to ensure the non-singular behaviour of the gauge in the perturbed spacetime.

Summary

The results of the analysis of the last two sections are summarised in the following:

Lemma 15.2 (***de Sitter spacetime as a solution to the extended conformal Einstein field equations***) *The fields*

$$(\bar{\Theta}, \bar{d}_{\boldsymbol{a}}, \bar{e}_{\boldsymbol{a}}{}^{\boldsymbol{b}}, \bar{\Gamma}_{\boldsymbol{a}}{}^{\boldsymbol{b}}{}_{\boldsymbol{c}}, \bar{L}_{\boldsymbol{ab}}, \bar{d}^{\boldsymbol{a}}{}_{\boldsymbol{bcd}})$$

as given by Equations (15.8)–(15.11) *or, equivalently, their spinorial counterparts*

$$(\bar{\Theta}, \bar{d}_{\boldsymbol{AA'}}, \bar{e}_{\boldsymbol{AA'}}{}^{\boldsymbol{b}}, \bar{\Gamma}_{\boldsymbol{AA'BC}}, \bar{L}_{\boldsymbol{AA'BB'}}, \bar{\phi}_{\boldsymbol{ABCD}})$$

defined over $\mathbb{R}\times\mathbb{S}^3$ *constitute a solution to the extended conformal Einstein field Equations* (8.46) *and the associated gauge constraints* (8.48) *and, respectively, the spinorial vacuum conformal Einstein field Equations* (8.54a) *and* (8.54b) *and* (8.55).

15.2 Perturbations of initial data for the de Sitter spacetime

This section clarifies the notion of perturbations of initial data for the de Sitter spacetime. In what follows, let $\mathcal{S}$ denote a three-dimensional manifold with the topology of $\mathbb{S}^3$. On $\mathcal{S}$ one considers a solution to the vacuum conformal Hamiltonian and momentum constraint equations $(\mathcal{S}, \boldsymbol{h}, \boldsymbol{K}, \Omega, \Sigma)$ with a de Sitter-like value of the cosmological constant, that is, Equations (11.15a) and (11.15b) with $\varrho = 0$ and $j_k = 0$.

Remark. For conceptual clarity it is often convenient to distinguish between the 3-manifold $\mathcal{S}$ and its embedding, $\mathcal{S}_\star$, in the spacetime arising as the development of the initial data set $(\mathcal{S}, \boldsymbol{h}, \boldsymbol{K}, \Omega, \Sigma)$.

15.2.1 Initial data on a standard initial hypersurface

Using the procedure described in Section 11.4.3, the tensor fields $\boldsymbol{h}$ and $\boldsymbol{K}$ can be used to construct a solution to the vacuum conformal constraint Equations (11.35a)–(11.35j). As the 3-manifold $\mathcal{S}$ is assumed to be compact, one can, without loss of generality, assume that $\Omega = 1$ and $\Sigma = 0$.

As $\mathcal{S} \approx \mathbb{S}^3$, there exists a diffeomorphism $\psi : \mathcal{S} \to \mathbb{S}^3$ which can be used to pull back coordinates $\underline{x} = (x^\alpha)$ in $\mathbb{S}^3$ to $\mathcal{S}$. In this way one obtains a system of coordinates $\underline{x}' \equiv \underline{x} \circ \psi$ on $\mathcal{S}$ and can write $\underline{x}' = (x'^\alpha)$. The diffeomorphism ψ can be used to push forward the vector fields $\{\boldsymbol{c}_i\}$ on $T(\mathbb{S}^3)$ to vector fields $\{\psi_*^{-1}\boldsymbol{c}_i\}$ on $T(\mathcal{S})$ and to pull back their dual covectors $\{\boldsymbol{\alpha}^i\}$ on $T^*(\mathbb{S}^3)$ to covectors $\{\psi_*\boldsymbol{\alpha}^i\}$ on $T^*(\mathcal{S})$. *For simplicity of the presentation, in a slight abuse of notation, the vectors and covectors* $\{\psi_*^{-1}\boldsymbol{c}_i\}$ *and* $\{\psi_*\boldsymbol{\alpha}^i\}$ *will be written (except for the next subsection) as* $\{\boldsymbol{c}_i\}$ *and* $\{\boldsymbol{\alpha}^i\}$, *respectively.*

Gauge fixing

The construction described in the previous paragraph depends strongly on the particular choice of the diffeomorphism ψ. This ***gauge freedom*** can be fixed by considerations similar to those used in the discussion of the *coordinate gauge source functions* of Section 13.2.1.

Given an $\boldsymbol{h}$-orthonormal frame $\{\boldsymbol{e}_i\}$ on $\mathcal{S}$, one can write $\boldsymbol{e}_i = e_i{}^j(\psi_*^{-1}\boldsymbol{c}_j)$ and use the frame coefficients $e_i{}^j$ to introduce a ***spatial coordinate gauge source function*** $F^i(\underline{x}')$ via the relation

$$D^j e_j{}^i = F^i(\underline{x}'),$$

where $\boldsymbol{D}$ denotes the Levi-Civita covariant derivative of $\boldsymbol{h}$. Writing

$$\psi_*\boldsymbol{\alpha}^i = (\psi_*\boldsymbol{\alpha}^i)_\alpha \mathbf{d}x'^\alpha$$

and noticing that $e_i{}^j = \langle \psi_*\boldsymbol{\alpha}^j, \boldsymbol{e}_i\rangle$ one finds that $D^j e_j{}^i = D^\beta(\psi_*\boldsymbol{\alpha}^i)_\beta$. Expressing the coordinates $\underline{x}$ in $\mathbb{S}^3$ in terms of the coordinates $\underline{x}'$ on $\mathcal{S}$ in the form $x^\alpha = x^\alpha(\underline{x}')$ one finds, by a calculation similar to the one discussed in Section 13.2.1, that the diffeomorphism $\psi : \mathcal{S} \rightarrow \mathbb{S}^3$ is a ***harmonic map***. That is, one has that

$$D^\beta D_\beta x^\alpha = 0,$$

if

$$h^{\alpha\beta}\not{D}_\gamma \alpha^i{}_\delta \frac{\partial x^\gamma}{\partial x'^\alpha}\frac{\partial x^\delta}{\partial x'^\beta} = F^i(\underline{x}'),$$

where $\not{D}$ denotes the Levi-Civita connection of the metric $\boldsymbol{\hbar}$ on $\mathbb{S}^3$ and $\boldsymbol{\alpha}^i = \alpha^i{}_\alpha \mathbf{d}x^\alpha$. Finally, if one lets $x'^\alpha = x'^\alpha(\underline{x})$ be the identity map so that $\underline{x}' = \underline{x}$, one concludes that

$$F^i(\underline{x}) = \delta^{jk}\mathring{\gamma}_j{}^i{}_k = 0,$$

where the last equality follows from (15.4). This construction and the resulting spatial gauge source function fixes the gauge freedom in the diffeomorphism ψ; see Figure 15.2.

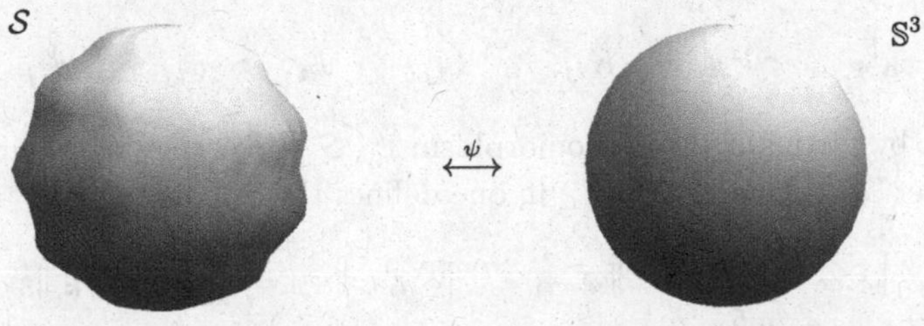

Figure 15.2 Construction of coordinates on a compact three-dimensional manifold describing perturbations of standard de Sitter initial data. The identification of the 3-manifolds $\mathcal{S}$ and $\mathbb{S}^3$ is realised through a harmonic map; see main text for further details.

Parametrising the perturbation data

While the frame $\{\boldsymbol{c}_i\}$ is orthonormal with respect to the standard metric $\boldsymbol{\hbar}$ of $\mathbb{S}^3$, in general, this will not be the case with respect to the 3-metric $\boldsymbol{h}$ on $\mathcal{S}$. Now, let $\{\boldsymbol{e}_i\}$ denote an $\boldsymbol{h}$-orthonormal frame over $T(\mathcal{S})$ and let $\{\boldsymbol{\omega}^i\}$ denote its corresponding cobasis. In what follows, it will be assumed that one can write

$$\boldsymbol{e}_i = \boldsymbol{c}_i + \breve{\boldsymbol{e}}_i, \tag{15.15}$$

for some vectors $\{\breve{\boldsymbol{e}}_i\}$. This is essentially equivalent to saying that one has introduced coordinates $\underline{x} = (x^\alpha)$ on $\mathcal{S}$ such that

$$\boldsymbol{h} = \boldsymbol{\hbar} + \breve{\boldsymbol{h}}.$$

It is important to emphasise that the above statement depends on the gauge.

From the split in Equation (15.15), it follows that the solution to the conformal constraint equations implied by $(\Omega = 1, \Sigma = 0, \boldsymbol{h}, \boldsymbol{K})$ on $\mathcal{S}$ can be written as

$$\begin{aligned}
e_{\boldsymbol{a}}{}^{\boldsymbol{b}} &= \delta_{\boldsymbol{a}}{}^{\boldsymbol{b}} + \breve{e}_{\boldsymbol{a}}{}^{\boldsymbol{b}}, \\
\gamma_{\boldsymbol{i}}{}^{\boldsymbol{j}}{}_{\boldsymbol{k}} &= \epsilon_{\boldsymbol{i}}{}^{\boldsymbol{j}}{}_{\boldsymbol{k}} + \breve{\gamma}_{\boldsymbol{i}}{}^{\boldsymbol{j}}{}_{\boldsymbol{k}}, \qquad \chi_{\boldsymbol{ij}} = \breve{K}_{\boldsymbol{ij}}, \\
L_{\boldsymbol{ij}} &= \delta_{\boldsymbol{ij}} + \breve{L}_{\boldsymbol{ij}}, \qquad L_{\boldsymbol{i}} = \delta_{\boldsymbol{i}}{}^{\boldsymbol{0}} + \breve{L}_{\boldsymbol{i}}, \\
d_{\boldsymbol{ij}} &= \breve{d}_{\boldsymbol{ij}}, \qquad d^*_{\boldsymbol{ij}} = \breve{d}^*_{\boldsymbol{ij}},
\end{aligned}$$

where the components of the various fields are expressed as components with respect to the frame $\{\boldsymbol{e}_i\}$ as given in (15.15) and one has

$$\breve{e}_{\boldsymbol{a}}{}^{\boldsymbol{b}} = 0, \qquad \breve{\gamma}_{\boldsymbol{i}}{}^{\boldsymbol{j}}{}_{\boldsymbol{k}} = 0, \qquad \breve{L}_{\boldsymbol{ij}} = 0, \qquad \breve{L}_{\boldsymbol{i}} = 0, \qquad \breve{d}_{\boldsymbol{ij}} = 0, \qquad \breve{d}^*_{\boldsymbol{ij}} = 0,$$

if and only if

$$\breve{\boldsymbol{e}}_i = 0, \qquad \boldsymbol{K} = 0.$$

Accordingly, the fields topped with a $\breve{}$ together with $K_{\boldsymbol{ij}}$ describe the deviation of a solution to the conformal constraint equations from data for the exact de Sitter spacetime. It is important to observe that as

$$\breve{e}_{\boldsymbol{a}}{}^{\boldsymbol{b}}, \qquad \breve{\gamma}_{\boldsymbol{i}}{}^{\boldsymbol{j}}{}_{\boldsymbol{k}}, \qquad \breve{K}_{\boldsymbol{ij}}, \qquad \breve{L}_{\boldsymbol{ij}}, \qquad \breve{L}_{\boldsymbol{i}}, \qquad \breve{d}_{\boldsymbol{ij}}, \qquad \breve{d}^*_{\boldsymbol{ij}}$$

are scalars, by virtue of the diffeomorphism $\psi : \mathcal{S} \to \mathbb{S}^3$, they can be considered as fields over $\mathbb{S}^3$. As such, for $m \geq 0$, one defines the Sobolev norms

$$\| \breve{e}_{\boldsymbol{a}}{}^{\boldsymbol{b}} \|_{\mathcal{S},m} \equiv \sum_{\boldsymbol{a},\boldsymbol{b}} \| \breve{e}_{\boldsymbol{a}}{}^{\boldsymbol{b}} \|_{\mathbb{S}^3,m}, \qquad \| \breve{\gamma}_{\boldsymbol{i}}{}^{\boldsymbol{j}}{}_{\boldsymbol{k}} \|_{\mathcal{S},m} \equiv \sum_{\boldsymbol{i},\boldsymbol{j},\boldsymbol{k}} \| \breve{\gamma}_{\boldsymbol{i}}{}^{\boldsymbol{j}}{}_{\boldsymbol{k}} \|_{\mathbb{S}^3,m},$$

and, similarly, for the other fields – the sums in the previous expressions are carried out over the independent components of the particular field under consideration. In terms of these norms, it will be said that the initial data for the conformal field equations are ***ε-close*** in the norm $\| \quad \|_{\mathcal{S},m}$ to initial data for the de Sitter spacetime if

$$\| \breve{e}_a{}^b \|_{\mathcal{S},m} + \| \breve{\gamma}_i{}^j{}_k \|_{\mathcal{S},m} + \| \breve{K}_{ij} \|_{\mathcal{S},m} + \| \breve{L}_{ij} \|_{\mathcal{S},m} + \| \breve{L}_i \|_{\mathcal{S},m} + \| \breve{d}_{ij} \|_{\mathcal{S},m} + \| \breve{d}^*_{ij} \|_{\mathcal{S},m} < \varepsilon. \tag{15.16}$$

This notion of closeness to initial data is gauge dependent. Nevertheless, it is the appropriate one to exploit the existence and stability theorems of Chapter 12.

15.2.2 Initial data on the conformal boundary

An important property of de Sitter-like spacetimes is that the individual components of the conformal boundary can serve as Cauchy hypersurfaces of the unphysical spacetimes. Accordingly, it is possible to formulate for these spacetimes an ***asymptotic initial value problem*** where initial data are prescribed on a 3-manifold corresponding to, say, $\mathscr{I}^-$.

The solutions to the conformal constraint equations at the conformal boundary have been discussed in Section 11.4.4. In particular, it has been shown that one needs to prescribe on $\mathscr{I}^-$ a 3-metric $\boldsymbol{h}$, a symmetric trace-free and divergence-free tensor corresponding to the initial value of the electric part of the rescaled Weyl tensor and a function $\varkappa$. From these free data it is possible to compute the values of the remaining conformal fields. In the particular case of the exact de Sitter spacetime it can be verified that the asymptotic free data are given by

$$\boldsymbol{h} \simeq \mathring{\boldsymbol{h}}, \qquad d_{ij} \simeq 0, \qquad \varkappa \simeq 0,$$

where components are expressed with respect to the $\mathring{\boldsymbol{h}}$-orthonormal frame $\{\boldsymbol{c}_i\}$. From the above one finds

$$e_i{}^j \simeq \delta_i{}^j, \quad \gamma_i{}^j{}_k \simeq \epsilon_i{}^j{}_k, \quad K_{ij} \simeq 0, \quad L_i \simeq 0, \quad L_{ij} \simeq \frac{1}{2}\delta_{ij}, \quad d^*_{ij} \simeq 0.$$

Perturbations of the above asymptotic initial data for the de Sitter spacetime are discussed in a manner similar to that of perturbations of standard Cauchy data. Accordingly, assuming that $\mathscr{I}^- \approx \mathbb{S}^3$, one can make use of diffeomorphisms $\psi : \mathscr{I}^- \to \mathbb{S}^3$ to introduce coordinates on the conformal boundary and to pull back the components of the various conformal fields to $\mathbb{S}^3$. Initial data corresponding to perturbations of asymptotic de Sitter initial data will then be described in terms of fields

$$\boldsymbol{h} = \mathring{\boldsymbol{h}} + \breve{\boldsymbol{h}}, \qquad d_{ij} = \breve{d}_{ij}, \qquad \breve{\varkappa},$$

where $\breve{d}_{ij}$ are the components of a symmetric, $\boldsymbol{h}$-trace-free and $\boldsymbol{h}$-divergence-free tensor expressed in terms of the components of the $\boldsymbol{h}$-orthonormal frame $\{\boldsymbol{e}_i\} = \{\boldsymbol{c}_i + \breve{\boldsymbol{e}}_i\}$. Mimicking the standard Cauchy case, the perturbation of asymptotic data for the de Sitter spacetime will be said to be ***ε-close to exact asymptotic de Sitter data*** in the $\| \quad \|_m$-norm if the various conformal fields on $\mathscr{I}^-$ satisfy an inequality of the form of (15.16). In principle, it is possible to express this smallness requirement in terms of a smallness condition on the *basic perturbation data* $\breve{e}_i{}^j$, $\breve{d}_{ij}$ and $\breve{\varkappa}$; this idea will not be further pursued here.

15.3 Global existence and stability using gauge source functions

In this section a first proof of the global existence and stability of de Sitter-like spacetimes is provided. This proof makes use of the hyperbolic reduction of the spinorial conformal field equations using gauge source functions as discussed in Section 13.2 and of the conformal representation of the de Sitter spacetime discussed in Section 15.1.1. This approach can be readily generalised to include trace-free matter. The discussion presented here follows the seminal work by Friedrich (1986b, 1991).

15.3.1 Gauge considerations

The first step in the construction of de Sitter-like spacetimes consists of the fixing of the gauge in the evolution equations. This *gauge fixing* allows one to relate, in an unambiguous manner, fields in the background de Sitter spacetime with fields in the perturbed spacetime; see Figure 15.3.

As the (unphysical) spacetime $(\mathcal{M}, \boldsymbol{g})$ to be constructed will be of the form $\mathcal{M} \approx [a,b] \times \mathbb{S}^3 \subset \mathbb{R} \times \mathbb{S}^3$ with a, $b \in \mathbb{R}$, it is natural to make use of the coordinates and frames in the background spacetime $(\mathbb{R} \times \mathbb{S}^3, \boldsymbol{g}_{\mathscr{E}})$ to coordinatise and construct a suitable gauge in the perturbed spacetime. Following the discussion of Section 13.2.1, coordinates $x = (\tau, x^\alpha)$ on the Einstein cylinder $\mathcal{M}_{\mathscr{E}} = \mathbb{R} \times \mathbb{S}^3$ can be regarded as coordinates on a perturbed spacetime $(\mathcal{M}, \boldsymbol{g})$ if one identifies the manifolds $\mathcal{M}$ and $\mathcal{M}_{\mathscr{E}}$. This coordinatisation is equivalent to the coordinate gauge source choice

$$F^{\boldsymbol{a}}(x) = -\eta^{\boldsymbol{bc}} \mathring{\Gamma}_{\boldsymbol{b}}{}^{\boldsymbol{a}}{}_{\boldsymbol{c}} = 0,$$

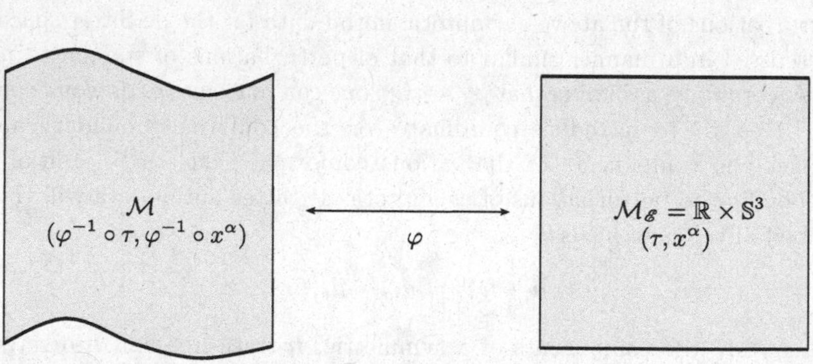

Figure 15.3 Schematic representation of the construction of coordinates on a perturbation of the de Sitter spacetime $(\mathcal{M}, \boldsymbol{g})$ using coordinates on the exact de Sitter spacetime $(\mathcal{M}_{\mathscr{E}}, \boldsymbol{g}_{\mathscr{E}})$ and a diffeomorphism $\varphi : \mathcal{M} \to \mathring{\mathcal{M}}$ as described in the main text. The particular realisation of the diffeomorphism identifies the manifolds $\mathcal{M}$ and $\mathcal{M}_{\mathscr{E}}$ in such a way that φ is a wave map.

where the last equality follows from the discussion leading to Lemma 15.1. This particular choice of coordinate gauge source function makes the identification between $\mathcal{M}$ and $\mathcal{M}_{\mathscr{E}}$ a wave map; see Section 13.2.1.

By similar considerations, the vectors $\{\mathring{\boldsymbol{c}}_{\boldsymbol{a}}\} = \{\boldsymbol{\partial}_\tau, \mathring{\boldsymbol{c}}_{\boldsymbol{i}}\}$ originally defined on $\mathcal{M}_{\mathscr{E}}$ can be regarded as vectors on the perturbed spacetime $(\mathcal{M}, \boldsymbol{g})$ in terms of which the $\boldsymbol{g}$-orthonormal frame $\{\boldsymbol{e}_{\boldsymbol{a}}\}$ can be expanded by writing $\boldsymbol{e}_{\boldsymbol{a}} = e_{\boldsymbol{a}}{}^{\boldsymbol{b}} \mathring{\boldsymbol{c}}_{\boldsymbol{b}}$. In an analogous manner, the fields

$$\mathring{\mathbf{u}} = (\mathring{\Xi}, \mathring{\Sigma}, \mathring{\Sigma}_{\boldsymbol{AA'}}, \mathring{e}_{\boldsymbol{AA'}}{}^{\boldsymbol{a}}, \mathring{\Gamma}_{\boldsymbol{AA'}}{}^{\boldsymbol{B}}{}_{\boldsymbol{C}}, \mathring{\Phi}_{\boldsymbol{AA'BB'}}, \mathring{\phi}_{\boldsymbol{ABCD}}),$$

as given by Lemma 15.1, can be regarded as fields over $\mathcal{M}$. It is important to emphasise that all of the above fields (except for $\mathring{e}_{\boldsymbol{AA'}}{}^{\boldsymbol{a}}$) are, in fact, components of tensors with respect to the background frame $\{\mathring{\boldsymbol{e}}_{\boldsymbol{a}}\} = \{\delta_{\boldsymbol{a}}{}^{\boldsymbol{b}} \mathring{\boldsymbol{c}}_{\boldsymbol{b}}\}$.

The gauge fixing is completed by setting the frame gauge source function $F_{\boldsymbol{AB}}(x)$ and the conformal gauge source function $R(x)$ equal to their values in the background spacetime $(\mathbb{R} \times \mathbb{S}^3, \boldsymbol{g}_{\mathscr{E}})$. That is, one sets

$$F_{\boldsymbol{AB}}(x) = 0, \qquad R(x) = -6;$$

compare Lemma 15.1.

15.3.2 The evolution system

The hyperbolic reduction procedure discussed in Section 13.2 and summarised in Proposition 13.1, leads to an evolution system which, in terms of local coordinates $x = (\tau, x^\alpha)$ of an open domain $\mathcal{U} \subset \mathbb{R} \times \mathbb{S}^3$, takes the form

$$\partial_\tau \boldsymbol{\sigma} = \mathbf{G}(\boldsymbol{\Gamma})\boldsymbol{\sigma} + \mathbf{H}(\boldsymbol{\sigma}, \boldsymbol{\upsilon}), \tag{15.17a}$$

$$\big(\mathbf{I} + \mathbf{D}^0(e)\big)\partial_\tau \boldsymbol{\upsilon} + \mathbf{D}^\alpha(e)\partial_\alpha \boldsymbol{\upsilon} = \mathbf{E}(\boldsymbol{\Gamma})\boldsymbol{\upsilon} + \mathbf{F}(\boldsymbol{\sigma}, \boldsymbol{\upsilon}, \boldsymbol{\phi}), \tag{15.17b}$$

$$\big(\mathbf{I} + \mathbf{A}^0(e)\big)\partial_\tau \boldsymbol{\phi} + \mathbf{A}^\alpha(e)\partial_\alpha \boldsymbol{\phi} = \mathbf{B}(\boldsymbol{\Gamma})\boldsymbol{\phi}, \tag{15.17c}$$

where $\boldsymbol{\sigma}$ encodes the conformal factor Θ and the independent components of its concomitants; $\boldsymbol{\upsilon}$ collects the independent components of the frame components, the connection coefficients and the trace-free Ricci spinor and $\boldsymbol{\phi}$ groups the independent components of the rescaled Weyl spinor.

To apply the methods of the theory of hyperbolic partial differential equations (PDEs) discussed in Chapter 12 it is convenient to split the various field unknowns into a ***background part*** and a ***perturbation part***. More precisely, one sets

$$\Xi = \mathring{\Xi} + \breve{\Xi}, \quad \Sigma = \mathring{\Sigma} + \breve{\Sigma}, \quad \Sigma_{\boldsymbol{AB}} = \breve{\Sigma}_{\boldsymbol{AB}}, \quad s = \mathring{s} + \breve{s}, \tag{15.18a}$$

$$e^{\mathbf{0}} = \mathring{e}^{\mathbf{0}} + \breve{e}^{\mathbf{0}}, \quad e_{\boldsymbol{AB}}{}^{\mathbf{0}} = \breve{e}_{\boldsymbol{AB}}{}^{\mathbf{0}}, \tag{15.18b}$$

$$e^{\boldsymbol{i}} = \breve{e}^{\boldsymbol{i}}, \quad e_{\boldsymbol{AB}}{}^{\boldsymbol{i}} = \mathring{e}_{\boldsymbol{AB}}{}^{\boldsymbol{i}} + \breve{e}_{\boldsymbol{AB}}{}^{\boldsymbol{i}}, \tag{15.18c}$$

$$\Gamma_{\boldsymbol{AB}} = \breve{\Gamma}_{\boldsymbol{AB}}, \quad \Gamma_{(\boldsymbol{AB})\boldsymbol{CD}} = \mathring{\Gamma}_{(\boldsymbol{AB})\boldsymbol{CD}} + \breve{\Gamma}_{(\boldsymbol{AB})\boldsymbol{CD}}, \tag{15.18d}$$

$$\Phi_{(\boldsymbol{ABCD})} = \breve{\Phi}_{(\boldsymbol{ABCD})}, \quad \Phi_{\boldsymbol{AB}} = \breve{\Phi}_{(\boldsymbol{AB})}, \quad \Phi = \mathring{\Phi} + \breve{\Phi}, \tag{15.18e}$$

$$\phi_{\boldsymbol{ABCD}} = \breve{\phi}_{\boldsymbol{ABCD}}, \tag{15.18f}$$

where

$$\mathring{\Xi}, \quad \mathring{\Sigma}, \quad \mathring{s}, \quad \mathring{e}^{\mathbf{0}}, \quad \mathring{e}_{\boldsymbol{AB}}{}^{\boldsymbol{i}}, \quad \mathring{\Gamma}_{(\boldsymbol{AB})\boldsymbol{CD}}, \quad \mathring{\Phi}$$

are the non-vanishing components of the fields describing the *background de Sitter solution* as discussed in Section 15.1.1, while

$$\breve{\Xi}, \quad \breve{\Sigma}, \quad \breve{\Sigma}_{\boldsymbol{AB}}, \quad \breve{s}, \quad \breve{e}^{\boldsymbol{a}}, \quad \breve{e}_{\boldsymbol{AB}}{}^{\boldsymbol{a}}, \quad \breve{\Gamma}_{\boldsymbol{AB}}, \quad \breve{\Gamma}_{(\boldsymbol{AB})\boldsymbol{CD}}, \tag{15.19a}$$

$$\breve{\Phi}_{(\boldsymbol{ABCD})}, \quad \breve{\Phi}_{(\boldsymbol{AB})}, \quad \breve{\Phi}, \quad \breve{\phi}_{\boldsymbol{ABCD}} \tag{15.19b}$$

describe the *perturbations away from the de Sitter solution.* The split between background and perturbations given by Equations (15.17a)–(15.17c) depends strongly on the choice of gauge.

By construction, the background fields are a solution to the conformal evolution Equations (15.17a)–(15.17c). Consequently, one has

$$\partial_\tau \mathring{\boldsymbol{\sigma}} = \mathbf{G}(\mathring{\boldsymbol{\Gamma}})\mathring{\boldsymbol{\sigma}} + \mathbf{H}(\mathring{\boldsymbol{\sigma}}, \mathring{\boldsymbol{v}}),$$

$$(\mathbf{I} + \mathbf{D}^0(\mathring{\boldsymbol{e}}))\partial_\tau \mathring{\boldsymbol{v}} + \mathbf{D}^\alpha(\mathring{\boldsymbol{e}})\partial_\alpha \mathring{\boldsymbol{v}} = \mathbf{E}(\mathring{\boldsymbol{\Gamma}})\mathring{\boldsymbol{v}}.$$

Accordingly, substituting now the ansatz (15.18a)–(15.18f) in the evolution system (15.17a)–(15.17c), one obtains equations for the independent components of the perturbation fields (15.19a) and (15.19b):

$$\begin{aligned}\partial_\tau \breve{\boldsymbol{\sigma}} &= \mathbf{G}(\mathring{\boldsymbol{\Gamma}})\breve{\boldsymbol{\sigma}} + \mathbf{G}(\breve{\boldsymbol{\Gamma}})\mathring{\boldsymbol{\sigma}} + \mathbf{G}(\breve{\boldsymbol{\Gamma}})\breve{\boldsymbol{\sigma}}\\ &\quad + \mathbf{H}(\mathring{\boldsymbol{\sigma}}, \breve{\boldsymbol{v}}) + \mathbf{H}(\breve{\boldsymbol{\sigma}}, \mathring{\boldsymbol{v}}) + \mathbf{H}(\breve{\boldsymbol{\sigma}}, \breve{\boldsymbol{v}}),\end{aligned} \tag{15.20a}$$

$$\begin{aligned}(\mathbf{I} + \mathbf{D}^0(\mathring{\boldsymbol{e}} + \breve{\boldsymbol{e}}))\partial_\tau \breve{\boldsymbol{v}} + \mathbf{D}^\alpha(\mathring{\boldsymbol{e}} + \breve{\boldsymbol{e}})\partial_\alpha \breve{\boldsymbol{v}} &= \mathbf{E}(\breve{\boldsymbol{\Gamma}})\mathring{\boldsymbol{v}} + \mathbf{E}(\mathring{\boldsymbol{\Gamma}})\breve{\boldsymbol{v}} + \mathbf{E}(\breve{\boldsymbol{\Gamma}})\breve{\boldsymbol{v}} + \mathbf{F}(\mathring{\boldsymbol{\sigma}}, \mathring{\boldsymbol{v}}, \breve{\boldsymbol{\phi}})\\ &\quad + \mathbf{F}(\mathring{\boldsymbol{\sigma}}, \breve{\boldsymbol{v}}, \breve{\boldsymbol{\phi}}) + \mathbf{F}(\breve{\boldsymbol{\sigma}}, \mathring{\boldsymbol{v}}, \breve{\boldsymbol{\phi}}) + \mathbf{F}(\breve{\boldsymbol{\sigma}}, \breve{\boldsymbol{v}}, \breve{\boldsymbol{\phi}})\\ &\quad - (\mathbf{I} + \mathbf{D}^0(\breve{\boldsymbol{e}}))\partial_\tau \mathring{\boldsymbol{v}},\end{aligned} \tag{15.20b}$$

$$(\mathbf{I} + \mathbf{A}^0(\mathring{\boldsymbol{e}} + \breve{\boldsymbol{e}}))\partial_\tau \breve{\boldsymbol{\phi}} + \mathbf{A}^\alpha(\mathring{\boldsymbol{e}} + \breve{\boldsymbol{e}})\partial_\alpha \breve{\boldsymbol{\phi}} = \mathbf{B}(\mathring{\boldsymbol{\Gamma}} + \breve{\boldsymbol{\Gamma}})\breve{\boldsymbol{\phi}}. \tag{15.20c}$$

In view of the properties of the original conformal evolution Equation (15.17a)–(15.17c) the above equations constitute a symmetric hyperbolic evolution system for the components of $\breve{\mathbf{u}} = (\breve{\boldsymbol{\sigma}}, \breve{\boldsymbol{v}}, \breve{\boldsymbol{\phi}})$. Accordingly, the theory of hyperbolic PDEs, as discussed in Chapter 12, can be applied in domains of the form $[0, \tau_\bullet] \times \mathbb{S}^3$ with $\tau_\bullet > 0$, to guarantee the existence of solutions and to assert Cauchy stability. In particular, as the background solution $(\mathring{\boldsymbol{\sigma}}, \mathring{\boldsymbol{v}}, \mathring{\boldsymbol{\phi}})$ is well defined on the whole of $\mathbb{R} \times \mathbb{S}^3$ one obtains the following existence, uniqueness and Cauchy stability result:

Proposition 15.1 (***existence of solutions to the standard conformal evolution equations***) *Let* $\mathbf{u}_\star = \mathring{\mathbf{u}}_\star + \breve{\mathbf{u}}_\star$ *denote de Sitter-like initial data for the conformal field equations prescribed on a 3-manifold* $\mathcal{S} \approx \mathbb{S}^3$*. Given* $m \geq 4$ *and* $\tau_\bullet > \frac{3}{4}\pi$*, then:*

(i) There exists $\varepsilon > 0$ *such that if*

$$\| \breve{\mathbf{u}}_\star \|_m < \varepsilon,$$

then there exists a C^{m-2} *unique solution to the conformal evolution Equations* (15.20a)–(15.20c) *defined on* $[0, \tau_\bullet] \times \mathbb{S}^3$.

(ii) Given a sequence of initial data $\mathbf{u}_\star^{(n)} = \mathring{\mathbf{u}}_\star^{(n)} + \breve{\mathbf{u}}_\star^{(n)}$ *such that*

$$\| \breve{\mathbf{u}}_\star^{(n)} \|_m < \varepsilon \qquad \textit{and} \qquad \| \breve{\mathbf{u}}_\star^{(n)} \|_m \to 0 \qquad \textit{as} \qquad n \to \infty,$$

then for the corresponding solutions $\breve{\mathbf{u}}^{(n)} \in C^{m-2}([0, \tau_\bullet] \times \mathbb{S}^3)$ *one has that* $\| \breve{\mathbf{u}}^{(n)} \|_m \to 0$ *uniformly in* $\tau \in [0, \tau_\bullet]$ *as* $n \to \infty$.

Proof The above proposition is a direct consequence of Theorem 12.4. To apply this theorem it is necessary to ensure that both

$$\mathbf{I} + \mathbf{A}^0(\mathring{e}_\star + \breve{e}_\star) \qquad \text{and} \qquad \mathbf{I} + \mathbf{D}^0(\mathring{e}_\star + \breve{e}_\star) \tag{15.21}$$

are both positive definite away from zero in a uniform manner over $\mathbb{S}^3$. An explicit calculation shows that

$$\mathbf{I} + \mathbf{A}^0(\mathring{e}_\star) \qquad \text{and} \qquad \mathbf{I} + \mathbf{D}^0(\mathring{e}_\star)$$

are positive definite away from zero. Thus, by setting ε sufficiently small, condition (15.21) can be guaranteed. By further reducing ε, if necessary, one can ensure that all solutions with $\| \breve{\mathbf{u}}_\star \|_m < \varepsilon$ have a minimum existence time $\tau_\bullet > \frac{3}{4}\pi$. Taking into account the above, point (i) follows from points (i)–(iii) of Theorem 12.4 while point (ii) follows from point (iv) in the same theorem. □

Remark 1. The purpose of point (i) in Proposition 15.1 is to guarantee a minimum existence time of solutions to the evolution system (15.20a)–(15.20c) containing the conformal boundary of the perturbed solution.

Remark 2. Point (ii) in Proposition 15.1 is a statement of Cauchy stability. It ensures that data sufficiently close to data for the de Sitter spacetime give rise to solutions with an existence time similar to that of the background solution. Moreover, within the established existence time, the solutions are suitably close to the background solution. Observe, however, that this result makes no statement about whether a particular solution converges in time to the background solution. Thus, one has obtained only an ***orbital stability result*** for the conformal evolution Equations (15.20a)–(15.20c).

The solutions $\breve{\mathbf{u}}$ provided by Proposition 15.1 give rise, in turn, to a solution to the conformal field equations. More precisely, one has:

Proposition 15.2 (***propagation of the constraints for the standard conformal evolution system***) *Given a solution* $\mathbf{u}_\star = \mathring{\mathbf{u}}_\star + \breve{\mathbf{u}}$ *to the conformal* *evolution Equations* (15.20a)–(15.20c) *on* $[0,\tau_\bullet]\times\mathbb{S}^3$ *such that the conformal constraint equations are satisfied on* $\mathcal{S}_\star$, *then*

$$Z_{\boldsymbol{a}}=0,\qquad Z_{\boldsymbol{ab}}=0,\qquad \Sigma_{\boldsymbol{a}}{}^{\boldsymbol{c}}{}_{\boldsymbol{b}}=0,\qquad \Xi^{\boldsymbol{c}}{}_{\boldsymbol{dab}}=0,$$
$$\Delta_{\boldsymbol{cdb}}=0,\qquad \Lambda_{\boldsymbol{bcd}}=0,$$

on $[0,\tau_\bullet]\times\mathbb{S}^3$.

Proof From the discussion in Chapters 11 and 13 it follows that if the conformal constraint equations and the conformal evolution equations are satisfied on the initial hypersurface $\mathcal{S}_\star$, then one obtains

$$Z_{\boldsymbol{a}}|_{\mathcal{S}_\star}=0,\qquad Z_{\boldsymbol{ab}}|_{\mathcal{S}_\star}=0,\qquad \Sigma_{\boldsymbol{a}}{}^{\boldsymbol{c}}{}_{\boldsymbol{b}}|_{\mathcal{S}_\star}=0,\qquad \Xi^{\boldsymbol{c}}{}_{\boldsymbol{dab}}|_{\mathcal{S}_\star}=0,$$
$$\Delta_{\boldsymbol{cdb}}|_{\mathcal{S}_\star}=0,\qquad \Lambda_{\boldsymbol{bcd}}|_{\mathcal{S}_\star}=0.$$

Now, from Proposition 13.2 it follows that the above zero quantities satisfy a symmetric hyperbolic subsidiary evolution system. As the initial data for this evolution system vanish and the evolution system is homogeneous in the zero quantities, it follows from Corollary 12.1 that the zero quantities must vanish on $[0,\tau_\bullet)\times\mathbb{S}^3$ so that the result follows. □

Locating the conformal boundary

The existence of solutions to the evolution Equations (15.20a)–(15.20c) for a minimum existence interval $[0,\tau_\bullet)\supset[0,\frac{3}{4}\pi)$ provides room enough for the development of the conformal boundary. That this does indeed happen is crucial for the interpretation of the solution to the conformal evolution equations as a *global solution to the Einstein field equations*. This property is ensured by the following:

Lemma 15.3 (***structure of the conformal boundary***) *Given a solution* $\check{\mathbf{u}}$, *as given by Proposition 15.1, with* $\parallel\check{\mathbf{u}}_\star\parallel_m<\varepsilon$ *sufficiently small, there exists a function* $\tau_+=\tau_+(\underline{x})$, $\underline{x}\in\mathbb{S}^3$ *such that* $0<\tau_+(\underline{x})<\tau_\bullet$ *and*

$$\Xi>0\qquad on\ \tilde{\mathcal{M}}\equiv\{(\tau,\underline{x})\in\mathbb{R}^3\,|\,0\le\tau<\tau_+(\underline{x})\},$$
$$\Xi=0\quad and\quad \Sigma_{\boldsymbol{a}}\Sigma^{\boldsymbol{a}}=-\frac{1}{3}\lambda<0\qquad on\ \mathscr{I}^+\equiv\{(\tau_+(\underline{x}),\underline{x})\in\mathbb{R}\times\mathbb{S}^3\}.$$

Remark. The above lemma ensures the existence, at least for sufficiently small perturbations, of a complete spacelike component of the conformal boundary. Observe also that the function $\tau_+(\underline{x})$ provided by Lemma 15.3 defines a diffeomorphism between $\mathbb{S}^3$ and $\mathscr{I}^+$. Consequently $\mathscr{I}^+\approx\mathbb{S}^3$.

Proof The key observation to prove this result is that $\mathring{\Xi}|_{\tau=3\pi/4}<0$. Using Proposition 15.1 (ii), for sufficiently small $\varepsilon>0$ one has $(\mathring{\Xi}+\breve{\Xi})|_{\tau=3\pi/4}<0$. As $(\mathring{\Xi}+\breve{\Xi})|_{\tau=0}>0$ there must exist a τ_+ for which $\Xi=0$. By reducing ε further – if necessary – one has that τ is unique, and, hence, the function $\tau_+(\underline{x})$ is well defined. Now, from the conformal Equation (8.28c) it follows that

$$\nabla^a \Xi \nabla_a \Xi = -\frac{1}{3}\lambda > 0, \qquad \text{if } \Xi = 0.$$

Accordingly, $\tau = \tau_+(\underline{x})$ defines a regular spacelike hypersurface $\mathscr{I}^+$. □

The last step in the present analysis is to show that the obtained solutions to the conformal evolution Equations (15.20a)–(15.20c) give rise to a global solution to the vacuum Einstein field equations. One has the following:

Theorem 15.1 (***global existence and stability of de Sitter-like spacetimes: gauge source functions version***) *Given $m \geq 4$, a solution $\mathbf{u}_\star = \mathring{\mathbf{u}}_\star + \breve{\mathbf{u}}_\star$ to the conformal constraint equations with de Sitter-like cosmological constant such that $\| \breve{\mathbf{u}}_\star \|_m < \varepsilon$ for $\varepsilon > 0$ suitably small gives rise to a unique C^{m-2} solution to the conformal Einstein field equations on*

$$\mathcal{M} \equiv \tilde{\mathcal{M}} \cup \mathscr{I}^+$$

with $\tilde{\mathcal{M}}$ and $\mathscr{I}^+$ as defined in Lemma 15.3. This solution implies, in turn, a solution $(\tilde{\mathcal{M}}, \tilde{\mathbf{g}})$, to the Einstein field equations with de Sitter-like cosmological constant for which $\mathscr{I}^+$ represents conformal infinity.

Remark. The above theorem together with Propositions 15.1 and 15.2 and Lemma 15.3 constitute a technical version of the main theorem of this chapter. As the component of the conformal boundary obtained by this procedure is a spacelike hypersurface with the topology of $\mathbb{S}^3$, one concludes that the solution $(\tilde{\mathcal{M}}, \tilde{\mathbf{g}})$ to the Einstein field equations has the same global structure as the exact de Sitter spacetime; see Figure 15.4.

Proof From Proposition 8.2, it follows that a solution to the conformal Einstein field equations implies the existence of a metric $\tilde{\mathbf{g}}$ satisfying the Einstein field

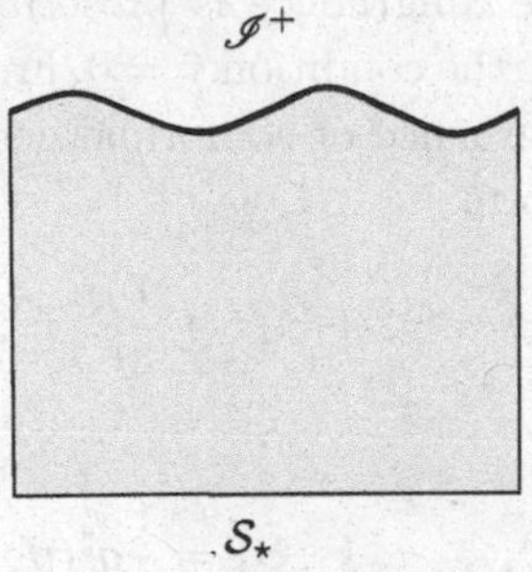

Figure 15.4 Penrose diagram of a perturbation of the de Sitter spacetime given by Theorem 15.1. The spacetime is obtained as a result of an initial value problem on the Cauchy hypersurface $\mathcal{S}_\star$.

equations wherever $\Xi \neq 0$. The statement about the interpretation of the conformal boundary follows from Lemma 15.3. □

15.4 Global existence and stability using conformal Gaussian systems

This section provides an alternative proof of the main theorem of this chapter using the extended conformal Einstein field equations expressed in terms of a conformal Gaussian system. This alternative proof allows one to contrast the strengths and weaknesses of the two different hyperbolic reduction methods discussed in Chapter 13. As will be seen in the following, the use of properties of conformal geodesics greatly simplifies the analysis of the conformal boundary of the spacetime. Generalising this approach to include matter fields is, however, more complicated than if one were to use gauge source functions.

The details of the construction of a *conformal Gaussian system* for the extended conformal field equations have already been discussed in Section 13.4.1. To apply this general discussion to the analysis of perturbations of the de Sitter spacetime, one needs to specify the particular form of the conformal factor Θ and the covector $\boldsymbol{d}$ associated to the congruence of conformal geodesics. This is done in the following subsection.

15.4.1 A priori analysis of the structure of the conformal boundary of perturbations of the de Sitter spacetime

One of the advantages of the formulation of the conformal evolution equations in terms of conformal Gaussian systems is that it provides an a priori knowledge of the location and structure of the conformal boundary; that is, one has an explicit description of the locus of its points, even before knowing that a solution actually exists. This a priori knowledge provides valuable insight into the nature of the underlying initial value problem.

In what follows, let $(\tilde{\mathcal{M}}, \tilde{\boldsymbol{g}})$ denote a vacuum spacetime with de Sitter-like cosmological constant. It will be assumed that $(\tilde{\mathcal{M}}, \tilde{\boldsymbol{g}})$ can be covered with a non-intersecting congruence of conformal geodesics $(x(\bar{\tau}), \tilde{\boldsymbol{\beta}}(\bar{\tau}))$ with affine parameter $\bar{\tau}$ and that the data for the congruence is prescribed on a fiduciary spacelike hypersurface $\mathcal{S}_\star$ described by the condition $\bar{\tau} = 0$. From Proposition 5.1 it follows that, associated to this congruence of conformal geodesics, one has a *canonical conformal factor* Θ of the form

$$\Theta = \Theta_\star + \dot{\Theta}_\star \bar{\tau} + \frac{1}{2}\ddot{\Theta}_\star \bar{\tau}^2, \tag{15.22}$$

with the constraints

$$\dot{\Theta}_\star = \langle \boldsymbol{d}_\star, \dot{\boldsymbol{x}}_\star \rangle, \qquad \Theta_\star \ddot{\Theta}_\star = \frac{1}{2}\boldsymbol{g}^\sharp(\boldsymbol{d}_\star, \boldsymbol{d}_\star) + \frac{1}{6}\lambda, \tag{15.23}$$

where the coefficients $\Theta_\star$, $\dot{\Theta}_\star$ and $\ddot{\Theta}_\star$ are constant along a given conformal geodesic. The conformal factor Θ allows one to obtain a conformal extension

$(\bar{\mathcal{M}}, \bar{\boldsymbol{g}})$ of the physical spacetime $(\tilde{\mathcal{M}}, \tilde{\boldsymbol{g}})$ with $\bar{\boldsymbol{g}} \equiv \Theta^2 \tilde{\boldsymbol{g}}$. The specific details of the conformal factor Θ depend on the location of the hypersurface $\mathcal{S}_\star$ with respect to the conformal boundary and give rise to two different initial value problems.

Standard Cauchy problem

First, consider a situation where the initial hypersurface $\mathcal{S}_\star$ does not coincide with the conformal boundary. As one is interested in the construction of spacetimes whose spatial sections have the topology of $\mathbb{S}^3$ it is natural to set, without loss of generality, that

$$\Theta_\star = 1, \qquad \dot{\Theta}_\star = 0,$$

so that no further distortion is introduced in the 3-manifold and the congruence of conformal geodesics is symmetric with respect to the initial hypersurface. Moreover, one can set

$$\tilde{\boldsymbol{\beta}}_\star = 0,$$

so that $\boldsymbol{d}_\star = \Theta_\star \tilde{\boldsymbol{\beta}}_\star = 0$ and the general expression for the conformal factor reduces to

$$\Theta = 1 + \frac{1}{12}\lambda \bar{\tau}^2.$$

Now, using Proposition 5.1 one finds that the components $d_{\boldsymbol{a}}$ of the covector $\boldsymbol{d}$ respect to a Weyl propagated frame $\{\boldsymbol{e}_{\boldsymbol{a}}\}$ along the congruence of conformal geodesics and such that $\boldsymbol{e}_{\boldsymbol{0}} = \dot{\boldsymbol{x}}$ are given by

$$d_{\boldsymbol{0}} = \dot{\Theta}, \qquad d_{\boldsymbol{i}} = 0.$$

A direct computation shows that the conformal factor Θ vanishes for

$$\bar{\tau}_{\pm} = \pm\sqrt{\frac{12}{|\lambda|}}.$$

The above expression gives the location of the conformal boundary. Accordingly, it is natural to define

$$\mathscr{I}^{\pm} \equiv \{\bar{\tau}_{\pm}\} \times \mathcal{S},$$

and one has $\mathscr{I}^{\pm} \approx \mathbb{S}^3$. Finally, recalling the constraint $\boldsymbol{d} = \Theta \boldsymbol{f} + \mathbf{d}\Theta$ and assuming that $\boldsymbol{f}$ is regular at $\mathscr{I}^{\pm}$ one finds

$$\boldsymbol{g}(\mathbf{d}\Theta, \mathbf{d}\Theta) = \eta^{\boldsymbol{ab}} d_{\boldsymbol{a}} d_{\boldsymbol{b}} = \dot{\Theta}^2 > 0 \qquad \text{at } \mathscr{I}^{\pm}.$$

Thus, if the conformal boundary exists and is regular, it must be a spacelike hypersurface.

Asymptotic Cauchy problem at $\mathscr{I}^-$

The conformal field equations allow the formulation of an alternative initial value problem in which initial data are prescribed on a spacelike hypersurface $\mathcal{S} \approx \mathbb{S}^3$ representing one of the components of the conformal boundary, say, $\mathscr{I}^-$ – an ***asymptotic initial value problem***. In this spirit, it is natural to prescribe the initial data for the congruence of conformal geodesics directly at the conformal boundary. This is made possible by the conformal invariance of the conformal geodesic equations.

By assumption, on an asymptotic initial value problem one has that $\Theta = 0$ on $\mathscr{I}^-$. Thus, one necessarily has that $\Theta_\star = 0$ and the conformal factor takes the form

$$\Theta = \dot{\Theta}_\star \bar{\tau} + \frac{1}{2}\ddot{\Theta}_\star \bar{\tau}^2.$$

The second expression in Equation (15.23) implies that $\boldsymbol{g}^\sharp(\boldsymbol{d}_\star, \boldsymbol{d}_\star) = -\lambda/3 > 0$ so that $\boldsymbol{d}_\star$ must be timelike. Now, taking into account the further constraint $\boldsymbol{d} = \Theta \boldsymbol{f} + \mathbf{d}\Theta$ and requiring $\dot{\boldsymbol{x}}_\star$ to be normal to $\mathscr{I}^-$, it follows that

$$d_{\boldsymbol{0}\star} = \dot{\Theta}_\star = \sqrt{\frac{|\lambda|}{3}}, \qquad d_{\boldsymbol{i}\star} = 0,$$

where the positive root has been chosen so that Θ is positive in the future of $\mathscr{I}^-$. Accordingly, off $\mathscr{I}^-$ one has

$$d_{\boldsymbol{0}} = \dot{\Theta} = \left(\sqrt{\frac{|\lambda|}{3}} + \ddot{\Theta}_\star \bar{\tau}\right), \qquad d_{\boldsymbol{i}} = 0.$$

So far, the coefficient $\ddot{\Theta}_\star$ has remained unspecified. Accordingly, it will be considered as free data. These data are, in fact, related to value of the Friedrich scalar

$$s \equiv \frac{1}{4}\nabla_a \nabla^a \Theta + \frac{1}{24} R\Theta$$

on $\mathscr{I}^-$. On the one hand, a calculation gives

$$\begin{aligned} s &\simeq \frac{1}{4}\left(\boldsymbol{e}_{\boldsymbol{a}}\Sigma^{\boldsymbol{a}} + \Gamma_{\boldsymbol{a}}{}^{\boldsymbol{a}}{}_{\boldsymbol{b}}\Sigma^{\boldsymbol{b}}\right) \\ &\simeq \frac{1}{4}\left(\ddot{\Theta}_\star + \dot{\Theta}_\star \Gamma_{\boldsymbol{a}}{}^{\boldsymbol{a}}{}_{\boldsymbol{0}}\right), \end{aligned}$$

where the last equality follows from the fact that $\Sigma_{\boldsymbol{i}}$ vanishes at $\mathscr{I}^-$. On the other hand, the solution to the conformal constraints at the conformal boundary, as discussed in Section 11.4.4, shows that $s_\star \simeq \dot{\Theta}_\star \varkappa$ where $\varkappa$ is a scalar field over $\mathscr{I}^-$; compare Equation (11.40). A further calculation using a frame adapted to $\mathscr{I}^-$ readily yields

$$\Gamma_{\boldsymbol{a}}{}^{\boldsymbol{a}}{}_{\boldsymbol{0}} \simeq \Gamma_{\boldsymbol{i}}{}^{\boldsymbol{i}}{}_{\boldsymbol{0}} \simeq \chi_{\boldsymbol{i}}{}^{\boldsymbol{i}} \simeq \varkappa \delta_{\boldsymbol{i}}{}^{\boldsymbol{i}} \simeq 3\varkappa.$$

One thus concludes that

$$\ddot{\Theta}_\star = \varkappa \dot{\Theta}_\star.$$

In practice, it is convenient to set $\varkappa$ to be constant on $\mathscr{I}^-$. The choice $\varkappa = 0$ gives a representation in which $\mathscr{I}^-$ is a hypersurface with vanishing extrinsic curvature; see Equation (11.41). This representation does not involve a second component of the conformal boundary.

To have a second component of the conformal boundary one hence needs $\varkappa \neq 0$. Adopting the simple choice $\ddot{\Theta}_\star = -1/2$, the conformal factor vanishes at

$$\bar{\tau}_- = 0, \qquad \bar{\tau}_+ = 4\sqrt{\frac{|\lambda|}{3}}.$$

In this conformal representation, the two components of the conformal boundary of the de Sitter-like spacetime $(\tilde{\mathcal{M}}, \tilde{g})$ are given by the sets

$$\mathscr{I}^- = \{0\} \times \mathcal{S}, \qquad \mathscr{I}^+ = \left\{ 4\sqrt{\frac{|\lambda|}{3}} \right\} \times \mathcal{S}.$$

More generally, keeping $\ddot{\Theta}_\star$ unspecified, one finds that the location of $\mathscr{I}^+$ is determined by the free data $\ddot{\Theta}_\star$. Finally, on $\mathscr{I}^+$ one has

$$\boldsymbol{g}(\mathbf{d}\Theta, \mathbf{d}\Theta) = \eta^{\boldsymbol{ab}} d_{\boldsymbol{a}} d_{\boldsymbol{b}} = -\frac{\lambda}{3} > 0,$$

so that both $\mathscr{I}^\pm$, if they exist, are spacelike hypersurfaces.

Remark. In what follows, the analysis of both the standard and the asymptotic initial value problems will be simplified by making use of the choice $\lambda = -3$ for the cosmological constant.

15.4.2 The extended conformal evolution system

Once the conformal factor Θ and the covector $\boldsymbol{d}$ associated to the conformal Gaussian system have been specified, one can proceed to the formulation of an initial value problem. In Proposition 13.3 it has been shown that the extended conformal Einstein field equations expressed in terms of a conformal Gaussian system imply a symmetric hyperbolic evolution system of the form

$$\partial_{\bar{\tau}} \hat{\boldsymbol{\upsilon}} = \mathbf{K}\hat{\boldsymbol{\upsilon}} + \mathbf{Q}(\hat{\boldsymbol{\Gamma}})\hat{\boldsymbol{\upsilon}} + \mathbf{L}(x)\boldsymbol{\phi}, \tag{15.24a}$$

$$(\mathbf{I} + \mathbf{A}^0(e))\partial_{\bar{\tau}}\boldsymbol{\phi} + \mathbf{A}^\alpha(e)\partial_\alpha \boldsymbol{\phi} = \mathbf{B}(\hat{\boldsymbol{\Gamma}})\boldsymbol{\phi}, \tag{15.24b}$$

for $\hat{\mathbf{u}} = (\hat{\boldsymbol{\upsilon}}, \boldsymbol{\phi})$ where $\hat{\boldsymbol{\upsilon}}$ encodes the independent components of the frame, the connection coefficients and the Schouten tensor, while $\boldsymbol{\phi}$ contains the components of the rescaled Weyl spinor.

Mimicking the analysis of Section 15.3, one considers solutions of the form

$$e_{AB}{}^{0} = \breve{e}_{AB}{}^{0}, \quad e_{AB}{}^{\alpha} = \bar{e}_{AB}{}^{\alpha} + \breve{e}_{AB}{}^{\alpha},$$
$$\xi_{ABCD} = \bar{\xi}_{ABCD} + \breve{\xi}_{ABCD}, \quad \chi_{ABCD} = \bar{\chi}_{ABCD} + \breve{\chi}_{ABCD}, \quad f_{AB} = \breve{f}_{AB},$$
$$\Theta_{ABCD} = \bar{\Theta}_{ABCD} + \breve{\Theta}_{ABCD}, \quad \phi_{ABCD} = \breve{\phi}_{ABCD},$$

where

$$\bar{e}_{AB}{}^{\mu}, \qquad \bar{\xi}_{ABCD}, \qquad \bar{\chi}_{ABCD}, \qquad \bar{\Theta}_{ABCD}$$

are the values of the *exact de Sitter solution* expressed in a conformal Gaussian system as discussed in Section 15.1.2; see, in particular, Proposition 15.2. Accordingly, one can write

$$\hat{\boldsymbol{\upsilon}} = \bar{\boldsymbol{\upsilon}} + \breve{\boldsymbol{\upsilon}}, \qquad \boldsymbol{\phi} = \breve{\boldsymbol{\phi}}, \tag{15.25a}$$
$$\boldsymbol{e} = \bar{\boldsymbol{e}} + \breve{\boldsymbol{e}}, \qquad \hat{\boldsymbol{\Gamma}} = \bar{\boldsymbol{\Gamma}} + \breve{\boldsymbol{\Gamma}}. \tag{15.25b}$$

On the initial hypersurface $\mathcal{S}_\star$ one has

$$\hat{\boldsymbol{\upsilon}}_\star = \bar{\boldsymbol{\upsilon}}_\star + \breve{\boldsymbol{\upsilon}}_\star, \qquad \boldsymbol{\phi} = \breve{\boldsymbol{\phi}}_\star,$$

where $\mathring{\mathbf{u}}_\star = (\bar{\boldsymbol{\upsilon}}_\star, \mathbf{0})$ is the exact de Sitter data discussed in Section 15.2 and $\breve{\mathbf{u}}_\star = (\breve{\boldsymbol{\upsilon}}_\star, \breve{\boldsymbol{\phi}}_\star)$.

As the conformal factor Θ and the covector $\boldsymbol{d}$ are universal – that is, they possess the same form for either the exact de Sitter data or the perturbations thereof – one has

$$\partial_{\bar{\tau}}\bar{\boldsymbol{\upsilon}} = \mathbf{K}\bar{\boldsymbol{\upsilon}} + \mathbf{Q}(\bar{\boldsymbol{\Gamma}})\bar{\boldsymbol{\upsilon}}.$$

Substituting the ansatz (15.25a) and (15.25b) into Equations (15.24a) and (15.24b) yields the following evolution equations for $\breve{\mathbf{u}} = (\breve{\boldsymbol{\upsilon}}, \breve{\boldsymbol{\phi}})$:

$$\partial_{\bar{\tau}}\breve{\boldsymbol{\upsilon}} = \mathbf{K}\breve{\boldsymbol{\upsilon}} + \mathbf{Q}(\bar{\boldsymbol{\Gamma}} + \breve{\boldsymbol{\Gamma}})\breve{\boldsymbol{\upsilon}} + \mathbf{Q}(\breve{\boldsymbol{\Gamma}})\bar{\boldsymbol{\upsilon}} + \mathbf{L}(x)\breve{\boldsymbol{\phi}}, \tag{15.26a}$$
$$(\mathbf{I} + \mathbf{A}^0(\bar{\boldsymbol{e}} + \breve{\boldsymbol{e}}))\partial_{\bar{\tau}}\breve{\boldsymbol{\phi}} + \mathbf{A}^\alpha(\bar{\boldsymbol{e}} + \breve{\boldsymbol{e}})\partial_\alpha\breve{\boldsymbol{\phi}} = \mathbf{B}(\bar{\boldsymbol{\Gamma}} + \breve{\boldsymbol{\Gamma}})\breve{\boldsymbol{\phi}}. \tag{15.26b}$$

The above equations are already in a form where the theory of hyperbolic PDEs, as discussed in Chapter 12, can be applied. In particular, existence and Cauchy stability of Equations (15.26a) and (15.26b) are given by Theorem 12.4. The natural domains for solutions to these equations are of the form

$$\mathcal{M}_{\bar{\tau}_\bullet} \equiv [0, \bar{\tau}_\bullet] \times \mathcal{S}, \qquad \mathcal{S} \approx \mathbb{S}^3,$$

for some $\bar{\tau}_\bullet > 0$. The analogue of Propositions 15.1 and 15.2 for the conformal evolution system (15.26a) and (15.26b) is given by:

Proposition 15.3 (***existence of de Sitter-like solutions to the extended conformal evolution equations and propagation of the constraints***) *Let* $\hat{\mathbf{u}}_\star = \bar{\mathbf{u}}_\star + \breve{\mathbf{u}}_\star$ *denote de Sitter-like (standard or asymptotic) initial data for the conformal field equations prescribed on a 3-manifold* $\mathcal{S} \approx \mathbb{S}^3$. *Given* $m \geq 4$*:*

(i) *There exists* $\varepsilon > 0$ *such that if*

$$\| \breve{\mathbf{u}}_\star \|_m < \varepsilon,$$

then there exists a C^{m-2} *unique solution* $\breve{\mathbf{u}}$ *to the conformal evolution equations* (15.26a) *and* (15.26b) *defined on* $(-\frac{5}{2}, \frac{5}{2}) \times \mathbb{S}^3$ *for the standard Cauchy problem and on* $[0, \frac{9}{2}) \times \mathbb{S}^3$ *for the asymptotic Cauchy problem.*

(ii) *If*

$$\hat{\Sigma}_a{}^c{}_b\big|_{\mathcal{S}_\star} = 0, \qquad \hat{\Xi}^c{}_{dab}\big|_{\mathcal{S}_\star} = 0, \qquad \hat{\Delta}_{abc}\big|_{\mathcal{S}_\star} = 0, \qquad \Lambda_{abc}\big|_{\mathcal{S}_\star} = 0,$$

and

$$\delta_a\big|_{\mathcal{S}_\star} = 0, \qquad \gamma_{ab}\big|_{\mathcal{S}_\star} = 0, \qquad \varsigma_{ab}\big|_{\mathcal{S}_\star} = 0,$$

then the solution $\breve{\mathbf{u}}$ *to the conformal evolution equations given by (i) implies, by reducing* ε *if necessary, a* C^{m-2} *solution* $\hat{\mathbf{u}} = \bar{\mathbf{u}} + \breve{\mathbf{u}}$ *to the extended conformal field equations on* $(-\frac{5}{2}, \frac{5}{2}) \times \mathbb{S}^3$ *and, respectively, on* $[0, \frac{9}{2}) \times \mathbb{S}^3$.

(iii) *Given a sequence of initial data* $\hat{\mathbf{u}}_\star^{(n)} = \bar{\mathbf{u}}_\star^{(n)} + \breve{\mathbf{u}}_\star^{(n)}$ *such that*

$$\| \breve{\mathbf{u}}_\star^{(n)} \|_m < \varepsilon, \quad \text{and} \quad \| \breve{\mathbf{u}}_\star^{(n)} \|_m \to 0 \qquad \text{as } n \to \infty,$$

then for the corresponding solutions $\breve{\mathbf{u}}^{(n)} \in C^{m-2}((-\frac{5}{2}, \frac{5}{2}) \times \mathbb{S}^3)$ *and, respectively,* $C^{m-2}([0, \frac{9}{2}) \times \mathbb{S}^3)$, *one has* $\| \breve{\mathbf{u}}^{(n)} \|_m \to 0$ *uniformly in* $\bar{\tau} \in (-\frac{5}{2}, \frac{5}{2})$ *and, respectively* $\bar{\tau} \in [0, \frac{9}{2})$, *as* $n \to \infty$.

Proof The proof of points (i) and (iii) of the above proposition is, again, a direct application of Theorem 12.4 along the lines of Proposition 15.1. The proof of point (ii) concerning the existence of an actual solution of the extended conformal field equations follows from the homogeneity of the subsidiary evolution system as given in Proposition 13.4 together with Corollary 12.1 by an argument identical to that used in Proposition 15.2. □

Remark. By an argument similar to the one leading to Proposition 15.3, using the expression (15.14) for a separation vector in the background de Sitter spacetime, it can be shown that if ε is sufficiently small, then the separation fields for the perturbed spacetime remain non-zero in $(-\frac{5}{2}, \frac{5}{2})$ and, respectively, $[0, \frac{9}{2})$. Thus, the conformal Gaussian system used in the hyperbolic reduction remains well behaved throughout.

Constructing solutions to the Einstein field equations

The discussion of this section can be summarised in the following two technical versions of the main theorem of this chapter:

Theorem 15.2 (***global existence and stability of de Sitter-like spacetimes: conformal Gaussian systems version***) *Given $m \geq 4$, a solution $\mathbf{u}_\star = \bar{\mathbf{u}}_\star + \breve{\mathbf{u}}_\star$ to the conformal constraint equations with $\lambda = -3$ on a standard Cauchy hypersurface $\mathcal{S}_\star \approx \mathbb{S}^3$ such that $\| \breve{\mathbf{u}}_\star \|_m < \varepsilon$, for $\varepsilon > 0$ suitably small, gives rise to a solution $\mathbf{u}$ to the conformal field equations on*

$$\mathcal{M} \equiv [-2, 2] \times \mathbb{S}^3.$$

This solution implies, in turn, a solution $(\tilde{\mathcal{M}}, \tilde{\boldsymbol{g}})$ to the Einstein field equations with cosmological constant $\lambda = -3$ where

$$\tilde{\mathcal{M}} \equiv (-2, 2) \times \mathbb{S}^3,$$

for which

$$\mathscr{I}^\pm \equiv \{\pm 2\} \times \mathbb{S}^3,$$

represent future and past conformal infinity, respectively.

In the case of asymptotic Cauchy data one obtains a similar statement:

Theorem 15.3 (***global existence and stability for the asymptotic initial value problem***) *Given $m \geq 4$, a solution $\mathbf{u}_\star = \bar{\mathbf{u}}_\star + \breve{\mathbf{u}}_\star$ to the conformal constraint equations with $\lambda = -3$ on a 3-manifold $\mathcal{S} \approx \mathbb{S}^3$ representing the past component of the conformal boundary such that $\| \breve{\mathbf{u}}_\star \|_m < \varepsilon$, for $\varepsilon > 0$ suitably small, gives rise to a solution $\mathbf{u}$ to the conformal field equations on*

$$\mathcal{M} \equiv [0, 4] \times \mathbb{S}^3.$$

This solution implies, in turn, a solution $(\tilde{\mathcal{M}}, \tilde{\boldsymbol{g}})$ to the Einstein field equations with cosmological constant $\lambda = -3$ where

$$\tilde{\mathcal{M}} \equiv (0, 4) \times \mathbb{S}^3,$$

for which

$$\mathscr{I}^- \equiv \{0\} \times \mathbb{S}^3, \qquad \mathscr{I}^+ \equiv \{4\} \times \mathbb{S}^3,$$

represent future and past conformal infinity, respectively.

The proofs of Theorems 15.2 and 15.3 are identical to that of Theorem 15.1. Penrose diagrams of the spacetimes thus obtained can be seen in Figure 15.5. Observe that in the gauge being considered, the Penrose diagrams for the exact de Sitter spacetime and the perturbations are identical!

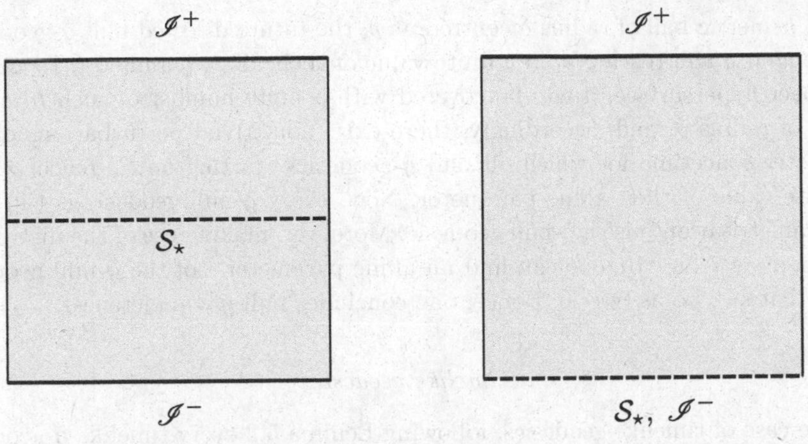

Figure 15.5 Penrose diagrams of de Sitter-like spacetimes obtained from Theorems 15.2 and 15.3: on the left is the spacetime obtained from a standard Cauchy initial value problem; on the right is the spacetime obtained from the asymptotic initial value problem.

15.4.3 Geodesic completeness and asymptotic analysis

The analysis of the existence and stability of de Sitter-like spacetimes developed in Sections 15.3 and 15.4 can be refined to include geodesic completeness. As the exact de Sitter spacetime is geodesically complete, it is to be expected that suitably small perturbations thereof will also share this property. More precisely:

Proposition 15.4 (***geodesic completeness of de Sitter-like spacetimes***) *Suitably small perturbations $(\tilde{\mathcal{M}}, \tilde{\mathbf{g}})$ of the de Sitter spacetime are null and timelike $\tilde{\mathbf{g}}$-geodesically complete.*

In particular, the above proposition together with the existence and stability results obtained in the previous sections show that suitably small perturbations of the de Sitter spacetime are *asymptotically simple spacetimes.*

It is convenient to divide the analysis of Proposition 15.4 into two cases.

Null geodesics

The key observation required to prove null geodesic completeness is the following: *given the conformal representation $(\mathbb{R} \times \mathbb{S}^3, \bar{\mathbf{g}}_{\mathscr{E}})$ any null $\bar{\mathbf{g}}_{\mathscr{E}}$-geodesic starting within the unphysical spacetime reaches the conformal boundary for a finite value of its affine parameter.*

In what follows, let $(\bar{\mathcal{M}}, \bar{\mathbf{g}})$ be one of the de Sitter-like spacetimes obtained from, say, a standard Cauchy initial value problem with data prescribed on a hypersurface $\mathcal{S}_\star$. Making use of a perturbative argument similar to the ones employed in Propositions 15.1 and 15.3 and by reducing ε, if necessary, it can be shown that given a point $p \in \mathcal{S}_\star$ and a fixed $\delta > 0$, for all points $q \in \mathcal{S}_\star$ lying in an $\boldsymbol{h}$-metric ball of radius δ centred at p, the future directed null $\bar{\boldsymbol{g}}$-geodesics starting at q will reach $\mathscr{I}^+$ in a finite value of their affine parameter. As $\mathcal{S}_\star$ is a compact hypersurface, it can be covered with a finite number of such $\boldsymbol{h}$-metric balls of radius δ, and, accordingly, there exist non-trivial perturbations of the de Sitter spacetime for which all null $\bar{\boldsymbol{g}}$-geodesics starting on $\mathcal{S}_\star$ reach $\mathscr{I}^+$ in a finite value of their affine parameter. Now, every $\bar{\boldsymbol{g}}$-null geodesic is (up to a reparametrisation) also a $\tilde{\boldsymbol{g}}$-null geodesic. Moreover, making use of the discussion in Chapters 7 and 10, one can find an affine parameter s of the $\tilde{\boldsymbol{g}}$-null geodesic such that s $\to \infty$ as $\Theta \to 0$. Hence, one concludes null $\tilde{\boldsymbol{g}}$-completeness.

Timelike geodesics

In the case of timelike geodesics, following Lemma 5.2 every timelike $\tilde{\boldsymbol{g}}$-geodesic is, up to a reparametrisation, a timelike $\bar{\boldsymbol{g}}_{\mathscr{E}}$-conformal geodesic. It can be explicitly checked – starting, for example, from the general solution to the conformal geodesic equations in the Minkowski spacetime as discussed in Section 6.2.3 – that every $\bar{\boldsymbol{g}}_{\mathscr{E}}$-conformal geodesic starting inside the region of the Einstein cylinder associated to the conformal de Sitter spacetime reaches the conformal boundary of the spacetime in a finite amount of its unphysical proper time $\bar{\tau}$. Using, as in the case of the null geodesics, a perturbative argument, this property is seen to be preserved for suitably small perturbations of the de Sitter spacetime. Of course, not every $\bar{\boldsymbol{g}}$-conformal geodesic can be reparametrised to a $\tilde{\boldsymbol{g}}$-geodesic. This is the case only for those curves reaching the conformally boundary orthogonally – as can be checked using Lemma 5.3. Finally, using the properties of conformal geodesics in Einstein spaces as discussed in Section 5.5.6, the physical proper time of $\tilde{\boldsymbol{g}}$ satisfies $\tilde{\tau} \to \infty$ as $\Theta \to 0$. This implies geodesic completeness.

15.5 Extensions

The results of this chapter can be extended to the case of the Einstein equations coupled with suitable trace-free matter; see Chapter 9.

For simplicity, the subsequent discussion will be restricted to the standard conformal field equations. One of the main difficulties when attempting a direct extension to include matter is the presence of the rescaled Cotton tensor T_{cdb} in the Cotton and Bianchi equations. As discussed in Chapter 9 this tensor involves derivatives of the matter fields. As the Cotton and the Bianchi equations are interpreted as differential conditions on the components of the Schouten tensor and the rescaled Weyl tensor, the inclusion of matter in the analysis brings further terms into the principal part of the conformal evolution equations which, in principle, destroy the symmetric hyperbolicity. In general, these derivatives cannot be eliminated using the field equations for the matter model. Thus, one introduces the derivatives of the matter variables as new unknowns into the problem. Equations for these ***auxiliary variables*** can be obtained by applying a covariant derivative to the matter equation and commuting derivatives. This procedure has been described, for the Maxwell field, in Section 9.2. For suitable matter models – such as the Maxwell field, the conformally invariant scalar field and radiation fluids – the resulting field equations for the auxiliary field admit a symmetric hyperbolic reduction without the need of introducing further gauge source functions.

The construction of suitable symmetric hyperbolic evolution equations for the auxiliary fields needs to be supplemented with their associated subsidiary evolution equations and a further subsidiary equation for the definition of the auxiliary variable. This procedure is similar in spirit to the construction of subsidiary equations for the geometric fields described in Sections 13.3 and 13.4.5.

The procedure briefly described in the previous paragraph has been implemented by Friedrich (1991) for the Einstein-Yang-Mills system, using the standard conformal field equations and a hyperbolic reduction involving gauge source functions to obtain a generalisation of the existence and stability result given in the main theorem of this chapter. The same basic ideas can be used to obtain a future global existence and stability result for perturbations of radiation perfect fluid Friedman-Robertson-Walker cosmological models; see Lübbe and Valiente Kroon (2013b).

Matter and the extended conformal field equations

The implementation of the ideas discussed in the previous paragraphs to the extended conformal field equations requires further consideration. The matter field equations are usually expressed in terms of the Levi-Civita connection $\boldsymbol{\nabla}$ of the unphysical metric $\boldsymbol{g}$. However, the conformal field equations provide direct access only to the Riemann tensor of the Weyl connection $\hat{\boldsymbol{\nabla}}$. Equation (5.30a), relating the Riemann tensors of the connections $\boldsymbol{\nabla}$ and $\hat{\boldsymbol{\nabla}}$, involves the covariant derivatives of the covector $f_{\boldsymbol{a}}$. Thus, further derivatives of the conformal fields enter the principal part of the evolution system in a way which destroys the symmetric hyperbolicity. The antisymmetric part of the derivative $\hat{\nabla}_{[\boldsymbol{a}}f_{\boldsymbol{b}]}$ can be replaced by terms not containing derivatives using the equation

$$\hat{\nabla}_{\boldsymbol{a}}f_{\boldsymbol{b}} - \hat{\nabla}_{\boldsymbol{b}}f_{\boldsymbol{a}} = \hat{L}_{\boldsymbol{ab}} - \hat{L}_{\boldsymbol{ba}};$$

compare Equation (8.45). However, a similar substitution is not possible for the symmetric part $\hat{\nabla}_{(\boldsymbol{a}}f_{\boldsymbol{b})}$. In order to obtain a suitable symmetric hyperbolic system one needs to introduce $\hat{\nabla}_{(\boldsymbol{a}}f_{\boldsymbol{b})}$ as an unknown of the system – or, alternatively, the components $L_{\boldsymbol{ab}}$ of the Schouten tensor of the unphysical Levi-Civita connection $\boldsymbol{\nabla}$. In the case of the Einstein-Maxwell equations it is possible to find suitable evolution equations for the auxiliary field $\psi_{\boldsymbol{AA'BC}} \equiv \nabla_{\boldsymbol{AA'}}\phi_{\boldsymbol{BC}}$ which do not contain the symmetrised derivative $\hat{\nabla}_{(\boldsymbol{a}}f_{\boldsymbol{b})}$; see Lübbe and Valiente Kroon (2012). This, however, is an exceptional situation.

15.6 Further reading

The results discussed in this chapter were first obtained in Friedrich (1986b). Similar results starting from asymptotic Cauchy data were first discussed in Friedrich (1986a). These results have been extended to the case of the Einstein equations coupled to a Yang-Mills field in Friedrich (1991). Alternative proofs, which make use of the extended conformal field equations and conformal gauge systems, in the vacuum and Einstein-Maxwell case, have been given in Lübbe and Valiente Kroon (2009, 2012).

A different way of generalising the global existence and stability results discussed in this chapter is to consider higher dimensions. In this case one cannot make use of the conformal Einstein field equations of Chapter 8, which are valid only for four-dimensional spacetimes. Alternative field equations are required. Global existence and stability results for de Sitter-like vacuum spacetimes of arbitrary dimension have been given in Anderson (2005a) and Anderson and Chruściel (2005) using the *conformal equations* implied by the Fefferman-Graham obstruction tensor.

The methods of this chapter can be adapted to analyse perturbations of cosmological models with radiation perfect fluids and an asymptotic structure similar to that of de Sitter spacetime. An example of this type of analysis can be found in Lübbe and Valiente Kroon (2013b).

16

Minkowski-like spacetimes

This chapter studies the existence and stability of Minkowski-like spacetimes, that is, solutions to the vacuum Einstein field equations with vanishing cosmological constant. The main result of this chapter is very similar in spirit to the main result concerning the global existence and stability of de Sitter-like spacetimes of Chapter 15. There is, however, a key difference: while the results in Chapter 15 are global in nature, the ones in the present chapter are ***semi-global***. More precisely, the spacetimes to be discussed arise as the development of suitable initial data on hyperboloidal hypersurfaces – an examination of the Penrose diagram of the Minkowski spacetime in Figure 16.1 reveals that these hypersurfaces are not Cauchy hypersurfaces of the spacetime. Accordingly, only a portion of the whole spacetime can be recovered from this type of initial value problem. The main result of this chapter can be formulated as follows:

Theorem *(semiglobal existence and stability of Minkowski-like spacetimes).* *Small enough perturbations of hyperboloidal initial data for the Minkowski spacetime give rise to solutions to the vacuum Einstein field equations which exist globally towards the future and have an asymptotic structure similar to that of the Minkowski spacetime.*

This result was first proved in Friedrich (1986b) and subsequently extended to the Einstein-Yang-Mills equations in Friedrich (1991). The original proof of the result made use of the standard conformal Einstein field equations and is similar to the argument given for the de Sitter spacetime in Section 15.3. In this chapter a proof of the theorem is given which makes use of the extended conformal field equations and conformal Gaussian systems following the ideas in Lübbe and Valiente Kroon (2009). This approach allows for a more detailed and explicit discussion of the structure of the conformal boundary.

The restriction of the analysis of the present chapter to the hyperboloidal initial value problem may seem mysterious at first sight. As will be discussed in some detail in Chapter 20, the initial data for the conformal Einstein field

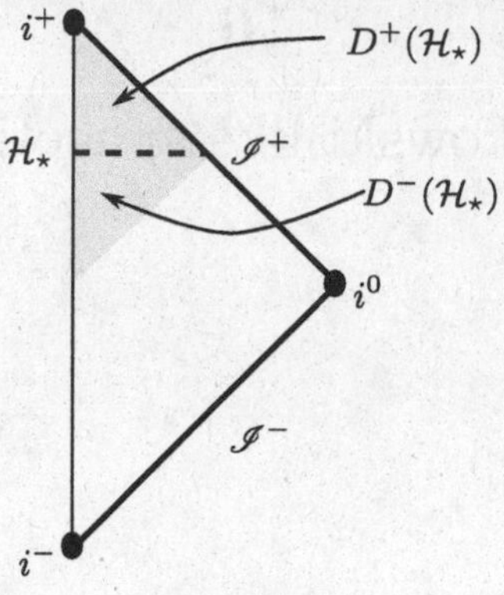

Figure 16.1 Penrose diagram of the Minkowski spacetime and the regions that can be recovered from data on the standard hyperboloid $\mathcal{H}_\star$. The future and past domains of dependence $D^+(\mathcal{H}_\star)$ and $D^-(\mathcal{H}_\star)$ are depicted in grey shading. Observe that $\mathcal{H}_\star$ is not a Cauchy hypersurface as there are portions of the conformal diagram that cannot be recovered from the data on $\mathcal{H}_\star$.

equations on an asymptotically Euclidean (Cauchy) hypersurface is generically singular at spatial infinity – the various issues associated to this singular behaviour are usually known as the ***problem of spatial infinity.***

Despite the above limitation, hyperboloidal initial value problems arise naturally in evolution problems in which the behaviour of gravitational radiation is the main concern; see, for example, Rinne and Moncrief (2013) or Zenginoglu (2008). While the ADM mass – which is computed on asymptotically flat hypersurfaces (see Section 11.6.1) – is a conserved quantity, the notion of mass associated to hyperboloidal hypersurfaces, the so-called ***Bondi mass***, shows a monotonic behaviour, and so it describes the process of mass loss due to gravitational radiation; see Section 10.4.

16.1 The Minkowski spacetime and the conformal field equations

The first step of the stability analysis is a study of the Minkowski spacetime in the gauge used to deduce the conformal evolution equations of Proposition 13.3.

16.1.1 The basic representation

As discussed in Section 6.2, the Minkowski spacetime $(\mathbb{R}^4, \tilde{\boldsymbol{\eta}})$ can be conformally embedded into the *expanding Einstein cylinder* $(\mathbb{R}\times\mathbb{S}^3, \bar{\boldsymbol{g}}_{\mathscr{E}})$ where

$$\bar{\boldsymbol{g}}_{\mathscr{E}} \equiv \mathbf{d}\bar{\tau}\otimes\mathbf{d}\bar{\tau} - \left(1+\frac{\bar{\tau}^2}{4}\right)^2 \boldsymbol{\hbar},$$

by means of the conformal rescaling

$$\bar{\boldsymbol{g}}_{\mathscr{E}} = \Theta^2_{\mathscr{M}}\tilde{\boldsymbol{\eta}}, \qquad \Theta_{\mathscr{M}} \equiv 2\cos^2\frac{\psi}{2}\left(1-\frac{1}{4}\tan^2\frac{\psi}{2}\,\bar{\tau}^2\right). \tag{16.1}$$

The coordinate $\bar{\tau}$ is an affine parameter of the conformal geodesics $(x_{\mathscr{G}}(\bar{\tau}), \bar{\boldsymbol{\beta}}_{\mathscr{G}}(\bar{\tau}))$ with

$$x_{\mathscr{G}}(\bar{\tau}) = (\bar{\tau}, x_\star^\alpha), \qquad \boldsymbol{x}'_{\mathscr{G}}(\bar{\tau}) = \boldsymbol{\partial}_{\bar{\tau}}, \qquad \bar{\boldsymbol{\beta}}_{\mathscr{G}}(\bar{\tau}) = \frac{2\bar{\tau}}{4+\bar{\tau}^2}\mathbf{d}\bar{\tau}.$$

The underlying geometry of the conformal representation of the Minkowski spacetime described in the previous paragraph is that of the expanding cylinder. Accordingly, the *geometric fields for this conformal representation of the Minkowski spacetime coincide with those of the conformal representation of the de Sitter spacetime* discussed in Section 15.1.2. That is, one has

$$\bar{e}^{\mathbf{0}} = 1, \qquad \bar{e}_{(\boldsymbol{AB})}{}^{\mathbf{0}} = 0, \tag{16.2a}$$

$$\bar{e}^{\boldsymbol{i}} = 0, \qquad \bar{e}_{(\boldsymbol{AB})}{}^{\boldsymbol{i}} = \frac{4}{4+\bar{\tau}^2}\sigma_{\boldsymbol{AB}}{}^{\boldsymbol{i}}, \tag{16.2b}$$

$$f_{\boldsymbol{AB}} = 0, \tag{16.2c}$$

$$\xi_{\boldsymbol{ABCD}} = -\frac{4\mathrm{i}}{4+\bar{\tau}^2}h_{\boldsymbol{ABCD}}, \tag{16.2d}$$

$$\bar{\chi}_{(\boldsymbol{AB})\boldsymbol{CD}} = \frac{2\mathrm{i}}{4+\bar{\tau}^2}h_{\boldsymbol{ABCD}}, \tag{16.2e}$$

$$\Theta_{\boldsymbol{AB}} = 0, \qquad \Theta_{\boldsymbol{ABCD}} = -\frac{2}{4+\bar{\tau}^2}h_{\boldsymbol{ABCD}}, \tag{16.2f}$$

$$\phi_{\boldsymbol{ABCD}} = 0. \tag{16.2g}$$

It is important to emphasise, however, that the *conformal gauge fields* $\Theta_{\mathscr{M}}$ and $\boldsymbol{d}_{\mathscr{M}}$, relating the conformal representation to the physical Minkowski spacetime, are different from those of the de Sitter spacetime.

A schematic representation of the conformal boundary associated to the above conformal representation of the Minkowski spacetime is given in Figure 16.2. It is observed that, as a consequence of the explicit time symmetry of the

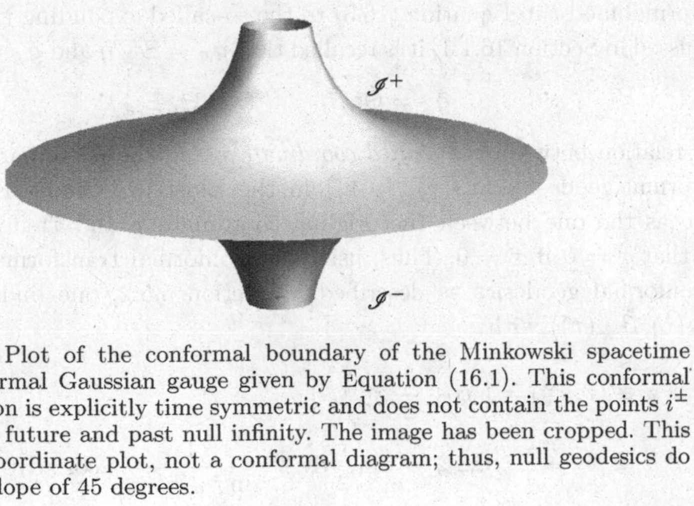

Figure 16.2 Plot of the conformal boundary of the Minkowski spacetime in the conformal Gaussian gauge given by Equation (16.1). This conformal representation is explicitly time symmetric and does not contain the points $i^\pm$ representing future and past null infinity. The image has been cropped. This figure is a coordinate plot, not a conformal diagram; thus, null geodesics do not have a slope of 45 degrees.

representation, the points $i^{\pm}$ representing future and past timelike infinity are not included. This representation is not the most convenient one to use in analysing a hyperboloidal initial value problem. A related, more convenient representation is given in the next subsection.

16.1.2 A conformal representation adapted to the standard hyperboloid

As discussed in Section 6.2, the Minkowski spacetime can be embedded into the Einstein cylinder using the conformal factor

$$\Xi_{\mathscr{M}} \equiv \cos\tau + \cos\psi.$$

In the following it will be convenient to shift the above standard embedding by $\pi/2$ to the past with the replacement $\tau \mapsto \check{\tau} + \pi/2$, so that the ***standard Minkowski hyperboloid*** $\mathcal{H}_\star$ which is given by the condition $\tau = \pi/2$ is now located at $\check{\tau} = 0$; see Equation (6.26). Accordingly, one obtains the shifted conformal factor

$$\check{\Xi}_{\mathscr{M}} \equiv \cos\left(\check{\tau} + \frac{\pi}{2}\right) + \cos\psi = \cos\psi - \sin\check{\tau}. \tag{16.3}$$

In particular, the conformal factor embedding the hyperboloidal 3-manifold into $\mathbb{S}^3$ is given by

$$\bar{\Omega} \equiv \cos\psi.$$

One has that $\bar{\Omega} = 0$ at $\psi = \pi/2$. Hence, it is natural to define

$$\partial\mathcal{H}_\star \equiv \left\{ p \in \mathbb{S}^3 \mid \psi(p) = \frac{\pi}{2} \right\}.$$

Observe that $\mathbf{d}\bar{\Omega} \neq 0$ at $\partial\mathcal{H}_\star$.

To relate the conformal representation of the Minkowski spacetime given by the conformal factor in Equation (16.3) to the so-called expanding Einstein cylinder discussed in Section 16.1.1, it is recalled that $\boldsymbol{g}_{\mathscr{E}} = \check{\Xi}^2_{\mathscr{M}}\tilde{\boldsymbol{\eta}}$ and $\bar{\boldsymbol{g}}_{\mathscr{E}} = \bar{\Theta}^2_{\mathscr{E}}\boldsymbol{g}_{\mathscr{E}}$ so that

$$\bar{\boldsymbol{g}}_{\mathscr{E}} = \check{\Theta}^2_{\mathscr{M}}\tilde{\boldsymbol{\eta}}, \qquad \check{\Theta} \equiv \bar{\Theta}_{\mathscr{E}}\check{\Xi}_{\mathscr{M}}.$$

The relation between the *shifted coordinate* $\check{\tau}$ and the affine parameter $\bar{\tau}$ of the conformal geodesics $\big(x_{\mathscr{E}}(\bar{\tau}), \boldsymbol{\beta}_{\mathscr{E}}(\bar{\tau})\big)$ in the Einstein cylinder is, *formally*, the same as the one between the original coordinate τ and $\bar{\tau}$; in particular, one has that $\bar{\tau} = 0$ if $\check{\tau} = 0$. Thus, using the conformal transformation properties of conformal geodesics as described in Section 5.5.2, one finds that the pair $\big(\check{x}_{\mathscr{M}}(\check{\tau}), \check{\boldsymbol{\beta}}_{\mathscr{M}}(\check{\tau})\big)$ with

$$\check{x}_{\mathscr{M}} \equiv (\bar{\tau}, x^\alpha_\star) = \left(\cos^2\frac{\check{\tau}}{2}, x^\alpha_\star\right),$$

$$\check{\boldsymbol{\beta}}_{\mathscr{M}} \equiv \boldsymbol{\beta}_{\mathscr{E}} + \check{\Xi}^{-1}_{\mathscr{M}}\mathbf{d}\check{\Xi}_{\mathscr{M}} = \tan\frac{\check{\tau}}{2}\,\mathbf{d}\check{\tau} + \frac{1}{\sin\check{\tau} - \cos\psi}\big(\cos\check{\tau}\,\mathbf{d}\check{\tau} + \sin\psi\,\mathbf{d}\psi\big)$$

and

$$\check{\tau} = 2\arctan\frac{\bar{\tau}}{2} \tag{16.4}$$

gives rise to a congruence of conformal geodesics in the Minkowski spacetime adapted to the conformal factor $\breve{\Xi}_{\mathscr{M}}$ in Equation (16.3). This congruence can be used to construct a conformal Gaussian gauge system for the Minkowski spacetime.

A calculation using standard trigonometric identities and the relation (16.4) between the parameters $\check{\tau}$ and $\bar{\tau}$ yields the expression

$$\breve{\Theta} = \cos\psi\left(1 - \sec\psi\bar{\tau} + \frac{\bar{\tau}^2}{4}\right) \tag{16.5}$$

for the conformal factor associated to the new conformal Gaussian gauge system. This conformal factor vanishes whenever

$$\bar{\tau} = \frac{2 \pm \sin\psi}{\cos\psi}.$$

A plot of this conformal factor can be seen in Figure 16.3. Moreover, the components of the covector $\breve{\boldsymbol{d}}_{\mathscr{M}} \equiv \breve{\Theta}\breve{\boldsymbol{\beta}}_{\mathscr{M}}$ with respect to a Weyl propagated frame $\{\bar{\boldsymbol{e}}_a\}$ such that $\bar{\boldsymbol{e}}_{\boldsymbol{0}} = \dot{\bar{\boldsymbol{x}}}_{\mathscr{M}}$ are given by

$$\breve{\beta}_{\boldsymbol{0}} = \partial_{\bar{\tau}}\breve{\Theta}_{\mathscr{M}}, \qquad \breve{\beta}_{\boldsymbol{i}} = \bar{\boldsymbol{e}}_{\boldsymbol{i}}(\bar{\Omega}).$$

Finally, it follows from the discussion from the previous paragraphs that the geometry of the conformal representation of the Minkowski spacetime given by the conformal factor (16.5) is described by the fields (16.2a)–(16.2g); that is, *the geometry of this representation coincides with that of the expanding Einstein*

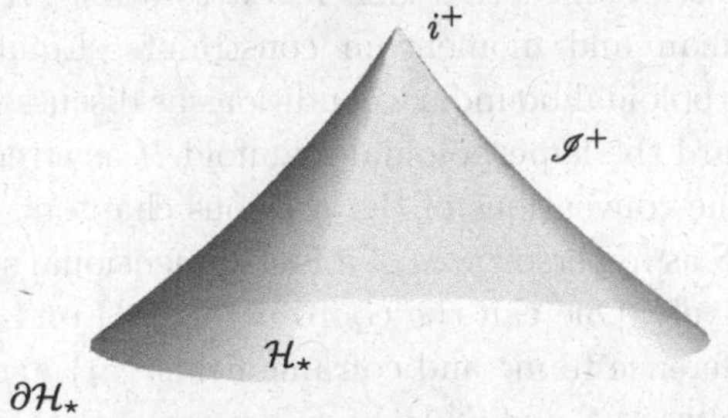

Figure 16.3 Plot of the conformal boundary for the Minkowski spacetime in a conformal Gaussian gauge adapted to the standard hyperboloid. In this particular representation timelike infinity i^+ is at a finite location. The set $\partial\mathcal{H}_\star$ denotes the intersection of the conformal boundary $\mathscr{I}^+$ with the initial hyperboloid $\mathcal{H}_\star$. As in the case of Figure 16.2, this plot is not a conformal diagram.

cylinder. In particular, a suitable Jacobi field $z_{\boldsymbol{AB}}$ measuring the deviation of the curves of the congruence of conformal geodesics is given by

$$z = 0, \qquad z_{(\boldsymbol{AB})} = \left(1 + \frac{\bar{\tau}^2}{4}\right) z_{\star(\boldsymbol{AB})},$$

where $z_{\star(\boldsymbol{AB})}$ is some fiduciary initial value at the standard hyperboloid $\mathcal{H}_\star$; compare Equation (15.14).

16.1.3 Initial data for the Minkowski spacetime on the standard hyperboloid

The congruence of conformal geodesics giving rise to the conformal Gaussian system is specified on the standard hyperboloid $\mathcal{H}_\star$ by the data

$$\breve{\Theta}_\star = \cos\psi, \qquad \breve{d}_{\boldsymbol{0}\star} = -1, \qquad \breve{d}_{\boldsymbol{i}\star} = \boldsymbol{c}_{\boldsymbol{i}}(\bar{\Omega}).$$

On $\mathcal{H}_\star$, the conformal fields satisfy the conditions

$$\bar{e}^{\boldsymbol{0}} = 1, \qquad \bar{e}_{(\boldsymbol{AB})}{}^{\boldsymbol{0}} = 0, \tag{16.6a}$$

$$\bar{e}^{\boldsymbol{i}} = 0, \qquad \bar{e}_{(\boldsymbol{AB})}{}^{\boldsymbol{i}} = \sigma_{\boldsymbol{AB}}{}^{\boldsymbol{i}}, \tag{16.6b}$$

$$f_{\boldsymbol{AB}} = 0, \qquad \bar{\xi}_{\boldsymbol{ABCD}} = -\mathrm{i}h_{\boldsymbol{ABCD}}, \qquad \bar{\chi}_{(\boldsymbol{AB})\boldsymbol{CD}} = 0, \tag{16.6c}$$

$$\bar{\Theta}_{\boldsymbol{AB}} = 0, \qquad \bar{\Theta}_{\boldsymbol{ABCD}} = -\frac{1}{2}h_{\boldsymbol{ABCD}}, \tag{16.6d}$$

$$\bar{\phi}_{\boldsymbol{ABCD}} = 0. \tag{16.6e}$$

16.2 Perturbations of hyperboloidal data for the Minkowski spacetime

In what follows, it is assumed that one has a solution $(\mathcal{H}, \boldsymbol{h}, \boldsymbol{K}, \Omega, \Sigma)$ to the conformal Hamiltonian and momentum constraints, Equations (11.15a) and (11.15b), with hyperboloidal boundary conditions as discussed in Section 11.7. It is convenient to regard the hyperboloidal manifold $\mathcal{H}$ as a region of a 3-manifold $\mathcal{S} \approx \mathbb{S}^3$. Following the conventions of the previous chapters, when regarding the 3-manifolds $\mathcal{H}$ and $\mathcal{S}$ as hypersurfaces of a four-dimensional spacetime one writes $\mathcal{H}_\star$ and $\mathcal{S}_\star$, respectively. One can use coordinates (x^α) on $\mathbb{S}^3$ as coordinates on $\mathcal{H}$ and introduce reference frame and coframe fields $\{\boldsymbol{c}_{\boldsymbol{i}}\}$ and $\{\boldsymbol{\alpha}^{\boldsymbol{i}}\}$ by requiring the identification between $\mathcal{S}$ and $\mathbb{S}^3$ to be a harmonic map; see the discussion in Section 15.2.1.

Initial data for the conformal evolution equations can be obtained from the basic initial data $(\mathcal{H}, \boldsymbol{h}, \boldsymbol{K}, \Omega, \Sigma)$ using the procedure described in Section 11.4.3. It will be assumed that the data can be written on the initial hyperboloid $\mathcal{H}_\star$ in the form

$$e^{\mathbf{0}} = 1, \qquad e_{(\boldsymbol{AB})}{}^{\mathbf{0}} = 0, \tag{16.7a}$$
$$e^{\boldsymbol{i}} = 0, \qquad e_{(\boldsymbol{AB})}{}^{\boldsymbol{i}} = \sigma_{\boldsymbol{AB}}{}^{\boldsymbol{i}} + \breve{e}_{(\boldsymbol{AB})}{}^{\boldsymbol{i}}, \tag{16.7b}$$
$$f_{\boldsymbol{AB}} = 0, \tag{16.7c}$$
$$\xi_{\boldsymbol{ABCD}} = \bar{\xi}_{\boldsymbol{ABCD}} + \breve{\xi}_{\boldsymbol{ABCD}}, \tag{16.7d}$$
$$\chi_{(\boldsymbol{AB})\boldsymbol{CD}} = \bar{\chi}_{(\boldsymbol{AB})\boldsymbol{CD}}, \tag{16.7e}$$
$$\Theta_{\boldsymbol{AB}} = \breve{\Theta}_{\boldsymbol{AB}}, \qquad \Theta_{\boldsymbol{ABCD}} = \bar{\Theta}_{\boldsymbol{ABCD}} + \breve{\Theta}_{\boldsymbol{ABCD}}, \tag{16.7f}$$
$$\phi_{\boldsymbol{ABCD}} = \breve{\phi}_{\boldsymbol{ABCD}}, \tag{16.7g}$$

with

$$\bar{\xi}_{\boldsymbol{ABCD}}, \quad \bar{\chi}_{(\boldsymbol{AB})\boldsymbol{CD}}, \quad \bar{\Theta}_{\boldsymbol{ABCD}}$$

as given by Equations (16.6a)–(16.6e), while the fields

$$\breve{e}_{(\boldsymbol{AB})}{}^{\boldsymbol{i}}, \quad \breve{\xi}_{\boldsymbol{ABCD}}, \quad \breve{\chi}_{(\boldsymbol{AB})\boldsymbol{CD}}, \quad \breve{\Theta}_{\boldsymbol{AB}}, \quad \breve{\Theta}_{\boldsymbol{ABCD}}, \quad \breve{\phi}_{\boldsymbol{ABCD}}$$

describe the perturbation from standard hyperboloidal Minkowski data and $\sigma_{\boldsymbol{AB}}{}^{\boldsymbol{i}}$ are the spatial Infeld-van der Waerden symbols.

While the ***background fields*** $\sigma_{\boldsymbol{AB}}{}^{\boldsymbol{i}}$, $\bar{\xi}_{\boldsymbol{ABCD}}$, $\bar{\chi}_{(\boldsymbol{AB})\boldsymbol{CD}}$, $\bar{\Theta}_{\boldsymbol{ABCD}}$ are defined on the whole of $\mathcal{S} \approx \mathbb{S}^3$, the perturbation fields $\breve{e}_{(\boldsymbol{AB})}{}^{\boldsymbol{i}}$, $\breve{\xi}_{\boldsymbol{ABCD}}$, $\breve{\chi}_{(\boldsymbol{AB})\boldsymbol{CD}}$, $\breve{\Theta}_{\boldsymbol{AB}}$, $\breve{\Theta}_{\boldsymbol{ABCD}}$, $\breve{\phi}_{\boldsymbol{ABCD}}$ are defined only on $\mathcal{H}$. To apply the basic existence and stability result, Theorem 12.4, to the present situation one extends the hyperboloidal initial data set on $\mathcal{H}$ to data on $\mathcal{S}$. Using the *extension theorem*, Proposition 12.2, and given $m \geq 4$ there exists a linear operator $\mathscr{E} : H^m(\mathcal{H}, \mathbb{C}^N) \to H^m(\mathcal{S}, \mathbb{C}^N)$ such that for $\breve{\mathbf{u}}_\star \in H^m(\mathcal{H}, \mathbb{C}^N)$ then $(\mathscr{E}\breve{\mathbf{u}}_\star)(x) = \breve{\mathbf{u}}_\star(x)$ almost everywhere in $\mathcal{H}$ and

$$\| \mathscr{E}\breve{\mathbf{u}}_\star \|_{m,\mathcal{S}} \leq K \| \breve{\mathbf{u}}_\star \|_{m,\mathcal{H}},$$

with K a universal constant for fixed m. As in the case of the de Sitter spacetime, the background initial data $\mathring{\mathbf{u}}_\star$ is defined on the whole of $\mathcal{S}$ so that the ***extended data***

$$\mathbf{u}_\star = \mathring{\mathbf{u}}_\star + \mathscr{E}\breve{\mathbf{u}}_\star \tag{16.8}$$

is a well-defined function in $H^m(\mathcal{S}, \mathbb{C}^N)$. The extension of the hyperboloidal data given by (16.8) is non-unique and, in general, will not satisfy the conformal constraint equations on $\mathcal{S} \setminus \mathcal{H}$. As the norm $\| \mathscr{E}\breve{\mathbf{u}}_\star \|_{m,\mathcal{S}}$ is dominated by the norm $\| \breve{\mathbf{u}}_\star \|_{m,\mathcal{H}}$, then $\| \mathscr{E}\breve{\mathbf{u}}_\star \|_{m,\mathcal{S}}$ can be made as small as necessary by making $\| \mathbf{u}_\star \|_{m,\mathcal{H}}$ suitably small. In complete analogy to the case of the de Sitter spacetime, one says that a hyperboloidal initial data set of the form (16.7a)–(16.7g) is ***ε-small*** (in the norm $\| \;\; \|_{\mathcal{S},m}$) if

$$\begin{aligned}\| \breve{e}_{(\boldsymbol{AB})}{}^{\boldsymbol{i}} \|_{\mathcal{S},m} &+ \| \breve{\xi}_{\boldsymbol{ABCD}} \|_{\mathcal{S},m} + \| \breve{\chi}_{\boldsymbol{ABCD}} \|_{\mathcal{S},m} \\ &+ \| \breve{\Theta}_{\boldsymbol{AB}} \|_{\mathcal{S},m} + \| \breve{\Theta}_{\boldsymbol{ABCD}} \|_{\mathcal{S},m} + \| \breve{\phi}_{\boldsymbol{ABCD}} \|_{\mathcal{S},m} < \varepsilon,\end{aligned}$$

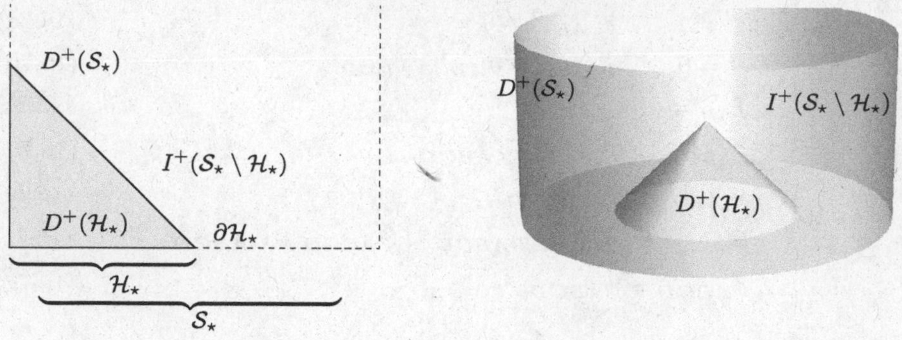

Figure 16.4 Domain of dependence $D^+(\mathcal{H}_\star)$ of data for the conformal evolution equations on a hyperboloid $\mathcal{H}_\star$: on the left, a schematic representation of the setup; on the right, a three-dimensional depiction. To make use of Kato's existence theorem, the data have to be extended to $\mathcal{S}_\star \setminus \mathcal{H}_\star$ where $\mathcal{S}_\star \approx \mathbb{S}^3$. The domain of dependence of the extended data $D^+(\mathcal{S}_\star)$ corresponds, in principle, to the cylinder $[0,\infty)\times\mathbb{S}^3$. The chronological future of the extension on $\mathcal{S}_\star \setminus \mathcal{H}_\star$, denoted by $I^+(\mathcal{S}_\star \setminus \mathcal{H}_\star)$, does not intersect the domain of dependence of the hyperboloidal data $D^+(\mathcal{H}_\star)$, and, thus, it is independent of the particular extension being used.

where it is understood that each of the terms in the above expression comprises a sum over all the independent components of the spinorial field under consideration.

The extended data (16.8) is non-unique. This non-uniqueness does not pose any problem for the considerations of this chapter. While it is true that the development $D^+(\mathcal{S}_\star)$ is clearly dependent on the particular extension of the initial data, one has

$$D^+(\mathcal{H}_\star) \cap I^+(\mathcal{S}_\star \setminus \mathcal{H}_\star) = \emptyset;$$

compare the Remark at the end of Section 14.2. Thus, the particular choice of extension of the data on $\mathcal{H}_\star$ has no effect on $D^+(\mathcal{H}_\star)$; see Figure 16.4 for further details.

16.3 A priori structure of the conformal boundary

This section discusses the available a priori knowledge of the structure of the conformal boundary of the development of hyperboloidal initial data.

In what follows, assume that $\mathcal{H}_\star$ can be regarded as an open subset of a compact manifold $\mathcal{S}_\star \approx \mathbb{S}^3$. Moreover, assume that $\partial\mathcal{H}_\star \approx \mathbb{S}^2$. On $\mathcal{S}_\star$ one considers a conformal factor Ω such that $\Omega > 0$ in the interior of $\mathcal{H}_\star$ and $\Omega = 0$ on $\partial\mathcal{H}_\star$; that is, the conformal factor Ω can be thought of as a boundary-defining function. Consistent with the hyperbolic reduction procedure for the extended conformal field equations as described in Section 13.4, it is assumed that the domain of dependence $D^+(\mathcal{S}_\star)$ can be covered by a non-singular congruence of conformal geodesics with data prescribed on $\mathcal{S}_\star$. In particular, it is required that the conformal geodesics are initially orthogonal to $\mathcal{S}_\star$.

Determining the conformal factor

From Proposition 5.1, it follows that the general form of the conformal factor associated to the congruence of conformal geodesics is given by

$$\Theta = \Theta_\star + \dot{\Theta}_\star \bar{\tau} + \frac{1}{2}\ddot{\Theta}_\star \bar{\tau}^2, \tag{16.9}$$

where the coefficients $\Theta_\star$, $\dot{\Theta}_\star$ and $\ddot{\Theta}_\star$ are functions of the spatial coordinates and are subject to the constraints

$$\dot{\Theta}_\star = \langle \boldsymbol{d}_\star, \dot{\boldsymbol{x}}_\star \rangle, \qquad \Theta_\star \ddot{\Theta}_\star = \frac{1}{2}\boldsymbol{g}^\sharp(\boldsymbol{d}_\star, \boldsymbol{d}_\star). \tag{16.10}$$

It is convenient to set

$$\Theta_\star = \Omega$$

and to make the spatial part of $\boldsymbol{d}_\star$ equal to $\mathbf{d}\Omega$. Accordingly, one finds that

$$d_{\boldsymbol{0}\star} = \dot{\Theta}_\star, \qquad d_{\boldsymbol{i}\star} = D_{\boldsymbol{i}}\Omega.$$

Now, letting

$$\alpha \equiv \Omega^{-1}\dot{\Theta}_\star,$$

one finds from the constraints in (16.10) that

$$2\Omega\ddot{\Theta}_\star = \boldsymbol{h}^\sharp(\mathbf{d}\Omega, \mathbf{d}\Omega) + \alpha^2,$$

where it is recalled that $\boldsymbol{h}^\sharp(\mathbf{d}\Omega, \mathbf{d}\Omega) < 0$ as a consequence of the signature convention. Whenever $\Omega = 0$, it follows from the constraints (16.10) that $\boldsymbol{d}_\star$ must be a null covector as $\mathbf{d}\Theta \neq 0$.

Making use of the above observations, one finds that Equation (16.9) takes the particular form

$$\Theta = \Omega\left(1 + \alpha\bar{\tau} + \left(\frac{1}{4}\alpha^2 - \frac{1}{\omega^2}\right)\bar{\tau}^2\right),$$

where

$$\omega \equiv \frac{2\Omega}{\sqrt{|\boldsymbol{h}^\sharp(\mathbf{d}\Omega, \mathbf{d}\Omega)|}}.$$

A calculation shows that $\Theta = 0$ for

$$\bar{\tau}_\pm \equiv \frac{2\alpha\omega^2 \pm 4\omega}{4 - \alpha^2\omega^2}. \tag{16.11}$$

Accordingly, it is natural to define the ***future (and, respectively, past) null infinity*** of the development associated to the hyperboloid $\mathcal{H}_\star$ as

$$\mathscr{I}^{\pm} \equiv \Big\{ (\bar{\tau}, x) \in \mathbb{R} \times \mathbb{S}^3 \,|\, \bar{\tau} = \bar{\tau}_{\pm}(x) \Big\}.$$

This expression shows how the location of the conformal boundary is predetermined by the initial data Ω and $\boldsymbol{d}_\star$ as long as the underlying congruence of conformal geodesics remains non-singular. As $\Omega \to 0$, one has that either $\bar{\tau}_{\pm} \to 0$ or $\bar{\tau}_{\pm} \to -2\dot{\Theta}_\star/\ddot{\Theta}_\star$. It follows that $\mathscr{I}^+$ and $\mathscr{I}^-$ are smooth hypersurfaces whenever $\mathbf{d}\Theta \neq 0$. Moreover, $\partial\mathcal{H}_\star$ is the intersection of $\mathscr{I}^{\pm}$ with $\mathcal{H}_\star = \{0\} \times \mathcal{H}$ as is to be expected for hyperboloidal data. In analogy to the model case of the hyperboloids in the Minkowski spacetime, *the development of generic hyperboloidal data has a conformal boundary which corresponds to either $\mathscr{I}^+$ or $\mathscr{I}^-$, but not both*; see Figure 16.5. This information is contained in the sign of the free datum $\dot{\Theta}_\star$. By convention, the conformal factor is positive in the region corresponding to the physical spacetime $(\tilde{\mathcal{M}}, \tilde{\boldsymbol{g}})$. Accordingly, if $\dot{\Theta}_\star > 0$ on $\partial\mathcal{H}_\star$, then $\tilde{\mathcal{M}}$ lies to the future of the conformal boundary and one speaks of a hyperboloid which intersects past null infinity, and, thus, the conformal boundary is identified with $\mathscr{I}^-$. If, by contrast, $\dot{\Theta}_\star < 0$ on $\partial\mathcal{H}_\star$, then $\tilde{\mathcal{M}}$ lies to the past of the conformal boundary. In this case, the hyperboloid intersects future null infinity and $\mathscr{I}^+$ gives the conformal boundary. *Without loss of generality, in the following, attention will be restricted to hyperboloids intersecting future null infinity so that $\dot{\Theta} < 0$ on $\partial\mathcal{H}_\star$.*

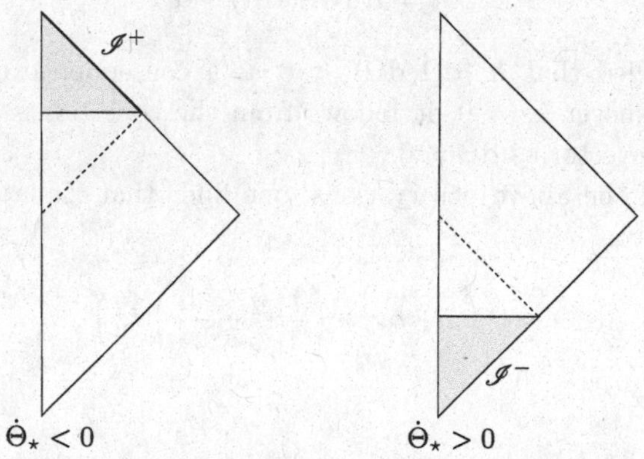

Figure 16.5 The two possible configurations of the conformal boundary for hyperboloidal data as discussed in the main text: on the left, one has the case $\dot{\Theta}_\star < 0$ at $\partial\mathcal{H}_\star$ where the conformal boundary given by the conformal Gaussian gauge system corresponds to $\mathscr{I}^+$; on the right, one has the situation corresponding to $\dot{\Theta}_\star > 0$ at $\partial\mathcal{H}_\star$ so the realised component of the conformal boundary is given by $\mathscr{I}^-$.

Timelike infinity

To identify the points which can be regarded as representing timelike infinity, one needs to investigate the critical points of Θ on the conformal boundary, that is, the points where $\mathbf{d}\Theta = 0$ and $\bar{\tau} = \bar{\tau}_{\pm}(x)$. A calculation shows that

$$\mathbf{d}\Theta = \left(1 + \alpha\bar{\tau} + \left(\frac{1}{4}\alpha^2 - \frac{1}{\omega^2}\right)\bar{\tau}^2\right)\mathbf{d}\Omega + \Omega\left(\alpha + 2\bar{\tau}\left(\frac{1}{4}\alpha^2 - \frac{1}{\omega^2}\right)\right)\mathbf{d}\bar{\tau}$$
$$+ \Omega\bar{\tau}\,\mathbf{d}\alpha + \Omega\bar{\tau}^2\left(\frac{1}{2}\alpha\mathbf{d}\alpha + \frac{2}{\omega^3}\mathbf{d}\omega\right).$$

Thus, a necessary condition for having a critical point of Θ on $\mathscr{I}^{\pm}$ is

$$\alpha + 2\bar{\tau}_{\pm}\left(\frac{1}{4}\alpha^2 - \frac{1}{\omega^2}\right) = 0.$$

A short computation shows that the above is equivalent to $\boldsymbol{h}(\mathbf{d}\Omega, \mathbf{d}\Omega) = 0$. That is, the critical points of Θ can occur only along conformal geodesics for which $\mathbf{d}\Omega = 0$ on the initial hypersurface $\mathcal{S}_{\star}$. The standard hyperboloid in the Minkowski spacetime contains precisely one such point. *By continuity, suitably small perturbations of this data will have only one point for which* $\mathbf{d}\Omega = 0$.

Now, for points lying along a conformal geodesic for which $\mathbf{d}\Omega = 0$, Equation (16.11) yields $\bar{\tau}_{\pm} = -2/\alpha$. Note that $\bar{\tau}_{\pm} > 0$ if $\alpha < 0$, that is, if $\dot{\Theta}_{\star} < 0$. To obtain a conformal representation which includes timelike infinity one needs to set $\alpha \neq 0$. *This condition will be assumed in the remainder of this chapter.* Moreover, one defines

$$\bar{\tau}_{i^+} \equiv -2/\alpha.$$

In particular, for the conformal representation of the Minkowski spacetime given by the conformal factor of Equation (16.5) one finds that $\bar{\tau}_{i^+} = 2$.

To conclude the discussion of timelike infinity, it is necessary to analyse the Hessian of the conformal factor Θ. *In what follows it is assumed that one has obtained a solution to the conformal field equations and that the associated unphysical metric* $\boldsymbol{g}$ *has been determined.*

From the general discussion of the conformal field equations in Chapter 8 it follows that Θ satisfies the equations

$$\nabla_{\boldsymbol{a}}\Theta = \Sigma_{\boldsymbol{a}}, \tag{16.12a}$$
$$\nabla_{\boldsymbol{a}}\Sigma_{\boldsymbol{b}} = s\eta_{\boldsymbol{ab}} - \Theta L_{\boldsymbol{ab}}, \tag{16.12b}$$
$$\nabla_{\boldsymbol{a}} s = -L_{\boldsymbol{ac}}\Sigma^{\boldsymbol{c}}, \tag{16.12c}$$

where s denotes the Friedrich scalar, $\boldsymbol{\nabla}$ is the Levi-Civita connection of the unphysical metric $\boldsymbol{g} \equiv \Theta^2\tilde{\boldsymbol{g}}$ and $L_{\boldsymbol{ab}}$ are the components of the Schouten tensor of $\boldsymbol{\nabla}$ with respect to the Weyl propagated frame $\{\boldsymbol{e_a}\}$. If s and $L_{\boldsymbol{ab}}$ are *regular* at the points for which $\bar{\tau} = \bar{\tau}_{i^+}$, one finds that

$$\boldsymbol{Hess}\,\Theta|_{i^+} = s|_{i^+}\boldsymbol{g}|_{i^+}.$$

If, in addition, $s|_{i^+} \neq 0$ – which, as will be seen, is the case for perturbations of the Minkowski spacetime – one concludes that the Hessian of the conformal factor Θ is non-degenerate at i^+, and, consequently, the point i^+ can be rightfully regarded as the timelike infinity of the development of the hyperboloidal initial data prescribed on $\mathcal{H}_\star$.

The Cauchy horizon of the hyperboloidal data and the conformal boundary

The discussion in the previous two subsections can be further refined to show that the conformal boundary $\mathscr{I}^+$ coincides with the ***Cauchy horizon*** $H^+(\mathcal{H}_\star)$ of the initial data prescribed on $\mathcal{H}_\star$. General results of ***Lorentzian causal theory*** as described in Chapter 14 imply that the Cauchy horizon $H^+(\mathcal{H}_\star)$ is generated by null geodesic segments with endpoints on $\partial\mathcal{H}_\star$; see Proposition 14.4. Since $\partial\mathcal{H}_\star$ is assumed to be a smooth two-dimensional manifold, it follows that $H^+(\mathcal{H}_\star)$ is, in a neighbourhood of $\partial\mathcal{H}_\star$, a $\boldsymbol{g}$-null hypersurface.

Setting $\Sigma_{\boldsymbol{a}} \equiv \nabla_{\boldsymbol{a}}\Theta$ it follows from the initial data on $\mathcal{H}_\star$ that

$$\Omega = 0, \qquad \text{and} \qquad \Sigma_{\boldsymbol{a}}\Sigma^{\boldsymbol{a}} = \eta^{\boldsymbol{ab}} d_{\boldsymbol{a}} d_{\boldsymbol{b}} = 0, \qquad \text{on} \quad \partial\mathcal{H}_\star,$$

where the various fields are expressed in terms of their components with respect to the Weyl propagated frame $\{\boldsymbol{e_a}\}$. Accordingly, the null directions tangent to $H^+(\mathcal{H}_\star)$ – the so-called ***null generators of null infinity*** – are given on $\partial\mathcal{H}_\star$ by $\Sigma_{\boldsymbol{a}}$. On $\partial\mathcal{H}_\star$ one can define $\boldsymbol{g}$-null vectors $\boldsymbol{l}$ and $\boldsymbol{n}$ by requiring

$$l_{\boldsymbol{a}} = \Sigma_{\boldsymbol{a}}, \qquad \boldsymbol{n} \perp \partial\mathcal{H}_\star, \qquad \boldsymbol{g}(\boldsymbol{l},\boldsymbol{n}) = 1, \qquad \text{on} \quad \partial\mathcal{H}_\star.$$

Moreover, on suitable open sets $\mathcal{O} \subset \partial\mathcal{H}_\star$ one can supplement $\boldsymbol{l}$ and $\boldsymbol{n}$ with complex vectors $\boldsymbol{m}$ and $\bar{\boldsymbol{m}}$ tangent to $\partial\mathcal{H}_\star$ with $\boldsymbol{g}(\boldsymbol{m},\bar{\boldsymbol{m}}) = -1$. The resulting Newman-Penrose frame $\{\boldsymbol{l}, \boldsymbol{n}, \boldsymbol{m}, \bar{\boldsymbol{m}}\}$ can be propagated along the null generators of $H^+(\mathcal{H}_\star)$ which terminate on $\mathcal{O} \subset \partial\mathcal{H}_\star$ by parallel transport in the direction of $\boldsymbol{l}$; that is, one has

$$l^{\boldsymbol{a}}\nabla_{\boldsymbol{a}} l^{\boldsymbol{b}} = 0, \qquad l^{\boldsymbol{a}}\nabla_{\boldsymbol{a}} n^{\boldsymbol{b}} = 0, \qquad l^{\boldsymbol{a}}\nabla_{\boldsymbol{a}} m^{\boldsymbol{b}} = 0.$$

Assuming now that the conformal field equations are satisfied on $H^+(\mathcal{H}_\star)$, it follows from transvecting Equations (16.12a) and (16.12b) with $l^{\boldsymbol{a}}$ and $m^{\boldsymbol{a}}$ that

$$\begin{aligned} l^{\boldsymbol{a}}\nabla_{\boldsymbol{a}}\Theta &= l^{\boldsymbol{a}}\Sigma_{\boldsymbol{a}}, \\ l^{\boldsymbol{a}}\nabla_{\boldsymbol{a}}(l^{\boldsymbol{b}}\Sigma_{\boldsymbol{b}}) &= -\Theta\left(L_{\boldsymbol{ab}}\right), \\ l^{\boldsymbol{a}}\nabla_{\boldsymbol{a}}(m^{\boldsymbol{b}}\Sigma_{\boldsymbol{b}}) &= -\Theta\big(L_{\boldsymbol{ab}} l^{\boldsymbol{a}} m^{\boldsymbol{b}}\big). \end{aligned}$$

These equations can be regarded as *ordinary differential equations* for the scalars Θ, $l^{\boldsymbol{b}}\Sigma_{\boldsymbol{b}}$ and $m^{\boldsymbol{b}}\Sigma_{\boldsymbol{b}}$ along the generators of $\mathscr{I}^+$. By construction, these fields vanish on $\mathcal{O} \subset \partial\mathcal{H}_\star$. Therefore, following a generator on $H^+(\mathcal{H}_\star)$ off $\partial\mathcal{H}_\star$ one finds that $\Theta = 0$, $l^{\boldsymbol{b}}\Sigma_{\boldsymbol{b}} = 0$, $m^{\boldsymbol{b}}\Sigma_{\boldsymbol{b}} = 0$ until, possibly, one arrives at a caustic point. Consequently, there is at least a portion of $H^+(\mathcal{H}_\star)$ where the conformal factor vanishes. It follows from the above that on $\mathcal{O}$ the field Σ_a must be proportional to l_a – more precisely, one can write

$$\Sigma^a = (n^b \Sigma_b)\, l^a \qquad \text{on} \quad \mathcal{O} \subset \partial\mathcal{H}_\star. \tag{16.13}$$

The portion of the Cauchy horizon where Θ vanishes can be identified with a portion of $\mathscr{I}^+$ as given by Equation (16.11). On this part of $H^+(\mathcal{H}_\star)$, the conformal field Equations (16.12b) and (16.12c) imply that

$$l^a \nabla_a (n^b \Sigma_b) = s,$$
$$l^a \nabla_a s = -(n^c \Sigma_c) L_{ab} l^a l^b.$$

Since $n^a \Sigma_a = 1$ on $\partial\mathcal{H}_\star$ it follows from the homogeneity of the above equations that s and $n^b \Sigma_b$ cannot vanish simultaneously. Moreover, contracting $\nabla_a \Theta = \Sigma_a$ with m^a, $\bar{m}^a$ one obtains

$$s = -(n^a \Sigma_a)\, \rho, \tag{16.14}$$

where $\rho \equiv m^a \bar{m}^b \nabla_b l_a$ is the Newman-Penrose spin coefficient associated to the expansion of the congruence of null generators. Thus, ρ is a measure of its convergence; see, for example, Stewart (1991), section 2.7. It follows from Equation (16.14) that $\rho \to \infty$ if $\mathbf{d}\Theta = 0$ at some point $p \in H^+(\mathcal{H}_\star)$; see Figure 16.6.

The discussion of the previous subsection shows that the development of hyperboloidal data suitably close to Minkowski data will contain an isolated point i^+ on the conformal boundary for which $\mathbf{d}\Theta = 0$. As $\mathscr{I}^+$ and $H^+(\mathcal{H}_\star)$ coincide wherever there are no caustics, it follows that the null geodesics on $H^+(\mathcal{H}_\star)$ must converge to i^+. Accordingly, $H^+(\mathcal{H}_\star)$ is the past light cone of i^+, and the causal past $J^-(i^+)$ and the future domain of dependence $D^+(\mathcal{H}_\star)$ coincide.

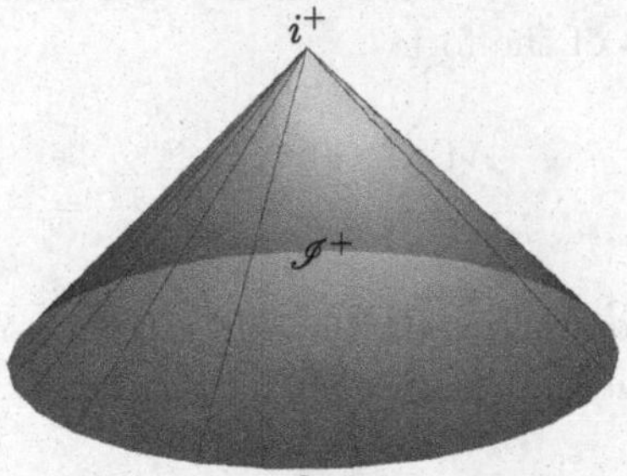

Figure 16.6 Null generators of $\mathscr{I}^+$ meeting at i^+, as discussed in the main text. The causal past of the caustic point i^+ corresponds to the future domain of dependence of hyperboloidal data; that is, $J^-(i^+) = D^+(\mathcal{H}^+_\star)$.

16.4 The proof of the main existence and stability result

Once the background Minkowski spacetime has been analysed in terms of a conformal Gaussian system adapted to the standard hyperboloid, a proof of semiglobal existence and stability is obtained by a procedure almost identical to the one used for the de Sitter spacetime in Section 15.4.

As in the case of the analysis of the de Sitter spacetime, it is convenient to consider an ansatz for the solutions to the conformal evolution equations of the form

$$e_{\boldsymbol{AB}}{}^{0} = \breve{e}_{\boldsymbol{AB}}{}^{0}, \quad e_{\boldsymbol{AB}}{}^{\alpha} = \bar{e}_{\boldsymbol{AB}}{}^{\alpha} + \breve{e}_{\boldsymbol{AB}}{}^{\alpha},$$
$$\xi_{\boldsymbol{ABCD}} = \bar{\xi}_{\boldsymbol{ABCD}} + \breve{\xi}_{\boldsymbol{ABCD}}, \quad \chi_{\boldsymbol{ABCD}} = \bar{\chi}_{\boldsymbol{ABCD}} + \breve{\chi}_{\boldsymbol{ABCD}}, \quad f_{\boldsymbol{AB}} = \breve{f}_{\boldsymbol{AB}},$$
$$\Theta_{\boldsymbol{ABCD}} = \bar{\Theta}_{\boldsymbol{ABCD}} + \breve{\Theta}_{\boldsymbol{ABCD}}, \quad \phi_{\boldsymbol{ABCD}} = \breve{\phi}_{\boldsymbol{ABCD}},$$

where

$$\bar{e}_{\boldsymbol{AB}}{}^{\mu}, \qquad \bar{\xi}_{\boldsymbol{ABCD}}, \qquad \bar{\chi}_{\boldsymbol{ABCD}}, \qquad \bar{\Theta}_{\boldsymbol{ABCD}}$$

are the values of the exact conformal Minkowski spacetime as discussed in Section 16.1. For conciseness, the above ansatz will be written schematically as $\mathbf{u} = \bar{\mathbf{u}} + \breve{\mathbf{u}}$. Taking into account that the background fields are also a solution to the conformal evolution equations and writing the (explicitly known) conformal gauge fields Θ and $d_{\boldsymbol{a}}$ in the form

$$\Theta = \breve{\Theta} + \breve{\Theta}, \qquad d_{\boldsymbol{a}} = \breve{d}_{\boldsymbol{a}} + \breve{d}_{\boldsymbol{a}},$$

one finds evolution equations for the perturbation fields of the form

$$\partial_{\bar{\tau}}\breve{\boldsymbol{\upsilon}} = \mathbf{K}\breve{\boldsymbol{\upsilon}} + \mathbf{Q}(\bar{\boldsymbol{\Gamma}} + \breve{\boldsymbol{\Gamma}})\breve{\boldsymbol{\upsilon}} + \mathbf{Q}(\breve{\boldsymbol{\Gamma}})\bar{\boldsymbol{\upsilon}} + \mathbf{L}(x)\breve{\boldsymbol{\phi}}, \tag{16.15a}$$
$$\big(\mathbf{I} + \mathbf{A}^{0}(\bar{e} + \breve{e})\big)\partial_{\bar{\tau}}\breve{\boldsymbol{\phi}} + \mathbf{A}^{\alpha}(\bar{e} + \breve{e})\partial_{\alpha}\breve{\boldsymbol{\phi}} = \mathbf{B}(\bar{\boldsymbol{\Gamma}} + \breve{\boldsymbol{\Gamma}})\breve{\boldsymbol{\phi}}, \tag{16.15b}$$

in the conventions of Proposition 13.3. The natural domains for solutions to the above equations are sets of the form

$$\mathcal{M}_{\bar{\tau}_\bullet} \equiv [0, \bar{\tau}_\bullet] \times \mathbb{S}^3$$

for some $\bar{\tau}_\bullet > 0$.

Using the evolution Equations (16.15a) and (16.15b) one obtains the following technical version of the main theorem of this chapter:

Theorem 16.1 (***semiglobal existence and stability for perturbations of hyperboloidal data***) *Let* $\mathbf{u}_\star = \bar{\mathbf{u}}_\star + \breve{\mathbf{u}}_\star$ *be hyperboloidal initial data for the conformal Einstein field equations given on a hyperboloidal manifold* $\mathcal{H}$. *Given* $m \geq 4$ *and* $\bar{\tau}_\bullet > 2$ *there exists* $\varepsilon > 0$ *such that:*

(i) For $\| \breve{\mathbf{u}}_\star \|_m < \varepsilon$ *there exists a solution* $\mathbf{u} = \bar{\mathbf{u}} + \breve{\mathbf{u}}$ *to the conformal propagation equations with a minimal existence interval* $[0, \bar{\tau}_\bullet]$ *such that*

$$\mathbf{u} \in C^{m-2}([0, \bar{\tau}_\bullet] \times \mathbb{S}^3),$$

and the associated congruence of conformal geodesics contains no conjugate points in $[0, \bar{\tau}_\bullet]$.

(ii) For every $\breve{\mathbf{u}}_\star$ *with* $\| \breve{\mathbf{u}}_\star \|_m < \varepsilon$ *there is a unique point* p_+ *in the interior of* $\mathcal{H}$ *such that* $\mathbf{d}\Omega = 0$ *with* $\tau_{i^+} \equiv \bar{\tau}_+(p_+) \in [0, \bar{\tau}_\bullet]$.

The solution $\mathbf{u} = \bar{\mathbf{u}} + \breve{\mathbf{u}}$ *is unique on* $D^+(\mathcal{H}_\star)$ *and implies, wherever* $\Theta \neq 0$*, a* C^{m-2} *solution to the vacuum Einstein field equations with a vanishing cosmological constant for which the set* $\mathscr{I}^+$*, as defined by*

$$\mathscr{I}^+ \equiv \{(\tau, p) \in \mathbb{R} \times \mathbb{S}^3 \mid \tau = \tau_\pm(p_+)\},$$

represents null infinity, while the point $i^+ \equiv (\bar{\tau}_{i^+}, x^\alpha(p_+))$ *represents timelike infinity. Moreover, one has*

$$D^+(\mathcal{H}_\star) = J^-(i^+).$$

Proof The assertion in (i) follows from the general existence result from symmetric hyperbolic systems in Theorem 12.4 along lines similar to the ones used in the proofs of Propositions 15.1 and 15.3. The key observation in this respect is that as $(\mathbf{I} + \mathbf{A}^0(\bar{e}))|_\star$ is positive definite and bounded away from zero, then $(\mathbf{I} + \mathbf{A}^0(\bar{e} + \breve{e}))|_\star$ can also be made positive definite and bounded away from zero by choosing $\varepsilon > 0$ small enough. This observation and the general structure of the evolution Equations (16.15a) and (16.15b) ensure the existence of C^{m-2} solutions $\breve{\mathbf{u}}$ with $\| \breve{\mathbf{u}}_\star \|_m < \varepsilon$ on $[0, \bar{\tau}_\bullet] \times \mathbb{S}^3$ with $\bar{\tau}_\bullet > 2$. The regularity of the congruence of conformal geodesics defining the gauge is obtained by supplementing the conformal evolution equations with evolution equations for the conformal deviation fields, Equations (13.67a) and (13.67b), and recalling that the deviation fields for the expanding Einstein cylinder are given by Equation (15.14).

The proof of point (ii) follows from the discussion in Section 16.3 and by observing that the spatial conformal factor $\bar{\Omega}$ for the exact (background) hyperboloidal data has an isolated critical point (in fact, a maximum) at $\psi = 0$. Accordingly, by continuity, any suitably small perturbations of this data will also have a unique isolated critical point of its spatial conformal factor. Again, choosing $\varepsilon > 0$ sufficiently small, one can ensure that $\bar{\tau}_+ < \bar{\tau}_\bullet$.

The final remarks in Theorem 16.1 follow from a *propagation of the constraints argument* using the properties of the subsidiary evolution system as given by Proposition 13.4 and the assumption that the initial data satisfy the conformal constraint equations on $\mathcal{H}_\star$. The solution to the conformal field equations obtained by the above argument implies a solution to the vacuum Einstein field equations whenever $\Theta \neq 0$ as a consequence of Proposition 8.3. Finally, the statements about the interpretation of $\mathscr{I}^+$ as the conformal boundary and the structure of i^+ follow from the analysis in Section 16.3. □

Remarks

(i) For conciseness, Theorem 16.1 is restricted to perturbations of the data implied by the Minkowski spacetime on the *standard hyperboloid*. An inspection of the argument, however, shows that this simplifying assumption is non-essential and that an analogous result can be obtained, at the expense of some further technical details, for perturbations of Minkowski data on *arbitrary hyperboloids*. In other words, the location of the initial hyperboloid within null infinity is irrelevant. A more subtle consequence of this observation is that it is, in principle, hard to quantify how far away a given hyperboloidal initial data set lies from spatial infinity or even whether there is any (asymptotically Euclidean) Cauchy initial data for the Einstein field equations whose development contains the hyperboloidal data.

(ii) Theorem 16.1 can be combined with the *method of exterior gluing* discussed in Section 11.8.2 to show the existence of a large class of asymptotically simple spacetimes with a complete conformal boundary, that is, whose null generators are inextendible geodesics starting at i^0 and ending at i^+ and, respectively, i^-. These ideas are discussed in more detail in Section 20.5.

(iii) The future domain of dependence $D^+(\mathcal{H}_\star)$ as given by Theorem 16.1 provides an infinite portion of spacetime where the framework of ***asymptopia***, as discussed in Chapter 10, can be applied; see also, for example, chapter 3 of Stewart (1991). In particular, if the hyperboloidal initial data are constructed using the methods of Theorem 11.2, one can obtain a development which has any desired degree of smoothness and, accordingly, satisfies the ***peeling behaviour***; see the discussion in Section 10.2.

16.5 Extensions and further reading

The first semiglobal existence and stability result for hyperboloidal vacuum data of Minkowski-like spacetimes was obtained in the seminal work by Friedrich (1986b). This analysis used the standard vacuum conformal field equations and gauge source functions. The approach adopted in this chapter, employing the extended conformal field equations and a gauge based on the properties of conformal geodesics, is adapted from the discussion given in Lübbe and Valiente Kroon (2009). Similar semiglobal existence and stability results have been obtained in Anderson and Chruściel (2005) for arbitrary even-dimensional spacetimes using the conformal equations given by the Graham-Fefferman obstruction tensor.

The main result of this chapter can be extended to the case of the Einstein-Maxwell and Einstein-Yang-Mills equations. This was done in Friedrich (1991) where the standard conformal field equations and a hyperbolic reduction procedure based on gauge source functions were used. An alternative proof of the semi-global existence and stability result for the Einstein-Maxwell equations has been obtained in Lübbe and Valiente Kroon (2012) using an approach similar in spirit to the one used in this chapter, that is, employing the extended conformal field equations and a conformal gauge based on the properties of conformal curves. Conformal curves were preferred in this analysis as they provide an explicit expression for the conformal factor. In the presence of matter, a standard conformal Gaussian system does not provide an explicit expression for the conformal factor. There is, however, no reason why a semi-global result of the type discussed in this chapter cannot be obtained using a gauge based on conformal geodesics. Another way of generalising the main result of this chapter is to consider the Einstein-conformally invariant scalar field system; see Hübner (1995).

The methods in this chapter can be adapted to analyse semiglobal existence and stability of asymptotically simple spacetimes with vanishing cosmological constant which are neither the Minkowski spacetime nor perturbations thereof – so-called ***purely radiative spacetimes***. These vacuum spacetimes consist of gravitational radiation (hence the name) which is not necessarily weak, but still tame enough to not form a black hole; see, for example, Friedrich (1986c) and the discussion in Chapter 19. Stability of these types of spacetimes from the perspective of a hyperboloidal initial value problem has been analysed, for the vacuum case, in Lübbe and Valiente Kroon (2010) and, for the Einstein-Maxwell case, in Lübbe and Valiente Kroon (2012).

The main theorem of this chapter has been beautifully verified in numerical simulations in Hübner (2001a). In particular, the numerical results show how the null generators of the conformal boundary converge, to machine precision, at timelike infinity. These numerical simulations are further discussed in Section 21.3.

17

Anti-de Sitter-like spacetimes

This chapter discusses the construction of ***anti-de Sitter-like spacetimes***, that is, solutions to the vacuum Einstein field equations with an anti-de Sitter-like value of the cosmological constant λ. Following the general discussion in Chapter 10, an anti-de Sitter-like value of the cosmological constant implies a timelike conformal boundary. This feature of anti-de Sitter-like spacetimes marks the essential difference between the analysis contained in this chapter and the ones given in Chapters 15 and 16 for de Sitter-like and Minkowski-like spacetimes, respectively.

While the de Sitter and Minkowski spacetimes are both globally hyperbolic, and, accordingly, perturbations thereof can be constructed by means of suitable initial value problems, *the anti-de Sitter spacetime is not-globally hyperbolic*; see the discussion in Section 14.5. Consequently, anti-de Sitter-like spacetimes cannot be solely reconstructed from initial data. One needs to prescribe some boundary data on the conformal boundary. Thus, the proper setting for the construction of anti-de Sitter-like spacetimes is that of an *initial boundary value problem.* In this spirit, one of the key objectives of this chapter is to identify suitable boundary data for the conformal Einstein field equations.

For both the de Sitter and Minkowski spacetimes it is possible to obtain conformal representations which are compact in time so that global existence of perturbations can be analysed in terms of problems which are local in time. However, the conformal representations of the anti-de Sitter spacetime discussed in Chapter 6 involve an infinite range of time. As a consequence, the main result of this chapter is local in time and makes no assertions about the stability of the anti-de Sitter spacetime. The main result of this chapter can be formulated as follows:

Theorem ***(local existence of anti-de Sitter-like spacetimes).*** *Given smooth anti-de Sitter-like initial data for the Einstein field equations on a three-dimensional manifold $\mathcal{S}$ with boundary and a smooth three-dimensional Lorentzian metric $\boldsymbol{\ell}$ on a cylinder $[0,\tau_\bullet)\times\partial\mathcal{S}$ for some $\tau_\bullet>0$, and assuming that* *these data satisfy certain corner conditions, there exists a local-in-time solution to the Einstein field equations with an anti-de Sitter-like cosmological constant such that on $\{0\} \times \mathcal{S}$ it implies the given anti-de Sitter-like initial data. Moreover, this solution to the Einstein field equations admits a conformal completion such that the intrinsic metric of the resulting (timelike) conformal boundary belongs to the conformal class $[\boldsymbol{\ell}]$.*

Thus, the conformal class of the intrinsic metric of the conformal boundary constitutes suitable boundary data for the construction of anti-de Sitter spacetimes. This insight was first obtained in Friedrich (1995).

17.1 General properties of anti-de Sitter-like spacetimes

In what follows, by an ***anti-de Sitter-like spacetime*** it will be understood an asymptotically simple spacetime $(\tilde{\mathcal{M}}, \tilde{\boldsymbol{g}})$ with *positive* (i.e. anti-de Sitter-like) cosmological constant. The basic intuition on this type of spacetimes is obtained from the paradigmatic example discussed in Section 6.4. In particular, it has been shown that making use of the conformal factor

$$\Xi_{adS} = a\cos\psi, \qquad a \text{ a constant},$$

the anti-de Sitter spacetime $(\mathbb{R}^4, \tilde{\boldsymbol{g}}_{adS})$ is conformal to the region

$$\tilde{\mathcal{M}}_{adS} \equiv \left\{ p \in \mathbb{R} \times \mathbb{S}^3 \;\middle|\; 0 \leq \psi(p) < \frac{\pi}{2} \right\}$$

of the Einstein cylinder $\mathbb{R} \times \mathbb{S}^3$ described in *standard coordinates* $(T, \psi, \theta, \varphi)$. Moreover, the conformal boundary of the spacetime is given by

$$\mathscr{I} \equiv \left\{ p \in \mathbb{R} \times \mathbb{S}^3 \;\middle|\; \psi(p) = \frac{\pi}{2} \right\},$$

which can be verified to be timelike.

17.1.1 General setting for the construction of anti-de Sitter-like spacetimes

Let $(\mathcal{M}, \boldsymbol{g}, \Xi)$ denote a conformal extension of an anti-de Sitter-like spacetime $(\tilde{\mathcal{M}}, \tilde{\boldsymbol{g}})$ with $\boldsymbol{g} = \Xi^2 \tilde{\boldsymbol{g}}$. It will be assumed that the spacetime is causal (i.e. it contains no closed timelike curves) and that it contains a smooth, oriented and compact spacelike hypersurface $\mathcal{S}_\star$ with boundary $\partial\mathcal{S}_\star$ which intersects the conformal boundary $\mathscr{I}$ in such a way that $\mathcal{S}_\star \cap \mathscr{I} = \partial\mathcal{S}_\star$. It is convenient to define $\tilde{\mathcal{S}}_\star \equiv \mathcal{S}_\star \setminus \partial\mathcal{S}_\star$. The portion of $\mathscr{I}$ in the future of $\mathcal{S}_\star$ will be denoted by $\mathscr{I}^+$. Furthermore, it will be assumed that the causal future $J^+(\mathcal{S}_\star)$ coincides with the future domain of dependence[1] $D^+(\mathcal{S}_\star \cup \mathscr{I}^+)$ and that $\mathcal{S}_\star \cup \mathscr{I}^+ \approx [0,1) \times \mathcal{S}_\star$

[1] In Chapter 14 the domain of dependence has been defined for achronal sets. However, that $\mathcal{S}_\star \cup \mathscr{I}^+$ is not achronal. This feature will not play a role in the subsequent discussion.

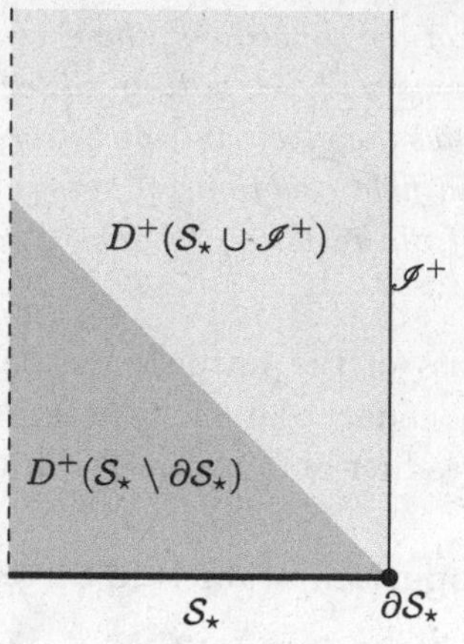

Figure 17.1 Penrose diagram of the set up for the construction of anti-de Sitter-like spacetimes as described in the main text. Initial data prescribed on $\mathcal{S}_\star \setminus \partial\mathcal{S}_\star$ allow one to recover the dark shaded region $D^+(\mathcal{S}_\star \setminus \partial\mathcal{S}_\star)$. In order to recover $D^+(\mathcal{S}_\star \cup \mathscr{I}^+)$ it is necessary to prescribe boundary data on $\mathscr{I}^+$. Notice that $D^+(\mathcal{S}_\star \cup \mathscr{I}^+) = J^+(\mathcal{S}_\star)$.

so that, in particular, $\mathscr{I}^+ \approx [0,1) \times \partial\mathcal{S}_\star$. A schematic depiction of the above setting is given in Figure 17.1. One of the key objectives of the present chapter is to address the question: *what data on $\mathcal{S}_\star \cup \mathscr{I}^+$ are needed to reconstruct the anti-de Sitter-like spacetime $(\tilde{\mathcal{M}}, \tilde{\boldsymbol{g}})$ in a neighbourhood $\mathcal{U} \subset J^+(\mathcal{S}_\star)$ of $\mathcal{S}_\star$?*

As a consequence of the properties of the standard Cauchy problem and the *localisation property of hyperbolic equations*, the solutions to the conformal Einstein field equations on $D^+(\tilde{\mathcal{S}}_\star)$ are determined, up to diffeomorphisms, in a unique manner by solutions to the constraint equations on $\mathcal{S}_\star$. To recover $J^+(\mathcal{S}_\star) \setminus D^+(\tilde{\mathcal{S}}_\star)$ one needs to prescribe suitable data on the conformal boundary $\mathscr{I}$. The analysis of the suitable boundary data requires the prescription of some appropriate gauge near $\mathscr{I}$. As will be seen, conformal geodesics are ideally suited to provide such a gauge.

The conformal constraints at the conformal boundary

Because for anti-de Sitter-like spacetimes the conformal boundary is a $\boldsymbol{g}$-timelike hypersurface, it follows that the metric $\boldsymbol{g}$ induces on $\mathscr{I}$ a three-dimensional Lorentzian metric $\boldsymbol{\ell}$. As discussed in Section 11.4.4, the conformal Einstein field equations satisfied by the (unphysical) spacetime $(\mathcal{M}, \boldsymbol{g})$ imply on $\mathscr{I}$ a *simplified* set of interior (constraint) equations. It is recalled that a solution to these conformal constraints at the conformal boundary can be computed from the metric $\boldsymbol{\ell}$, a smooth scalar function $\varkappa$ and a symmetric $\boldsymbol{\ell}$-tracefree three-dimensional tensor on $\mathscr{I}$; see Proposition 11.1. The scalar function is, in particular, a conformal gauge-dependent quantity which can be set to zero by considering a different metric in $[\boldsymbol{\ell}]$.

17.1.2 Conformal geodesics at the conformal boundary

In Section 6.4.2 it has been shown that the anti-de Sitter spacetime can be covered by a congruence of (non-intersecting) conformal geodesics. In this congruence, curves which for some value of their affine parameter $\bar{\tau}$ are tangent to $\mathscr{I}$ remain on $\mathscr{I}$ for all values of $\bar{\tau}$. It will be shown that this observation is, in fact, a generic property of anti-de Sitter-like spacetimes.

On the conformal boundary of an anti-de Sitter-like spacetime consider an *adapted* $\boldsymbol{g}$-orthonormal frame $\{\boldsymbol{e_a}\}$ such that $\boldsymbol{e_3}$ is inward pointing and orthogonal to $\mathscr{I}$. This frame can then be extended to a neighbourhood $\mathcal{U}$ of $\mathscr{I}$ by requiring the frame to be parallely propagated in the direction of $\boldsymbol{e}_3$. It follows that the connection coefficients of $\boldsymbol{\nabla}$ associated to this frame satisfy

$$\Gamma_{\boldsymbol{3}}{}^{\boldsymbol{a}}{}_{\boldsymbol{b}} = 0 \qquad \text{on } \mathcal{U}.$$

If one uses ***Gaussian coordinates*** $x = (x^\mu)$ based on $\mathscr{I}$ such that

$$\mathscr{I} = \{p \in \mathcal{U} \,|\, x^3(p) = 0\},$$

it follows from writing $\boldsymbol{e_a} = e_{\boldsymbol{a}}{}^{\mu}\boldsymbol{\partial}_\mu$ that

$$e_{\boldsymbol{3}}{}^{\mu} = \delta_3{}^{\mu}, \qquad e_{\boldsymbol{a}}{}^{3} = 0.$$

To analyse the behaviour of conformal geodesics at the conformal boundary it is convenient to consider the equations for these curves expressed in terms of the connection $\boldsymbol{\nabla}$. These equations can be decomposed in components using the adapted frame discussed in the previous paragraph. One writes

$$\dot{\boldsymbol{x}} = z^{\boldsymbol{a}}\boldsymbol{e_a}, \qquad \boldsymbol{\beta} = \beta_{\boldsymbol{a}}\boldsymbol{\omega}^{\boldsymbol{a}}.$$

The conformal curve equations split into two groups. Firstly, one has the ***normal equations***:

$$\begin{aligned}
\dot{x}^3 &= z^{\boldsymbol{a}}e_{\boldsymbol{a}}{}^3 = z^{\boldsymbol{3}},\\
\dot{z}^{\boldsymbol{3}} &= -\Gamma_{\boldsymbol{a}}{}^{\boldsymbol{3}}{}_{\boldsymbol{b}}z^{\boldsymbol{a}}z^{\boldsymbol{b}} - 2(\beta_{\boldsymbol{c}}z^{\boldsymbol{c}})z^{\boldsymbol{3}} + (z_{\boldsymbol{c}}z^{\boldsymbol{c}})\beta^{\boldsymbol{3}},\\
\dot{\beta}_{\boldsymbol{3}} &= \Gamma_{\boldsymbol{a}}{}^{\boldsymbol{c}}{}_{\boldsymbol{3}}z^{\boldsymbol{a}}\beta_{\boldsymbol{c}} + (\beta_{\boldsymbol{c}}z^{\boldsymbol{c}})\beta_{\boldsymbol{3}} - \frac{1}{2}(\beta_{\boldsymbol{c}}\beta^{\boldsymbol{c}})z_{\boldsymbol{3}} + L_{\boldsymbol{33}}z^{\boldsymbol{3}} + L_{\boldsymbol{i3}}z^{\boldsymbol{i}}.
\end{aligned}$$

Secondly, for $\boldsymbol{i}$, $\alpha = 0, 1, 2$ one has the ***intrinsic equations***:

$$\begin{aligned}
\dot{x}^\alpha &= e_{\boldsymbol{a}}{}^{\alpha}z^{\boldsymbol{a}},\\
\dot{z}^{\boldsymbol{i}} &= -\Gamma_{\boldsymbol{c}}{}^{\boldsymbol{i}}{}_{\boldsymbol{b}}z^{\boldsymbol{c}}z^{\boldsymbol{b}} - 2(\beta_{\boldsymbol{c}}z^{\boldsymbol{c}})z^{\boldsymbol{i}} + (z_{\boldsymbol{c}}z^{\boldsymbol{c}})\beta^{\boldsymbol{i}},\\
\dot{\beta}_{\boldsymbol{i}} &= \Gamma_{\boldsymbol{b}}{}^{\boldsymbol{c}}{}_{\boldsymbol{i}}\beta_{\boldsymbol{c}}z^{\boldsymbol{b}} + (\beta_{\boldsymbol{c}}z^{\boldsymbol{c}})\beta_{\boldsymbol{i}} - \frac{1}{2}(\beta_{\boldsymbol{c}}\beta^{\boldsymbol{c}})z_{\boldsymbol{i}} + L_{\boldsymbol{3i}}z^{\boldsymbol{3}} + L_{\boldsymbol{ci}}z^{\boldsymbol{c}}.
\end{aligned}$$

To simplify the analysis of the above equations one can exploit the conformal freedom and choose an element of the conformal class of the intrinsic 3-metric $\boldsymbol{\ell}$ of $\mathscr{I}$ for which

$$s = \frac{1}{4}\nabla^c\nabla_c\Xi + \frac{1}{24}R\Xi = 0.$$

Following the discussion of Section 11.4.4, this can always be done locally. Under this choice of conformal gauge, the solution of the conformal constraint equations on $\mathscr{I}$ implies that

$$\Gamma_{\boldsymbol{a}}{}^{\boldsymbol{3}}{}_{\boldsymbol{b}} = 0, \qquad \Gamma_{\boldsymbol{a}}{}^{\boldsymbol{c}}{}_{\boldsymbol{3}} = 0, \qquad L_{\boldsymbol{3a}} = 0.$$

Moreover, one has

$$L_{\boldsymbol{3a}} = 0, \qquad L_{\boldsymbol{ij}} = l_{\boldsymbol{ij}}.$$

That is, the spacetime (unphysical) Schouten tensor on $\mathscr{I}$ is determined by the Schouten tensor of the intrinsic metric $\boldsymbol{\ell}$.

From the previous discussion it follows that the normal subset of the conformal geodesic equations reduces to:

$$\begin{aligned}
\dot{x}^{\boldsymbol{3}} &= z^{\boldsymbol{3}}, \\
\dot{z}^{\boldsymbol{3}} &= -2(\beta_{\boldsymbol{c}}\beta^{\boldsymbol{c}})z^{\boldsymbol{3}} + (z_{\boldsymbol{c}}z^{\boldsymbol{c}})\beta_{\boldsymbol{3}}, \\
\dot{\beta}_{\boldsymbol{3}} &= (\beta_{\boldsymbol{c}}z^{\boldsymbol{c}})\beta_{\boldsymbol{3}} - \frac{1}{2}(\beta_{\boldsymbol{c}}\beta^{\boldsymbol{c}})z^{\boldsymbol{3}} + l_{\boldsymbol{33}}z^{\boldsymbol{3}}.
\end{aligned}$$

These equations are homogeneous in the unknowns $(x^3, z^{\boldsymbol{3}}, \beta_{\boldsymbol{3}})$. Thus, by choosing initial data such that

$$x^3_\star = 0, \qquad \dot{x}^{\boldsymbol{3}}_\star = 0, \qquad \beta_{\boldsymbol{3}\star} = 0, \tag{17.1}$$

one obtains that

$$x^3(\tau) = 0, \qquad z^{\boldsymbol{3}}(\tau) = 0, \qquad \beta_{\boldsymbol{3}}(\tau) = 0$$

for later times. Accordingly, conformal curves with initial data given by (17.1) will remain on $\mathscr{I}$. Looking now at the intrinsic part of the conformal geodesic equations one observes that the equations reduce to

$$\begin{aligned}
\dot{x}^{\alpha} &= z^{\boldsymbol{i}}e_{\boldsymbol{i}}{}^{\alpha}, \\
\dot{z}^{\boldsymbol{i}} &= -\Gamma_{\boldsymbol{k}}{}^{\boldsymbol{i}}{}_{\boldsymbol{j}}z^{\boldsymbol{k}}z^{\boldsymbol{j}} - 2(\beta_{\boldsymbol{k}}z^{\boldsymbol{k}})z^{\boldsymbol{i}} + (z_{\boldsymbol{k}}z^{\boldsymbol{k}})\beta^{\boldsymbol{i}}, \\
\dot{\beta}_{\boldsymbol{i}} &= \Gamma_{\boldsymbol{j}}{}^{\boldsymbol{k}}{}_{\boldsymbol{i}}z^{\boldsymbol{j}}\beta_{\boldsymbol{k}} + (\beta_{\boldsymbol{k}}z^{\boldsymbol{k}})\beta_{\boldsymbol{i}} - \frac{1}{2}(\beta_{\boldsymbol{k}}\beta^{\boldsymbol{k}})z_{\boldsymbol{i}} + l_{\boldsymbol{ki}}z^{\boldsymbol{k}}.
\end{aligned}$$

These are the *conformal geodesic equations for the 3-metric* $\boldsymbol{\ell}$ on $\mathscr{I}$.

To verify the consistency between the construction described in the previous paragraphs and the adapted $\boldsymbol{g}$-orthonormal frame $\{\boldsymbol{e_a}\}$, consider a vector $\boldsymbol{v}$ satisfying the Weyl propagation equation

$$\nabla_{\dot{\boldsymbol{x}}}\boldsymbol{v} = -\langle\boldsymbol{\beta}, \boldsymbol{v}\rangle\dot{\boldsymbol{x}} - \langle\boldsymbol{\beta}, \dot{\boldsymbol{x}}\rangle\boldsymbol{v} + \boldsymbol{g}(\boldsymbol{v}, \dot{\boldsymbol{x}})\boldsymbol{\beta}^\sharp,$$

along $\mathscr{I}$. Making the ansatz $\boldsymbol{v} = \alpha\boldsymbol{e}_3$, where α denotes a scalar function on $\mathscr{I}$, one finds the equation $\dot{\alpha} = -\langle\boldsymbol{\beta}, \dot{\boldsymbol{x}}\rangle\alpha$. Thus, if initially one has $\alpha_\star \neq 0$, then $\alpha \neq 0$ at later times. Accordingly, if one prescribes at some point of the conformal geodesic in $\mathscr{I}$ an orthonormal frame $\{\boldsymbol{e_a}\}$ containing a vector which is normal to $\mathscr{I}$, one finds that the solution to the Weyl propagation equations will be a frame

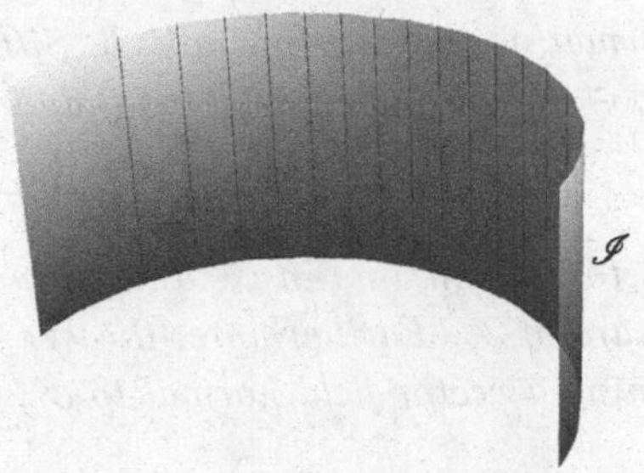

Figure 17.2 Representation of conformal geodesics on the conformal boundary of an anti-de Sitter-like spacetime: those curves that at some point are tangent to $\mathscr{I}$ remain in the conformal boundary and are conformal geodesics for the conformal structure implied by the intrinsic metric $\boldsymbol{\ell}$; see Lemma 17.1. The conformal geodesics are depicted by black lines.

along the conformal geodesic which contains a vector normal to $\mathscr{I}$. Moreover, as the Weyl propagation preserves the orthogonality of vectors, it follows that the elements of the frame which are at some point intrinsic to $\mathscr{I}$ will remain so at later times; see Figure 17.2.

A more general result

The results obtained in the previous paragraphs make use of a particular metric in the conformal class $[\boldsymbol{\ell}]$. Thus, it is of interest to reformulate them in an arbitrary conformal gauge. As in Chapter 10, the symbol $\simeq$ denotes equality on $\mathscr{I}$. Now, consider on $\mathcal{M}$ a conformal factor $\vartheta > 0$ such that $\vartheta \simeq 1$ to perform a rescaling of the form $\boldsymbol{g}' \equiv \vartheta^2 \boldsymbol{g}$. This rescaling leaves the metric $\boldsymbol{\ell}$ unchanged in the sense that $\boldsymbol{\ell}' \simeq \vartheta^2 \boldsymbol{\ell} \simeq \boldsymbol{\ell}$. Furthermore, one finds that

$$s' \simeq (\nabla^a \Xi \nabla_a \vartheta) \simeq \boldsymbol{e_3}(\vartheta),$$

with $\boldsymbol{e_3} = (\mathbf{d}\Xi)^\sharp$ as $\Xi = x^3$ in local coordinates. The comparison of the above expression with the solution to the conformal constraints at the conformal boundary as given in Section 11.4.4 suggests defining

$$\varkappa \equiv \sqrt{3/\lambda}\, \boldsymbol{e_3}(\vartheta)\big|_{\mathscr{I}}.$$

Defining the covector $\boldsymbol{k} \equiv \vartheta^{-1}\mathbf{d}\vartheta$, and taking into account the transformation properties of conformal geodesics as given in Section 5.5.2, it follows that

$$(x(\tau), \boldsymbol{\beta}'(\tau)), \qquad \text{with } \boldsymbol{\beta}' \equiv \boldsymbol{\beta} - \boldsymbol{k},$$

is a solution to the conformal geodesic equations associated to the connection $\boldsymbol{\nabla}' \equiv \boldsymbol{\nabla} + \boldsymbol{S}(\boldsymbol{k})$. From the definition of $\boldsymbol{k}$ it follows that $\boldsymbol{\nabla}'$ is the Levi-Civita connection of the metric $\boldsymbol{g}' = \vartheta^2 \boldsymbol{g}$. Observe, in particular, that

$$\beta'_{\boldsymbol{3}}(\tau) \simeq -k_{\boldsymbol{3}}(\tau) \simeq -\boldsymbol{e_3}(\vartheta) \simeq -s'.$$

The discussion of this section can be summarised as follows:

Lemma 17.1 *A conformal geodesic in an anti-de Sitter-like spacetime which passes through a point $p \in \mathscr{I}$, is tangent to $\mathscr{I}$ at p and which satisfies*

$$\langle \boldsymbol{\beta}, \boldsymbol{\nu} \rangle|_p = -s,$$

with $\boldsymbol{\nu}$ the unit normal to $\mathscr{I}$, remains in $\mathscr{I}$ and defines a conformal geodesic for the conformal structure of $\mathscr{I}$. Furthermore, the Weyl propagation equations admit a solution containing a vector field normal to $\mathscr{I}$.

17.2 The formulation of an initial boundary value problem

The properties of conformal geodesics in anti-de Sitter-like spacetimes will now be exploited to construct a conformal Gaussian system for the extended conformal Einstein field equations. As will be seen, the hyperbolic reduction associated to this gauge leads to an initial boundary value problem for the conformal evolution equations.

17.2.1 Construction of a boundary adapted gauge

Following the discussion of Chapter 14, the solution to the Einstein field equations on the domain of dependence $D^+(\tilde{\mathcal{S}}_\star) = D^+(\mathcal{S}_\star \setminus \partial\mathcal{S}_\star)$ is determined in a unique manner, up to diffeomorphisms, by a pair of tensors $(\tilde{\boldsymbol{h}}, \tilde{\boldsymbol{K}})$ satisfying the Einstein constraint equations on $\tilde{\mathcal{S}}_\star$. On $\mathcal{S}_\star$, let

$$\Omega \equiv \Theta|_{\tilde{\mathcal{S}}_\star}, \qquad \tilde{\Sigma}_\star \equiv \tilde{\boldsymbol{\nu}}(\Theta)|_{\tilde{\mathcal{S}}_\star},$$

with $\tilde{\boldsymbol{\nu}}$ the future-directed unit normal field of $\tilde{\mathcal{S}}_\star$ with respect to $\tilde{\boldsymbol{g}}$. In addition to the usual smoothness and positivity assumptions, the fields Ω and $\tilde{\Sigma}_\star$ are restricted by their behaviour near $\partial\mathcal{S}_\star$ where one requires that $\Sigma_\star \equiv \boldsymbol{\nu}(\Theta)\big|_{\mathcal{S}_\star} = \Omega^{-1}\tilde{\Sigma}_\star$, with $\boldsymbol{\nu}$ the future-directed $\boldsymbol{g}$-unit normal, to be smooth. Using the above fields one can use Equations (11.1a) and (11.1b) to compute the unphysical fields $(\boldsymbol{h}, \boldsymbol{K})$.

To simplify the subsequent discussion, it is assumed that the initial hypersurface $\mathcal{S}_\star$ is such that the unit normal $\boldsymbol{\nu}$ is tangent to $\mathscr{I}$ on $\partial\mathcal{S}_\star$. Accordingly, one has

$$\Sigma_\star \equiv \boldsymbol{\nu}(\Theta)\big|_{\mathcal{S}_\star} = 0 \qquad \text{on } \partial\mathcal{S}_\star.$$

Moreover, recalling that at the conformal boundary s can be made to vanish by a convenient choice of conformal gauge, it is assumed that

$$s = 0, \qquad \text{on } \partial\mathcal{S}_\star.$$

In what follows, each $p \in \mathcal{S}_\star$ will be considered as the starting point of a future-directed conformal geodesic $(x(\tau), \boldsymbol{\beta}(\tau))$ and an associated Weyl propagated frame $\{\boldsymbol{e}_{\boldsymbol{a}}\}$. The parametrisation of the curves is naturally chosen so that $\tau = 0$ on $\mathcal{S}_\star$. For points $p \in \tilde{\mathcal{S}}_\star$, the data for these curves are set in terms of $\tilde{\boldsymbol{g}}$ and its Levi-Civita connection $\tilde{\boldsymbol{\nabla}}$ by the conditions:

(i) $\dot{\boldsymbol{x}}$ is future directed, orthogonal to $\tilde{\mathcal{S}}_\star$ and satisfies the normalisation condition

$$\tilde{\boldsymbol{g}}(\dot{\boldsymbol{x}}, \dot{\boldsymbol{x}})_\star = \Theta_\star^{-2}.$$

(ii) $\boldsymbol{\beta}_\star = \Omega^{-1}\mathbf{d}\Omega$ so that $\langle \boldsymbol{\beta}_\star, \dot{\boldsymbol{x}}_\star \rangle = 0$ —as $\Sigma_\star = 0$ by assumption.
(iii) $\boldsymbol{e}_{\boldsymbol{0}\star} = \dot{\boldsymbol{x}}_\star$ and $\tilde{\boldsymbol{g}}(\boldsymbol{e}_{\boldsymbol{a}}, \boldsymbol{e}_{\boldsymbol{b}})_\star = \Theta_\star^{-2}\eta_{\boldsymbol{ab}}$.

On suitable neighbourhoods $\mathcal{W} \subset J^+(\mathcal{S}_\star)$ of $\mathcal{S}_\star$, the conformal geodesics $x(\tau)$ define a smooth timelike congruence in $\mathcal{W}$, $\{\boldsymbol{e}_{\boldsymbol{a}}\}$ a smooth frame field and $\boldsymbol{\beta}$, a smooth covector. The conformal geodesics can be used to fix a conformal Gaussian coordinate system on $\mathcal{W}$ by setting $x^0 = \tau$ and then extending local coordinates $\underline{x} = (x^\alpha)$ on $\mathcal{S}_\star$ by requiring them to remain constant along conformal geodesics. The coefficients $e_{\boldsymbol{a}}{}^\mu = \langle \mathbf{d}x^\mu, \boldsymbol{e}_{\boldsymbol{a}} \rangle$ of the frame $\{\boldsymbol{e}_{\boldsymbol{a}}\}$ with respect to the Gaussian coordinates satisfy on $\mathcal{W}$ the condition $e_{\boldsymbol{0}}{}^\mu = \delta_0{}^\mu$. Observe, however, that in general $e_{\boldsymbol{a}}{}^0 = 0$ only on $\mathcal{S}_\star$. The conformal factor Θ is then fixed on $\mathcal{W}$ by requiring

$$\boldsymbol{g}(\boldsymbol{e}_{\boldsymbol{a}}, \boldsymbol{e}_{\boldsymbol{b}}) = \eta_{\boldsymbol{ab}}.$$

The discussion of the conformal geodesics in the conformal boundary needs to be done in terms of the metric $\boldsymbol{g}$ and its Levi-Civita connection $\boldsymbol{\nabla}$. In terms of these, the conformal geodesics are represented by a pair $(x(\tau), \boldsymbol{f}(\tau))$ with $\boldsymbol{f} \equiv \boldsymbol{\beta} - \Theta^{-1}\mathbf{d}\Theta$. Accordingly, one has

$$\boldsymbol{f} = 0, \qquad \text{on } \mathcal{S}_\star.$$

As a result of Lemma 17.1, conformal geodesics which start on $\partial\mathcal{S}_\star$ remain on $\mathscr{I}$. As $s = 0$ on $\partial\mathcal{S}_\star$ one can write

$$s_\star = \Omega\varsigma_\star, \tag{17.2}$$

with $\varsigma_\star$ a smooth function on $\partial\mathcal{S}_\star$. It follows from Proposition 5.1 that

$$\Theta = \Omega\left(1 - \frac{1}{2}\varsigma_\star\tau^2\right), \tag{17.3}$$

while for $d_{\boldsymbol{a}} \equiv \langle \boldsymbol{d}, \boldsymbol{e}_{\boldsymbol{a}} \rangle$ one obtains the explicit expression

$$d_{\boldsymbol{a}} = (\dot{\Theta}, \boldsymbol{e}_{\boldsymbol{i}}(\Omega)_\star), \qquad \boldsymbol{e}_{\boldsymbol{i}}(\Omega)_\star \equiv (e_{\boldsymbol{i}}{}^\alpha\partial_\alpha\Omega)_\star, \tag{17.4}$$

where the functions Ω, $\varsigma_\star$ and $\boldsymbol{e}_{\boldsymbol{i}}(\Omega)_\star$ defined initially on $\mathcal{S}_\star$ are extended to $\mathcal{W}$ so that they are constant along conformal geodesics.

Remark. Insight into the behaviour of the conformal factor (17.3) can be obtained from the constraint Equation (11.35c). Using Equation (17.2), exploiting that in an adapted gauge $(\mathbf{d}\Omega)^\sharp = -\boldsymbol{e}_{\boldsymbol{3}}$ and evaluating at $\partial\mathcal{S}_\star$ one concludes

$$\varsigma_\star \simeq -L_{\boldsymbol{03}}\Sigma - L_{\boldsymbol{33}}.$$

Finally, from Equations (11.40) and (11.41) it follows that in a conformal gauge for which $s \simeq 0$ one also necessarily has $L_{\mathbf{03}} \simeq 0$. Thus, one obtains the simple expression

$$\varsigma_\star \simeq -L_{\mathbf{33}}.$$

In particular, if $L_{\mathbf{33}} > 0$, then from Equation (17.3) the conformal factor Θ vanishes only if Ω vanishes. This observation is consistent with the discussion of Section 17.1.2 – conformal geodesics which start normal to $\mathcal{S}_\star$ and away from $\partial\mathcal{S}_\star$ cannot enter the conformal boundary. Ideally, one would like to deduce the property $L_{\mathbf{33}} > 0$ from an analysis of the conformal constraint equations. For data for the exact de Sitter spacetime, Equation (6.8b) implies $L_{\mathbf{33}} = \frac{1}{2}$ on $\partial\mathcal{S}_\star$. Suitable perturbations of data for the anti-de Sitter spacetime should preserve this property.

17.2.2 The conformal evolution system

Combining the gauge construction with the hyperbolic reduction for the extended conformal field equations discussed in Section 13.4 one obtains an evolution system of the form

$$\partial_\tau \hat{\boldsymbol{\upsilon}} = \mathbf{K}\hat{\boldsymbol{\upsilon}} + \mathbf{Q}(\hat{\boldsymbol{\Gamma}})\hat{\boldsymbol{\upsilon}} + \mathbf{L}(x)\boldsymbol{\phi}, \qquad (17.5a)$$

$$\big(\mathbf{I} + \mathbf{A}^0(\boldsymbol{e})\big)\partial_\tau \boldsymbol{\phi} + \mathbf{A}^\alpha(\boldsymbol{e})\partial_\alpha \boldsymbol{\phi} = \mathbf{B}(\hat{\boldsymbol{\Gamma}})\boldsymbol{\phi}, \qquad (17.5b)$$

where the notation of Proposition 13.3 is retained and the matrix-valued function $\boldsymbol{L}(x)$ is given explicitly in terms of the conformal gauge fields Θ and $d_{\boldsymbol{a}}$ as given by Equations (17.3) and (17.4). In the above system, Equation (17.5b) is understood to correspond to the boundary-adapted Bianchi evolution system (13.60a) and (13.60b) in Chapter 13. The evolution system (17.5a) and (17.5b) is ideally suited to the formulation of a boundary value problem, as the equations described by the subsystem (17.5a) are mere ***transport equations*** along the conformal boundary which do not need to be supplemented by boundary conditions. Hence, all the boundary conditions arise from the subsystem (17.5b) associated to the evolution of the Weyl tensor.

Following the discussion of initial boundary value problems for symmetric hyperbolic equations as described in Section 12.4, the identification of suitable boundary conditions for Equation (17.5b) stems from an analysis of the ***normal matrix*** $\mathbf{A}^3$ at the conformal boundary. Making use of the explicit expression for the principal part of the boundary-adapted Bianchi system given in Equation (13.61) and taking into account that, in the *boundary adapted conformal Gaussian gauge*, one has

$$e_{\mathbf{00}}{}^3 \simeq 0, \qquad e_{\mathbf{11}}{}^3 \simeq 0,$$

it follows that

$$\mathbf{A}^3 \simeq 2e_{\mathbf{01}}{}^3\big|_{\mathscr{I}} \begin{pmatrix} -1 & 0 & 0 & 0 & 0 \\ 0 & 0 & 0 & 0 & 0 \\ 0 & 0 & 0 & 0 & 0 \\ 0 & 0 & 0 & 0 & 0 \\ 0 & 0 & 0 & 0 & 1 \end{pmatrix}.$$

This normal matrix is almost in the form required by the theory of Chapter 12. It needs only to be verified that the evolution of the frame coefficient $e_{\mathbf{01}}{}^3$ on $\mathscr{I}$ can be decoupled from that of the components of the Weyl tensor. An inspection of the conformal evolution Equations (13.59a)–(13.59g) – of which Equation (17.5a) above is a schematic representation – shows that whenever $\Theta = 0$, the evolution equations for *certain components* of the fields $e_{\boldsymbol{AB}}{}^{\alpha}$, $\chi_{(\boldsymbol{AB})\boldsymbol{CD}}$, $\Theta_{\boldsymbol{CD}(\boldsymbol{AB})}$ decouple from the evolution of $\phi_{\boldsymbol{ABCD}}$. Thus, it is possible to determine the frame coefficient $e_{\mathbf{01}}{}^3$ directly from the initial data at $\partial\mathcal{S}_\star$ – hence, it is independent of any boundary value prescriptions on $\mathscr{I}$. This observation will be discussed in some detail in the following subsection.

Remark. The normal matrix for the *standard Bianchi system* is given by

$$\mathbf{A}^3 \simeq 2e_{\mathbf{01}}{}^3\big|_{\mathscr{I}} \begin{pmatrix} -1 & 0 & 0 & 0 & 0 \\ 0 & -2 & 0 & 0 & 0 \\ 0 & 0 & 0 & 0 & 0 \\ 0 & 0 & 0 & 2 & 0 \\ 0 & 0 & 0 & 0 & 1 \end{pmatrix},$$

so that this normal matrix leads to a much more complicated analysis of boundary conditions.

17.2.3 Behaviour of the frame at the conformal boundary

In this section, the discussion is restricted to a suitable open neighbourhood $\mathcal{W}$ of a point on $\partial\mathcal{S}_\star$ such that the intersection with conformal geodesics is connected. Consistent with the discussion in Section 17.2.1, one introduces on $\mathcal{S}_\star \cap \mathcal{W}$ an adapted three-dimensional spatial frame $\{\boldsymbol{e}_{\boldsymbol{i}}\}$ such that $\boldsymbol{e}_{\mathbf{3}}$ is orthogonal and inward directed at $\partial\mathcal{S}_\star$ and such that $\nabla_{\mathbf{3}}\boldsymbol{e}_{\boldsymbol{a}} = 0$ on $\mathcal{S}_\star \cap \mathcal{W}$. One introduces coordinates $\underline{x} = (x^\alpha)$ on $\mathcal{S}_\star \cap \mathcal{W}$ so that x^3 vanishes on $\partial\mathcal{S}_\star$ and $\langle \mathbf{d}x^\alpha, \boldsymbol{e}_{\mathbf{3}}\rangle = \delta_{\mathbf{3}}{}^\alpha$ on $\mathcal{S}_\star \cap \mathcal{W}$. A conformal Gaussian gauge system satisfying the above assumptions near $\partial\mathcal{S}_\star$ will be called a ***boundary adapted gauge***.

For future reference it is observed that the conformal evolution Equations (13.59b), (13.59e) and (13.59f) reduce, on the conformal boundary, to

$$\partial_\tau e_{\boldsymbol{AB}}{}^\alpha \simeq -\chi_{(\boldsymbol{AB})}{}^{\boldsymbol{PQ}} e_{\boldsymbol{PQ}}{}^\alpha, \tag{17.6a}$$

$$\partial_\tau \chi_{(\boldsymbol{AB})\boldsymbol{CD}} \simeq -\chi_{(\boldsymbol{AB})}{}^{\boldsymbol{PQ}} \chi_{\boldsymbol{PQCD}} - \Theta_{\boldsymbol{AB}(\boldsymbol{CD})}, \tag{17.6b}$$

$$\partial_\tau \Theta_{\boldsymbol{CD}(\boldsymbol{AB})} \simeq -\chi_{(\boldsymbol{CD})}{}^{\boldsymbol{PQ}} \Theta_{\boldsymbol{PQ}(\boldsymbol{AB})} + \mathrm{i}\sqrt{2} d^{\boldsymbol{P}}{}_{(\boldsymbol{A}}{}^{\mu}{}_{\boldsymbol{B})\boldsymbol{CDP}}. \tag{17.6c}$$

The above evolution equations at the conformal boundary are conveniently analysed in terms of a ***1+1+2 spinorial formalism***. Given a spinorial basis $\{\epsilon_{\boldsymbol{A}}{}^{A}\}$ such that

$$\tau^{\boldsymbol{AA}'} = \delta_{\boldsymbol{0}}{}^{\boldsymbol{A}}\delta_{\boldsymbol{0}'}{}^{\boldsymbol{A}'} + \delta_{\boldsymbol{1}}{}^{\boldsymbol{A}}\delta_{\boldsymbol{1}'}{}^{\boldsymbol{A}'},$$

it is convenient to introduce a spatial spinor $\rho^{AA'}$ with components with respect to the basis $\{\epsilon_{\boldsymbol{A}}{}^{A}\}$ given by

$$\rho^{\boldsymbol{AA}'} \equiv \delta_{\boldsymbol{0}}{}^{\boldsymbol{A}}\delta_{\boldsymbol{0}'}{}^{\boldsymbol{A}'} - \delta_{\boldsymbol{1}}{}^{\boldsymbol{A}}\delta_{\boldsymbol{1}'}{}^{\boldsymbol{A}'}.$$

The space spinor counterpart of $\rho^{AA'}$ is given by

$$\rho_{\boldsymbol{AB}} \equiv \tau_{\boldsymbol{B}}{}^{\boldsymbol{A}'}\rho_{\boldsymbol{AA}'} = -2\delta_{(\boldsymbol{A}}{}^{\boldsymbol{0}}\delta_{\boldsymbol{B})}{}^{\boldsymbol{1}}.$$

It can be verified that, in addition to the condition $\sqrt{2}\boldsymbol{e}_0 = \tau^{\boldsymbol{AA}'}\boldsymbol{e}_{\boldsymbol{AA}'}$, one has

$$\sqrt{2}\boldsymbol{e}_3 = \rho^{\boldsymbol{AA}'}\boldsymbol{e}_{\boldsymbol{AA}'} = \rho^{\boldsymbol{AB}}\boldsymbol{e}_{\boldsymbol{AB}} = 2\boldsymbol{e}_{\boldsymbol{01}}, \qquad \text{on } \mathcal{S}_\star \cap \mathcal{W}. \tag{17.7}$$

In particular, one has

$$\boldsymbol{e}_{\boldsymbol{AB}}(\Theta) = d_{\boldsymbol{AB}} = -\sqrt{\lambda/6}\,\rho_{\boldsymbol{AB}} \qquad \text{on } \partial\mathcal{S}_\star.$$

The spinor $\rho^{\boldsymbol{AB}}$ will be used to split space spinor fields into parts orthogonal and tangent to $\mathscr{I}$. Accordingly, one defines

$$\begin{aligned}
e^{3\perp} &\equiv \rho^{\boldsymbol{AB}}e_{\boldsymbol{AB}}{}^{3}, & e_{\boldsymbol{AB}}{}^{3\|} &\equiv \rho_{(\boldsymbol{A}}{}^{\boldsymbol{C}}e_{\boldsymbol{B})\boldsymbol{C}}{}^{3}\\
\chi^{\perp\perp} &\equiv \rho^{\boldsymbol{AB}}\rho^{\boldsymbol{CD}}\hat{\chi}_{\boldsymbol{ABCD}}, & \chi^{\|\perp}{}_{\boldsymbol{AB}} &\equiv \rho_{(\boldsymbol{A}}{}^{\boldsymbol{E}}\hat{\chi}_{\boldsymbol{B})\boldsymbol{ECD}}\rho^{\boldsymbol{CD}},\\
\chi^{\perp\|}{}_{\boldsymbol{CD}} &\equiv \rho^{\boldsymbol{AB}}\hat{\chi}_{\boldsymbol{ABE}(\boldsymbol{C}}\rho^{\boldsymbol{E}}{}_{\boldsymbol{D})}, & \chi^{\|\|}{}_{\boldsymbol{ABCD}} &\equiv \rho_{(\boldsymbol{A}}{}^{\boldsymbol{E}}\hat{\chi}_{\boldsymbol{B})\boldsymbol{EF}(\boldsymbol{C}}\rho^{\boldsymbol{F}}{}_{\boldsymbol{D})},\\
\Theta^{\perp\perp} &\equiv \rho^{\boldsymbol{AB}}\rho^{\boldsymbol{CD}}\hat{\Theta}_{\boldsymbol{ABCD}}, & \Theta^{\|\perp}{}_{\boldsymbol{AB}} &\equiv \rho_{(\boldsymbol{A}}{}^{\boldsymbol{E}}\hat{\Theta}_{\boldsymbol{B})\boldsymbol{ECD}}\rho^{\boldsymbol{CD}},
\end{aligned}$$

where

$$\hat{\chi}_{\boldsymbol{ABCD}} \equiv \chi_{(\boldsymbol{AB})\boldsymbol{CD}}, \qquad \hat{\Theta}_{\boldsymbol{ABCD}} \equiv \Theta_{\boldsymbol{AB}(\boldsymbol{CD})}.$$

Observing that $\partial_\tau\rho_{\boldsymbol{AB}} = 0$, it follows from Equations (17.6a)–(17.6c) that

$$\begin{aligned}
\partial_\tau e_{\boldsymbol{AB}}{}^{3} &\simeq -\hat{\chi}_{\boldsymbol{AB}}{}^{\boldsymbol{PQ}}e_{\boldsymbol{PQ}}{}^{3},\\
\partial_\tau\big(\hat{\chi}_{\boldsymbol{ABCD}}\rho^{\boldsymbol{CD}}\big) &\simeq -\hat{\chi}_{\boldsymbol{AB}}{}^{\boldsymbol{PQ}}\hat{\chi}_{\boldsymbol{PQCD}}\rho^{\boldsymbol{CD}} - \hat{\Theta}_{\boldsymbol{ABCD}}\rho^{\boldsymbol{CD}},\\
\partial_\tau\big(\hat{\Theta}_{\boldsymbol{CDAB}}\rho^{\boldsymbol{AB}}\big) &\simeq -\hat{\chi}_{\boldsymbol{AB}}{}^{\boldsymbol{PQ}}\hat{\Theta}_{\boldsymbol{PQCD}}\rho^{\boldsymbol{AB}},
\end{aligned}$$

where it has been used that $d^{\boldsymbol{P}}{}_{(\boldsymbol{A}\mu\boldsymbol{B})\boldsymbol{CDP}}\rho^{\boldsymbol{AB}} = 0$ as $d_{\boldsymbol{AB}}$ and $\rho_{\boldsymbol{AB}}$ are proportional to each other. By further contractions with $\rho^{\boldsymbol{AB}}$ one finds that the above equations split into the subsystems

$$\partial_\tau e_{\boldsymbol{AB}}{}^{3\|} \simeq \frac{1}{2}\chi^{\|\perp}{}_{\boldsymbol{AB}} e^{3\perp} + \chi^{\|\|} e_{\boldsymbol{AB}}{}^{3\|}, \tag{17.8a}$$

$$\partial_\tau \chi^{\|\perp}{}_{\boldsymbol{AB}} \simeq \frac{1}{2}\chi^{\|\perp}{}_{\boldsymbol{AB}}\chi^{\perp\perp} + \chi^{\|\|}{}_{\boldsymbol{ABPQ}}\chi^{\|\perp\boldsymbol{PQ}} - \Theta^{\|\perp}{}_{\boldsymbol{AB}}, \tag{17.8b}$$

$$\partial_\tau \Theta^{\|\perp}{}_{\boldsymbol{AB}} \simeq \frac{1}{2}\chi^{\|\perp}{}_{\boldsymbol{AB}}\Theta^{\perp\perp} + \chi^{\|\|}{}_{\boldsymbol{ABPQ}}\Theta^{\|\perp\boldsymbol{PQ}}, \tag{17.8c}$$

and

$$\partial_\tau e^{3\perp} \simeq \frac{1}{2}\chi^{\perp\perp} e^{3\perp} + \chi^{\perp\|}{}_{\boldsymbol{PQ}} e^{3\|\boldsymbol{PQ}}, \tag{17.9a}$$

$$\partial_\tau \chi^{\perp\perp} \simeq \frac{1}{2}(\chi^{\perp\perp})^2 + \chi^{\perp\|}{}_{\boldsymbol{PQ}}\chi^{\|\perp\boldsymbol{PQ}} - \Theta^{\perp\perp}, \tag{17.9b}$$

$$\partial_\tau \Theta^{\perp\perp} \simeq \frac{1}{2}\Theta^{\perp\perp}\chi^{\perp\perp} + \chi^{\perp\|}{}_{\boldsymbol{PQ}}\Theta^{\|\perp\boldsymbol{PQ}}. \tag{17.9c}$$

Initial data for $e^{3\perp}$ and $e_{\boldsymbol{AB}}{}^{3\|}$ at $\partial\mathcal{S}_\star$ follow directly from (17.7). Namely, one has

$$e^{3\perp}|_{\partial\mathcal{S}_\star} = \sqrt{2}, \qquad e_{\boldsymbol{AB}}{}^{3\|}|_{\partial\mathcal{S}_\star} = 0. \tag{17.10}$$

For $\chi^{\|\perp}{}_{\boldsymbol{AB}}$ and $\chi^{\perp\perp}$, initial data can be extracted from the conformal constraint Equation (11.35b) which, taking into account that by assumption $\Sigma = 0$ and $L_{\boldsymbol{a}} = 0$ on $\mathcal{S}_\star$, takes the form $\chi_{\boldsymbol{a}}{}^{\boldsymbol{c}}D_{\boldsymbol{c}}\Omega = 0$ on $\partial\mathcal{S}_\star$. It follows then that

$$\chi^{\perp\perp} = 0, \qquad \chi^{\|\perp}{}_{\boldsymbol{AB}} = 0, \qquad \text{on } \partial\mathcal{S}_\star. \tag{17.11}$$

Finally, to compute the data for $\Theta^{\|\perp}{}_{\boldsymbol{AB}}$ and $\Theta^{\perp\perp}$ one considers the conformal constraint (11.35c) which, in the present context, takes the form

$$D_{\boldsymbol{3}}s = -D^{\boldsymbol{b}}\Omega L_{\boldsymbol{b3}}.$$

Recalling that $s = \Omega\varsigma_\star$ and that, in local Gaussian coordinates, $\Omega = x^3$ one concludes that

$$\Theta^{\perp\perp} = 2\varsigma_\star, \qquad \Theta^{\|\perp}{}_{\boldsymbol{AB}} = 0, \qquad \text{on } \partial\mathcal{S}_\star. \tag{17.12}$$

Using the initial conditions (17.10), (17.11) and (17.12) together with the homogeneity of the subsystem (17.8a)–(17.8c), it follows directly that

$$e_{\boldsymbol{AB}}{}^{3\|} \simeq 0, \qquad \chi^{\|\perp}{}_{\boldsymbol{AB}} \simeq 0, \qquad \Theta^{\|\perp}{}_{\boldsymbol{AB}} \simeq 0.$$

The solution to the subsystem (17.9a)–(17.9c) is given by

$$e^{3\perp} = -\frac{2\sqrt{2}}{2+\tau^2\varsigma_\star}, \qquad \chi^{\perp\perp} = -\frac{4\tau\varsigma_\star}{2+\tau^2\varsigma_\star}, \qquad \Theta^{\perp\perp} = \frac{4\varsigma_\star}{2+\tau^2\varsigma_\star}.$$

The discussion in this section is summarised in the following:

Lemma 17.2 *For any solution to the conformal evolution Equations* (17.5a) *and* (17.5b) *satisfying on* $\partial\mathcal{S}_\star$ *the conditions* (17.10), (17.11) *and* (17.12), *one has that the normal matrix* $\mathbf{A}^3|_{\mathscr{I}}$ *of the boundary adapted Bianchi system is given by*

$$\mathbf{A}^3 \simeq \frac{2\sqrt{2}}{2+\tau^2\varsigma_\star}\begin{pmatrix} -1 & 0 & 0 & 0 & 0 \\ 0 & 0 & 0 & 0 & 0 \\ 0 & 0 & 0 & 0 & 0 \\ 0 & 0 & 0 & 0 & 0 \\ 0 & 0 & 0 & 0 & 1 \end{pmatrix},$$

irrespectively of the value of $\phi_{\boldsymbol{ABCD}}$ *on* $\mathcal{W}\cap\mathscr{I}$.

17.2.4 Identification of boundary conditions

The results of the previous paragraphs allow the identification of maximally dissipative boundary conditions for the conformal evolution equations. Following the discussion in Section 12.4, the basic condition to be satisfied by the normal matrix is the inequality

$$\langle \boldsymbol{\phi}, \mathbf{A}^3|_{\mathscr{I}}\boldsymbol{\phi}\rangle \leq 0,$$

which, assuming that $2+\tau^2\varsigma_\star > 0$, implies that

$$|\phi_4|^2 - |\phi_0|^2 \leq 0. \tag{17.13}$$

To characterise the subspaces of $\mathbb{C}^5$ satisfying the above condition consider two smooth complex-valued functions c_1 and c_2 on $\mathscr{I}$ and let

$$\phi_4 = c_1\phi_0 + c_2\bar{\phi}_0.$$

Exploiting that $(c_1\phi_0 - c_2\bar{\phi}_0)(\bar{c}_1\bar{\phi}_0 - \bar{c}_2\phi_0) \geq 0$ one finds that

$$|\phi_4|^2 - |\phi_0|^2 \leq (|c_1|^2 + |c_2|^2 - 1)|\phi_0|^2.$$

Thus, condition (17.13) is satisfied if one requires

$$|c_1|^2 + |c_2|^2 \leq 1.$$

The above discussion shows that suitable ***inhomogeneous maximally dissipative*** boundary conditions for the conformal evolution equations are given by

$$\phi_4 - c_1\phi_0 - c_2\bar{\phi}_0 = q, \qquad |c_1|^2 + |c_2|^2 \leq 1, \tag{17.14}$$

with c_1, c_2, q smooth complex-valued functions on $\mathscr{I}$.

Corner conditions

As seen in Section 12.4, the smoothness of a solution to an initial boundary value problem requires certain compatibility conditions between the initial data and the boundary conditions at the edge $\partial\mathcal{S}_\star$ – so-called ***corner conditions.*** Following the general discussion given in Section 12.4, one can use the boundary-adapted Bianchi system (17.5b) to determine a formal expansion in terms of τ of the vector $\boldsymbol{\phi}$ on $\mathscr{I}$ near $\partial\mathcal{S}_\star$. This expansion implies, in turn, an expansion for $\phi_4 - c_1\phi_0 - c_2\bar{\phi}_0$ and must be consistent with the prescription of the freely specifiable function q. The explicit form of these corner conditions is rather cumbersome. In what follows, it will be assumed that these corner conditions are satisfied to any order.

17.2.5 The local existence result

The analysis of the boundary conditions leads to a local existence result for an initial boundary value problem for the conformal evolution system (17.5a) and (17.5b) with boundary conditions of the form (17.14). This result is a direct application of Theorem 12.6. More precisely, one has the following:

Proposition 17.1 (***local existence for the initial boundary value problem***) *Given an initial boundary value problem for Equations* (17.5a) *and* (17.5b) *with smooth initial data*

$$(\hat{\boldsymbol{\upsilon}}_\star(\underline{x}), \boldsymbol{\phi}_\star(\underline{x})), \qquad \textit{on } \mathcal{S}_\star,$$

and inhomogeneous maximally dissipative boundary data

$$\phi_4 - c_1\phi_0 - c_2\bar{\phi}_0 = q, \qquad |c_1|^2 + |c_2|^2 \leq 1, \qquad \textit{on } \mathscr{I},$$

with c_1, c_2, q smooth complex-valued functions on $\mathscr{I}$ and assuming that the required corner conditions at $\partial\mathcal{S}_\star$ between initial and boundary data are satisfied to any order, there exists $\tau_\bullet > 0$ such that the initial boundary value problem has a unique smooth solution $(\hat{\boldsymbol{\upsilon}}(\tau,\underline{x}), \boldsymbol{\phi}(\tau,\underline{x}))$ defined on

$$\mathcal{M}_{\tau_\bullet} \equiv [0, \tau_\bullet) \times \mathcal{S}.$$

Remark. Although the above result is local in time, it is nevertheless global in space. As already mentioned, existence on $D^+(\mathcal{S}_\star \setminus \partial\mathcal{S}_\star)$ follows from the standard Cauchy problem. The solutions away from the boundary and those close to the boundary are then patched together to render the full solution.

17.2.6 Propagation of the constraints

In order to transform the existence result given by Proposition 17.1 into an assertion about the Einstein field equations it is necessary to provide an analysis of the propagation of the constraints.

The subsidiary evolution system associated to the conformal evolution Equations (17.5a) and (17.5b) has been discussed in Proposition 13.4. The key structural feature of these subsidiary equations is that they are homogeneous in the zero quantities. A further crucial feature is that the equations for the zero quantities

$$\hat{\Sigma}_{\boldsymbol{a}}{}^{\boldsymbol{c}}{}_{\boldsymbol{b}}, \quad \hat{\Xi}^{\boldsymbol{c}}{}_{\boldsymbol{dab}}, \quad \hat{\Delta}_{\boldsymbol{abc}}, \quad \delta_{\boldsymbol{a}}, \quad \gamma_{\boldsymbol{ab}}, \quad \varsigma_{\boldsymbol{ab}}$$

are all *transport equations*, and, accordingly, they do not give rise to boundary conditions on $\mathscr{I}$. For the zero quantity $\Lambda_{\boldsymbol{abc}}$ associated to the Bianchi identity, the subsidiary system implied by the boundary-adapted system contains no derivatives with respect to the coordinate x^3 and, thus, has a vanishing normal matrix; compare Equations (13.66a)–(13.66c). It follows that the subsidiary evolution equations require no boundary condition on $\mathscr{I}$. From the uniqueness result for initial boundary value problems, Theorem 12.5, if the conformal Einstein equations are satisfied on $\mathcal{S}$ – that is, the zero quantities vanish – then they are also satisfied on $\mathcal{M}_{\tau_\bullet}$. Combining this discussion with Proposition 8.3 one obtains the following existence result for the Einstein field equations:

Theorem 17.1 (***propagation of the constraints for the initial boundary value problem***) *Consider smooth anti-de Sitter-like initial data for the extended conformal Einstein field equations on a three-dimensional manifold $\mathcal{S}$ and boundary initial data of the form (17.14) on $\mathscr{I}$. Assume that the above data satisfy the required corner conditions to all orders on $\partial\mathcal{S}_\star = \mathcal{S}_\star \cap \mathscr{I}$. Then the solution of the initial boundary value problem given by Proposition 17.1 implies a solution to the extended conformal Einstein field equations on $\mathcal{M}_{\tau_\bullet}$. This solution, in turn, implies an anti-de Sitter-like solution to the vacuum Einstein field equations on*

$$\tilde{\mathcal{M}}_{\tau_\bullet} \equiv \mathcal{M}_{\tau_\bullet} \setminus \mathscr{I},$$

for which $\mathscr{I}$ represents the conformal boundary.

Remark. For an ***anti-de Sitter-like initial data set*** it is understood a collection of conformal fields satisfying the conformal constraint equations with the required anti-de Sitter asymptotic behaviour; see Section 11.7.

17.3 Covariant formulation of the boundary conditions

From a geometric point of view, the formulation of the boundary conditions in Proposition 17.1 is not satisfactory. The fields appearing in the maximally dissipative boundary conditions (17.14) are expressed with respect to a certain *boundary adapted gauge*. This gauge specification is an integral part of the boundary conditions: changes on the adapted boundary imply changes in the data. It is therefore important to recast the conditions (17.14), or at least a subclass thereof, in a covariant manner. In what follows, attention will be restricted to the subclass

$$\phi_4 - c\bar{\phi}_0 = q, \qquad c \text{ constant}, \ |c| \leq 1. \tag{17.15}$$

17.3.1 Space spinor split of the boundary data

To recast the boundary condition (17.15) in a covariant manner, it is first necessary to express the fields in terms of objects intrinsic to the conformal boundary $\mathscr{I}$. It is convenient to make use of a ***timelike spinor formalism*** based on the spacelike spinor

$$\rho^{\boldsymbol{AA}'} = \delta_{\boldsymbol{0}}{}^{\boldsymbol{A}}\delta_{\boldsymbol{0}'}{}^{\boldsymbol{A}'} - \delta_{\boldsymbol{1}}{}^{\boldsymbol{A}}\delta_{\boldsymbol{1}'}{}^{\boldsymbol{A}'},$$

as defined in Section 17.2.3, to project spinorial fields into $\mathscr{I}$ in analogy to the space spinor splits with respect to $\tau^{\boldsymbol{AA}'}$. The spinor $\rho^{\boldsymbol{AA}'}$ is the spinorial counterpart of the inward-pointing normal $\boldsymbol{\nu} = \boldsymbol{e_3}$ to $\mathscr{I}$. Notice, however, the normalisation $\rho_{\boldsymbol{AA}'}\rho^{\boldsymbol{AA}'} = -2$. Define the *space spinor version* $\tau_{\boldsymbol{AB}}$ of $\tau_{\boldsymbol{AA}'}$ as

$$\tau_{\boldsymbol{AB}} = \rho_{\boldsymbol{B}}{}^{\boldsymbol{B}'}\tau_{\boldsymbol{AB}'} = 2\delta_{(\boldsymbol{A}}{}^{\boldsymbol{0}}\delta_{\boldsymbol{B})}{}^{\boldsymbol{1}}.$$

Now, taking into account the decomposition of the spinorial counterpart of the Weyl spinor one can compute its ***electric*** and ***magnetic parts*** with respect to $\rho^{\boldsymbol{AA}'}$ as

$$E_{\boldsymbol{ABCD}} \equiv \frac{1}{2}\rho_{\boldsymbol{B}}{}^{\boldsymbol{A}'}\rho^{\boldsymbol{EE}'}\rho_{\boldsymbol{D}}{}^{\boldsymbol{C}'}\rho^{\boldsymbol{FF}'}d_{\boldsymbol{AA}'\boldsymbol{EE}'\boldsymbol{CC}'\boldsymbol{FF}'} = \frac{1}{2}(\phi_{\boldsymbol{ABCD}} + \phi^{\ddagger}_{\boldsymbol{ABCD}}),$$

$$B_{\boldsymbol{ABCD}} \equiv \frac{1}{2}\rho_{\boldsymbol{B}}{}^{\boldsymbol{A}'}\rho^{\boldsymbol{EE}'}\rho_{\boldsymbol{D}}{}^{\boldsymbol{C}'}\rho^{\boldsymbol{FF}'}d^{*}_{\boldsymbol{AA}'\boldsymbol{EE}'\boldsymbol{CC}'\boldsymbol{FF}'} = -\frac{\mathrm{i}}{2}(\phi_{\boldsymbol{ABCD}} - \phi^{\ddagger}_{\boldsymbol{ABCD}}),$$

with

$$\phi^{\ddagger}_{\boldsymbol{ABCD}} \equiv \rho_{\boldsymbol{A}}{}^{\boldsymbol{A}'}\rho_{\boldsymbol{B}}{}^{\boldsymbol{B}'}\rho_{\boldsymbol{C}}{}^{\boldsymbol{C}'}\rho_{\boldsymbol{D}}{}^{\boldsymbol{D}'}\bar{\phi}_{\boldsymbol{A}'\boldsymbol{B}'\boldsymbol{C}'\boldsymbol{D}'}.$$

By construction $E_{\boldsymbol{ABCD}} = E_{(\boldsymbol{ABCD})}$ and $B_{\boldsymbol{ABCD}} = B_{(\boldsymbol{ABCD})}$.

The spinors $E_{\boldsymbol{ABCD}}$ and $B_{\boldsymbol{ABCD}}$ can be decomposed in a $1+2$ manner with respect to the spinor $\tau_{\boldsymbol{AB}}$. The subsequent discussion will be restricted to $B_{\boldsymbol{ABCD}}$, but an identical analysis can be carried out for $E_{\boldsymbol{ABCD}}$. This decomposition is best carried out using tensor frame components and then translating the result into spinors. One obtains

$$B_{\boldsymbol{ABCD}} = \mu_{\boldsymbol{ABCD}} + \mu_{\boldsymbol{AB}}\tau_{\boldsymbol{CD}} + \tau_{\boldsymbol{AB}}\mu_{\boldsymbol{CD}} + \frac{1}{4}\mu(3\tau_{\boldsymbol{AB}}\tau_{\boldsymbol{CD}} - 2\epsilon_{\boldsymbol{A}(\boldsymbol{C}}\epsilon_{\boldsymbol{D})\boldsymbol{B}}), \tag{17.16}$$

with the fields

$$\mu_{\boldsymbol{ABCD}} = \mu_{(\boldsymbol{ABCD})}, \qquad \mu_{\boldsymbol{AB}} = \mu_{(\boldsymbol{AB})}, \qquad \mu = \bar{\mu},$$

satisfying

$$\tau^{AB}\mu_{ABCD} = 0, \qquad \tau^{AB}\mu_{AB} = 0.$$

The geometric interpretation of the various spinors follows from the above properties. By inspection, it can be shown that the only non-vanishing components of the spinor μ_{ABCD} are given by $\overline{\mu_{1111}} = \mu_{0000}$. Similarly, for the rank-2 spinor μ_{AB} one has the non-vanishing components and $\mu_{00} = \overline{\mu_{11}}$. From the definitions of the magnetic parts of ϕ_{ABCD} it follows that

$$\mu_{1111} = -\frac{\mathrm{i}}{2}\big(\phi_{1111} - \bar{\phi}_{0'0'0'0'}\big) \qquad \mu_{11} = -\frac{\mathrm{i}}{2}\big(\phi_{0111} - \bar{\phi}_{1'0'0'0'}\big),$$
$$\mu = -\mathrm{i}\big(\phi_{0011} - \bar{\phi}_{1'1'0'0'}\big).$$

It follows from the above expressions and their analogues for E_{ABCD} that the boundary condition (17.15) can be rewritten in terms of the components of the spinors E_{ABCD} and B_{ABCD}. Of particular interest are the cases

$$c = 1: \qquad B_{1111} = q, \tag{17.17a}$$
$$c = -1: \qquad E_{1111} = q. \tag{17.17b}$$

The Bianchi constraints at the conformal boundary

Now, assume that one is provided with boundary data in the form (17.17a) or (17.17b). A natural question is whether it is possible to recover the full spinor E_{ABCD} and, respectively, B_{ABCD}. It is recalled that the conformal field equation

$$\nabla^{A}{}_{A'}\phi_{ABCD} = 0$$

implies on $\mathscr{I}$ the constraint equations

$$\mathcal{D}^{PQ}\eta_{PQAB} = 0, \qquad \mathcal{D}^{PQ}\mu_{PQAB} = 0, \tag{17.18}$$

with $\mathcal{D}_{AB} \equiv \rho_{(A}{}^{A'}\nabla_{B)A'}$; see Section 11.4. The above equations are the spinorial versions of the conformal constraints (11.39f) and (11.39g). They can be decomposed by introducing the directional derivatives

$$\mathcal{P} \equiv \tau^{AA'}\nabla_{AA'}, \qquad \delta_{AB} \equiv \tau_{(A}{}^{Q}\mathcal{D}_{B)Q},$$

along, respectively, the direction dictated by the conformal geodesics threading the conformal boundary and the direction orthogonal to them. A direct computation gives

$$\mathcal{D}_{AB} = \frac{1}{2}\tau_{AB}\mathcal{P} + \delta_{AB}.$$

Combining this split with the decomposition (17.16) of the spinor B_{ABCD} one finds that the constraint Equations (17.18) imply the system

$$2\mathcal{P}\mu + 4\delta^{AB}\mu_{AB} = 2\mu_{AB}\mathcal{P}\tau^{AB} - 3\mu\mathcal{D}^{AB}\tau_{AB} + 2\tau^{EF}\mathcal{D}^{AB}\mu_{ABEF}, \tag{17.19a}$$

$$4\mathcal{P}\mu_{CD} + 2\delta_{CD}\mu = 4(\mu_{CD}\mathcal{D}^{EF}\tau_{EF} + \mu_{EF}\mathcal{D}^{EF}\tau_{CD}) - 3\mu\mathcal{P}\tau_{CD} + 4(\delta_C{}^E\delta_D{}^F + \tau_{CD}\tau^{EF})\mathcal{D}^{AB}\mu_{ABEF}. \tag{17.19b}$$

A similar system is satisfied by the components of E_{ABCD}. Direct inspection reveals that the above equations constitute a *linear symmetric hyperbolic system (intrinsic to $\mathscr{I}$)* for the fields μ and μ_{AB} if the field μ_{ABCD} is provided; that is, μ_{ABCD} plays the role of *source terms*. The terms involving derivatives with respect to the spinor field τ_{AB} appearing in the right-hand sides of the above equations can be simplified if one assumes a boundary-adapted gauge on $\mathscr{I}$.

The discussion of the previous paragraphs can be summarised in the following manner: suppose one is given boundary data on $\mathscr{I}$ of the form (17.17a) and suppose one knows the values of the fields μ and μ_{AB} on $\partial\mathcal{S}_\star$; then, at least in a neighbourhood of the edge $\partial\mathcal{S}_\star$, it is possible to determine the components μ and μ_{AB} by solving the hyperbolic system (17.19a) and (17.19b). A similar discussion holds for the electric part.

17.3.2 Prescribing the Cotton tensor of the conformal boundary

Despite the formal symmetry between the boundary conditions (17.17a) and (17.17b), the former condition possesses a much stronger geometric content. As a consequence of Equation (11.42), the magnetic part of the rescaled Weyl tensor corresponds, essentially, to the components of the Cotton tensor y_{ijk} of the intrinsic Lorentzian metric $\boldsymbol{\ell}$ of $\mathscr{I}$. Thus, one can ask whether, given the components y_{ijk} of a tensor on $\mathscr{I}$ with the symmetries of the Cotton tensor, it is possible to find a Lorentzian metric $\boldsymbol{\ell}$ on $\mathscr{I}$ such that y_{ijk} are the components, with respect to a *boundary-adapted frame*, of the Cotton tensor of $\boldsymbol{\ell}$. If this is possible, then, as a consequence of its conformal transformation properties, *one has obtained a way of reexpressing a subset of the general maximally dissipative boundary conditions for the conformal field equations in terms of the conformal structure on $\mathscr{I}$*. One has the following result, adapted from lemma 7.1 in Friedrich (1995):

Proposition 17.2 (*geometric formulation of boundary conditions*) *Suppose one has a solution to the extended conformal field equations with anti-de Sitter-like cosmological constant on $\mathcal{M}_{\tau_\bullet} = [0,\tau_\bullet)\times\mathcal{S}$ for $\tau_\bullet > 0$ for which $\mathscr{I} = [0,\tau_\bullet)\times\partial\mathcal{S}$ represents the conformal boundary. Let $\boldsymbol{g}$ denote the metric on $\mathcal{M}_{\tau_\bullet}$ obtained from the solution to the conformal field equations and let $\boldsymbol{\ell}$ denote the 3-metric induced on $\mathscr{I}$ by $\boldsymbol{g}$. Assume that the boundary-adapted conformal Gaussian gauge system can be extended to all of $\mathcal{M}_{\tau_\bullet}$. One then has:*

(i) *Given the restriction to $\partial\mathcal{S}_\star$ of the data for the conformal Einstein field equations in the boundary-adapted gauge and given the conformal class $[\boldsymbol{\ell}]$, it is possible to compute the function q appearing in the boundary condition* (17.17a).

(ii) *Conversely, given on $\partial\mathcal{S}_\star$ the restriction of the data for the conformal Einstein field equations in the boundary-adapted gauge and the boundary condition* (17.17a), *it is possible to determine, in a unique manner, the conformal class $[\boldsymbol{\ell}]$.*

Proof To prove (i) it is observed that as a consequence of Lemma 17.1, the boundary-adapted conformal Gaussian gauge at the conformal boundary can be constructed by solving the conformal geodesic equations for the metric $\boldsymbol{\ell}$. Once the associated Weyl-propagated frame $\{\boldsymbol{e}_i\}$ has been obtained, one can directly compute the components y_{ijk} of the Cotton tensor. Using the discussion of the previous subsection one can, in turn, compute the function q appearing in the boundary condition (17.17a).

The proof of (ii) is much more involved and only a sketch of the main ideas will be provided. Here, one has to verify whether a given three-dimensional tensor is the Cotton tensor of a three-dimensional Lorentzian metric. In view of the Lorentzian nature of this problem, one can address this question by formulating a suitable initial value problem on $\mathscr{I}$ with data on $\partial\mathcal{S}_\star$ for the evolution equations implied by the structural equations on $\mathscr{I}$. Formulated in this manner one has a situation which is very similar to the Cauchy problem for the extended conformal Einstein field equations.

In what follows, let $\boldsymbol{D}$ denote the Levi-Civita covariant derivative of the metric $\boldsymbol{\ell}$, and let $\hat{\boldsymbol{D}}$ denote a Weyl connection in the conformal class of $\boldsymbol{\ell}$. As in the four-dimensional case, the connections are related to each other via a relation of the form $\hat{\boldsymbol{D}} - \boldsymbol{D} = \boldsymbol{S}(\boldsymbol{f})$, with $\boldsymbol{f}$ representing a three-dimensional covector and $\boldsymbol{S}$ the three-dimensional version of the transition tensor discussed in Section 5.2.1. Let $\{\boldsymbol{e}_i\}$ denote an $\boldsymbol{\ell}$-orthogonal frame on $\mathscr{I}$, and let $\hat{\gamma}_i{}^j{}_k$ be the associated connection coefficients of the connection $\hat{\boldsymbol{D}}$. Moreover, let $\hat{l}_{ij}$ denote the components of the Schouten tensor of the connection $\hat{\boldsymbol{D}}$. In analogy to the discussion of the conformal field equations, it is convenient to introduce a number of zero quantities encoding the ***structure equations*** to be satisfied by the various geometric fields:

$$
\begin{aligned}
\hat{\Sigma}_i{}^k{}_j \boldsymbol{e}_k &\equiv [\boldsymbol{e}_i, \boldsymbol{e}_j] - (\hat{\gamma}_i{}^k{}_j - \hat{\gamma}_j{}^k{}_i)\boldsymbol{e}_k, \\
\hat{\Xi}^k{}_{lij} &\equiv \boldsymbol{e}_i(\hat{\gamma}_j{}^k{}_l) - \boldsymbol{e}_j(\hat{\gamma}_i{}^k{}_l) + \hat{\gamma}_m{}^k{}_l(\hat{\gamma}_j{}^m{}_i - \hat{\gamma}_i{}^m{}_j) \\
&\qquad + \hat{\gamma}_j{}^m{}_l\hat{\gamma}_i{}^k{}_m - \hat{\gamma}_i{}^m{}_l\hat{\gamma}_j{}^k{}_m - 2S_{l[i}{}^{km}\hat{l}_{j]m}, \\
\hat{\Delta}_{ijk} &\equiv \hat{D}_i\hat{l}_{jk} - \hat{D}_j\hat{l}_{ik} - y_{ijk}, \\
\Lambda_j &\equiv D^i y_{ij},
\end{aligned}
$$

where

$$y_{ij} \equiv -\frac{1}{2}\epsilon_j{}^{kl} y_{ikl}, \qquad y_i{}^i = 0, \qquad y_{ij} = y_{ji},$$

is the so-called ***Bach tensor***. The zero quantity $\hat{\Sigma}_i{}^k{}_j$ encodes the vanishing of the torsion of the connection $\hat{\boldsymbol{D}}$, $\hat{\Xi}^k{}_{lij}$ contains the relation between the geometric and algebraic curvatures (the Ricci identities), $\hat{\Delta}_{ijk}$ describes the second Bianchi identity for $\hat{\boldsymbol{D}}$ while Λ_j corresponds to the so-called ***third Bianchi identity*** – the differential identity satisfied by the Bach tensor.

To obtain a hyperbolic reduction of the above equations one considers the conformal Gaussian system implied by the conformal geodesics on $\mathscr{I}$. Using arguments similar to the ones in the four-dimensional case one has

$$e_i{}^\alpha = \delta_i^\alpha, \qquad \hat{\gamma}_0{}^k{}_j = 0, \qquad \hat{l}_{0j} = 0, \tag{17.20}$$

and one considers the *evolution equations*

$$\hat{\Sigma}_0{}^k{}_j e_k = 0, \qquad \hat{\Xi}^k{}_{l0j} = 0, \qquad \hat{\Delta}_{0jk} = 0, \qquad \Lambda_j = 0. \tag{17.21}$$

Taking into consideration the gauge conditions (17.20), it can be verified that the first three equations in (17.21) are transport equations on $\mathscr{I}$. The fourth equation requires a more careful discussion: using the solution to the conformal constraint equations as given by Equation (11.42) some components of y_{ij} can be expressed in terms of the boundary conditions; for the remaining components one has that Equations (17.19a) and (17.19b) imply a symmetric hyperbolic system. Thus, one has obtained a symmetric hyperbolic system for the fields $e_i{}^\alpha$, $\hat{\gamma}_i{}^k{}_j$, $\hat{l}_{ij}$ and for the components of y_{ij} not determined by the boundary conditions. Initial data on $\partial\mathcal{S}_\star$ for these fields can be computed from the restriction to $\partial\mathcal{S}_\star$ of the initial data for the conformal evolution equations. Hence, using the general theory of symmetric hyperbolic systems as discussed in Chapter 12, one obtains a solution to Equations (17.21) in a neighbourhood $\mathcal{U}$ in $\mathscr{I}$ of $\partial\mathcal{S}_\star$. To show that this solution implies, in turn, a solution to the equations

$$\hat{\Sigma}_i{}^k{}_j e_k = 0, \qquad \hat{\Xi}^k{}_{lij} = 0, \qquad \hat{\Delta}_{ijk} = 0, \qquad \Lambda_j = 0, \qquad \text{on } \mathcal{U}$$

provided that they are satisfied at $\partial\mathcal{S}$, one needs to discuss the *propagation of the constraints* along the lines of Section 13.4.5. The resulting frame $\{\boldsymbol{e}_i\}$ can be used to construct on $\mathcal{U} \subset \mathscr{I}$ a Lorentzian metric $\boldsymbol{\ell}$. This metric characterises the conformal class of the intrinsic metric of the conformal boundary. □

Reflective boundary conditions

An important class of boundary conditions covered by the prescription (17.17a) is that of the so-called ***reflective boundary conditions***. These correspond to the particular choice of $q = 0$ so that one has

$$\phi_{11111} = \bar{\phi}_{0'0'0'0'}, \qquad \text{on } \mathscr{I}.$$

In what follows, this boundary condition will be supplemented by the conditions

$$\phi_{0111} = \bar{\phi}_{0'0'0'1'}, \qquad \phi_{0011} = \bar{\phi}_{0'0'1'1'}, \qquad \text{on } \partial\mathcal{S}_\star.$$

Accordingly, from the discussion in Section 17.3.1 it follows that $B_{ABCD} = 0$ on $\partial\mathcal{S}_\star$. Furthermore, using the interior evolution system (17.19a) and (17.19b) one has

$$B_{ABCD} = 0, \qquad \text{on } \mathscr{I}.$$

As B_{ABCD} corresponds to the Cotton tensor of $\mathscr{I}$, it follows that *reflective boundary conditions together with some supplementary conditions at the edge imply that the intrinsic metric on $\mathscr{I}$ is conformally flat.*

As pointed out in Friedrich (2014a), despite the above neat geometric characterisation of reflective boundary conditions, if one wants to construct a smooth solution to the initial boundary value problem, one still needs to satisfy an infinite hierarchy of corner conditions. Whether this requirement is compatible with the known procedures for constructing anti-de Sitter-like initial data remains an open question.

Comparison with other initial boundary value problems for the Einstein field equations

Initial boundary value problems in general relativity arise in a natural manner in numerical applications. There exists a number of treatments of the well-posedness of this type of partial differential equation problem for the Einstein field equations; see, for example, Friedrich and Nagy (1999) and Kreiss et al. (2009). The approach and formulation of the Einstein equations considered in the former reference are similar to the ones discussed in this book.

The analysis in Friedrich and Nagy (1999) makes use of a frame formulation of the Einstein field equations. The equations employed in this reference can be obtained from the standard conformal Einstein field equations discussed in Section 8.3.1 by setting $\Xi = 1$. Given these equations, the question is what type of boundary data need to be prescribed on a, in principle, arbitrary, timelike hypersurface to obtain a well-posed initial boundary value problem and to ensure the propagation of the constraints. It turns out that the allowed boundary data are essentially expressed as a combination of components of the Weyl tensor (with respect to a boundary adapted frame) of the form given in Equation (17.14).

Despite these parallels, the situation of the initial boundary value problem analysed in Friedrich and Nagy (1999) and the one discussed in this chapter differ in a key aspect: the boundary hypersurface in anti-de Sitter spacetimes has a *canonic* character. As a consequence, it is possible to formulate covariant boundary conditions, and one ends up with a setting where ***geometric uniqueness*** of the solutions can be ensured. In Friedrich and Nagy (1999) it was not possible to obtain a geometric formulation of the boundary conditions on the timelike hypersurface. Thus, they remain tied to the prescription of the boundary-adapted gauge. As geometric uniqueness cannot be asserted, it is, in principle, not possible to determine whether two seemingly different boundary conditions will lead to the same spacetime, modulo diffeomorphisms. A further discussion can be found in Friedrich (2009).

17.4 Other approaches to the construction of anti-de Sitter-like spacetimes

The analysis of this section has been concerned with the construction of four-dimensional anti-de Sitter-like spacetimes by means of an initial boundary value problem for the conformal Einstein field equations. There are, however, other approaches to this problem if, for example, one assumes the existence of a *static Killing vector* on the spacetime. The assumption of staticity is a strong one and renders results of a global nature. As an example of this type of statement one has the following theorem from Anderson et al. (2002):

Theorem 17.2 (***existence of static anti-de Sitter-like spacetimes***) *Let $\boldsymbol{\ell}$ denote a smooth strictly globally static Lorentzian metric of non-negative scalar curvature on $\mathbb{R} \times \mathbb{S}^2$. Then $(\mathbb{R} \times \mathbb{S}^2, \boldsymbol{\ell})$ is the conformal boundary of a complete strictly globally static vacuum Lorentzian metric on $\mathbb{R}^4$ with anti-de Sitter-like cosmological constant.*

A ***strictly globally static spacetime*** is a spacetime containing an everywhere timelike vector which is orthogonal to the level sets of a globally defined time function. The proof of this result relies on the use of the Fefferman-Graham obstruction tensor; see Fefferman and Graham (1985, 2012). Related to the above theorem is the *rigidity result* given in Anderson (2006), in which it is shown that complete non-singular anti-de Sitter-like spacetimes with a globally stationary conformal infinity and an asymptotically stationary *bulk* must be globally stationary. This result seems to suggest the instability of anti-de Sitter-like spacetimes, at least for certain types of boundary conditions. This expectation has been reinforced by the evidence of turbulent instability observed in numerical simulations of spherically symmetric solutions of the Einstein-scalar field system with anti-de Sitter-like boundary conditions reported by Bizon and Rostworowski (2011).

17.5 Further reading

The approach to the construction of anti-de Sitter-like spacetimes discussed in this chapter has been adapted from the seminal analysis in Friedrich (1995). Boundary conditions for a range of test fields in the anti-de Sitter spacetime have been studied in Ishibashi and Wald (2004). General properties of the exact anti-de Sitter spacetime are examined in detail in Griffiths and Podolský (2009), while properties of anti-de Sitter-like spacetimes are discussed in Henneaux and Teitelboim (1985) and Frances (2005). An issue which has not been touched on in this chapter is that of the definition of the mass for anti-de Sitter-like spacetimes. Conformal approaches to this question have been discussed, for example, in Ashtekar and Magnon (1984) and Ashtekar and Das (2000). Readers interested in a discussion of the issue of the stability/instability of the anti-de Sitter spacetime are referred to the reviews by Bizon (2013) and Maliborski and Rostworowski (2013) and references within.

A considerable part of the interest on anti-de Sitter-like spacetime stems from the so-called AdS/CFT correspondence; see, for example, Maldacena (1998), Witten (1998) and Witten and Yau (1999). A good discussion of the issues involved from a mathematician's point of view are presented in Anderson (2005b).

18

Characteristic problems for the conformal field equations

This chapter discusses the basic theory of characteristic problems for the conformal field equations. Characteristic problems have been of great conceptual value in the development of the modern theory of gravitational radiation. Indeed, the seminal works by Bondi et al. (1962) and Sachs (1962b), in which the modern understanding of gravitational waves was established, were carried out in a setting based on a characteristic initial value problem; see also Sachs (1962c) and Newman and Penrose (1962). The connection between characteristic problems and the notion of asymptotic flatness, already present in the seminal work by Penrose (1963), was further elaborated in Penrose (1965, 1980). From a mathematical point of view, the realisation that the characteristic initial value problem for the Einstein field equations leads to a symmetric hyperbolic evolution system for which the machinery of the theory of partial differential equations (PDEs) is available was first established in Friedrich (1981b). In Friedrich (1981a, 1982) these ideas were subsequently extended to a situation in which part of the data is prescribed at null infinity –a so-called ***asymptotic characteristic initial value problem***, the subject of this chapter. These results established the local existence of *analytic solutions* and were later extended to the smooth case by Kánnár (1996b) using the method of reduction to a standard Cauchy problem by Rendall (1990); see Section 12.5.3.

There are two basic types of asymptotic characteristic problem for the conformal Einstein field equations. The first type is the so-called ***standard asymptotic characteristic problem*** – introduced in Friedrich (1981b) – where initial data are prescribed on null infinity and a null hypersurface intersecting null infinity in a two-dimensional surface with the topology of a 2-sphere; see Figure 18.1, left. In the second type – the so-called ***characteristic problem on a cone***, first discussed in Friedrich (1986c) – one prescribes information on a null cone down to its vertex; see Figure 18.1, right. For reasons discussed in Section 12.5, characteristic problems on a cone are more technically involved. Existence results have been obtained in Chruściel and Paetz (2013).

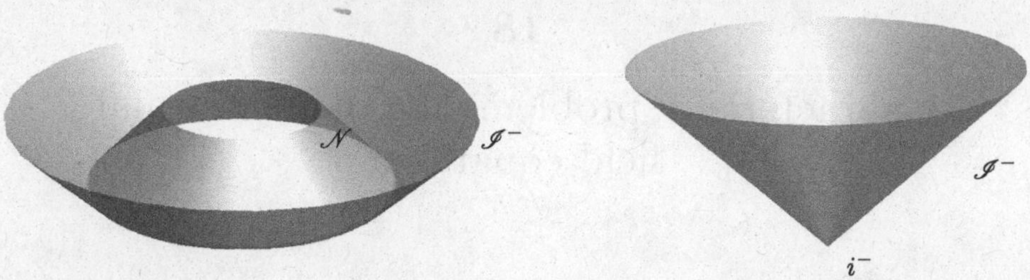

Figure 18.1 Two possible asymptotic characteristic problems for the conformal field equations: on the left, initial data are prescribed on an outgoing null hypersurface $\mathscr{N}$ and null infinity $\mathscr{I}^-$; on the right, data are prescribed on a null cone representing past null infinity $\mathscr{I}^-$. The vertex of the cone corresponds to past timelike infinity, i^-.

The standard and characteristic initial value problems have several structural properties in common. Moreover, the characteristic problem on a cone can be regarded as a limiting case of the standard characteristic problem. In both cases, the Einstein field equations on the initial hypersurfaces split into a set of ***interior (or intrinsic) equations*** and a set of ***transverse equations***. The interior equations split, in turn, into ***constraint equations*** which need to be satisfied only on some subsets of the initial hypersurface (the intersection of the null hypersurfaces or the vertex of the cone) and ***transport equations*** which propagate information along the generators of the null hypersurfaces. The transverse equations dictate the evolution off the initial hypersurfaces. One of the key aspects of the analysis of asymptotic characteristic problems is the identification of ***freely specifiable data*** from which the full data for the evolution equations can be derived. An appealing feature of this type of setting is the natural interpretation of the free data in terms of radiation fields so that a clear-cut connection with the theory of asymptotics as discussed in Chapter 10 can be established.

The discussion in the present chapter is mostly concerned with standard characteristic problems. Certain aspects of the characteristic problem on a cone are briefly considered. The existence results discussed are local in nature. That is, one obtains existence of solutions in a neighbourhood of the intersection of the null hypersurfaces or the vertex of the initial cone. From the perspective of the physical spacetime these local neighbourhoods represent unbounded domains in the asymptotic region.

18.1 Geometric and gauge aspects of the standard characteristic initial value problem

This section provides a discussion of the geometric setting and the gauge fixing procedure for the standard asymptotic characteristic problem. Taking into account the general theory of characteristic problems described in Section 12.5.1 one can consider two possible configurations (see Figure 18.2): (i) that of a

Figure 18.2 The two possible *standard* asymptotic characteristic problems for the conformal Einstein field equations. Case (i) where data are prescribed on a future-oriented (*outgoing*) null hypersurface $\mathscr{N}'$ and future null infinity $\mathscr{I}^+$, and case (ii) where data are prescribed on a past-oriented (*incoming*) null hypersurface $\mathscr{N}$ and past null infinity $\mathscr{I}^-$.

future-oriented (i.e. *outgoing*) null hypersurface intersecting future null infinity or (ii) a past-oriented (i.e. *incoming*) null hypersurface intersecting past null infinity. In order to compare with the characteristic problem on a cone, the present discussion focuses in the latter case. A careful inspection of the setting discussed here leads to the formulation of case (i).

18.1.1 Geometric setting

In what follows, let $(\mathcal{M}, \boldsymbol{g}, \Xi)$ denote a conformal extension of an asymptotically simple spacetime $(\tilde{\mathcal{M}}, \tilde{\boldsymbol{g}})$ satisfying $\boldsymbol{Ric}[\tilde{\boldsymbol{g}}] = 0$ which contains past null infinity $\mathscr{I}^-$. Let $\mathcal{W}$ denote a region of $\mathcal{M}$ with $\mathcal{W} \approx \mathbb{R}^+ \times \mathbb{R}^+ \times \mathbb{S}^2$ bounded by an *incoming null hypersurface* $\mathscr{N}$ and past null infinity $\mathscr{I}^-$. It will be assumed that both $\mathscr{N}$ and $\mathscr{I}^-$ have the topology of $\mathbb{R}^+ \times \mathbb{S}^2$. Let $\mathscr{Z} \equiv \mathscr{N} \cap \mathscr{I}^-$ with $\mathscr{Z} \approx \mathbb{S}^2$. One has that $\mathcal{W} \subset J^+(\mathscr{Z})$. A schematic representation of the geometric setting can be seen in Figure 18.3.

An adapted coordinate system (x^μ) and an associated null tetrad $\{\boldsymbol{e}_{\boldsymbol{AA'}}\}$will be used to describe the geometry of the region $\mathcal{W}$. Let $\{\boldsymbol{\omega}^{\boldsymbol{AA'}}\}$ denote the associated coframe and require that

$$\boldsymbol{g}(\boldsymbol{e}_{\boldsymbol{AA'}}, \boldsymbol{e}_{\boldsymbol{BB'}}) = \epsilon_{\boldsymbol{AB}}\epsilon_{\boldsymbol{A'B'}}. \tag{18.1}$$

On $\mathscr{Z}$ one considers some coordinate system $(x^{\mathcal{A}})$ where $\mathcal{A} = 2, 3$. The complex vectors $\boldsymbol{e}_{\boldsymbol{01'}}$ and $\boldsymbol{e}_{\boldsymbol{10'}} = \overline{\boldsymbol{e}_{\boldsymbol{01'}}}$ of the null tetrad $\{\boldsymbol{e}_{\boldsymbol{AA'}}\}$ will be chosen so that they span the tangent bundle $T(\mathscr{Z})$ – recall that in standard Newman-Penrose notation the vectors $\boldsymbol{e}_{\boldsymbol{01'}}$ and $\boldsymbol{e}_{\boldsymbol{10'}}$ correspond to $\boldsymbol{m}$ and $\overline{\boldsymbol{m}}$.

Now, choose $\boldsymbol{e}_{\boldsymbol{00'}}$ so that, on $\mathscr{I}^-$, it is tangent to the null generators of the conformal boundary – in standard Newman-Penrose notation this vector corresponds to $\boldsymbol{l}$. Let v denote an affine parameter of these generators with the property that $v|_{\mathscr{Z}} = 0$. Thus, one has that $\boldsymbol{e}_{\boldsymbol{00'}} \simeq \boldsymbol{\partial}_v$ where, following the conventions of Chapter 10, the symbol $\simeq$ denotes equality at $\mathscr{I}^-$. The vectors $\boldsymbol{e}_{\boldsymbol{01'}}$ and $\boldsymbol{e}_{\boldsymbol{10'}}$ can be extended to the rest of $\mathscr{I}^-$ by parallel propagation along the null generators. Accordingly, one has

$$\nabla_{\boldsymbol{00'}}\boldsymbol{e}_{\boldsymbol{00'}} \simeq 0, \quad \nabla_{\boldsymbol{00'}}\boldsymbol{e}_{\boldsymbol{01'}} \simeq 0, \quad \nabla_{\boldsymbol{00'}}\boldsymbol{e}_{\boldsymbol{10'}} \simeq 0, \qquad \text{on } \mathscr{I}^-, \tag{18.2}$$

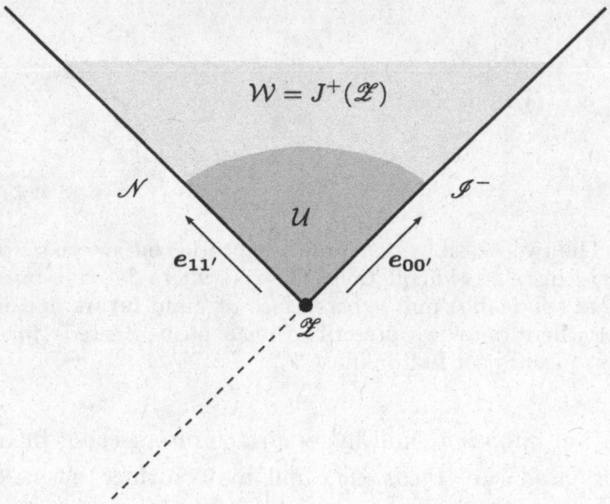

Figure 18.3 Schematic representation of the set up for the standard asymptotic characteristic problem. The existence results are restricted to a neighbourhood $\mathcal{U}$ of $\mathscr{Z}$ in $J^+(\mathscr{Z})$.

where $\nabla_{00'} \equiv \boldsymbol{e}_{00'}{}^a \nabla_a$ is the directional derivative in the direction of $\boldsymbol{e}_{00'}$. Given $v_\bullet \in [0, \infty)$, let $\mathscr{Z}_{v_\bullet} \subset \mathscr{I}^-$ denote the two-dimensional surfaces given by

$$\mathscr{Z}_{v_\bullet} \equiv \{p \in \mathscr{I}^- \mid v(p) = v_\bullet\}.$$

As a result of their parallel propagation, the vectors $\boldsymbol{e}_{01'}$ and $\boldsymbol{e}_{10'}$ span $T(\mathscr{Z}_{v_\bullet})$. Having fixed the vectors $\boldsymbol{e}_{00'}$, $\boldsymbol{e}_{01'}$ and $\boldsymbol{e}_{10'}$ on $\mathscr{I}^-$, regarding the conformal boundary as a submanifold of $\mathcal{M}$, and given that the spacetime metric $\boldsymbol{g}$ is assumed to be known, it follows that at every point $p \in \mathscr{I}^-$, there exists a unique future-pointing null vector linearly independent to $\{\boldsymbol{e}_{00'}, \boldsymbol{e}_{01'}, \boldsymbol{e}_{10'}\}$. This vector is used to complete the null frame $\{\boldsymbol{e}_{\boldsymbol{AA}'}\}$ on $\mathscr{I}^-$ – accordingly, it will be denoted by $\boldsymbol{e}_{11'}$, or $\boldsymbol{n}$ in Newman-Penrose notation. The vector $\boldsymbol{e}_{11'}$ is fixed by the four conditions

$$\boldsymbol{g}(\boldsymbol{e}_{11'}, \boldsymbol{e}_{\boldsymbol{BB}'}) = \epsilon_{\boldsymbol{1B}}\epsilon_{\boldsymbol{1}'\boldsymbol{B}'}.$$

Now, for fixed $v_\bullet$, there exists (at least locally) a unique null hypersurface $\mathscr{N}_{v_\bullet}$ in $\mathcal{M}$ satisfying $\mathscr{N}_{v_\bullet} \cap \mathscr{I}^- = \mathscr{Z}_{v_\bullet}$ such that at $\mathscr{Z}_{v_\bullet}$ the vector $\boldsymbol{e}_{11'}$ is tangent to $\mathscr{N}_{v_\bullet}$ – this involves solving the eikonal equation $\boldsymbol{g}(\mathbf{d}\Phi, \mathbf{d}\Phi) = 0$ for some scalar $\Phi \in \mathcal{X}(\mathcal{W})$ near $\mathscr{I}^-$ with the appropriate initial conditions; for further details see, for example, Stewart (1991), section 4.3. By varying $v_\bullet$ one thus obtains (at least locally) a foliation of null hypersurfaces intersecting $\mathscr{I}^-$. Thence, the affine parameter v along the null generators of $\mathscr{I}^-$ can be used as a coordinate on $\mathcal{W}$. Accordingly, one sets $x^0 = v$, and has

$$\mathscr{N}_{v_\bullet} \equiv \{p \in \mathcal{W} \,|\, v(p) = v_\bullet\},$$

so that the normal to $\mathscr{N}_{v_\bullet}$ is given by $\mathbf{d}v$. The vector $\boldsymbol{e}_{11'}$ can now be extended into $\mathcal{W}$ by requiring it to be tangent to the generators of these hypersurfaces; that is, one has

$$\boldsymbol{e}_{11'} = \boldsymbol{g}^\sharp(\mathbf{d}x^0, \cdot). \tag{18.3}$$

Let r denote an affine parameter of the integral curves of $\boldsymbol{e}_{11'}$ so that one can write $\boldsymbol{e}_{11'} = \boldsymbol{\partial}_r$. Without loss of generality one can choose $r \simeq 0$. The coordinate system (x^μ) on $\mathcal{W}$ is then completed by setting $x^1 = r$ and by extending the coordinates $(x^{\mathcal{A}})$ on $\mathscr{Z}_v$ so that they are constant along the integral curves of $\boldsymbol{e}_{00'}$ and $\boldsymbol{e}_{11'}$. As a consequence of this construction one has

$$\mathscr{N} \equiv \{p \in \mathcal{W} \mid x^0(p) = 0\}, \qquad \mathscr{I}^- \equiv \{p \in \mathcal{W} \mid x^1(p) = 0\}.$$

The vectors $\boldsymbol{e}_{00'}$, $\boldsymbol{e}_{01'}$ and $\boldsymbol{e}_{10'}$ can be extended off $\mathscr{I}^-$ by parallel propagation along the direction of $\boldsymbol{e}_{11'}$. Accordingly, one has

$$\nabla_{11'}\boldsymbol{e}_{11'} = 0, \qquad \nabla_{11'}\boldsymbol{e}_{01'} = 0, \qquad \nabla_{11'}\boldsymbol{e}_{10'} = 0, \qquad \text{on } \mathcal{W}. \tag{18.4}$$

To obtain an explicit expression for the frame $\{\boldsymbol{e}_{\boldsymbol{A}\boldsymbol{A}'}\}$ in the coordinates $(x^\mu) = (v, r, x^{\mathcal{A}})$, it is observed that from Equation (18.3) – rewritten in the form $\boldsymbol{g}(\boldsymbol{\partial}_r, \cdot) = \langle \mathbf{d}v, \cdot\rangle$ – one obtains the pairings

$$\boldsymbol{g}(\boldsymbol{\partial}_r, \boldsymbol{\partial}_v) = 1, \qquad \boldsymbol{g}(\boldsymbol{\partial}_r, \boldsymbol{\partial}_r) = 0, \qquad \boldsymbol{g}(\boldsymbol{\partial}_r, \boldsymbol{\partial}_{\mathcal{A}}) = 0. \tag{18.5}$$

Taking into account the above, the most general form for the frame $\{\boldsymbol{e}_{\boldsymbol{A}\boldsymbol{A}'}\}$ consistent with Equations (18.1) and (18.3) is given by

$$\begin{aligned}
\boldsymbol{e}_{00'} &= \boldsymbol{\partial}_v + U\boldsymbol{\partial}_r + X^{\mathcal{A}}\boldsymbol{\partial}_{\mathcal{A}},\\
\boldsymbol{e}_{11'} &= \boldsymbol{\partial}_r,\\
\boldsymbol{e}_{01'} &= \omega\boldsymbol{\partial}_r + \xi^{\mathcal{A}}\boldsymbol{\partial}_{\mathcal{A}},\\
\boldsymbol{e}_{10'} &= \bar{\omega}\boldsymbol{\partial}_r + \bar{\xi}^{\mathcal{A}}\boldsymbol{\partial}_{\mathcal{A}},
\end{aligned}$$

where U and $X^{\mathcal{A}}$ are real functions and ω and $\xi^{\mathcal{A}}$ are complex functions. Observe, in particular, that because of the conditions in (18.5), $\boldsymbol{e}_{01'}$ and $\boldsymbol{e}_{10'}$ cannot have a v-component. Using, again, relation (18.1) one finds that the components $g^{\mu\nu} = \boldsymbol{g}^\sharp(\mathbf{d}x^\mu, \mathbf{d}x^\nu)$ are of the form

$$(g^{\mu\nu}) = \begin{pmatrix} 0 & 1 & 0 \\ 1 & g^{11} & g^{1\mathcal{A}} \\ 0 & g^{\mathcal{A}1} & g^{\mathcal{A}\mathcal{B}} \end{pmatrix},$$

where

$$g^{11} = 2(U - \omega\bar{\omega}), \qquad g^{1\mathcal{A}} = X^{\mathcal{A}} - (\xi^{\mathcal{A}}\bar{\omega} + \bar{\xi}^{\mathcal{A}}\omega), \qquad g^{\mathcal{A}\mathcal{B}} = -(\xi^{\mathcal{A}}\bar{\xi}^{\mathcal{B}} + \bar{\xi}^{\mathcal{A}}\xi^{\mathcal{B}}).$$

In particular, one has that $U \simeq 0$, $X^{\mathcal{A}} \simeq 0$, $\omega \simeq 0$, consistent with the fact that $\boldsymbol{e}_{\mathbf{11}'}$ is tangent to the generators of null infinity and that v is an affine parameter; hence, $\boldsymbol{e}_{\mathbf{11}'} \simeq \boldsymbol{\partial}_v$. Observe also that $\boldsymbol{e}_{\mathbf{01}'} \simeq \xi^{\mathcal{A}}\boldsymbol{\partial}_{\mathcal{A}}$. Thus, the pull-back to $\mathscr{Z}_v$ of

$$g^{\mathcal{AB}}\boldsymbol{\partial}_{\mathcal{A}} \otimes \boldsymbol{\partial}_{\mathcal{B}} = -(\xi^{\mathcal{A}}\bar{\xi}^{\mathcal{B}} + \bar{\xi}^{\mathcal{A}}\xi^{\mathcal{B}})\boldsymbol{\partial}_{\mathcal{A}} \otimes \boldsymbol{\partial}_{\mathcal{B}},$$

to be denoted by $\boldsymbol{\varsigma}^{\sharp}$, corresponds to the two-dimensional (contravariant) metric of the sections of null infinity. Now, by assumption $\mathscr{Z}_v \approx \mathbb{S}^2$ so that $\boldsymbol{\varsigma}$ is conformal to the standard metric of $\mathbb{S}^2$.

Finally, combining the propagation conditions (18.2) and (18.4) with the definition of the spin connection coefficients – see Equations (3.31) and (3.33) – in the form

$$\Gamma_{\boldsymbol{AA'BC}} = \frac{1}{2}\epsilon_{\boldsymbol{BP}}\langle \boldsymbol{\omega}^{\boldsymbol{PQ'}}, \nabla_{\boldsymbol{AA'}}\boldsymbol{e}_{\boldsymbol{CQ'}}\rangle,$$

one finds

$$\Gamma_{\mathbf{00'10}} \simeq 0, \qquad \Gamma_{\mathbf{00'00}} \simeq 0,$$

and

$$\Gamma_{\mathbf{00'11}} = \bar{\Gamma}_{\mathbf{10'1'0'}} + \Gamma_{\mathbf{10'10}}, \quad \Gamma_{\mathbf{01'11}} = \bar{\Gamma}_{\mathbf{10'1'1'}}, \quad \Gamma_{\mathbf{11'}\boldsymbol{AB}} = 0, \quad \text{on } \mathcal{W} \subset \mathcal{M}.$$

The discussion of this section is summarised in the following

Lemma 18.1 (***frame gauge conditions for the standard characteristic problem***) *Let* $(\tilde{\mathcal{M}}, \tilde{\boldsymbol{g}})$ *denote an asymptotically simple spacetime satisfying* $\boldsymbol{Ric}[\tilde{\boldsymbol{g}}] = 0$ *and let* $(\mathcal{M}, \boldsymbol{g}, \Xi)$ *with* $\boldsymbol{g} = \Xi^2\tilde{\boldsymbol{g}}$ *be a conformal extension thereof for which the condition* $\Xi = 0$ *describes past null infinity* $\mathscr{I}^-$. *The frame* $\{\boldsymbol{e}_{\boldsymbol{AA'}}\}$ *can be chosen so that, given a null hypersurface* $\mathscr{N}$ *intersecting* $\mathscr{I}^-$ *on* $\mathscr{Z} \approx \mathbb{S}^2$, *one has*

$$\Gamma_{\mathbf{00'11}} = \bar{\Gamma}_{\mathbf{10'1'0'}} + \Gamma_{\mathbf{10'10}},$$
$$\Gamma_{\mathbf{01'11}} = \bar{\Gamma}_{\mathbf{10'1'1'}}, \quad \Gamma_{\mathbf{11'}\boldsymbol{AB}} = 0, \quad \text{on } \mathcal{W} \subset \mathcal{M}.$$

In addition, one has that

$$\Gamma_{\mathbf{00'01}} = \Gamma_{\mathbf{00'10}} = \Gamma_{\mathbf{00'00}} = U = X^{\mathcal{A}} = \omega = 0, \qquad \text{on } \mathscr{I}^-.$$

Remark. The conventions used here for the vectors $\boldsymbol{e}_{\mathbf{00}'}$ and $\boldsymbol{e}_{\mathbf{11}'}$ are the opposite of those used in Kánnár (1996b). They have been chosen to agree with the standard conventions in the treatment of asymptotics as given in Penrose and Rindler (1986) and Stewart (1991) and to ease the comparison with the characteristic problem on a cone.

18.1.2 The choice of conformal gauge

The geometric setting discussed in the previous section has an inherent conformal gauge freedom which can be exploited to simplify the analysis.

As discussed in Section 8.2.5, the Ricci scalar $R[\boldsymbol{g}]$ plays the role of a ***conformal gauge source function*** for the conformal field equations. A possible choice in the present setting is to fix the conformal factor Ξ linking the metrics $\tilde{\boldsymbol{g}}$ and $\boldsymbol{g}$ in such a manner that $R[\boldsymbol{g}] = 0$. To see that this can always be done, consider first a situation involving a generic conformal factor Ξ for which $R[\boldsymbol{g}] \neq 0$, and let

$$\boldsymbol{g}' \equiv \vartheta^2 \boldsymbol{g}, \tag{18.6}$$

with ϑ a positive function on $\mathcal{W}$. Defining $\Xi' \equiv \vartheta\Xi$ one finds that $\boldsymbol{g}' = \Xi'^2 \tilde{\boldsymbol{g}}$. Consistent with the above conformal rescaling one considers the following transformation behaviour for the $\boldsymbol{g}$-orthonormal frame $\{\boldsymbol{e}_{\boldsymbol{AA}'}\}$:

$$\boldsymbol{e}'_{\boldsymbol{00}'} = \boldsymbol{e}_{\boldsymbol{00}'}, \qquad \boldsymbol{e}'_{\boldsymbol{11}'} = \vartheta^{-2}\boldsymbol{e}_{\boldsymbol{11}'}, \qquad \boldsymbol{e}'_{\boldsymbol{01}'} = \vartheta^{-1}\boldsymbol{e}_{\boldsymbol{01}'}, \qquad \boldsymbol{e}'_{\boldsymbol{10}'} = \vartheta^{-1}\boldsymbol{e}_{\boldsymbol{10}'}.$$

Using the transformation law under conformal rescalings for the Ricci scalar, Equation (5.6c), one finds that the requirement $R[\boldsymbol{g}'] = 0$ is equivalent to the wave equation

$$\nabla^a \nabla_a \vartheta = \frac{1}{6} R[\boldsymbol{g}]; \tag{18.7}$$

see also Equation (8.30). The general theory of the characteristic problem for wave equations ensures the existence of a unique solution to this equation in a neighbourhood $\mathcal{U}$ of $\mathscr{Z}$ in $J^+(\mathscr{Z})$ if some suitable data are prescribed on $\mathscr{N} \cup \mathscr{I}^-$; see, for example, Rendall (1990). A natural requirement on the initial data for Equation (18.7) is to have $\boldsymbol{\omega}'^{\boldsymbol{11}'} = \mathbf{d}\Xi'$ on $\mathscr{Z}$ where $\{\boldsymbol{\omega}'^{\boldsymbol{AA}'}\}$ denotes the coframe dual to $\{\boldsymbol{e}'_{\boldsymbol{AA}'}\}$. This is equivalent to setting

$$\boldsymbol{\omega}'^{\boldsymbol{11}'} = \vartheta \mathbf{d}\Xi \qquad \text{on } \mathscr{Z}.$$

By choosing $\vartheta^{-1}|_{\mathscr{Z}} = \boldsymbol{e}_{\boldsymbol{11}'}(\Xi)|_{\mathscr{Z}} = \langle \mathbf{d}\Xi, \boldsymbol{e}_{\boldsymbol{11}'}\rangle|_{\mathscr{Z}}$ one can, in fact, ensure that

$$\boldsymbol{e}'_{\boldsymbol{11}'}(\Xi') = \langle \mathbf{d}\Xi', \boldsymbol{e}'_{\boldsymbol{11}'}\rangle = 1 \qquad \text{on } \mathscr{Z}.$$

The principal part of the wave Equation (18.7), expressed in terms of frame derivatives, is given by

$$\boldsymbol{e}_{\boldsymbol{00}'}(\boldsymbol{e}_{\boldsymbol{11}'}(\vartheta)) + \boldsymbol{e}_{\boldsymbol{11}'}(\boldsymbol{e}_{\boldsymbol{00}'}(\vartheta)) - \boldsymbol{e}_{\boldsymbol{01}'}(\boldsymbol{e}_{\boldsymbol{10}'}(\vartheta)) - \boldsymbol{e}_{\boldsymbol{10}'}(\boldsymbol{e}_{\boldsymbol{01}'}(\vartheta)).$$

Thus, Equation (18.7) implies an *intrinsic propagation equation on* $\mathscr{N}$ for $\boldsymbol{e}_{\boldsymbol{00}'}(\vartheta)$ if $\boldsymbol{e}_{\boldsymbol{11}'}(\vartheta)$ is known on $\mathscr{N}$. Analogously, one has an *intrinsic propagation equation on* $\mathscr{I}^-$ for $\boldsymbol{e}_{\boldsymbol{11}'}(\vartheta)$ if $\boldsymbol{e}_{\boldsymbol{00}'}(\vartheta)$ is known on $\mathscr{I}^-$. The freedom in the specification of characteristic data can be exploited by observing that under the conformal rescaling (18.6) one obtains the transformation rules

$$\Gamma'_{\mathbf{0}\mathbf{1}'\mathbf{1}\mathbf{1}} = \Gamma_{\mathbf{0}\mathbf{1}'\mathbf{1}\mathbf{1}} - \vartheta^{-1}\boldsymbol{e}_{\mathbf{1}\mathbf{1}'}(\vartheta), \qquad \Gamma'_{\mathbf{1}\mathbf{0}'\mathbf{0}\mathbf{0}} = \vartheta^{-2}\Gamma_{\mathbf{1}\mathbf{0}'\mathbf{0}\mathbf{0}} - \vartheta^{-3}\boldsymbol{e}_{\mathbf{0}\mathbf{0}'}(\vartheta).$$

Accordingly, by setting

$$\boldsymbol{e}_{\mathbf{1}\mathbf{1}'}(\vartheta) = \vartheta\Gamma_{\mathbf{0}\mathbf{1}'\mathbf{1}\mathbf{1}}, \qquad \boldsymbol{e}_{\mathbf{0}\mathbf{0}'}(\vartheta) = \vartheta\Gamma_{\mathbf{1}\mathbf{0}'\mathbf{0}\mathbf{0}} \qquad \text{on } \mathscr{Z},$$

one obtains

$$\Gamma'_{\mathbf{0}\mathbf{1}'\mathbf{1}\mathbf{1}} = 0, \qquad \Gamma'_{\mathbf{1}\mathbf{0}'\mathbf{0}\mathbf{0}} = 0, \qquad \text{on } \mathscr{Z}.$$

To propagate the freely specifiable components of $\nabla_{\boldsymbol{AA}'}\vartheta$ along $\mathscr{N}$ and $\mathscr{I}^-$ it is convenient to consider the transformation law under conformal rescalings of the trace-free part of the Ricci tensor

$$\Phi'_{ab} - \Phi_{ab} = -2\vartheta^{-1}\bigg(\nabla_a\nabla_b\vartheta - 2\vartheta^{-1}\nabla_a\vartheta\nabla_b\vartheta \\ - \frac{1}{4}g_{ab}(\nabla^c\nabla_c\vartheta - 2\vartheta^{-1}\nabla_c\vartheta\nabla^c\vartheta)\bigg). \tag{18.8}$$

Now, recalling that

$$\Phi_{\boldsymbol{AA}'\boldsymbol{BB}'} = e_{\boldsymbol{AA}'}{}^a e_{\boldsymbol{BB}'}{}^b\Phi_{ab}, \qquad \Phi'_{\boldsymbol{AA}'\boldsymbol{BB}'} = e'_{\boldsymbol{AA}'}{}^a e'_{\boldsymbol{BB}'}{}^b\Phi'_{ab},$$

one can consider the propagation equations

$$\boldsymbol{e}_{\mathbf{1}\mathbf{1}'}(\boldsymbol{e}_{\mathbf{1}\mathbf{1}'}(\vartheta)) - 2\vartheta^{-1}\big(\boldsymbol{e}_{\mathbf{1}\mathbf{1}'}(\vartheta)\big)^2 = \vartheta\Phi_{22} \qquad \text{on } \mathscr{N}, \tag{18.9a}$$

$$\boldsymbol{e}_{\mathbf{0}\mathbf{0}'}(\boldsymbol{e}_{\mathbf{0}\mathbf{0}'}(\vartheta)) - 2\vartheta^{-1}\big(\boldsymbol{e}_{\mathbf{0}\mathbf{0}'}(\vartheta)\big)^2 = \vartheta\Phi_{00} \qquad \text{on } \mathscr{I}^-. \tag{18.9b}$$

These two equations can be read as ordinary differential equations along the generators of $\mathscr{N}$ and $\mathscr{I}^-$ for $\boldsymbol{e}_{\mathbf{1}\mathbf{1}'}(\vartheta)$ and $\boldsymbol{e}_{\mathbf{0}\mathbf{0}'}(\vartheta)$, respectively. Accordingly, a solution exists in a neighbourhood of $\mathscr{Z}$ on $\mathscr{N}$ and, respectively, on $\mathscr{I}^-$. Comparing with Equation (18.8), one sees that these solutions ensure

$$\Phi'_{22} = 0 \qquad \text{on } \mathscr{N}, \tag{18.10a}$$

$$\Phi'_{00} = 0 \qquad \text{on } \mathscr{I}^-. \tag{18.10b}$$

Once the solutions $\boldsymbol{e}_{\mathbf{1}\mathbf{1}'}(\vartheta)$ and $\boldsymbol{e}_{\mathbf{0}\mathbf{0}'}(\vartheta)$ to the propagation conditions (18.9a) and (18.9b) have been obtained, one can use the intrinsic equations implied by (18.7) on $\mathscr{N} \cup \mathscr{I}^-$ to obtain $\boldsymbol{e}_{\mathbf{0}\mathbf{0}'}(\vartheta)$ on $\mathscr{N}$ and $\boldsymbol{e}_{\mathbf{1}\mathbf{1}'}(\vartheta)$ on $\mathscr{I}^-$.

The analysis of this section can be summarised in the following:

Lemma 18.2 (***conformal gauge conditions for the standard characteristic problem***) *Let $(\tilde{\mathcal{M}}, \tilde{\boldsymbol{g}})$ denote an asymptotically simple spacetime satisfying $\boldsymbol{Ric}[\tilde{\boldsymbol{g}}] = 0$ and let $(\mathcal{M}, \boldsymbol{g}, \Xi)$ with $\boldsymbol{g} = \Xi^2\tilde{\boldsymbol{g}}$ be a conformal extension thereof for which the condition $\Xi = 0$ describes past null infinity $\mathscr{I}^-$. Given the frame*

$\{e_{AA'}\}$ of Lemma 18.1, the conformal factor Ξ can be chosen so that given a null hypersurface $\mathscr{N}$ intersecting $\mathscr{I}^-$ on $\mathscr{Z} \approx \mathbb{S}^2$ one has

$$R[\boldsymbol{g}] = 0, \qquad \textit{in a neighbourhood } \mathcal{W} \textit{ of } \mathscr{Z} \textit{ on } J^+(\mathscr{Z}).$$

Moreover, one has the additional gauge conditions

$$e_{11'}(\Xi) = 1, \qquad \Gamma_{01'11} = \Gamma_{10'00} = 0, \qquad \textit{on } \mathscr{Z},$$
$$\Phi_{22} = 0 \qquad \textit{on } \mathscr{N},$$
$$\Sigma_{AA'} = e_{11'}(\Xi)\delta_A{}^1\delta_{A'}{}^{1'}, \qquad \Phi_{00} = 0, \qquad \textit{on } \mathscr{I}^-.$$

Remark. In the gauge given by Lemma 18.2 one has that $L_{AA'BB'} = \Phi_{AA'BB'}$. This fact will be used repeatedly in the following without any further mention.

18.2 The conformal evolution equations in the standard characteristic initial value problem

This section analyses general aspects of the standard characteristic initial value problem for the conformal Einstein field equations with data prescribed on the null hypersurfaces $\mathscr{N}$ and $\mathscr{I}^-$. The spinorial conformal field equations, as discussed in Section 8.3.2, will be used to formulate this problem. Accordingly, on $\mathcal{W}$ it will be required that

$$\Sigma_{AA'BB'} = 0, \qquad \Xi^C{}_{DAA'BB'} = 0, \tag{18.11a}$$
$$\Xi_{AA'} = 0, \qquad Z_{AA'BB'} = 0, \qquad Z_{AA'} = 0, \qquad Z = 0, \tag{18.11b}$$
$$\Delta_{CDBB'} = 0, \qquad \Lambda_{BB'CD} = 0, \tag{18.11c}$$

where, for convenience, one defines

$$\Xi_{AA'} \equiv \Sigma_{AA'} - \nabla_{AA'}\Xi.$$

Following the conventions of Chapter 13 let $\mathbf{u}$ denote the collection of independent components of the unknowns appearing in the conformal field Equations (18.11a)–(18.11c) and let $\mathbf{u}_\star$ be its value on $\mathscr{N} \cup \mathscr{I}^+$.

Strictly speaking, as no hyperbolic reduction procedure has yet been applied to equations (18.11a)–(18.11c) – that is, the equations do not constitute a symmetric hyperbolic system – one does not directly obtain a characteristic problem in the sense described in Section 12.1.2. Nevertheless, the structure of the conformal evolution equations can be used to obtain a symmetric hyperbolic system for which the theory of Section 12.5 can be applied. Thus, it is necessary to analyse the properties of the conformal field equations on the hypersurfaces $\mathscr{N}$ and $\mathscr{I}^+$. When evaluated on $\mathscr{N} \cup \mathscr{I}^+$ the system (18.11a)–(18.11c) splits into a set of ***interior*** and a set of ***transverse*** equations. As the name suggests, interior equations contain only derivatives which are intrinsic to the null hypersurfaces. The interior equations divide, in turn, into ***transport equations*** containing the directional derivative along the generators of the hypersurface and ***constraint***

equations which do not contain this derivative. In the transverse equations one deals with the directional derivative transverse to the surface.

To see how this split comes about, it is convenient to recall some aspects of the hyperbolic reduction procedure for the Equations (18.11a)–(18.11c). Given a timelike vector τ^μ and a suitable set of gauge source functions $F^{\boldsymbol{a}}(x)$ and $F_{\boldsymbol{AB}}(x)$ on $\mathcal{W}$, one obtains a symmetric hyperbolic system for the independent components of the various conformal fields. As discussed in Proposition 13.1, the characteristic polynomial of this system contains factors of the form $g^{\mu\nu}\xi_\mu\xi_\nu$. Accordingly, the combined null hypersurface $\mathscr{N}\cap\mathscr{I}^+$ is a null hypersurface of the reduced evolution system. Following the discussion of Section 12.1.2, it follows that the reduced system contains equations which are intrinsic to $\mathscr{N}\cap\mathscr{I}^+$ and equations which are transverse to the initial hypersurface. In the following, it is shown how this observation can be extended to the full conformal field equations.

The interior equations on $\mathscr{N}$

The interior equations on the null hypersurface $\mathscr{N}$ should contain only the directional derivatives along the directions given by $\boldsymbol{e}_{\boldsymbol{11'}}$, $\boldsymbol{e}_{\boldsymbol{01'}}$ and $\boldsymbol{e}_{\boldsymbol{10'}}$. Inspection shows that the subset of (18.11a)–(18.11c) with this property is given by the equations

$$\Xi_{\boldsymbol{11'}}=0,\quad Z_{\boldsymbol{11'AA'}}=0,\quad Z_{\boldsymbol{11'}}=0, \tag{18.12a}$$

$$\Sigma_{\boldsymbol{11'BB'}}=0,\quad \Xi^{\boldsymbol{C}}{}_{\boldsymbol{D11'BB'}}=0, \tag{18.12b}$$

$$\Delta_{\boldsymbol{1DBB'}}=0,\quad \Lambda_{\boldsymbol{B1'CD}}=0. \tag{18.12c}$$

More explicitly, taking into account the gauge conditions given by Lemmas 18.1 and 18.2 one has the equations

$$\boldsymbol{e}_{\boldsymbol{11'}}(\Xi)=\Sigma_{\boldsymbol{11'}}, \tag{18.13a}$$

$$\boldsymbol{e}_{\boldsymbol{11'}}(\Sigma_{\boldsymbol{00'}})=-\Xi\Phi_{11}-s,\quad \boldsymbol{e}_{\boldsymbol{11'}}(\Sigma_{\boldsymbol{01'}})=-\Xi\Phi_{12},\quad \boldsymbol{e}_{\boldsymbol{11'}}(\Sigma_{\boldsymbol{11'}})=0, \tag{18.13b}$$

$$\boldsymbol{e}_{\boldsymbol{11'}}(s)=-\Phi_{11}\Sigma_{\boldsymbol{11'}}+2\Phi_{12}\Sigma_{\boldsymbol{01'}}, \tag{18.13c}$$

$$\boldsymbol{e}_{\boldsymbol{11'}}(e_{\boldsymbol{BB'}}{}^{\mu})=-\Gamma_{\boldsymbol{BB'}}{}^{\boldsymbol{C}}{}_{\boldsymbol{1}}e_{\boldsymbol{C1'}}{}^{\mu}-\bar{\Gamma}_{\boldsymbol{B'B}}{}^{\boldsymbol{C'}}{}_{\boldsymbol{1'}}e_{\boldsymbol{1C'}}{}^{\mu}, \tag{18.13d}$$

$$\begin{aligned}\boldsymbol{e}_{\boldsymbol{11'}}(\Gamma_{\boldsymbol{BB'CD}})&=-\Gamma_{\boldsymbol{F1'CD}}\Gamma_{\boldsymbol{BB'}}{}^{\boldsymbol{F}}{}_{\boldsymbol{1}}-\Gamma_{\boldsymbol{1F'CD}}\bar{\Gamma}_{\boldsymbol{B'B}}{}^{\boldsymbol{F'}}{}_{\boldsymbol{1'}}\\&\quad-\Xi\phi_{\boldsymbol{BCD1}}\epsilon_{\boldsymbol{1'B'}}-\Phi_{\boldsymbol{C1'DB'}}\epsilon_{\boldsymbol{1B}},\end{aligned} \tag{18.13e}$$

$$\begin{aligned}\boldsymbol{e}_{\boldsymbol{11'}}(\Phi_{\boldsymbol{D0'BB'}})&=\nabla_{\boldsymbol{10'}}\Phi_{\boldsymbol{D1'BB'}}-\nabla_{\boldsymbol{D1'}}\Phi_{\boldsymbol{10'BB'}}+\nabla_{\boldsymbol{D0'}}\Phi_{\boldsymbol{11'BB'}}\\&\quad-2\Sigma_{\boldsymbol{1B'}}\phi_{\boldsymbol{BD01}}+2\Sigma_{\boldsymbol{0B'}}\phi_{\boldsymbol{BD11}},\end{aligned} \tag{18.13f}$$

$$\boldsymbol{e}_{\boldsymbol{11'}}(\phi_{\boldsymbol{ABC0}})=\nabla_{\boldsymbol{01'}}\phi_{\boldsymbol{ABC1}}, \tag{18.13g}$$

with the understanding that equations for quantities already determined by gauge conditions are dropped from the list. Despite their apparent complexity, the above equations possess a delicate hierarchical structure which allows one to solve them sequentially from some basic data on $\mathscr{Z}$ and $\mathscr{N}$. This structure is briefly described in the following paragraphs.

One starts by combining Equation (18.13a) with the third equation in (18.13b) and then using that $\boldsymbol{c}_{11'} = \boldsymbol{\partial}_r$ to find that $\partial_r^2 \Xi = 0$. Hence, taking into account Lemma 18.2 one concludes that $\Xi = r$ along $\mathscr{N}$. Next, one can consider Equation (18.13e) for $\Gamma_{\mathbf{0}\mathbf{1}'\mathbf{1}\mathbf{1}}$ and $\Gamma_{\mathbf{1}\mathbf{0}'\mathbf{1}\mathbf{1}}$ (in standard Newman-Penrose (NP) notation γ and λ) which, in view of the gauge conditions, gives the subsystem

$$\partial_r \Gamma_{\mathbf{0}\mathbf{1}'\mathbf{1}\mathbf{1}} = -(\Gamma_{\mathbf{0}\mathbf{1}'\mathbf{1}\mathbf{1}})^2 - \Gamma_{\mathbf{1}\mathbf{0}'\mathbf{1}\mathbf{1}}\bar{\Gamma}_{\mathbf{1}'\mathbf{0}\mathbf{1}'\mathbf{1}'},$$
$$\partial_r \Gamma_{\mathbf{1}\mathbf{0}'\mathbf{1}\mathbf{1}} = -2\Gamma_{\mathbf{0}\mathbf{1}'\mathbf{1}\mathbf{1}}\Gamma_{\mathbf{1}\mathbf{0}'\mathbf{1}\mathbf{1}} + \Xi\phi_4.$$

The above ***Riccati system*** can be solved if ϕ_4 is known along $\mathscr{N}$. With $\Gamma_{\mathbf{0}\mathbf{1}'\mathbf{1}\mathbf{1}}$ and $\Gamma_{\mathbf{1}\mathbf{0}'\mathbf{1}\mathbf{1}}$ known, one can then make use of Equation (18.13d) for $e_{\mathbf{0}\mathbf{1}'}{}^{\mathcal{A}} = \xi^{\mathcal{A}}$ which takes the form

$$\partial_r \xi^{\mathcal{A}} = -\Gamma_{\mathbf{0}\mathbf{1}'\mathbf{1}\mathbf{1}}\xi^{\mathcal{A}} - \bar{\Gamma}_{\mathbf{1}'\mathbf{0}\mathbf{1}'\mathbf{1}'}\bar{\xi}^{\mathcal{A}}.$$

This equation together with its complex conjugate constitute a system of ordinary differential equations for $\xi^{\mathcal{A}}$ and $\bar{\xi}^{\mathcal{A}}$ which can be solved with the information already available. To determine the frame coefficient ω one considers Equation (18.13d) for $e_{\mathbf{0}\mathbf{1}'}{}^{1} = \omega$ so that

$$\partial_r \omega = -\Gamma_{\mathbf{0}\mathbf{1}'\mathbf{1}\mathbf{1}}\omega - \bar{\Gamma}_{\mathbf{1}'\mathbf{0}\mathbf{1}'\mathbf{1}'}\bar{\omega} + \Gamma_{\mathbf{0}\mathbf{1}'\mathbf{0}\mathbf{1}} + \bar{\Gamma}_{\mathbf{1}'\mathbf{0}\mathbf{0}'\mathbf{1}'}.$$

Accordingly, one also needs to consider the equations for $\Gamma_{\mathbf{0}\mathbf{1}'\mathbf{0}\mathbf{1}}$ and $\Gamma_{\mathbf{1}\mathbf{0}'\mathbf{0}\mathbf{1}}$ (β and α in NP notation), namely,

$$\partial_r \Gamma_{\mathbf{0}\mathbf{1}'\mathbf{0}\mathbf{1}} = -\Gamma_{\mathbf{0}\mathbf{1}'\mathbf{0}\mathbf{1}}\Gamma_{\mathbf{1}\mathbf{0}'\mathbf{1}\mathbf{1}} - \Gamma_{\mathbf{1}\mathbf{0}'\mathbf{0}\mathbf{1}}\bar{\Gamma}_{\mathbf{1}'\mathbf{0}\mathbf{1}'\mathbf{1}'} + \Phi_{12},$$
$$\partial_r \Gamma_{\mathbf{1}\mathbf{0}'\mathbf{0}\mathbf{1}} = -\Gamma_{\mathbf{0}\mathbf{1}'\mathbf{0}\mathbf{1}}\Gamma_{\mathbf{1}\mathbf{0}'\mathbf{1}\mathbf{1}} - \Gamma_{\mathbf{1}\mathbf{0}'\mathbf{0}\mathbf{1}}\bar{\Gamma}_{\mathbf{0}'\mathbf{1}\mathbf{1}'\mathbf{1}'} + \Xi\phi_3,$$

so that, in addition, one requires equations for ϕ_3 and Φ_{12}. These can be found to be given by

$$\partial_r \phi_3 = \omega\partial_r\phi_4 + \xi^{\mathcal{A}}\partial_{\mathcal{A}}\phi_4 - 4\Gamma_{\mathbf{0}\mathbf{1}'\mathbf{1}\mathbf{1}}\phi_3 + 4\Gamma_{\mathbf{0}\mathbf{1}'\mathbf{0}\mathbf{1}}\phi_4,$$
$$\partial_r \Phi_{12} = \Sigma_{\mathbf{0}\mathbf{1}'}\phi_4 - \Sigma_{\mathbf{1}\mathbf{1}'}\phi_3.$$

Thus, to close the system one considers the third equation in (18.13b). The key observation is that for a given choice of ϕ_4 on $\mathscr{N}$ and with the knowledge of $\Gamma_{\mathbf{0}\mathbf{1}'\mathbf{1}\mathbf{1}}$ and $\Gamma_{\mathbf{1}\mathbf{0}'\mathbf{1}\mathbf{1}}$ from a previous integration one obtains a system of ordinary differential equations along the generators of $\mathscr{N}$ for the unknowns ω, $\xi^{\mathcal{A}}$, $\Gamma_{\mathbf{0}\mathbf{1}'\mathbf{0}\mathbf{1}}$, $\Gamma_{\mathbf{1}\mathbf{0}'\mathbf{0}\mathbf{1}}$, ϕ_3, Φ_{12} and $\Sigma_{\mathbf{0}\mathbf{1}'}$.

At this point, one considers Equation (18.13d) for $e_{\mathbf{1}\mathbf{1}'}{}^{\mathcal{A}}$. One has

$$\partial_r X^{\mathcal{A}} = -\Gamma_{\mathbf{0}\mathbf{0}'\mathbf{1}\mathbf{1}}\xi^{\mathcal{A}} - \bar{\Gamma}_{\mathbf{0}'\mathbf{0}\mathbf{1}'\mathbf{1}'}\bar{\xi}^{\mathcal{A}}.$$

Recalling the gauge condition $\Gamma_{\mathbf{0}\mathbf{0}'\mathbf{1}\mathbf{1}} = \bar{\Gamma}_{\mathbf{1}\mathbf{0}'\mathbf{1}'\mathbf{0}'} + \Gamma_{\mathbf{1}\mathbf{0}'\mathbf{1}\mathbf{0}}$ one has enough information to integrate along the generators of $\mathscr{N}$. Next, one considers the equations for $\Gamma_{\mathbf{0}\mathbf{1}'\mathbf{0}\mathbf{0}}$ and $\Gamma_{\mathbf{1}\mathbf{0}'\mathbf{0}\mathbf{0}}$ (σ and ρ in NP notation):

$$\partial_r \Gamma_{\mathbf{0}\mathbf{1}'\mathbf{0}\mathbf{0}} = -\Gamma_{\mathbf{0}\mathbf{1}'\mathbf{0}\mathbf{0}}\Gamma_{\mathbf{0}\mathbf{1}'\mathbf{1}\mathbf{1}} - \Gamma_{\mathbf{1}\mathbf{0}'\mathbf{0}\mathbf{0}}\bar{\Gamma}_{\mathbf{0}\mathbf{1}'\mathbf{1}'\mathbf{1}'} + \Phi_{02},$$
$$\partial_r \Gamma_{\mathbf{1}\mathbf{0}'\mathbf{0}\mathbf{0}} = -\Gamma_{\mathbf{0}\mathbf{1}'\mathbf{0}\mathbf{0}}\Gamma_{\mathbf{1}\mathbf{0}'\mathbf{1}\mathbf{1}} - \Gamma_{\mathbf{1}\mathbf{0}'\mathbf{0}\mathbf{0}}\bar{\Gamma}_{\mathbf{1}\mathbf{0}'\mathbf{1}'\mathbf{1}'} + \Xi\phi_2.$$

Hence, one has to couple the above to the equations for ϕ_2 and Φ_{02}:

$$\begin{aligned}\partial_r\phi_2 &= \omega\partial_r\phi_3 + \xi^{\mathcal{A}}\partial_{\mathcal{A}}\phi_3 + \Gamma_{\mathbf{0}\mathbf{1}'\mathbf{0}\mathbf{0}}\phi_4 + 2\Gamma_{\mathbf{0}\mathbf{1}'\mathbf{0}\mathbf{1}}\phi_3 - 3\Gamma_{\mathbf{0}\mathbf{1}'\mathbf{1}\mathbf{1}}\phi_2,\\ \partial_r\Phi_{02} &= \nabla_{\mathbf{1}\mathbf{0}'}\Phi_{12} + \Sigma_{\mathbf{0}\mathbf{1}'}\phi_3 - \Sigma_{\mathbf{0}\mathbf{0}'}\phi_4.\end{aligned}$$

Thus, it is then necessary to consider simultaneously the first equation in (18.13b) and Equation (18.13c) to determine $\Sigma_{\mathbf{0}\mathbf{0}'}$ and s – notice that at this stage one already knows all the frame and connection coefficients appearing in $\nabla_{\mathbf{1}\mathbf{0}'}$. In turn, this forces the coupling with the equation for Φ_{11} obtained from (18.13f):

$$\partial_r\Phi_{11} = \nabla_{\mathbf{1}\mathbf{0}'}\Phi_{12} - \Sigma_{\mathbf{1}\mathbf{1}'}\phi_2 + \Sigma_{\mathbf{0}\mathbf{1}'}\phi_3.$$

Recapitulating, one has obtained a further closed subsystem of ordinary differential equations along the generators of $\mathscr{N}$ for the fields $\Gamma_{\mathbf{0}\mathbf{1}'\mathbf{0}\mathbf{0}}$, $\Gamma_{\mathbf{1}\mathbf{0}'\mathbf{0}\mathbf{0}}$, ϕ_2, Φ_{02}, Φ_{11}, s and $\Sigma_{\mathbf{0}\mathbf{0}'}$. With the information obtained from the solution to this system, one can also solve for the frame coefficient U and the connection coefficient $\Gamma_{\mathbf{0}\mathbf{0}'\mathbf{0}\mathbf{1}}$ (ϵ in NP notation) via the equations

$$\begin{aligned}\partial_r U &= -\Gamma_{\mathbf{0}\mathbf{0}'\mathbf{1}\mathbf{1}}\omega - \bar{\Gamma}_{\mathbf{0}'\mathbf{0}\mathbf{1}'\mathbf{1}'}\bar{\omega} + \Gamma_{\mathbf{0}\mathbf{0}'\mathbf{0}\mathbf{1}} + \bar{\Gamma}_{\mathbf{0}'\mathbf{0}\mathbf{0}'\mathbf{1}'},\\ \partial_r\Gamma_{\mathbf{0}\mathbf{0}'\mathbf{0}\mathbf{1}} &= -\Gamma_{\mathbf{0}\mathbf{1}'\mathbf{0}\mathbf{1}}\Gamma_{\mathbf{0}\mathbf{0}'\mathbf{1}\mathbf{1}} - \Gamma_{\mathbf{1}\mathbf{0}'\mathbf{0}\mathbf{1}}\bar{\Gamma}_{\mathbf{0}'\mathbf{0}\mathbf{1}'\mathbf{1}'} + \Xi\phi_2 + \Phi_{11}.\end{aligned}$$

The integration of the connection coefficients can now be completed with the equation for $\Gamma_{\mathbf{0}\mathbf{0}'\mathbf{0}\mathbf{0}}$ (κ in NP notation) dictated by (18.13e), that is,

$$\partial_r\Gamma_{\mathbf{0}\mathbf{0}'\mathbf{0}\mathbf{0}} = -\Gamma_{\mathbf{0}\mathbf{1}'\mathbf{0}\mathbf{0}}\Gamma_{\mathbf{0}\mathbf{0}'\mathbf{1}\mathbf{1}} - \Gamma_{\mathbf{1}\mathbf{0}'\mathbf{0}\mathbf{0}}\bar{\Gamma}_{\mathbf{0}'\mathbf{0}\mathbf{1}'\mathbf{1}'} + \Xi\phi_1 + \Phi_{01},$$

which needs to be supplemented by the equations for ϕ_1 and Φ_{01}:

$$\begin{aligned}\partial_r\phi_1 &= \nabla_{\mathbf{0}\mathbf{1}'}\phi_2,\\ \partial_r\Phi_{01} &= \nabla_{\mathbf{1}\mathbf{0}'}\Phi_{11} - \Sigma_{\mathbf{0}\mathbf{1}'}\phi_2 + \Sigma_{\mathbf{0}\mathbf{0}'}\phi_3.\end{aligned}$$

Again, one has a subsystem of ordinary differential equations along the generators of $\mathscr{N}$. The integration of the interior equations on $\mathscr{N}$ is completed by considering the equation for the rescaled Weyl spinor component ϕ_0

$$\partial_r\phi_0 = \nabla_{\mathbf{0}\mathbf{1}'}\phi_1,$$

which, too, is an ordinary differential equation, and by that for Φ_{00}:

$$\partial_r\Phi_{00} = \nabla_{\mathbf{1}\mathbf{0}'}\Phi_{01} - \nabla_{\mathbf{0}\mathbf{1}'}\Phi_{01} + 2\Sigma_{\mathbf{0}\mathbf{0}'}\phi_2 - 2\Sigma_{\mathbf{0}\mathbf{1}'}\phi_1 + \nabla_{\mathbf{0}\mathbf{0}'}\Phi_{11}.$$

This last equation is different from the other ones in the hierarchy as its last term in the right-hand side (i.e. $\nabla_{\mathbf{0}\mathbf{0}'}\Phi_{11}$) contains transverse derivatives. However, using the evolution equations in Section 18.2.2, this term can be formally computed on $\mathscr{N}$ from the available data.

The interior equations on $\mathscr{I}^-$

On $\mathscr{I}^-$ the intrinsic equations should contain only the derivatives along the directions given by $e_{00'}$, $e_{10'}$ and $e_{01'}$. The relevant subset of (18.11a)–(18.11c) is, in this case, given by

$$\Xi_{AA'} \simeq 0, \quad Z_{AA'} \simeq 0, \quad Z_{AA'BB'} \simeq 0, \quad \text{for } AA' \neq 11', \tag{18.14a}$$

$$\Sigma_{AA'BB'} \simeq 0, \quad \Xi^{C}{}_{DAA'BB'} \simeq 0, \quad \text{for } AA',\ BB' \neq 11', \tag{18.14b}$$

$$\Delta_{0DBB'} \simeq 0, \quad \Lambda_{B0'CD} \simeq 0. \tag{18.14c}$$

More explicitly, taking into account the gauge conditions given by Lemmas 18.1 and 18.2 the above equations encode the following ***transport equations:***

$$e_{00'}(\Xi) \simeq 0, \tag{18.15a}$$

$$e_{00'}(\Sigma_{11'}) \simeq -s, \tag{18.15b}$$

$$e_{00'}(s) \simeq -\Phi_{11}\Sigma_{11'}, \tag{18.15c}$$

$$e_{00'}(e_{01'}{}^{\mu}) \simeq \Gamma_{00'}{}^{CC'}{}_{01'}e_{CC'}{}^{\mu} - \Gamma_{01'}{}^{CC'}{}_{00'}e_{CC'}{}^{\mu}, \tag{18.15d}$$

$$P_{CD00'BB'} \simeq -\Xi\phi_{BCD0}\epsilon_{0'B'} - \Phi_{C0'DB'}\epsilon_{0B} \quad BB' \neq 11', \tag{18.15e}$$

$$\nabla_{00'}\phi_{ABC1} \simeq \nabla_{10'}\phi_{ABC0}, \tag{18.15f}$$

$$\nabla_{00'}\Phi_{D1'BB'} + \nabla_{D0'}\Phi_{01'BB'} - \nabla_{01'}\Phi_{D0'BB'} - \nabla_{D1'}\Phi_{00'BB'} \simeq 2\Sigma_{1B'}\phi_{0DB1}, \tag{18.15g}$$

where, following the notation of Chapter 8, the field $P_{CDAA'BB'}$ denotes the geometric curvature. In addition to the above, Equations (18.14a)–(18.14c) also contain the ***constraint equations***

$$e_{01'}(\Xi) \simeq 0, \qquad e_{01'}(\Sigma_{11'}) \simeq 0, \qquad e_{01'}(s) \simeq -\Phi_{01}\Sigma_{11'}, \tag{18.16a}$$

$$e_{01'}(e_{10'}{}^{\mu}) - e_{10'}(e_{01'}{}^{\mu}) \simeq \Gamma_{01'}{}^{CC'}{}_{10'}e_{CC'}{}^{\mu} - \Gamma_{10'}{}^{CC'}{}_{01'}e_{CC'}{}^{\mu}, \tag{18.16b}$$

$$P_{CD01'10'} \simeq \Xi\phi_{1CD0}\epsilon_{1'0'} - \Phi_{C1'D0'}. \tag{18.16c}$$

If Equations (18.16a)–(18.16c) hold in a certain section of $\mathscr{I}^-$, then using an argument similar to that of the propagation of the constraints in the standard Cauchy problem, it can be shown that they will hold everywhere else on null infinity by virtue of the transport Equations (18.15a)–(18.15g). Thus, they need to be solved only on $\mathscr{Z}$.

In analogy to the transport equations on $\mathscr{N}$, the transport Equations (18.15a)–(18.15g) can be solved along the generators of $\mathscr{I}^-$ exploiting a hierarchical structure if some basic data are provided. Some inspection reveals that the basic data are given by either the connection coefficient $\Gamma_{10'11}$ or the rescaled Weyl spinor component ϕ_0. The details of this construction will not be further elaborated.

18.2.1 The freely specifiable data

The discussion of the hierarchical structure of the interior equations on $\mathscr{N} \cup \mathscr{I}^-$ allows the identification of the basic ***reduced initial data set*** $\mathbf{r}_\star$ from which the full initial data $\mathbf{u}_\star$ on $\mathscr{N} \cup \mathscr{I}^-$ for the conformal Einstein field equations can be computed. As already observed, the choice of reduced initial data sets is not unique. Two possible ways of specifying the reduced data are given in the following:

Lemma 18.3 (***freely specifiable data for the standard characteristic problem***) *Assume that the gauge conditions given by Lemmas 18.1 and 18.2 are satisfied in a neighbourhood $\mathscr{U}$ of $\mathscr{Z}$ on $\mathscr{N} \cup \mathscr{I}^-$. Initial data $\mathbf{u}_\star$ for the conformal Einstein field equations on $\mathscr{N} \cup \mathscr{I}^-$ can be computed from either of the two following reduced initial data sets:*

(i) $\mathbf{r}_{1\star}$ *consisting of*

$$\Gamma_{\mathbf{10'11}} \quad \text{on } \mathscr{I}^-,$$
$$\phi_4 \quad \text{on } \mathscr{N},$$
$$\phi_3, \quad \phi_2 + \bar{\phi}_2, \quad \xi^{\mathcal{A}}, \quad \text{on } \mathscr{Z};$$

(ii) $\mathbf{r}_{2\star}$ *consisting of*

$$\phi_0 \quad \text{on } \mathscr{I}^-,$$
$$\phi_4 \quad \text{on } \mathscr{N},$$
$$\Gamma_{\mathbf{10'11}}, \quad \Phi_{20}, \quad \phi_3, \quad \phi_2 + \bar{\phi}_2, \quad \xi^{\mathcal{A}}, \quad \text{on } \mathscr{Z}.$$

In both cases the field $\xi^{\mathcal{A}}$ is chosen so that $-(\xi^{\mathcal{A}}\bar{\xi}^{\mathcal{B}} + \bar{\xi}^{\mathcal{A}}\xi^{\mathcal{B}})\boldsymbol{\partial}_{\mathcal{A}} \otimes \boldsymbol{\partial}_{\mathcal{B}}$ is conformal to the standard (contravariant) metric on $\mathbb{S}^2$.

Remark. The reduced set $\mathbf{r}_{2\star}$ in (ii) has the advantage of being symmetric with respect to $\mathscr{N}$ and $\mathscr{I}^-$.

Proof The proof of this lemma follows from the discussion in the previous subsection. Further discussion can be found in Friedrich (1981a). □

18.2.2 The reduced conformal field equations

To apply the theory on the characteristic initial value problem discussed in Section 12.5 one has to extract a suitable symmetric hyperbolic system out of the conformal field Equations (18.11a)–(18.11c). Given the split between intrinsic and transverse equations, a hyperbolic reduction procedure such as the one discussed in Chapter 13 is not required. Instead, a suitable choice of ***reduced conformal field equations*** is given by the combinations

$$\Xi_{\mathbf{11'}} = 0, \qquad Z_{\mathbf{11'}} = 0, \qquad Z_{\mathbf{11'AB'}} = 0, \tag{18.17a}$$
$$\Sigma_{\mathbf{11'BB'}} = 0, \qquad \Xi^{\boldsymbol{C}}{}_{\mathbf{D11'BB'}} = 0, \tag{18.17b}$$

$$-\Delta_{\mathbf{1}BC\mathbf{0}'} = 0, \quad \Delta_{\mathbf{0}BC\mathbf{0}'} - \Delta_{\mathbf{1}BC\mathbf{1}'} = 0, \quad \Delta_{\mathbf{0}BC'\mathbf{1}'} = 0, \tag{18.17c}$$

$$-\Lambda_{\mathbf{0}\mathbf{1}'\mathbf{0}\mathbf{0}} = 0, \quad \Lambda_{\mathbf{0}\mathbf{0}'BC} - \Lambda_{\mathbf{1}\mathbf{1}'BC} = 0, \quad \Lambda_{\mathbf{1}\mathbf{0}'\mathbf{1}\mathbf{1}} = 0. \tag{18.17d}$$

A more explicit form of the equations is discussed in Section 18.3. From these expressions, adopting the matricial notation of Chapter 12 and considering *suitable multiples* of the equations, the reduced conformal field equations can be written schematically in the form

$$\mathbf{A}^\mu(x,\mathbf{u})\partial_\mu \mathbf{u} + \mathbf{B}(x,\mathbf{u}) = 0, \tag{18.18}$$

with $\mathbf{A}^\mu$ Hermitian matrices and

$$\mathbf{A}^\mu(\omega^{\mathbf{0}\mathbf{0}'}{}_\mu + \omega^{\mathbf{1}\mathbf{1}'}{}_\mu) \qquad \text{positive definite.} \tag{18.19}$$

Thus, one obtains a symmetric hyperbolic system for the components of $\mathbf{u}$. Using the expressions for the principal part of the system (18.17a)–(18.17d), a computation shows that the characteristic polynomial of the reduced system contains factors of the form $g^{\mu\nu}\xi_\mu\xi_\nu$ so that the null hypersurfaces $\mathscr{N}$ and $\mathscr{I}^-$ are indeed characteristics of the system. It follows from (18.19) that the surfaces with normal $\boldsymbol{\omega}^{\mathbf{0}\mathbf{0}'} + \boldsymbol{\omega}^{\mathbf{1}\mathbf{1}'}$ are spacelike for the symmetric hyperbolic system. Although the coordinates $x^0 = v$ and $x^1 = r$ have been constructed so that they have non-negative values, the reduced Equations (18.17a)–(18.17d) also hold for negative values of the coordinates. It follows that the hypersurface

$$\mathcal{S}_\star \equiv \{p \in \mathbb{R} \times \mathbb{R} \times \mathbb{S}^2 \mid x^0(p) + x^1(p) = 0\} \tag{18.20}$$

is spacelike for Equation (18.18) in a neighbourhood of $\mathscr{Z}$.

18.3 A local existence result for characteristic problems

As discussed in Section 12.5, the existence and uniqueness of solutions to a characteristic initial value problem can be obtained via an *auxiliary Cauchy initial value problem* on a spacelike hypersurface – in the present case the hypersurface $\mathcal{S}_\star$ defined by (18.20). The formulation of this auxiliary Cauchy problem crucially depends on Whitney's extension theorem so that initial data on $\mathscr{N} \cup \mathscr{I}^-$ can be extended to a spacetime neighbourhood $\mathcal{U}$ of $\mathscr{Z}$. In turn, the application of Whitney's theorem depends on being able to evaluate all (interior and transverse) derivatives of the initial data on $\mathscr{N} \cup \mathscr{I}^-$.

18.3.1 Computation of the formal derivatives on $\mathscr{N} \cup \mathscr{I}^-$

To verify that one can compute all derivatives of the initial data on $\mathscr{N} \cup \mathscr{I}^-$ one needs to inspect the principal part of the reduced Equations (18.17a)–(18.17d).

Borrowing the notation of Proposition 13.1, the reduced Equations (18.17a)–(18.17b) take the form

$$\partial_r \boldsymbol{\sigma} = \mathbf{G}(\boldsymbol{\sigma}, \boldsymbol{\Gamma}, \boldsymbol{\Phi}, \boldsymbol{\phi}), \tag{18.21a}$$

$$\partial_r \boldsymbol{e} = \mathbf{H}(\boldsymbol{e}, \boldsymbol{\Gamma}), \tag{18.21b}$$

$$\partial_r \boldsymbol{\Gamma} = \mathbf{K}(\boldsymbol{\Gamma}, \boldsymbol{\Phi}, \boldsymbol{\phi}); \tag{18.21c}$$

that is, they are transport equations along the direction given by $\boldsymbol{e}_{00'}$. For the equations in (18.17c) one has

$$\partial_r \Phi_{20} - \bar{\omega}\partial_r \Phi_{21} - \bar{\xi}^{\mathcal{A}}\partial_{\mathcal{A}}\Phi_{21} = L_{20}(\boldsymbol{\Gamma}, \boldsymbol{\Phi}, \boldsymbol{\phi}), \tag{18.22a}$$

$$\partial_r \Phi_{10} - \bar{\omega}\partial_r \Phi_{11} - \bar{\xi}^{\mathcal{A}}\partial_{\mathcal{A}}\Phi_{11} = L_{10}(\boldsymbol{\Gamma}, \boldsymbol{\Phi}, \boldsymbol{\phi}), \tag{18.22b}$$

$$\partial_r \Phi_{00} - \bar{\omega}\partial_r \Phi_{01} - \bar{\xi}^{\mathcal{A}}\partial_{\mathcal{A}}\Phi_{01} = L_{00}(\boldsymbol{\Gamma}, \boldsymbol{\Phi}, \boldsymbol{\phi}), \tag{18.22c}$$

$$\begin{aligned}&\partial_r \Phi_{21} + \partial_v \Phi_{21} + U\partial_r \Phi_{21} + X^{\mathcal{A}}\partial_{\mathcal{A}}\Phi_{21}\\ &\quad - \omega\partial_r \Phi_{20} - \xi^{\mathcal{A}}\partial_{\mathcal{A}}\Phi_{20} - \bar{\omega}\partial_r \Phi_{22} - \bar{\xi}^{\mathcal{A}}\partial_{\mathcal{A}}\Phi_{22} = M_{21}(\boldsymbol{\Gamma}, \boldsymbol{\Phi}, \boldsymbol{\phi}),\end{aligned} \tag{18.22d}$$

$$\begin{aligned}&\partial_r \Phi_{11} + \partial_v \Phi_{11} + U\partial_r \Phi_{11} + X^{\mathcal{A}}\partial_{\mathcal{A}}\Phi_{11}\\ &\quad - \omega\partial_r \Phi_{10} - \xi^{\mathcal{A}}\partial_{\mathcal{A}}\Phi_{10} - \bar{\omega}\partial_r \Phi_{12} - \bar{\xi}^{\mathcal{A}}\partial_{\mathcal{A}}\Phi_{12} = M_{11}(\boldsymbol{\Gamma}, \boldsymbol{\Phi}, \boldsymbol{\phi}),\end{aligned} \tag{18.22e}$$

$$\begin{aligned}&\partial_r \Phi_{01} + \partial_v \Phi_{01} + U\partial_r \Phi_{01} + X^{\mathcal{A}}\partial_{\mathcal{A}}\Phi_{01}\\ &\quad - \omega\partial_r \Phi_{00} - \xi^{\mathcal{A}}\partial_{\mathcal{A}}\Phi_{00} - \bar{\omega}\partial_r \Phi_{02} - \bar{\xi}^{\mathcal{A}}\partial_{\mathcal{A}}\Phi_{02} = M_{01}(\boldsymbol{\Gamma}, \boldsymbol{\Phi}, \boldsymbol{\phi}),\end{aligned} \tag{18.22f}$$

$$\partial_v \Phi_{22} + U\partial_r \Phi_{22} + X^{\mathcal{A}}\partial_{\mathcal{A}}\Phi_{22} - \omega\partial_r \Phi_{21} - \xi^{\mathcal{A}}\partial_{\mathcal{A}}\Phi_{21} = N_{22}(\boldsymbol{\Gamma}, \boldsymbol{\Phi}, \boldsymbol{\phi}), \tag{18.22g}$$

$$\partial_v \Phi_{12} + U\partial_r \Phi_{12} + X^{\mathcal{A}}\partial_{\mathcal{A}}\Phi_{12} - \omega\partial_r \Phi_{11} - \bar{\xi}^{\mathcal{A}}\partial_{\mathcal{A}}\Phi_{11} = N_{12}(\boldsymbol{\Gamma}, \boldsymbol{\Phi}, \boldsymbol{\phi}), \tag{18.22h}$$

$$\partial_v \Phi_{02} + U\partial_r \Phi_{02} + X^{\mathcal{A}}\partial_{\mathcal{A}}\Phi_{02} - \bar{\omega}\partial_r \Phi_{01} - \bar{\xi}^{\mathcal{A}}\partial_{\mathcal{A}}\Phi_{01} = N_{02}(\boldsymbol{\Gamma}, \boldsymbol{\Phi}, \boldsymbol{\phi}), \tag{18.22i}$$

where L_{20}, L_{10}, L_{00}, M_{21}, M_{11}, M_{01}, N_{22}, N_{12} and N_{02} are smooth functions of their arguments – their explicit form will not be required. Finally, for the Equations (18.17d) involving the components of the rescaled Weyl tensor one has

$$\partial_r \phi_0 - \omega\partial_r \phi_1 - \xi^{\mathcal{A}}\partial_{\mathcal{A}}\phi_1 = W_0(\boldsymbol{\Gamma}, \boldsymbol{\phi}), \tag{18.23a}$$

$$\partial_r \phi_1 + \partial_v \phi_1 + U\partial_r \phi_1 + X^{\mathcal{A}}\partial_{\mathcal{A}}\phi_1 \tag{18.23b}$$

$$- \bar{\omega}\partial_r \phi_0 - \bar{\xi}^{\mathcal{A}}\partial_{\mathcal{A}}\phi_0 - \omega\partial_r \phi_2 - \xi^{\mathcal{A}}\partial_{\mathcal{A}}\phi_2 = W_1(\boldsymbol{\Gamma}, \boldsymbol{\phi}), \tag{18.23c}$$

$$\partial_r \phi_2 + \partial_v \phi_2 + U\partial_r \phi_2 + X^{\mathcal{A}}\partial_{\mathcal{A}}\phi_2 \tag{18.23d}$$

$$- \bar{\omega}\partial_r \phi_1 - \bar{\xi}^{\mathcal{A}}\partial_{\mathcal{A}}\phi_1 - \omega\partial_r \phi_3 - \xi^{\mathcal{A}}\partial_{\mathcal{A}}\phi_3 = W_2(\boldsymbol{\Gamma}, \boldsymbol{\phi}), \tag{18.23e}$$

$$\partial_r \phi_3 + \partial_v \phi_3 + U\partial_r \phi_3 + X^{\mathcal{A}}\partial_{\mathcal{A}}\phi_3 \tag{18.23f}$$

$$- \bar{\omega}\partial_r \phi_2 - \bar{\xi}^{\mathcal{A}}\partial_{\mathcal{A}}\phi_2 - \omega\partial_r \phi_4 - \xi^{\mathcal{A}}\partial_{\mathcal{A}}\phi_4 = W_3(\boldsymbol{\Gamma}, \boldsymbol{\phi}), \tag{18.23g}$$

$$\partial_v \phi_4 + U\partial_r \phi_4 + X^{\mathcal{A}}\partial_{\mathcal{A}}\phi_4 - \bar{\omega}\partial_r \phi_3 - \bar{\xi}^{\mathcal{A}}\partial_{\mathcal{A}}\phi_3 = W_4(\boldsymbol{\Gamma}, \boldsymbol{\phi}), \tag{18.23h}$$

with W_0, W_1, W_2, W_3 and W_4 smooth functions of their arguments – again, their explicit form will not be required.

In what follows, it is shown that all *formal partial derivatives* on $\mathscr{N} \cup \mathscr{I}^-$ can indeed be computed from the above equations.

Computation of formal derivatives on $\mathscr{I}^-$

To compute the formal derivatives on $\mathscr{I}^-$ one first observes that the partial derivatives ∂_v, ∂_2, ∂_3 are interior, while ∂_r is transverse. In this case, direct inspection shows that except for

$$\partial_r\phi_4, \quad \partial_r\Phi_{22}, \quad \partial_r\Phi_{12}, \quad \partial_r\Phi_{02},$$

all ∂_r-derivatives of the unknown $\mathbf{u}$ can be computed using Equations (18.21a)–(18.21c), (18.22a)–(18.22f) and (18.23a)–(18.23g). The exceptional cases shown above arise due to the fact that $\omega = U = 0$ on $\mathscr{I}^-$ so that Equations (18.22g)–(18.22i) and (18.23h) evaluated at $\mathscr{I}^-$ do not, in fact, contain ∂_r-derivatives. To get around this problem one computes the ∂_r-derivative of (18.22g)–(18.22i) and (18.23h) and then evaluates on $\mathscr{I}^-$ to obtain the system

$$\begin{aligned}
&\partial_v(\partial_r\Phi_{22}) + \partial_r U\partial_r\Phi_{22} + \partial_r X^{\mathcal{A}}\partial_{\mathcal{A}}\Phi_{22} - \partial_r\omega\partial_r\Phi_{12}\\
&\qquad\qquad - \partial_r\xi^{\mathcal{A}}\partial_{\mathcal{A}}\Phi_{12} - \xi^{\mathcal{A}}\partial_{\mathcal{A}}\partial_r\Phi_{12} \simeq \partial_r N_{22},\\
&\partial_v(\partial_r\Phi_{12}) + \partial_r U\partial_r\Phi_{12} + \partial_r X^{\mathcal{A}}\partial_{\mathcal{A}}\Phi_{12} - \partial_r\omega\partial_r\Phi_{11}\\
&\qquad\qquad - \partial_r\xi^{\mathcal{A}}\partial_{\mathcal{A}}\Phi_{11} - \xi^{\mathcal{A}}\partial_{\mathcal{A}}\partial_r\Phi_{11} \simeq \partial_r N_{12},\\
&\partial_v(\partial_r\Phi_{02}) + \partial_r U\partial_r\Phi_{02} + \partial_r X^{\mathcal{A}}\partial_{\mathcal{A}}\Phi_{02} - \partial_r\omega\partial_r\Phi_{01}\\
&\qquad\qquad - \partial_r\xi^{\mathcal{A}}\partial_{\mathcal{A}}\Phi_{01} - \xi^{\mathcal{A}}\partial_{\mathcal{A}}\partial_r\Phi_{01} \simeq \partial_r N_{02},\\
&\partial_v(\partial_r\phi_4) + \partial_r U\partial_r\phi_4 + \partial_r X^{\mathcal{A}}\partial_{\mathcal{A}}\phi_4 - \partial_r\bar{\omega}\partial_r\phi_3\\
&\qquad\qquad - \partial_r\bar{\xi}^{\mathcal{A}}\partial_{\mathcal{A}}\phi_3 - \bar{\xi}^{\mathcal{A}}\partial_{\mathcal{A}}\partial_r\phi_3 \simeq \partial_r W_4.
\end{aligned}$$

The latter can be interpreted as a system of first-order linear ordinary differential equations for $\partial_r\phi_4$, $\partial_r\Phi_{22}$, $\partial_r\Phi_{12}$, $\partial_r\Phi_{02}$. The initial data on $\mathscr{Z}$ for these equations can be computed from the data on $\mathscr{N} \cup \mathscr{I}^-$. General results of the theory of ordinary differential equations ensures that this system of equations can be solved in a neighbourhood of $\mathscr{Z}$ on $\mathscr{I}^-$. Accordingly, all the first transverse derivatives on $\mathscr{I}^-$ can be explicitly computed. The argument described in this paragraph can be generalised, by repeatedly differentiating the reduced equations with respect to ∂_r, to iteratively compute higher order ∂_r-derivatives as the solution to a system of algebraic equations and linear PDEs.

Computation of formal derivatives on $\mathscr{N}$

The analysis of the formal derivatives on $\mathscr{N}$ is almost the mirror image of that on $\mathscr{I}^-$. In this case ∂_r, ∂_2, ∂_3 are interior derivatives, while ∂_v is transverse. After an inspection of the list of Equations (18.21a)–(18.21c), (18.22a)–(18.22i) and (18.23a)–(18.23h) one finds that *only*

$$\partial_v\phi_4, \quad \partial_v\phi_3, \quad \partial_v\phi_2, \quad \partial_v\phi_1,$$
$$\partial_v\Phi_{22}, \quad \partial_v\Phi_{12}, \quad \partial_v\Phi_{11}, \quad \partial_v\Phi_{02}, \quad \partial_v\Phi_{01}$$

are algebraically determined by the initial data on $\mathscr{N}$. To obtain the remaining transverse derivatives, one computes the ∂_v-derivatives of Equations (18.21a)–(18.21c), (18.22a)–(18.22c) and (18.23a) and evaluates them on $\mathscr{N}$ to obtain a first-order system of ordinary differential equations along the generators of $\mathscr{N}$ for

$$\partial_v \boldsymbol{\sigma}, \quad \partial_v \boldsymbol{e}, \quad \partial_v \boldsymbol{\Gamma}, \quad \partial_v \Phi_{02}, \quad \partial_v \Phi_{01}, \quad \partial_v \Phi_{00}, \quad \partial_v \phi_0.$$

Supplementing this system with the information on $\mathscr{Z}$ implied by the initial data for the reduced equations, one finds that the general theory of ordinary differential equations ensures the existence of solutions in a neighbourhood of $\mathscr{Z}$ on $\mathscr{N}$. In this manner one obtains a complete set of first-order transverse derivatives on $\mathscr{N}$. Higher order transverse derivatives can be obtained iteratively by computing higher order ∂_v-derivatives of the reduced conformal field equations as required.

The analysis described in the previous paragraphs can be summarised in the following:

Lemma 18.4 (***computation of formal derivatives***) *Any arbitrary formal derivatives* $(\partial^{\alpha}\mathbf{u})_\star$ *of the vector unknown* $\mathbf{u}$ *on* $\mathscr{N} \cup \mathscr{I}^-$ *can be computed from the prescribed initial data* $\mathbf{u}_\star$ *for the reduced conformal field equations on* $\mathscr{N} \cup \mathscr{I}^-$.

18.3.2 The subsidiary system

To show that the solutions of the reduced equations imply a solution to the full conformal field equations if initial data satisfying the constraints on $\mathscr{N}$ and $\mathscr{I}^-$ are prescribed, it is necessary to obtain a suitable subsidiary system for the zero quantities encoding the conformal field equations. The *propagation of the constraints* is ensured by the following:

Proposition 18.1 (***propagation of the constraints***) *A solution* $\mathbf{u}$ *of the reduced conformal field Equations* (18.17a)–(18.17d) *on a neighbourhood* $\mathcal{U}$ *of* $\mathscr{Z}$ *on* $J^+(\mathscr{Z})$ *that coincides with initial data on* $\mathscr{N} \cup \mathscr{I}^-$ *satisfying the conformal equations is a solution to the conformal field Equations* (18.11a)–(18.11c) *on* $\mathcal{U}$.

A subsidiary system adapted to the geometry of the characteristic problem described in the previous sections is obtained from the following derivatives of the zero quantities associated to the conformal field equations:

$$\nabla_{\mathbf{11'}}\Xi_{\boldsymbol{AA'}}, \qquad \nabla_{\mathbf{11'}}Z_{\boldsymbol{AA'}}, \qquad \nabla_{\mathbf{11'}}Z_{\boldsymbol{AA'BB'}},$$

$$\nabla_{\mathbf{11'}}\Sigma_{\boldsymbol{AA'BB'}}, \qquad \nabla_{\mathbf{11'}}\Xi_{\boldsymbol{CDAA'BB'}}$$

$$(\nabla_{\mathbf{00'}} + \nabla_{\mathbf{11'}})\Delta_{\boldsymbol{CDBB'}}, \qquad (\nabla_{\mathbf{00'}} + \nabla_{\mathbf{11'}})\Lambda_{\boldsymbol{BB'CD}}.$$

Using arguments similar to those employed in Sections 13.3 and 13.4.5 one rewrites the above derivatives as homogeneous expressions in the zero quantities. Further details of these lengthy calculations can be found in Friedrich (1981a).

Once a subsidiary system of the required form has been obtained, the propagation of the constraints follows from the uniqueness of solutions to the characteristic problem.

In addition to Proposition 18.1 one has the following:

Corollary 18.1 (***preservation of the conformal gauge***) *Let* $\mathbf{u}$ *denote a solution to the characteristic problem for the conformal field equations on a neighbourhood* $\mathcal{U}$ *of* $\mathscr{Z}$ *on* $J^+(\mathscr{Z})$ *which satisfies the gauge conditions given in Lemmas 18.1 and 18.2. Then the metric* $\boldsymbol{g}$ *constructed from the components of the solution* $\mathbf{u}$ *satisfies the vacuum Einstein field equations* $R[\boldsymbol{g}] = 0$.

This result follows from an argument similar to the one used to prove the propagation of the *algebraic* conformal field equation encoding the transformation rule for the Ricci scalar in Lemma 8.1. Here one considers the derivative

$$\nabla_{\boldsymbol{11}'}(\Xi\nabla^{\boldsymbol{AA}'}\nabla_{\boldsymbol{AA}'}\Xi - 2\nabla_{\boldsymbol{AA}'}\Xi\nabla^{\boldsymbol{AA}'}\Xi)$$

and makes use of the conformal field equations to rewrite it as a homogeneous expression in zero quantities. In view of the transformation law of the Ricci scalar under conformal rescalings, the term in brackets coincides with $R[\boldsymbol{g}]$. Now, from the discussion leading to Lemma 18.2 one concludes that $R[\boldsymbol{g}] = 0$ on $\mathscr{N} \cup \mathscr{I}^-$. The corollary then follows from the uniqueness of the solutions to the characteristic problem.

18.3.3 The existence result

Combining the analysis developed in the previous subsections with the theory of characteristic initial value problems for symmetric hyperbolic systems of Section 12.5, one obtains the following existence result:

Theorem 18.1 (***existence and uniqueness to the standard asymptotic characteristic problem***) *Given a smooth reduced initial data set* $\mathbf{r}_\star$ *for the conformal Einstein field equations on* $\mathscr{N} \cup \mathscr{I}^-$, *there exists a unique smooth solution of the conformal field equations in a neighbourhood* $\mathcal{U}$ *of* $\mathscr{Z}$ *in* $J^+(\mathscr{Z})$ *which implies the prescribed initial data on* $\mathscr{N} \cup \mathscr{I}^-$.

Proof It follows from Lemma 18.4 that the formal derivatives of $\mathbf{u}$ can be computed to any arbitrary order from the reduced data $\mathbf{r}_\star$ on $\mathscr{N} \cup \mathscr{I}^-$. Hence, it is possible to formulate an auxiliary Cauchy problem for the reduced conformal field Equations (18.17a)–(18.17d) with data implied by the extension to a neighbourhood of $\mathscr{Z}$ given by Whitney's theorem. Thus, using Theorem 12.7 and the discussion in Section 12.5.3 there is a neighbourhood $\mathcal{W}$ of $\mathscr{Z}$ in $J^+(\mathscr{Z})$ in which there exists a unique solution $\mathbf{u}$ to the reduced conformal field equations which on $\mathscr{N} \cup \mathscr{I}^-$ coincides with the data $\mathbf{u}_\star$ implied by the prescribed reduced initial data – as $\mathscr{Z} \approx \mathbb{S}^2$, it is necessary to combine solutions in two different patches. Finally Proposition 18.1 and Corollary 18.1 imply that the solution to the reduced equations is, in fact, a solution to the full conformal field equations. □

The characteristic problem on $\mathscr{N}' \cup \mathscr{I}^+$

The analysis leading to Theorem 18.1 can be adapted to analyse the dual asymptotic characteristic problem with data on $\mathscr{N}' \cup \mathscr{I}^+$ where $\mathscr{N}'$ is a future-oriented null hypersurface. In this case one endeavours to find a solution in a neighbourhood $\mathcal{U}'$ of $\mathscr{Z}' = \mathscr{N}' \cap \mathscr{I}^+$ in $J^-(\mathscr{Z}')$. All the relevant expressions can be obtained from those for the characteristic problem on $\mathscr{N} \cup \mathscr{I}^-$ through the replacements $_\mathbf{0} \mapsto {_\mathbf{1}}$, $_\mathbf{1} \mapsto {_\mathbf{0}}$ in the spinorial frame indices so that

$$\boldsymbol{e}_{\mathbf{00'}} \mapsto \boldsymbol{e}_{\mathbf{11'}}, \quad \boldsymbol{e}_{\mathbf{11'}} \mapsto \boldsymbol{e}_{\mathbf{00'}}, \quad \boldsymbol{e}_{\mathbf{01'}} \mapsto \boldsymbol{e}_{\mathbf{10'}}, \quad \boldsymbol{e}_{\mathbf{10'}} \mapsto \boldsymbol{e}_{\mathbf{01'}}.$$

In particular, one has

$$\phi_0 \mapsto \phi_4, \quad \phi_1 \mapsto \phi_3, \quad \phi_2 \mapsto \phi_2, \quad \phi_3 \mapsto \phi_1, \quad \phi_4 \mapsto \phi_0$$

and

$$\omega \mapsto \bar{\omega}, \qquad \xi^{\mathcal{A}} \mapsto \bar{\xi}^{\mathcal{A}}.$$

Similarly, for the connection coefficients and the components of the trace-free Ricci spinor one has

$$\Gamma_{\mathbf{01'00}} \mapsto \Gamma_{\mathbf{10'11}}, \qquad \Phi_{12} \mapsto \Phi_{10} = \Phi_{01}, \qquad \text{and so on.}$$

For consistency, one should replace the coordinate v along the generators of $\mathscr{I}^-$ with a coordinate u along the generators of $\mathscr{I}^+$.

18.4 The asymptotic characteristic problem on a cone

As discussed in the introduction, an alternative characteristic problem for the conformal Einstein field equations consists of a configuration where initial data is prescribed in a neighbourhood of the vertex of a cone representing the timelike infinity of a Minkowski-like spacetime; see Figure 18.1, right. This type of geometric setup for a characteristic initial value problem was originally introduced in Friedrich (1986c) and is intended to model ***purely radiative spacetimes***, that is, a system describing gravitational radiation from past null infinity which interacts non-linearly with itself and eventually escapes to future null infinity. Intuitively, one would expect this type of solution to the Einstein field equations to have a smooth structure at null infinity. To ensure that the gravitational field consists only of gravitational radiation one requires that the generators of null infinity are complete and that past timelike infinity is represented by a point i^- which is *regular from the point of view of the conformal completion.*

To discuss the geometric setting in a more precise manner it is convenient to introduce some definitions.

Definition 18.1 (***spacetimes with a cone past boundary***) *A spacetime $(\mathcal{M}, \boldsymbol{g})$ is said to have a* **cone past boundary** *if:*

(i) There exists a causal, oriented and time-oriented spacetime $(\mathcal{M}', \boldsymbol{g}')$ (the **ambient manifold***).*
(ii) There exists a point $o \in \mathcal{M}'$ such that the set consisting of o and all points of $\mathcal{M}'$ which can be joined to o by a causal curve in $\mathcal{M}'$ – to be denoted by $J^+(o, \mathcal{M}')$ – is closed in $\mathcal{M}'$.
(iii) Given $\mathscr{N}_o \equiv \partial J^+(o, \mathcal{M}')$, then $\mathscr{N}_o \setminus \{o\}$ is a smooth null hypersurface of $\mathcal{M}'$.
(iv) The set $\mathcal{M}$ corresponds to $J^+(o, \mathcal{M}')$ together with the structures it inherits from $(\mathcal{M}', \boldsymbol{g}')$ – in particular, $\boldsymbol{g}$ is the pull-back of $\boldsymbol{g}'$ to $\mathcal{M}$.

Given $p \in \mathcal{M}$, the set $\mathscr{N}_p \subset \mathcal{M}$ is called the **future null cone of** *p.*

In terms of the above notions one introduces the further notion:

Definition 18.2 (***spacetimes with a complete past null infinity cone***) *A vacuum spacetime $(\tilde{\mathcal{M}}, \tilde{\boldsymbol{g}})$ is said to be a* **solution to the Einstein field equations with complete null cone at past timelike infinity** *i^- if there exists a conformal extension $(\mathcal{M}, \boldsymbol{g}, \Xi)$ with cone-like past boundary $\mathscr{N}_{i^-}$ such that the conformal factor satisfies*

$$\Xi > 0 \quad \text{on } \mathcal{M} \setminus \mathscr{N}_{i^-}, \tag{18.24a}$$

$$\Xi = 0 \quad \text{on } \mathscr{N}_{i^-}, \tag{18.24b}$$

$$\mathbf{d}\Xi \neq 0 \quad \text{on } \mathscr{N}_{i^-} \setminus \{i^-\}, \tag{18.24c}$$

$$\mathbf{d}\Xi = 0, \quad \boldsymbol{Hess}\,\Xi \text{ non-degenerate at } i^-, \tag{18.24d}$$

and there is a diffeomorphism by means of which the manifolds $\tilde{\mathcal{M}}$ and $\mathcal{M} \setminus \mathscr{N}_{i^-}$ can be identified so that $\boldsymbol{g} = \Xi^2 \tilde{\boldsymbol{g}}$ on $\mathcal{M} \setminus \mathscr{N}_{i^-}$. The set $\mathscr{N}_{i^-} \setminus \{i^-\}$ is swept by the future-directed null geodesics through i^- and represents the past null infinity $\mathscr{I}^-$ of the spacetime.

Equipped with the above definitions, one can formulate a ***pure radiation problem*** in which one asks: *given data on a cone $\mathscr{N}_o$, is there a unique solution to the Einstein field equations with complete past null infinity implying fields on $\mathscr{I}^-$ which can be identified with the data prescribed on $\mathscr{N}_o$ and such that the point o can be identified with i^-?*

18.4.1 Gauge conditions

This section gives a brief discussion of the gauge specification process for the characteristic initial value problem on a cone. As is the case in all initial value problems concerning the conformal field equations, one has to consider three different types of gauges: conformal, coordinate and frame gauges. These are analysed in turn.

The conformal gauge

Given a null cone $\mathscr{N}_o$ with vertex o, let $\boldsymbol{l}$ denote the vector tangent to the null generators of $\mathscr{N}_o$. Consistent with conditions (18.24a)–(18.24d), it is assumed that one has a conformal factor Ξ such that

$$\Xi = 0, \qquad \mathbf{d}\Xi = 0, \qquad s \neq 0 \qquad \text{at } o.$$

Mimicking the discussion of Section 16.3, one can transvect the conformal field equations

$$\nabla_a \nabla_b \Xi = -\Xi L_{ab} + s g_{ab}, \qquad \nabla_a s = -\nabla^b \Xi L_{ba}, \tag{18.25}$$

with $\boldsymbol{l}$ to find that $\Xi = 0$ and $s \neq 0$ on $\mathscr{N}_o$ and, moreover, that $\mathbf{d}\Xi \neq 0$ on $\mathscr{N}_o \setminus \{o\}$. It is also observed that if $s|_o = 0$, then $\mathbf{d}\Xi = 0$ on $\mathscr{N}_o$. The behaviour of the conformal gauge at o can be refined by considering a rescaling as in Equation (18.6) with $\vartheta > 0$. Making use of the transformation formula for the Friedrich scalar s, Equation (8.29b), one finds that $s'|_o = (s\vartheta^{-1})|_o$. Let $\gamma(\varsigma)$ with $\varsigma \in \mathbb{R}$ denote a future-directed null geodesic on $\mathscr{N}_o$ with $\gamma(0) = o$ such that $\boldsymbol{l} = \dot{\gamma}$ and, consequently, $\nabla_{\boldsymbol{l}} \boldsymbol{l} = 0$. Setting $\boldsymbol{l}' \equiv \vartheta^{-1}\boldsymbol{l}$, one finds that $\boldsymbol{g}'(\boldsymbol{l}', \boldsymbol{l}') = 0$ and $\nabla'_{\boldsymbol{l}'} \boldsymbol{l}' = 0$ as well. Using the transformation formula for the trace-free Ricci tensor Φ_{ab}, Equation (18.8), one finds that along γ it holds that

$$\vartheta^3 l'^a l'^b \Phi'_{ab} = \vartheta^{-1} l^a l^b \Phi_{ab} + 2 l^b \nabla_b \big(l^a \nabla_a (\vartheta^{-1}) \big).$$

Thus, if the value of the component $l'^a l'^b \Phi'_{ab}$ is prescribed, the above equation can be read as an ordinary differential equation for ϑ along the null geodesic γ. The initial value of ϑ can be fixed through the specification of $s|_o$. Using the first of the equations in (18.25) one finds that

$$s|_o \boldsymbol{g}'(\boldsymbol{l}', \boldsymbol{l}')|_o = \nabla_{l'} \nabla_{l'} \Xi'|_o.$$

In order to have a local minimum of Ξ at o, one needs that $\nabla_{l'} \nabla_{l'} \Xi'|_o > 0$ forcing $s|_o > 0$ – in the signature $(+---)$. Without loss of generality, one can then set

$$s = 2 \qquad \text{at } o, \tag{18.26}$$

and

$$l^a l^b \Phi_{ab} = 0 \qquad \text{on } \mathscr{N}_o \text{ near } o. \tag{18.27}$$

In this construction there is still the freedom of specifying the value of $\mathbf{d}\vartheta$ at o. Adapting the arguments of Section 18.1.2 one can set a characteristic initial value problem on $\mathscr{N}_o$ for the wave Equation (18.7) in such a way that

$$R[\boldsymbol{g}] = 0 \qquad \text{on } J^+(\mathscr{N}_o) \text{ near } o. \tag{18.28}$$

The coordinates and the frame near o

A convenient four-dimensional description of the null cone $\mathscr{N}_o$ is obtained using $\boldsymbol{g}$-normal coordinates $y = (y^\mu)$ centred at o; see Sections 2.4.5 and 11.6.2. Accordingly, one has that $y^\mu(o)=0$, $g_{\mu\nu}(o)=\eta_{\mu\nu}$, $\partial_\lambda g_{\mu\nu}(o)=0$ and $\Gamma_\mu{}^\nu{}_\lambda(o)=0$. These properties can be more concisely summarised in the expression

$$y^\mu g_{\mu\nu} = y^\mu \eta_{\mu\nu} \qquad \text{in a neighbourhood of } o. \tag{18.29}$$

In these coordinates, for fixed $(y^\mu) \neq 0$ one has that the curve $\gamma : \varsigma \to \varsigma y^\mu$ is a geodesic through o and that

$$\mathscr{N}_o = \{y^\mu \in \mathbb{R}^4 \,|\, \eta_{\mu\nu} y^\mu y^\nu = 0,\ y^0 \geq 0\}.$$

Thus, in *these coordinates the null cone $\mathscr{N}_o$ can be thought of as being the null cone through the origin in Minkowski spacetime.*

Associated to the $\boldsymbol{g}$-normal coordinates, it is natural to consider a ***normal frame centred at*** o, that is, a frame $\{\boldsymbol{e_a}\}$ which, in a neighbourhood $\mathcal{U}$ of o, satisfies $\boldsymbol{g}(\boldsymbol{e_a}, \boldsymbol{e_b}) = \eta_{\boldsymbol{ab}}$ and $\nabla_{\dot\gamma}\boldsymbol{e_a} = 0$ for any geodesic passing through o. Without loss of generality, one can assume that the frame coefficients in $\boldsymbol{e_a} = e_{\boldsymbol{a}}{}^\mu \boldsymbol{\partial}_\mu$ satisfy $e_{\boldsymbol{a}}{}^\mu(o) = \delta_{\boldsymbol{a}}{}^\mu$. Using the properties of the exponential function, it can be shown that the frame coefficients $e_{\boldsymbol{a}}{}^\mu$ depend smoothly on the coordinates (y^μ). It can then be verified that $\boldsymbol{g}(\dot\gamma, \boldsymbol{e_a})$ is constant along γ. Moreover, using that $g_{\mu\nu} = \eta_{\boldsymbol{ab}}\omega^{\boldsymbol{a}}{}_\mu \omega^{\boldsymbol{b}}{}_\nu$, it can be shown that

$$y^\mu \delta_\mu{}^{\boldsymbol{a}} e_{\boldsymbol{a}}{}^\nu(y) = y^\nu, \qquad y^\mu \eta_{\mu\nu} e_{\boldsymbol{a}}{}^\nu(y) = y^\mu \eta_{\mu\nu}\delta_{\boldsymbol{a}}{}^\nu. \tag{18.30}$$

The above conditions can be regarded as an alternative definition of normal coordinates. More precisely, if a set of coordinates $y = (y^\mu)$ and frame coefficients $\{e_{\boldsymbol{a}}{}^\mu\}$ satisfy the conditions in (18.30) the metric components $g_{\mu\nu}$ will satisfy condition (18.29).

To complete the discussion, it is convenient to introduce the vector field $\boldsymbol{y}(y) = y^\mu \boldsymbol{\partial}_\mu$ tangent to the geodesics through o. One then has

$$\boldsymbol{y}(o) = 0, \qquad (\nabla_\mu y^\nu)|_o = \delta_\mu{}^\nu, \qquad \nabla_{\boldsymbol{y}}\boldsymbol{y} = \boldsymbol{y}.$$

Writing $\boldsymbol{y}$ in terms of a $\boldsymbol{g}$-normal frame one has that $\boldsymbol{y} = y^{\boldsymbol{a}}\boldsymbol{e_a}$ where $y^{\boldsymbol{a}}(y) = \delta_\nu{}^{\boldsymbol{a}} y^\nu$. Furthermore, using $\nabla_{\boldsymbol{y}}\boldsymbol{e_a} = 0$ one concludes that

$$y^{\boldsymbol{a}}(y)\Gamma_{\boldsymbol{a}}{}^{\boldsymbol{b}}{}_{\boldsymbol{c}}(y) = \delta_\nu{}^{\boldsymbol{a}} y^\nu \Gamma_{\boldsymbol{a}}{}^{\boldsymbol{b}}{}_{\boldsymbol{c}}(y) = 0, \qquad \text{close to } o.$$

The coordinates $y = (y^\mu)$ and the frame $\{\boldsymbol{e_a}\}$ satisfying the conditions discussed in the previous paragraphs will be collectively known as a ***normal gauge***. This gauge system is supplemented by a normalised spin frame $\{\boldsymbol{\epsilon_A}{}^A\}$ satisfying $y^{AA'}\nabla_{AA'}\boldsymbol{\epsilon_A}{}^B = 0$ such that $\{\boldsymbol{e_{AA'}}\} = \{\boldsymbol{\epsilon_A}\bar{\boldsymbol{\epsilon}}_{\boldsymbol{A'}}\}$ with $\boldsymbol{e_{AA'}} = \sigma_{\boldsymbol{AA'}}{}^{\boldsymbol{a}}\boldsymbol{e_a}$ – here $y^{AA'}$ is the spinorial counterpart of the vector $\boldsymbol{y}$. In what follows, *all spinors will be expressed in components with respect to this type of frame.*

Adapted coordinates on $\mathscr{N}_o$

The coordinates $y = (y^\mu)$ introduced in the previous subsections provide a convenient spacetime description of $\mathscr{N}_o$. However, to analyse the intrinsic geometry of the cone, one needs *adapted coordinates*. The construction of these coordinates is similar to that of the coordinates $(v, r, x^{\mathcal{A}})$ used in the analysis of the characteristic problem on $\mathscr{N} \cup \mathscr{I}^-$ in Section 18.1.1. The fundamental difference is that, in the case of a cone, these adapted coordinates degenerate at the vertex o. More precisely, one can consider adapted coordinates $x = (x^\mu)$ such that $\mathscr{N}_o$ is given as a level surface by the condition $r \equiv x^1 = 0$ and $v \equiv x^0$ is a parameter along the generators with tangent $\boldsymbol{l}$ – thus, $\boldsymbol{l} = \boldsymbol{\partial}_v$. The two-dimensional spacelike surfaces $\mathscr{Z}_{v_\bullet} \equiv \{p \in \mathscr{N} \mid v(p) = v_\bullet\}$ satisfy $\mathscr{Z}_{v_\bullet} \approx \mathbb{S}^2$, except for the limit case $\mathscr{Z}_0 = \{o\}$ which is a point. As in Section 18.1.1, $(x^{\mathcal{A}})$ denote local coordinates on $\mathscr{Z}_{v_\bullet}$. On $\mathscr{N}_o$ the covector $\boldsymbol{n}^\flat \equiv \mathbf{d}r$ is null and normal to $\mathscr{N}_o$. The coordinate r can be chosen so that one has the usual normalisation $\boldsymbol{g}(\boldsymbol{l}, \boldsymbol{n}) = 1$. Finally, the vectors $\boldsymbol{l}$ and $\boldsymbol{n}$ can be completed to a frame by choosing a pair of complex conjugate vectors $\boldsymbol{m}, \bar{\boldsymbol{m}} \in T(\mathscr{Z}_{v_\bullet})$, for $v_\bullet \neq 0$, such that $\boldsymbol{g}(\boldsymbol{m}, \bar{\boldsymbol{m}}) = -1$. As in Section 18.1.1 the vectors $\boldsymbol{m}$ and $\bar{\boldsymbol{m}}$ can be parallelly propagated along the generators of $\mathscr{N}_o$ off some fiduciary section $\mathscr{Z}_{v_\bullet}$.

18.4.2 Null data on the cone

As in the case of the characteristic problem on $\mathscr{N} \cup \mathscr{I}^-$, there are several ways of prescribing the free data. The most physically meaningful specification consists of the so-called ***radiation field*** encoding information on the two components of the Weyl tensor with the slowest fall-off at null infinity, and can be thought of as describing *the two polarisation states of incoming radiation.*

To describe the null data, let, as in previous sections, $\boldsymbol{l}$ denote the vector tangent to the generators of the null cone $\mathscr{N}_o$. As $\boldsymbol{l}$ is a null vector, there exists a spinor κ^A such that $l^{AA'} = \kappa^A \bar{\kappa}^{A'}$ with $l^{AA'}$ the spinorial counterpart of $\boldsymbol{l}$. The spinor κ^A is defined up to a phase $\kappa^A \mapsto e^{i\vartheta}\kappa^A$ with $\vartheta \in \mathbb{R}$ constant along the null generators. The ***radiation field*** is then defined as the component

$$\phi_0 \equiv \kappa^A \kappa^B \kappa^C \kappa^D \phi_{ABCD}$$

of the rescaled Weyl spinor. Due to the phase ambiguity in κ^A, the radiation field is a spin-weighted quantity. The information encoded in the radiation field is equivalent to information on the pull-back of $d_{abcd} l^a l^c$ to $\mathscr{N}_o$. More precisely, if $\boldsymbol{m}$ and $\bar{\boldsymbol{m}}$ are complex vectors tangent to the sections of $\mathscr{N}_o$ such that $\boldsymbol{g}(\boldsymbol{l}, \boldsymbol{m}) = 0$, then it follows from the symmetries of the Weyl tensor that $\phi_0 = d_{abcd} l^a m^b l^c m^d$.

Solving the constraints on $\mathscr{N}_o$

In analogy to the characteristic problem on $\mathscr{N} \cup \mathscr{I}^-$, and making use of the adapted coordinates $x = (v, r, x^{\mathcal{A}})$ and of the frame $\{\boldsymbol{l}, \boldsymbol{n}, \boldsymbol{m}, \bar{\boldsymbol{m}}\}$, the conformal

Einstein field equations split into equations transverse and intrinsic to $\mathcal{N}_o$. The intrinsic equations divide, in turn, into propagation equations (i.e. ordinary differential equations) along the generators of the cone and constraints which need to be solved only at a particular cut. Assuming the conformal gauge discussed in Section 18.4.1, the knowledge of the radiation field ϕ_0 on $\mathcal{N}_o$ allows one to compute the value of the remaining conformal fields in a neighbourhood of o on $\mathcal{N}_o$. More precisely, one has the following:

Proposition 18.2 (***reduced initial data for the asymptotic characteristic problem on a cone***) *In the conformal gauge given by conditions* (18.26), (18.27) *and* (18.28), *the transport equations induced by the conformal Einstein field equations and the structure equations on $\mathcal{N}_o$ uniquely determine the fields Ξ, s, $\Phi_{AA'BB'}$ and ϕ_{ABCD} on $\mathcal{N}_o$ once the radiation field ϕ_0 has been prescribed. The resulting fields satisfy the constraint equations on $\mathcal{N}_o$.*

Details on this result can be found in Friedrich (2014b).

Evaluating formal derivatives on $\mathcal{N}_o$

In addition to solving the constraint equations on $\mathcal{N}_o$, and in order to apply the theory of characteristic problems on a cone, given a choice of radiation field, it is necessary to show that the (formal) derivatives of any order of the conformal fields can be determined on the null cone along the generators of $\mathcal{N}_o$. This analysis is analogous to the one discussed in Section 18.3.1 for the characteristic problem on $\mathcal{N} \cup \mathscr{I}^-$. In the present case, however, the analysis is more delicate as the set $\mathscr{Z} = \mathcal{N} \cap \mathscr{I}^-$ shrinks to a point, so that the information for the integration along the generators has to be extracted solely from the null data. The key result is the following (see Friedrich (2014b)):

Proposition 18.3 (***computation of formal derivatives at the vertex***) *In a neighbourhood of the point o let the fields Ξ, s, $\Phi_{AA'BB'}$ ϕ_{ABCD}, $e_{AA'}{}^{\mu}$, $\Gamma_{AA'BC}$ be smooth and be expressed in an o-centred normal gauge and a conformal gauge satisfying Equations* (18.26), (18.27) *and* (18.28). *If the above fields satisfy the conformal field equations, then the Taylor expansions of the fields Ξ, s, $\Phi_{AA'BB'}$ and ϕ_{ABCD} in a suitable neighbourhood of o are determined by the null datum ϕ_0.*

Remark. In the above proposition, the neighbourhoods of o are spacetime neighbourhoods in the ambient manifold $\mathcal{M}'$ containing the cone $\mathcal{N}_o$.

18.4.3 The existence result

The setting described in the previous paragraphs leads to the following existence result, adapted from Chruściel and Paetz (2013):

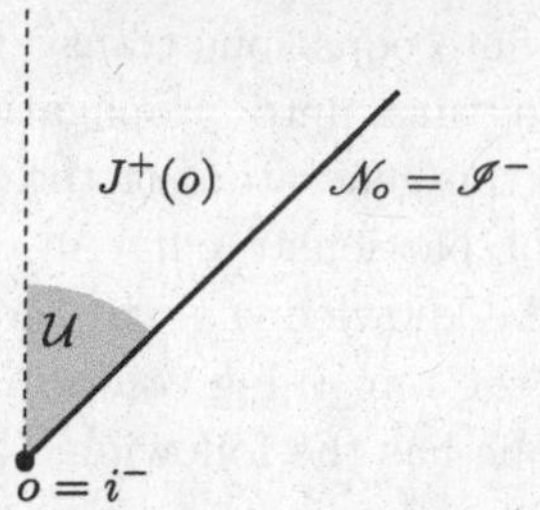

Figure 18.4 Schematic representation of the set up for the asymptotic characteristic problem on a cone. The existence results are restricted to a neighbourhood $\mathcal{U}$ of o in $J^+(o)$.

Theorem 18.2 (***local existence for the asymptotic characteristic problem on a cone***) *For any smooth prescription of the radiation field ϕ_0 on the null cone at the origin of the Minkowski spacetime, $\mathscr{N}_o$, there exists a neighbourhood $\mathcal{U} \subset J^+(o)$ of o, a smooth metric $\boldsymbol{g}$ and a smooth function Ξ such that:*

(i) $\mathscr{N}_o$ is the light cone of o for $\boldsymbol{g}$.
(ii) $\Xi = 0$ on $\mathscr{N}_o$.
(iii) $\mathbf{d}\Xi = 0$, $\boldsymbol{Hess}\,\Xi \neq 0$ on o.
(iv) $\mathbf{d}\Xi \neq 0$ on $\partial J^+(o) \cap \mathcal{U} \setminus \{o\}$.
(v) The function Ξ has no zeros on $\mathcal{U} \cap I^+(o)$ and the metric $\tilde{\boldsymbol{g}} = \Xi^{-2}\boldsymbol{g}$ satisfies the vacuum Einstein field equations on $\mathcal{U} \cap I^+(o)$.

Moreover, the rescaled Weyl spinor ϕ_{ABCD} of the pair $(\boldsymbol{g}, \Xi)$ extends smoothly across $\mathscr{N}_o$ and the restriction of $\phi_{ABCD}\epsilon_{\boldsymbol{0}}{}^A\epsilon_{\boldsymbol{0}}{}^B\epsilon_{\boldsymbol{0}}{}^C\epsilon_{\boldsymbol{0}}{}^D$ to $\mathscr{N}_o \setminus \{o\}$ coincides with the prescribed radiation field ϕ_0. The solution is unique up to isometries.

Remark. It follows from points (ii), (iii) and (iv) that the set $\mathscr{N}_o \setminus \{o\}$ corresponds to the past null infinity $\mathscr{I}^-$ of the resulting spacetime, while the vertex o is its past timelike infinity i^-. A schematic representation of the set up of the above theorem is given in Figure 18.4.

The proof of the above theorem, as given in Chruściel and Paetz (2013), makes use of the metric version of the conformal field equations and the associated wave equations discussed in Paetz (2015); see also Section 13.5.2. The reason behind the use of a hyperbolic reduction based on wave equations – as opposed, say, to the first-order symmetric hyperbolic systems used throughout this book – lies in the fact that the available theory of characteristic problems on a cone is well understood for this type of equations; see Dossa (1986, 2002).

18.5 Further reading

Characteristic problems in general relativity have a long history. The first systematic discussion has been given in Sachs (1962c). Further classical discussions can be found in Penrose (1965, 1980) and Müller zu Hagen and Seifert (1977).

A review on the various approaches to the problem, including an analysis of the possible choices of free data, can be found in Chruściel and Paetz (2012); this reference provides a convenient point of entry to the literature on the subject.

The basic theory of asymptotic characteristic initial value problems for the conformal field equations has been developed in the articles by Friedrich (1981a,b). A version of Theorem 18.1 in the analytic setting was given in Friedrich (1982). This result has been extended to the smooth setting in Kánnár (1996b) using the reduction to an auxiliary Cauchy problem given in Rendall (1990). The geometric set up for the asymptotic characteristic problem on a cone has first been given in Friedrich (1986c). The relation between Taylor expansions at the vertex of the null cone and the interior equations implied by the conformal Einstein field equations has been examined in Friedrich (2014b). The existence result for the characteristic problem in the cone has been given in Chruściel and Paetz (2013). Characteristic problems on a cone are less studied than those on intersecting null hypersurfaces. A good point of entry to the literature is Choquet-Bruhat et al. (2011).

Characteristic problems provide a natural approach to the construction of solutions to the Einstein equations by means of numerical methods. An advantage of this formulation is its clear-cut connection with the notion of gravitational radiation; see, for example, Damour and Schmidt (1990). A review on the subject can be found in Winicour (2012).

The characteristic initial value problem has been used in the seminal work by Christodoulou on the collapse of a spherically symmetric self-gravitating scalar field; see Christodoulou (1986).

19
Static solutions

In this chapter a study of static solutions to the vacuum Einstein field equations from the point of view of conformal methods is undertaken. Static and, more generally, stationary solutions provide valuable physical and mathematical intuition concerning the behaviour of solutions to the Einstein field equations. Static solutions describe the exterior region of time-independent, non-rotating, isolated bodies. Accordingly, they provide an interesting class of solutions to analyse the structure of spatial infinity; see Chapter 20. In addition, some particular static solutions (the Schwarzschild spacetime) are expected to describe the asymptotic state of the evolution dictated by the Einstein field equations. From a mathematical point of view, the results discussed in this chapter are of particular interest as they lie at the interface of classical potential theory, conformal geometry and general relativity. Throughout this chapter, the focus is restricted to the asymptotic region of an asymptotically flat static spacetime. Several of the key results for static spacetimes admit a suitable stationary counterpart; the interested reader is referred to the literature for further details. These generalisations of the theory are much more technically involved than the original static version and they will not be considered here.

19.1 The static field equations

For a ***static spacetime*** it will be understood a solution to the Einstein field equations $\boldsymbol{Ric}[\tilde{g}] = 0$ endowed with a hypersurface orthogonal Killing vector $\boldsymbol{\xi}$ which, in a suitable asymptotic region, is timelike. Using coordinates $(t, \underline{y}) = (t, y^\alpha)$ adapted to this Killing vector, one has that $\boldsymbol{\xi} = \boldsymbol{\partial}_t$. As $\boldsymbol{\xi}$ is hypersurface orthogonal, then there exists a function $v = v(\underline{y})$ such that

$$\boldsymbol{\xi}^\flat = \tilde{\boldsymbol{g}}(\boldsymbol{\xi}, \cdot) = v^2 \mathbf{d}t.$$

Thus, $v^2 = \tilde{\boldsymbol{g}}(\boldsymbol{\partial}_t, \boldsymbol{\partial}_t)$ is the square of the norm of $\boldsymbol{\xi}$. It follows that the hypersurfaces of constant coordinate t define a foliation of the spacetime. In what follows, it will be convenient to consider a frame $\{\boldsymbol{e_a}\}$ adapted to the static Killing vector and set $\boldsymbol{e_0}$ to be parallel to $\boldsymbol{\xi}$; that is, one has $\boldsymbol{\xi} = v\boldsymbol{e_0}$. The spatial part of the frame, $\{\boldsymbol{e_i}\}$, spans the tangent bundle of the hypersurfaces of constant t. Without loss of generality, the spatial frame can be parallely propagated along the direction of $\boldsymbol{e_0}$ so that using the definition of the connection coefficients one has that $\tilde{\Gamma}_{\boldsymbol{0}}{}^{\boldsymbol{a}}{}_{\boldsymbol{i}} = 0$. Let $\{\omega^{\boldsymbol{a}}\}$ be the associated coframe. One readily finds that $\boldsymbol{\omega^0} = v\mathbf{d}t$. It follows from this discussion that the metric $\tilde{\boldsymbol{g}}$ takes the form

$$\tilde{\boldsymbol{g}} = v^2\mathbf{d}t \otimes \mathbf{d}t + \tilde{\boldsymbol{h}}, \qquad v = v(\underline{y}) > 0, \qquad \tilde{\boldsymbol{h}} = \tilde{h}_{\alpha\beta}(\underline{y})\mathbf{d}y^\alpha \otimes \mathbf{d}y^\beta, \tag{19.1}$$

where $\tilde{\boldsymbol{h}}$ denotes a (negative definite) Lorentzian metric on the hypersurfaces of constant time coordinate.

Derivation of the static equations

The equations satisfied by the fields v and $\tilde{\boldsymbol{h}}$ appearing in the metric (19.1) can be deduced using the frame formalism introduced in the previous paragraphs. Observing that $\xi_{\boldsymbol{0}} \equiv \langle \boldsymbol{\xi}^\flat, \boldsymbol{e_0}\rangle = v$ one concludes that $\xi_{\boldsymbol{a}} = v\delta_{\boldsymbol{a}}{}^{\boldsymbol{0}}$. It follows that the ***Killing equation***

$$\tilde{\nabla}_{\boldsymbol{a}}\xi_{\boldsymbol{b}} + \tilde{\nabla}_{\boldsymbol{b}}\xi_{\boldsymbol{a}} = 0$$

takes the form

$$(\tilde{\nabla}_{\boldsymbol{a}}v\delta_{\boldsymbol{b}}{}^{\boldsymbol{0}} + \tilde{\nabla}_{\boldsymbol{b}}v\delta_{\boldsymbol{a}}{}^{\boldsymbol{0}}) + v(\tilde{\Gamma}_{\boldsymbol{a}}{}^{\boldsymbol{0}}{}_{\boldsymbol{b}} + \tilde{\Gamma}_{\boldsymbol{b}}{}^{\boldsymbol{0}}{}_{\boldsymbol{a}}) = 0.$$

As v is time independent, one concludes from setting $\boldsymbol{a}, \boldsymbol{b} = \boldsymbol{0}$ that $\tilde{\Gamma}_{\boldsymbol{0}}{}^{\boldsymbol{0}}{}_{\boldsymbol{0}} = 0$. Setting $\boldsymbol{a} = \boldsymbol{i}$ and $\boldsymbol{b} = \boldsymbol{j}$ one finds that $\tilde{\Gamma}_{\boldsymbol{i}}{}^{\boldsymbol{0}}{}_{\boldsymbol{j}} + \tilde{\Gamma}_{\boldsymbol{j}}{}^{\boldsymbol{0}}{}_{\boldsymbol{i}} = 0$ so that from the definition of the extrinsic curvature, Equation (2.45), one concludes that $K_{\boldsymbol{ij}} = 0$; that is, the surfaces of constant coordinate t are *time symmetric*. Accordingly, the Einstein constraint Equations (11.13a) and (11.13b) reduce to the condition

$$r[\tilde{\boldsymbol{h}}] = 0. \tag{19.2}$$

A further condition can be obtained from the equation

$$\tilde{\nabla}_{\boldsymbol{0}}(\tilde{\nabla}_{\boldsymbol{a}}\xi_{\boldsymbol{b}} + \tilde{\nabla}_{\boldsymbol{b}}\xi_{\boldsymbol{a}}) = 0.$$

Commuting covariant derivatives and using that the Killing equation implies

$$\tilde{\nabla}_{\boldsymbol{0}}\xi_{\boldsymbol{b}} = -\tilde{\nabla}_{\boldsymbol{b}}\xi_{\boldsymbol{0}} = -\tilde{\nabla}_{\boldsymbol{b}}v$$

one finds that

$$\tilde{\nabla}_{\boldsymbol{a}}\tilde{\nabla}_{\boldsymbol{b}}v + v\tilde{R}_{\boldsymbol{0a0b}} = 0.$$

From the last equation, using the Gauss-Codazzi identity, Equation (2.47), one concludes that

$$\tilde{\nabla}_{\boldsymbol{i}}\tilde{\nabla}_{\boldsymbol{j}}v = v\tilde{r}_{\boldsymbol{ij}} \tag{19.3}$$

where r_{ij} denotes the components with respect to $\{\boldsymbol{e}_i\}$ of the Ricci tensor of the 3-metric $\tilde{\boldsymbol{h}}$.

Equations (19.2) and (19.3) provide the required ***static Einstein field equations*** for the fields v and $\tilde{\boldsymbol{h}}$. After some further slight manipulations they can be rewritten in tensorial form as

$$\Delta_{\tilde{\boldsymbol{h}}} v = 0, \tag{19.4a}$$

$$\tilde{r}_{ij} = \frac{1}{v} \tilde{D}_i \tilde{D}_j v, \tag{19.4b}$$

where $\tilde{D}_i$ and $\tilde{r}_{ij}$ denote, respectively, the Levi-Civita connection and the Ricci tensor of the 3-metric $\tilde{\boldsymbol{h}}$. In what follows, a pair $(v, \tilde{\boldsymbol{h}})$ solving the ***static equations*** (19.4a) and (19.4b) will be called a ***static solution***. A static solution, expressed in terms of $\tilde{\boldsymbol{h}}$-harmonic coordinates is analytic; see Müller zu Hagen (1970).

Observe that discarding the field v, a solution to the static equations gives rise to a solution to the time-symmetric Einstein constraints. This dual perspective of static solutions as a spacetime and as time-symmetric initial data for a spacetime will be used often. The context will dictate the appropriate point of view. Equations (19.4a) and (19.4b) can be regarded as a three-dimensional analogue of the Einstein field equations in which the curvature is coupled to a *fictitious matter field* described by v. This interpretation also holds for other *symmetry reductions* of the vacuum Einstein field equations, say, axial symmetry; see, for example, Geroch (1971a, 1972a).

Asymptotic conditions and the Licnerowicz theorem

Of special interest are static solutions describing the asymptotic region of isolated systems. For simplicity, it will be assumed that $\tilde{\mathcal{S}}$ has a single asymptotic region in which coordinates $\underline{y} = (y^\alpha)$ can be found such that

$$v = 1 - \frac{m}{|y|} + O_k(|y|^{-(1+\varepsilon)}), \tag{19.5a}$$

$$\tilde{h}_{\alpha\beta} = -\left(1 + \frac{2m}{|y|}\right)\delta_{\alpha\beta} + O_k(|y|^{-(1+\varepsilon)}), \tag{19.5b}$$

as $|y| \to \infty$ where $m \neq 0$ denotes the Arnowitt-Deser-Miser (ADM) mass and $\varepsilon > 0$. The notation O_k has been described in the Appendix to Chapter 11. The above decay conditions can be deduced from more primitive assumptions which make no reference to asymptotic flatness; see Reiris (2014a,b). In order to describe an isolated system – say, the exterior of a star – Equations (19.4a) and (19.4b) need to be supplemented with suitable boundary conditions at an interior boundary $\partial\tilde{\mathcal{S}}$ – say, the surface of a star. An analysis of this type has been carried out by Reula (1989) and Miao (2003). The role played by boundary conditions in the determination of static solutions is nicely exhibited in the case where $\tilde{\mathcal{S}} \approx \mathbb{R}^3$. In this case, it follows from Equation (19.4a) by integration by parts that

$$0 = \int_{\tilde{\mathcal{S}}} v \Delta_{\tilde{\boldsymbol{h}}} v \mathrm{d}\mu = -\int_{\tilde{\mathcal{S}}} \tilde{D}^i v \tilde{D}_i v \mathrm{d}\mu,$$

so that $\tilde{D}_i v = 0$ on $\tilde{\mathcal{S}}$. This implies that v has to be constant on $\tilde{\mathcal{S}}$. Moreover, using (19.5a) one concludes that $v = 1$. Substituting into Equation (19.4a) one finds that $\tilde{r}_{ij} = 0$ so that $\tilde{\boldsymbol{h}}$ must be flat – recall that in three dimensions the curvature is fully determined by the Ricci tensor. Consequently, in order to have static solutions other than the Minkowski solution one needs hypersurfaces $\tilde{\mathcal{S}}$ with a non-trivial topology or with some inner boundary $\partial\tilde{\mathcal{S}}$. This result is usually known as ***Licnerowicz's theorem***.

19.1.1 The conformal static field equations

In the remainder of this chapter, *the discussion of static solutions will be restricted to a suitable asymptotic region where the decay conditions* (19.5a) *and* (19.5b) *hold.* Accordingly, it is convenient to make use of the definition of *asymptotically Euclidean and regular manifolds* given in Section 11.6.2. Hence, one considers a function Ω on $\mathcal{S} \equiv \tilde{\mathcal{S}} \cup \{i\}$ with $\Omega \in C^2(\tilde{\mathcal{S}}) \cap C^\infty(\tilde{\mathcal{S}})$, $\Omega > 0$ on $\tilde{\mathcal{S}}$ which conformally extends $\tilde{\boldsymbol{h}}$ to a smooth metric

$$\boldsymbol{h} \equiv \Omega^2 \tilde{\boldsymbol{h}} \qquad \text{on } \mathcal{S},$$

in such a way that

$$\Omega = 0, \qquad D_i\Omega = 0, \qquad D_iD_j\Omega = -2h_{ij}, \qquad \text{at } i. \tag{19.6}$$

In order to exploit the above conformal setting, it is convenient to rewrite the static Equations (19.4a) and (19.4b) in terms of fields satisfying regular equations in a neighbourhood of i. The procedure of constructing a system of ***regular conformal static equations*** is similar in spirit to the one carried out in Chapter 8 to obtain the conformal field equations. The key idea is to identify quantities which in the conformally rescaled picture are both suitably regular and which satisfy equations that are formally regular at i. In this spirit, the equation obtained from combining the static Equation (19.4b) with the transformation law of the three-dimensional Ricci tensor, Equation (5.16a), should be read not as a differential condition for the components of a conformally rescaled metric but rather as differential equations involving second derivatives of a quantity associated to the conformal factor. Similar considerations need to be taken into account when attempting to construct a *conformal equation* for the scalar field v. Using the transformation law for the Yamabe operator, Equation (11.23), one obtains

$$\left(\Delta_{\boldsymbol{h}} - \frac{1}{8} r[\boldsymbol{h}]\right)(\Omega^{-1/2} v) = 0.$$

This equation is formally singular at i unless it is possible to tie the behaviour of Ω with that of v. Alternatively, one could try to find a regular equation for a quantity which indirectly allows one to gain knowledge about v. These ideas are explored in the following subsections.

Fixing the conformal gauge

The standard approach to obtain a set of regular conformal static field equations relies on a specific choice of conformal gauge which explicitly prescribes the conformal factor Ω in terms of the norm of the static Killing vector v; see, for example, Beig and Simon (1980a) and Friedrich (1988, 2004, 2007). In the following, the approach taken in the last two references will be followed. A general version of the conformal static equations which retains the whole conformal freedom has been given in Friedrich (2013).

It can be verified that the conditions (19.6) expressed in terms of physical coordinates $\underline{y} = (y^\alpha)$ require Ω to behave like $1/|y|^2$ as $|y| \to 0$. This observation suggests, in turn, considering a conformal factor of the form

$$\Omega = \left(\frac{1-v}{m}\right)^2. \tag{19.7}$$

As will be seen in Section 19.2, this is not the only possible way of fixing the conformal freedom. The choice in Equation (19.7) fixes the value of the Ricci scalar of the conformal metric $\boldsymbol{h}$. This can be seen from the transformation law of the Yamabe operator, Equation (11.23), by setting $u = \Omega^{1/2}$ and making the replacements $\phi \mapsto \Omega^{1/2}$, $\boldsymbol{h}' \mapsto \boldsymbol{h}$, $\boldsymbol{h} \mapsto \tilde{\boldsymbol{h}}$ so that, on the one hand, one has

$$\mathbf{L}_{\boldsymbol{h}}[1] = \Omega^{-5/2}\mathbf{L}_{\tilde{\boldsymbol{h}}}\left(\frac{1-v}{m}\right) = -\frac{1}{m}\Omega^{-5/2}\Delta_{\tilde{\boldsymbol{h}}}v = 0,$$

while, on the other hand,

$$\mathbf{L}_{\boldsymbol{h}}[1] = \left(\Delta_{\boldsymbol{h}} - \frac{1}{8}r[\boldsymbol{h}]\right)[1] = -\frac{1}{8}r[\boldsymbol{h}].$$

Hence, one concludes that $r[\boldsymbol{h}] = 0$.

A decomposition of the conformal factor

Following the general discussion of Section 11.6.3, one has that the conformal factor Ω satisfies

$$\left(\Delta_{\boldsymbol{h}} - \frac{1}{8}r[\boldsymbol{h}]\right)(\Omega^{-1/2}) = 0, \qquad \text{on } \tilde{\mathcal{S}},$$

and $|x|\Omega^{-1/2} \to 1$ as $|x| \to 0$. Here, and in what follows, let $\underline{x} = (x^\alpha)$ denote some coordinates in a neighbourhood $\mathcal{U} \subset \mathcal{S}$ with $x^\alpha(i) = 0$. Close to i one has the representation

$$\Omega^{-1/2} = \zeta^{-1/2} + W, \tag{19.8}$$

with ζ, W smooth – confront this decomposition with the discussion in Section 11.6.4. In particular, one has

$$\left(\Delta_{\boldsymbol{h}} - \frac{1}{8}r[\boldsymbol{h}]\right)W = 0, \qquad W(i) = \frac{m}{2}, \tag{19.9}$$

and

$$\left(\Delta_{\boldsymbol{h}} - \frac{1}{8}r[\boldsymbol{h}]\right)(\zeta^{-1/2}) = 4\pi\delta[i]. \tag{19.10}$$

From this last equation it follows that

$$\zeta(i) = 0, \qquad D_i\zeta(i) = 0, \qquad D_iD_j\zeta(i) = -2h_{ij}(i). \tag{19.11}$$

One has that ζ is essentially the Green function of the Yamabe operator and describes the local geometry in a neighbourhood of i, while W encodes information of a global nature – in particular, its ADM mass m. Accordingly, the functions ζ and W will be called, respectively, the ***massless part*** and ***mass part*** of the conformal factor Ω. Given a conformal metric $\boldsymbol{h}$, *the decomposition* (19.8) *is unique.* Moreover, using the so-called ***Hadamard's parametrix construction***, it can be shown that ζ and W are analytic if $\boldsymbol{h}$ is analytic; see Friedrich (1998c, 2004) for further details about this last statement and Garabedian (1986) for the underlying PDE theory. In particular, for the choice of the conformal factor (19.7) it follows that the parametrisation (19.8) takes the form

$$\zeta = \frac{1}{\mu}\left(\frac{1-v}{1+v}\right)^2, \qquad W = \frac{m}{2}, \qquad \mu \equiv \frac{m^2}{4}. \tag{19.12}$$

It can be verified that the function ζ satisfies the asymptotic conditions (19.11) and that W is, indeed, a solution of Equation (19.9).

Using the chain rule to rewrite Equation (19.10) as an equation for $\Delta_{\boldsymbol{h}}\zeta$ and taking into account the asymptotic conditions (19.11), one finds that

$$2\zeta\varsigma = D_i\zeta D^i\zeta, \qquad \text{with } \varsigma \equiv \frac{1}{3}\Delta_{\boldsymbol{h}}\zeta, \tag{19.13}$$

which is a regular equation in a suitable neighbourhood of i. In particular, it can be verified that $\varsigma(i) = -2$. Equation (19.13) is the analogue of the conformal Einstein field Equation (8.24). It encodes a regularised version of the transformation equation for the Ricci scalar. As will be seen in the following, it can be interpreted as a constraint which is automatically satisfied if other equations hold.

Equations for the curvature

To exploit the fact that one is working with a gauge for which $r[\boldsymbol{h}] = 0$, it is convenient to introduce an $\boldsymbol{h}$-tracefree tensor s_{ij} such that

$$r_{ij}[\boldsymbol{h}] = s_{ij}.$$

Recalling that in three dimensions the Riemann curvature tensor $r^i{}_{jkl}$ is fully determined by the Ricci tensor, it is then natural to interpret the tensors r_{ij} and s_{ij} as describing, respectively, the ***geometric*** and ***algebraic three-dimensional curvatures***; see Section 8.3.1 for further discussion on these notions in the context of the conformal Einstein field equations. If the zero quantity

$$\Xi_{ij} \equiv r_{ij}[\boldsymbol{h}] - s_{ij}$$

vanishes, then the three-dimensional (contracted) Bianchi identity takes the form

$$D^i s_{ij} = 0. \tag{19.14}$$

The fields ζ, ς and s_{ij} can be used to obtain a regular version of the formally singular transformation law for the Ricci tensor; see Equation (5.16a). Rewriting derivatives of the conformal factor Ω as derivatives of ζ one obtains

$$S_{ij} \equiv D_i D_j \zeta - \varsigma h_{ij} + \zeta(1-\mu\zeta)s_{ij} = 0. \tag{19.15}$$

Equation (19.15) will be read as a differential equation for ζ. To close the system one needs differential equations for ς and s_{ij}. Suitable equations can be obtained from the integrability conditions

$$D^i S_{ij} = 0, \qquad D_{[k}S_{i]j} + \frac{1}{2}D^l S_{l[k}h_{i]j} = 0, \tag{19.16}$$

encoding the three-dimensional second Bianchi identity in contracted and uncontracted form, respectively. The identities (19.16) can be verified through a direct computation and introducing the *zero quantities*

$$\begin{aligned} S_i &\equiv D_i\varsigma + (1-\mu\zeta)s_{ij}D^j\zeta = 0, \\ H_{kij} &\equiv (1-\mu\zeta)D_{[k}s_{i]j} - \mu\big(2D_{[k}\zeta s_{i]j} + D^l\zeta h_{l[k}s_{i]j}\big). \end{aligned}$$

It follows from a further computation that $S_i = 0$ and $H_{kij} = 0$ are equivalent to the integrability conditions (19.16). The condition $H_{kij} = 0$ can be read as an expression for the *Cotton tensor*

$$b_{jki} \equiv D_{[k}r_{i]j} - \frac{1}{4}D_{[k}r h_{i]j} = 2D_{[k}l_{i]j},$$

where l_{ij} denotes the three-dimensional Schouten tensor. In the remainder of this chapter it is often more convenient to work with the dualised version of b_{jki}, $b_{ij} \equiv \frac{1}{2}b_{ikl}\epsilon_j{}^{kl}$. A computation shows that

$$b_{ij} = \frac{\mu}{1-\mu\zeta}\left(s_{li}\epsilon_j{}^{kl}D_k\zeta - \frac{1}{2}s_{lm}\epsilon_{ji}{}^l D^m\zeta\right).$$

Summary: the conformal static equations

In what follows, the conditions

$$\Xi_{ij} = 0, \qquad S_i = 0, \qquad S_{ij} = 0, \qquad H_{kij} = 0 \tag{19.17}$$

will be known as the ***conformal static equations***. They provide an overdetermined system of differential conditions for the fields h_{ij}, s_{ij}, ζ and ς. As will be seen, the equations in (19.17) imply an elliptic system for the components of the various ***conformal fields***.

Remark. A direct computation yields the identity

$$D_i\big(2\zeta\varsigma - D_k\zeta D^k\varsigma\big) = 2\zeta S_i - 2S_{ik}D^k\zeta.$$

Thus, if $S_i = 0$ and $S_{ij} = 0$, then $2\zeta\varsigma - D_k\zeta D^k\varsigma$ is a constant. Evaluating at i and using the known values of the various fields at this point, one concludes that the expression in brackets must vanish. This argument shows that Equation (19.13) plays the role of a constraint. Hence, it has not been included in the list (19.17).

19.1.2 Spinorial version of the equations

To write the spinorial version of the conformal static equations, let $s_{ABCD} = s_{(ABCD)}$ denote the spinorial counterpart of the trace-free tensor s_{ij}. The Bianchi identity (19.14) takes the form

$$D^{AB}s_{ABCD} = 0.$$

In terms of the spinor s_{ABCD} the equation $H_{kij} = 0$ takes, after exploiting the antisymmetry in the pair $_{ki}$, the simple form

$$D_A{}^Q s_{BCDQ} = \frac{2\mu}{1-\mu\zeta} s_{Q(ABC}D_{D)}{}^Q\zeta. \tag{19.18}$$

The spinorial transcription of equations $S_i = 0$ and $S_{ij} = 0$ is completely direct. When working with spinors, the equation $r_{ij} = s_{ij}$ is replaced by the Cartan structure equations, Equations (2.41) and (2.42), for the 3-geometry with an ***algebraic 3-curvature*** given by s_{ij}. These structure equations provide, respectively, differential conditions for the coefficients of a frame and for the associated spin connection coefficients; see, for example, Friedrich (2007). It will often be convenient to express the various spinorial fields and the associated equations in terms of their components (e.g. $s_{\boldsymbol{ABCD}}$) with respect to some spin dyad $\{\epsilon_{\boldsymbol{A}}{}^A\}$.

For later use, let b_{ABCDEF} denote the spinorial counterpart of the Cotton tensor b_{ijk}. Exploiting the antisymmetry in the indices $_{jk}$, one obtains the decomposition

$$b_{ABCDEF} = b_{ABCE}\epsilon_{DF} + b_{ABDF}\epsilon_{CE}, \qquad b_{ABCD} \equiv D^Q{}_{(A}s_{BCD)Q}. \tag{19.19}$$

Consequently, one has the symmetry $b_{ABCD} = b_{(ABCD)}$. Moreover, it can be verified that

$$D^{AB} b_{ABCD} = 0.$$

In what follows, b_{ABCD} will be referred to as the ***Cotton spinor***.

19.2 Analyticity at infinity

The conformal static Equations (19.17) allow one to show that, under some basic regularity assumptions, there exist coordinates in a neighbourhood of the point at infinity for which all the conformal fields are analytic. This result brings to the forefront the inherent ellipticity of the equations and constitutes the foundation of any further analysis of static solutions from a conformal perspective. The result was originally proven by Beig and Simon (1980a). In the following, an adaptation of this result will be given. One has:

Theorem 19.1 (***analyticity of static solutions at infinity***) *Let* $(v, \tilde{\boldsymbol{h}})$ *denote a solution to the static Equations* (19.4a) *and* (19.4b) *such that* Ω *as defined by Equation* (19.7) *satisfies the conditions* (19.6) *with* $h_{\alpha\beta} = \Omega^2 \tilde{h}_{\alpha\beta}$ *the components of a* $C^{4,\alpha}$ *metric for some coordinates* $\underline{x} = (x^\alpha)$ *in a neighbourhood of* i. *Then there exist coordinates* $\underline{x}' = (x'^\alpha)$ *defined in a neighbourhood of* i *such that* $h'_{\alpha\beta}$, ζ', ς' *and* $s'_{\alpha\beta}$ *are analytic.*

Remark. The regularity assumptions in this result are expressed in terms of Hölder spaces; see the Appendix to this chapter.

Proof The proof exploits the fact that the *Ricci operator of a Riemannian metric expressed in harmonic coordinates is elliptic* – the Lorentzian counterpart of this observation has been discussed in the Appendix to Chapter 13. The general theory of elliptic equations – see, for example, Garabedian (1986) – shows that it is always possible to find a neighbourhood of i in which the equations

$$\Delta_h x'^\alpha = 0 \tag{19.20}$$

have a solution $x'^\alpha = x'^\alpha(\underline{x})$. The coefficients of the differential operator in Equation (19.20) consist of $h_{\alpha\beta}$ and its derivatives so that they are of class $C^{3,\alpha}$. The general theory of elliptic partial differential equations (PDEs) shows that solutions of second-order elliptic equations gain two derivatives with respect to the coefficients of the equation. Accordingly, one has that $x'^\alpha = x'^\alpha(\underline{x})$ is $C^{5,\alpha}$. This regularity is sufficient to invert the coordinates. Taking into account the transformation law of the metric tensor under change of coordinates,

$$h'_{\alpha\beta} = \frac{\partial x^\gamma}{\partial x'^\alpha} \frac{\partial x^\delta}{\partial x'^\beta} h_{\gamma\delta},$$

it follows that $h'_{\alpha\beta}$ is $C^{4,\alpha}$. Similarly, the field ζ' can be verified to be $C^{4,\alpha}$, while ς' and $s'_{\alpha\beta}$ are $C^{2,\alpha}$. To conclude the proof, one needs to construct a system of elliptic equations for the various fields. In the remainder of the proof it is assumed that all the fields are expressed in terms of the coordinates $\underline{x}'$, and the primes will be dropped from the expressions. Let $\gamma_\alpha{}^\beta{}_\gamma$ denote the Christoffel symbols of the metric $\boldsymbol{h}$ and denote by $\gamma^\beta \equiv h^{\alpha\gamma}\gamma_\alpha{}^\beta{}_\gamma$ the associated ***contracted Christoffel symbols***. A discussion analogous to that of the hyperbolic reduction of the Einstein field equations in generalised wave coordinates – see the discussion in the Appendix of Chapter 13 – shows that

$$r_{\alpha\beta} - D_{(\alpha}\gamma_{\beta)} = s_{\alpha\beta}$$

is an elliptic equation for the components $h_{\alpha\beta}$ of the metric $\boldsymbol{h}$ in the coordinates (x^α) if $s_{\alpha\beta}$ are known. To close the system one considers the equations

$$S^\alpha{}_\alpha = 0, \qquad D^\alpha S_\alpha = 0, \qquad D^\gamma H_{\gamma\alpha\beta} = 0.$$

Using the Bianchi identity (19.14) and the conformal static field equations to remove all the second derivatives of the conformal fields which are not Laplacians, one obtains a system of the form

$$\Delta_{\boldsymbol{h}}\big(h_{\alpha\beta}, \zeta, \varsigma, s_{\alpha\beta}\big) = \mathbf{F}(h_{\alpha\beta}, \zeta, \varsigma, s_{\alpha\beta}, D_\gamma h_{\alpha\beta}, D_\alpha\zeta, D_\alpha\varsigma, D_\gamma s_{\alpha\beta}), \qquad (19.21)$$

with $\mathbf{F}$ an analytic vector-valued function of its entries. Despite having a Laplacian operator on the left-hand side, it is a priori not clear that the system (19.21) is elliptic as $\Delta_{\boldsymbol{h}}$ applied to $h_{\alpha\beta}$ and $s_{\alpha\beta}$ gives rise to further second-order derivatives of $h_{\alpha\beta}$ which come from derivatives of the Christoffel symbols. To verify ellipticity one needs to compute the determinant of the symbol of (19.21). A calculation shows that this determinant is, in fact, proportional to $(h^{\alpha\beta}\xi_\alpha\xi_\beta)^{13}$ so that one, indeed, has an elliptic system as $h_{\alpha\beta}$ are the components of a Riemannian metric; compare the definition of ellipticity in Section 11.2. The general theory of the regularity of solutions of elliptic systems shows that if one has a $C^{2,\alpha}$ solution to the above equation, then it must, in fact, be analytic in $\mathcal{U}$; a discussion of this result is given in Morrey (1958). □

Remarks

(i) The original proof in Beig and Simon (1980a) was carried out in a conformal gauge obtained from writing the static metric (19.1) in the form

$$\tilde{\boldsymbol{g}} = e^{2U}\mathbf{d}t \otimes \mathbf{d}t - e^{-2U}\hat{h}_{\alpha\beta}\mathbf{d}y^\alpha \otimes \mathbf{d}y^\beta,$$

where U is a scalar field and $\hat{h}_{\alpha\beta}$ denote the components of a Riemannian 3-metric. Their analysis shows that $\omega \equiv (U/m)^2$ and $\boldsymbol{h}' \equiv \omega^2\hat{\boldsymbol{h}}$ are analytic in $\boldsymbol{h}'$-harmonic coordinates. This gauge and the one used to prove Theorem 19.1 can be related by letting $\Omega' \equiv \omega e^U$. One has by analogy to Equation (19.8) the split $\Omega'^{-1/2} = \zeta'^{-1/2} + W'$ with

$$\zeta' = \frac{\omega}{\cosh^2 U/2}, \qquad W' = \frac{m \sinh U/2}{U}.$$

It can be verified that the conformal metrics $\boldsymbol{h}$ and $\boldsymbol{h}'$ are related to each other via $\boldsymbol{h} = \vartheta^4 \boldsymbol{h}'$ with $\vartheta \equiv 2W'/m$.

(ii) Kennefick and O'Murchadha (1995) have shown that the smoothness assumption on the conformal metric made in Theorem 19.1 can be deduced from weaker differentiability and decay conditions on the physical 3-metric $\tilde{\boldsymbol{h}}$.

(iii) Theorem 19.1 can be further strengthened by considering $\boldsymbol{h}$-normal coordinates based on i. It can be verified that the coordinate transformation relating the analytic coordinate system $\underline{x}'$ and normal coordinates is also analytic.

A remark concerning the notion of analyticity at i

As a consequence of the analytic behaviour ensured by Theorem 19.1 one has that, for example, the field ζ can be expressed in a suitably small neighbourhood $\mathcal{U}$ of i as a convergent series of the form

$$\zeta = \sum_{p=2}^{\infty} \zeta_{\alpha_2\cdots\alpha_p} x^{\alpha_2} \cdots x^{\alpha_p}, \qquad \zeta_{\alpha_2\cdots\alpha_p} \in \mathbb{R}. \tag{19.22}$$

The other conformal fields have similar expansions. An alternative description of the above expansion can be obtained by introducing ***polar coordinates***. Accordingly, one defines

$$\rho^2 \equiv |x|^2 = \delta_{\alpha\beta} x^\alpha x^\beta, \qquad \rho^\alpha \equiv \frac{x^\alpha}{|x|}. \tag{19.23}$$

The ***unit position vector*** ρ^α can be parametrised by means of some coordinates $\theta = (\theta^{\mathcal{A}})$ on the 2-sphere $\mathbb{S}^2$ so that one can write $\rho^\alpha = \rho^\alpha(\theta)$. In what follows it will be assumed, for convenience, that the coordinates $(\theta^{\mathcal{A}})$ are analytic functions of the original coordinates $\underline{x}$ – clearly, the coordinates $(\theta^{\mathcal{A}})$ cannot cover the whole of $\mathbb{S}^2$. This fact will not play a role in the sequel. In terms of the coordinates $(\rho, \theta^{\mathcal{A}})$ the expansion (19.22) takes the form

$$\zeta = \sum_{p=2}^{\infty} \zeta_{\alpha_2\cdots\alpha_p} \rho^{\alpha_2} \cdots \rho^{\alpha_p} \rho^p.$$

Accordingly, ζ is also an analytic function of the coordinates $(\rho, \theta^{\mathcal{A}})$. Decomposing the product $\rho^{\alpha_2} \cdots \rho^{\alpha_p}$ (which depends only on the angular coordinates) into symmetric, trace-free terms one obtains the usual expansion in terms of spherical harmonics Y_{lm}. This computation can be conveniently performed in space spinors; see, for example, Torres del Castillo (2003).

Remark. *Not every analytic function of* $(\rho, \theta^{\mathcal{A}})$ *is an analytic function of the associated Cartesian coordinates.* The standard counterexample for this is the radial coordinate ρ as defined by Equation (19.23), whose second derivative with respect to the coordinates (x^α) is singular at i. To have analyticity with respect to the coordinates (x^α) one needs the *right combination of spherical harmonics and powers of* ρ.

A particular case of the above discussion concerns the conformal factor Ω. From Equation (19.8) it follows that

$$\Omega = \frac{\zeta}{(1+\zeta^{1/2}W)^2}.$$

A direct computation taking into account the asymptotic conditions (19.11) shows that while Ω is C^2 at i, it will fail to be of class C^3 unless $W = 0$ – which, in the present gauge, means that $m = 0$. Thus, in general, the conformal factor Ω is not analytic in the harmonic coordinates (x^α) even if ζ is analytic. It is, nevertheless, analytic in ρ.

19.2.1 A spacetime conformal completion of static solutions

Theorem 19.1 is a statement about the conformal structure of hypersurfaces of a canonical foliation of a static spacetime. Thus, it is of natural interest to analyse the consequences of this property from a spacetime perspective. Intuitively, one expects that the nice conformal properties of the leaves of the foliation will lead to a good spacetime conformal behaviour. As the spatial conformal factor is not analytic with respect to the harmonic coordinates $\underline{x} = (x^\alpha)$, one cannot expect analyticity of a spacetime conformal extension in terms of these coordinates. Instead, one looks for extensions which are analytic in the associated radial coordinate.

Following Remark (iii) after Theorem 19.1, it is assumed that the harmonic coordinates $\underline{x} = (x^\alpha)$ are $\boldsymbol{h}$-normal and centred at i. Writing $\theta = (\theta^{\mathcal{A}})$, one has that for $\theta = \theta_\star$ fixed and $\mathrm{s} \in [0, \mathrm{s}_\star)$ for $\mathrm{s}_\star \geq 0$ suitably small, $x^\alpha(\mathrm{s}) = \mathrm{s}\rho^\alpha(\theta_\star)$ describes a geodesic passing through i. A function $f : \mathcal{U} \to \mathbb{R}$ evaluated along one of these geodesics will be denoted by $f(\mathrm{s}\rho^\alpha)$. From $x^\alpha = \rho\, \rho^\alpha$ it follows that

$$\mathbf{d}x^\alpha = \rho^\alpha \mathbf{d}\rho + \mathbf{d}\rho^\alpha = \rho^\alpha \mathbf{d}\rho + \rho \partial_{\mathcal{A}} \rho^\alpha \mathbf{d}\theta^{\mathcal{A}},$$

so that, using the normal coordinates condition $h_{\alpha\beta}x^\alpha = -\delta_{\alpha\beta}x^\alpha$, one concludes that

$$\boldsymbol{h} = -\mathbf{d}\rho \otimes \mathbf{d}\rho + \rho^2 \boldsymbol{k}, \tag{19.24}$$

where

$$\boldsymbol{k} \equiv h_{\alpha\beta}\mathbf{d}\rho^\alpha \otimes \mathbf{d}\rho^\beta = h_{\alpha\beta}\partial_{\mathcal{A}}\rho^\alpha \partial_{\mathcal{B}}\rho^\beta \mathbf{d}\theta^{\mathcal{A}} \otimes \mathbf{d}\theta^{\mathcal{B}}$$

corresponds to the metric of the two-dimensional surfaces of constant ρ. In particular, one has that $\boldsymbol{k}|_{\rho=0} = -\boldsymbol{\sigma}$ – the negative definite standard metric of $\mathbb{S}^2$.

Putting together the discussion of the previous paragraph and recalling that $\boldsymbol{h} = \Omega^2 \tilde{\boldsymbol{h}}$, one finds that the static metric (19.1) can be rewritten as

$$\tilde{\boldsymbol{g}} = v^2 \mathbf{d}t \otimes \mathbf{d}t - \Omega^{-2}\mathbf{d}\rho \otimes \mathbf{d}\rho + \Omega^{-2}\rho^2 \boldsymbol{k}. \tag{19.25}$$

The claim is now that the conformal metric $\boldsymbol{g} \equiv \Xi^2 \tilde{\boldsymbol{g}}$ with

$$\Xi = \Omega^{1/2}$$

gives rise to a conformal extension of the static spacetime which is *as regular as one can possibly expect*, that is, analytic in the coordinates $(\rho, \theta^{\mathcal{A}})$. From Equation (19.25) one has that

$$\boldsymbol{g} = \Omega v^2 \mathbf{d}t \otimes \mathbf{d}t - \Omega^{-1}\mathbf{d}\rho \otimes \mathbf{d}\rho + \rho^2 \Omega^{-1}\boldsymbol{k}. \tag{19.26}$$

Recalling that $\Omega = O(\rho^2)$ and $v = O(1)$, one finds that while the first and third terms of the above metric are regular, the second one is singular. This singularity is a coordinate artefact which can be removed by considering the *null coordinate*

$$u \equiv t + \int_\rho^{\rho_\star} \frac{\mathrm{d}s}{v(s\rho^\alpha)\Omega(s\rho^\alpha)},$$

for fixed ρ^α and $\rho_\star > 0$. Observe, in particular, that as a consequence of the behaviour of Ω near i one has that $u \to -\infty$ as $\rho \to 0$. The differential of the null coordinate u is given by

$$\mathbf{d}u = \mathbf{d}t - \frac{1}{v\Omega}\mathbf{d}\rho + \boldsymbol{\lambda},$$

where

$$\boldsymbol{\lambda} \equiv \lambda_{\mathcal{A}} \mathbf{d}\theta^{\mathcal{A}}, \qquad \lambda_{\mathcal{A}} \equiv \int_\rho^{\rho_\star} \partial_{\mathcal{A}} \left(\frac{1}{v(s\rho^\alpha)\Omega(s\rho^\alpha)} \right) \mathrm{d}s.$$

Substituting the above expressions into the conformal metric (19.26) yields

$$\begin{aligned} \boldsymbol{g} = \Omega v^2 \mathbf{d}u \otimes \mathbf{d}u + v(\mathbf{d}u \otimes \mathbf{d}\rho + \mathbf{d}\rho \otimes \mathbf{d}u) - \Omega v^2 (\mathbf{d}u \otimes \boldsymbol{\lambda} + \boldsymbol{\lambda} \otimes \mathbf{d}u) \\ - v(\boldsymbol{\lambda} \otimes \mathbf{d}\rho + \mathbf{d}\rho \otimes \boldsymbol{\lambda}) + \Omega v^2 \boldsymbol{\lambda} \otimes \boldsymbol{\lambda} + \rho^2 \Omega^{-2} \boldsymbol{k}, \end{aligned}$$

which is regular whenever $\Omega = 0$. Moreover, following the discussion of Section 19.2, the various metric coefficients are analytic in the coordinates $(\rho, \theta^{\mathcal{A}})$. The conformal representation of static spacetimes given above shows that *static spacetimes admit a smooth conformal extension which includes a portion of null infinity.* However, this description is not suitable for a spacetime discussion of spatial infinity. This issue will be elaborated in Chapter 20. The discussion of this section can be extended to include stationary spacetimes; for a discussion of the required considerations, see Dain (2001b).

19.3 A regularity condition

As an application of the results on the analyticity of solutions to the conformal static equations at i, in this section a proof is given of a property of the conformal structure of static solutions which plays a central role in the discussion of Chapter 20. The analysis of this section is best carried out in spinors and is adapted from Beig (1991).

Before stating the main result of this section it is convenient to discuss some ancillary consequences of the conformal static equations. In what follows, all the spinors are expressed in terms of their components with respect to some spin dyad $\{\epsilon_{\boldsymbol{A}}{}^{A}\}$ associated to the frame $\{\boldsymbol{e}_{\boldsymbol{i}}\}$ corresponding to the particular realisation of harmonic $\boldsymbol{h}$-normal coordinates at i.

Lemma 19.1 (*behaviour of the symmetrised derivatives of* ζ *at* i) *A solution to the conformal static equations satisfies*

$$D_{(\boldsymbol{A}_p\boldsymbol{B}_p}D_{\boldsymbol{A}_{p-1}\boldsymbol{B}_{p-1}}\cdots D_{\boldsymbol{A}_1)\boldsymbol{B}_1}\zeta(i)=0.$$

Proof For the cases $p=0,\,1,\,2$, the result follows from a direct computation using the conditions in (19.11), observing that $h_{(\boldsymbol{ABCD})}=0$. For higher order derivatives, the result follows by induction, using that the spinorial version of the equation associated to the zero quantity S_{ij} is given by

$$D_{\boldsymbol{AB}}D_{\boldsymbol{CD}}\zeta=\varsigma h_{\boldsymbol{ABCD}}+(\mu-1)\zeta s_{\boldsymbol{ABCD}},\tag{19.27}$$

and using that $D_{\boldsymbol{EF}}h_{\boldsymbol{ABCD}}=0$ and $h_{(\boldsymbol{ABC})\boldsymbol{D}}=0$. □

Remark. Lemma 19.1 implies, in particular, that

$$D_{(\boldsymbol{A}_p\boldsymbol{B}_p}D_{\boldsymbol{A}_{p-1}\boldsymbol{B}_{p-1}}\cdots D_{\boldsymbol{A}_1\boldsymbol{B}_1)}\zeta(i)=0.$$

The main result of this section is the following:

Proposition 19.1 (*behaviour of the derivatives of the Cotton spinor at* i) *A solution to the conformal static equations satisfies*

$$D_{(\boldsymbol{A}_p\boldsymbol{B}_p}D_{\boldsymbol{A}_{p-1}\boldsymbol{B}_{p-1}}\cdots D_{\boldsymbol{A}_1\boldsymbol{B}_1}b_{\boldsymbol{ABCD})}(i)=0,\qquad p=0,\,1,\,2,\ldots.\tag{19.28}$$

The original proofs of this result were given independently by Friedrich (1988) and Beig (1991).

Proof The proof of this result follows from considering Equation (19.18) in the form

$$(1-\mu\zeta)b_{\boldsymbol{ABCD}}=2\mu s_{\boldsymbol{Q}(\boldsymbol{ABC}}D_{\boldsymbol{D})}{}^{\boldsymbol{Q}}\zeta.\tag{19.29}$$

Using the conditions in (19.11) one obtains $b_{ABCD}(i) = 0$. Repeated differentiation and symmetrisation of Equation (19.29) yields

$$\begin{aligned}(1-\mu\zeta)D_{(A_pB_p}D_{A_{p-1}B_{p-1}}\cdots D_{A_1B_1}b_{ABCD)}\\ -p\mu D_{(A_pB_p}\zeta D_{A_{p-1}B_{p-1}}\cdots D_{A_1B_1}b_{ABCD)}\\ +\cdots-\mu D_{(A_pB_p}D_{A_{p-1}B_{p-1}}\cdots D_{A_1B_1}\zeta b_{ABCD)}\\ = D_{(A_pB_p}\cdots D_{A_1B_1}s_{|Q|ABC}D_{D)}{}^{Q}\zeta\\ +\cdots+s_{Q(ABC}D_{A_pB_p}\cdots D_{A_1B_1}D_{D)}{}^{Q}\zeta.\end{aligned}$$

Using Lemma 19.1 it follows that every term in the above expression, save for the first one in the left-hand side, vanishes when evaluated at i. This yields the desired result. □

Remark. Condition (19.28) has been called in Friedrich (1988), for reasons to be elaborated in Chapter 20, the ***radiativity condition***. In Friedrich (1998c) it has been given the name ***regularity condition***. In tensorial notation Equation (19.28) takes the form

$$D_{\{\alpha_p}D_{\alpha_{p-1}}\cdots D_{\alpha_1}b_{\beta\gamma\}}(i) = 0 \qquad p = 0, 1, 2, \ldots.$$

Conformal transformation properties

Let ω denote a smooth function defined in a neighbourhood of i satisfying $\omega(i) \neq 0$. From the conformal transformation properties of the Cotton tensor – see Equation (5.19) – it follows that under the rescaling $\boldsymbol{h} \mapsto \boldsymbol{h}' = \omega^2\boldsymbol{h}$ the Cotton spinor satisfies

$$b'_{ABCD} = \omega^{-1}b_{ABCD}.$$

Thus, $b'_{ABCD}(i) = 0$ if $b_{ABCD}(i) = 0$. Using the transformation law of the connection one finds that $D'_{A_1B_1}b'_{ABCD}(i) = D_{A_1B_1}b_{ABCD}(i)$ as the correction terms associated to the transition tensor involve $b_{ABCD}(i) = 0$. Hence, $D'_{(A_1B_1}b'_{ABCD)}(i) = 0$. Proceeding inductively one concludes that

$$D'_{(A_pB_p}D'_{A_{p-1}B_{p-1}}\cdots D'_{A_1B_1}b'_{ABCD)}(i) = 0, \qquad p = 0, 1, 2, \ldots.$$

Consequently, *the regularity condition* (19.28) *is conformally invariant*. This conformal invariance allows the following reading of Proposition 19.1: *the conformal class of a 3-metric satisfying the static equations cannot be arbitrary.* More precisely, condition (19.28) is a necessary condition for a metric $\boldsymbol{h}$ to belong to the conformal class of a static metric.

19.4 Multipole moments

In Newtonian gravity time-independent gravitational fields are characterised by a sequence of multipole moments. It is desirable to have a similar characterisation for time-independent solutions to the Einstein field equations describing isolated bodies. One of the advantages of the conformal approach to static spacetimes is that it allows a geometric formulation of the notion of multipole moments. Following the original treatment in Geroch (1970a,b) one defines a sequence of tensor fields $\{P,\, P_i,\, P_{i_1 i_2},\, \dots\}$ in a neighbourhood $\mathcal{U}$ of i via the recursive relations

$$
\begin{aligned}
&P \equiv \Omega^{-1/2}(1-v),\\
&P_i \equiv D_i P,\\
&P_{i_2 i_1} \equiv D_{\{i_2} P_{i_1\}} - \frac{1}{2} P r_{i_2 i_1},\\
&P_{i_{p+1}\cdots i_1} \equiv D_{\{i_{p+1}} P_{i_p \cdots i_1\}} - \frac{1}{2} p(2p-1) P_{\{i_{p+1}\cdots i_3} r_{i_2 i_1\}}, \qquad p = 2,\, 3, \dots .
\end{aligned}
$$

The particular form of the lower order correction terms in the definition of the tensors $P_{i_{p+1}\cdots i_1}$ has been chosen so as to ensure conformal invariance of the definition of multipole moments to be given below – this observation follows from a tedious computation which will not be further elaborated here. The multipole moments of a static solution are then obtained by evaluating the above tensors at i. To this end, choose a smooth coordinate system $\underline{x} = (x^\alpha)$ on $\mathcal{U}$ and denote by $P_\alpha,\, P_{\alpha_2\alpha_1},\, \dots$ the components of the tensors with respect to these coordinates and define the ***multipole moments of the static solution*** with respect to the coordinates $\underline{x}$ to be the sequence $\{m,\, m_\alpha,\, m_{\alpha_2\alpha_1},\, \cdots\}$ with

$$
m \equiv P(i), \qquad m_{\alpha_p\cdots\alpha_1} \equiv P_{\alpha_p\cdots\alpha_1}(i), \qquad p = 1,\, 2,\, 3 \dots .
$$

For a given p, the 2^p quantities $m_{\alpha_p\cdots\alpha_1}$ are called the ***2^p-poles***. In particular, m is the ***monopole*** (the mass) and m_α is the ***dipole moment***. As the multipole moments are expressed as the value of a tensor at a point, it follows that, under a coordinate transformation $\underline{x}' = (x'^\alpha(\underline{x}))$, the multipole moments transform as

$$
m' = m, \qquad m'_\alpha = A_\alpha{}^\beta m_\beta, \qquad m'_{\alpha_p\cdots\alpha_1} = A_{\alpha_p}{}^{\beta_p} \cdots A_{\alpha_1}{}^{\beta_1} m_{\beta_p\cdots\beta_1}, \tag{19.30}
$$

where $(A_\alpha{}^\beta)$ are the components of 3×3 invertible real matrices; that is, $(A_\alpha{}^\beta) \in GL(3,\mathbb{R})$. Observe that the monopole is invariant under a change of coordinates. For the particular choice of the conformal factor given by Equation (19.7) one has that $P = m$ so that $D_i P = 0$ and accordingly $m_\alpha = 0$; in other words, in the conformal gauge determined by (19.7) one is automatically in the *centre of mass*.

The properties of the multipole expansions in Newtonian gravity raise the question of to what extent the general relativistic multipole moments determine a solution to the static equations, and vice versa. The construction described in the previous paragraph can be thought of as mapping a static solution $(v, \tilde{\boldsymbol{h}})$ to the collection of multipoles $\{m,\, m_\alpha,\, m_{\alpha_2\alpha_1},\, \dots\}$. Now, two collections of multipoles $\{m,\, m_\alpha,\, m_{\alpha_2\alpha_1},\, \cdots\}$ and $\{m',\, m'_\alpha,\, m'_{\alpha_2\alpha_1},\, \dots\}$ are said to be ***equivalent*** if there exists $(A_\alpha{}^\beta) \in GL(3,\mathbb{R})$ such that the relations in (19.30) hold. In Beig and Simon (1980a) the following has been proved:

Theorem 19.2 (***multipole theorem***) *If two static solutions* $(v, \tilde{\boldsymbol{h}})$ *and* $(v', \tilde{\boldsymbol{h}}')$ *lead to multipole sequences which are equivalent, then the static solutions are isometric in a neighbourhood of* i.

Although a detailed proof of the above theorem will not be provided, it is of interest to discuss the basic underlying ideas. The fundamental problem is the following: given a sequence of multipoles $\{m, m_\alpha, m_{\alpha_2\alpha_1}, \ldots\}$, *how can one reconstruct the pair* $(v, \tilde{\boldsymbol{h}})$ *solving the static equations?* To answer this question one first employs an inductive argument which relies on the definition of the multipole moments, the conformal static equations and the commutator of covariant derivatives to show that the values of the fields ζ, s, $s_{\alpha\beta}$ and any of their covariant derivatives at the point i can be expressed in terms of the multipole moments. Thus, one can compute the Taylor expansions (in harmonic $\boldsymbol{h}$-normal coordinates) of these fields around i. From the general theory of Taylor expansions one knows that the expansions are unique. Moreover, it is a classical result of Riemannian geometry that the sequence

$$\{r_{\alpha\beta\gamma\delta}(i), D_\eta r_{\alpha\beta\gamma\delta}(i), D_{\eta_2\eta_1} r_{\alpha\beta\gamma\delta}(i), \ldots\}$$

determines, in a unique way, the Taylor expansion of the components of the metric $h_{\alpha\beta}$ – again, in $\boldsymbol{h}$-normal coordinates (x^α) centred at i; see, for example, Günther (1975). A final argument shows that applying the above procedure to two equivalent sequences of multipoles leads to two metrics which are isometric.

Now, given any set of multipole moments subject to the appropriate convergence condition, it is natural to expect that there exists a static solution having precisely those multipole moments. In other words, the sequence of multipoles characterises (in a suitable) unique manner the static spacetime. As a result of the analyses in Friedrich (2007) and Herberthson (2009) one has the following:

Theorem 19.3 (***sufficient conditions on the sequence of multipoles for the existence of a static solution***) *Let* $\{m, m_\alpha, m_{\alpha_2\alpha_1}, \ldots\}$ *denote the components of a sequence of real valued, totally symmetric trace-free tensors at the origin of* $\mathbb{R}^3$ *expressed in terms of Cartesian coordinates* $\underline{x} = (x^\alpha)$. *If constants* M, $C > 0$ *can be found such that*

$$|m_{\alpha_p \cdots \alpha_1}| \leq \frac{p! M}{C^p}, \tag{19.31}$$

then there exists a static, asymptotically flat spacetime having the multipole moments $\{m, m_\alpha, m_{\alpha_2\alpha_1}, \ldots\}$.

The proof of the above result goes beyond the scope of this book. Again, only the basic underlying ideas are briefly discussed. The starting point of the analysis is to exploit the analyticity of the solutions to the conformal static equations provided by Theorem 19.1 to implement a ***complex analytic extension*** of the whole setting. More precisely, the fields $h_{\alpha\beta}$, ζ, ς, $s_{\alpha\beta}$ can be extended near i by analyticity into the complex domain and regarded as *holomorphic* (i.e. complex analytic) fields on a complex analytic manifold $\mathcal{S}_{\mathbb{C}}$. Restricting $\mathcal{S}_{\mathbb{C}}$ to be a sufficiently small neighbourhood of i one can use similarly extended normal coordinates $\underline{x} = (x^\alpha)$ centred at i to define an analytic system of coordinates on $\mathcal{S}_{\mathbb{C}}$ which identifies the latter with an open neighbourhood of the origin in $\mathbb{C}^3$. The original manifold $\mathcal{S}$ is then a three-dimensional real analytic submanifold of $\mathcal{S}_{\mathbb{C}}$. Under the analytic extension the main differential geometric concepts and formulas remain valid. In particular, the extended fields, to be denoted again by $h_{\alpha\beta}$, ζ, s, $s_{\alpha\beta}$, satisfy the conformal static vacuum field equations on $\mathcal{S}_{\mathbb{C}}$. In order to provide a geometric perspective of the problem, one considers the function $\Gamma \equiv \delta_{\alpha\beta}x^\alpha x^\beta$ on $\mathcal{S}$ which extends to a holomorphic function on $\mathcal{S}_{\mathbb{C}}$ satisfying the equation $h^{\alpha\beta}D_\alpha\Gamma D_\beta\Gamma = -4\Gamma$. While restricted to $\mathcal{S}$, the function Γ vanishes only at i. On $\mathcal{S}_{\mathbb{C}}$ its set of zeros is a two-dimensional complex submanifold of $\mathcal{S}_{\mathbb{C}}$,

$$\mathcal{N}_i \equiv \{p \in \mathcal{S}_{\mathbb{C}} \,|\, \Gamma(p) = 0\},$$

the so-called ***complex null cone at*** i. This cone is generated by the complex null geodesics through i; see Figure 19.1. The analogy between the (spacetime) conformal field equations and the conformal static field equations discussed in Section 19.1.1 suggests the formulation of a characteristic initial value problem for the conformal static field equations on the null cone $\mathcal{N}_i$. The formulation of this characteristic initial value problem requires the determination of suitable initial data. An argument involving the idea of *exact sets of fields* – see Penrose and Rindler (1984) – allows one to show that the basic data for this characteristic problem are is given by the sequence of fields

$$\{s_{\alpha\beta}(i),\, D_{\{\alpha_1}s_{\alpha\beta\}}(i),\, \ldots,\, D_{\{\alpha_p\cdots\alpha_1}s_{\alpha\beta\}}(i), \ldots\}.$$

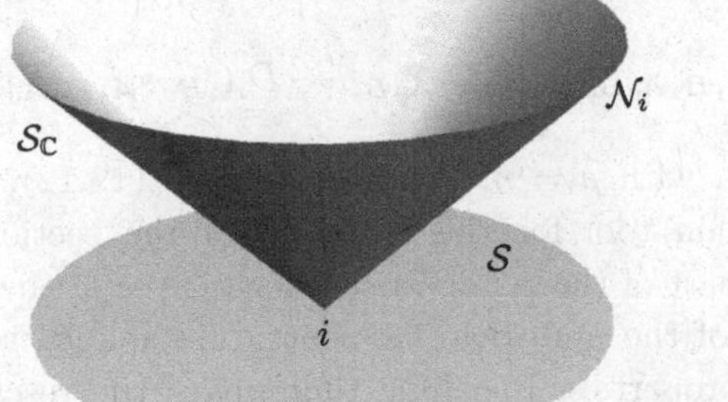

Figure 19.1 Schematic representation of the complex null cone through i, $\mathcal{N}_i$, as described in the main text.

The above ***null data*** can be obtained by repeated differentiation along the direction of the complex null generators of $\mathcal{N}_i$ of the components of $s_{\alpha\beta}$; see, for example, the discussion in Friedrich (2004, 2007).

Given the analyticity of the setting described in the previous paragraphs, one can make use of the *Cauchy-Kowalewskaya theorem* to discuss the existence of analytic solutions to this characteristic problem and to provide convergence conditions on the null data which ensure the existence of a solution; see the Appendix to this chapter. The convergence conditions thus obtained are similar to the ones in Equation (19.31) of Theorem 19.3 and, in particular, ensure the existence of a real static solution. This is the main result of Friedrich (2007). To obtain the convergence condition on the sequence of multipoles one needs to analyse the relation between the null data and the sequence of multipoles. Inspection shows that the null data and the multipole moments are in a one-to-one correspondence. This correspondence, however, is non-linear and implicit. The detailed analysis of this correspondence in Herberthson (2009) allows the transformation of the convergence conditions for the null data into convergence conditions for the sequence of multipoles.

19.5 Uniqueness of the conformal structure of static metrics

As a final application of the conformal static equations, the extent to which the conformal class of the 3-metric $\tilde{\boldsymbol{h}}$ determines a solution to the static equations will be analysed. This question was first analysed in Beig (1991) from where the main ideas of the analysis are adapted. An alternative discussion of some aspects of this problem is given in Friedrich (2008a,b).

The multipole Theorem 19.2 shows that a static solution is determined by its multipole moments. Thus, it is natural to try to relate the multipole moments to the conformal class of the metric $\boldsymbol{h}$. In what follows, for $p = 1, 2, 3, \ldots$ define

$$\beta_{\boldsymbol{A}_p\boldsymbol{B}_p\cdots\boldsymbol{A}_1\boldsymbol{B}_1\boldsymbol{A}_0\boldsymbol{B}_0} \equiv D^{\boldsymbol{Q}}{}_{(\boldsymbol{B}_p}D_{\boldsymbol{A}_{p-1}\boldsymbol{B}_{p-1}}\cdots D_{\boldsymbol{A}_1\boldsymbol{B}_1}b_{\boldsymbol{A}_0\boldsymbol{B}_0\boldsymbol{A}_p\boldsymbol{Q})}(i).$$

Using an inductive argument similar (albeit lengthier!) to the one leading to Proposition 19.1 one obtains the family of identities

$$\beta_{\boldsymbol{A}_p\boldsymbol{B}_p\cdots\boldsymbol{A}_1\boldsymbol{B}_1\boldsymbol{A}_0\boldsymbol{B}_0} = 6\mu D_{(\boldsymbol{A}_p\boldsymbol{B}_p}\cdots D_{\boldsymbol{A}_2\boldsymbol{B}_2}s_{\boldsymbol{A}_1\boldsymbol{B}_1\boldsymbol{A}_0\boldsymbol{B}_0)}(i), \tag{19.32}$$

for $p = 1, 2, 3, \ldots$ with $\mu = m^2/4$; see Equation (19.12). The above identities constitute the main tool for the reminder of the section. Observe that the tensorial counterpart of the symmetrised derivatives of the spinor enter directly in the definition of the multipole moments. The quantities $\beta_{\boldsymbol{A}_p\boldsymbol{B}_p\cdots\boldsymbol{A}_0\boldsymbol{B}_0}$ have good conformal properties. Recalling that under the rescaling $\boldsymbol{h}' = \omega^2\boldsymbol{h}$ with $\omega(i) \neq 0$, one has the transformation rules

$$b'_{\boldsymbol{ABCD}} = \omega^{-1}b_{\boldsymbol{ABCD}}, \qquad \epsilon'^{\boldsymbol{AB}} = \omega^{-1}\epsilon^{\boldsymbol{AB}}.$$

It follows that

$$\beta'_{A_pB_p\cdots A_0B_0} = \omega^{-2}(i)\beta_{A_pB_p\cdots A_0B_0}.$$

Consider now two solutions $(\boldsymbol{h}, s_{ABCD}, \zeta, \varsigma)$ and $(\boldsymbol{h}', s'_{ABCD}, \zeta', \varsigma')$ to the conformal static equations such that

$$\boldsymbol{h}' = \omega^2\boldsymbol{h},$$

and consider the question of under which circumstances will the above two solutions determine *the same physical static solution* $(v, \tilde{\boldsymbol{h}})$ – modulo isometries. The identities (19.32) show how the multipole moments of the two solutions are connected to each other. One then needs further conditions that allow one to constrain the relation between solutions further. In view of the conformal nature of the problem, the natural object to look for those extra conditions is the Cotton spinor. Proposition 19.1 and the identities (19.32) already provide information about some of the derivatives of b_{ABCD} at i. The only derivatives which have not yet been considered are *divergences* of the form $D^{PQ}D_{(PQ}D_{A_pB_p}\cdots D_{A_1B_1}b_{ABCD)}$. A direct computation using Equations (19.27) and (19.29) yields

$$D^{PQ}b_{PQCD}(i) = 0, \tag{19.33a}$$

$$D^{PQ}D_{(PQ}b_{ABCD)}(i) = 0, \tag{19.33b}$$

$$D^{PQ}D_{(PQ}D_{EF}b_{ABCD)}(i) = 0. \tag{19.33c}$$

However, a lengthy computation reveals that

$$D^{PQ}D_{(PQ}D_{GH}D_{EF}b_{ABCD)}(i) = -24\mu D_{(GH}s^{Q}{}_{EFA}(i)s_{BCD)Q}(i).$$

Defining, for convenience, $O_{GHEFABCD} \equiv D_{(GH}s^{Q}{}_{EFA}(i)s_{BCD)Q}(i)$, a computation using the definition of the quantities $\beta_{A_pB_p\cdots A_0B_0}$ allows the reexpression of this quantity in the form

$$O_{GHEFABCD} = \frac{1}{36}\beta^{Q}{}_{(GHFEA}\beta_{BCD)Q}. \tag{19.34}$$

The conformal transformation properties can be easily read from this last expression. Namely, one has that

$$O'_{GHEFABCD} = \omega^{-5}(i)O_{GHEFABCD}.$$

On the other hand, it can be checked that

$$D'^{PQ}D'_{(PQ}D'_{GH}D'_{EF}b'_{ABCD)}(i) = \omega^{-3}(i)D^{PQ}D_{(PQ}D_{GH}D_{EF}b_{ABCD)}(i).$$

From the above transformation rules, assuming that $O_{GHEFABCD} \neq 0$ one concludes that

$$\omega^2(i) = \mu'/\mu.$$

Observe that if $O_{GHEFABCD} = 0$, no conclusion can be extracted from the analysis. As a consequence of Equation (19.34), the requirement $O_{GHEFABCD} \neq 0$ is a condition on the conformal structure of the static solutions under consideration. If it holds, then using the identities (19.32) one concludes that the two solutions to the conformal static equations will have the same multipole moments if they have the same mass. Moreover, as a consequence of the multipole Theorem 19.2, they are isometric. The analysis of this section is summarised in the following theorem, first proven in Beig (1991):

Theorem 19.4 (***uniqueness of the conformal structure of static solutions***) *Two solutions to the conformal static equations with the same mass, lying in the same conformal class and satisfying $O_{GHEFABCD} \neq 0$ are isometric.*

The condition $O_{GHEFABCD} \neq 0$ can be seen to be violated if the two static solutions are axially symmetric about a common axis; see, for example, Beig (1991). As discussed in Friedrich (2008a) this is, in fact, the only possibility. More precisely, a static solution which admits a non-trivial rescaling leading to a new static solution must be axially symmetric and admit a conformal Killing vector. There exists a three-parameter family of such solutions. These have been explicitly found in Friedrich (2008b).

The Schwarzschild solution

A case of particular interest is when $\tilde{\boldsymbol{h}}$ is conformally flat. It then follows that $\beta_{\boldsymbol{A}_p\boldsymbol{B}_p\cdots\boldsymbol{A}_0\boldsymbol{B}_0} = 0$ for all p, and the only non-vanishing mass multipole is the mass m. Invoking, again, the multipole Theorem 19.2 it follows that for a given value of m there exists only one solution, up to isometries, with this property – namely, the Schwarzschild solution. An alternative derivation of the *uniqueness of the Schwarzschild spacetime among the class of conformally flat static solutions* which makes no use of the multipole theorem has been given in Friedrich (2004). In this analysis the conformal static equations are explicitly integrated along geodesics starting at i.

19.6 Characterisation of static initial data

An issue related to the questions discussed in the previous section concerns the characterisation of initial data for a static spacetime – this question will be of relevance in Chapter 20. More precisely, one is interested in the following question: *given a 3-metric $\boldsymbol{h}$, under which circumstances does there exist in the conformal class $[\boldsymbol{h}]$ another metric $\tilde{\boldsymbol{h}}$ which, together with some scalar v, constitutes a solution to the static equations?*

As in the rest of the chapter, the above question is restricted to a suitable neighbourhood of infinity. Proposition 19.1 shows that not every conformal class will contain a static metric. In other words, condition (19.28) is a *necessary condition* for a metric $\boldsymbol{h}$ to be conformal to a static metric. Now, condition (19.28) is not sufficient. The relations (19.33b) and (19.33c) show that there exist further conditions (in fact, an infinite hierarchy of them) on the conformal class, algebraically independent from (19.28), which need to be satisfied by a metric $\boldsymbol{h}$ in order to be conformal to a static metric. The *gap* between a conformal class of 3-metrics satisfying the regularity condition (19.28) and a conformal class containing a static metric has been analysed in detail in Friedrich (2013).

The level of detail required to discuss the main result of Friedrich (2013) goes well beyond the scope of this chapter, and only the key ideas are briefly mentioned. If a metric $\boldsymbol{h}$ is conformal to a metric $\boldsymbol{h}'$ solving the conformal static equations, then writing $\boldsymbol{h}' = \omega^2 \boldsymbol{h}$ for some suitable conformal factor ω, it is possible to rewrite the conformal static equations as a highly overdetermined system of differential equations for ω. To analyse the solvability of the conditions one needs to consider the associated integrability conditions. As already anticipated by (19.33b) and (19.33c), these integrability conditions give rise, in addition to (19.28), to restrictions on the conformal structure which take the form of an infinite hierarchy of differential conditions on the Cotton tensor at i. These conditions can be expressed in terms of requirements on a covector constructed from the 3-metric $\boldsymbol{h}$. An interesting feature of the analysis is that the overdetermined system involving the conformal factor ω is highly singular at i. For this system to have a solution, a hierarchy of regularity conditions need to be imposed on the singular part of the equation so that it admits a smooth extension to a neighbourhood of i – this is reminiscent of a procedure which arises in the construction of *radiative initial data sets* in Section 20.2. Remarkably, the required *regularity conditions* turn out to be nothing else but the conditions (19.28).

19.7 Further reading

A systematic analysis of time-independent solutions to the Einstein field equations is provided in Beig and Schmidt (2000). This reference provides an excellent point of entry to the extensive literature on static and stationary solutions in general relativity. A survey of the various approaches to define multipole moments for time-independent solutions to the Einstein field equations can be found in Quevedo (1990). An analysis of global aspects of static and stationary spacetimes can be found in Anderson (2000).

Several of the results discussed in this chapter admit a generalisation to the case of stationary solutions. The definitions of multipole moments given by Geroch (1970a,b) have been extended to the stationary case in Hansen (1974). The analyticity of solutions of the conformal stationary field equations has been analysed in Beig and Simon (1980b, 1981); see also Kundu (1981). However, in this case the 3-metric $\boldsymbol{h}$ of a surface of constant time will not be analytic; see Dain (2001b). Instead, the analyticity refers to the 3-metric γ of the ***quotient space*** obtained from identifying points on the spacetime lying on the same orbit of the stationary Killing vector. The analysis of the convergence conditions for null data of static solutions in Friedrich (2007) has been extended to the case of stationary solutions in Aceña (2009). An alternative analysis of multipole expansions of static solutions with the aim of obtaining convergence conditions on sequences of multipoles has been given in Bäckdahl and Herberthson (2005a,b, 2006) and Bäckdahl (2007).

The analysis of the conformal static equations by means of the complex null cone through i was first introduced in Friedrich (1988). Further extensions of this method have been given in Friedrich (2004, 2007, 2013).

Appendix 1: Hölder conditions

Given $0 < \alpha \leq 1$, a real valued function f on an open set $\mathcal{U} \subset \mathbb{R}^n$ is said to satisfy the ***Hölder condition*** with exponent α on $\mathcal{U}$ if there exists a non-negative constant C such that

$$|f(x) - f(y)| \leq C|x - y|^\alpha, \qquad \text{for all } x,\, y \in \mathcal{U}.$$

If the above is the case, one writes $f \in C^{0,\alpha}(\mathcal{U})$. The Hölder condition is a stronger notion of continuity; that is, a function satisfying the Hölder condition is continuous, but not all continuous functions satisfy the Hölder condition for some α. More generally, one says that $f \in C^{k,\alpha}(\mathcal{U})$ if all its derivatives up to order k satisfy the Hölder condition for a given α. The Hölder condition plays an important role in the regularity of solutions to elliptic PDEs; see, for example, Evans (1998) for further details.

Appendix 2: the Cauchy-Kowalewskaya theorem

The ***Cauchy-Kowalewskaya*** theorem asserts the local existence, in a neighbourhood of $t = 0$, of a real analytic solution $\mathbf{u}(t, \underline{x})$ to the quasilinear first-order initial value problem

$$\begin{aligned}\partial_t \mathbf{u} &= \mathbf{A}^\alpha(t, \underline{x}, \mathbf{u})\partial_\alpha \mathbf{u} + \mathbf{B}(t, \underline{x}, \mathbf{u}),\\ \mathbf{u}(0, \underline{x}) &= \mathbf{u}_\star(\underline{x}),\end{aligned}$$

where $\mathbf{A}^\alpha(t, \underline{x}, \mathbf{u})$, $\mathbf{B}(t, \underline{x}, \mathbf{u})$ and $\mathbf{u}_\star(\underline{x})$ are real analytic functions of their arguments; see, for example, Evans (1998) for further details. A discussion of the various approaches to prove this result can be found in Shinbrot and Welland (1976).

20

Spatial infinity

This chapter discusses the properties of the conformal Einstein field equations and the behaviour of their solutions in a suitable neighbourhood of spatial infinity. This analysis is key in any attempt to extend the *semiglobal existence results* for Minkowski-like spacetimes of Chapter 16 to a truly *global problem* where initial data is prescribed on a *Cauchy hypersurface.* An interesting feature of the semiglobal existence Theorem 16.1 is that the location of the intersection of the initial hyperboloid with null infinity does not play any role in the formulation of the result. This observation suggests that the essential difficulty in formulating a Cauchy problem is concentrated in an arbitrary (spacetime) neighbourhood of spatial infinity. The subject of this chapter can be regarded, in some sense, as a natural extension of the study of static spacetimes in Chapter 19 to dynamic spacetimes – a considerable amount of the discussion of the present chapter is devoted to understanding why this is indeed the case. A further objective of this chapter is to understand the close relation between the behaviour of the gravitational field at spatial infinity and the so-called peeling behaviour discussed in Chapter 10. The main technical tool in this chapter is the construction of the so-called *cylinder at spatial infinity* – a conformal representation of spatial infinity allowing the formulation of a regular initial value problem by means of which it is possible to relate properties of the initial data on a Cauchy hypersurface with the behaviour of the gravitational field at null infinity.

Despite recent developments in the understanding of the behaviour of solutions to the Einstein field equations in a neighbourhood of spatial infinity, several key issues still remain open.

20.1 Cauchy data for the conformal field equations near spatial infinity

To begin to understand the difficulties behind the formulation of a *standard* initial value problem for a Minkowski-like spacetime, it is convenient to look at the behaviour of Cauchy data for the conformal equations near spatial infinity.

20.1.1 General set up

In what follows, initial data sets $(\tilde{\mathcal{S}}, \tilde{\boldsymbol{h}}, \tilde{\boldsymbol{K}})$ which are *asymptotically Euclidean and regular* in the sense of Definition 11.2 will be considered. As the discussion in this chapter will be mainly concerned with the behaviour of the data in a neighbourhood of spatial infinity, it will be assumed, without any loss of generality, that the manifold $\tilde{\mathcal{S}}$ has only one asymptotic end. The basic aspects of the analysis of spatial infinity are already present in *time-symmetric initial data sets.* Thus, attention is restricted to this type of configuration. Finally, it will be assumed, unless otherwise explicitly stated, that the *conformal metric* $\boldsymbol{h}$ *is analytic in a suitable neighbourhood of the point at infinity* i. This assumption allows the simplification of a number of arguments and calculations and allows one to analyse the solutions to the Einstein field equations under optimal regularity assumptions of the initial data – it is, however, non-essential.

Remark. Static initial data sets satisfy the assumptions described in the previous paragraph.

In Chapter 11 it has been seen that the conformal factor Ω linking a particular choice of conformal metric $\boldsymbol{h}$ with the physical metric $\tilde{\boldsymbol{h}}$ admits, in a suitable neighbourhood $\mathcal{U}$ of i and in terms of *normal coordinates* $\underline{x} = (x^\alpha)$ *centred at* i, the decomposition

$$\Omega = \frac{|x|^2}{(U + |x|W)^2}, \qquad |x|^2 = \delta_{\alpha\beta}x^\alpha x^\beta, \tag{20.1}$$

where $U/|x|^2$ is the *Green function* of the Yamabe operator and describes the local geometry around i, while W contains global information; see the discussion in Section 11.6.4. In particular, one has that $U = 1 + O(|x|^2)$ is analytic if $\boldsymbol{h}$ is analytic and, moreover, $W(i) = m/2$ where m denotes the Arnowitt-Deser-Misner (ADM) mass of the initial data set.

There is a certain amount of freedom in the choice of the conformal scaling of the metric $\boldsymbol{h}$. For the purposes of the present discussion, it is convenient to consider the scaling introduced in Section 11.6.2 for which

$$h_{\alpha\beta} = -\delta_{\alpha\beta} + O(|x|^3), \tag{20.2}$$

so that the curvature tensor of $\boldsymbol{h}$ satisfies, in this gauge, $r_{\alpha\beta\gamma\delta}(i) = 0$. This gauge construction is supplemented by an $\boldsymbol{h}$*-normal frame* $\{\boldsymbol{e}_i\}$ *centred at* i; that is, one has

$$h_{\boldsymbol{ij}} \equiv \boldsymbol{h}(\boldsymbol{e_i}, \boldsymbol{e_j}) = -\delta_{\boldsymbol{ij}}, \qquad D_{\dot\gamma}\boldsymbol{e_i} = 0,$$

where $\dot\gamma$ denotes the tangent vector to any geodesic passing through i; compare the discussion in Section 18.4.1. Consistent with Equation (20.2), the frame coefficients in $\boldsymbol{e_i} = e_{\boldsymbol{i}}{}^\alpha\partial_\alpha$ satisfy

$$e_{\boldsymbol{i}}{}^\alpha = \delta_{\boldsymbol{i}}{}^\alpha + O(|x|^3).$$

Moreover, the associated connection coefficients are of the form

$$\gamma_i{}^j{}_k = O(|x|^2),$$

and one has that

$$r_{ij} = O(|x|),$$

where $r_{ij} \equiv r_{\alpha\beta} e_i{}^\alpha e_j{}^\beta$ are the components of the Ricci tensor of $\boldsymbol{h}$ with respect to the frame $\{\boldsymbol{e}_i\}$.

20.1.2 The rescaled Weyl and Schouten tensors on $\mathcal{U}$

For time-symmetric initial data, the components of the electric part of the rescaled Weyl tensor, d_{ij}, and the Schouten tensor, L_{ij}, with respect to the frame $\{\boldsymbol{e}_i\}$ are given on $\mathcal{U}$, respectively, by

$$d_{ij} = \frac{1}{\Omega^2}\big(D_{\{i}D_{j\}}\Omega + \Omega s_{ij}\big), \qquad L_{ij} = -\frac{1}{\Omega}D_{\{i}D_{j\}}\Omega + \frac{1}{12} r h_{ij};$$

see Section 11.4.3. The spinorial version of the above expressions is readily found to be given by

$$\phi_{ABCD} = \frac{1}{\Omega^2}\big(D_{(AB}D_{CD)}\Omega + \Omega s_{ABCD}\big), \tag{20.3a}$$

$$L_{ABCD} = -\frac{1}{\Omega}D_{(AB}D_{CD)}\Omega + \frac{1}{12} r h_{ABCD}. \tag{20.3b}$$

The first of the above equations implies an expression for the Cotton spinor b_{ABCD}; see Equation (19.19). Rewriting Equation (20.3a) in the form

$$\Omega^2 \phi_{ABCD} = D_{(AB}D_{CD)}\Omega + \Omega s_{ABCD},$$

taking the spinorial curl of the latter, commuting covariant derivatives in the term with the triple derivatives of Ω and recalling that the Cotton spinor is given by $b_{ABCD} = D_{(A}{}^Q s_{BCD)Q}$, one concludes that

$$b_{ABCD} = 2D_{(A}{}^Q \Omega \phi_{BCD)Q} + \Omega D_{(A}{}^Q \phi_{BCD)Q}. \tag{20.4}$$

Behaviour close to infinity

As in the case of *hyperboloidal data* discussed in Section 11.7, the expressions (20.3a) and (20.3b) are formally singular whenever $\Omega = 0$. Accordingly, the discussion of the behaviour of d_{ij} and L_{ij} close to i requires some care.

In view of the decomposition (20.1) it is convenient to define the ***massless part of the conformal factor*** as $\grave{\Omega} \equiv |x|^2/U^2$. By construction one has

$$\grave{\Omega}(i) = 0, \qquad D_i\grave{\Omega}(i) = 0, \qquad D_iD_j\grave{\Omega}(i) = -2\delta_{ij}, \tag{20.5}$$

so that one obtains the expansion

$$\grave{\Omega} = \delta_{\alpha\beta}x^{\alpha}x^{\beta} + O(|x|^3).$$

In particular, it is observed that $D_{\{i}D_{j\}}\grave{\Omega} = O(|x|^3)$.

One can define a ***massive part of the conformal factor*** as $\check{\Omega} \equiv \Omega - \grave{\Omega}$. Rewriting Equation (20.1) as

$$\Omega = \grave{\Omega}\left(1 + \frac{|x|W}{U}\right)^{-2},$$

and using that

$$D_i|x| = \frac{x_i}{r} + O(|x|^2), \qquad D_iD_j|x| = \frac{1}{|x|^3}(|x|^2\delta_{ij} - x_ix_j) + O(|x|),$$

where $x_i \equiv \delta_i{}^{\beta}\delta_{\alpha\beta}x^{\alpha}$, one concludes, taking into account the boundary conditions (20.5), that

$$D_{\{i}D_{j\}}\Omega = -\frac{3mx_{\{i}x_{j\}}}{|x|} + O(|x|^2).$$

Finally, observing that, in the present gauge, $s_{ij} = O(|x|)$ and $r = O(|x|)$, one finds

$$d_{ij} = -\frac{3mx_{\{i}x_{j\}}}{|x|^5} + O(|x|^{-2}), \qquad L_{ij} = \frac{3mx_{\{i}x_{j\}}}{|x|^3} + O(|x|^0).$$

Accordingly, one concludes that both d_{ij} and L_{ij} are singular at i with

$$d_{ij} = O(|x|^{-3}), \qquad L_{ij} = O(|x|^{-1}), \qquad \text{as } |x| \to 0.$$

The analysis of the consequences of this singular behaviour and how to deal with it will be the central subject of the remainder of this chapter.

Remark. Even if the massive part of the conformal factor vanishes, one still has a potential source of singularities in the fields d_{ij} and L_{ij}. This can be seen from computing the ***massless part of the electric part of the Weyl tensor*** given by

$$\begin{aligned}\grave{d}_{ij} &\equiv \frac{1}{\grave{\Omega}^2}(D_{\{i}D_{j\}}\grave{\Omega} + \grave{\Omega}s_{ij}) \\ &= \frac{1}{|x|^4}(U^2D_{\{i}D_{j\}}|x|^2 - 4UD_{\{i}|x|^2D_{j\}}U \\ &\quad - 2|x|^2UD_{\{i}D_{j\}}U + 6|x|^2D_{\{i}UD_{j\}}U + |x|^2U^2s_{ij}). \qquad (20.6)\end{aligned}$$

A similar expression holds for $\grave{L}_{ij}$ – the massless part of the Schouten tensor. In the next section, it will be seen that under suitable assumptions on the metric $\boldsymbol{h}$, both $\grave{d}_{ij}$ and $\grave{L}_{ij}$ extend to analytic fields in a neighbourhood of i.

20.2 Massless and purely radiative spacetimes

Intuition on the behaviour of solutions to the conformal Einstein field equations in a neighbourhood of spatial infinity can be obtained from the analysis of ***massless initial data sets***, that is, data sets for which $\Omega = \grave{\Omega}$. In view of the ***mass positivity theorem*** – see Schoen and Yau (1979) – the idea of considering initial data sets which are massless might at first seem strange. The *rigidity part* of this theorem implies that if the ADM mass m of an initial data set vanishes, then one has, in fact, initial data for the Minkowski spacetime or the initial data set is singular somewhere. Since the present chapter is mainly focused on an analysis local to i (i.e. in a suitably small neighbourhood of i) the presence of singularities in the interior of the 3-manifold $\mathcal{S}$ can be disregarded.

20.2.1 Geometric setting

Given a massless initial data set for the conformal Einstein field equations in a neighbourhood $\mathcal{U}$ of the point at infinity i, the conformal evolution equations determine a (future and past) development $(\mathcal{M}, \boldsymbol{g}, \Xi)$ which is contained in $D(\mathcal{U}) = D^+(\mathcal{U}) \cup D^-(\mathcal{U})$. Following the notation of Chapter 14, let $I^+(i)$ and $I^-(i)$ denote the timelike future and timelike past of i in $(\mathcal{M}, \boldsymbol{g})$ and by $\mathscr{N}_i^{\,+}$ and $\mathscr{N}_i^{\,-}$ the null cones generated by the null geodesics passing through i. From the boundary conditions (20.5) satisfied by $\grave{\Omega}$ it follows that the spacetime conformal factor Ξ has a non-degenerate critical point at i which, for simplicity, is assumed to be the only critical point of Ξ. The locus of points for which $\Xi = 0$ coincides with $\mathscr{N}_i^{\,+} \cup \mathscr{N}_i^{\,-}$; see the discussion in Section 16.3.

As observed in Friedrich (1988), the development $(\mathcal{M}, \boldsymbol{g}, \Xi)$ of the conformal field equations can be regarded from a dual perspective:

(i) On the set $\mathcal{M}_i^c \equiv \mathcal{M} \setminus \big(I^+(i) \cup \mathscr{N}_i^{\,+} \cup I^-(i) \cup \mathscr{N}_i^{\,-}\big)$, corresponding to the exterior of the null cones, the metric $\tilde{\boldsymbol{g}} \equiv \Xi^{-2}\boldsymbol{g}$ is a solution to the Einstein field equations with vanishing mass for which i represents spatial infinity i^0.

(ii) On $I^+(i)$ the metric $\tilde{\boldsymbol{g}}$ represents a solution to the Einstein field equations for which the point i represents past timelike infinity i^- and the set $\mathscr{I}^- \equiv \mathscr{N}_i^{\,+} \setminus \{i\}$ past null infinity. For suitably smooth initial data, the solution thus obtained has a regular past timelike infinity and provides an example of ***purely radiative spacetimes***; see the discussion in Section 18.4.

A schematic depiction of the above geometric setting can be found in Figure 20.1.

20.2.2 A regularity condition at spatial infinity

Not all initial data sets lead to developments $(\mathcal{M}, \boldsymbol{g}, \Xi)$ such that $I^+(i)$ admits a regular past timelike infinity – as given in point (ii) of the previous section. The purpose of this section is to identify the initial data sets with this property.

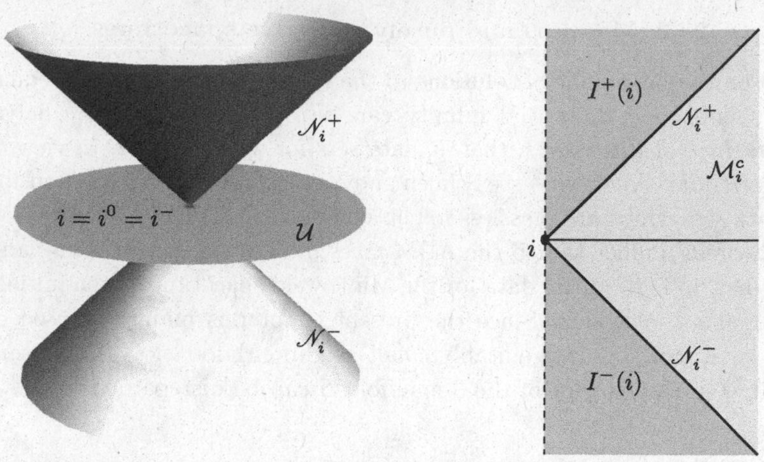

Figure 20.1 Schematic depiction of the geometric set up for massless spacetimes as described in the main text. The set $\mathcal{M}_i^c \equiv \mathcal{M} \setminus \big(I^+(i) \cup \mathcal{N}_i^+ \cup I^-(i) \cup \mathcal{N}_i^-\big)$ contains a solution to the vacuum Einstein field equations with vanishing mass for which i represents spatial infinity i^0, while on $I^+(i)$ one obtains a purely radiative solution where $\mathcal{N}_i^+$ represents past null infinity $\mathscr{I}^-$ and i corresponds to past timelike infinity i^-.

The arguments of this section are always carried out in a suitable neighbourhood of the vertex i.

As already discussed, direct inspection of expression (20.6) shows that although $\grave{\Omega} = |x|^2/U^2$ is a real analytic function in a suitable neighbourhood $\mathcal{U}$ of i fixed by the equation

$$2\grave{\Omega}\Delta_{\boldsymbol{h}}\grave{\Omega} = 3D^i\grave{\Omega}D_i\grave{\Omega} - \frac{1}{2}\grave{\Omega}^2 r[\boldsymbol{h}]$$

and the boundary conditions (20.5), in general, the corresponding fields $\grave{d}_{ij}$ and $\grave{L}_{ij}$ will not have the same degree of smoothness.

To identify conditions on $\boldsymbol{h}$ ensuring that the fields $\grave{d}_{ij}$ and $\grave{L}_{ij}$ are also analytic, it is convenient to consider a ***complex analytic extension*** of $\mathcal{U}$ similar to the one discussed in Section 19.4. To this end, one allows the normal coordinates $\underline{x} = (x^\alpha)$ in $\mathcal{U}$ to take values in a neighbourhood $\mathcal{U}_{\mathbb{C}}$ of the origin of $\mathbb{C}^3$ so that the original neighbourhood $\mathcal{U}$ is a real three-dimensional analytic submanifold of $\mathcal{U}_{\mathbb{C}}$. Accordingly, the fields $\Gamma \equiv |x|^2$, $\boldsymbol{h}$, $\boldsymbol{e}_i$, $\grave{\Omega}$, s_{ij} and $r[\boldsymbol{h}]$ are extended by analyticity into the complex domain and are regarded as *holomorphic* fields over $\mathcal{U}_{\mathbb{C}}$. Assuming that $i = \{p \in \mathcal{U}_{\mathbb{C}} \mid x^\alpha(p) = 0\}$ is the only critical point of $\grave{\Omega}$ in $\mathcal{U}_{\mathbb{C}}$, the complex null cone generated by the complex null geodesics through i is given by the two-dimensional complex submanifold

$$\mathcal{N}_{\mathbb{C}}(i) \equiv \{p \in \mathcal{U}_{\mathbb{C}} \mid \Gamma(p) = 0\}.$$

By construction, the set of points in $\mathcal{U}_{\mathbb{C}}$ where $\grave{\Omega}$ vanishes coincides with $\mathcal{N}_{\mathbb{C}}(i)$.

A first criterion ensuring the analyticity of $\grave{\phi}_{ABCD}$ – or, equivalently, $\grave{d}_{ij}$ – is given by the following:

Proposition 20.1 (***analyticity of the massless part of the Weyl spinor: first version***) *The analyticity of $\grave{\phi}_{ABCD}$ near i is equivalent to the requirement*

$$D_{(P_pQ_p} \cdots D_{P_1Q_1)} D_{EF}\big(D_{(AB}D_{CD)}\grave{\Omega} + \grave{\Omega} s_{ABCD}\big)(i) = 0 \tag{20.7}$$

for $p = 0, 1, 2, \ldots$.

Remark. As will be seen in the following, this condition is, in fact, a condition on the conformal class of $\boldsymbol{h}$. The conformal constraint equations imply that $\grave{L}_{ij}$ is analytic if $\grave{d}_{ij}$ is analytic.

Proof The proof of the lemma makes repeated use of a *factorisation lemma* for holomorphic functions, which is discussed in the Appendix to this chapter; see Lemma 19.2. The analyticity of $\boldsymbol{h}$ implies that the field $\grave{\phi}_{ABCD}$ on $\mathcal{U}$ extends to a holomorphic field on $\mathcal{U}_{\mathbb{C}} \setminus \mathcal{N}_{\mathbb{C}}(i)$ which satisfies

$$\grave{\Omega}^2 \grave{\phi}_{ABCD} = D_{(AB}D_{CD)}\grave{\Omega} + \grave{\Omega} s_{ABCD}. \tag{20.8}$$

Now, if $\grave{\phi}_{ABCD}$ is analytic at $\mathcal{N}_{\mathbb{C}}(i)$ one can take a derivative of the above expression and evaluate on $\mathcal{N}_{\mathbb{C}}(i)$ to find that

$$D_{EF}\big(D_{(AB}D_{CD)}\grave{\Omega} + \grave{\Omega} s_{ABCD}\big)\big|_{\mathcal{N}_{\mathbb{C}}(i)} = 0. \tag{20.9}$$

Conversely, given condition (20.9), one would like to verify that $\grave{\phi}_{ABCD}$ is analytic at $\mathcal{N}_{\mathbb{C}}(i)$. Using the factorisation Lemma 19.2 in the Appendix to this chapter, it follows that there is a holomorphic field f_{ABCDEF} such that, in a neighbourhood of $\mathcal{N}_{\mathbb{C}}(i)$, one has

$$D_{EF}\big(D_{(AB}D_{CD)}\grave{\Omega} + \grave{\Omega} s_{ABCD}\big) = \grave{\Omega} f_{ABCDEF}. \tag{20.10}$$

Defining $Z_{ABCD} \equiv D_{(AB}D_{CD)}\grave{\Omega} + \grave{\Omega} s_{ABCD}$, the last equation can be written as $D_{EF}Z_{ABCD} = \grave{\Omega} f_{ABCDEF}$. Moreover, transvecting Equation (20.10) with $D^{EF}\grave{\Omega}$ one obtains

$$D^{EF}\grave{\Omega} D_{EF} Z_{ABCD}\big|_{\mathcal{N}_{\mathbb{C}}(i)} = 0, \tag{20.11}$$

which can be read as an ordinary differential equation for Z_{ABCD} along the generators of $\mathcal{N}_{\mathbb{C}}(i)$. It follows that the field Z_{ABCD} is constant along the generators, so that evaluating Equation (20.11) at the vertex one concludes that

$$D_{(AB}D_{CD)}\grave{\Omega} + \grave{\Omega} s_{ABCD} = 0 \qquad \text{on } \mathcal{N}_{\mathbb{C}}(i).$$

Using again Lemma 19.2 again, one finds that there exists a further holomorphic field f_{ABCD} such that

$$D_{(AB}D_{CD)}\grave{\Omega} + \grave{\Omega} s_{ABCD} = \grave{\Omega} f_{ABCD} \qquad \text{in a neighbourhood of } \mathcal{N}_{\mathbb{C}}(i).$$

Taking a derivative of this expression and comparing the result with Equation (20.10) it follows that

$$\left(D_{\boldsymbol{EF}}\grave{\Omega} f_{\boldsymbol{ABCD}}\right)\big|_{\mathcal{N}_{\mathbb{C}}(i)} = 0.$$

One observes that $D_{\boldsymbol{EF}}\grave{\Omega} \neq 0$ on $\mathcal{U}_{\mathbb{C}} \setminus \{i\}$. It follows that there exists a holomorphic function $g_{\boldsymbol{ABCD}}$ such that $f_{\boldsymbol{ABCD}} = \grave{\Omega} g_{\boldsymbol{ABCD}}$ so that

$$\grave{\Omega}^2 g_{\boldsymbol{ABCD}} = D_{(\boldsymbol{AB}}D_{\boldsymbol{CD})}\grave{\Omega} + \grave{\Omega} s_{\boldsymbol{ABCD}}.$$

Comparing the latter with Equation (20.8) one concludes that $g_{\boldsymbol{ABCD}}$ coincides with $\grave{\phi}_{\boldsymbol{ABCD}}$ on $\mathcal{U}_{\mathbb{C}} \setminus \mathcal{N}_{\mathbb{C}}(i)$, and, thus, $\grave{\phi}_{\boldsymbol{ABCD}}$ is analytic near i as required.

Having encoded the analyticity of $\grave{\phi}_{\boldsymbol{ABCD}}$ in terms of the vanishing of a spinorial field at $\mathcal{N}_{\mathbb{C}}(i)$, one makes use of Lemma 19.3 in the Appendix to this chapter to express the latter condition as an equivalent series of conditions at the vertex i. □

An alternative way of imposing conditions on the metric $\boldsymbol{h}$ ensuring that $\grave{\phi}_{\boldsymbol{ABCD}}$ is analytic at i can be obtained using expression (20.4) for the Cotton spinor. One has the following:

Proposition 20.2 (***analyticity of the massless part of the Weyl spinor: second version***) *A necessary condition on the metric* $\boldsymbol{h}$ *for* $\grave{\phi}_{\boldsymbol{ABCD}}$ *to be analytic in a neighbourhood of* i *is given by the sequence of conditions*

$$D_{(\boldsymbol{P}_p\boldsymbol{Q}_p} \cdots D_{\boldsymbol{P}_1\boldsymbol{Q}_1} \grave{b}_{\boldsymbol{ABCD})}(i) = 0, \qquad p = 0,\, 1,\, 2, \ldots \tag{20.12}$$

Proof As in the proof of Proposition 20.1, one considers an arbitrary null geodesic $\gamma(\mathrm{s}) \in \mathcal{N}_{\mathbb{C}}(i)$, $\mathrm{s} \in \mathbb{C}$, such that $\gamma(0) = i$ with tangent vector having the spinorial counterpart $\kappa^{\boldsymbol{A}}\kappa^{\boldsymbol{B}}$ with $\kappa^{\boldsymbol{A}}$ parallely propagated along $\gamma(\mathrm{s})$. The latter implies that

$$\kappa^{\boldsymbol{A}} D_{\boldsymbol{AB}}\grave{\Omega} = 0. \tag{20.13}$$

For $\Omega = \grave{\Omega}$, relation (20.4) takes the form

$$\grave{b}_{\boldsymbol{ABCD}} = 2D_{(\boldsymbol{A}}{}^{\boldsymbol{Q}}\grave{\Omega}\grave{\phi}_{\boldsymbol{BCD})\boldsymbol{Q}} + \grave{\Omega} D_{(\boldsymbol{A}}{}^{\boldsymbol{Q}}\grave{\phi}_{\boldsymbol{BCD})\boldsymbol{Q}},$$

which can be extended to $\mathcal{U}_{\mathbb{C}}$ by analyticity. In particular, at $\mathcal{N}_{\mathbb{C}}(i)$, contracting the previous expression four times with $\kappa^{\boldsymbol{A}}$ one obtains, in view of (20.13), that

$$\left(\kappa^{\boldsymbol{A}}\kappa^{\boldsymbol{B}}\kappa^{\boldsymbol{C}}\kappa^{\boldsymbol{D}}\grave{b}_{\boldsymbol{ABCD}}\right)\big|_{\mathcal{N}_{\mathbb{C}}(i)} = 0.$$

Applying repeatedly $\kappa^{\boldsymbol{P}}\kappa^{\boldsymbol{Q}} D_{\boldsymbol{PQ}}$ one finds that

$$\left(\kappa^{\boldsymbol{P}_p}\kappa^{\boldsymbol{Q}_p} \cdots \kappa^{\boldsymbol{P}_1}\kappa^{\boldsymbol{Q}_1}\kappa^{\boldsymbol{A}}\kappa^{\boldsymbol{B}}\kappa^{\boldsymbol{C}}\kappa^{\boldsymbol{D}} D_{\boldsymbol{P}_p\boldsymbol{Q}_p} \cdots D_{\boldsymbol{P}_1\boldsymbol{Q}_1}\grave{b}_{\boldsymbol{ABCD}}\right)\big|_{\mathcal{N}_{\mathbb{C}}(i)} = 0,$$

for $p = 0,\, 1,\, 2, \ldots$ Restricting the previous expression to i and recalling that $\kappa^{\boldsymbol{A}}$ is arbitrary, one obtains (20.12). □

Remark. In Chapter 19 it has been shown that the Cotton spinor of the 3-metric of a static solution satisfies a condition that is identical to (20.12); see Proposition 19.1. The role it plays in ensuring the analyticity of the rescaled Weyl tensor at i motivates the alternative name ***regularity condition***. From the analysis of Section 19.3 it follows that the expression (20.12) is conformally invariant and, accordingly, is a condition on the conformal class $[\boldsymbol{h}]$.

In Friedrich (1998c) it has been proven that conditions (20.7) and (20.12) are, in fact, equivalent. The following proposition rounds out nicely the discussion of this section.

Proposition 20.3 (***equivalence between the conditions ensuring analyticity of the Weyl spinor***) *Conditions* (20.7) *and* (20.12) *are equivalent. Consequently, a necessary and sufficient condition on the conformal class* $[\boldsymbol{h}]$ *to ensure that the fields* $\grave{\phi}_{ABCC}$ *and* $\grave{L}_{ABCD}$ *extend analytically to* i *is given by*

$$D_{(P_pQ_p} \cdots D_{P_1Q_1}\grave{b}_{ABCD)}(i) = 0, \qquad p = 0, 1, 2, \ldots$$

The proof of the equivalence between (20.7) and (20.12) involves a lengthy computation that goes beyond the scope of this section. Interested readers are referred to Friedrich (1998c), theorem 4.2 and its proof, for full details.

20.2.3 Construction of massless data

In Friedrich (1988) it has been observed that asymptotically initial data sets can be used as *seeds* for the construction of massless initial data sets.

Let $(\boldsymbol{h}_{\circledast}, s_{ij}, \zeta, \varsigma)$ denote an asymptotically Euclidean solution to the conformal static Equations (19.17). The above fields are expressed in a conformal gauge for which $r[\boldsymbol{h}_{\circledast}] = 0$. Moreover, the conformal factor linking the conformal metric $\boldsymbol{h}_{\circledast}$ with the physical metric $\tilde{\boldsymbol{h}}_{\circledast}$ via $\boldsymbol{h}_{\circledast} \equiv \Omega_{\circledast}^2 \tilde{\boldsymbol{h}}_{\circledast}$ satisfies

$$\Omega_{\circledast}^{-1/2} = \zeta^{-1/2} + \frac{1}{2}m \tag{20.14}$$

with m the ADM mass of the static solution; compare Equation (19.8). One can then look for conformal factors Ω solving the conformal Hamiltonian constraint

$$2\Omega\Delta_{\boldsymbol{h}_{\circledast}}\Omega = 3|\boldsymbol{D}_{\circledast}\Omega|^2$$

– compare Equation (11.15a) – where $\boldsymbol{D}_{\circledast}$ and $\Delta_{\boldsymbol{h}_{\circledast}}$ denote, respectively, the Levi-Civita covariant derivative and Laplacian of the conformally static metric $\boldsymbol{h}_{\circledast}$. Making use of the ansatz $\Omega = \Omega(\zeta)$, observing that

$$D_i\Omega = \frac{\mathrm{d}\Omega}{\mathrm{d}\zeta}D_i\zeta, \qquad D_iD_j\Omega = \frac{\mathrm{d}^2\Omega}{\mathrm{d}\zeta^2}D_i\zeta D_j\zeta + \frac{\mathrm{d}\Omega}{\mathrm{d}\zeta}D_iD_j\zeta,$$

and taking into account the conformal static equations one arrives at the ordinary differential equation

$$2\zeta\Omega\frac{\mathrm{d}^2\Omega}{\mathrm{d}\zeta^2}+3\Omega\frac{\mathrm{d}\Omega}{\mathrm{d}\zeta}=3\zeta\left(\frac{\mathrm{d}\Omega}{\mathrm{d}\zeta}\right)^2.$$

The general solution to this equation is given by

$$\Omega=\frac{c_1\zeta}{(1+c_2\sqrt{\zeta})^2}, \qquad c_1, c_2 \text{ constants.}$$

The subclass of analytic solutions is given by $c_2 = 0$, so that – up to a constant factor – one has

$$\Omega=\zeta. \tag{20.15}$$

When $c_2 \neq 0$, that is, in the case of a non-analytic solution, one has a non-vanishing mass. In hindsight, the solution (20.15) could have been guessed directly from Equation (20.14) as ζ satisfies the boundary conditions (20.5); compare also Equation (19.11). Summarising, one has:

Proposition 20.4 (***massless initial data out of static data***) *Given a solution to the conformal static equations* $(\boldsymbol{h}_{\circledast}, s_{ij}, \zeta, \varsigma)$ *in a neighbourhood* $\mathcal{U}$ *of the point at infinity* i*, the metric*

$$\tilde{\boldsymbol{h}}=\zeta^{-2}\boldsymbol{h}_{\circledast},$$

defined in a suitable punctured neighbourhood of i*, satisfies the time-symmetric Hamiltonian constraint* $r[\tilde{\boldsymbol{h}}]=0$ *and has vanishing mass. Moreover, the rescaled Weyl and Schouten spinors obtained from Equations (20.3a) and* (20.3b) *by setting* $\Omega=\zeta$ *are analytic at* i*.*

The above result can be generalised to obtain massless initial data for the conformal Einstein-Maxwell field equations; see Simon (1992).

20.2.4 Evolution of massless data

Proposition 20.4 can be combined with the conformal evolution equations to obtain a development admitting the dual interpretation discussed in Section 20.1.1. The simplest way of implementing the construction is to make use of the extended conformal Einstein field equations expressed in terms of a conformal Gaussian gauge; see Section 13.4.

Initial data for the congruence of conformal geodesics $(x(\tau), \tilde{\boldsymbol{\beta}}(\tau))$ underlying the conformal Gaussian gauge can be set by the conditions

$$\tau=0, \quad \dot{\boldsymbol{x}}\perp\mathcal{U}, \quad \Theta_\star=\zeta, \quad \dot{\Theta}_\star=0, \quad \boldsymbol{d}_\star\equiv\Theta_\star\tilde{\boldsymbol{\beta}}_\star=\mathbf{d}\zeta, \qquad \text{on } \mathcal{U}.$$

The conformal factor associated to the congruence of conformal geodesics – see Proposition 5.1 – is then given by

$$\Theta=\zeta\left(1+\frac{\varsigma\tau^2}{2\zeta}\right), \qquad \varsigma\equiv\frac{1}{3}\Delta_{\boldsymbol{h}_{\circledast}}\zeta.$$

In the above expression ζ and ς are regarded as constant along a given conformal geodesic. Now, one has that $\varsigma(i) = -2$. Hence, by choosing $\mathcal{U}$ sufficiently small so that $\varsigma < 0$ in this neighbourhood, one can ensure that Θ has real roots at $\tau = \pm\sqrt{2\zeta/|\varsigma|}$. Observe, in addition, that $\Theta = 0$ at i.

The existence of a development for the massless data provided by Proposition 20.4 is given by the following result:

Theorem 20.1 (***existence of purely radiative spacetimes***) *Let $\mathbf{u}_\star$ denote initial data for the extended conformal Einstein field equations on a neighbourhood $\mathcal{U}$ of i constructed from a pair $(\boldsymbol{h}_\circledast, \zeta)$ as given by Proposition 20.4. Then there exists $\tau_\bullet > 0$ ensuring the existence of a smooth solution $\mathbf{u}$ to the conformal Einstein field equations on*

$$\mathcal{M}_{\tau_\bullet} \equiv D^+(\mathcal{U}) \cap \big([0, \tau_\bullet) \times \mathcal{U}\big),$$

such that the restriction of $\mathbf{u}$ to $\mathcal{U}$ coincides with $\mathbf{u}_\star$. Define

$$\mathscr{N}_{\tau_\bullet} \equiv \{p \in \mathcal{M}_{\tau_\bullet} \mid \Theta(p) = 0\},$$

and let $\boldsymbol{g}$ be the Lorentzian metric constructed from the solution $\mathbf{u}$. For this solution one has the following:

(i) On $\mathcal{M}_{\tau_\bullet} \setminus (\mathscr{N}_{\tau_\bullet} \cup (I^+(i) \cap \mathcal{M}_{\tau_\bullet}))$ the metric $\tilde{\boldsymbol{g}} \equiv \Theta^{-2}\boldsymbol{g}$ is a solution to the vacuum Einstein field equations with vanishing mass for which $\mathscr{N}_{\tau_\bullet} \setminus \{i\}$ represents future null infinity $\mathscr{I}^+$ and the point i corresponds to spatial infinity i^0.

(ii) On $I^+(i) \cap \mathcal{M}_{\tau_\bullet}$ the Lorentzian metric $\tilde{\boldsymbol{g}}$ is a purely radiative solution to the Einstein field equations for which $\mathscr{N}_{\tau_\bullet} \setminus \{i\}$ represents past null infinity and the point i corresponds to past timelike infinity.

A schematic depiction of the spacetimes constructed via the above result is given in Figure 20.1.

Proof The local existence of smooth solutions follows from the hyperbolic form of the evolution equations given in Proposition 13.3 together with the local existence for this type of evolution equations provided by Kato's Theorem 12.2. The existence of an actual solution to the full conformal Einstein field equations follows from the form of the associated subsidiary system – see Proposition 13.4 – while the existence of a solution to the Einstein field equations is obtained from Proposition 8.3 whenever $\Theta \neq 0$. The interpretation of the solutions in the regions where $\Theta > 0$ and $\Theta < 0$ follows from the discussion in Section 20.2.1. □

Remark. Although the result is, from the conformal perspective, purely local, from the physical point of view, it is nevertheless of a semi-global nature. It follows from the smoothness of the solution $\mathbf{u}$ to the conformal Einstein field equations provided by Theorem 20.1 on $\mathscr{N}_{\tau_\bullet} \setminus \{i\}$ and, in particular, of the field

ϕ_{ABCD}, that the Weyl tensor of both of the spacetimes in (i) and (ii) satisfy the *Peeling behaviour*; see Theorem 10.4.

20.3 A regular initial value problem at spatial infinity

The purpose of this section is to discuss the formulation of a regular asymptotic initial value problem for the conformal evolution equations for data with non-vanishing mass.

Consider a suitable neighbourhood $\mathcal{U}$ of the point at infinity i of the point-compactification $(\mathcal{S}, \boldsymbol{h}, \Omega)$ of an asymptotically Euclidean (time-symmetric) Cauchy hypersurface of a vacuum spacetime $(\tilde{\mathcal{M}}, \tilde{\boldsymbol{g}})$. Let $\{\boldsymbol{e}_i\}$ denote an $\boldsymbol{h}$-orthonormal frame and let $\{\epsilon_{\boldsymbol{A}}{}^{A}\}$ denote an associated spin frame. The key idea behind the formulation of a *regular asymptotic initial value problem* is based on the observation that a conformal rescaling of the form

$$\Omega \mapsto \Omega' \equiv \kappa^{-1}\Omega \qquad (20.16)$$

induces a rescaling of the frame of the form

$$\boldsymbol{e}_i \mapsto \boldsymbol{e}'_i \equiv \kappa \boldsymbol{e}_i, \qquad \epsilon_{\boldsymbol{A}}{}^{A} \mapsto \epsilon'_{\boldsymbol{A}}{}^{A} \equiv \kappa^{1/2}\epsilon_{\boldsymbol{A}}{}^{A}.$$

Accordingly, the components of the rescaled Weyl spinor with respect to the spin frame $\{\epsilon_{\boldsymbol{A}}{}^{A}\}$ rescale as

$$\phi_{ABCD} \mapsto \phi'_{ABCD} \equiv \kappa^3 \phi_{ABCD}.$$

Now, if one considers the rescaling (20.16) with a function κ of the form

$$\kappa = |x|\varkappa, \qquad \text{with } \varkappa \text{ smooth such that } \varkappa(i) \neq 0, \qquad (20.17)$$

one finds that $\phi'_{ABCD} = O(1)$. Thus, the components of the Weyl spinor with respect to the new frame are bounded at i.

20.3.1 Rescaling of the initial data for the conformal field equations

The discussion of the previous paragraph suggests that the rescaling (20.16) with κ given by (20.17) could be used to formulate a regular Cauchy problem on $\mathcal{U}$. Note, however, that while ϕ'_{ABCD} is bounded at i, there is no guarantee that it will be smooth since $|x|$ is not a smooth function of the normal coordinates $\underline{x} = (x^\alpha)$. Thus, one needs to resort to *polar coordinates* similar to the ones used in Section 19.2.1 to analyse the spacetime conformal extensions of static spacetimes. Letting

$$\rho^2 \equiv \delta_{\alpha\beta}x^\alpha x^\beta, \qquad \rho^\alpha \equiv \frac{x^\alpha}{|x|},$$

and using some coordinates $\theta = (\theta^{\mathcal{A}})$ on $\mathbb{S}^2$ to parametrise the position vector ρ^α, one has that the 3-metric $\boldsymbol{h}$ can be written as

$$\boldsymbol{h} = -\mathbf{d}\rho \otimes \mathbf{d}\rho + \rho^2 \boldsymbol{k}, \qquad \boldsymbol{k} \equiv h_{\alpha\beta}\partial_{\mathcal{A}}\rho^\alpha \partial_{\mathcal{B}}\rho^\beta \mathbf{d}\theta^{\mathcal{A}} \otimes \mathbf{d}\theta^{\mathcal{B}},$$

with $\boldsymbol{k}|_{\rho=0} = -\boldsymbol{\sigma}$; compare Equation (19.24). It is natural to consider an $\boldsymbol{h}$-orthonormal frame $\{\boldsymbol{e}_i\}$ with dual coframe $\{\boldsymbol{\omega}^i\}$ satisfying

$$\boldsymbol{\omega}^3 = \mathbf{d}\rho, \qquad \rho^2\boldsymbol{k} = -\boldsymbol{\omega}^1 \otimes \boldsymbol{\omega}^1 - \boldsymbol{\omega}^2 \otimes \boldsymbol{\omega}^2.$$

The indexing of the basis vectors has been chosen so as to match that of the spatial Infeld-van der Waerden symbols; see Equations (4.11a) and (4.11b). From the above expressions it follows, writing $\boldsymbol{\omega}^i = \omega^i{}_\alpha \mathbf{d}x^\alpha$, that $\omega^3{}_\alpha = O(1)$, $\omega^1{}_\alpha, \omega^2{}_\alpha = O(\rho)$, while for the frame coefficients in $\boldsymbol{e}_i = e_i{}^\alpha \boldsymbol{\partial}_\alpha$ one has that $e_3{}^\alpha = O(1)$, $e_1{}^\alpha$, $e_2{}^\alpha = O(\rho^{-1})$.

Consistent with the rescaling (20.16), let

$$\boldsymbol{e}'_i \equiv \kappa \boldsymbol{e}_i, \qquad \boldsymbol{\omega}'^i \equiv \kappa^{-1}\boldsymbol{\omega}^i, \tag{20.18}$$

and set $\boldsymbol{e}'_i = e'_i{}^\alpha \partial_\alpha$ and $\boldsymbol{\omega}^i = \omega'^i{}_\alpha \mathbf{d}x^\alpha$. It follows that if the function κ is chosen as in Equation (20.17), then

$$e'_3{}^\alpha = O(\rho), \qquad e'_1{}^\alpha,\ e'_2{}^\alpha = O(1),$$
$$\omega'^3{}_\alpha = O(\rho^{-1}), \qquad \omega'_1{}^\alpha,\ \omega'_2{}^\alpha = O(1),$$

and, moreover, that

$$\boldsymbol{h}' \equiv \kappa^{-2}\boldsymbol{h} = -\frac{1}{\rho^2}\mathbf{d}\rho \otimes \mathbf{d}\rho + \boldsymbol{k}. \tag{20.19}$$

Thus, the coframe coefficients and, consequently, also the metric coefficients are singular at ρ. *This singular behaviour is, however, not an obstacle for the construction of a regular initial value problem as these objects do not explicitly appear as unknowns in the spinorial conformal Einstein field equations in either their standard or their extended form.* Introducing the coordinate $r \equiv -\log\rho$ so that $r \to \infty$ as $\rho \to 0$ one obtains the line element

$$\boldsymbol{h}' = -\mathbf{d}r \otimes \mathbf{d}r + \boldsymbol{k}.$$

Hence, the locus of points for which $\rho = 0$ lies at infinity with respect to the metric $\boldsymbol{h}$ but has finite circumference – and is, in fact, a metric 2-sphere. Accordingly, the rescaling (20.19) *resolves (blows up) the point at infinity into a 2-sphere* which is described locally in terms of the coordinates $\theta = (\theta^{\mathcal{A}})$.

In the remainder of this chapter, for the ***blow up of*** i ***to*** $\mathbb{S}^2$ it will be understood the pair

$$(\mathcal{C}(\mathcal{U}), \{\boldsymbol{e}'_i\})$$

consisting of a 3-manifold $\mathcal{C}(\mathcal{U})$ with boundary $\partial\mathcal{C}(\mathcal{U}) \approx \mathbb{S}^2$ such that $\mathcal{C}(\mathcal{U}) \setminus \partial\mathcal{C}(\mathcal{U})$ can be identified with $\mathcal{U} \setminus \{i\}$ and where the frame $\{\boldsymbol{e}'_i\}$ is given as in Equation (20.18) with a choice of κ as in Equation (20.17). The set $\mathcal{I}^0 \equiv \partial\mathcal{C}(\mathcal{U})$ will be called the ***sphere at infinity.*** Observe that the definition of a blow up makes reference neither to the metric $\boldsymbol{h}'$ nor to the coframe $\{\boldsymbol{\omega}'^i\}$ which are singular as $\rho \to 0$.

The previous definition of the blow up of i has the purpose of simplifying the discussion in the remainder of the chapter. A precise and rigorous discussion of this construction requires the use of the language of fibre bundles. The interested reader is referred to Friedrich (1998c, 2004) for a detailed account; see also Aceña and Valiente Kroon (2011).

The rescaling of the conformal fields

The effects of the frame rescaling (20.18) on the connection coefficients can be analysed by means of the usual transformation formulae for the connection. One has

$$\begin{aligned}\gamma'_{\boldsymbol{i}}{}^{\boldsymbol{j}}{}_{\boldsymbol{k}} &= \omega'^{\boldsymbol{j}}{}_{k} e'_{\boldsymbol{i}}{}^{i} D'_{i} e'_{\boldsymbol{k}}{}^{k} \\ &= \omega'^{\boldsymbol{j}}{}_{k} e'_{\boldsymbol{i}}{}^{i} D_{i} e'_{\boldsymbol{k}}{}^{k} - \kappa^{-1} \omega'^{\boldsymbol{j}}{}_{k} e'_{\boldsymbol{i}}{}^{i} e'_{\boldsymbol{k}}{}^{l} S_{il}{}^{mk} D_{m}\kappa \\ &= \kappa \gamma_{\boldsymbol{i}}{}^{\boldsymbol{j}}{}_{\boldsymbol{k}} - (\delta_{\boldsymbol{k}}{}^{\boldsymbol{j}} D_{\boldsymbol{i}}\kappa + \delta_{\boldsymbol{i}}{}^{\boldsymbol{j}} D_{\boldsymbol{k}}\kappa + \delta_{\boldsymbol{ik}} D^{\boldsymbol{j}}\kappa);\end{aligned}$$

compare a similar computation in Section 15.1.2. The spinorial version of the above expression is given by

$$\gamma'_{\boldsymbol{ABCD}} = \kappa \gamma_{\boldsymbol{ABCD}} - \frac{1}{2}(\epsilon_{\boldsymbol{AC}} D_{\boldsymbol{BD}}\kappa + \epsilon_{\boldsymbol{BD}} D_{\boldsymbol{AC}}\kappa).$$

To complete the discussion of the connection, one needs to consider the rescaling of the covector $\boldsymbol{f}$. From the transformation rules of solutions to the conformal geodesics equations, Equation (5.40), it follows that $\boldsymbol{f}' = \boldsymbol{f} + \mathbf{d}\kappa$. Thus, if $\boldsymbol{f}_\star = 0$, it follows that

$$f'_{\boldsymbol{i}} = D_{\boldsymbol{i}}\kappa, \qquad f'_{\boldsymbol{AB}} = D_{\boldsymbol{AB}}\kappa.$$

Finally, it follows from the transformation rules of the components of the Schouten tensor under the transition to a Weyl connection that

$$\hat{L}_{\boldsymbol{ij}} = \kappa^2 L_{\boldsymbol{ij}}, \qquad \Theta'_{\boldsymbol{ABCD}} = \kappa^2 \Theta_{\boldsymbol{ABCD}}.$$

Comparing the above expressions with (20.3b), it follows that if κ is chosen as in (20.17), then $\Theta'_{\boldsymbol{ABCD}} = O(1)$.

A closer look at the frame

It is convenient to have a more detailed expression for the frame $\{\boldsymbol{e_i}\}$ or, alternatively, its frame spinorial index counterpart $\{\boldsymbol{e_{AB}}\}$ – recall that $\boldsymbol{e_{AB}} \equiv \sigma^{\boldsymbol{i}}{}_{\boldsymbol{AB}} \boldsymbol{e_i}$. Let $\{\boldsymbol{\partial}_+, \boldsymbol{\partial}_-\}$ denote a local basis of vectors on $\mathbb{S}^2$ with dual cobasis $\{\boldsymbol{\alpha}^+, \boldsymbol{\alpha}^-\}$ such that $\boldsymbol{\partial}_- = \overline{\boldsymbol{\partial}_+}$ and, furthermore,

$$\boldsymbol{\sigma} = 2(\boldsymbol{\alpha}^+ \otimes \boldsymbol{\alpha}^- + \boldsymbol{\alpha}^- \otimes \boldsymbol{\alpha}^+),$$

with $\boldsymbol{\sigma}$ denoting the standard metric of $\mathbb{S}^2$. The vectors can be expressed in terms of the local coordinates $\theta = (\theta^{\mathcal{A}})$, but the explicit correspondence will not be required. The vectors $\{\boldsymbol{\partial}_+, \boldsymbol{\partial}_-\}$ originally defined on $\mathbb{S}^2$ can be extended to $\mathcal{C}(\mathcal{U})$

by Lie propagation along the radial direction given by $\boldsymbol{\partial}_\rho$; that is, one requires that $[\boldsymbol{\partial}_\rho, \boldsymbol{\partial}_\pm] = 0$. Using the vector fields $\{\boldsymbol{\partial}_\rho, \boldsymbol{\partial}_\pm\}$ one can then locally write

$$\boldsymbol{e_{AB}} = e_{\boldsymbol{AB}}{}^{\boldsymbol{3}}\boldsymbol{\partial}_\rho + e_{\boldsymbol{AB}}{}^{+}\boldsymbol{\partial}_+ + e_{\boldsymbol{AB}}{}^{-}\boldsymbol{\partial}_-,$$

for suitable frame coefficients $e_{\boldsymbol{AB}}{}^{\boldsymbol{3}}$ and $e_{\boldsymbol{AB}}{}^{\pm}$. These coefficients can be expanded, in turn, in terms of the ***basic valence-2 symmetric spinors***

$$x_{\boldsymbol{AB}} \equiv \sqrt{2}\epsilon_{(\boldsymbol{A}}{}^{\boldsymbol{0}}\epsilon_{\boldsymbol{B})}{}^{\boldsymbol{1}}, \qquad y_{\boldsymbol{AB}} \equiv -\frac{1}{\sqrt{2}}\epsilon_{\boldsymbol{A}}{}^{\boldsymbol{1}}\epsilon_{\boldsymbol{B}}{}^{\boldsymbol{1}}, \qquad z_{\boldsymbol{AB}} \equiv \frac{1}{\sqrt{2}}\epsilon_{\boldsymbol{A}}{}^{\boldsymbol{0}}\epsilon_{\boldsymbol{B}}{}^{\boldsymbol{0}},$$

satisfying

$$x_{\boldsymbol{AB}}x^{\boldsymbol{AB}} = -1, \qquad x_{\boldsymbol{AB}}y^{\boldsymbol{AB}} = 0, \qquad x_{\boldsymbol{AB}}z^{\boldsymbol{AB}} = 0, \tag{20.20a}$$

$$y_{\boldsymbol{AB}}y^{\boldsymbol{AB}} = 0, \qquad y_{\boldsymbol{AB}}z^{\boldsymbol{AB}} = -\frac{1}{2}, \qquad z_{\boldsymbol{AB}}z^{\boldsymbol{AB}} = 0. \tag{20.20b}$$

Expressing the spinorial basis $\{\epsilon_{\boldsymbol{A}}{}^A\}$ in the form $\epsilon_{\boldsymbol{0}}{}^A = o^A$, $\epsilon_{\boldsymbol{1}}{}^A = \iota^A$ one finds that the fields $x_{\boldsymbol{AB}}$, $y_{\boldsymbol{AB}}$ and $z_{\boldsymbol{AB}}$ are, up to a normalisation, the components of the pairs $o_{(A}\iota_{B)}$, $o_A o_B$ and $\iota_A\iota_B$ with respect to the spin basis. Taking into account the contractions (20.20a) and (20.20b) and the line element (20.19) one finds the more detailed expression

$$\boldsymbol{e_{AB}} = x_{\boldsymbol{AB}}\boldsymbol{\partial}_\rho + e_{\boldsymbol{AB}}{}^{+}\boldsymbol{\partial}_+ + e_{\boldsymbol{AB}}{}^{-}\boldsymbol{\partial}_-, \qquad e_{\boldsymbol{01}}{}^{\pm} = 0.$$

20.3.2 The cylinder at spatial infinity

After providing regular initial data for the conformal field equations in the neighbourhood $\mathcal{U}$ of i, one can now specify in more detail the conformal Gaussian system underlying the hyperbolic reduction of the conformal Einstein field equations.

In what follows, the initial data for the congruence of conformal geodesics will be fixed, so that for $p \in \mathcal{U} \setminus \{i\}$ one has:

$$x_\star \equiv x(0) = p, \qquad \dot{\boldsymbol{x}}_\star \equiv \dot{\boldsymbol{x}}(0) = \boldsymbol{e_0} \quad \text{future directed and orthogonal to } \tilde{\mathcal{S}},$$

$$\Theta_\star \tilde{\boldsymbol{g}}(\boldsymbol{e_a}, \boldsymbol{e_b}) = \eta_{\boldsymbol{ab}}, \qquad \Theta_\star > 0,$$

$$\langle \tilde{\boldsymbol{\beta}}, \dot{\boldsymbol{x}} \rangle_\star = 0.$$

For the above data one further lets

$$\Theta_\star = \kappa^{-1}\Omega, \qquad \tilde{\boldsymbol{\beta}}_\star = \Omega^{-1}\mathbf{d}\Omega \quad \text{in } \mathcal{U} \setminus \{i\}, \tag{20.21}$$

where, in a slight abuse of notation, $\tilde{\boldsymbol{\beta}}_\star$ denotes the pull-back of $\tilde{\boldsymbol{\beta}}$ to $\mathcal{U} \setminus \{i\}$. While $\tilde{\boldsymbol{\beta}}_\star$ is singular at i, $\boldsymbol{d}_\star \equiv \Theta_\star\tilde{\boldsymbol{\beta}}_\star$ is regular under the present assumptions.

Using Proposition 5.1 in Chapter 5 it follows that

$$\Theta = \kappa^{-1}\Omega\left(1 - \frac{\kappa^2\tau^2}{\omega^2}\right), \qquad d_{\boldsymbol{a}} = \left(-\frac{2\kappa\Omega\tau}{\omega^2}, \kappa^{-1}\mathbf{d}\Omega\right), \tag{20.22}$$

where

$$\omega \equiv \frac{2\Omega}{\sqrt{|\mathbf{d}\Omega|^2}}.$$

Now, as $\Omega = O(\rho^2)$, it follows that $\omega = O(\rho)$ so that, together with the choice (20.17) for κ one finds that $\kappa/\omega \to 1$. Moreover, both Θ and $d_{\boldsymbol{a}}$ can be seen to have well-defined limits as $\rho \to 0$. Accordingly, the conformal Gaussian gauge can be extended to the set

$$\mathcal{I}^0 \equiv \{p \in \mathcal{U} \mid \rho(p) = 0\} \approx \mathbb{S}^2,$$

despite the fact that the second prescription in (20.21) is singular at the above set.

Assume now that the congruence of conformal geodesics underlying the gauge has no conjugate points on $D(\mathcal{U})$. A point $p \in D(\mathcal{U})$ is described by coordinates $(\tau, \underline{x}_\star)$ where $\underline{x}_\star$ denotes the *normal coordinates* of the intersection of the unique conformal geodesic passing through p with $\mathcal{U}$. Accordingly, a suitable region of $D(\mathcal{U})$ close to $\mathcal{U}$ can be thought of as a subset of $\mathbb{R} \times \mathcal{U}$. In the following it will be convenient to consider the sets

$$\mathcal{M}(\mathcal{U}) \equiv \left\{(\tau, q) \in \mathbb{R} \times \mathcal{U} \mid -\frac{\omega(q)}{\kappa(q)} \leq \tau \leq \frac{\omega(q)}{\kappa(q)}\right\}, \tag{20.23a}$$

$$\mathcal{I} \equiv \{(\tau, q) \in \mathcal{M}(\mathcal{U}) \mid q \in \mathcal{I}^0,\ |\tau| < 1\}, \tag{20.23b}$$

$$\mathcal{I}^\pm \equiv \{(\tau, q) \in \mathcal{M}(\mathcal{U}) \mid q \in \mathcal{I}^0,\ \tau \pm 1\}, \tag{20.23c}$$

$$\mathscr{I}^\pm \equiv \left\{(\tau, q) \in \mathcal{M}(\mathcal{U}) \mid \tau = \pm\frac{\omega(q)}{\kappa(q)}\right\}. \tag{20.23d}$$

If an existence result for solutions to the conformal evolution equations can be obtained, then the set $\mathcal{M}(\mathcal{U})$ gives rise to an extension of the physical spacetime manifold $\tilde{\mathcal{M}}$ in a neighbourhood of spatial infinity, while $\mathscr{I}^\pm$ describe the two components of null infinity. In this representation the point i^0 is replaced by an extended set $\mathcal{I}$, the ***cylinder at spatial infinity***, with both spatial and temporal dimensions. The sets $\mathcal{I}^\pm$ where *"null infinity touches spatial infinity"* will be called, for reasons which will become clearer in the subsequent discussion, the ***critical sets***.

The set up discussed in the previous paragraphs is fixed up to a specific choice of the function $\varkappa$ in (20.17). A convenient and simple choice of this function consists in setting $\kappa = \rho$ so that $\varkappa = 1$ – this choice will be called the ***basic representation***. A schematic depiction of the sets in (20.23a)–(20.23d) of the basic representation is given in Figure 20.2. An alternative choice consists of setting $\kappa = \omega$. In this case Θ vanishes at $\tau \pm 1$, and, accordingly, one calls this construction the ***horizontal representation***. A schematic depiction of the sets in (20.23a)–(20.23d) of the horizontal representation is given in Figure 20.3.

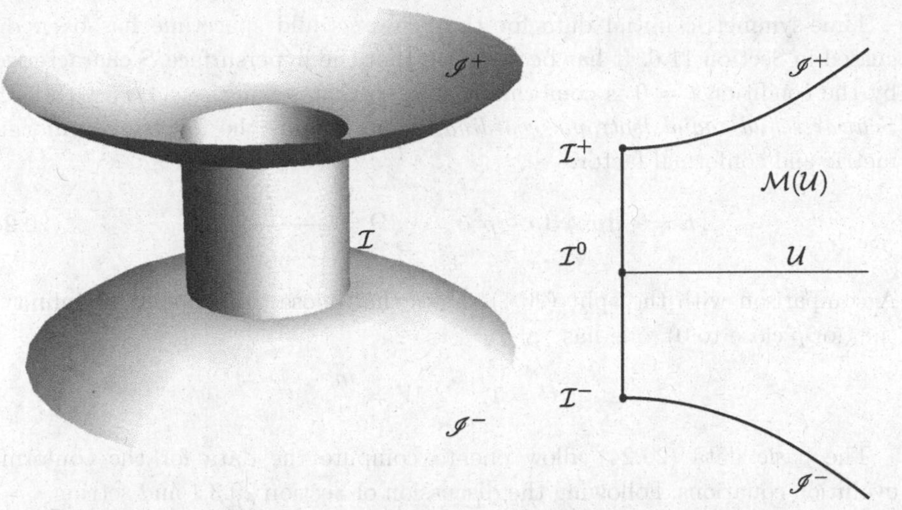

Figure 20.2 Schematic depiction of the basic representation of the set up of the cylinder at spatial infinity. Left, a three-dimensional diagram; right, a two-dimensional longitudinal section. See the main text for further details. Note that the diagram to the right is not a conformal diagram but a graph of the location of the conformal boundary in the conformal Gaussian coordinates.

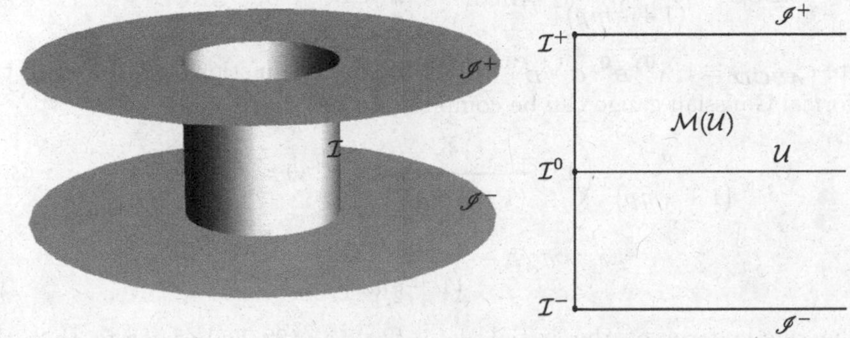

Figure 20.3 Schematic depiction of the horizontal representation of the set up of the cylinder at spatial infinity. Left, a three-dimensional diagram; right, a two-dimensional longitudinal section. See the main text for further details. Note that the diagram to the right is not a conformal diagram but a graph of the location of the conformal boundary in the conformal Gaussian coordinates.

20.3.3 The cylinder at spatial infinity for the Minkowski and Schwarzschild spacetimes

A good way of obtaining intuition about the properties of the conformal evolution equations in a neighbourhood of $\mathcal{I}$ is to consider the case of initial data for the Schwarzschild spacetime. The discussion in this section follows that of section 6 in Friedrich (1998c).

Time-symmetric initial data for the Schwarzschild spacetime has been discussed in Section 11.6. It has been shown that the hypersurface $\tilde{\mathcal{S}}$ characterised by the condition $t = 0$ is conformally flat, so that setting $\rho = 1/\bar{r}$, with $\bar{r}$ the *Schwarzschild radial isotropic coordinate*, one obtains the following conformal metric and conformal factor:

$$\boldsymbol{h} = -\mathbf{d}\rho \otimes \mathbf{d}\rho - \rho^2 \boldsymbol{\sigma}, \qquad \Omega = \frac{\rho^2}{\left(1 + \frac{1}{2}m\rho\right)^2}. \tag{20.24}$$

A comparison with the split (20.1) shows that, close to the point at infinity i (i.e. for ρ close to 0), one has

$$U = 1, \qquad W = \frac{m}{2}.$$

The basic data (20.24) allow one to compute the data for the conformal evolution equations. Following the discussion of Section 20.3.1 and setting $\kappa = \rho$ (i.e. using the standard representation) one finds that

$$e_{\boldsymbol{AB}}{}^{0} = 0, \quad e_{\boldsymbol{AB}}{}^{1} = \rho x_{\boldsymbol{AB}}, \quad e_{\boldsymbol{AB}}{}^{+} = z_{\boldsymbol{AB}}, \quad e_{\boldsymbol{AB}}{}^{-} = y_{\boldsymbol{AB}}, \tag{20.25a}$$

$$f_{\boldsymbol{AB}} = x_{\boldsymbol{AB}}, \quad \xi_{\boldsymbol{ABCD}} = 0, \quad \chi_{(\boldsymbol{AB})\boldsymbol{CD}} = 0, \tag{20.25b}$$

$$\Theta_{\boldsymbol{ABCD}} = \frac{6m\rho}{\left(1 + \frac{1}{2}m\rho\right)^2} \epsilon^2_{\boldsymbol{ABCD}}, \quad \phi_{\boldsymbol{ABCD}} = -6m\epsilon^2_{\boldsymbol{ABCD}}, \tag{20.25c}$$

where $\epsilon^2_{\boldsymbol{ABCD}} \equiv \epsilon_{(\boldsymbol{A}}{}^{\boldsymbol{0}}\epsilon_{\boldsymbol{B}}{}^{\boldsymbol{0}}\epsilon_{\boldsymbol{C}}{}^{\boldsymbol{1}}\epsilon_{\boldsymbol{D})}{}^{\boldsymbol{1}}$. In addition, the functions associated to the conformal Gaussian gauge can be computed to be

$$\Theta = \frac{\rho}{\left(1 + \frac{1}{2}m\rho\right)^2}\left(1 - \frac{\tau^2}{\left(1 + \frac{1}{2}m\rho\right)^2}\right), \qquad \dot{\Theta} = -\frac{2\tau\rho}{\left(1 + \frac{1}{2}m\rho\right)^4},$$

$$d_{\boldsymbol{AB}} = \frac{2\rho x_{\boldsymbol{AB}}}{\left(1 + \frac{1}{2}m\rho\right)^3}.$$

The simple form of the initial data (20.25a)–(20.25c) suggests that the discussion of the conformal evolution equations can be simplified by considering a specific ansatz for the solutions. Some experimentation shows that a *consistent ansatz* is given by

$$e_{\boldsymbol{AB}}{}^{0} = e^{0} x_{\boldsymbol{AB}}, \quad e_{\boldsymbol{AB}}{}^{1} = e^{1} x_{\boldsymbol{AB}}, \quad e_{\boldsymbol{AB}}{}^{+} = e^{+} z_{\boldsymbol{AB}}, \quad e_{\boldsymbol{AB}}{}^{-} = e^{-} y_{\boldsymbol{AB}},$$

$$f_{\boldsymbol{AB}} = f x_{\boldsymbol{AB}}, \quad \xi_{\boldsymbol{ABCD}} = \frac{1}{\sqrt{2}}\xi(\epsilon_{\boldsymbol{AC}}x_{\boldsymbol{BD}} + \epsilon_{\boldsymbol{BD}}x_{\boldsymbol{AC}}),$$

$$\chi_{(\boldsymbol{AB})\boldsymbol{CD}} = \chi_2 \epsilon^2_{\boldsymbol{ABCD}} + \frac{1}{3}\chi h_{\boldsymbol{ABCD}},$$

$$\Theta_{\boldsymbol{ABCD}} = \theta_2 \epsilon^2_{\boldsymbol{ABCD}} + \frac{1}{3}\theta_h h_{\boldsymbol{ABCD}} + \frac{1}{\sqrt{2}}\theta_x \epsilon_{\boldsymbol{AB}} x_{\boldsymbol{CD}},$$

$$\phi_{\boldsymbol{ABCD}} = \phi \epsilon^2_{\boldsymbol{ABCD}},$$

where the components of the vector-valued unknown

$$\mathbf{u} = (e^0, e^1, e^+, e^-, f, \xi, \chi_2, \chi, \theta_2, \theta_h, \theta_x, \phi)$$

are assumed to be real-valued functions of (τ, ρ). The ansatz allows one to reduce the spinorial evolution equations to a system of scalar equations. A lengthy computation renders

$$\partial_\tau e^0 = \frac{1}{3}(\chi_2 - \chi)e^0 - f, \qquad \partial_\tau e^1 = \frac{1}{3}(\chi_2 - \chi)e^1,$$

$$\partial_\tau e^\pm = -\frac{1}{6}(\chi_2 + 2\chi)e^\pm, \qquad \partial_\tau \xi = -\frac{1}{6}(\chi_2 + 2\chi)\xi - \frac{1}{2}\chi_2 f - \theta_x,$$

$$\partial_\tau f = \frac{1}{3}(\chi_2 - \chi)f + \theta_x, \qquad \partial_\tau \chi_2 = \frac{1}{6}\chi_2^2 - \frac{2}{3}\chi_2\chi - \theta_2 + \Theta\phi,$$

$$\partial_\tau \chi = -\frac{1}{6}\chi_2^2 - \frac{1}{3}\chi^2 - \theta_h, \qquad \partial_\tau \theta_2 = \frac{1}{6}\chi_2\theta_2 - \frac{1}{3}(\chi_2\theta_h + \chi\theta_2) - \dot{\Theta}\phi,$$

$$\partial_\tau \theta_h = -\frac{1}{6}\chi_2\theta_2 - \frac{1}{3}\chi\theta_h, \qquad \partial_\tau \theta_x = \frac{1}{3}(\chi_2 - \chi)\theta_x - \frac{2\rho}{3(1 + \frac{1}{2}m\rho)^3}\phi,$$

$$\partial_\tau \phi = -\frac{1}{2}(\chi_2 + 2\chi)\phi.$$

Initial data for these components can be obtained from a comparison of the ansatz with Equations (20.25a)–(20.25c). One concludes that

$$e^0 = 0, \quad e^1 = \rho, \quad e^+ = 1, \quad e^- = 1, \quad f = 1, \quad \xi = 0, \quad \chi_2 = 0, \quad \chi = 0,$$

$$\theta_2 = \frac{6m\rho}{(1 + \frac{1}{2}m\rho)^2}, \quad \theta_h = 0, \quad \theta_x = 0, \quad \phi = -6m.$$

The ***symmetry reduced system*** and associated initial data can be written in a schematic form as

$$\partial_\tau \mathbf{u} = F(\mathbf{u}, \tau, \rho; m), \qquad \mathbf{u}(0, \rho; m) = \mathbf{u}_\star(\rho; m), \tag{20.26}$$

where F and $\mathbf{u}_\star$ are analytic functions of their arguments.

The $m = 0$ case

The particular case $m = 0$ – that is, the Minkowski spacetime – can be solved explicitly with the only non-vanishing geometric fields given by

$$e^0 = -\tau, \qquad e^1 = \rho, \qquad e^\pm = 1, \qquad f = 1, \tag{20.27}$$

while the fields associated to the conformal gauge are

$$\Theta = \rho(1 - \tau^2), \qquad d_{\boldsymbol{AB}} = 2\rho x_{\boldsymbol{AB}}.$$

Consequently, this solution exists for all τ, $\rho \in \mathbb{R}$. From the expressions in (20.27) one finds that

$$\boldsymbol{\omega} \equiv \tau_{\boldsymbol{AA'}}\boldsymbol{\omega}^{\boldsymbol{AA'}} = \sqrt{2}\left(\mathbf{d}\tau + \frac{\tau}{\rho}\mathbf{d}\rho\right), \quad \boldsymbol{\omega}^{\boldsymbol{AB}} = -\frac{1}{\rho}x^{\boldsymbol{AB}}\mathbf{d}\rho - 2y^{\boldsymbol{AB}}\boldsymbol{\alpha}^+ - 2z^{\boldsymbol{AB}}\boldsymbol{\alpha}^-.$$

Using the above covectors one can recover the metric associated to the conformal representation of the Minkowski spacetime under consideration. From Equation (4.14) one finds

$$\boldsymbol{g} = \frac{1}{\rho^2}\big(\rho^2 \mathbf{d}\tau \otimes \mathbf{d}\tau + \tau\rho(\mathbf{d}\tau \otimes \mathbf{d}\rho + \mathbf{d}\rho \otimes \mathbf{d}\tau) - (1-\tau^2)\mathbf{d}\rho \otimes \mathbf{d}\rho - \boldsymbol{\sigma}\big).$$

Consistent with the discussion of the previous sections, this metric is singular at $\rho = 0$. Now, as

$$\boldsymbol{f} = f_{\boldsymbol{AB}}\boldsymbol{\omega}^{\boldsymbol{AB}} = \frac{1}{\rho}\mathbf{d}\rho$$

is a closed form, it follows that the Weyl connection $\hat{\boldsymbol{\nabla}}$ associated to this representation is, in fact, the Levi-Civita connection of the metric $\rho^2\boldsymbol{g}$. The standard Minkowski metric can be recovered by setting $x^0 = \tau\rho$ so that

$$\begin{aligned}\tilde{\boldsymbol{g}} = \Theta^{-1}\boldsymbol{g} &= \frac{1}{\big(\rho^2 - (x^0)^2\big)^2}\big(\mathbf{d}x^0 \otimes \mathbf{d}x^0 - \mathbf{d}\rho \otimes \mathbf{d}\rho - \rho^2\boldsymbol{\sigma}\big),\\ &= \frac{1}{(x_\lambda x^\lambda)^2}\eta_{\mu\nu}\mathbf{d}x^\mu \otimes \mathbf{d}x^\nu.\end{aligned}$$

Performing the inversion $x^\mu \mapsto -x^\mu/(x_\lambda x^\lambda)$ in the last line element one obtains the standard Minkowski metric; compare the discussion in Section 6.2.2.

Null geodesics orthogonal to the spheres of constant ρ are given by

$$\tau = \frac{\mathrm{s}}{1 \pm \mathrm{s}}, \qquad \rho = \rho_\star(1 \pm \mathrm{s}), \qquad \theta = (\theta^{\mathcal{A}}) = (\theta^{\mathcal{A}}_\star), \tag{20.28}$$

for constant $\rho_\star$, $\theta^{\mathcal{A}}_\star$ and s an affine parameter. A direct computation shows that outgoing null geodesics intersecting future null infinity $\mathscr{I}^+$ correspond to the choice of the minus sign in Equations (20.28) – the intersection occurring at $\mathrm{s} = \frac{1}{2}$ so that $\rho = \frac{1}{2}\rho_\star$. These outgoing null geodesics do not intersect past null infinity $\mathscr{I}^-$ for a finite value of s. As $\rho_\star \to 0$, the outgoing null geodesics approach in a non-uniform manner the set $\mathscr{I}^- \cup \mathcal{I} \cup \mathcal{I}^+ \cup \mathcal{I}^-$. An analogous discussion applies to the incoming null geodesics obtained from taking the plus sign in (20.28). Accordingly, the cylinder at spatial infinity can be regarded as a *limit set of outgoing and incoming null geodesics*; see Figure 20.4.

The $m \neq 0$ case

Now, returning to the case $m \neq 0$, it follows from the *Cauchy stability* of ordinary differential equations – see, for example, Hartman (1987) – that given $\tau_\bullet > 1$ there exist $m_\bullet > 0$, $\rho_\bullet > 0$ such that the solution $\mathbf{u}(\tau, \rho; m)$ is analytic in all variables and exists for

$$|\tau| \leq \tau_\bullet, \qquad \rho \leq \rho_\bullet, \qquad |m| \leq m_\bullet.$$

By choosing $\tau_\bullet$ sufficiently large and observing the properties of the reference $m = 0$ solution, one can ensure that for each conformal geodesic with

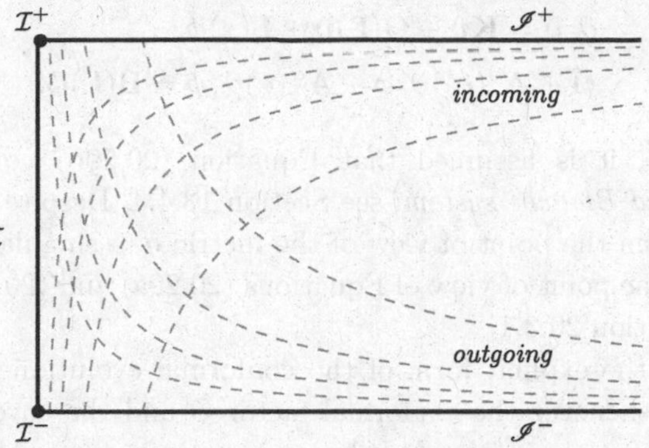

Figure 20.4 Schematic depiction of the null geodesics close to the cylinder at spatial infinity for the Minkowski spacetime as discussed in the main text; see the parametric equations in (20.28). The curves intersecting $\mathscr{I}^+$ are outgoing geodesics, while the ones intersecting $\mathscr{I}^-$ are incoming. The cylinder $\mathscr{I}$ can be seen as a limit set of the two classes of geodesics.

$0 < \rho < \rho_\bullet$ there exists a $\tau_{\mathscr{I}} < \tau_\bullet$ such that $\Theta|_{\pm\tau_{\mathscr{I}}} = 0$, $\mathbf{d}\Theta|_{\pm\tau_{\mathscr{I}}} \neq 0$. To obtain a statement that is valid for any value of m, it is observed that the symmetry-reduced evolution equations and the associated data are invariant under the rescaling

$$m \mapsto \frac{1}{\ell}m, \quad \rho \mapsto \ell\rho, \quad \phi \mapsto \frac{1}{\ell}\phi, \qquad e^1 \mapsto \ell e^1, \qquad \Theta \mapsto \ell\Theta,$$

for $\ell > 0$. Consequently, for any arbitrary m it is always possible to obtain a solution to the symmetry-reduced system (20.26) reaching null infinity if ρ is sufficiently small. Moreover, if ρ is sufficiently small, the underlying congruence of conformal geodesics is free of conjugate points on $\mathcal{M}(\mathcal{U})$. Null geodesics in the Schwarzschild spacetime behave more and more like null geodesics in the Minkowski spacetime as $\rho \to 0$. Numerically constructed solutions of the reduced spherically symmetric evolution system for the Schwarzschild spacetime can be found in Zenginoglu (2006, 2007).

20.3.4 Structural properties of the conformal evolution equations near the cylinder at spatial infinity

Having briefly analysed the regular initial value problem at spatial infinity for the Minkowski and the Schwarzschild spacetimes, one is now in the position of making some general remarks about this type of initial value problem.

The cylinder at spatial infinity as a total characteristic

Following Proposition 13.3, the hyperbolic reduction of the extended conformal Einstein field equations by means of a conformal Gaussian system leads to an evolution system which can be written as

$$\partial_\tau \hat{\boldsymbol{\upsilon}} = \mathbf{K}\hat{\boldsymbol{\upsilon}} + \mathbf{Q}(\hat{\boldsymbol{\Gamma}})\hat{\boldsymbol{\upsilon}} + \mathbf{L}(x)\boldsymbol{\phi}, \tag{20.29a}$$

$$\big(\mathbf{I} + \mathbf{A}^0(\boldsymbol{e})\big)\partial_\tau \boldsymbol{\phi} + \mathbf{A}^\alpha(\boldsymbol{e})\partial_\alpha \boldsymbol{\phi} = \mathbf{B}(\hat{\boldsymbol{\Gamma}})\boldsymbol{\phi}. \tag{20.29b}$$

For convenience, it is assumed that Equation (20.29b) corresponds to the *boundary adapted Bianchi system*; see Section 13.4.4. Despite the fact that the cylinder $\mathcal{I}$ is, from the point of view of the metric $\boldsymbol{g}$, a singular hypersurface, it is regular from the point of view of Equations (20.29a) and (20.29b) and its data $(\hat{\boldsymbol{\upsilon}}_\star, \boldsymbol{\phi}_\star)$; see Section 20.3.1.

Inspection of the explicit form of the conformal evolution equations reveals that $\mathbf{L}(x) = 0$ whenever the conformal factor Θ and the covector $\boldsymbol{d}$ vanish. It follows that, on $\mathcal{I}$, the conformal evolution equations decouple and one has

$$\partial_\tau \hat{\boldsymbol{\upsilon}}^{[0]} = \mathbf{K}\hat{\boldsymbol{\upsilon}}^{[0]} + \mathbf{Q}(\hat{\boldsymbol{\Gamma}}^{[0]})\hat{\boldsymbol{\upsilon}}^{[0]}, \qquad \hat{\boldsymbol{\upsilon}}^{[0]} \equiv \hat{\boldsymbol{\upsilon}}|_{\mathcal{I}}, \quad \hat{\boldsymbol{\Gamma}}^{[0]} \equiv \hat{\boldsymbol{\Gamma}}|_{\mathcal{I}}.$$

These ***transport equations*** can be integrated along the cylinder $\mathcal{I}$ from the observation that, irrespective of the particular choice of $\varkappa$, the restriction of the initial data $\hat{\boldsymbol{\upsilon}}_\star$ to $\mathcal{I}^0$ coincides with the restriction of initial data for the Minkowski spacetime as given in Section 20.3.3. Accordingly, the solution one obtains must also coincide with the Minkowskian one – namely,

$$(e^0_{\boldsymbol{AB}})^{[0]} = -\tau x_{\boldsymbol{AB}}, \quad (e^1_{\boldsymbol{AB}})^{[0]} = 0, \quad (e^+_{\boldsymbol{AB}})^{[0]} = y_{\boldsymbol{AB}}, \quad (e^-_{\boldsymbol{AB}})^{[0]} = z_{\boldsymbol{AB}},$$
$$(\xi_{\boldsymbol{ABCD}})^{[0]} = 0, \quad (\chi_{(\boldsymbol{AB})\boldsymbol{CD}})^{[0]} = 0, \quad (f_{\boldsymbol{AB}})^{[0]} = 0, \quad (\Theta_{\boldsymbol{ABCD}})^{[0]} = 0.$$

Substituting the above values in the restriction to $\mathcal{I}$ of the partial differential equation (PDE) (20.29b) one finds that the *normal matrix* $\mathbf{A}^3$ satisfies

$$\mathbf{A}^3|_{\mathcal{I}} = 0. \tag{20.30}$$

Hence, on $\mathscr{I}$ the restriction of Equation (20.29b) acquires the simplified form

$$(\mathbf{I} + \mathbf{A}^0(\boldsymbol{e}^{[0]}))\partial_\tau \boldsymbol{\phi}^{[0]} + \mathbf{A}^+(\boldsymbol{e}^{[0]})\boldsymbol{\partial}_+ \boldsymbol{\phi}^{[0]} + \mathbf{A}^-(\boldsymbol{e}^{[0]})\boldsymbol{\partial}_- \boldsymbol{\phi}^{[0]} = \mathbf{B}(\hat{\boldsymbol{\Gamma}}^{[0]})\boldsymbol{\phi}^{[0]} \tag{20.31}$$

where $\boldsymbol{\phi}^{[0]} \equiv \boldsymbol{\phi}|_{\mathcal{I}}$; that is, one obtains an interior system. It follows that the cylinder at spatial infinity $\mathcal{I}$ is a ***total characteristic*** of the conformal evolution Equations (20.29a) and (20.29b) and the restriction to $\mathcal{I}$ of all the conformal fields can be obtained from the restriction of the initial data to $\mathcal{I}^0$ by solving the resulting system of transport equations. Thus, although at first sight it seems that the construction of the cylinder at spatial infinity is introducing a set on which boundary data must be prescribed, the structural properties of the equations do not allow this: *no boundary conditions can be prescribed on* $\mathcal{I}$; compare the discussion in Section 12.4.

The solution to the interior equations for the Weyl tensor, Equation (20.31), can be obtained by observing that the restriction of the initial data for the Weyl tensor coincides with that of Schwarzschildean data so that the solution must be the Schwarzschild spacetime. From the symmetry-reduced conformal evolution equations it follows that ϕ_{ABCD} is constant along $\mathcal{I}$. Accordingly, one finds that

$$(\phi_{ABCD})^{[0]} = -6m\epsilon^2_{ABCD}.$$

The conformal evolution system and the critical sets

The analysis of the transport equations on $\mathcal{I}$ provides valuable insights into the hyperbolicity of the conformal evolution system (20.29a) and (20.29b). Observing that $(e^0_{AB})^{[0]} = -\tau x_{AB}$ it follows that

$$(\mathbf{I} + \mathbf{A}^0(e))|_{\mathcal{I}} = \begin{pmatrix} 1-\tau & 0 & 0 & 0 & 0 \\ 0 & 1 & 0 & 0 & 0 \\ 0 & 0 & 1 & 0 & 0 \\ 0 & 0 & 0 & 1 & 0 \\ 0 & 0 & 0 & 0 & 1+\tau \end{pmatrix};$$

compare Equation (13.61). Accordingly, the matrix $\mathbf{A}^0$ loses rank at the critical sets $\mathcal{I}^{\pm}$ and is no longer positive definite. Thus, the standard theory of hyperbolic PDEs as discussed in Chapter 12 cannot be employed to make assertions about the existence and uniqueness of solutions of the evolution system (20.29a) and (20.29b) up to $\mathcal{I}^{\pm}$. This *degeneracy* of the conformal evolution system is the essential source of difficulties in the analysis of the so-called ***problem of spatial infinity*** and requires the development of tailor-made techniques in order for one to be able to make assertions about the behaviour of its solutions.

Expansions in a neighbourhood of the cylinder at spatial infinity

On an intuitive level, one would expect the degeneracy of the conformal evolution system at the critical sets $\mathcal{I}^{\pm}$ to manifest itself through a loss of smoothness of its solutions. The discussion of Section 20.3.3 shows that this potential loss of regularity does not occur for all initial data. This observation hints that the precise algebraic structure of the evolution equations plays a decisive role in the nature of the solutions. In Friedrich (1998c) a procedure to analyse in detail the properties of the solutions to the conformal evolution equations has been put forward. Exploiting the total characteristic nature of the cylinder at spatial infinity one can repeatedly differentiate the evolution Equations (20.29a) and (20.29b) with respect to ∂_ρ and then evaluate on $\mathcal{I}$. In view of condition (20.30) one obtains a *hierarchy of transport equations* for the fields

$$\hat{\boldsymbol{v}}^{[p]} \equiv \partial^p_\rho \hat{\boldsymbol{v}}|_{\mathcal{I}}, \qquad \boldsymbol{\phi}^{[p]} \equiv \partial^p_\rho \boldsymbol{\phi}|_{\mathcal{I}}, \qquad p = 1, 2, 3, \ldots,$$

of the form

$$\partial_\tau \hat{\boldsymbol{v}}^{[p]} = \mathbf{K}\hat{\boldsymbol{v}}^{[p]} + \mathbf{Q}(\hat{\boldsymbol{\Gamma}}^{[0]})\hat{\boldsymbol{v}}^{[p]} + \mathbf{Q}(\hat{\boldsymbol{\Gamma}}^{[p]})\hat{\boldsymbol{v}}^{[0]} + \sum_{j=1}^{p-1}\binom{p}{j}\left(\mathbf{Q}(\hat{\Gamma}^{[j]})\hat{\boldsymbol{v}}^{[p-j]} + \mathbf{L}^{[j]}\boldsymbol{\phi}^{[p-j]}\right) + \mathbf{L}^{[p]}\boldsymbol{\phi}^{[0]},$$

$$
\begin{aligned}
(\mathbf{I} + \mathbf{A}^0(\mathbf{e}^{[0]}))\partial_\tau \boldsymbol{\phi}^{[p]} &+ \mathbf{A}^+(\mathbf{e}^{[0]})\boldsymbol{\partial}_+\boldsymbol{\phi}^{[p]} + \mathbf{A}^-(\mathbf{e}^{[0]})\boldsymbol{\partial}_-\boldsymbol{\phi}^{[p]} \\
&= \mathbf{B}(\hat{\Gamma}^{[0]})\boldsymbol{\phi}^{[p]} + \sum_{j=1}^{p}\binom{p}{j}\Big(\mathbf{B}(\hat{\Gamma}^{[j]})\boldsymbol{\phi}^{[p-j]} - \mathbf{A}^+(\mathbf{e}^{[j]})\boldsymbol{\partial}_+\boldsymbol{\phi}^{[p-j]} \\
&\quad - \mathbf{A}^-(\mathbf{e}^{[j]})\boldsymbol{\partial}_-\boldsymbol{\phi}^{[p-j]}\Big).
\end{aligned}
$$

The above equations will be called the ***transport equations of order*** p. The non-homogeneous terms depend on $\hat{\boldsymbol{\upsilon}}^{[j]}$ and $\boldsymbol{\phi}^{[j]}$ for $0 \leq j < p$. Thus, if these lower order solutions are known, the above pair of equations constitutes an interior system for $\hat{\boldsymbol{\upsilon}}^{[p]}$ and $\boldsymbol{\phi}^{[p]}$ on $\mathcal{I}$. The principal part of these equations is *universal* – in the sense that it is independent of the value of p. Initial data for these transport equations can be obtained from repeated ρ-differentiation and evaluation on $\mathcal{I}^0$ of the initial data $\hat{\boldsymbol{\upsilon}}_\star$, $\boldsymbol{\phi}_\star$ on $\mathcal{U}$. The coefficients obtained from this integration can, in turn, be collected in *formal expansions* of the form

$$
\hat{\boldsymbol{\upsilon}} = \sum_{p=0}^{\infty} \frac{1}{p!}\hat{\boldsymbol{\upsilon}}^{[p]}\rho^p, \qquad \hat{\boldsymbol{\phi}} = \sum_{p=0}^{\infty}\frac{1}{p!}\boldsymbol{\phi}^{[p]}\rho^p. \tag{20.32}
$$

At the time of writing, the analysis of the convergence of these formal expansions and the way they relate to actual solutions to the conformal Einstein field equations is an outstanding open aspect in the understanding of the problem of spatial infinity. Some ideas on how this problem could be addressed can be found in, for example, Friedrich (2003b) and Valiente Kroon (2009).

The structure of the hierarchy of transport equations for $\hat{\boldsymbol{\upsilon}}^{[p]}$ and $\boldsymbol{\phi}^{[p]}$ makes them amenable to a treatment by means of computer algebra methods. This approach has been pursued in Valiente Kroon (2004a,b) where solutions up to order $p = 8$ have been obtained. As is to be expected, the algebraic complexity of the solutions increases as p increases, eventually making the evaluation of further orders in the expansion no longer feasible due to computer limitations. The solutions to the transport equations obtained in this manner provide a valuable insight into the behaviour of the conformal field equations at spatial infinity.

As first observed in Friedrich (1998c), quite remarkably, there is a non-trivial relation between the regularity condition for the Cotton tensor, Equation (20.12), and the smoothness of the solutions to the transport equations:

Theorem 20.2 (***necessary conditions for the regularity of solutions to the conformal field equations at the critical sets***) *Given a vacuum time-symmetric initial data set with a conformal metric which is analytic in a neighbourhood of infinity, the solution to the regular finite initial value problem at spatial infinity is smooth through $\mathcal{I}^\pm$ only if the regularity condition*

$$
D_{(\boldsymbol{E}_p\boldsymbol{F}_p}\cdots D_{\boldsymbol{E}_1\boldsymbol{F}_1}b_{\boldsymbol{ABCD})}(i) = 0 \tag{20.33}
$$

holds for $p = 0, 1, 2, \ldots$ If this condition is violated at some order p', the solutions to the transport equations at order p' will develop logarithmic singularities at $\mathcal{I}^\pm$.

The analysis leading to the above result requires only the homogeneous part of the transport equations.

A toy model: the spin-2 massless field

A way of gaining insight into the behaviour of the solutions to the conformal evolution equations on the cylinder $\mathcal{I}$ is to consider an analogous discussion for a test ***spin-2 massless field on the Minkowski spacetime***. Accordingly, let ζ_{ABCD} denote the components of a totally symmetric rank-4 spinorial field satisfying the equation

$$\nabla^{Q}{}_{A'}\zeta_{ABCQ} = 0. \tag{20.34}$$

The principal part of the evolution equations implied by (20.34) along the cylinder $\mathcal{I}$ is identical to that of the Bianchi evolution equations. Several aspects of this toy model have been considered in Valiente Kroon (2002), Friedrich (2003b) and Beyer et al. (2012), and the following discussion is adapted from various parts of these references.

The background Minkowski geometry has already been obtained in Section 20.3.3; see Equation (20.27). From these expressions, a computation shows that the evolution equations implied by the spin-2 massless field equation can be explicitly written as

$$(1-\tau)\partial_\tau\zeta_0 + \rho\partial_\rho\zeta_0 - \eth\zeta_1 - 2\zeta_0 = 0, \tag{20.35a}$$
$$\partial_\tau\zeta_1 - \frac{1}{2}(\eth\zeta_2 + \bar{\eth}\zeta_0) - \zeta_1 = 0, \tag{20.35b}$$
$$\partial_\tau\zeta_2 - \frac{1}{2}(\eth\zeta_3 + \bar{\eth}\zeta_1) = 0, \tag{20.35c}$$
$$\partial_\tau\zeta_3 - \frac{1}{2}(\eth\zeta_4 + \bar{\eth}\zeta_2) + \zeta_3 = 0, \tag{20.35d}$$
$$(1+\tau)\partial_\tau\zeta_4 - \rho\partial_\rho\zeta_4 - \bar{\eth}\zeta_3 + 2\zeta_4 = 0, \tag{20.35e}$$

where $\zeta_0 \equiv \zeta_{\mathbf{0000}}$, $\zeta_1 \equiv \zeta_{\mathbf{0001}}, \ldots$, and where for convenience of the subsequent discussion, the connection coefficients associated to $\mathbb{S}^2$ (i.e. $\Gamma_{\mathbf{00}CD}$ and $\Gamma_{\mathbf{11}CD}$) have been absorbed in the differential operators $\eth$ and $\bar{\eth}$; see the Appendix to Chapter 10. The subsequent analysis will also require the *constraint equations* implied by Equation (20.34). These are given by

$$\tau\partial_\tau\zeta_1 - \rho\partial_\rho\zeta_1 - \frac{1}{2}(\eth\zeta_0 - \bar{\eth}\zeta_2) = 0, \tag{20.36a}$$
$$\tau\partial_\tau\zeta_2 - \rho\partial_\rho\zeta_2 - \frac{1}{2}(\eth\zeta_1 - \bar{\eth}\zeta_2) = 0, \tag{20.36b}$$
$$\tau\partial_\tau\zeta_3 - \rho\partial_\rho\zeta_3 - \frac{1}{2}(\eth\zeta_2 - \bar{\eth}\zeta_4) = 0. \tag{20.36c}$$

Differentiating Equations (20.35a)–(20.35e) and (20.36a)–(20.36c) repeatedly with respect to ∂_ρ and evaluating at $\mathcal{I}$ one obtains the transport equations

$$(1-\tau)\partial_\tau\zeta_0^{[p]} - \bar{\eth}\zeta_1^{[p]} + (p-2)\zeta_0^{[p]} = 0, \tag{20.37a}$$

$$\partial_\tau\zeta_1^{[p]} - \frac{1}{2}(\eth\zeta_0^{[p]} + \bar{\eth}\zeta_2^{[p]}) - \zeta_1^{[p]} = 0, \tag{20.37b}$$

$$\partial_\tau\zeta_2^{[p]} - \frac{1}{2}(\eth\zeta_1^{[p]} + \bar{\eth}\zeta_3^{[p]}) = 0, \tag{20.37c}$$

$$\partial_\tau\zeta_3^{[p]} - \frac{1}{2}(\eth\zeta_2^{[p]} + \bar{\eth}\zeta_4^{[p]}) + \zeta_3^{[p]} = 0, \tag{20.37d}$$

$$(1+\tau)\partial_\tau\zeta_4^{[p]} - \eth\zeta_3^{[p]} - (p-2)\zeta_4^{[p]} = 0, \tag{20.37e}$$

and

$$\tau\partial_\tau\zeta_1^{[p]} - \frac{1}{2}(\eth\zeta_4^{[p]} + \bar{\eth}\zeta_2^{[p]}) - p\zeta_1^{[p]} = 0, \tag{20.38a}$$

$$\tau\partial_\tau\zeta_2^{[p]} - \frac{1}{2}(\eth\zeta_3^{[p]} - \bar{\eth}\zeta_1^{[p]}) - p\zeta_2^{[p]} = 0, \tag{20.38b}$$

$$\tau\partial_\tau\zeta_3^{[p]} - \frac{1}{2}(\eth\zeta_2^{[p]} - \bar{\eth}\zeta_0^{[p]}) - p\zeta_3^{[p]} = 0. \tag{20.38c}$$

The linearity of the above equations suggests eliminating the angular dependence of the solutions through an expansion in terms of spin-weighted spherical harmonics. Consistent with the spin weight of the various components of ζ_{ABCD} one considers the ansatz

$$\zeta_k^{[p]} = \sum_{l=|k-2|}^{p} \sum_{m=-l}^{l} z_{k,p;l,m}(\tau)\, {}_{k-2}Y_{lm}.$$

Observe, in particular, that the number of l-modes is bounded by the differentiation order p. This ansatz can be shown to be the most general possible. Taking into account the action of the operators $\eth$ and $\bar{\eth}$ on the spin-weighted spherical harmonics, a calculation combining Equations (20.37a)–(20.37e) and (20.38a)–(20.38c) shows that the coefficients $z_{k,p;l,m}(\tau)$ satisfy the ***Jacobi ordinary differential equation***

$$(1-\tau^2)\ddot{z}_{k,p;l,m} + \big(2(k-2) + 2(p-1)\tau\big)\dot{z}_{k,p;l,m} + \big(l(l+1) - p(p-1)\big)z_{k,p;l,m} = 0,$$

where $\dot{}$ denotes differentiation with respect to τ. The solutions to this equation are well understood; see, for example, Szegö (1978). For $|k-2| \leq l < p$ the solutions are a linear combination of the polynomials

$$P_{p-l-1}^{(-p-6+k,-p+k-2)}(\tau), \qquad \left(\frac{1-\tau}{2}\right)^{p+k-2} P_{l-2}^{(p+k-2,-p+k-2)}(\tau),$$

where $P_n^{(\alpha,\beta)}(\tau)$ denotes the ***Jacobi polynomial*** of degree n with integer parameters (α,β) given by

$$P_n^{(\alpha,\beta)}(\tau) \equiv \sum_{s=0}^{n} \binom{n+\alpha}{s}\binom{n+\beta}{n-s}\left(\frac{\tau-1}{2}\right)^{n-s}\left(\frac{\tau+1}{2}\right)^{s}.$$

The case $l = p$ is the one of most interest as the general solution can be found to be a linear combination of

$$\left(\frac{1-\tau}{2}\right)^{p+k-2}\left(\frac{1+\tau}{2}\right)^{p+2-k},$$

$$\left(\frac{1-\tau}{2}\right)^{p+k-2}\left(\frac{1+\tau}{2}\right)^{p+2-k}\int_0^\tau \frac{\mathrm{d}s}{(1-s)^{p-1+k}(1+s)^{p+3-k}}.$$

Using partial fractions one finds that the integral in the second solution can be expressed as

$$\begin{aligned}\int_0^\tau \frac{\mathrm{d}s}{(1-s)^{p-1+k}(1+s)^{p+3-k}} &= a_\bullet \ln(1+\tau) + \frac{a_{p+2-k}}{(1+\tau)^{p+2-k}} + \cdots + \frac{a_1}{1+\tau}\\ &\quad + b_\bullet \ln(1-\tau) + \frac{b_{p-2+k}}{(1-\tau)^{p-2+k}} + \cdots + \frac{b_1}{1-\tau} + b_0,\end{aligned}$$

where $a_\bullet$, $b_\bullet$, $a_{p+2-k}, \ldots, a_1$, $b_{p-2+k}, \ldots, b_0$ are some constants. Thus, generically, the solutions for the $l = p$ modes will be non-smooth and develop logarithmic singularities at the critical sets $\mathcal{I}^\pm$ even in the case where the initial data is as smooth as it can be. Direct inspection of the above expressions shows that at $\tau = 1$ the most singular component of ζ_{ABCD} is ζ_0, while at $\tau = -1$ it is ζ_4.

The singular behaviour can be avoided if the initial data is fine tuned. Indeed, a lengthy analysis renders the following (see Valiente Kroon (2002)):

Proposition 20.5 (***regularity of solutions to the massless spin-2 field equations at the critical sets***) *The solutions to the transport equations implied on the cylinder at spatial infinity $\mathcal{I}$ of the Minkowski spacetime by the spin-2 massless field Equation (20.34) extends analytically to the critical sets $\mathcal{I}^\pm$ if and only if the regularity condition*

$$D_{(E_pF_p}\cdots D_{E_1F_1}\breve{b}_{ABCD)}(i) = 0, \qquad p = 0, 1, 2, \ldots,$$

where

$$\breve{b}_{ABCD} \equiv 2D_{P(A}\Omega\zeta_{BCD)}{}^P + \Omega D_{P(A}\zeta_{BCD)}{}^P$$

denotes the linearisation of the Cotton spinor around Minkowski data.

This result is the spin-2 field version of Theorem 20.2 for the full conformal Einstein field equations.

20.3.5 The cylinder at spatial infinity and static solutions

The analysis of static solutions provides deeper insights into the behaviour of the solutions to the transport equations on $\mathcal{I}$. The discussion in Section 19.2.1 shows that static solutions admit a smooth conformal completion at null infinity. Thus, it is natural to conjecture that they also extend smoothly through the critical sets $\mathcal{I}^{\pm}$. The analysis of the conformal evolutions for the Schwarzschild spacetime provides further support to this idea – this evidence, however, must be taken with care as the spherical symmetry of the spacetime gives rise to a number of non-generic simplifications.

A lengthy computation which combines the ideas of Sections 19.2.1 and 20.3.3 yields the following satisfactory result:

Proposition 20.6 (***regularity of static solutions at the critical sets***) *The solutions to the transport equations at $\mathcal{I}$ for static data extend smoothly (and, in fact, analytically) through the critical sets $\mathcal{I}^{\pm}$.*

A proof of this result can be found in Friedrich (2004). A generalisation of the analysis to the stationary case is given in Aceña and Valiente Kroon (2011).

20.4 Spatial infinity and peeling

At the time of writing, one of the outstanding challenges in the analysis of the problem of spatial infinity is to obtain a satisfactory understanding of the connection between the solutions to the transport equations on $\mathcal{I}$ and the peeling (or lack thereof) of the Weyl tensor at $\mathscr{I}$.

The key hypothesis in the ***peeling theorem***, Theorem 10.4, is the smoothness of the rescaled Weyl tensor on null infinity. Direct inspection allows one to relax this assumption to a certain minimum regularity threshold. Now, it has been seen in the previous section that generic solutions to the transport equations on the cylinder $\mathcal{I}$ have logarithmic singularities at the critical sets $\mathcal{I}^{\pm}$. In view of the hyperbolic character of the conformal evolution equations, it is to be expected that this singular behaviour will spread along the conformal boundary, thus destroying the smoothness of the rescaled Weyl tensor along the conformal boundary. These singularities may lead to a *restricted peeling behaviour*; see, for example, Chruściel et al. (1995) and Valiente Kroon (1998, 1999a,b) for a discussion of more general types of peeling. A detailed and rigorous treatment of these ideas is not yet available; some heuristic discussions can be found in Valiente Kroon (2002, 2003, 2005, 2007a).

The most promising avenue to obtain a link between the generic singular behaviour at the critical sets and the peeling behaviour at null infinity consists of computing the formal expansions (20.32) up to a certain order N. Letting $\hat{\boldsymbol{\upsilon}}$ and $\boldsymbol{\phi}$ denote the actual solutions (if any) to the conformal evolution equations one defines the ***remainders***

$$\mathbf{R}_N[\hat{\boldsymbol{\upsilon}}] \equiv \hat{\boldsymbol{\upsilon}} - \sum_{p=0}^{N} \frac{1}{p!} \hat{\boldsymbol{\upsilon}}^{[p]} \rho^p, \qquad \mathbf{R}_N[\boldsymbol{\phi}] \equiv \boldsymbol{\phi} - \sum_{p=0}^{N} \frac{1}{p!} \boldsymbol{\phi}^{[p]} \rho^p.$$

If the expansion order N is sufficiently high, it may be possible to use the conformal evolution equations to obtain estimates on the remainders $\mathbf{R}_N[\hat{\boldsymbol{\upsilon}}]$ and $\mathbf{R}_N[\boldsymbol{\phi}]$. The idea behind this approach is that the expansion terms should contain the most singular part of the solution, thus leaving a remainder which is more regular and, accordingly, more amenable to an analytic treatment. This strategy has been implemented with success for the model problems of the spin-2 massless field in the Minkowski spacetime in Friedrich (2003b) and for the spinorial Maxwell equations (i.e. the spin-1 massless field) on the Schwarzschild spacetime in Valiente Kroon (2009).

20.5 Existence of asymptotically simple spacetimes

The regularity of static solutions at spatial infinity provides a procedure to construct a wide class of asymptotically simple solutions to the Einstein field equations from a Cauchy initial value problem: the so-called ***Cutler-Wald-Chruściel-Delay construction***; see Cutler and Wald (1989); Chruściel and Delay (2002) and Corvino (2007). The key idea behind this construction is to consider time-symmetric initial data sets $(\tilde{\mathcal{S}}, \tilde{\boldsymbol{h}})$ for the Einstein field equations which are exactly Schwarzschildean in a suitable exterior region $\tilde{\mathcal{E}}$ of the asymptotic end but otherwise arbitrary in a compact region $\mathcal{B}$ in the interior. The existence of such initial data sets is ensured by the ***exterior asymptotic gluing construction***; see Theorems 11.3 and 11.4. Denote by $(\mathcal{S}, \boldsymbol{h})$ a suitable point compactification of the data $(\tilde{\mathcal{S}}, \tilde{\boldsymbol{h}})$ and let $\mathcal{E}$ denote the neighbourhood of i corresponding to the exterior region $\tilde{\mathcal{E}}$. *As a consequence of the causal properties of general relativity, the development of* $(\tilde{\mathcal{S}}, \tilde{\boldsymbol{h}})$ *is such that* $D^+(\mathcal{E})$ *coincides with a suitable spacetime neighbourhood of the spatial infinity of* $\mathcal{S}$. In a slight abuse of terminology one can say that these data have *compact support*. Accordingly, $D^+(\mathcal{S})$ will contain hyperboloidal hypersurfaces $\mathcal{H}$ which coincide with $\mathcal{S} \setminus \mathcal{E}$ on $D^+(\mathcal{S} \setminus \mathcal{E})$. On $\mathcal{H} \cap D^+(\mathcal{E})$, the initial data for the conformal field equations implied by the development on $\mathcal{H}$ will be Schwarzschildean hyperboloidal data – and, thus, smooth at $\mathscr{I} \cap \mathcal{H}$. An important technical aspect of this construction is to ensure that the gluing region does not *drift away* into the asymptotic region as one considers a sequence of data tending to data for the Minkowski spacetime. This is ensured by Theorem 11.4. Now, if the data on $\mathcal{B}$ are sufficiently close to data for the Minkowski spacetime, one can apply the semi-global existence Theorem 16.1 to the data on $\mathcal{H}$ to obtain a development $D^+(\mathcal{H})$ which is asymptotically simple. As the Schwarzschild spacetime is asymptotically simple, one concludes that $D^+(\mathcal{S})$ is asymptotically simple; see Figure 20.5. In view of the time symmetry of the initial data, one, in fact, obtains a spacetime where the two components of null infinity $\mathscr{I}^-$ and $\mathscr{I}^+$ are complete. While the development $D^+(\mathcal{S})$ is static in $D^+(\mathcal{E})$, in general, radiation will be registered on $\mathscr{I}^+ \cap J^+(\mathcal{S} \setminus \mathcal{E})$ and $\mathscr{I}^- \cap J^-(\mathcal{S} \setminus \mathcal{E})$.

For further details on the construction described in the previous paragraph, see Chruściel and Delay (2002).

Remark. The original version of the above construction was carried out for solutions to the Einstein-Maxwell equations. Remarkably, it is possible to

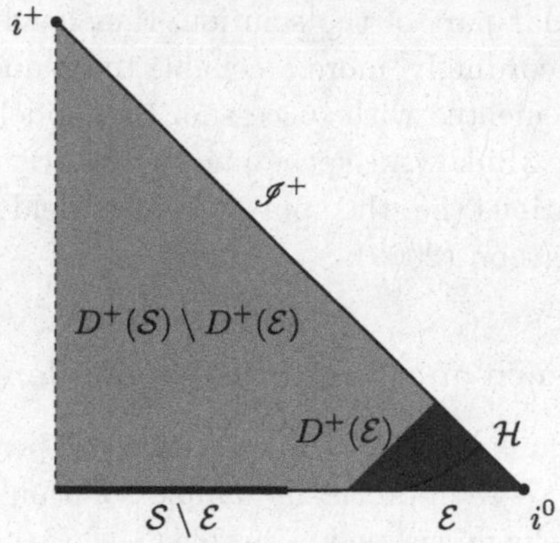

Figure 20.5 Schematic depiction of the Cutler-Wald-Chruściel-Delay construction. The spacetime is asymptotically simple and coincides with the Schwarzschild spacetime on $D^+(\mathcal{H})$. Generically, radiation is registered on $\mathscr{I}^+ \cap J^+(\mathcal{S}\setminus\mathcal{E})$.

construct initial data with *compact support* for the Einstein-Maxwell equations without the need of a gluing construction; see Cutler and Wald (1989).

20.6 Obstructions to the smoothness of null infinity

The spacetimes obtained from the Cutler-Wald-Chruściel-Delay construction are very special. Thus, it is natural to ask whether it is possible to construct asymptotically simple spacetimes which do not have such a rigid behaviour in a neighbourhood of spatial infinity. Insight into this question can be obtained from the analysis of the transport equations on the cylinder at spatial infinity.

The systematic analysis of the transport equations on $\mathcal{I}$ has shown that two different types of ***obstructions to the smoothness of null infinity*** arise in the development of time-symmetric data $(\tilde{\mathcal{S}}, \tilde{\boldsymbol{h}})$ admitting a smooth point compactification $(\mathcal{S}, \boldsymbol{h})$ at spatial infinity. These are briefly discussed in the following.

Obstructions associated to the conformal class $[\tilde{\boldsymbol{h}}]$

As already discussed, obstructions to the smoothness of null infinity associated to the conformal class $[\tilde{\boldsymbol{h}}]$ can be removed by requiring that the Cotton tensor of the conformal metric $\boldsymbol{h}$ satisfies the regularity condition (20.33).

Obstructions associated to the scaling of the conformal metric

To discuss the obstructions to the smoothness of null infinity associated to the particular scaling of the conformal metric, suppose that the conformal metric $\boldsymbol{h}_{\circledast}$ is a solution to the conformal static equations with associated conformal factor Ω; see Chapter 19. Now, restricting the subsequent considerations to a suitable small neighbourhood $\mathcal{U}$ of i consider another conformal factor Ω' satisfying the boundary conditions of a point compactification and such that the metric $\tilde{\boldsymbol{h}}' \equiv \Omega'^{-2}\boldsymbol{h}_{\circledast}$ satisfies the time-symmetric Hamiltonian constraint on $\tilde{\mathcal{U}} \equiv \mathcal{U}\setminus\{i\}$; that is, $r[\tilde{\boldsymbol{h}}'] = 0$. It follows that there exists $\vartheta \in C^2(\mathcal{U}) \cap C^\infty(\tilde{\mathcal{U}})$ such that $\Omega' \equiv \vartheta\Omega$. Now, assume that $\vartheta(i) = 1$, $\mathbf{d}\vartheta(i) = 0$ and $\boldsymbol{Hess}\,\vartheta(i) = 0$ so that the metrics $\tilde{\boldsymbol{h}}$ and $\tilde{\boldsymbol{h}}' = \vartheta^{-2}\tilde{\boldsymbol{h}}$ have the same mass. As the conformal metric $\boldsymbol{h}_{\circledast}$ is static it satisfies the regularity condition (20.33). Moreover, as $\boldsymbol{h}' = \vartheta^2\boldsymbol{h}_{\circledast} \in [\boldsymbol{h}_{\circledast}]$ it also satisfies the regularity condition. After a lengthy inductive argument one obtains the following:

Theorem 20.3 (***obstructions to the smoothness of null infinity associated to the scaling of the conformal metric***) *Given time-symmetric initial data with an analytic conformal metric* $\boldsymbol{h}$*, the solution to the regular finite initial value problem at spatial infinity for the conformal Einstein field equations is smooth through the critical sets* $\mathcal{I}^{\pm}$ *(and, in particular, free of logarithmic singularities) if and only if* $\vartheta - 1$ *vanishes at i at all orders.*

The proof of this result can be found in Valiente Kroon (2010, 2011). The analysis leading to the above theorem assumed the analyticity of the metric in $\mathcal{U}$. However, the result also holds if one assumes smoothness. This result provides strong indication that static initial data play a privileged role among the class of time-symmetric data which extend smoothly through the critical sets. A precise clarification of this role is one of the outstanding challenges in the analysis. Despite the insights obtained so far, at the time of writing, it cannot be excluded that there exist data which are not asymptotically static at i and for which the solutions to the transport equations on $\mathcal{I}$ extend smoothly through the critical sets. To address this point, it is necessary to identify the gap between initial data satisfying the regularity condition (20.33) and static data. The further conditions required to single out static data have been analysed in Friedrich (2013). It has been found that a *sufficient condition* for the staticity of the data satisfying condition (20.33) and the *non-degeneracy* requirement associated to the hypothesis of Theorem 19.4 (concerning the uniqueness of the conformal structure of a static solution) can be expressed in terms of a covector with conformally invariant differential. The challenge is now to analyse whether data violating this sufficient condition develop singularities at the critical sets.

20.7 Further reading

Although it has long been recognised that, for spacetimes with a non-vanishing mass, spatial infinity is a singular point of the conformal structure – see, for example, Penrose (1963, 1965) – systematic attempts to understand the behaviour of the geometry of spacetime in a neighbourhood of this point in the light of the Einstein field equations took time to get started. Early analyses of the behaviour of the Einstein field equations in a neighbourhood of a suitable representation of spatial infinity have been given in Schmidt (1981), Beig and Schmidt (1982), Beig (1984) and Schmidt (1987). The approach to the analysis of spatial infinity discussed in this chapter started in Friedrich (1988). The construction of the cylinder at spatial infinity was presented in Friedrich (1998c) which to date remains the most comprehensive reference in the matter. A useful discussion which overlaps with the previous reference but also expands in certain aspects not covered in the original work is given in Friedrich (2004); this reference provides, in particular, a detailed discussion of the construction of the cylinder at spatial infinity for static solutions. The extension of the later analysis to stationary solutions has been carried out in Aceña and Valiente Kroon (2011). A programme to analyse the solutions to the transport equations on $\mathcal{I}$ was started in Friedrich and Kánnár (2000a); see also Friedrich and Kánnár (2000b). Expansions to a sufficiently high order to observe the first obstructions to the smoothness of null infinity have been carried out in Valiente Kroon (2004a,b,c, 2005). General results concerning these expansions showing the special role played by static solutions (in a time-symmetric setting) are given in Valiente Kroon (2010, 2011). An account of the state of the art concerning the problem of spatial infinity is provided in Friedrich (2013) where the gap between data satisfying the regularity condition on the Cotton tensor and static data is analysed in detail. A discussion of general aspects of the behaviour of the massless spin-2 field in a neighbourhood of spatial infinity of the Minkowski spacetime can be found in Valiente Kroon (2002); see also Beyer et al. (2012). A method for the construction of estimates for the massless spin-2 field which remain regular at the critical sets of the Minkowski spacetime has been provided in Friedrich (2003b). These ideas have been adapted to the case of the Maxwell equations on a Schwarzschild background in Valiente Kroon (2007b, 2009).

Appendix: properties of functions on the complex null cone

The following result of complex analysis is used repeatedly in the main text of this chapter.

Lemma 19.2 (***factorisation lemma***) *Let f denote a holomorphic function on a neighbourhood $\mathcal{U}_{\mathbb{C}}$ of the origin of $\mathbb{C}^3$, and let $\mathcal{N}_{\mathbb{C}}(0)$ denote the complex null cone through the origin. If $f|_{\mathcal{N}_{\mathbb{C}}(0)} = 0$, then there exists a holomorphic function g defined on a neighbourhood of the origin of $\mathbb{C}^3$ such that $f = \Gamma g$ where $\Gamma = |x|^2$.*

The proof of this result can be found in Kodaira (1986). Recall that $\mathcal{N}_{\mathbb{C}}(0)$ coincides with the locus of points in $\mathbb{C}^3$ for which Γ vanishes. One also has the following:

Lemma 19.3 (***characterisation of functions vanishing on the null cone***) *A holomorphic spinorial field $\zeta_{\boldsymbol{A}\cdots\boldsymbol{D}}$ in some neighbourhood $\mathcal{U}_{\mathbb{C}}$ of the origin in $\mathbb{C}^3$ vanishes on $\mathcal{N}_{\mathbb{C}}(0)$ if and only if it satisfies the sequence of conditions*

$$D_{(\boldsymbol{P}_p\boldsymbol{Q}_p}\cdots D_{\boldsymbol{P}_1\boldsymbol{Q}_1)}\zeta_{\boldsymbol{A}\cdots\boldsymbol{D}}(0)=0, \qquad p=0,\,1,\,2,\ldots \tag{20.39}$$

The proof of this result is based on the observation that the conditions (20.39) can be used to construct a Taylor-like expansion of the field $\zeta_{\boldsymbol{A}\cdots\boldsymbol{D}}$ of the form

$$\zeta_{\boldsymbol{A}\cdots\boldsymbol{D}}(\gamma(\mathrm{s}))=\sum_{p=0}^{\infty}\frac{1}{p!}\mathrm{s}^p\kappa^{\boldsymbol{P}_p}\kappa^{\boldsymbol{Q}_p}\cdots\kappa^{\boldsymbol{P}_1}\kappa^{\boldsymbol{Q}_1}D_{\boldsymbol{P}_p\boldsymbol{Q}_p}\cdots D_{\boldsymbol{P}_1\boldsymbol{Q}_1}\zeta_{\boldsymbol{A}\cdots\boldsymbol{D}}(0)$$

along the generators $\gamma(\mathrm{s})$ of $\mathcal{N}_{\mathbb{C}}(0)$ for s an affine parameter sufficiently close to 0. As a consequence of the analyticity of the set up, the above expansion uniquely determines the function $\zeta_{\boldsymbol{A}\cdots\boldsymbol{D}}$ in a neighbourhood of 0 on $\mathcal{N}_{\mathbb{C}}(0)$. A more detailed discussion of the proof can be found in Friedrich (2013), lemma 6.1.

21
Perspectives

And it seemed as though in a little while the solution would be found, and then a new and splendid life would begin; and it was clear to both of them that they had still a long, long road before them, and that the most complicated and difficult part was only just beginning.

– A. Chekhov, *The lady with the dog*

Conformal notions provide valuable tools for the analysis of global properties of spacetimes. In Part IV of this book it has been shown how a *conformal point of view* leads to proofs of the global existence and non-linear stability of de Sitter-like spacetimes, of the semiglobal existence and non-linear stability of Minkowski-like spacetimes, and how they provide a systematic procedure for the construction of anti-de Sitter-like spacetimes. Moreover, conformal methods provide a robust framework for the analysis of the gravitational field of isolated systems in a neighbourhood of both null and spatial infinity.

The application of conformal methods in general relativity is a mature area of research with a considerable number of open problems. Several of these have been discussed in various places of this book. Unavoidably, there are other problems and aspects of the subject which, for reasons of space, could not be covered in the main text. This last chapter presents a list of ideas and problems which, in the opinion of the author, may play a role in the future development of the subject.

21.1 Stability of cosmological models

The global non-linear stability of the de Sitter spacetime was discussed in Chapter 15. This exact solution can be regarded as a basic cosmological model. The analysis of Chapter 15 can be extended to include a non-vacuum matter content with good conformal properties: for example, the Maxwell, Yang-Mills and conformally coupled scalar field; see Friedrich (1991) and Lübbe and Valiente

Kroon (2012). More recently, the ideas behind these proofs have been adapted in Lübbe and Valiente Kroon (2013b) to provide an analysis of the future stability of Friedman-Robertson-Walker cosmological models with a perfect fluid with the equation of state of radiation; see Section 9.4. A natural question is whether conformal methods could be adapted to more general matter models, that is, matter models with a non-vanishing trace. That this may be possible is suggested by the analysis in Friedrich (2015b) where it is shown that massive scalar fields for which the mass parameter is related to the cosmological constant by the condition $3m^2 = -2\lambda$ give rise to a set of regular conformal evolution equations for the Einstein-massive scalar field system. A further indication that conformal methods may be applicable to more general matter models is provided by the observation that for a large class of equations of state, perfect fluid Friedman-Lemaître-Robertson-Walker (FLRW) cosmological models can be smoothly conformally compactified – see, for example, Griffiths and Podolský (2009), section 6.4 – this despite the fact that the "natural" conformal evolution equations for these models are not conformally regular.

An important motivation behind the analysis of the future non-linear stability of cosmological models is the so-called ***cosmic no-hair conjecture*** – the expectation that for a large class of models the late-time evolution approximates, in some sense, a de Sitter state; see, for example, Wald (1983). As the analysis of Lübbe and Valiente Kroon (2013b) exemplifies, conformal methods provide a convenient setting for this discussion – at least for some suitable matter models.

Conformal methods provide a natural tool for the analysis of so-called ***isotropic cosmological singularities***. These are singularities of the physical spacetime that can be removed by means of a conformal rescaling of the metric. The singularity of the rescaled metric is assumed to occur on a spacelike surface. Accordingly, the conformal structure can be extended through the singularity and one can study the Cauchy problem for the cosmological model with data at the location of the singularity; see, for example, Tod (2002) for an introduction into the subject and Anguige and Tod (1999a,b) and Tod (2003) for further details. The Big Bang singularity in FLRW models provides the prototypical example of this type of singularity: as these spacetimes are conformally flat, any curvature singularity must be restricted to the (physical) Ricci tensor. In view of the highly symmetric nature of FLRW spacetimes, the Ricci tensor has only one essential component; combining this observation with the fact that under conformal rescalings the Ricci tensor satisfies a transformation law which is non-homogeneous, one can then see that in FLRW spacetimes the conformal factor can be chosen so as to absorb the singular behaviour of the curvature.

In the analysis of isotropic singularities, the role of the conformal factor is different from the one in the study of asymptotics: the conformal factor diverges at the singularity rather than going to zero – thus, it "blows up" the shrinking physical metric to make it finite. This type of behaviour is not expected to be a general feature of cosmological solutions to the Einstein field equations. This observation is related to Penrose's ***Weyl curvature hypothesis***: the idea that the early geometry of the universe should be such that the Weyl tensor vanishes, singling out a state of *low gravitational entropy*; see Penrose (1979).

In the discussion of isotropic cosmological singularities one pursues conformal rescalings of the form $\tilde{g} = \mho^2 g$ where $\tilde{g}$ denotes the physical metric, while the unphysical metric g extends the conformal structure through the singularity characterised by the condition $\mho \to 0$. Under these conventions the Einstein field equations, written in a suitable gauge, lead to *conformal evolution equations* having a well-understood singular behaviour at the Big Bang. These evolution equations are an example of so-called ***Fuchsian differential equations*** – a class of equations with a well-defined theory. Using this theory, a number of statements concerning isotropic singularities can be obtained; see again Tod (2002) and references within for further details. More recently, it has been shown that a duality property of the conformally coupled scalar field equation allows one to analyse isotropic singularities in a framework involving the conformal Einstein field equations; see Lübbe (2014). It is of interest to see whether these ideas can be pursued further and extended to more general contexts.

21.2 Stability of black hole spacetimes

One of the outstanding open problems in mathematical general relativity is the question of the ***non-linear stability of the Kerr spacetime***; see, for example, Dafermos and Rodnianski (2010) for an entry point into the literature of the subject. The expectation associated with this question is that perturbations of a Kerr metric should dynamically approach a member of the Kerr family of solutions in the exterior of the black hole region. This problem involves both an orbital and an asymptotic stability analysis; see the discussion in Section 14.4. The non-linear stability of the Kerr spacetime poses both technical and conceptual challenges. On the technical side, it requires the development of robust partial differential equation (PDE) techniques to control the behaviour of the Einstein field equations in the strong gravitational field regime of a black hole. Current efforts in this direction have involved a detailed analysis of linear wave equations whose solutions propagate on the domain of outer communication of a Kerr background. This analysis makes systematic use of so-called *vector field methods*. The expectation is that these wave equations provide a suitable model for the Einstein equations written, say, in harmonic coordinates; see again Dafermos and Rodnianski (2010) for an account of this approach. On the conceptual side, the problem needs a detailed specification, compatible with the needs of PDE theory, of what is meant by the statement that a given spacetime is *close* to the Kerr spacetime; some ideas on how to address this issue are discussed in Bäckdahl and Valiente Kroon (2010a,b).

Given that conformal methods, as discussed in this book, provide a tool for the analysis of the non-linear stability of asymptotically simple spacetimes – compare

Chapters 15 and 16 – it is natural to wonder whether they could also provide an avenue for the analysis of the non-linear stability of black hole spacetimes. The stability proofs discussed in this book start from the premise that a detailed understanding of the conformal geometry of a background solution is key to the analysis. Once this has been achieved, the existence and stability results follow by means of general results of the theory of PDEs – namely, the Cauchy stability guaranteed by Theorem 12.2. From the perspective of conformal geometry, the essential difference between the basic asymptotically simple spacetimes and the exact solutions describing black hole spacetimes is that while the former are conformally regular, the latter have a conformal structure with singular regions. This observation rules out the possibility of directly using arguments based solely on the notion of Cauchy stability to prove global existence and stability of black hole spacetimes. In order to go any further, it seems necessary to analyse the structure of the singularities in the conformal structure of the background solutions so as to obtain, if possible, conformal representations of the black hole spacetimes for which the conformal Einstein field equations acquire a form which is amenable to a PDE analysis. An example of the *regularisation* of singularities in the conformal structure is provided by the analysis of the ***problem of spatial infinity*** in Section 20.3 where a detailed knowledge of the singular behaviour of the various conformal fields led to the construction of a regular Cauchy problem for the conformal field equations. It is possible that some singular regions in the conformal structure of black hole spacetimes – such as neighbourhoods of $i^{\pm}$ in the extreme Reissner-Nordström and extreme Kerr spacetimes – are amenable to an analogous discussion; see, for example, Lübbe and Valiente Kroon (2014).

A systematic approach to the analysis of the conformal structure of black holes is through the study of suitable congruences of conformal geodesics. In Friedrich (2003a) it is shown that it is possible to construct a non-intersecting congruence of conformal geodesics that covers the whole of the Kruskal extension of the Schwarzschild spacetime. This congruence is prescribed by initial data on the time symmetric hypersurface of the spacetime, and it provides a preferred conformal representation of the spacetime as well as a global conformal Gaussian gauge system from which, say, a global numerical evaluation of the spacetime can be undertaken; see the discussion in the next section. In addition, this type of construction sheds some light on the singular behaviour of the conformal structure at the timelike infinity; see Friedrich (2002), section 1.4.4. A similar analysis has been carried out in the Reissner-Nordström spacetime (including the extremal case) using so-called ***conformal curves*** in Lübbe and Valiente Kroon (2013a) and the Schwarzschild-de Sitter and Schwarzschild-anti de Sitter spacetime with conformal geodesics in García-Parrado et al. (2014). It would be of great interest extend this type of analysis to stationary black holes, that is, the Kerr spacetime.

The expectation driving the constructions described in the previous paragraph is that they will lead to a suitable conformal representation of the background black hole spacetimes which, in turn, lends itself to the formulation of an initial value problem allowing the analysis of the non-linear stability of black hole spacetimes. Nevertheless, the presence of singular points of the conformal structure of the background solution will require considerations of *asymptotic stability* – rather than just *orbital stability* as in the case of the proofs of stability of the de Sitter and Minkowski spacetimes given in Chapters 15 and 16. The development of methods that allow this type of analysis for the conformal field equations is an interesting and challenging problem.

Finally, it should be mentioned that the notion of conformal compactification of spacetimes, as introduced in Chapter 7, has been used as the starting point of a programme to construct a *theory of peeling and scattering of fields* (including gravity) on black hole spacetimes; see Nicolas (2015) and references within. It would be of great interest to combine this approach to the asymptotic analysis of spacetimes with the methods for the conformal Einstein field equations developed in this book.

21.3 Conformal methods and numerics

Numerical relativity, the study of the Einstein field equations by means of numerical methods, has undergone a great development in recent years. Extended numerical simulations of coalescing black holes have become almost routine; see, for example, Alcubierre (2008), Pretorious (2009) and Baumgarte and Shapiro (2010). To a great extent, these numerical simulations have been concerned with astrophysical aspects of black holes – most notably the extraction of gravitational wave forms; see, for example, Lehner and Pretorious (2014). In addition to this important application aimed at the detection of gravitational waves, numerical relativity offers a powerful tool in mathematical investigations of the equations of general relativity. Some promising areas for this type of interaction have been described in, for example, Jaramillo et al. (2008); for an alternative perspective, see Andersson (2006).

The conformal field equations suggest the possibility of performing *global numerical evaluations of spacetimes*, that is, evaluations which are not limited in their spatial and temporal dimensions by the finiteness of the computational grids. In addition, one would expect such evaluations to be free, to some extent, of the problems posed by the presence of *unphysical* boundary conditions required to obtain a discretisation in a finite grid without periodic boundary conditions.

There have been a number of efforts geared towards the construction of global numerical evaluations of spacetimes using the conformal Einstein field equations. An early implementation of these ideas for the spherically symmetric Einstein-conformally invariant scalar field system can be found in Hübner (1995). A programme to construct a computer code for numerical simulations using the metric vacuum conformal field equations has been reported in Hübner (1999a,b, 2001b) and culminated in Hübner (2001a) where a numerical demonstration of the semi-global existence result for hyperboloidal data discussed in Chapter 16

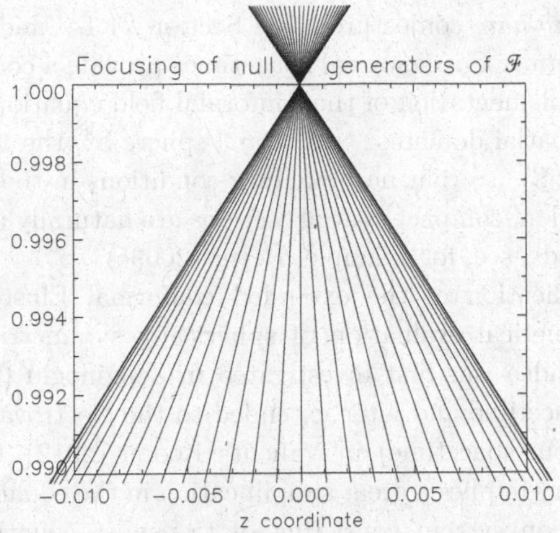

Figure 21.1 Focusing of the generators of null infinity in a numerical evaluation of hyperboloidal initial data close to Minkowski initial data. From Figure 2 in P. Hübner (2001), *From now to timelike infinity in a finite grid*, Class. Quantum Grav. **18**, 1871–1884. © IOP Publishing. Reproduced by permission of IOP Publishing. All rights reserved.

has been provided. Remarkably, the numerical simulations obtained by means of this code show how the generators of null infinity intersect, up to numerical precision at a single point, future timelike infinity i^+; see Figure 21.1. An alternative approach based on the frame version of the standard vacuum conformal field equations has been described in Frauendiener (1998a,b, 2002) and implemented in Frauendiener and Hein (2002); see also the review by Frauendiener (2004). A critical discussion of the numerical implementation of the standard conformal Einstein field equation can be found in Husa (2002).

Conformal Gaussian gauge systems provide an alternative approach to the numerical implementation of the conformal Einstein field equations. As shown in Chapter 13, the evolution equations implied by the extended conformal Einstein equations in this type of gauge splits into a subsystem of transport equations for the components of the frame, connection and Schouten tensor and a symmetric hyperbolic system for the components of the rescaled Weyl tensor. This remarkable structure, highlighting the special role of the Weyl tensor as describing the *free gravitational field*, may facilitate the numerical implementation of the system. An added advantage of this formulation of the conformal field equations is, in the vacuum case, the a priori knowledge of the conformal factor linking the unphysical spacetime with the physical spacetime for which the Einstein equations hold.

A programme to analyse the global dynamics of cosmological spacetimes by numerical methods using the extended conformal field equations has been pursued in Beyer (2007, 2008, 2009a,b). This work has provided valuable insights into the *cosmic no-hair* conjecture – see Section 21.1 – and the role of the so-called *Nariai solution*. Cosmological spacetimes provide a convenient testbed for the numerical implementation of the conformal field equations as they allow the use of compact spatial domains – say, the 3-sphere $\mathbb{S}^3$, the 3-torus $\mathbb{S} \times \mathbb{S} \times \mathbb{S}$ or the 3-handle $\mathbb{S}^2 \times \mathbb{S}$ – so that no boundary conditions in the spatial domain are required. In addition, compact spatial sections are naturally amenable to the use of spectral methods; see, for example, Beyer (2009c).

A further application of the extended conformal Einstein field equations is the global numerical evaluation of spherically symmetric static black hole spacetimes. This idea was first investigated in Zenginoglu (2006, 2007) for the Schwarzschild spacetime and later extended to the electrovacuum case (i.e. the Reissner-Nordström spacetime) in Valiente Kroon (2012). The assumption of spherical symmetry implies a great simplification in the equations so as to render a reduced evolution system consisting of transport equations solely. Notice, however, that the conformal gauge in terms of which the evolution equations are expressed is not adapted to the orbits of the static Killing vectors, and, thus, one has non-trivial *gauge dynamics*. An important property of these spherically symmetric reduced equations is that their essential dynamics is governed by a *core system* consisting of three equations in the vacuum case (for a component of the connection, a component of the Schouten tensor and the non-vanishing component of the rescaled Weyl tensor) and four equations in the electrovacuum case (connection, Schouten tensor, rescaled Weyl tensor and the single non-vanishing component of the Faraday tensor). These equations can be easily implemented and numerically solved with present-day desktop computers and allow the global computation of a privileged conformal representation of the black hole spacetime from an initial hypersurface to either the singularity or null infinity and beyond; see Figure 21.2. These small-scale numerical simulations could be used, in the future, as the first step in the global numerical evaluation of dynamic, non-spherically symmetric spacetimes.

More recently, there have been efforts aimed at the numerical implementation of the construction of the cylinder at spatial infinity described in Section 20.3.2. The ultimate goal of this programme is the numerical computation of hyperboloidal data from Cauchy data and to obtain insight into the numerical consequences of the obstructions to the smoothness of null infinity discussed in the later sections of Chapter 20. At the time of writing, the analysis has been restricted to the analysis of the spin-2 field equation on a Minkowski background – in the spirit of Section 20.3.4 – with the expectation that this situation contains the essential difficulties in the implementation; see Beyer et al. (2012) and Frauendiener and Hennig (2014).

Foliations of spacetimes by means of hyperboloidal hypersurfaces have been used in numerical simulations aimed at analysing radiative processes in gravitational collapse and perturbations of black hole spacetimes; see, for example, Zenginoglu (2008, 2011a,b), Rinne (2010, 2014), Zenginoglu and Kidder (2010)

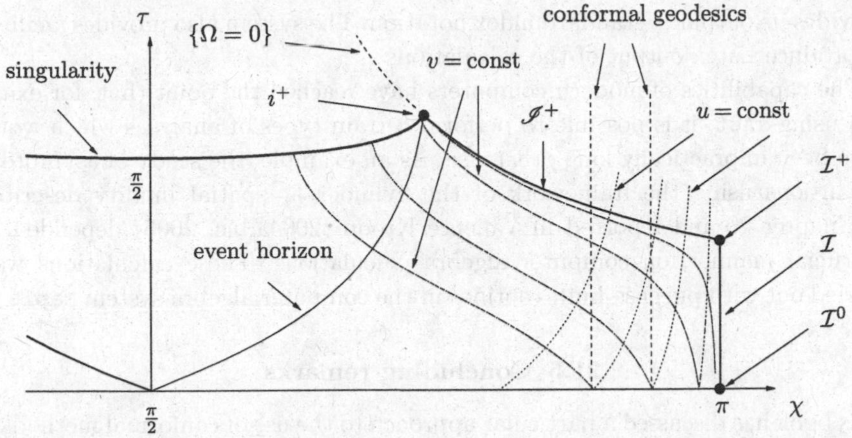

Figure 21.2 The Schwarzschild spacetime in a conformal Gaussian gauge system; see the discussion in the main text and compare also Section 20.3.3. From Figure 3.3, page 57, of A. Zenginoglu, *A conformal approach to numerical computations of asymptotically simple spacetimes*, PhD thesis, University of Postdam (2006). Reproduced courtesy of the author.

and Rinne and Moncrief (2013). These numerical investigations make use of formulations of the Einstein field equations alternative to the ones discussed in this book. Finally, hyperboloidal foliations have also been used in the implementation of fully spectral (i.e. in time and space) evolution schemes for various fields; see Hennig and Ansorg (2009) and Macedo and Ansorg (2014).

21.4 Computer algebra

The analysis of hyperbolic reductions in Chapter 13 shows that despite their elegance and appealing geometric nature, the study of the consequences of the conformal field equations requires a considerable amount of algebraic manipulations. Modern computer algebra systems provide a natural way of performing these manipulations in an effective and efficient way. At the time of writing, the suite of packages `xAct` for `Mathematica` provides a robust and versatile framework for the type of tensorial and spinorial manipulations discussed in this book; see Martín-García (2014). The packages in the suite `xAct` allow one to perform tensorial and spinorial abstract index computations on generic tensors as well as explicit component computations for a given metric. At the core of `xAct` is a *canonicalisation routine* which allows one to simplify large tensorial and spinorial expressions by identifying "dummy" indices and exploiting the symmetries of the various objects involved. In addition, `xAct` allows one to carry out cumbersome operations such as the decomposition of spinors into irreducible parts. An additional appeal of this system is that it provides its output in standard index notation. The system also provides facilities to produce Latex output of the calculations

The capabilities of modern computers have reached the point that, for example, using `xAct`, it is possible to perform certain types of analyses which would have been impractically long otherwise. As an example, the study of asymptotic expansions using the framework of the cylinder at spatial infinity described in Chapter 20 and reported in Valiente Kroon (2004a,b,c, 2005) depended, in a crucial manner, on computer algebra calculations. These calculations were carried out with purpose-built routines in the computer algebra system `Maple V`.

21.5 Concluding remarks

This book has discussed a particular approach to the use of conformal methods in mathematical general relativity. Clearly, the approach presented is not the only one possible nor are the potential applications restricted to the ones discussed in these pages. It constitutes a body of work extending over a period of more than 30 years starting with the work of H. Friedrich in the early 1980s – or 50 years if one considers the seminal work by R. Penrose in the 1960s. This extended period of time is proof of the vitality of the subject. Nevertheless, a more exacting assessment of its vitality and relevance should come from its influence in the whole of mathematical general relativity and its ability to foster new ideas and research problems. Time will be the ultimate judge on this matter.

This book is an attempt to bring to the fore the relevance of conformal methods in modern research in general relativity and to make the subject as accessible as possible to those interested in using these ideas in their own research. The reader is left to decide whether this goal has been achieved.

References

Abraham, R., and Robbin, J. 1967. *Transversal mapping and flows.* W. A. Benjamin.

Aceña, A. 2009. Convergent null data expansions at spacelike infinity of stationary vacuum solutions. *Ann. Henri Poincaré*, **10**, 275.

Aceña, A., and Valiente Kroon, J. A. 2011. Conformal extensions for stationary spacetimes. *Class. Quantum Grav.*, **28**, 225023.

Alcubierre, M. 2008. *Introduction to* 3+1 *numerical relativity.* Oxford University Press.

Ambrosetti, A., and Prodi, G. 1995. *A primer of nonlinear analysis.* Cambridge University Press.

Anderson, M. T. 2000. On stationary vacuum solutions to the Einstein equations. *Ann. Henri Poincaré* **1**, 977.

Anderson, M. T. 2005a. Existence and stability of even dimensional asymptotically de Sitter spaces. *Ann. Henri Poincaré*, **6**, 801.

Anderson, M. T. 2005b. Geometric aspects of the AdS/CFT correspondence. Page 1 of: Biquard, O. (ed), *AdS/CFT correspondence: Einstein metrics and their conformal boundaries.* Euro. Math. Soc. Zürich.

Anderson, M. T. 2006. On the uniqueness and global dynamics of AdS spacetimes. *Class. Quantum Grav.*, **23**, 6935.

Anderson, M. T., and Chruściel, P. T. 2005. Asymptotically simple solutions of the vacuum Einstein equations in even dimensions. *Comm. Math. Phys.*, **260**, 557.

Anderson, M. T., Chruściel, P. T., and Delay, E. 2002. Non-trivial static, geodesically complete vacuum spacetimes with a negative cosmological constant. *JHEP*, **10**, 6.

Andersson, L. 2006. On the relation between mathematical and numerical relativity. *Class. Quantum Grav.*, **23**, S307.

Andersson, L., and Chruściel, P. T. 1993. Hyperboloidal Cauchy data for vacuum Einstein equations and obstructions to smoothness of null infinity. *Phys. Rev. Lett.*, **70**, 2829.

Andersson, L., and Chruściel, P. T. 1994. On "hyperboloidal" Cauchy data for vacuum Einstein equations and obstructions to smoothness of scri. *Comm. Math. Phys.*, **161**, 533.

Andersson, L., Chruściel, P. T., and Friedrich, H. 1992. On the regularity of solutions to the Yamabe equation and the existence of smooth hyperboloidal initial data for Einstein's field equations. *Comm. Math. Phys.*, **149**, 587.

Anguige, K., and Tod, K. P. 1999a. Isotropic cosmological singularities I. Polytropic perfect fluid spacetimes. *Ann. Phys.*, **276**, 257.

Anguige, K., and Tod, K. P. 1999b. Isotropic cosmological singularities II. The Einstein-Vlasov system. *Ann. Phys.*, **276**, 294.

Appel, W. 2007. *Mathematics for physics and physicists.* Princeton University Press.

Arnowitt, R., Deser, S., and Misner, C. W. 1962. The dynamics of general relativity. Page 227 of: Witten, L. (ed), *Gravitation: an introduction to current research.* John Wiley & Sons.

Arnowitt, R., Deser, S., and Misner, C. 2008. Reprint of: The dynamics of general relativity. *Gen. Rel. Grav.*, **40**, 1997.

Ashtekar, A. 1980. Asymptotic structure of the gravitational field at spatial infinity. Page 37 of: Held, A. (ed), *General relativity and gravitation: one hundred years after the birth of Albert Einstein*, vol. 2. Plenum Press.

Ashtekar, A. 1984. Asymptotic properties of isolated systems: recent developments. Page 37 of: Bertotti, B., de Felice, F., and Pascolini, A. (eds), *General relativity and gravitation*. D. Reidel Publishing Company.

Ashtekar, A. 1987. *Asymptotic quantization*. Bibliopolis.

Ashtekar, A. 1991. *Lectures on non-perturbative canonical gravity*. World Scientific.

Ashtekar, A. 2014. Geometry and physis of null infinity. Page 99 of: Bieri, L., and Yau, S.-T. (eds), *One hundred years of general relativity*. Surveys in Differential Geometry, vol. 20. International Press.

Ashtekar, A., and Das, S. 2000. Asymptotically anti-de Sitter spacetimes: conserved quantities. *Class. Quantum Grav.*, **17**, L17.

Ashtekar, A., and Dray, T. 1981. On the existence of solutions to Einstein's field equations with non-zero Bondi news. *Comm. Math. Phys.*, **79**, 581.

Ashtekar, A., and Hansen, R. O. 1978. A unified treatment of null and spatial infinity in general relativity. I. Universal structure, asymptotic symmetries, and conserved quantities at spatial infinity. *J. Math. Phys.*, **19**, 1542.

Ashtekar, A., and Magnon, A. 1984. Asymptotically anti-de Sitter space-times. *Class. Quantum Grav.*, **1**, L39.

Ashtekar, A., and Schmidt, B. G. 1980. Null infinity and Killing fields. *J. Math. Phys.*, **21**, 862.

Ashtekar, A., and Xanthopoulos, B. C. 1978. Isometries compatible with asymptotic flatness at null infinity: a complete description. *J. Math. Phys.*, **19**, 2216.

Ashtekar, A., Horowitz, G. T., and Magnon-Ashtekar, A. 1982. A generalization of tensor calculus and its applications to physics. *Gen. Rel. Grav.*, **14**, 411.

Aubin, T. 1976. Équations différentielles non linéaires et le problème de Yamabe concernant la courbure scalaire. *J. Math. Pures Appl.*, **55**, 269.

Bäckdahl, T. 2007. Axisymmetric stationary solutions with arbitrary multipole moments. *Class. Quantum Grav.*, **24**, 2205.

Bäckdahl, T. 2009. Relating the Newman-Penrose constants to the Geroch-Hansen multipole moments. *Class. Quantum Grav.*, **26**, 231102.

Bäckdahl, T., and Herberthson, M. 2005a. Explicit multipole moments of stationary axisymmetric spacetimes. *Class. Quantum Grav.*, **22**, 3585.

Bäckdahl, T., and Herberthson, M. 2005b. Static axisymmetric space-times with prescribed multipole moments. *Class. Quantum Grav.*, **22**, 1607.

Bäckdahl, T., and Herberthson, M. 2006. Calculations of, and bounds for, the multipole moments of stationary spacetimes. *Class. Quantum Grav.*, **23**, 5997.

Bäckdahl, T., and Valiente Kroon, J. A. 2010a. Geometric invariant measuring the deviation from Kerr data. *Phys. Rev. Lett.*, **104**, 231102.

Bäckdahl, T., and Valiente Kroon, J. A. 2010b. On the construction of a geometric invariant measuring the deviation from Kerr data. *Ann. Henri Poincaré*, **11**, 1225.

Bartnik, R. 1986. The mass of an asymptotically flat manifold. *Comm. Pure Appl. Math.*, 661.

Bartnik, R., and Isenberg, J. 2004. The constraint equations. Page 1 of: Chruściel, P. T., and Friedrich, H. (eds), *The Einstein equations and the large scale behaviour of gravitational fields*. Birkhauser.

Baston, R. J., and Mason, L. J. 1987. Conformal gravity, the Einstein equation and spaces of complex null geodesics. *Class. Quantum Grav.*, **4**, 815.

Baumgarte, T. W., and Shapiro, S. L. 2010. *Numerical relativity: solving Einstein's equations on the computer*. Cambridge University Press.

Beig, R. 1984. Integration of Einstein's equations near spatial infinity. *Proc. Roy. Soc. Lond. A*, **391**, 295.

Beig, R. 1985. A remarkable property of spherical harmonics. *J. Math. Phys.*, **26**, 769.

Beig, R. 1991. Conformal properties of static spacetimes. *Class. Quantum Grav.*, **8**, 263.

Beig, R., and Heinzle, J. M. 2005. CMC-slicings of Kottler-Schwarzschild-de Sitter Cosmologies. *Comm. Math. Phys.*, **260**, 673.

Beig, R., and Husa, S. 1994. Initial data for general relativity with toroidal conformal symmetry. *Phys. Rev. D*, **50**, R7116.

Beig, R., and O'Murchadha, N. 1991. Trapped surfaces due to concentration of gravitational radiation. *Phys. Rev. Lett.*, **66**, 2421.

Beig, R., and O'Murchadha, N. 1994. Trapped surfaces in vacuum spacetimes. *Class. Quantum Grav.*, **11**, 419.

Beig, R., and O'Murchadha, N. 1996. The momentum constraints of general relativity and spatial conformal isometries. *Comm. Math. Phys.*, **176**, 723.

Beig, R., and O'Murchadha, N. 1998. Late time behaviour of the maximal slicing of the Schwarzschild black hole. *Phys. Rev. D*, **57**, 4728.

Beig, R., and Schmidt, B. G. 1982. Einstein's equation near spatial infinity. *Comm. Math. Phys.*, **87**, 65.

Beig, R., and Schmidt, B. G. 2000. Time-independent gravitational fields. Page 325 of: Schmidt, B. G. (ed), *Einstein's field equations and their physical implications*. Springer.

Beig, R., and Simon, W. 1980a. Proof of a multipole conjecture due to Geroch. *Comm. Math. Phys.*, **75**, 79.

Beig, R., and Simon, W. 1980b. The stationary gravitational field near spatial infinity. *Gen. Rel. Grav.*, **12**, 1003.

Beig, R., and Simon, W. 1981. On the multipole expansions for stationary space-times. *Proc. Roy. Soc. Lond. A*, **376**, 333.

Beig, R., and Szabados, L. B. 1997. On a global invariant of initial data sets. *Class. Quantum Grav.*, **14**, 3091.

Bekenstein, J. 1974. Exact solutions of Einstein-conformal scalar equations. *Ann. Phys.*, **82**, 535.

Bernal, A. N., and Sánchez, M. 2007. Globally hyperbolic spacetimes can be defined as 'causal' instead of strongly causal. *Class. Quantum Grav.*, **24**, 745.

Besse, A. L. 2008. *Einstein manifolds*. Springer Verlag.

Beyer, F. 2007. Asymptotics and singularities in cosmological models with positive cosmological constant. PhD thesis, University of Potsdam.

Beyer, F. 2008. Investigations of solutions of Einstein's field equations close to lambda-Taub-NUT. *Class. Quantum Grav.*, **25**, 235005.

Beyer, F. 2009a. Non-genericity of the Nariai solutions: I. Asymptotics and spatially homogeneous perturbations. *Class. Quantum Grav.*, **26**, 235015.

Beyer, F. 2009b. Non-genericity of the Nariai solutions: II. Investigations within the Gowdy class. *Class. Quantum Grav.*, **26**, 235016.

Beyer, F. 2009c. A spectral solver for evolution problems with spatial S3-topology. *J. Comput. Phys.*, **228**, 6496.

Beyer, F., Doulis, G., Frauendiener, J., and Whale, B. 2012. Numerical space-times near space-like and null infinity. The spin-2 system on Minkowski space. *Class. Quantum Grav.*, **29**, 245013.

Bičák, J. 2000. Selected solutions of Einstein's field equations: their role in general relativity and astrophysics. In: Schmidt, B. G. (ed), *Einstein field equations and their physical implications (selected essays in honour of Juergen Ehlers).* Springer Verlag.

Bičák, J., and Krtouš, P. 2001. Accelerated sources in de Sitter spacetime and the insufficiency of retarded fields. *Phys. Rev. D*, **64**, 124020.

Bičák, J., and Krtouš, P. 2002. The fields of uniformly accelerated charges in de Sitter spacetime. *Phys. Rev. Lett.*, **88**, 211101.

Bičák, J., and Schmidt, B. G. 1989. Asymptotically flat radiative space-times with boost-rotation symmetry: the general structure. *Phys. Rev. D*, **40**, 1827.

Bičák, J., Scholtz, M., and Tod, K. P. 2010. On asymptotically flat solutions of Einstein's equations periodic in time: II. Spacetimes with scalar-field sources. *Class. Quantum Grav.*, **27**, 175011.

Bizon, P. 2013. Is AdS stable? *Gen. Rel. Grav.*, **46**, 1724.

Bizon, P., and Friedrich, H. 2012. A remark about wave equations on the extreme Reissner-Nordström black hole exterior. *Class. Quantum Grav.*, **30**, 065001.

Bizon, P., and Rostworowski, A. 2011. Weakly turbulent instability of anti-de Sitter spacetime. *Phys. Rev. Lett.*, **107**, 031102.

Bondi, H., van der Burg, M. G. J., and Metzner, A. W. K. 1962. Gravitational waves in general relativity VII. Waves from axi-symmetric isolated systems. *Proc. Roy. Soc. Lond. A*, **269**, 21.

Brännlund, J. 2004. Conformal isometry of the Reissner-Nordström-de Sitter Black Hole. *Gen. Rel. Grav.*, **36**, 883.

Branson, T., Eastwood, M., and Wang, M. 2004. Conformal geometry. In www.birs.ca/workshops/2004/04w5006/report04w5006.pdf.

Brill, D.R., and Hayward, S. A. 1994. Global structure of a black hole cosmos and its extremes. *Class. Quantum Grav.*, **11**, 359.

Buchdahl, H. A. 1959. Reciprocal static metrics and scalar fields in the general theory of relativity. *Phys. Rev.*, **115**, 1325.

Butscher, A. 2002. Exploring the conformal constraint equations. Page 195 of: Frauendiener, J., and Friedrich, H. (eds), *The conformal structure of spacetime: geometry, analysis, numerics.* Lect. Notes. Phys. Springer.

Butscher, A. 2007. Perturbative solutions of the extended constraint equations in general relativity. *Comm. Math. Phys.*, **272**, 1.

Cagnac, F. 1981. Problème de Cauchy sur un conoïde caractéristique pour des equations quasi-lineaires. *Ann. Mat. Pure Appl.*, **129**, 13.

Carmeli, M. 1977. *Group theory and general relativity: representations of the Lorentz group and their applications to the gravitational field.* McGraw-Hill.

Carter, B. 1966a. The complete analytic extension of the Reissner-Nordström metric in the special case $e^2 = m^2$. *Phys. Rev. Lett.*, **21**, 423.

Carter, B. 1966b. Complete analytic extension of the symmetry axis of Kerr's solution of Einstein's equations. *Phys. Rev.*, **141**, 1242.

Carter, B. 1968. Global structure of the Kerr family of gravitational fields. *Phys. Rev.*, **174**, 1559.

Carter, B. 1971. Causal structure in space-time. *Gen. Rel. Grav.*, **1**, 349.

Carter, B. 1973. Black hole equilibrium states. Page 61 of: DeWitt, C., and DeWitt, B.S. (eds), *Black holes: les astres occlus.* Gordon and Breach.

Chaljub-Simon, A. 1982. Decomposition of the space of covariant two-tensors on $\mathbb{R}^3$. *Gen. Rel. Grav.*, **14**, 743.

Chaljub-Simon, A., and Choquet-Bruhat, Y. 1980. Global solutions of the Licnerowicz equation in general relativity on an asymptotically Euclidean manifold. *Gen. Rel. Grav.*, **12**, 175.

Chang, S.-Y. A., Qing, J., and Yang, P. 2007. Some progress in conformal geometry. *Sigma*, **3**, 122.

Choptuik, M. W. 1993. Universality and scaling in gravitational collapse of a massless scalar field. *Phys. Rev. Lett.*, **70**, 9.

Choquet-Bruhat, I., Isenberg, J., and York, J. W. Jr. 2000. Einstein constraints on asymptotically Euclidean manifolds. *Phys. Rev. D*, **61**, 084034.

Choquet-Bruhat, Y. 2007. Results and open problems in mathematical general relativity. *Milan J. Math.*, **75**, 273.

Choquet-Bruhat, Y. 2008. *General relativity and the Einstein equations.* Oxford University Press.

Choquet-Bruhat, Y., and Christodoulou, D. 1981. Existence of global solutions of the Yang-Mills, Higgs and spinor field equations in 3+1 dimensions. *Ann. Sci. l'E.N.S.*, **14**, 481.

Choquet-Bruhat, Y., and Geroch, R. 1969. Global aspects of the Cauchy problem in general relativity. *Comm. Math. Phys.*, **14**, 329.

Choquet-Bruhat, Y., and York, J. W. Jr. 1980. The Cauchy problem. In: Held, A. (ed), *General relativity and gravitation*, Vol. I. Plenum Press.

Choquet-Bruhat, Y., Dewitt-Morette, C., and Dillard-Bleck, M. 1982. *Analysis, manifolds and physics: part I.* North Holland Publishing Company.

Choquet-Bruhat, Y., Chruściel, P. T., and Martín-García, J. M. 2011. The Cauchy problem on a characteristic cone for the Einstein equations in arbitrary dimensions. *Ann. Henri Poincaré*, **12**, 419.

Christodoulou, D. 1986. The problem of a self-gravitating scalar field. *Comm. Math. Phys.*, **105**, 337.

Christodoulou, D., and Klainerman, S. 1993. *The global nonlinear stability of the Minkowski space.* Princeton University Press.

Christodoulou, D., and O'Murchadha, N. 1981. The boost problem in general relativity. *Comm. Math. Phys.*, **80**, 271.

Chruściel, P. T. 1991. *On the uniqueness in the large of solutions of Einstein's equations ("Strong Cosmic Censorship").* Centre for Mathematics and Its Applications, Australian National University.

Chruściel, P. T., and Delay, E. 2002. Existence of non-trivial, vacuum, asymptotically simple spacetimes. *Class. Quantum Grav.*, **19**, L71.

Chruściel, P. T., and Delay, E. 2003. On mapping properties of the general relativistic constraint operator in weighted function spaces, with applications. *Mem. Soc. Math. France*, **94**, 1.

Chruściel, P. T., and Delay, E. 2009. Gluing constructions for asymptotically hyperbolic manifolds with constant scalar curvature. *Comm. Anal. Geom.*, **17**, 343.

Chruściel, P. T., and Paetz, T.-T. 2012. The many ways of the characteristic Cauchy problem. *Class. Quantum Grav.*, **29**, 145006.

Chruściel, P. T., and Paetz, T.-T. 2013. Solutions of the vacuum Einstein equations with initial data on past null infinity. *Class. Quantum Grav.*, **30**, 235037.

Chruściel, P. T., MacCallum, M. A. H., and Singleton, D. B. 1995. Gravitational waves in general relativity XIV. Bondi expansions and the "polyhomogeneity" of $\mathscr{I}$. *Phil. Trans. Roy. Soc. Lond. A*, **350**, 113.

Chruściel, P. T., Ölz, C. R., and Szybka, S. J. 2012a. Space-time diagrammatics. *Phys. Rev. D*, **86**, 124041.

Chruściel, P. T., and Costa, J. L., and Heusler, M. 2012b. Stationary black holes: uniqueness and beyond. *Living Rev. Relativity* **15**, 7. URL (cited on 24 May 2016): www.livingreviews.org/lrr-2012-7.

Cocke, W. J. 1989. Table for constructing spin coefficients in general relativity. *Phys. Rev. D*, **40**, 650.

Cook, G. B. 2000. Initial data for numerical relativity. *Living Rev. Relativity*, **3**, 5. URL (cited on 24 May 2016): www.livingreviews.org/lrr-2000-5.

Corvino, J. 2000. Scalar curvature deformations and a gluing construction for the Einstein constraint equations. *Comm. Math. Phys.*, **214**, 137.

Corvino, J. 2007. On the existence and stability of the Penrose compactification. *Ann. Henri Poincaré*, **8**, 597.

Corvino, J., and Schoen, R. 2006. On the asymptotics for the Einstein constraint vacuum equations. *J. Diff. Geom.*, **73**, 185.

Cotton, É. 1899. Sur les variétés à trois dimensions. *Ann. Fac. Sci. Toulouse 2e série*, **1**, 385.

Couch, W. E., and Torrence, R. J. 1984. Conformal invariance under spatial inversion of extreme Reissner-Nordström black holes. *Gen. Rel. Grav.*, **16**, 789.

Courant, R., and Hilbert, D. 1962. *Methods of mathematical physics.* Vol. II. John Wiley & Sons.

Courant, R., and John, F. 1989. *Introduction to calculus and analysis II/1.* Springer.

Cutler, C. 1989. Properties of spacetimes that are asymptotically flat at timelike infinity. *Class. Quantum Grav.*, **6**, 1075.

Cutler, C., and Wald, R. M. 1989. Existence of radiating Einstein-Maxwell solutions which are C^∞ on all of $\mathscr{I}^+$ and $\mathscr{I}^-$. *Class. Quantum Grav.*, **6**, 453.

Dafermos, M. 2003. Stability and instability of the Cauchy horizon for the spherically symmetric Einstein-Maxwell-scalar field equations. *Ann. Math.*, **158**, 875.

Dafermos, M. 2005. The interior of charged black holes and the problem of uniqueness in general relativity. *Comm. Pure Appl. Math.*, **58**, 0445.

Dafermos, M., and Rodnianski, I. 2005. A proof of Price's law for the collapse of a self-gravitating scalar field. *Invent. Math.*, **162**, 381.

Dafermos, M., and Rodnianski, I. 2010. Lectures on black holes and linear waves. Page 97 of: Ellwood, D., Rodnianski, I., Staffilani, G., and Wunsch, J. (eds), *Evolution equations.* Clay Mathematics Proceedings, Vol. 17. American Mathematical Society-Clay Mathematics Institute.

Dain, S. 2001a. Initial data for a head-on collision of two Kerr-like black holes with close limit. *Phys. Rev. D*, **64**, 124002.

Dain, S. 2001b. Initial data for stationary spacetimes near spacelike infinity. *Class. Quantum Grav.*, **18**, 4329.

Dain, S. 2001c. Initial data for two Kerr-like black holes. *Phys. Rev. Lett.*, **87**, 121102.

Dain, S. 2006. Elliptic systems. Page 117 of: Frauendiener, J., Giulini, D., and Perlick, V. (eds), *Analytical and numerical approaches to general relativity.* Lect. Notes. Phys., vol. 692. Springer Verlag.

Dain, S., and Friedrich, H. 2001. Asymptotically flat initial data with prescribed regularity at infinity. *Comm. Math. Phys.*, **222**, 569.

Dain, S., and Gabach-Clement, M. E. 2011. Small deformations of extreme Kerr black hole initial data. *Class. Quantum Grav.*, **28**, 075003.

Damour, T., and Schmidt, B. 1990. Reliability of perturbation theory in general relativity. *J. Math. Phys.*, **31**, 2441.

Dossa, M. 1986. Solution globale d'un probléme de Cauchy caractéristique non linéare. *Comptes Rendus de l'Academie des Sciences. Series I*, **303**, 795.

Dossa, M. 1997. Espaces de Sobolev non isotropes à poids et problémes de Cauchy quasi-linéaires sur un conoïde caractéristique. *Ann. Inst. H. Poincaré Phys. Théor.*, **1**, 37.

Dossa, M. 2002. Solutions C^∞ d'une classe de problèmes de Cauchy quasi-linéaires hyperboliques du second ordre sur un conoïde caractéristique. *Annales de la faculté des sciences de Toulouse Sér. 6*, **11**, 351.

Eastwood, M. 1996. Notes on conformal differential geometry. Page 57 of: Slovák, J. (ed), *Proceedings of the 15th Winter School "Geometry and Physics"*. Rendiconti del Circolo Matematico de Palermo, Serie II, vol. Supplemento No. 43. Circolo Matematico di Palermo.

Ehlers, J. 1973. Spherically symmetric spacetimes. Page 114 of: Israel, W. (ed), *Relativity, astrophysics and cosmology*. D. Reidel Publishing Company.

Ellis, G. F. R. 1984. Relativistic cosmology: its nature, aims and problems. In: Bertotti, B., de Felice, F., and Pascolini, A. (eds), *General relativity and gravitation*. D. Reidel Publishing Company.

Ellis, G. F. R. 2002. The state of cosmology 2001: two views and a middle way. In: Bishop, N. T., and Maharaj, S. D. (eds), *General relativity and gravitation*. World Scientific.

Ellis, G. F. R., and van Elst, H. 1998. Cosmological models: Cargese lectures 1998. *NATO Adv. Study Inst. Ser. C. Math. Phys. Sci.*, **541**, 1.

Ellis, G. F. R., Maartens, R., and MacCallum, M. A. H. 2012. *Relativistic cosmology*. Cambridge University Press.

Estabrook, F., Wahlquist, H., Christensen, S., DeWitt, B., Smarr, L., and Tsiang, E. 1973. Maximally slicing a black hole. *Phys. Rev. D*, **7**, 2814.

Evans, L. C. 1998. *Partial differential equations*. American Mathematical Society.

Exton, A. R., Newman, E. T., and Penrose, R. 1969. Conserved quantites in the Einstein-Maxwell theory. *J. Math. Phys.*, **10**, 1566.

Fefferman, C., and Graham, C. R. 1985. Conformal invariants. Page 95 of: *Élie Cartan et les mathématiques d'aujord'hui. The mathematical heritage of Élie Cartan, Sémin. Lyon 1984*. Astérisque, No. Hors Sér.

Fefferman, C., and Graham, C. R. 2012. *The ambient metric*. Annals of Mathematical Studies, vol. 178. Princeton University Press.

Fischer, A. E., and Marsden, J. E. 1972. The Einstein evolution equations as a first-order quasi-linear symmetric hyperbolic system. I. *Comm. Math. Phys.*, **28**, 1.

Fourès-Bruhat, Y. 1952. Théorème d'existence pour certains systèmes d'équations aux derivées partielles non linéaires. *Acta Mathematica*, **88**, 141.

Frances, C. 2005. The conformal boundary of anti de Sitter spacetimes. Page 205 of: Biquard, O. (ed), *AdS/CFT correspondence: Einstein metrics and their conformal boundaries*. Euro. Math. Soc. Zürich.

Frankel, T. 2003. *The geometry of physics*. Cambridge University Press.

Frauendiener, J. 1998a. Numerical treatment of the hyperboloidal initial value problem for the vacuum Einstein equations. I. The conformal field equations. *Phys. Rev. D*, **58**, 064002.

Frauendiener, J. 1998b. Numerical treatment of the hyperboloidal initial value problem for the vacuum Einstein equations. II. The evolution equations. *Phys. Rev. D*, **58**, 064003.

Frauendiener, J. 2002. Some aspects of the numerical treatment of the conformal field equations. Page 261 of: Frauendiener, J., and Friedrich, H. (eds), *The conformal structure of space-time: geometry, analysis, numerics*. Springer.

Frauendiener, J. 2004. Conformal infinity. *Living Rev. Relativity*, **7**, 1. URL (cited on 24 May 2016): www.livingreviews.org/lrr-2004-1.

Frauendiener, J., and Hein, M. 2002. Numerical evolution of axisymmetric, isolated systems in general relativity. *Phys. Rev. D*, **66**, 124004.

Frauendiener, J., and Hennig, J. 2014. Fully pseudospectral solution of the conformally invariant wave equation near the cylinder at spacelike infinity. *Class. Quantum Grav.*, **31**, 085010.

Frauendiener, J., and Sparling, G. A. 2000. Local twistors and the conformal field equations. *J. Math. Phys.*, **41**, 437.

Frauendiener, J., and Szabados, L. B. 2001. The kernel of the edth operators on higher-genus spacelike 2-surfaces. *Class. Quantum Grav.*, **18**, 1003.

Friedrich, H. 1981a. The asymptotic characteristic initial value problem for Einstein's vacuum field equations as an initial value problem for a first-order quasilinear symmetric hyperbolic system. *Proc. Roy. Soc. Lond. A*, **378**, 401.

Friedrich, H. 1981b. On the regular and the asymptotic characteristic initial value problem for Einstein's vacuum field equations. *Proc. Roy. Soc. Lond. A*, **375**, 169.

Friedrich, H. 1982. On the existence of analytic null asymptotically flat solutions of Einstein's vacuum field equations. *Proc. Roy. Soc. Lond. A*, **381**, 361.

Friedrich, H. 1983. Cauchy problems for the conformal vacuum field equations in General Relativity. *Comm. Math. Phys.*, **91**, 445.

Friedrich, H. 1984. Some (con-)formal properties of Einstein's field equations and consequences. In: Flaherty, F. J. (ed), *Asymptotic behaviour of mass and spacetime geometry*. Lecture notes in physics 202. Springer Verlag.

Friedrich, H. 1985. On the hyperbolicity of Einstein's and other gauge field equations. *Comm. Math. Phys.*, **100**, 525.

Friedrich, H. 1986a. Existence and structure of past asymptotically simple solutions of Einstein's field equations with positive cosmological constant. *J. Geom. Phys.*, **3**, 101.

Friedrich, H. 1986b. On the existence of n-geodesically complete or future complete solutions of Einstein's field equations with smooth asymptotic structure. *Comm. Math. Phys.*, **107**, 587.

Friedrich, H. 1986c. On purely radiative space-times. *Comm. Math. Phys.*, **103**, 35.

Friedrich, H. 1988. On static and radiative space-times. *Comm. Math. Phys.*, **119**, 51.

Friedrich, H. 1991. On the global existence and the asymptotic behaviour of solutions to the Einstein-Maxwell-Yang-Mills equations. *J. Diff. Geom.*, **34**, 275.

Friedrich, H. 1992. Asymptotic structure of space-time. Page 147 of: Janis, A. I., and Porter, J. R. (eds), *Recent advances in general relativity*. Einstein Studies, vol. 4. Birkhauser.

Friedrich, H. 1995. Einstein equations and conformal structure: existence of anti-de Sitter-type space-times. *J. Geom. Phys.*, **17**, 125.

Friedrich, H. 1996. Hyperbolic reductions for Einstein's equations. *Class. Quantum Grav.*, **13**, 1451.

Friedrich, H. 1998a. Einstein's equation and geometric asymptotics. Page 153 of: Dadhich, N., and Narlinkar, J. (eds), *Proceedings of the GR-15 conference*. Inter-University Centre for Astronomy and Astrophysics.

Friedrich, H. 1998b. Evolution equations for gravitating ideal fluid bodies in general relativity. *Phys. Rev. D*, **57**, 2317.

Friedrich, H. 1998c. Gravitational fields near space-like and null infinity. *J. Geom. Phys.*, **24**, 83.

Friedrich, H. 1999. Einstein's equation and conformal structure. Page 81 of: Huggett, S. A., Mason, L. J., Tod, K. P., Tsou, S. T., and Woodhouse, N. M. J. (eds), *The geometric universe: science, geometry and the work of Roger Penrose*. Oxford University Press.

Friedrich, H. 2002. Conformal Einstein evolution. Page 1 of: Frauendiener, J., and Friedrich, H. (eds), *The conformal structure of spacetime: geometry, analysis, numerics*. Lecture Notes in Physics. Springer.

Friedrich, H. 2003a. Conformal geodesics on vacuum spacetimes. *Comm. Math. Phys.*, **235**, 513.

Friedrich, H. 2003b. Spin-2 fields on Minkowski space near space-like and null infinity. *Class. Quantum Grav.*, **20**, 101.

Friedrich, H. 2004. Smoothness at null infinity and the structure of initial data. In: Chruściel, P. T., and Friedrich, H. (eds), *50 years of the Cauchy problem in general relativity*. Birkhausser.

Friedrich, H. 2005. On the non-linearity of the subsidiary systems. *Class. Quantum Grav.*, **22**, L77.

Friedrich, H. 2007. Static vacuum solutions from convergent null data expansions at space-like infinity. *Ann. Henri Poincaré*, **8**, 817.

Friedrich, H. 2008a. Conformal classes of asymptotically flat, static vacuum data. *Class. Quantum Grav.*, **25**, 065012.

Friedrich, H. 2008b. One-parameter families of conformally related asymptotically flat, static vacuum data. *Class. Quantum Grav.*, **25**, 135012.

Friedrich, H. 2009. Initial boundary value problems for Einstein's field equations and geometric uniqueness. *Gen. Rel. Grav.*, **41**, 1947.

Friedrich, H. 2011. Yamabe numbers and the Brill-Cantor criterion. *Ann. Inst. H. Poincaré*, **12**, 1019.

Friedrich, H. 2013. Conformal structure of static vacuum data. *Comm. Math. Phys.*, **321**, 419.

Friedrich, H. 2014a. On the AdS stability problem. *Class. Quantum Grav.*, **31**, 105001.

Friedrich, H. 2014b. The Taylor expansion at past time-like infinity. *Comm. Math. Phys.*, **324**, 263.

Friedrich, H. 2015a. Geometric asymptotics and beyond. Page 37 of: Bieri, L., and Yau, S.-T. (eds), *One hundred years of general relativity*. Surveys in Differential Geometry, vol. 20. International Press.

Friedrich, H. 2015b. Smooth non-zero rest-mass evolution across time-like infinity. *Ann. Henri Poincaré*, **16**, 2215.

Friedrich, H., and Kánnár, J. 2000a. Bondi-type systems near space-like infinity and the calculation of the NP-constants. *J. Math. Phys.*, **41**, 2195.

Friedrich, H., and Kánnár, J. 2000b. Calculating asymptotic quantities near space-like infinity and null infinity from Cauchy data. *Annalen Phys.*, **9**, 321.

Friedrich, H., and Nagy, G. 1999. The initial boundary value problem for Einstein's vacuum field equation. *Comm. Math. Phys.*, **201**, 619.

Friedrich, H., and Rendall, A. D. 2000. The Cauchy problem for the Einstein equations. *Lect. Notes. Phys.*, **540**, 127.

Friedrich, H., and Schmidt, B.G. 1987. Conformal geodesics in general relativity. *Proc. Roy. Soc. Lond. A*, **414**, 171.

Friedrich, H., and Stewart, J. 1983. Characteristic initial data and wavefront singularities in general relativity. *Proc. Roy. Soc. Lond. A*, **385**, 345.

Garabedian, P. R. 1986. *Partial differential equations*. AMS Chelsea Publishing.

García, A. A., Hehl, F. W., Heinicke, C., and Macias, A. 2004. The Cotton tensor in Riemannian spacetimes. *Class. Quantum Grav.*, **21**, 1099.

García-Parrado, A., Gasperín, E., and Valiente Kroon, J.A. 2014. Conformal geodesics in the Schwarzshild-de Sitter and Schwarzschild anti-de Sitter spacetimes. In preparation.

Gasperín, E., and Valiente Kroon, J. A. 2015. Spinorial wave equations and the stability of the Milne universe. *Class. Quantum Grav.*, **32**, 185021.

Geroch, R. 1968. Spinor structure of spacetimes in general relativity I. *J. Math. Phys.*, **9**, 1739.

Geroch, R. 1970a. Multipole moments. I. Flat space. *J. Math. Phys.*, **11**, 1955.

Geroch, R. 1970b. Multipole moments. II. Curved space. *J. Math. Phys.*, **11**, 2580.

Geroch, R. 1970c. Spinor structure of spacetimes in general relativity II. *J. Math. Phys.*, **11**, 343.

Geroch, R. 1971a. A method for generating new solutions of Einstein's equations. *J. Math. Phys.*, **12**, 918.

Geroch, R. 1971b. Space-time structure from a global view point. Page 71 of: Sachs, R. K. (ed), *General relativity and cosmology. Proceedings of the International School in Physics "Enrico Fermi", Course 48.* Academic Press.

Geroch, R. 1972a. A method for generating new solutions of Einstein's equations. II. *J. Math. Phys.*, **13**, 394.

Geroch, R. 1972b. Structure of the gravitational field at spatial infinity. *J. Math. Phys.*, **13**, 956.

Geroch, R. 1976. Asymptotic structure of space-time. In: Esposito, E. P., and Witten, L. (eds), *Asymptotic structure of spacetime.* Plenum Press.

Geroch, R., Held, A., and Penrose, R. 1973. A space-time calculus based on pairs of null directions. *J. Math. Phys.*, **14**, 874.

Gourgoulhon, E. 2012. 3 + 1 *Formalism in general relativity: bases of numerical relativity.* Lect. Notes. Phys. **846**. Springer Verlag.

Gowers, T. (ed). 2008. *The Princeton companion to mathematics.* Princeton University Press.

Graham, C. R., and Hirachi, K. 2005. The ambient obstruction tensor and Q-curvature. Page 59 of: *AdS/CFT correspondence: Einstein metrics and their conformal boundaries.* IRMA Lect. Math. Theor. Phys., vol. 8. Eur. Math. Soc. Zürich.

Griffiths, J. B., and Podolský, J. 2009. *Exact space-times in Einstein's general relativity.* Cambridge University Press.

Guès, O. 1990. Problème mixte hyperbolique quasi-linéaire caracteristique. *Comm. Part. Diff. Eqns.*, **15**, 595.

Gundlach, C., and Martín-García, J. M. 2007. Critical phenomena in gravitational collapse. *Living Rev. Relativity,* **10**, 5. URL (cited on 24 May 2016): www.livingreviews.org/lrr-2007-5.

Günther, P. 1975. Spinorkalkül und Normalkoordinaten. *Z. Angew. Math. Mech.*, **55**, 205.

Guven, J., and O'Murchadha, N. 1995. Constraints in spherically symmetric classical general relativity. I. Optical scalars, foliations, bounds on the configuration space variables, and the positivity of the quasilocal mass. *Phys. Rev. D*, **52**, 758.

Hamilton, R. 1982. The inverse function theorem of Nash and Moser. *Bull. Am. Math. Soc.*, **7**, 65.

Hansen, R. O. 1974. Multipole moments of stationary spacetimes. *J. Math. Phys.*, **15**, 46.

Hartman, P. 1987. *Ordinary differential equations.* SIAM.

Hawking, S. W., and Ellis, G. F. R. 1973. *The large scale structure of space-time.* Cambridge University Press.

Henneaux, M., and Teitelboim, C. 1985. Asymptotically anti-de Sitter spaces. *Comm. Math. Phys.*, **98**, 391.

Hennig, J., and Ansorg, M. 2009. A fully pseudospectral scheme for solving singular hyperbolic equations. *J. Hyp. Diff. Eqns.*, **6**, 161.

Herberthson, H. 2009. Static spacetimes with prescribed multipole moments: a proof of a conjecture by Geroch. *Class. Quantum Grav.*, **26**, 215009.

Herberthson, M., and Ludwig, G. 1994. Time-like infinity and direction-dependent metrics. *Class. Quantum Grav.*, **11**, 187.

Holst, M., Nagy, G., and Tsogtgerel, G. 2008a. Far-from-constant mean curvature solutions of Einstein's contraint equations with positive Yamabe metrics. *Phys. Rev. Lett.*, **100**, 161101.

Holst, M., Nagy, G., and Tsogtgerel, G. 2008b. Rough solutions of the Einstein constraints on closed manifolds without near-CMC conditions. *Comm. Math. Phys.*, **288**, 547.

Hübner, P. 1995. General relativistic scalar-field models and asymptotic flatness. *Class. Quantum Grav.*, **12**, 791.

Hübner, P. 1999a. How to avoid artificial boundaries in the numerical calculation of black hole spacetimes. *Class. Quantum Grav.*, **16**, 2145.

Hübner, P. 1999b. A scheme to numerically evolve data for the conformal Einstein equation. *Class. Quantum Grav.*, **16**, 2823.

Hübner, P. 2001a. From now to timelike infinity on a finite grid. *Class. Quantum Grav.*, **18**, 1871.

Hübner, P. 2001b. Numerical calculation of conformally smooth hyperboloidal data. *Class. Quantum Grav.*, **18**, 1421.

Hughes, T. J. R., Kato, T., and Marsden, J. E. 1977. Well-posed quasi-linear second-order hyperbolic systems with applications to nonlinear elastodynamics and general relativity. *Arch. Ration. Mech. Anal.*, **63**, 273.

Husa, S. 2002. Problems and successes in the numerical approach to the conformal field equations. Page 239 of: Frauendiener, J., and Friedrich, H. (eds), *The conformal structure of space-time: geometry, analysis, numerics.* Springer.

Ionescu, A. D., and Klainerman, S. 2009a. On the uniqueness of smooth, stationary black holes in vacuum. *Inventiones mathematicae*, **175**, 35.

Ionescu, A. D., and Klainerman, S. 2009b. Uniqueness results for ill-posed characteristic problems in curved spacetimes. *Comm. Math. Phys.*, **285**, 873.

Isenberg, J. 1995. Constant mean curvature solutions of the Einstein constraint equations on closed manifolds. *Class. Quantum Grav.*, **12**, 2249.

Isenberg, J. 2013. Initial value problem in general relativity. In: Ashtekar, A., and Petkov, V. (eds), *The Springer handbook of spacetime.* Springer Verlag.

Ishibashi, A., and Wald, R. M. 2004. Dynamics in non-globally-hyperbolic spacetimes III: anti de Sitter spacetime. *Class. Quantum Grav.*, **21**, 2981.

Jaramillo, J. L., Valiente Kroon, J.A., and Gourgoulhon, E. 2008. From geometry to numerics: interdisciplinary aspects in mathematical and numerical relativity. *Class. Quantum Grav.*, **25**, 093001.

Kánnár, J. 1996a. Hyperboloidal initial data for the vacuum field equations with cosmological constant. *Class. Quantum Grav.*, **13**, 3075.

Kánnár, J. 1996b. On the existence of C^∞ solutions to the asymptotic characteristic initial value problem in general relativity. *Proc. Roy. Soc. Lond. A*, **452**, 945.

Kato, T. 1975a. The Cauchy problem for quasi-linear symmetric hyperbolic systems. *Arch. Ration. Mech. Anal.*, **58**, 181.

Kato, T. 1975b. Quasi-linear equations of evolution, with applications to partial differential equations. *Lect. Notes Math.*, **448**, 25.

Kennefick, D. 2007. *Traveling at the speed of thought: Einstein and the quest for gravitational waves.* Princeton University Press.

Kennefick, D., and O'Murchadha, N. 1995. Weakly decaying asymptotically flat static and stationary solutions to the Einstein equations. *Class. Quantum Grav.*, **12**, 149.

Klainerman, S. 1984. The null condition and global existence to nonlinear wave equations. *Lect. Appl. Math.* **23**, 293.

Klainerman, S. 2008. Partial differential equations. Page 455 of: Gowers, T., Barrow-Green, J., and Leader, I. (eds), *The Princeton companion to mathematics*. Princeton University Press.

Kobayashi, S. 1995. *Transformation groups in differential geometry*. Springer Verlag.

Kobayashi, S., and Nomizu, K. 2009. *Foundations of differential geometry*. Vol. 1. Wiley.

Kodaira, K. 1986. *Complex manifolds and deformation of complex structures*. Springer Verlag.

Kozameh, C., Newman, E. T., and Tod, K. P. 1985. Conformal Einstein spaces. *Gen. Rel. Grav.*, **17**, 343.

Krantz, S. G. 2006. *Geometric function theory*. Birkhäuser.

Kreiss, H.-O., and Lorenz, J. 1998. Stability for time-dependent differential equations. *Acta Numerica*, **7**(203).

Kreiss, H.-O., Reula, O., Sarbach, O., and Winicour, J. 2009. Boundary conditions for coupled quasilinear wave equations with applications to isolated systems. *Comm. Math. Phys.*, **289**, 1099.

Kulkarni, R. S., and Pinkall, U. (eds). 1988. *Conformal geometry*. Aspects of Mathematics. Friedrich Vieweg & Sohn.

Kundu, P. 1981. On the analyticity of stationary gravitational fields at spacial infinity. *J. Math. Phys.*, **22**, 2006.

Künzle, H. P. 1967. Construction of singularity free spherically symmetric spacetime manifolds. *Proc. Roy. Soc. Lond. A*, **297**, 244.

Lee, J. M. 1997. *Riemannian manifolds: an introduction to curvature*. Springer Verlag.

Lee, J. M. 2000. *Introduction to topological manifolds*. Springer Verlag.

Lee, J. M. 2002. *Introduction to smooth manifolds*. Springer Verlag.

Lee, J. M., and Parker, T. H. 1987. The Yamabe problem. *Bull. Am. Math. Soc.*, **17**, 37.

Lehner, L., and Pretorious, F. 2014. Numerical relativity and astrophysics. *Ann. Rev. Astron. Astrophys.*, **52**, 661.

Lübbe, C. 2014. *Conformal scalar fields, isotropic singularities and conformal cyclic cosmologies*. In `arXIv:1312.2059`.

Lübbe, C., and Valiente Kroon, J. A. 2009. On de Sitter-like and Minkowski-like spacetimes. *Class. Quantum Grav.*, **26**, 145012.

Lübbe, C., and Valiente Kroon, J. A. 2010. A stability result for purely radiative spacetimes. *J. Hyp. Diff. Eqns.*, **7**, 545.

Lübbe, C., and Valiente Kroon, J. A. 2012. The extended conformal Einstein field equations with matter: the Einstein-Maxwell system. *J. Geom. Phys.*, **62**, 1548.

Lübbe, C., and Valiente Kroon, J. A. 2013a. A class of conformal curves in the Reissner-Nordström spacetime. *Ann. Henri Poincaré*, **15**, 1327.

Lübbe, C., and Valiente Kroon, J. A. 2013b. A conformal approach for the analysis of the non-linear stability of pure radiation cosmologies. *Ann. Phys.*, **328**, 1.

Lübbe, C., and Valiente Kroon, J. A. 2014. On the conformal structure of the extremal Reissner-Nordström spacetime. *Class. Quantum Grav.*, **31**, 175015.

Ludvigsen, M., and Vickers, J. A. G. 1981. The positivity of the Bondi mass. *J. Phys. A*, **14**, L389.

Ludvigsen, M., and Vickers, J. A. G. 1982. A simple proof of the positivity of the Bondi mass. *J. Phys. A*, **15**, L67.

Macedo, R. P., and Ansorg, M. 2014. Axisymmetric fully spectral code for hyperbolic equations. In `arXiv:1402.7343`.

Machado, M. P., and Vickers, J. A. G. 1995. A space-time calculus invariant under null rotations. *Proc. Roy. Soc. Lond. A*, **450**, 1.

Machado, M. P., and Vickers, J. A. G. 1996. Invariant differential operators and the Karlhede classification of type N vacuum solutions. *Class. Quantum Grav.*, **13**, 1589.

Maldacena, J. 1998. The large N limit of superconformal field theories and supergravity. *Adv. Theor. Math. Phys.*, **2**, 231.

Maliborski, M., and Rostworowski, A. 2013. Lecture notes on turbulent instability of anti-de Sitter spacetime. *J. Mod. Phys. A*, **28**, 1340020.

Martín-García, J. M. 2014. www.xact.es.

Mason, L. J. 1986. The conformal Einstein equations. *Twistor Newsletter*, **22**, 41.

Mason, L. J. 1995. The vacuum and Bach equations in terms of light cone cuts. *J. Math. Phys.*, **36**, 3704.

Miao, P. 2003. On the existence of static metric extensions in general relativity. *Comm. Math. Phys.*, **241**, 27.

Misner, C. W., Thorne, K. S., and Wheeler, J. A. 1973. *Gravitation.* W. H. Freeman.

Morrey, C. B. 1958. On the analyticity of the solutions of analytic non-linear elliptic systems of partial differential equations. *Am. J. Math.*, **80**, 198.

Morris, M. S., and Thorne, K. S. 1988. Wormholes in spacetime and their use for interstellar travel: a tool for teaching general relativity. *Am. J. Phys.*, **56**, 395.

Müller zu Hagen, H. 1970. On the analyticity of stationary vacuum solutions of Einstein's equation. *Proc. Camb. Phil. Soc.*, **68**, 199.

Müller zu Hagen, H., and Seifert, H.-J. 1977. On characteristic initial-value and mixed problems. *Gen. Rel. Grav.*, **8**, 259.

Newman, E. T., and Penrose, R. 1962. An approach to gravitational radiation by a method of spin coefficients. *J. Math. Phys.*, **3**, 566.

Newman, E. T., and Penrose, R. 1963. Errata: an approach to gravitational radiation by a method of spin coefficients. *J. Math. Phys.*, **4**, 998.

Newman, E. T., and Penrose, R. 1965. 10 exact gravitationally-conserved quantities. *Phys. Rev. Lett.*, **15**, 231.

Newman, E. T., and Penrose, R. 1966. Note on the Bondi-Metzner-Sachs group. *J. Math. Phys.*, **7**, 863.

Newman, E. T., and Penrose, R. 1968. New conservation laws for zero rest-mass fields in asymptotically flat space-time. *Proc. Roy. Soc. Lond. A*, **305**, 175.

Newman, E. T., and Tod, K. P. 1980. Asymptotically flat space-times. In: Held, A. (ed), *General relativity and gravitation: one hundred years after the birth of Albert Einstein.* Plenum.

Newman, R. P. A. C. 1989. The global structure of simple space-times. *Comm. Math. Phys.*, **123**, 17.

Nicolas, J.-P. 2015. The conformal approach to asymptotic analysis. In `arXiv:1508.02592`.

O'Donnell, P. 2003. *Introduction to 2-spinors in general relativity.* World Scientific.

Ogiue, K. 1967. Theory of conformal connections. *Kodai Math. Sem. Rep.*, **19**, 193.

O'Murchadha, N. 1988. The Yamabe problem and general relativity. *Proc. Centre Math. Anal. (A.N.U.)*, **19**(137).

O'Neill, B. 1983. *Semi-Riemannian geometry with applications to relativity.* Academic Press.

O'Neill, B. 1995. *The geometry of Kerr black holes.* A. K. Peters.

Paetz, T.-T. 2015. Conformally covariant systems of wave equations and their equivalence to Einstein's field equations. *Ann. Henri Poincaré*, **16**, 2059.

Penrose, R. 1960. A spinor approach to general relativity. *Ann. Phys. (New York)*, **10**, 171.

Penrose, R. 1963. Asymptotic properties of fields and space-times. *Phys. Rev. Lett.*, **10**, 66.

Penrose, R. 1964. Conformal approach to infinity. In: DeWitt, B. S., and DeWitt, C. M. (eds), *Relativity, groups and topology: the 1963 Les Houches lectures*. Gordon and Breach.

Penrose, R. 1965. Zero rest-mass fields including gravitation: asymptotic behaviour. *Proc. Roy. Soc. Lond. A*, **284**, 159.

Penrose, R. 1967. Structure of space-time. Page 121 of: DeWitt, C. M., and Wheeler, J. A. (eds), *Battelle rencontres: 1967 lectures in mathematics and physics*. W. A. Benjamin.

Penrose, R. 1969. Gravitational collapse: the role of general relativity. *Rev. Nuovo Cimento*, **I**, 257.

Penrose, R. 1979. Singularities and time asymmetry. Page 581 of: Hawking, S. W., and Israel, W. (eds), *General relativity: an Einstein centenary survey*. Cambridge University Press.

Penrose, R. 1980. Null hypersurface initial data for classical fields of arbitrary spin and for general relativity. *Gen. Rel. Grav.*, **12**, 225.

Penrose, R. 1983. Spinors and torsion in general relativity. *Found. Phys.*, **13**, 325.

Penrose, R. 2002. Gravitational collapse: the role of general relativity. *Gen. Rel. Grav.*, **34**, 1141.

Penrose, R. 2011. Reprint of: Conformal treatment of infinity. *Gen. Rel. Grav.*, **43**, 901.

Penrose, R., and Rindler, W. 1984. *Spinors and space-time*. Vol. 1: *Two-spinor calculus and relativistic fields*. Cambridge University Press.

Penrose, R., and Rindler, W. 1986. *Spinors and space-time*. Vol. 2: *Spinor and twistor methods in space-time geometry*. Cambridge University Press.

Persides, S. 1979. A definition of asymptotically Minkowskian space-times. *J. Math. Phys.*, **20**, 1731.

Persides, S. 1980. Structure of the gravitational field at spatial infinity. II. Asymptotically Minkowski spaces. *J. Math. Phys.*, **21**, 142.

Persides, S. 1982a. Timelike infinity. *J. Math. Phys.*, **23**, 283.

Persides, S. 1982b. A unified formulation of timelike, null and spatial infinity. *J. Math. Phys.*, **23**, 289.

Petersen, P. 1991. *Riemannian geometry*. Graduate Texts in Mathematics, vol. 171. Springer.

Poisson, E., and Will, C. M. 2015. *Gravity: Newtonian, post-Newtonian, relativistic*. Cambridge University Press.

Porrill, J. 1982. The structure of timelike infinity for isolated systems. *Proc. Roy. Soc. Lond. A*, **381**, 323.

Pretorious, F. 2009. Binary black hole coalescence. Page 305 of: Colpi, M., Casella, P., Gorini, V., Moschella, U., and Possenti, A. (eds), *Physics of relativistic objects in compact binaries: from birth to coalescence*. Springer.

Pugliese, D., and Valiente Kroon, J. A. 2012. On the evolution equations for ideal magnetohydrodynamics in curved spacetime. *Gen. Rel. Grav.*, **44**, 2785.

Pugliese, D., and Valiente Kroon, J. A. 2013. On the evolution equations for a self-gravitating charged scalar field. *Gen. Rel. Grav.*, **45**, 1247.

Quevedo, H. 1990. Multipole moments in general relativity: static and stationary vacuum solutions. *Fortsch. der Physik*, **38**, 733.

Reinhart, B. L. 1973. Maximal foliations of extended Schwarzschild space. *J. Math. Phys.*, **14**, 719.

Reiris, M. 2014a. Stationary solutions and asymptotic flatness I. *Class. Quantum Grav.*, **31**, 155012.

Reiris, M. 2014b. Stationary solutions and asymptotic flatness II. *Class. Quantum Grav.*, **31**, 155013.

Rendall, A. D. 1990. Reduction of the characteristic initial value problem to the Cauchy problem and its application to the Einstein equations. *Proc. Roy. Soc. Lond. A*, **427**, 221.

Rendall, A. D. 2005. Theorems on existence and global dynamics for the Einstein equations. *Living Rev. Relativity*, **8**, 6. URL (cited on 24 May 2016): www.livingreviews.org/lrr-2005-6.

Rendall, A. D. 2008. *Partial differential equations in general relativity.* Oxford University Press.

Reula, O. 1989. On existence and behaviour of asymptotically flat solutions to the stationary Einstein equations. *Comm. Math. Phys.*, **122**, 615.

Reula, O. 1998. Hyperbolic methods for Einstein's equations. *Living Rev. Rel.*, **3**, 1.

Ringström, H. 2009. *The Cauchy problem in general relativity.* Eur. Math. Soc. Zürich.

Rinne, O. 2010. An axisymmetric evolution code for the Einstein equations on hyperboloidal slices. *Class. Quantum Grav.*, **27**, 035014.

Rinne, O. 2014. Formation and decay of Einstein-Yang-Mills black holes. *Phys. Rev. D*, **90**, 124084.

Rinne, O., and Moncrief, V. 2013. Hyperboloidal Einstein-matter evolution and tails for scalar and Yang–Mills fields. *Class. Quantum Grav.*, **30**, 095009.

Rodnianski, I., and Speck, J. 2013. The nonlinear future-stability of the FLRW family of solutions to the Euler-Einstein system with a positive cosmological constant. *J. Eur. Math. Soc.*, **15**, 2369.

Sachs, R. K. 1962a. Asymptotic symmetries in gravitational theory. *Phys. Rev.*, **128**, 2851.

Sachs, R. K. 1962b. Gravitational waves in general relativity VIII. Waves in asymptotically flat space-time. *Proc. Roy. Soc. Lond. A*, **270**, 103.

Sachs, R. K. 1962c. On the characteristic initial value problem in gravitational theory. *J. Math. Phys.*, **3**, 908.

Sbierski, J. 2013. On the existence of a maximal Cauchy development for the Einstein equations: a dezornification. *Ann. Henri Poincaré: Online First.*

Schmidt, B., Walker, M., and Sommers, P. 1975. A characterization of the Bondi-Metzner-Sachs group. *Gen. Rel. Grav.*, **5**, 489.

Schmidt, B. G. 1978. Asymptotic structure of isolated systems. Page 11 of: Ehlers, J. (ed), *Isolated systems in general relativity, Proceedings of the International School of Physics "Enrico Fermi", Course 67.* North Holland Publishing Company.

Schmidt, B. G. 1981. The decay of the gravitational field. *Comm. Math. Phys.*, **78**, 447.

Schmidt, B. G. 1986. Conformal geodesics. *Lect. Notes. Phys.*, **261**, 135.

Schmidt, B. G. 1987. Gravitational radiation near spatial and null infinity. *Proc. Roy. Soc. Lond. A*, **410**, 201.

Schmidt, B. G. 1996. Vacuum space-times with toroidal null infinities. *Class. Quantum Grav.*, **13**, 2811.

Schmidt, B. G., and Walker, M. 1983. Analytic conformal extensions of asymptotically flat spacetimes. *J. Phys. A: Math. Gen.*, **16**, 2187.

Schoen, R. 1984. Conformal deformation of a Riemannian metric to constant scalar curvature. *J. Diff. Geom.*, **20**, 479.

Schoen, R., and Yau, S. T. 1979. On the proof of the positive mass conjecture in general relativity. *Comm. Math. Phys.*, **65**, 45.

Schouten, J. A. 1921. Über die konforme Abbildung n-dimensionaler Mannigfaltigkeiten mit quadratischer Massbestimmung auf eine Mannigfaltigkeit mit euklidischer Massbestimmung. *Math. Zeitschrift*, **11**, 58.

Schwarzschild, K. 1916. Über das Gravitationsfeld eines Massenpunktes nach der Einsteinschen Theorie. *Sitz. Preuss. Akad. Wiss. Berlin*, **7**, 189–196.

Schwarzschild, K. 2003. "Golden oldie": on the gravitational field of a mass point according to Einstein's theory. *Gen. Rel. Grav.*, **35**, 951.

Sen, A. 1981. On the existence of neutrino "zero-modes" in vacuum spacetimes. *J. Math. Phys.*, **22**, 1781.

Sexl, R. U., and Urbantke, H. K. 2000. *Relativity, groups, particles: special relativity and relativistic symmetry in field and particle physics.* Springer.

Shapiro, I. 1999. A century of relativity. *Rev. Mod. Phys.*, **71**, S41.

Shinbrot, M., and Welland, R. R. 1976. The Cauchy-Kowaleskaya. *J. Math. An. App.*, **55**, 757.

Simon, W. 1992. Radiative Einstein-Maxwell spacetimes and "no-hair" theorems. *Class. Quantum Grav.*, **9**, 241.

Sommers, P. 1980. Space spinors. *J. Math. Phys.*, **21**, 2567.

Speck, J. 2012. The nonlinear future-stability of the FLRW family of solutions to the Euler-Einstein system with a positive cosmological constant. *Selecta Mathematica*, **18**, 633.

Spivak, M. 1970. *A comprehensive introduction to differential geometry.* Vol. I. Publish or Perish.

Stephani, H., Kramer, D., MacCallum, M. A. H., Hoenselaers, C., and Herlt, E. 2003. *Exact solutions of Einstein's field equations*, 2nd edition. Cambridge University Press.

Stewart, J. 1991. *Advanced general relativity.* Cambridge University Press.

Synge, J. L. 1960. *Relativity: the general theory.* North Holland Publishing Company.

Szabados, L. B. 1994. Two-dimensional Sen connections in general relativity. *Class. Quantum Grav.*, **11**, 1833.

Szabados, L. B. 2009. Quasi-local energy-momentum and angular momentum in General Relativity. *Living Rev. Relativity*, **12**, 4. URL (cited on 24 May 2016): www.livingreviews.org/lrr-2009-4.

Szegö, G. 1978. *Orthogonal polynomials.* AMS Colloq. Pub., vol. 23. AMS.

Szekeres, P. 1965. The gravitational compass. *J. Math. Phys.*, **6**, 1387.

Tataru, D. 2004. The wave maps equation. *Bull. Am. Math. Soc.*, **41**, 185.

Taubes, C. H. 2011. *Differential geometry.* Oxford University Press.

Taylor, M. E. 1996a. *Partial differential equations I: basic theory.* Springer Verlag.

Taylor, M. E. 1996b. *Partial differential equations II: qualitative studies of linear equations.* Springer Verlag.

Taylor, M. E. 1996c. *Partial differential equations III: nonlinear equations.* Springer Verlag.

Tod, K. P. 1984. Three-surface twistors and conformal embedding. *Gen. Rel. Grav.*, **16**, 435.

Tod, K. P. 2002. Isotropic cosmological singularities. Page 123 of: Frauendiener, J., and Friedrich, H. (eds), *The conformal structure of space-time: geometry, analysis, numerics.* Lect. Notes. Phys. **604** Springer.

Tod, K. P. 2003. Isotropic cosmological singularities: other matter models. *Class. Quantum Grav.*, **20**, 251.

Tod, K. P. 2012. Some examples of the behaviour of conformal geodesics. *J. Geom. Phys.*, **62**, 1778.

Torres del Castillo, G. F. 2003. *3-D spinors, spin-weighted functions and their applications.* Birkhäuser.

Trautman, A. 1958. Radiation and boundary conditions in the theory of gravitation. *Bull. Academ. Pol. Sciences: Series des Sciences Math. Astron. Phys.*, **6**, 407.

Trudinger, N. 1968. Remarks concerning the conformal deformation of Riemannian structures on compact manifolds. *Ann. Sc. Norm. Sup. Pisa*, **22**, 265.

Valiente Kroon, J. A. 1998. Conserved quantities for polyhomogeneous spacetimes. *Class. Quantum Grav.*, **15**, 2479.

Valiente Kroon, J. A. 1999a. A comment on the outgoing radiation condition and the peeling theorem. *Gen. Rel. Grav.*, **31**, 1219.

Valiente Kroon, J. A. 1999b. Logarithmic Newman-Penrose constants for arbitrary polyhomogeneous spacetimes. *Class. Quantum Grav.*, **16**, 1653.

Valiente Kroon, J. A. 2002. Polyhomogeneous expansions close to null and spatial infinity. Page 135 of: Frauendiner, J., and Friedrich, H. (eds), *The conformal structure of spacetimes: geometry, numerics, analysis*. Lecture Notes in Physics. Springer.

Valiente Kroon, J. A. 2003. Early radiative properties of the developments of time symmetric conformally flat initial data. *Class. Quantum Grav.*, **20**, L53.

Valiente Kroon, J. A. 2004a. Does asymptotic simplicity allow for radiation near spatial infinity? *Comm. Math. Phys.*, **251**, 211.

Valiente Kroon, J. A. 2004b. A new class of obstructions to the smoothness of null infinity. *Comm. Math. Phys.*, **244**, 133.

Valiente Kroon, J. A. 2004c. Time asymmetric spacetimes near null and spatial infinity. I. Expansions of developments of conformally flat data. *Class. Quantum Grav.*, **23**, 5457.

Valiente Kroon, J. A. 2005. Time asymmetric spacetimes near null and spatial infinity. II. Expansions of developments of initial data sets with non-smooth conformal metrics. *Class. Quantum Grav.*, **22**, 1683.

Valiente Kroon, J. A. 2007a. Asymptotic properties of the development of conformally flat data near spatial infinity. *Class. Quantum Grav.*, **24**, 3037.

Valiente Kroon, J. A. 2007b. The Maxwell field on the Schwarzschild spacetime: behaviour near spatial infinity. *Proc. Roy. Soc. Lond. A*, **463**, 2609.

Valiente Kroon, J. A. 2009. Estimates for the Maxwell field near the spatial and null infinity of the Schwarzschild spacetime. *J. Hyp. Diff. Eqns.*, **6**, 229.

Valiente Kroon, J. A. 2010. A rigidity property of asymptotically simple spacetimes arising from conformally flat data. *Comm. Math. Phys.*, **298**, 673.

Valiente Kroon, J. A. 2011. Asymptotic simplicity and static data. *Ann. Henri Poincaré*, **13**, 363.

Valiente Kroon, J. A. 2012. *Global evaluations of static black hole spacetimes*. In preparation.

Wald, R. M. 1983. Asymptotic behaviour of homogeneous cosmological models in the presence of a positive cosmological constant. *Phys. Rev. D*, **28**, 2118.

Wald, R. M. 1984. *General relativity*. University of Chicago Press.

Walker, M. 1970. Block diagrams and the extension of timelike two-surfaces. *J. Math. Phys.*, **11**, 2280.

Weyl, H. 1918. Reine Infinitesimalgeometrie. *Math. Zeitschrift*, **2**, 404.

Weyl, H. 1968. Zur Infinitesimal Geometrie: Einordnung der projektiven und der konformen Auffassung. Page 195 of: *Gesammelte Abhandlungen.*, vol. II. Springer Verlag.

Will, C. M. 2014. The confrontation between general relativity and experiment. *Living. Rev. Relativity*, **17**.

Willmore, T. J. 1993. *Riemannian geometry*. Oxford University Press.

Winicour, J. 2012. Characteristic evolution and matching. *Living. Rev. Relativity*, **14**.

Witten, E. 1998. Anti de Sitter space and holography. *Adv. Theor. Math. Phys.*, **2**, 253.

Witten, E., and Yau, S.-T. 1999. Connectedness of the boundary in the AdS/CFT correspondence. *Adv. Theor. Math. Phys.*, 1635.

Wong, W. W.-Y. 2013. A comment on the construction of the maximal globally hyperbolic Cauchy development. *J. Math. Phys.*, **54**, 113511.

Yamabe, H. 1960. On a deformation of Riemannian structures on compact manifolds. *Ōsaka J. Mathematics*, **12**, 6126.

York, J. W. Jr. 1971. Gravitational degrees of freedom and the initial-value problem. *Phys. Rev. Lett.*, **26**, 1656.

York, J. W. Jr. 1972. Role of conformal three-geometry in the dynamics of gravitation. *Phys. Rev. Lett.*, **28**, 1082.

York, J. W. Jr. 1973. Conformally covariant orthogonal decomposition of symmetric tensor on Riemannian manifolds and the initial value problem of general relativity. *J. Math. Phys.*, **14**, 456.

Zenginoglu, A. 2006. A conformal approach to numerical calculations of asymptotically flat spacetimes. PhD thesis, Max-Planck Institute for Gravitational Physics (AEI) and University of Potsdam.

Zenginoglu, A. 2007. Numerical calculations near spatial infinity. *J. Phys. Conf. Ser.*, **66**, 012027.

Zenginoglu, A. 2008. A hyperboloidal study of tail decay rates for scalar and Yang-Mills fields. *Class. Quantum Grav.*, **25**, 195025.

Zenginoglu, A. 2011a. A geometric framework for black hole perturbations. *Phys. Rev. D*, **83**, 127502.

Zenginoglu, A. 2011b. Hyperboloidal layers for hyperbolic equations on unbounded domains. *J. Comput. Phys.*, **230**, 2286.

Zenginoglu, A., and Kidder, L. E. 2010. Hyperboloidal evolution of test fields in three spatial dimensions. *Phys. Rev. D*, **81**, 124010.

Zhiren, J. 1988. A counterexample to the Yamabe problem for complete non-compact manifolds. *Lect. Notes Math.*, **1306**, 93.

Index

Printed in the United States
by Baker & Taylor Publisher Services